T0315024

Classical Electromagnetism in a Nutshell

Classical Electromagnetism in a Nutshell

Anupam Garg

PRINCETON UNIVERSITY PRESS · PRINCETON AND OXFORD

Copyright © 2012 by Princeton University Press

Published by Princeton University Press, 41 William Street, Princeton,
New Jersey 08540

In the United Kingdom: Princeton University Press, 6 Oxford Street,
Woodstock, Oxfordshire OX20 1TW

press.princeton.edu

Library of Congress Cataloging-in-Publication Data

Garg, Anupam Kumar, 1956-
  Classical electromagnetism in a nutshell / Anupam Garg.
      p. cm.
  Includes bibliographical references and index.
  ISBN-13: 978-0-691-13018-7 (cloth : alk. paper)
  ISBN-10: 0-691-13018-3 (cloth : alk. paper)   1. Electromagnetism–Textbooks.   I. Title.
  QC760.G37 2012
  537–dc23
                    2011037153
British Library Cataloging-in-Publication Data is available

This book has been composed in Scala

Printed on acid-free paper. ∞

Typeset by S R Nova Pvt Ltd, Bangalore, India
Printed in the United States of America

10 9 8 7 6 5 4 3 2 1

To my parents, and to Neerja, Gaurav, and Arjun

Contents

Preface

In the United States, a graduate course in classical electromagnetism is often seen as a vehicle for learning mathematical methods of physics. Students mostly perceive the subject as hand-to-hand combat with the ∇ symbol, with endless nights spent gradding, divvying, and curling. Instructors have a more benign view, but many of them too regard the subject as a way of teaching the classical repertoire of Green functions, Sturm-Liouville theory, special functions, and boundary value problems. Perhaps this viewpoint is traditional. In the 1907 preface to the first edition of his classic text *Electricity and Magnetism*, James Jeans writes:

> Maxwell's treatise was written for the fully equipped mathematician: the present book is written more especially for the student, and for the physicist of more limited mathematical attainments.

The emphasis on mathematics continues throughout the preface: words derived from *mathematics* appear ten times in the single page, while derivatives of *physics* appear just five times.

To me, this point of view is both a puzzle and a pity, for while the mathematics of electrodynamics is indeed very beautiful, the physics is much more so. And so, while this book does not shy away from mathematical technique, the emphasis is very much on the physics, and I have striven to provide physical context and motivation for every topic in the book. Naturally, greater mastery of the mathematics will lead to greater enjoyment of the physics. I try to model good and efficient methods of calculation, but this is never an end in itself, and I do not hesitate to use the physicist's quick and dirty methods when appropriate. I urge readers, students in particular, to approach the subject in the same spirit.

This book started out many years ago as lecture notes for students in a graduate course. Over the years, to keep my own enthusiasm from waning, I varied the topics in the latter part of the course. A few others that I never got to teach are also included in the book.

As a result, there is more material here than can be covered in a single full-year course. I hope that all instructors will find something to interest them among the more advanced topics and applications, and that the book will serve as a reference for students after the course is complete.

The sequence of topics is as follows. After a very brief survey [1] and a longish math review [2],[1] we proceed to electrostatics [3] and magnetostatics [4] in vacuum, Faraday's law of induction [5], and electromagnetic waves in vacuum [7]. Chapter 6 is on the symmetries of the laws of electromagnetism, and it is here that the concepts of field energy and momentum are introduced. These chapters are part of the core of the subject and belong in all courses.

Chapter 8 deals with interference and diffraction. This material is somewhat nonstandard in an electromagnetism course and may be skipped without loss of continuity. On the other hand, the phenomena of interference and diffraction are the most striking manifestations of the wave aspects of light, and the related topics of coherence and intensity interferometry (the Hanbury-Brown and Twiss effect) are among the more subtle aspects of light that were understood in the twentieth century. They also connect to our understanding of quantum mechanics, and similar ideas are active research topics even today in other areas of physics. A section on the Pancharatnam phase and how the polarization of light affects interference rounds out the chapter.

Next we come to radiation from accelerated charges [9, 10]. Although the field of a moving charge is found in full generality [9], only nonrelativistic sources are considered for the present [10], and relativistic sources are treated later [25]. While this is also core material, the more advanced sections, such as those on higher multipoles, radiation of angular momentum, and radiation reaction, can be omitted in a first reading.

The motion of charges and magnetic moments in external fields is studied next [11]. In addition to charges in uniform and static fields, and the betatron, I have included a moderately complete account of Alfven's guiding center method for charges in inhomogeneous magnetic fields. This method is indispensable for the study of plasmas. I also discuss Larmor precession of magnetic moments, in both static and time-dependent magnetic fields. The concept of adiabatic invariants is developed both for charges and moments in magnetic fields. Again, subsequent chapters do not depend on this development in an essential way and many sections in this chapter may be skipped in a first pass.

Chapter 12 puts electromagnetism within the framework of the action principle. It is thus an essential launching point for the study of *quantum* electrodynamics, and even within the classical theory it is a deep unifying principle. For the study of electromagnetic *phenomena*, however, it is inessential, and less formally minded readers may choose to skip it.

Chapters 3 through 12 constitute a fairly complete treatment (except for relativity, for which see later) of the basic laws and phenomena of electromagnetic fields in vacuum. With these fundamentals over, we turn to electromagnetic fields in matter [13–21]. This

---

[1] Chapter numbers are in square brackets.

separation of the study of fields in vacuum and matter is a little nonstandard,[2] and it is more common to study the electrostatics of conductors [14] and dielectrics [15] along with electrostatics in vacuum, and electromagnetic waves in matter [20, 21] along with those in vacuum. I believe that no good can come of this fusion. It appears to me to be motivated by the incidental similarity of the mathematics required. However, the interesting physical questions are quite different in the two situations, and the basic electromagnetic equations in vacuum and in macroscopic materials have rather different meaning and ontological status. Thus, studying the propagation of light in glass and vacuum as common instances of the solution of the wave equation obscures the physical content of the dielectric constant, or the Kramers-Kronig relations, or the interplay between electromagnetic energy, mechanical work, and heat dissipation in a medium. A mathematically oriented approach also misses out on the beautiful physics involved in modeling constitutive relations. With specific reference to electrostatics, most students have studied it with conductors and in vacuum as one topic as undergraduates, so they already know that the field is strong near sharp points and edges and other similar facts, and there is little compelling reason to follow the same path again. Nevertheless, instructors who wish to stress the solution of Laplace's equation through methods such as separation of variables and orthogonal functions, can cover chapter 14 earlier.

In more detail, the treatment of electromagnetism in matter is organized as follows. A short survey chapter [13] introduces the concepts of spatially averaged or macroscopic fields, magnetization and polarization fields, and the vital role of constitutive relations. It is particularly stressed that the latter are largely phenomenological, but no pejorative connotation is attached to this word. It is also stressed that the similarity of form of the macroscopic or material Maxwell equations with those in vacuum is deceptive and sweeps important physics under the rug. This chapter is very important, and readers are urged not to skip any of it.

Chapter 14 covers electrostatics with conductors. My treatment is quite traditional, and it is with some ambivalence that so much detail is given. The prevalence of numerical methods has rather reduced the importance of the classical mathematical techniques, but they provide invaluable insight, and even the numerical solver would be unwise to ignore them. I have also included a section on the work function and contact potentials, if for no other reason than to show the reader that there is subtle physics in this otherwise mathematical territory. The coverage of electrostatics with dielectrics [15] is also traditional. In both chapters, there are several advanced sections that may be skipped without loss of continuity.

Magnetostatics in matter comes next [16]. The central point of physics here is to understand the constitutional differences (constitutive relations, if you will) between para-, dia-, and ferromagnets. Much of the chapter is devoted to ferromagnets, and relevant phenomena, such as domain walls, hysteresis, and demagnetization, ancient topics that are astonishingly current for the design of hard disks and other computer elements. The dangers of uncritical use of equations such as $\mathbf{B} = \mu\mathbf{H}$ are dwelt on. I have

---

[2] The same approach is taken in the Landau and Lifshitz *Course of Theoretical Physics*, vols. 2 and 8.

tried hard to obey Paul Muzikar's exhortation to explain the difference between **B** and **H** carefully, and I am sure I will hear from him if I have let him down! The chapter also includes a short section on superconductors and the Meissner effect.

The next chapter is on Ohm's law, emf, and electrical circuits [17]. These topics are of great practical importance, but they are often regarded as belonging to the realm of engineering. My focus is on understanding of the meaning of emf, and of the physics of the lumped circuit approximation. I also discuss the distribution of current in extended conductors, and van der Pauw's method of measuring resistivities.

With this, we turn to dynamic phenomena, beginning with the general issue of frequency-dependent response or constitutive relations [18]. This chapter covers the Drude and Drude-Lorentz models for the frequency-dependent conductivity and dielectric functions, which are of great pedagogical value because they fix several key physical concepts in the student's mind. I also discuss Kramers-Kronig relations and electromagnetic energy in material media. None of this material should be skipped.

Quasistatic phenomena [19], such as the skin effect, eddy currents, and maglev, come next, followed by electromagnetic waves in insulators [20] and conductors [21]. The former chapter covers dispersion, and reflection and refraction at interfaces, while the latter includes plasma oscillations, ultraviolet transparency and metallic reflection, waveguides, and resonant cavities.

This ends the discussion of macroscopic electromagnetism. Completeness would demand that I include topics such as spin waves and Walker modes in ferromagnets, light in anisotropic media, some magnetohydrodynamics, and nonlinear optics. But life is short, and the interested reader must seek these elsewhere.

Scattering of light [22] is a vast subject, and the selection of subtopics reflects my personal tastes. I do not discuss scattering of charged particles, collision radiation, virtual quanta, or the energy loss formula, as there are excellent treatments of these topics by other authors, and my own knowledge of them is limited.

The formalism of special relativity [23], its role in electromagnetism [24], and radiation from relativistic particles [25] bring up the rear, but they could in fact be covered any time after chapter 12. I have chosen not to present electromagnetism as an outgrowth of relativity as many other authors do, partly because that is ahistorical, but mainly because I have found that students find it hard to relate the formal relativistic underpinnings to actual electromagnetic phenomena, especially in matter, and the machinery of four-vectors and four-tensors seems unconnected to concepts such as the multipole expansion and mutual inductance. This machinery also requires more sophistication to master, so I have found it helpful to delay introducing it. Nevertheless, relativity is used to provide insight throughout the book, the first-order transformation of **E** and **B** fields is found early on, and much of the treatment of radiation and charged particle motion is fully relativistically correct.

Several appendixes provide mathematical reference material (Bessel and Airy functions, the Wiener-Khinchine theorem, etc.), which need not be covered explicitly. Appendix A on spherical harmonics is an exception, as these are used throughout the book, and an easy familiarity with them is a must. I have also included two appendixes on caustics (the rainbow and the teacup nephroid) and the motion of charged particles in the earth's magnetic field as specialized applications of matter in the main text.

Several colleagues have asked me if my book would include numerical methods. It does, but in a very limited way. My experience has been that good use of numerical methods requires so much attention to basic calculus and linear algebra that it reduces the time students can devote to physical concepts. Further, solving problems numerically requires much trial and error, making the exercise closer to experimental than theoretical physics, and thus ill suited to a traditional lecture-based course. Secondly, the peripheral aspects of computer use—the operating system, the precise computer language, the graphics package—are so varied and riddled with minutiae that they often end up dominating the learning of the numerical methods themselves. I do not advocate the use of canned black-box packages, as the student learns neither much physics nor much numerical analysis from them.

The book includes over three hundred exercises. These appear in each section rather than at the ends of chapters, because in my view the immediacy of the problems makes them more relevant. Some exercises are quite simple and designed to develop technical skills, while others may require significant extension of concepts in the text. The latter are often accompanied by short solutions or hints. I have tried to make sure that all exercises have a point and to explain that point, often by adding explicit commentary, and that none are just make-work for the sake of having a large number. A few exercises *are* purely mathematical or formal, but these are generally used to develop results used elsewhere in the book.

The book does not require in-depth understanding of quantum mechanics or thermo-dynamics, and though quantal and thermodynamic concepts are used or mentioned in a matter-of-fact way when needed, they are not critical to the central development, and an acquaintance with them at the undergraduate level will suffice.

I have benefited from encouragement and discussions with so many colleagues and students over the years that I am sure I cannot remember them all. I am indebted, among others, to Greg Anderson, Mike Bedzyk, Carol Braun, Paul Cadden-Zimansky, Pulak Dutta, Shyamsunder Erramilli, André de Gouvêa, Nick Giordano, C. Jayaprakash, John Ketterson, Mike Peskin, Mohit Randeria, Wayne Saslow, Surendra Singh, Mike Stone, and Tony Zee. I am materially indebted, in addition, to David Mermin and Saul Teukolsky for lectures at Cornell University, which I relied on in writing appendix A on spherical harmonics and chapters 23 and 24 on relativity. I am likewise indebted to Jens Koch for his invaluable help with some of the figures. I am particularly indebted to Onuttam Narayan and Paul Sievert for reading large sections of the text, and their valuable feedback and criticisms. None of these people are, of course, responsible for any errors or omissions in the book.

I also wish to thank Karen Fortgang and Ingrid Gnerlich at Princeton University Press for their gentle prodding and guidance through the production process.

I owe a different kind of debt to the authors of the texts from which I learned this subject. These include the delightful *Electricity and Magnetism* by Purcell; *The Feynman Lectures* (all three volumes) by Feynman, Leighton, and Sands; *Classical Electrodynamics* by Jackson; Vols. 2 and 8 of the *Course of Theoretical Physics* by Landau and Lifshitz; and *Classical Electrodynamics* by Schwinger, DeRaad, Milton, and Tsai. I have consulted these books so often that I am sure I have internalized and come to regard as my own the ideas and the very language that they use, and I hope I have not failed to give them credit on any occasion. To all these authors, I am deeply grateful.

I cannot adequately acknowledge my wife, Neerja, for her unwavering encouragement and support and uncomplaining acceptance of the domestic burdens created by my absorption with this project.

Lastly, I thank you, the readers, in advance. Your comments, suggestions, and notifications of mistakes are welcome and will be gratefully received.

Anupam Garg
Wilmette, Illinois
December 2010

List of symbols

Professor, to student on oral exam: *Was ist diese Gleichung, $E = h\nu$? (What is this equation, $E = h\nu$?)*
Student, after a while: *Das ist das Ohm'sche Gesetz in der phantastichen Darstellung. (That is Ohm's law in fancy notation.)*

The following is a list of the principal symbols used in this book, along with key sections where their meanings are discussed or where they are first introduced. Many symbols do double and even triple duty—they mean more than one thing. It is usually clear which meaning is intended from the context. In such cases, the table lists all the different usages. Some symbols are used generically, e.g., a, b, ℓ, etc., to denote lengths, $\mathbf{v}$ to denote a velocity, and so on. Such usages are not listed. Other symbols not listed are used in specific, local contexts, and they are defined in that context.

Symbol(s)	Meaning	Key Sections/First Mention
a_0	Bohr radius, interatomic spacing	14, 81
a_1, a_2, a_3	Stokes parameters	45
$\mathbf{a}_a$	apparent acceleration of moving charge	56
$\mathbf{a}_r$	retarded acceleration of moving charge	56
$\mathbf{a}_{co}$	acceleration in comoving frame	154
a^μ	acceleration four-vector	154
$\mathbf{A}$	vector potential	24

Symbol(s)	Meaning	Key Sections/First Mention
$\mathbf{A}_k$	Fourier component of $\mathbf{A}$	78
A^μ	four-vector potential	160
$A_{\mu,\nu}$, $A_{\mu\nu,\sigma}$	covariant four-derivatives	156
Ai, Bi	Airy functions	Appendix E
$\mathbf{b}$	microscopic magnetic field in medium	81
$\mathbf{B}$	magnetic field,	
	in vacuum	21
	macroscopic, in material medium	81
B_n, $\mathbf{B}_t$	normal and tangential components of $\mathbf{B}$	83
c	speed of light	Chap. 1
C, C_{ab}	capacitance, electrostatic induction coefficients	88, 116
$\mathbf{d}$	dipole moment	19
$d\ell$, $d\boldsymbol{\ell}$	scalar and vectorial line elements	7
ds, $d\mathbf{s}$	scalar and vectorial surface elements	7
$d\Omega$	element of solid angle	19
$\mathbf{D}$	electric displacement	82, 94
D_{ij}	quadrupole moment tensor	19, 63
e	electron charge magnitude	Chap. 1
$\mathbf{e}$	microscopic electric field in medium	81
$\mathbf{E}$	electric field,	
	in vacuum	12
	macroscopic, in material medium	81
$\mathbf{E}_d$	depolarization field	96
E_n, $\mathbf{E}_t$	normal and tangential components of $\mathbf{E}$	86
E_{ms}	magnetostatic energy in ferromagnets	105
$\mathcal{E}$	circulation of $\mathbf{E}$	28
$\mathcal{E}$	emf	114, 115

Symbol(s)	Meaning	Key Sections/First Mention
$\mathcal{E}$	energy of a particle/system/radiation	35, Chap. 10, Chap. 11
f_i	oscillator strength	123, 147
F, $\tilde{F}$, F_{bod}	various free energies	97, 98, 101
$\mathbf{F}$	force	2, Chap. 11
$F^{\mu\nu}$	electromagnetic field tensor	161
g	g-factor	75
$\mathbf{g}$	momentum density	36
$g_{\mu\nu}$	metric tensor	153
G	Green function (for various equations)	48, 54
$\mathbf{G}$	total momentum in electromagnetic field	46
$G^{\perp}_{jj}$	transverse current–current autocorrelation	61, Appendix F
H_{ex}	exchange field in ferromagnets	105
$\mathbf{H}$	magnetizing field	83, 100
$\mathbf{H}_d$	demagnetizing field	83, 105, 108
$\mathcal{H}$	Hamiltonian	Chap. 12
$\hbar$	Planck's constant	
I	current in a wire	22, 31
I	intensity of light	44, Chap. 7
I_n	modified Bessel functions	Appendix B
$\mathbf{j}$	volume current density,	
	in vacuum	23
	macroscopic, in material medium	83
$\mathbf{j}_{\text{free}}$	free (mobile) current density	83, 100
$\mathbf{j}_{\text{mag}}$	magnetization current density	83
$\mathbf{j}^{L,T}$, $\mathbf{j}^{\parallel,\perp}$	longitudinal and transverse currents	33, 60
j^{μ}	four-current	159
J_n	Bessel functions	Appendix B

Symbol(s)	Meaning	Key Sections/First Mention
$J^{\mu\nu}$	four-angular momentum of composite system	165
k, $\mathbf{k}$	wave vector, spatial FT variable	9, Chap. 7
k_B	Boltzmann's constant	
K	undulator/wiggler parameter	175
K_n	modified Bessel functions	Appendix B
$\mathbf{K}$	surface current density	83
$\mathbf{K}_{\mathrm{mag}}$	surface magnetization current density	83
$\mathbf{K}_s$	effective surface current in superconductors/ skin effect	109, 126
L, L_{ab}	inductance, mutual inductance	31, 116
L	Lagrangian	Chap. 12
$\mathbf{L}$	angular momentum	37, 47, 66
$\mathbf{L}_{\mathrm{op}}$, $L_{\pm}$, L_z	angular momentum operators	Appendix A
$\mathcal{L}$	angular momentum density	37
m	mass of a particle, often electron	
$\mathbf{m}$, $\mathcal{M}$	magnetic dipole moment	21, 75
$\mathbf{M}$	magnetization	83, 100
M_s	spontaneous magnetization of ferromagnets	104
M_{ij}	angular momentum flux	37
$\mathbf{M}_{\ell m}$	magnetic multipole	27
$\mathcal{M}_0$	magnetic dipole moment of the earth	Appendix G
n	number density of electrons/atoms/molecules	23, 82, 99
n	real part of refractive index	133
$\tilde{n}$	complex refractive index	133
n_b	density of bound electrons	123
n_f	density of free electrons	119, 123
n_{ij}	depolarization/demagnetization tensor	96

Symbol(s)	Meaning	Key Sections/First Mention
$\hat{\mathbf{n}}$	outward normal to surface, generic unit vector	7
	direction from radiation source to observer	Chap. 10
N	electron or molecular number density	23, 149
$\mathbf{N}$	torque	21, 75
p	pressure	139
$\mathbf{p}$	momentum of a particle	2, Chap. 11
	canonical momentum	77
$\mathbf{p}_{kin}$	kinetic momentum	77
p^{μ}	momentum four-vector	154
P	degree of polarization of light	45
P	power of radiation field	Chap. 10
P^{μ}	total four-momentum of composite system	165
$P_{\ell}(z)$	Legendre polynomials	Appendix A
$\mathbf{P}$	electric polarization	82, 94
$\mathbf{P_k}$	conjugate momentum to $\mathbf{A_k}$	78
$\mathcal{P}$	induced dipole moment of body	89
$dP/d\Omega$	angular power distribution	58
$d^2P/d\omega d\Omega$	frequency and angular power distribution	58
q, Q	charge	12
$q, \mathbf{q}$	wave vector, spatial FT variable	Chap. 7
Q	quality factor of cavity	144
$q_{\ell m}$	electrostatic 2^{ℓ}-pole moment	19
$\dot{Q}$	rate of heat evolution	121, 122
$r_{\perp}$	radial distance in cylindrical coordinates	24
r_c	cyclotron radius	70
r_e	classical electron radius	18, 146
$\mathbf{r}_{12}$	vector from point 2 to point 1	21

Symbol(s)	Meaning	Key Sections/First Mention
$\mathbf{r}_a$, $\mathbf{R}_a$	apparent position of accelerating charge	56
R, R_s	resistance, sheet resistance	110, 112, 116
R_{TE}, R_{TM}	reflectivity of surface in TE, TM modes	135–136, 141
s, Δs	interval in special relativity	152
S	action	Chap. 12
S	entropy	97, 101
$\mathbf{S}$, S	Poynting vector, magnitude thereof	
	in vacuum	35
	in material medium	85, 122
S^{α}	spin four-vector	165
t_r, t_e	retarded time, time of emission	56, 170
Δt_{obs}	elapsed time in inertial observer's frame	170
T	temperature	97, 101
T_{ij}	stress tensor	36
$T^{\mu\nu}$	stress–energy tensor	163
u	energy density of EM wave	41
u^{μ}	velocity four-vector	154
U	energy,	
	electrostatic	14, 88
	magnetostatic	21, 29, 31
	in general electromagnetic field	35
U, $\tilde{U}$	thermodynamic internal energy	97, 101
U^{μ}	four-velocity of center of inertia	165
$\mathbf{v}$	velocity of a particle	21, Chap. 11
v_d	drift/guiding center velocity	71, 73
v_g, v_p, v_s	group, phase, and signal velocities	134
v_F	Fermi velocity	139

Symbol(s)	Meaning	Key Sections/First Mention
$\mathbf{v}_a$	apparent velocity of accelerating charge	56
$\mathbf{v}_r$	retarded velocity of accelerating charge	56
V	voltage	110, 116
V_{AB}	contact potential	93
W	work function	93
W	work received by a system	97, 101
$W(\omega)$	power spectrum	51, Appendix F
x^μ	four-vector of position coordinates	153
$Y_{\ell m}$	spherical harmonics	Appendix A
Y_n	Bessel functions of second kind	Appendix B
Z	total number of electrons per atom	123, 147
Z_0	impedance of the vacuum, 376.7 Ω (SI)	61
Z_s	effective surface impedance (skin effect)	126
$Z(\omega)$	impedance of circuit	116
$\mathbf{Z}(\mathbf{r}, t), \mathbf{Z}_\omega$	displacement field at high frequency	120
α	power attenuation coefficient in waveguides	143
α, α_{ij}	polarizability, tensor of	89, 99, 149
$\boldsymbol{\beta}, \beta$	dimensionless velocity, $\mathbf{v}/c$, v/c	56
β	extinction or absorption coefficient	149
γ	$(1 - v^2/c^2)^{-1/2}$, Lorentz or time dilation factor	152
γ	gyromagnetic ratio	75
γ	damping rate in Lorentz model	123, 147
δ	skin depth	126
$\delta_{ij}, \delta(x)$	Kronecker, Dirac delta function	5, 9
∇^2_Ω	angular part of Laplacian	6
ϵ, ϵ_{ij}	static dielectric constant, tensor of	94
ϵ_0	permittivity of the vacuum (SI)	12

Symbol(s)	Meaning	Key Sections/First Mention
ϵ_{ijk}, $\epsilon_{\alpha\beta\gamma\delta}$	Levi-Civita symbol/tensor	5, 155
$\epsilon(\omega)$	frequency-dependent dielectric constant	119
$\epsilon_b(\omega)$	bound electron contribution to $\epsilon(\omega)$	120, 123
$\epsilon'(\omega)$, $\epsilon''(\omega)$	real and imaginary parts of $\epsilon(\omega)$	121, 122
ζ	inverse of pulse compression factor	56
$\zeta(\omega)$	electric propensity or full dielectric function	120
κ	complex skin effect wave vector	126
κ	imaginary part of refractive index	133
κ, κ_T	compressibility, isothermal	139
λ	wavelength of light	48
$\Lambda^{\mu}{}_{\nu}$	Lorentz transformation matrix	153
μ_0	magnetic permeability of the vacuum (SI)	4, 21
μ	magnetic permeability of medium	84, 100
μ	(electro)chemical potential	93
$\mu(\omega)$	frequency-dependent susceptibility	122
$\mu'(\omega)$, $\mu''(\omega)$	real and imaginary parts of $\mu(\omega)$	122
μ_B	Bohr magneton	103
ρ	charge density,	
	in vacuum	12
	macroscopic, in material medium	82
ρ	resistivity	110
ρ_n	number density of molecules	149, 151
ρ_{pol}, ρ_b	polarization or bound charge density	82
ρ_{free}	free (mobile) charge density	82, 94
ρ_{ij}	polarization tensor	45
σ	surface charge density	86
σ	static conductivity	110

Symbol(s)	Meaning	Key Sections/First Mention
σ	scattering cross section	145
σ_{free}	free surface charge density	82
σ_{pol}	polarization surface charge density	82
$\sigma(\omega)$	ac or frequency-dependent conductivity	119
$\sigma'(\omega), \sigma''(\omega)$	real and imaginary parts of $\sigma(\omega)$	121
Σ_s	surface charge density	111, 125
$d\sigma/d\Omega$	differential scattering cross section	Chap. 22
τ	collision time in conductors	119
τ	proper time in relativity	152
τ_d	quasistatic diffusion time	125
τ_e	classical electron time	18, 67
ϕ	potential, electrostatic and scalar	13, 33
φ	azimuthal angular coordinate	
Φ	magnetic flux	25, 28, 31
χ_e	electric susceptibility	94
χ_m	magnetic susceptibility	100
ψ	eikonal function	43
ψ	scalar wave field	Chap. 8
$\boldsymbol{\omega}, \omega$	angular velocity	75
ω	frequency (angular), temporal FT variable	Chap. 7
ω_c	cyclotron frequency	70
ω_c	characteristic frequency of synchrotron radiation	171, 173
ω_p	plasma frequency	138
ω_L	Larmor frequency	75
$\Omega, d\Omega$	solid angle, element thereof	19

Suggestions for using this book

As stated in the preface, there is more material in the text than can be covered comfortably in a one-year course, but I believe that all essential topics are treated, so there should be enough to accommodate many different instructors' tastes. Further, not everyone will agree with my order of presentation of topics. I have already commented in the preface on what I think the essential topics are. However, I have also added two further aids to navigation. First, sections that present advanced applications or topics are identified by an asterisk. These sections can be skipped in a first reading or in a course where time is short. Second, in the table below, I show what the prerequisites for each chapter are. Naturally, if a prerequisite chapter itself has prerequisites, then the earlier prerequisites must also be studied. A chapter of which only a small part is required or that is not required in an essential way is given in parentheses. Where the prerequisite is given as "UG," it means that a familiarity with the topic at an undergraduate level is necessary.

Readers will note that chapter 2 is not listed as a prerequisite for any of the later chapters for the simple reason that it (or rather the mathematical content that is summarized there) is essential for *all* of them. I strongly urge readers, students, and instructors alike not to skip this chapter. It will help the students strengthen their mathematical technique, so that the math is not in the way when they need to focus on the physics. Equally importantly, it aids in separating the mathematical ideas from the physical ones and helps the student to see which are which. I particularly recommend investing the time on Fourier methods, as the book makes heavy use of them, and not only are they ubiquitous in all other areas of physics, but they always lead to deeper physical insight.

	Chapter	Prerequisites
1.	Survey	None/UG
2.	Math review	UG
3.	Vacuum electrostatics	UG
4.	Vacuum magnetostatics	UG, 3
5.	Induced fields	4
6.	Symmetries	1 (3–5)
7.	EM waves	1 (6)
8.	Interference	UG (7)
9.	Retarded fields	5 (6, 7)
10.	Radiation	9
11.	Particle trajectories	1 (4, 5)
12.	Action formalism	1, 7 (4–6)
13.	Material media	1 (3–6)
14.	Electrostatics with conductors	3, 13
15.	Dielectrics	13 (3)
16.	Magnetostatics in matter	4, 13 (14, 15)
17.	Ohm's law	13 (3, 7)
18.	Response functions	13 (6, 15–17)
19.	Quasistatic fields	13, 14 (3–5, 18)
20.	EM waves in insulators	18 (7, 13)
21.	EM waves in conductors	18–20
22.	Scattering	7, 10 (18)
23.	Formalism of relativity	UG
24.	Relativistic EM	5, 6, 12, 23 (3, 4, 7)
25.	Relativistic radiation	10, 23, 24

Classical Electromagnetism in a Nutshell

1 Introduction

1 The field concept

The central concept in the modern theory of electromagnetism is that of the electromagnetic *field*. The forces that electrical charges, currents, and magnets exert on each other were believed by early thinkers to be of the action-at-a-distance type, i.e., the forces acted instantaneously over arbitrarily large distances. Experiments have shown, however, that this is not true. A radio signal, for example, can be sent by moving electrons back and forth in a metallic antenna. This motion will cause electrons in a distant piece of metal to move back and forth in response—this is how the signal is picked up in a radio or cell phone receiver. We know that the electrons in the receiver cannot respond in a time less than that required by light to travel the distance between transmitter and receiver. Indeed, radio waves, or electromagnetic waves more generally, are a form of light.

Facts such as these have led us to abandon the notion of action at a distance. Instead, our present understanding is that electrical charges and currents produce physical entities called *fields*, which permeate the space around them and which in turn act on other charges and currents. When a charge moves, the fields that it creates change, but this change is not instantaneous at every point in space. For a complete description, one must introduce two *vector* fields, $\mathbf{E}(\mathbf{r}, t)$, and $\mathbf{B}(\mathbf{r}, t)$, which we will call the electric and magnetic fields, respectively. In other words, at every time t, and at every point in space $\mathbf{r}$, we picture the existence of two vectors, $\mathbf{E}$ and $\mathbf{B}$. This picture is highly abstract, and early physicists had great trouble in coming to grips with it. Because the fields did not describe particulate matter and could exist in vacuum, they seemed very intangible, and early physicists were reluctant to endow them with physical reality. The modern view is quite different. Not only do these fields allow us to describe the interaction of charges and currents with each other in the mathematically simplest and cleanest way, we now believe them to be absolutely real physical entities, as real as a rhinoceros. Light is believed to be nothing but a jumble of wiggling $\mathbf{E}$ and $\mathbf{B}$ vectors everywhere, which implies that these fields can exist independently of charges and currents. Secondly, these fields carry such concrete physical properties as energy, momentum, and angular momentum. When one gets to a quantum mechanical description, these three attributes become properties of a particle called the *photon*, a quantum of light. At sufficiently high energies, two of these particles can spontaneously change into an electron and a positron, in a process called *pair production*. Thus, there is no longer any reason for regarding the **E** and **B** fields as adjuncts, or aids to understanding, or to picture the interactions of charges through lines of force or flux. Indeed, it is the latter concepts that are now regarded as secondary, and the fields as primary.

The impossibility of action at a distance is codified into the modern theory of relativity. The principle of relativity as enunciated by Galileo states that the laws of physics are identical in all inertial reference frames.[1] One goes from Galilean relativity to the modern theory by recognizing that there is a maximum speed at which physical influences or signals may propagate, and since this is a law of physics, the maximum speed must then be the same in all inertial frames.[2] This speed immediately acquires the status of a fundamental constant of nature and is none other than the speed of light in vacuum. Needless to say, this law, and the many dramatic conclusions that follow from considering it in conjunction with the principle of relativity, are amply verified by experiment.

The application of the principle of relativity also leads us to discover that **E** and **B** are two aspects of the same thing. A static set of charges creates a time-independent electric field, and a steady current creates a time-independent magnetic field. Since a current can be regarded as a charge distribution in motion, it follows that **E** and **B** will, in general, transform into one another when we change reference frames. In fact, the relativistic invariance of the laws of electrodynamics is best expressed in terms of a single tensor field, generally denoted F. The fields **E** and **B** are obtained as different components of F. At low speeds, however, these two different components have so many dissimilar aspects that greater physical understanding is obtained by thinking of them as separate vector fields. This is what we shall do in this book.

2 The equations of electrodynamics

The full range of electromagnetic phenomena is very wide and can be very complicated. It is somewhat remarkable that it can be captured in a small number of equations of

---

[1] That such frames exist is a matter of physical experience, and actual frames can be made to approximate an ideal inertial reference frame as closely as we wish.

[2] Einstein took the frame invariance of the speed of light as a postulate in addition to the principle of relativity. It was recognized fairly soon after, however, that this postulate was not strictly necessary: the relativity principle alone was enough to show that the most general form of the velocity addition law was that derived by Einstein, with some undetermined but finite limiting speed that any object could attain. That this speed is that of light is, then, a wonderful fact, but not of essential importance to the theory. Some works that explore this issue are W. V. Ignatowsky, *Arch. Math. Phys.* **17**, 1 (1911); **18**, 17 (1911); V. Mitavalsky, *Am. J. Phys.* **34**, 825 (1966); Y. P. Terletskii (1968); A. R. Lee and T. M. Kalotas, *Am. J. Phys.* **43**, 434 (1975); N. D. Mermin, *Am. J. Phys.* **52**, 119 (1984); A. Sen, *Am. J. Phys.* **62**, 157 (1994).

relatively simple form:

Law	Equation (Gaussian)	Equation (SI)
Gauss's law	$\nabla \cdot \mathbf{E} = 4\pi\rho$	$\nabla \cdot \mathbf{E} = \dfrac{\rho}{\epsilon_0}$
Ampere-Maxwell law	$\nabla \times \mathbf{B} = \dfrac{4\pi}{c}\mathbf{j} + \dfrac{1}{c}\dfrac{\partial \mathbf{E}}{\partial t}$	$\nabla \times \mathbf{B} = \mu_0\mathbf{j} + \mu_0\epsilon_0\dfrac{\partial \mathbf{E}}{\partial t}$
Faraday's law	$\nabla \times \mathbf{E} + \dfrac{1}{c}\dfrac{\partial \mathbf{B}}{\partial t} = 0$	$\nabla \times \mathbf{E} + \dfrac{\partial \mathbf{B}}{\partial t} = 0$
No magnetic monopoles	$\nabla \cdot \mathbf{B} = 0$	$\nabla \cdot \mathbf{B} = 0$
Lorentz force law	$\mathbf{F} = q\left(\mathbf{E} + \dfrac{1}{c}\mathbf{v} \times \mathbf{B}\right)$	$\mathbf{F} = q(\mathbf{E} + \mathbf{v} \times \mathbf{B})$

$$(2.1)$$

These laws are confirmed by extensive experience and the demands of consistency with general principles of symmetry and relativistic invariance, although their full content can be appreciated only after detailed study. We have written them in the two most widespread systems of units in use today and given them the names commonly used in the Western literature. The first four equations are also collectively known as the *Maxwell equations*, after James Clerk Maxwell, who discovered the last term on the right-hand side of the Ampere-Maxwell law in 1865 and thereby synthesized the, till then, separate subjects of electricity and magnetism into one.[3]

We assume that readers have at least some familiarity with these laws and are aware of some of their more basic consequences. A brief survey is still useful, however. We begin by discussing the symbols. The parameter c is the speed of light, and ϵ_0 and μ_0 are constant scale factors or conversion factors used in the SI system. The quantity ρ is a scalar field $\rho(\mathbf{r}, t)$, denoting the *charge distribution* or density. Likewise, $\mathbf{j}$ is a vector field $\mathbf{j}(\mathbf{r}, t)$, denoting the *current distribution*. This means that the total charge inside any closed region of space is the integral of $\rho(\mathbf{r}, t)$ over that space, and the current flowing across any surface is the integral of the normal component of $\mathbf{j}(\mathbf{r}, t)$ over the surface. This may seem a roundabout way of specifying the position and velocity of all the charges, which we know, after all, to be made of discrete objects such as electrons and protons.[4] But, it is in these terms that the equations for $\mathbf{E}$ and $\mathbf{B}$ are simplest. Further, in most macroscopic

---

[3] Although modern practice attaches the names of particular scientists to these laws, it should be remembered that they distill the collective efforts of several hundred individuals over the eighteenth and nineteenth centuries, if not more. A survey of the history may be found in E. M. Whittaker (1951). For a more modern history covering a more limited period, see O. Darrigol (2000).

[4] In fact, in dealing with discrete point charges, or idealized current loops of zero thickness, the distributions $\rho(\mathbf{r}, t)$ and $\mathbf{j}(\mathbf{r}, t)$ must be given in terms of the Dirac delta function. A certain amount of mathematical quick-stepping is then necessary, which we shall learn how to do in chapter 2.

situations, one does not know where each charge is and how fast it is moving, so that, at least in such situations, this description is the more natural one anyway.

The four Maxwell equations allow one to find $\mathbf{E}$ and $\mathbf{B}$ if ρ and $\mathbf{j}$ are known. For this reason, the terms involving ρ and $\mathbf{j}$ are sometimes known as *source terms*, and the $\mathbf{E}$ and $\mathbf{B}$ fields are said to be "due to" the charges and currents. However, we began by talking of the forces exerted by charges on one another, and of this there is no mention in the Maxwell equations. This deficiency is filled by the last law in our table—the Lorentz force law—which gives the rule for how the fields acts on charges. According to this law, the force on a particle with charge q at a point $\mathbf{r}$ and moving with a velocity $\mathbf{v}$ depends only on the instantaneous value of the fields at the point $\mathbf{r}$, which makes it a local law. Along with Newton's second law,

$$\frac{d\mathbf{p}}{dt} = \mathbf{F}, \tag{2.2}$$

equating force to the rate of change of momentum,[5] it allows us to calculate, in principle, the complete motion of the charges.

Let us now discuss some of the more salient features of the equations written above. First, the Maxwell equations are linear in $\mathbf{E}$ and $\mathbf{B}$, and in ρ and $\mathbf{j}$. This leads immediately to the *superposition principle*. If one set of charges and currents produces fields $\mathbf{E}_1$ and $\mathbf{B}_1$, and another set produces fields $\mathbf{E}_2$ and $\mathbf{B}_2$, then if both sets of charges and currents are simultaneously present, the fields produced will be given by $\mathbf{E}_1 + \mathbf{E}_2$, and $\mathbf{B}_1 + \mathbf{B}_2$. This fact enables one to simplify the calculation of the fields in many circumstances. In principle, one need only know the fields produced by a single moving charge, and the fields due to any distribution may be obtained by addition. In practice, the problem of addition is often not easy, and one is better off trying to solve the differential equations directly.[6] A large part of electromagnetic theory is devoted to developing the classical mathematical machinery for this purpose. This includes the theorems named after Gauss, Stokes, and Green, and Fourier analysis and expansions in complete sets of orthogonal functions. With modern-day computers, direct numerical solution is the method of choice in many cases, but a sound grasp of the analytic techniques and concepts is essential if one is to make efficient use of computational resources.

The second point is that the equations respect the symmetries of nature. We discuss these in considerably greater detail in chapter 6, and here we only list the symmetries. The first of these is invariance with respect to space and time translations, i.e., the equivalence of two frames with different origins or zeros of time. As in mechanics, this symmetry is connected with the conservation of momentum and energy. The fact that it holds

---

[5] In the form (2.2) the equation remains relativistically correct. This is not so if we write $\mathbf{F} = m\mathbf{a}$, with m and $\mathbf{a}$ being the mass and the acceleration, respectively. The reason is that for particles with speeds close to c, $\mathbf{p} \neq m\mathbf{v}$.

[6] Supplemented, one might add, by boundary conditions. Note though, that not all boundary conditions that lead to a well-posed mathematical problem are physically sensible. The physically acceptable boundary conditions are that in static problems, the fields die off at infinity, and in dynamic problems, they represent outgoing solutions, i.e., that there be no flow of energy from infinity into the region of interest, unless such irradiation is specifically known to be present.

for Maxwell's equations automatically leads us to assign energy and momentum to the electromagnetic field itself. The second symmetry is rotational invariance, or the isotropy of space. That this holds can be seen directly from the vector nature of **E** and **B**, and the properties of the divergence and curl. It is connected with the conservation of angular momentum.[7] The third symmetry is spatial inversion, or parity, which in conjunction with rotations is the same as mirror symmetry.[8] We shall find that under inversion, **E** → −**E**, in the same way that a "normal" vector like the velocity **v** behaves, but **B** → **B**. One therefore says that **E** is a *polar vector*, or just a vector, while **B** is a *pseudovector* or *axial vector*. The fourth symmetry is time reversal, or what might be better called motion reversal. This is the symmetry that says that if one could make a motion picture of the world and run it backward, one would not be able to tell that it was running backward.[9] The fifth symmetry is the already mentioned equivalence of reference frames, also known as *relativistic invariance* or *Lorentz invariance*.[10] This symmetry is extremely special and, in contrast to the first three, is the essential way in which electromagnetism differs from Newtonian or pre-Einsteinian classical mechanics. We shall devote chapter 23 to its study. Historically, electromagnetism laid the seed for modern (Einsteinian) relativity. The problem was that the Maxwell equations are not Galilean invariant. This fact is mostly clearly seen by noting that light propagation, which is a consequence of the Maxwell equations, is described by a wave equation of the form

$$\nabla^2 f - \frac{1}{c^2} \frac{\partial^2 f}{\partial t^2} = 0. \tag{2.3}$$

Here, f stands for any Cartesian component of **E** or **B**. As is well known, classical wave phenomena are *not* Galilean invariant. Sound, e.g., requires a material medium for its propagation, and the frame in which this medium is at rest is clearly special. The lack of Galilean invariance of Maxwell's equations was well known to physicists around the year 1900, but experimental support for the most commonly proposed cure, namely, that there was a special frame for light as well, and a special medium (the *ether*) filling empty space, through which light traveled, failed to materialize. Finally, in 1905, Einstein saw that Galilean invariance itself had to be given up. Although rooted in electromagnetism, this proposal has far-reaching consequences for all branches of physics. In mechanics, we mention the nonabsolute nature of time, the equivalence of mass and energy, and

[7] The connection between space translation invariance and the conservation of momentum, time translation invariance and the conservation of energy, and rotational invariance and the conservation of angular momentum is a general consequence of Noether's theorem, which states that any continuous symmetry leads to a conservation law and also gives the form of the conserved quantity. Noether's theorem is proved in almost all texts on mechanics. See, e.g., Jose and Saletan (1998), secs. 3.2.2., 9.2.

[8] The weak interactions do not respect this symmetry, but they lie outside the realm of classical physics. The same comment applies to time reversal.

[9] Anyone who has seen the Charlie Chaplin gag where he rises from his bed, stiff as a corpse, while his heels stay glued in one spot, will disagree with this statement. In fact, the laws of physics possess only *microscopic* reversibility. How one goes from this to *macroscopic* irreversibility and the second law of thermodynamics is a profound problem in statistical mechanics, and continues to be a matter of debate.

[10] This term is sometimes expanded to include the previous four symmetries also, and one then speaks of *full* or *general* Lorentz invariance.

the impossibility of the existence of rigid bodies and elementary particles with finite dimensions. Today, relativity is not regarded as a theory of a particular phenomenon but as a framework into which all of physics must fit. Much of particle physics in the twentieth century can be seen as an outcome of this idea in conjunction with quantum mechanics.

Another feature of the Maxwell equations that may be described as a symmetry is that they imply charge conservation. If we add the time derivative of the first equation, Gauss's law, to the divergence of the second, the Ampere-Maxwell law, we obtain the continuity equation for charge,

$$\frac{\partial \rho}{\partial t} = -\nabla \cdot \mathbf{j}. \tag{2.4}$$

If we integrate this equation over any closed region of space, and any finite interval of time, the left-hand side gives the net increase in charge inside the region, while, by Gauss's theorem, the right-hand side gives the inflow of charge through the surface bounding the region. Thus, eq. (2.4) states that charge is *locally* conserved. This conservation law is intimately connected with a symmetry known as *gauge invariance*. We shall say more about this in chapter 12.

The last symmetry to be discussed is a certain duality between $\mathbf{E}$ and $\mathbf{B}$. Let us consider the second and third Maxwell equations and temporarily ignore the current source term. The equations would then transform into one another under the replacements $\mathbf{E} \to \mathbf{B}$, $\mathbf{B} \to -\mathbf{E}$. The same is true of the remaining pair of equations if the charge source term is ignored. This makes it natural to ask whether we should not modify the equations for $\nabla \cdot \mathbf{B}$ and $\nabla \times \mathbf{E}$ to include magnetic charge and current densities ρ_m and $\mathbf{j}_m$, in other words, to write (in the Gaussian system),

$$\nabla \times \mathbf{E} = -\frac{1}{c} \frac{\partial \mathbf{B}}{\partial t} + \frac{4\pi}{c} \mathbf{j}_m, \tag{2.5}$$

$$\nabla \cdot \mathbf{B} = 4\pi \rho_m. \tag{2.6}$$

All the existing experimental evidence to date, however, indicates that free magnetic charges or monopoles do not exist.[11]

In the same connection, we should note that there is another source of magnetic field besides currents caused by charges in motion. All the charged elementary particles, the electron (and the other leptons, the muon and the taon) and the quarks, possess an intrinsic or spin magnetic moment. This moment cannot be understood as arising from a classical spinning charged object, however. The question then arises whether we should

---

[11] For extremely precise-minded readers, we should note that there is a certain convention implicit in the making of this statement. By adopting a larger set of duality transformations, one could, in fact, modify Maxwell's equations as per eqs. (2.5) and (2.6). Instead of asserting the absence of magnetic monopoles, one would then say that the ratio ρ_m/ρ_e (ρ_e being the electric charge) was the same for all known particles. There is little to be gained from this point of view, however, and it is simpler to pick a fixed representation for $\mathbf{E}$ and $\mathbf{B}$ and write Maxwell's equations in the usual form. See Jackson (1999), sec. 6.12, for more on this point.

not add a source term to the equation for $\nabla \cdot \mathbf{B}$ to take account of this magnetic moment. If we are interested only in describing the field classically, however, we can do equally well by thinking of these moments as idealized current loops of zero spatial extent and including this current in the source term proportional to $\mathbf{j}$ in the Ampere-Maxwell law. The integral of the divergence of this current over any finite volume is always zero, so the equation of continuity is unaffected, and we need never think of the charge distribution carried by these loops separately. In fact, the alternative of putting all or some of the source terms into the equation for $\nabla \cdot \mathbf{B}$ is not an option, for it leads to unacceptable properties for the vector potential. We discuss this point further in section 26.

3 A lightspeed survey of electromagnetic phenomena

Having surveyed the essential properties of the equations of electrodynamics, let us now mention some of the most prominent phenomena implied by them. First, let us consider a set of static charges. This is the subject of electrostatics. Then $\mathbf{j} = 0$, and $\rho(\mathbf{r})$ is time independent. The simplest solution is then to take $\mathbf{B} = 0$, and the E-field, which is also time independent, is given by Gauss's law. In particular, we can find $\mathbf{E}(\mathbf{r})$ for a point charge, and then, in combination with the Lorentz force law, we obtain Coulomb's force law—namely, that the force between two charges is proportional to the product of the charges, to the inverse square of their separation, and acts along the line joining the charges. We study electrostatics further in chapter 3.

Similarly, suppose we have a time-independent current density $\mathbf{j}(\mathbf{r})$, and $\rho = 0$. (The current density must be divergenceless to have a well-posed problem, for otherwise the equation of continuity would be violated.) This makes up the subject of magnetostatics. The simplest solution now is $\mathbf{E} = 0$, and a time-independent $\mathbf{B}$, which is given by the Ampere-Maxwell equation (now known as just *Ampere's law*) and the equation $\nabla \cdot \mathbf{B} = 0$. There is now no analog of Coulomb's law, but several simple setups can be considered. One can, for example, calculate the $\mathbf{B}$ field produced by a straight infinite current-carrying wire. A second wire parallel to the first will experience a force which is given by the Lorentz force law. The force per unit length on any wire is proportional to the product of the currents, is inversely proportional to the distance between the wires, and lies in the plane of the wires, perpendicular to the wires themselves. This relationship is the basis of the definition of the unit of current in the SI system, the ampere. We study magnetostatics in detail in chapter 4.

The simplest time-dependent phenomena are described by Faraday's law. This law says that a changing magnetic field, which could be created in several ways—a time-dependent current $\mathbf{j}(\mathbf{r}, t)$, or a moving magnet—produces an electric field. If a metallic wire loop is placed in the region of the electric field, a current and an emf will be induced in the loop. This phenomenon, known as *induction*, is the basis of transformers, generators, and motors, and therefore of the unfathomable technological revolution wrought by these devices. We study this in chapter 5. A related phenomenon is seen when a wire loop, or, more generally, any extended conductor, moves in a static magnetic field. The induced electric field can then drive currents through the conductor. This effect is exploited in dynamos and is believed to lie behind the earth's magnetic field, as we shall see in section 131.

The term $\partial \mathbf{E}/\partial t$ in the Ampere-Maxwell law is needed to make the equations consistent with charge conservation. Its greatest consequence, however, is seen by considering the equations in the absence of any currents or charges. If we take the curl of the Faraday equation, for example, and use the Ampere-Maxwell and Coulomb's laws, we obtain

$$\nabla^2 \mathbf{E} - \frac{1}{c^2}\frac{\partial^2 \mathbf{E}}{\partial t^2} = 0. \tag{3.1}$$

The same equation is obtained for $\mathbf{B}$ if we take the curl of the equation for $\nabla \times \mathbf{B}$. These two equations have nonzero solutions that are consistent with the first-order equations coupling together $\mathbf{E}$ and $\mathbf{B}$, and with $\nabla \cdot \mathbf{E} = \nabla \cdot \mathbf{B} = 0$. These solutions describe electromagnetic waves or light, and we study them in chapter 7, except for certain observer-dependent properties, such as the Doppler effect, which are covered in chapter 24. We have already commented on the implications of the existence of these solutions for the reality of the electromagnetic field.

Maxwell's equations also describe the production of electromagnetic waves via the phenomenon of radiation. We shall see this in chapters 9 and 10, when we consider the fields produced by moving charges. We shall see that an accelerating charge emits fields that die away only inversely with distance from the charge at large distances and that locally look like plane electromagnetic waves everywhere. These radiated fields carry energy and momentum. This phenomenon underlies radio, TV, cell phones, and all other wireless communications. If the charges are moving at speeds close to that of light, the properties of the radiation change dramatically. This is illustrated by the phenomenon of synchrotron radiation, which we study in chapter 25 after we have discussed special relativity in chapters 23 and 24.

The interaction of radiation or light with matter opens a whole new set of phenomena, which can be divided into large subclasses. First, when the matter is microscopic— individual charges, atoms, and molecules—interest attaches to scattering, i.e., the acceleration of the charges by the incident radiation, and reradiation of an electromagnetic field due to this acceleration. One now obtains the phenomena of Compton scattering, atomic and molecular spectra, etc. A proper treatment of these must be quantum mechanical. Nevertheless, much can be learned even in a classical approach, and we do this in chapter 22 using phenomenological models of atoms and molecules. Second, when the matter is in the form of a bulk medium, the most striking fact is that at certain wavelengths, light can propagate through matter, e.g., visible light goes through window glass. How this happens is examined in chapter 20. We also examine the attendant phenomena of reflection and refraction at interfaces between different media. If the medium is inhomogeneous, then, in addition to propagation, one also gets some scattering of the light. We see this phenomenon every day in the sky, and it also occurs when the medium is denser, such as a liquid. These topics are also discussed in chapter 22. Third, when the matter is in the form of opaque obstacles, large on the scale of the wavelength, application of the superposition principle to light fields leads to distinctive phenomena known as interference and diffraction. We take these up in chapter 8.

Next, let us turn to the behavior of charges in external fields. This is described by the Lorentz force equation. A large variety of motions is obtained, especially in inhomogeneous magnetic fields. We discuss these in chapter 11. Motion of charges in the earth's magnetic field is discussed in appendix G. The motion of magnetic moments in a magnetic field is also discussed in chapter 11.

We have already touched on the phenomena encountered when light interacts with bulk matter, without indicating how these are to be understood. For that, one must first tackle the larger problem of describing electromagnetic fields in matter more generally, not just for radiation fields. This is a very complex problem, as evidenced by the huge variety in the types of matter: conductors, insulators, magnets, and so on. Indeed, matter is itself held together largely by electromagnetic forces, and much of the distinction between the broadly different types of matter we have mentioned above is based on the response of these types to electromagnetic fields. Thus, it would seem that one first needs to develop a theory of matter, so that one may understand how some materials can be, say, conductors, and other materials insulators. Fortunately, one can make substantial progress by relying on intuitive and simplified notions of these terms. The key property that helps us is that matter is neutral on a very short distance scale, essentially a few atomic spacings. Thus, coarse-grained or *macroscopic* electromagnetic fields may be defined by spatially averaging over this length scale. These fields obey equations that resemble those for the fields in vacuum. The resemblance is only skin deep, however. The response of the medium cannot be trivialized. It is modeled through so-called constitutive relations that differ from medium to medium. In conductors, e.g., we have Ohm's law, which says that an internal electric field is accompanied by a proportionally large transport current. In insulators (also known as *dielectrics*), it relates the polarization of the matter to the internal electric field. It is in these constitutive relations that the complexity of the material is buried. Finding them from "first principles" is the province of condensed matter physics and statistical mechanics, which we shall not enter. Instead, we will work with semiempirical and phenomenological constitutive laws. Essentially all phenomena can be understood in this way. The coarse-graining procedure is discussed in chapter 13, and Ohm's law and the related topics of emf and electrical circuits in chapter 17. A simple but widely applicable constitutive model for time-dependent phenomena in many materials is developed in chapter 18.

The simplest kinds of phenomena involving matter are static. Electrostatic fields in the presence of conductors and insulators (or dielectrics) are discussed in chapters 14 and 15. In the first case, the central phenomenon is the expulsion of the electrical field from the interior of the conductors and from any hollow cavity inside a conductor. In the second case, the field is not expelled entirely, but is reduced, and the concern shifts to understanding why and estimating the reduction.

In contrast to the electric case, the response of most materials to static magnetic fields is rather tame. Permanent or ferromagnets are a notable exception. Unfortunately, the most interesting phenomena that they exhibit, such as hysteresis, domain formation, etc., are rather difficult to analyze or even formulate, since the particulars are dominated by material properties and even the shape of the body because of long-range dipole–dipole interactions. Still, several general aspects can be studied, and we do so in chapter 16.

When matter is subjected to time-dependent fields, even more phenomena emerge. In conductors, when the frequency is low, one gets eddy currents, and the expulsion of electric fields that was perfect in the static case is only slightly weakened. We discuss this in chapter 19. When the frequency is high, we get plasma oscillations and waves, as discussed in chapter 21. The near-perfect reflectivity of metals is also discussed in this chapter, along with the waveguides and resonant cavities that this property makes possible.

Needless to say, in attempting to understand such a vast array of phenomena, one must develop and draw upon many general concepts. These include conservation laws, relativistic invariance, thermodynamics and statistical mechanics, causality, stochasticity, the action principle, and the Lagrangian and Hamiltonian formulations of mechanics. The discussion of these concepts is woven into the entire text intimately, for it is in this way that we see the relation between electromagnetism and other branches of physics. The two exceptions are the action formalism, to which we devote an entire chapter, chapter 12, and the formalism of special relativity, which is covered in chapter 23. We shall, by and large, stay away from quantum mechanics, although a knowledge of some elementary quantum mechanical ideas is presumed in a few places.

4 SI versus Gaussian

Two common systems of units and dimensions are used today in electromagnetism. These are the Gaussian and the SI or rationalized MKSA systems. The Gaussian system is designed for use with the cgs (centimeter–gram–second) system of mechanical units, and the SI is designed for use with the MKS (meter–kilogram–second) system. Unfortunately, converting between Gaussian and SI is not as easy as converting between dynes and newtons, or even foot-pounds and newtons, as physical quantities do not even have the same engineering dimensions in the two systems. In the Gaussian system, $\mathbf{E}$ and $\mathbf{B}$ have the same dimensions, while in the SI system, $\mathbf{E}$ has dimensions of velocity times $\mathbf{B}$. This means that equations intended for use in the two systems do not have the same form, and one must convert not only amounts but also equations. For example, in the SI system, the factor $1/c$ does not appear in Faraday's law or the Lorentz force law. Additional differences are present in the relations for macroscopic fields $\mathbf{D}$ and $\mathbf{H}$ that arise in the discussion of material bodies. Further, the SI system entails two dimensional constants, ϵ_0 and μ_0, known as the *permittivity* and *permeability* of the vacuum, respectively. The net result is that converting between the two systems is almost invariably irritating, but there seems to be little that can be done to bring about a standardization. A prominent twentieth-century magnetician captures the frustration perfectly:

> Devotees of the Giorgi system will not be happy with my units; but I can assure them that the unhappiness that my system inflicts upon them will be no greater than the unhappiness that their system over the last thirty years has inflicted on me.[12]

In this book, we shall give the most important formulas in both systems, but intermediate calculational steps will be given in Gaussian only. From the point of view of physics and conceptual understanding, the Gaussian system is better. From the point of view of practical application, on the other hand, the SI system is better, as it gives currents in amperes, voltages in volts, etc. Thus, a hard insistence on using one system or the other gets one nowhere, and it is necessary for everyone who works with electricity and magnetism to understand *both* systems and to have an efficient and reliable method of going back and forth between them. Conversion tables that achieve this end can be found in almost all textbooks.[13] We too give such tables (see tables 1.1 and 1.2, pages 16 and 17). However, we also show how to derive these conversion factors. For now, we limit ourselves to the basic quantities **E**, **B**, charge, etc. Relationships for the macroscopic fields **D**, **H**, etc., and related quantities are discussed in chapter 13.

The scheme given here requires knowing (i) that the symbols for all mechanical quantities—mass, length, time, force, energy, power, etc.—are the same in the two systems, and (ii) formulas for three mechanical quantities in both systems. Other choices for this set of three are possible, but the one that we find most easy to remember is tabulated below.

Quantity	Formula (Gaussian)	Formula (SI)	
Coulomb force	$\dfrac{q^2}{r^2}$	$\dfrac{q^2}{4\pi\epsilon_0 r^2}$	(4.1)
Energy density	$\dfrac{1}{8\pi}(\mathbf{E}^2 + \mathbf{B}^2)$	$\dfrac{1}{2}(\epsilon_0\mathbf{E}^2 + \mu_0^{-1}\mathbf{B}^2)$	
Lorentz force	$q\left(\mathbf{E} + \dfrac{1}{c}\mathbf{v}\times\mathbf{B}\right)$	$q(\mathbf{E} + \mathbf{v}\times\mathbf{B})$	

The first formula is for the Coulomb force between two equal charges q separated by a distance r. Since the symbols for force and distance are the same, it follows that

$$q_{\mathrm{SI}} = \sqrt{4\pi\epsilon_0}\, q_{\mathrm{Gau}}, \tag{4.2}$$

---

[12] Brown (1966). By the Giorgi system, Brown means SI. He himself uses a mixed system he calls Gaussian mks, which allows for conversion between SI and Gaussian at the expense of introducing a multiplicative factor in Coulomb's law whose value is different depending on the unit system, and replacement rules for the current and emf.

[13] See, e.g., Jackson (1999), appendix, or Pugh and Pugh (1970), chap. 1.

where the suffix "Gau" is short for Gaussian. The same conversion applies to charge density ρ, current I, and current density $\mathbf{j}$. (Recall that current is the amount of charge flowing through a surface per unit time.)

The second formula is for the energy density in the electromagnetic field. Since we can vary $\mathbf{E}$ and $\mathbf{B}$ independently, this is a "twofer"—it gives us two conversions for the price of one. Since energy and volume are the same in the two systems, we see that

$$\mathbf{E}_{SI} = \frac{\mathbf{E}_{Gau}}{\sqrt{4\pi\epsilon_0}}, \tag{4.3}$$

$$\mathbf{B}_{SI} = \sqrt{\frac{\mu_0}{4\pi}} \mathbf{B}_{Gau}. \tag{4.4}$$

The Lorentz force formula is the third one. It too is a "twofer." Consider a situation in which there is only an electric field. Since the symbol for force is unchanged in going from one system to the other, so must be the product qE:

$$(q\mathbf{E})_{SI} = (q\mathbf{E})_{Gau}. \tag{4.5}$$

But, this is exactly what we get from eqs. (4.2) and (4.3), so we already knew this. Something new is learned when we apply the same reasoning to the magnetic field term. We get

$$(q\mathbf{B})_{SI} = \frac{1}{c}(q\mathbf{B})_{Gau}. \tag{4.6}$$

Changing q and B to the Gaussian system using eqs. (4.2) and (4.4), we get

$$\sqrt{\epsilon_0\mu_0} = \frac{1}{c}. \tag{4.7}$$

These relationships are enough to convert any formula in the SI system to the Gaussian, or vice versa. Take, for example, the magnetic field of an infinite current-carrying wire. In the Gaussian system this is given by the formula

$$B = \frac{2I}{cr_\perp}, \tag{4.8}$$

where $r_\perp$ is the distance from the wire to the point where the field is desired. To get the SI formula, we replace B by $(4\pi/\mu_0)^{1/2} B$ and I by $(4\pi\epsilon_0)^{-1/2} I$. This yields

$$B = \frac{1}{4\pi}\sqrt{\frac{\mu_0}{\epsilon_0}}\frac{2I}{cr_\perp}$$

$$= \frac{\mu_0 I}{2\pi r_\perp}, \tag{4.9}$$

where we have used eq. (4.7) to eliminate c.

As another example, let us take the formula for the power radiated by an electric dipole oscillator. In the SI system, this is

$$P = \frac{c^2 Z_0 k^4}{12\pi}|\mathbf{d}|^2; \quad Z_0 = \sqrt{\frac{\mu_0}{\epsilon_0}}. \tag{4.10}$$

Here k is the wave number of the radiation, and $\mathbf{d}$ is the dipole moment. Since the dipole moment for a charge distribution is the volume integral of $\mathbf{r}\rho(\mathbf{r})$, its conversion is the same as that for charge. The quantities P and k are evidently unchanged, so the Gaussian formula is

$$P = c^2 \sqrt{\frac{\mu_0}{\epsilon_0}} 4\pi\epsilon_0 \frac{k^4}{12\pi} |\mathbf{d}|^2 = \frac{ck^4}{3} |\mathbf{d}|^2. \tag{4.11}$$

One check that this is correct is that it is free of ϵ_0 and μ_0.

As the third example, let us change the Ampere-Maxwell law from its SI to the Gaussian form. In the SI system, the law reads

$$\nabla \times \mathbf{B} = \mu_0 \mathbf{j} + \mu_0\epsilon_0 \frac{\partial \mathbf{E}}{\partial t}. \tag{4.12}$$

Using eqs. (4.2)–(4.4), we see that the Gaussian system form is

$$\sqrt{\frac{\mu_0}{4\pi}} \nabla \times \mathbf{B} = \mu_0 \sqrt{4\pi\epsilon_0} \mathbf{j} + \frac{\mu_0\epsilon_0}{\sqrt{4\pi\epsilon_0}} \frac{\partial \mathbf{E}}{\partial t}, \tag{4.13}$$

or, dividing by $\sqrt{\mu_0/4\pi}$ and using eq. (4.7),

$$\nabla \times \mathbf{B} = \frac{4\pi}{c} \mathbf{j} + \frac{1}{c} \frac{\partial \mathbf{E}}{\partial t}, \tag{4.14}$$

as given in the table on page 3. The reader should carry out the same exercise for the remaining Maxwell equations.

The rules for converting capacitance, inductance, and conductance, and related quantities such as resistance and impedance will be found later when these quantities are defined.

Finally, let us see how to carry out ordinary or "engineering" dimensional analysis. We will denote the dimensions of a quantity by putting square brackets around it: $[\mathbf{E}]$ will denote the dimensions of $\mathbf{E}$, and so on.

In the Gaussian system, all quantities have dimensions that can be expressed in terms of M, L, and T, the dimensions of mass, length, and time. However, these quantities often have to be raised to fractional exponents. Let us see how this happens, starting with charge. From the Coulomb force formula, we have $[q] = [F L^2]^{1/2}$, and since $[F] = MLT^{-2}$, we have

$$[q] = M^{1/2} L^{3/2} T^{-1}. \tag{4.15}$$

The dimensions of $\mathbf{E}$ now follow from a formula such as $E = q/r^2$ for the electric field magnitude due to a point charge. We get

$$[\mathbf{E}] = M^{1/2} L^{-1/2} T^{-1}. \tag{4.16}$$

As a check, we examine the dimensions of the product qE:

$$[qE] = MLT^{-2}, \tag{4.17}$$

which are the same as those of force, as they should be. Similarly, E^2 has dimensions of $ML^{-1}T^{-2}$, which are the same as those of energy density [Energy (ML^2T^{-2})/Volume (L^3)].

In the Gaussian system, the dimensions of **B** and **E** are the same. This can be seen either from the expression for the energy density, or the Lorentz force law. Thus,

$$[\mathbf{B}] = M^{1/2}L^{-1/2}T^{-1}. \tag{4.18}$$

Exercise 4.1 Obtain the dimensions of I, ρ, and **d** in the Gaussian system, and verify the dimensional correctness of all formulas given in this chapter in the Gaussian system.

In the SI system, fractional exponents are avoided by including current (I) as a fourth basic unit. The dimensions of all electromagnetic quantities, including the constants ϵ_0 and μ_0, are given in terms of M, L, T, and I.

As the starting point, we again consider the dimensions of charge. This is now very simple. By the definition of current, we have

$$[q] = TI. \tag{4.19}$$

The Lorentz force formula now gives us [**E**] and [**B**]:

$$[\mathbf{E}] = MLT^{-3}I^{-1}, \tag{4.20}$$

$$[\mathbf{B}] = MT^{-2}I^{-1}. \tag{4.21}$$

Note that **E** and **B** do not have the same dimensions in SI; rather **E** has dimensions of velocity times **B**, as already stated.

The dimensions of ϵ_0 and μ_0 can now be obtained from the formula for energy density:

$$[\epsilon_0] = M^{-1}L^{-3}T^4I^2, \tag{4.22}$$

$$[\mu_0] = MLT^{-2}I^{-2}. \tag{4.23}$$

Exercise 4.2 Verify the dimensional correctness of all formulas given in this chapter in the SI system.

Exercise 4.3 A famous text in quantum mechanics states that "in atomic units," the probability per unit time of ionization of a hydrogen atom in its ground state in an external electric field $\mathcal{E}$ (also in atomic units) is given by

$$w = \frac{4}{\mathcal{E}}e^{-(2/3\mathcal{E})}. \tag{4.24}$$

Atomic units are such that $\hbar$, m (electron mass), and a_0 (Bohr radius) all have the value 1. Rewrite the above formula in the Gaussian and SI systems, and find the value of the ionization rate for a field of 10^{10} V/m.

Answer: 10^4 sec^{-1}.

We conclude with a brief history of the two systems of units. Knowing this helps in keeping an open mind about the benefits of one versus the other. In the early 1800s, with the cgs system for mechanical quantities (force, energy, mass, etc.), Coulomb's law provided the natural unit of charge. Likewise, the law for the force between two magnetic poles gave the unit of magnetic pole strength.[14] With Oersted's discovery that currents also produce magnetic fields, and the precise formulation of this discovery via the Biot-Savart law, current could be defined in terms of the magnetic pole strength. All other quantities, such as capacitance, resistance, magnetic flux, etc., could also be connected to the pole strength. This led to the so-called electromagnetic or cgs-emu units. However, current is also the rate of charge flow, so the magnetic field and all other electromagnetic quantities could be related to the unit of charge. This led to the electrostatic or cgs-esu units. It was then noticed by many workers that the ratio of the numerical value of any quantity in cgs-emu units to that in cgs-esu units was very close to 3×10^{10}, or its reciprocal, or the square of one of these numbers,[15] and that this number coincided with the speed of light in cgs units. Gauss saw that by putting a quantity with dimensions of velocity in the denominator of the Biot-Savart law, the cgs-emu and cgs-esu systems could be replaced by a single system; this is how the Gaussian system came to be. Further, this appearance of the speed of light was a key factor behind Maxwell's proposal of the displacement current in 1865, and the idea that light was an electromagnetic wave. All this while, many workers had adopted a "practical" system of units based on the same idea as the cgs-emu, but with units of length and mass equal to 10^9 cm and 10^{-11} g, respectively. In 1901, Giorgi adjusted the constants μ_0 and ϵ_0 to make the practical units compatible with the mks system; this is essentially the SI system in use today. It may surprise some readers that the joule and the newton were not created until the 1930s!

---

[14] Since magnetic monopoles do not exist, this sentence requires some explanation. The only sources of magnetism known in the eighteenth century were permanent magnets. It was well known that the poles of a magnet could not be separated and that breaking a bar magnet produced new poles at the broken ends. By careful torsion balance experiments with long magnetic needles, however, Coulomb was able to establish in 1785 that they behaved as if there were a force between the poles at the ends of the needles that varied as the inverse square of the separation and the product of the pole strengths.

[15] For example, the ratio for charge was measured by Weber and Kohlrausch to be 3.107×10^{10} in 1856; that for resistance was found to be $(2.842 \times 10^{10})^2$ by Maxwell in 1868, and $(2.808 \times 10^{10})^2$ by W. Thomson in 1869.

Table 1.1. If Gaussian makes you groan

Physical quantity	Replace	By	1 Gaussian unit (symbol)	Equals
Charge	q	$(4\pi\epsilon_0)^{-1/2}q$	esu or statcoulomb	$10^9/3$ C
Current	I	$(4\pi\epsilon_0)^{-1/2}I$	esu/sec or statampere	$10^9/3$ A
Electric field	$\mathbf{E}$	$(4\pi\epsilon_0)^{1/2}\mathbf{E}$	statvolt/cm	3×10^4 V/m
Potential, voltage	ϕ	$(4\pi\epsilon_0)^{1/2}\phi$	statvolt	300 V
Electric displacement	$\mathbf{D}$	$(4\pi/\epsilon_0)^{1/2}\mathbf{D}$	statvolt/cm	$10^{-5}/(3\times4\pi)$ C/m^2
Electric polarization	$\mathbf{P}$	$(4\pi\epsilon_0)^{-1/2}\mathbf{P}$	esu/cm^3	$10^5/3$ C/m^2
Magnetic field	$\mathbf{B}$	$(4\pi/\mu_0)^{1/2}\mathbf{B}$	gauss (G)	10^{-4} T
Magnetic flux	Φ	$(4\pi/\mu_0)^{1/2}\Phi$	gauss cm^2 or maxwell	10^{-8} Wb
Magnetizing field	$\mathbf{H}$	$(4\pi\mu_0)^{1/2}\mathbf{H}$	oersted (Oe)	$10^3/4\pi$ A/m
Magnetization	$\mathbf{M}$	$(\mu_0/4\pi)^{1/2}\mathbf{M}$	emu/cm^3	10^3 A/m
Conductivity	σ	$\sigma/4\pi\epsilon_0$	sec^{-1}	$10^{-9}/3\times3$ mho/m
Resistance	R	$(4\pi/c\,Z_0)\,R$	sec/cm	$3\times3\times10^{11}$ Ω
Conductance	G	$G/4\pi\epsilon_0$	cm/sec	$10^{-11}/3\times3$ mho
Capacitance	C	$C/4\pi\epsilon_0$	cm	$10^{-11}/3\times3$ F
Inductance	L	$4\pi\epsilon_0 L$	esu, or stathenry, or Gaussian unit	$3\times3\times10^{11}$ H
Dielectric constant	ϵ	ϵ/ϵ_0	dimensionless	same
Magnetic permeability	μ	μ/μ_0	dimensionless	same
Electric polarizability	α	$\alpha/4\pi$	dimensionless	$4\pi\times$ Gaussian value
Magnetic susceptibility	χ	$\chi/4\pi$	dimensionless	$4\pi\times$ Gaussian value

This table is designed to convert equations and numerical values of quantities from the Gaussian to the SI system. Conversions for purely mechanical quantities (force, power, etc.) are not listed. The following should be noted:

Factors of 3 should be replaced by 2.997 924 58 to be exact.

$c = 2.997\,924\,58 \times 10^8$ m/sec (speed of light by fiat; 1 meter = distance covered by light in so many seconds)

$\mu_0 = 4\pi \times 10^{-7}$ H/m $\simeq 1.257 \times 10^{-6}$ H/m

$\epsilon_0 = 1/(\mu_0 c^2) \simeq 8.854 \times 10^{-12}$ F/m

$Z_0 \equiv (\mu_0/\epsilon_0)^{1/2} \simeq 376.7$ Ω

Table 1.2. If SI makes you sigh

Physical quantity	Replace	By	1 SI unit (symbol)	Equals
Charge	q	$(4\pi\epsilon_0)^{1/2}q$	coulomb (C)	3×10^9 esu or statcoul
Current	I	$(4\pi\epsilon_0)^{1/2}I$	ampere (A)	3×10^9 esu/sec or statamp
Electric field	$\mathbf{E}$	$(4\pi\epsilon_0)^{-1/2}\mathbf{E}$	volt/m	$10^{-4}/3$ statvolt/cm
Potential, voltage	ϕ	$(4\pi\epsilon_0)^{-1/2}\phi$	volt (V)	$1/300$ statvolt
Electric displacement	$\mathbf{D}$	$(\epsilon_0/4\pi)^{1/2}\mathbf{D}$	coulomb/m^2	$3 \times 4\pi \times 10^5$ statvolt/cm
Electric polarization	$\mathbf{P}$	$(4\pi\epsilon_0)^{1/2}\mathbf{P}$	coulomb/m^2	3×10^5 esu/cm^3
Magnetic field	$\mathbf{B}$	$(\mu_0/4\pi)^{1/2}\mathbf{B}$	tesla (T)	10^4 G
Magnetic flux	Φ	$(\mu_0/4\pi)^{1/2}\Phi$	weber (Wb)	10^8 G cm^2
Magnetizing field	$\mathbf{H}$	$(4\pi\mu_0)^{-1/2}\mathbf{H}$	ampere/m	$4\pi \times 10^{-3}$ Oe
Magnetization	$\mathbf{M}$	$(4\pi/\mu_0)^{1/2}\mathbf{M}$	ampere/m	10^{-3} emu/cm^3
Conductivity	σ	$4\pi\epsilon_0\sigma$	mho/m	$3 \times 3 \times 10^9$ sec^{-1}
Resistance	R	$(c\,Z_0/4\pi)\,R$	ohm (Ω)	$10^{-11}/3 \times 3$ sec/cm
Conductance	G	$4\pi\epsilon_0\,G$	mho	$3 \times 3 \times 10^{11}$ cm/sec
Capacitance	C	$4\pi\epsilon_0 C$	farad (F)	$3 \times 3 \times 10^{11}$ cm
Inductance	L	$L/4\pi\epsilon_0$	henry (H)	$10^{-11}/3 \times 3$ esu
Dielectric constant	ϵ	$\epsilon_0\epsilon$	dimensionless	same
Magnetic permeability	μ	$\mu_0\mu$	dimensionless	same
Electric polarizability	α	$4\pi\alpha$	dimensionless	(SI value)/4π
Magnetic susceptibility	χ	$4\pi\chi$	dimensionless	(SI value)/4π

This table is designed to convert equations and numerical values of quantities from the SI to the Gaussian system. Conversions for purely mechanical quantities (force, power, etc.) are not listed. The following should be noted:

Factors of 3 should be replaced by 2.997 924 58 to be exact.

$c = 2.997\,924\,58 \times 10^8$ m/sec (speed of light by fiat; 1 meter = distance covered by light in so many seconds)

$\mu_0 = 4\pi \times 10^{-7}$ H/m $\simeq 1.257 \times 10^{-6}$ H/m

$\epsilon_0 = 1/(\mu_0 c^2) \simeq 8.854 \times 10^{-12}$ F/m

$Z_0 \equiv (\mu_0/\epsilon_0)^{1/2} \simeq 376.7\,\Omega$

2 | Review of mathematical concepts

We review in this chapter some mathematical tools that are of use in the study of electromagnetism. The most common of these concern vectors, vector calculus, and the integral theorems of Gauss, Stokes, and Green. We will carry out vector manipulations both via direct use of vector identities and via tensor notation and the Einstein summation convention. Experience shows that it is highly useful to master both techniques, and that very often a mix of the two is needed.

We also briefly discuss Fourier transforms, the Dirac delta function and its derivatives, and generalized functions more generally. These concepts are of direct value in solving many boundary value problems, in plane wave decompositions of the fields, and also provide additional insight into general properties of vector fields.

For completeness, we also include a discussion of rotation matrices and the transformations of vectors and tensors under rotations. This material is not needed for most of the mathematical manipulations that arise in the study of electromagnetism, but an understanding of the concepts is especially useful when one goes on to study relativity and four-vectors, four-tensors, etc.

Lastly, we discuss orthogonal curvilinear coordinates. Particular examples are cylindrical and spherical polar coordinates. We show how the formulas for gradient, curl, etc., arise naturally from the geometrical meanings of these operations.

We stress that our discussion here is only a review, and that the reader is expected to have seen most of this material before, at least at an elementary level.[1]

5 Vector algebra

We begin by recalling some rules for vectors in ordinary three-dimensional Euclidean space. We denote generic vectors by lowercase bold letters, **a**, **b**, **c**, etc., and scalars by

---

[1] There are many excellent texts that meet this need. Some that the author particularly likes are Hardy (1952) (basic one-variable calculus); Boas (2005) or Kreyszig (2005) (more advanced and multivariable calculus, linear algebra, vectors); Copson (1970) (complex analysis); Bender and Orszag (1978) (asymptotic analysis); Courant and Hilbert (1953) (classical mathematical physics).

italic letters f, g, h, etc. The Cartesian components[2] of a vector with respect to some suitably chosen set of axes will be denoted interchangeably by subscripts x, y, and z, or 1, 2, and 3, whichever is convenient. Thus, a general component of a vector **a** is denoted a_i, where the index i stands for any of the three values 1, 2, and 3.

Components of vectors are manipulated in the same way as ordinary numbers, i.e., with the rules of simple arithmetic. The ith component of the sum of two vectors is given by adding their corresponding components. Thus, if $\mathbf{c} = \mathbf{a} + \mathbf{b}$, then

$$c_i = (\mathbf{a} + \mathbf{b})_i = a_i + b_i. \tag{5.1}$$

We now recall the various ways of multiplying vectors. First, vectors may be multiplied by scalars to obtain new vectors, which may be added as before. The resulting algebra is straightforward, and one has identities such as

$$(f\mathbf{a} + g\mathbf{b})_i = f a_i + g b_i. \tag{5.2}$$

Second, the *length* or *magnitude* of a vector **a**, denoted $|\mathbf{a}|$, or a, is given by

$$a = |\mathbf{a}| = \sqrt{a_x^2 + a_y^2 + a_z^2}. \tag{5.3}$$

This is a special case of the *dot product* or *scalar product* of two vectors, which may be defined in terms of components as

$$\mathbf{a} \cdot \mathbf{b} = a_x b_x + a_y b_y + a_z b_z. \tag{5.4}$$

The squared length of a vector (often known as just the square) is then the dot product of the vector with itself:

$$a^2 = \mathbf{a} \cdot \mathbf{a}. \tag{5.5}$$

By considering the square of the sum $\mathbf{a} + \mathbf{b}$ and making use of the cosine formula from trigonometry, we are led to the geometrical meaning of the dot product, namely,

$$\mathbf{a} \cdot \mathbf{b} = ab \cos\theta, \tag{5.6}$$

where θ is the angle between **a** and **b**. For this reason, we often refer to $\mathbf{a} \cdot \mathbf{b}$ as the projection of **a** onto **b**, or vice versa.

Third, the *cross product* or *vector product* of two vectors **a** and **b** is a third vector **c**. We write

$$\mathbf{c} = \mathbf{a} \times \mathbf{b}. \tag{5.7}$$

The components of **c** are given by

$$\begin{aligned}
c_x &= a_y b_z - a_z b_y, \\
c_y &= a_z b_x - a_x b_z, \\
c_z &= a_x b_y - a_y b_x.
\end{aligned} \tag{5.8}$$

---

[2] We shall mean Cartesian coordinates or components whenever the words *coordinate* and *component* are used without qualification.

We remind readers that the second and third lines can be obtained from the first by cyclic permutations of the subscripts, and that the cross product is antisymmetric, i.e.,

$$\mathbf{b} \times \mathbf{a} = -\mathbf{a} \times \mathbf{b}. \tag{5.9}$$

The cross product of a vector with itself or of any two parallel vectors is zero. Further, the geometrical meaning of the cross product is that the vector $\mathbf{c}$ is perpendicular to both $\mathbf{a}$ and $\mathbf{b}$, pointing along the direction in which a right-handed screw would advance when turned from $\mathbf{a}$ to $\mathbf{b}$, and its magnitude is given by

$$c = |\mathbf{a} \times \mathbf{b}| = ab \sin \theta, \tag{5.10}$$

where θ is, as before, the angle between $\mathbf{a}$ and $\mathbf{b}$. Thus, c is equal to the area of the parallelogram formed by the vectors $\mathbf{a}$ and $\mathbf{b}$.

Fourth, the *triple* or *scalar* product of three vectors is given by

$$(\mathbf{a} \times \mathbf{b}) \cdot \mathbf{c}. \tag{5.11}$$

This quantity has the geometrical meaning of the volume of the (oriented) parallelpiped formed by the vectors $\mathbf{a}$, $\mathbf{b}$, and $\mathbf{c}$. This product has the important property that one may permute the three vectors cyclically without affecting its value:

$$(\mathbf{a} \times \mathbf{b}) \cdot \mathbf{c} = (\mathbf{b} \times \mathbf{c}) \cdot \mathbf{a} = (\mathbf{c} \times \mathbf{a}) \cdot \mathbf{b}. \tag{5.12}$$

This rule, combined with the symmetry of the dot product, leads to the rule

$$(\mathbf{a} \times \mathbf{b}) \cdot \mathbf{c} = \mathbf{a} \cdot (\mathbf{b} \times \mathbf{c}). \tag{5.13}$$

In other words, one may interchange the position of the dot and the cross in a triple product. Because of this, one often encounters specialized notations such as $[\mathbf{a}, \mathbf{b}, \mathbf{c}]$ or $\langle \mathbf{a}, \mathbf{b}, \mathbf{c} \rangle$ for this product. We shall not use these notations, but we shall occasionally omit the parentheses in eq. (5.11), as there can be no ambiguity about the meaning. On the other hand, the expressions $\mathbf{a} \times (\mathbf{b} \times \mathbf{c})$ and $(\mathbf{a} \times \mathbf{b}) \times \mathbf{c}$ are different, and parentheses must be used to make clear which one is meant.

Many vector manipulations can be made easier by using unit vectors, i.e., vectors of unit length, $\hat{\mathbf{x}}$, $\hat{\mathbf{y}}$, and $\hat{\mathbf{z}}$, or equivalently $\mathbf{e}_1$, $\mathbf{e}_2$, $\mathbf{e}_3$, that point along the three Cartesian axes. The following results are immediate consequences of the definitions of the dot and cross products:

$$\hat{\mathbf{x}} \cdot \hat{\mathbf{x}} = \hat{\mathbf{y}} \cdot \hat{\mathbf{y}} = \hat{\mathbf{z}} \cdot \hat{\mathbf{z}} = 1,$$
$$\hat{\mathbf{x}} \cdot \hat{\mathbf{y}} = \hat{\mathbf{y}} \cdot \hat{\mathbf{z}} = \hat{\mathbf{z}} \cdot \hat{\mathbf{x}} = 0, \tag{5.14}$$
$$\hat{\mathbf{x}} \times \hat{\mathbf{y}} = \hat{\mathbf{z}}, \quad \hat{\mathbf{y}} \times \hat{\mathbf{z}} = \hat{\mathbf{x}}, \quad \hat{\mathbf{z}} \times \hat{\mathbf{x}} = \hat{\mathbf{y}}.$$

Since $\hat{\mathbf{x}}$, $\hat{\mathbf{y}}$, and $\hat{\mathbf{z}}$ are a complete and orthonormal triad, a general vector $\mathbf{a}$ can be expanded as $a_x \hat{\mathbf{x}} + a_y \hat{\mathbf{y}} + a_z \hat{\mathbf{z}}$, where $a_x = \mathbf{a} \cdot \hat{\mathbf{x}}$ is the projection of $\mathbf{a}$ onto the x direction, and likewise for a_y and a_z. Using such an expansion for both $\mathbf{a}$ and $\mathbf{b}$, e.g., we immediately obtain eq. (5.4) by using the rules eq. (5.14) and the distributivity of the dot product over addition.

As a less trivial example, by using the distributivity of the cross product over addition, we can obtain the important formulas

$$\mathbf{a} \times (\mathbf{b} \times \mathbf{c}) = (\mathbf{a} \cdot \mathbf{c})\mathbf{b} - (\mathbf{a} \cdot \mathbf{b})\mathbf{c}, \qquad (5.15)$$

$$(\mathbf{a} \times \mathbf{b}) \times \mathbf{c} = (\mathbf{a} \cdot \mathbf{c})\mathbf{b} - (\mathbf{b} \cdot \mathbf{c})\mathbf{a}. \qquad (5.16)$$

The second formula is equivalent to the first, since we may write the left-hand side as $-\mathbf{c} \times (\mathbf{a} \times \mathbf{b})$, and then rename the vectors. It is quite useful to remember both formulas independently, however.[3]

Exercise 5.1 Prove eqs. (5.15) and (5.16).

Solution: It is enough to consider the first formula. We first establish its truth for all possible triples of $\hat{\mathbf{x}}$, $\hat{\mathbf{y}}$, and $\hat{\mathbf{z}}$. Explicit calculation gives

$$\hat{\mathbf{x}} \times (\hat{\mathbf{y}} \times \hat{\mathbf{z}}) = \hat{\mathbf{x}} \times \hat{\mathbf{x}} = 0,$$

$$\hat{\mathbf{x}} \times (\hat{\mathbf{x}} \times \hat{\mathbf{y}}) = \hat{\mathbf{x}} \times \hat{\mathbf{z}} = -\hat{\mathbf{y}}, \qquad (5.17)$$

and similar results obtainable by cyclic permutations of x, y, and z, and antisymmetry of the inner cross product. Equations (5.17) are exactly what the formula gives. We then note that in the formula, the expression on the left is separately linear in the vectors $\mathbf{a}$, $\mathbf{b}$, and $\mathbf{c}$. In other words,

$$(\mathbf{a}_1 + \mathbf{a}_2) \times (\mathbf{b} \times \mathbf{c}) = \mathbf{a}_1 \times (\mathbf{b} \times \mathbf{c}) + \mathbf{a}_2 \times (\mathbf{b} \times \mathbf{c}),$$

$$\mathbf{a} \times ((\mathbf{b}_1 + \mathbf{b}_2) \times \mathbf{c}) = \mathbf{a} \times (\mathbf{b}_1 \times \mathbf{c}) + \mathbf{a} \times (\mathbf{b}_2 \times \mathbf{c}),$$

etc. Since each vector may be expanded in terms of the unit vectors, the desired formula is proved.

---

[3] One commonly suggested mnemonic for this purpose is based on writing the right-hand side of eq. (5.15) as $\mathbf{b}(\mathbf{a} \cdot \mathbf{c}) - \mathbf{c}(\mathbf{a} \cdot \mathbf{b})$, and remembering this as "back cab." This is unsatisfactory, as it doesn't work for eq. (5.16) without relabeling the vectors. I prefer the following. Think of the vectors in eq. (5.15) as

$\mathbf{a}$: outer, $\mathbf{b}$: near, $\mathbf{c}$: far,

in the sense that $\mathbf{a}$ is in the outer cross product, $\mathbf{b}$ is near it, and $\mathbf{c}$ is far away. Then the mnemonic is "dot outer with far, minus dot outer with near." In other words, we write

$(\mathbf{a} \cdot \mathbf{c})_{\sqcup} - (\mathbf{a} \cdot \mathbf{b})_{\sqcup}$.

But this is not yet right, since the expression has to be a vector. So, we must multiply each dot product with the vector that is left over by filling it in the empty spaces ($\sqcup$):

$(\mathbf{a} \cdot \mathbf{c})\mathbf{b} - (\mathbf{a} \cdot \mathbf{b})\mathbf{c}$.

Since this mnemonic also works for eq. (5.16), with outer $= \mathbf{c}$, near $= \mathbf{b}$, and far $= \mathbf{a}$, you can send the cab back. Of course, one now needs a way to remember the "outer far, outer near" (OFON) rule, but this is easy if you recall the old Confucian saying "odiferous feet offend nose."

Exercise 5.2 Show that

$$(a \times b) \cdot (c \times d) = (a \cdot c)(b \cdot d) - (a \cdot d)(b \cdot c). \tag{5.18}$$

Exercise 5.3 Show that

$$a \times (b \times c) + b \times (c \times a) + c \times (a \times b) = 0. \tag{5.19}$$

All the above manipulations can be carried out very compactly by using a notation due to Einstein. It is now more convenient to label the components by 1, 2, and 3 rather than x, y, and z, and to use Latin letters i, j, k, etc., to denote an arbitrary component. To see the basic idea, let us rewrite eq. (5.4) for the dot product as

$$a \cdot b = \sum_{i=1}^{3} a_i b_i. \tag{5.20}$$

Einstein's innovation was to note that an index would appear twice in a sum every time a dot product arose and to use this fact to simplify the writing. That an index is repeated is enough to indicate a sum, and there is no need to show the summation sign explicitly. Thus, we write

$$a \cdot b = a_i b_i. \tag{5.21}$$

Likewise, consider one of the terms on the right-hand side of eq. (5.18). We have

$$(a \cdot c)(b \cdot d) = \sum_{i=1}^{3} \sum_{j=1}^{3} a_i c_i b_j d_j = a_i c_i b_j d_j, \tag{5.22}$$

where the last expression is written using the implicit summation convention. Note that we could have written $b_k d_k$ instead of $b_j d_j$. The indices are known as *dummy* indices for this reason.

It is evident that the summation convention enhances the readability of formulas considerably. With practice one can learn to recognize patterns and common structures quite readily. There is one pitfall that must be studiously avoided, however, which is that in any expression, a dummy index should not be repeated more than once. If, e.g., we write

$$(a \cdot c)(b \cdot d) = a_k c_k b_k d_k, \quad \text{(Wrong!)} \tag{5.23}$$

the right-hand side is ambiguous. Which vector is to be dotted with which? One cannot say.[4] One might wonder what is to be done if one really has an expression like $\sum_i a_i b_i c_i d_i$. In practice, such expressions almost never occur. When they do, one can always revert to explicitly displaying the summation. Likewise, if one encounters a product $a_i b_i$ without a sum on i, it is easy to make a note of this fact in the calculation.

---

[4] As the joke goes, the indices have now made dummies out of us.

To make full use of the summation convention, we introduce two special symbols. The *Kronecker delta symbol* is defined by

$$\delta_{ij} = \begin{cases} 1, & \text{if } i = j; \\ 0, & \text{otherwise.} \end{cases} \tag{5.24}$$

Note that this is symmetric, i.e., $\delta_{ji} = \delta_{ij}$. The second symbol is the *Levi-Civita symbol*, which has three indices and is defined to be completely antisymmetric under interchange of any two. In other words,

$$\epsilon_{ijk} = -\epsilon_{jik} = -\epsilon_{ikj} = -\epsilon_{kji}. \tag{5.25}$$

It follows that the symbol is zero if any two of the indices are equal; e.g., ϵ_{112}, ϵ_{121}, and ϵ_{233} are all zero. Second, since the indices can take on only values 1, 2, and 3, it follows that the only nonzero combinations of the indices that need be considered are $(1, 2, 3)$, $(1, 3, 2)$, etc.—six in all, corresponding to the number of distinct permutations. By successively interchanging the first two and last two indices, we have, however,

$$\epsilon_{kij} = -\epsilon_{ikj} = \epsilon_{ijk}, \tag{5.26}$$

which shows that the symbol is unchanged under *cyclic* permutation of its indices and changes sign under *anticyclic* permutation. Therefore, we need only give its value for one choice of indices for which it is nonzero. The standard choice is

$$\epsilon_{123} = 1. \tag{5.27}$$

This, along with eq. (5.25), specifies the Levi-Civita symbol completely.[5]

The Kronecker delta symbol is naturally associated with the dot product. For example, we can write

$$\mathbf{a} \cdot \mathbf{b} = \delta_{ij} a_i b_j. \tag{5.28}$$

On the right-hand side, we have an implicit double sum. But, all terms with $i \neq j$ in this sum vanish, and the only ones that survive are the products of a_i with b_i. These have unit coefficient, and the sum on i gives the dot product. This example may seem overly simple, but such combinations arise frequently.

Exercise 5.4 What is the value of the expression δ_{ii}? (**Answer:** 3.)

The Levi-Civita or epsilon symbol is naturally associated with the cross product. Thus, as is easily verified, the definition (5.8) of the cross product may be written as

$$(\mathbf{a} \times \mathbf{b})_i = \epsilon_{ijk} a_j b_k. \tag{5.29}$$

---

[5] We deliberately refer to δ_{ij} and ϵ_{ijk} as only *symbols* at this stage. In fact, when used in formulas such as (5.28) and (5.29), they have a natural interpretation as *tensors*. We will see this below. δ_{ij} is called the *unit tensor*, and ϵ_{ijk} is called the Levi-Civita tensor, or the *completely antisymmetric tensor of the third rank in three dimensions*.

For example, let $i = 1$ in this equation. The only nonzero terms in the implicit sum on j and k on the right-hand side are those with $j = 2$, $k = 3$, or $j = 3$, $k = 2$, since ϵ_{ijk} vanishes otherwise. Finally, since $\epsilon_{123} = 1$ and $\epsilon_{132} = -1$,

$$(\mathbf{a} \times \mathbf{b})_1 = a_2 b_3 - a_3 b_2, \tag{5.30}$$

as required.

In terms of the Levi-Civita symbol, the triple product is written as

$$\mathbf{a} \cdot (\mathbf{b} \times \mathbf{c}) = \epsilon_{ijk} a_i b_j c_k. \tag{5.31}$$

The invariance properties (5.12) and (5.13) now follow from the symmetries of ϵ_{ijk}.

Next, let us consider the following basic combination of two Levi-Civita symbols:

$$\epsilon_{ijk} \epsilon_{ilm}. \tag{5.32}$$

Consider the terms in this sum where $i = 1$. The only terms that contribute are those where j and k belong to the set $(2, 3)$, and likewise for the indices l and m. Considering the other values of i in the same fashion, we conclude that the only nonzero terms in this sum are those where the ordered pair (j, k) (with $j \neq k$) coincides either with (l, m) or with (m, l). In the first case, both ϵ's have the same sign, while in the second, they have opposite signs, so the product equals $+1$ and -1, respectively. These results can be summarized in the formula

$$\epsilon_{ijk} \epsilon_{ilm} = \delta_{jl} \delta_{km} - \delta_{jm} \delta_{kl}. \tag{5.33}$$

We can use this result to derive the formulas (5.15) and (5.16) for a triple vector product. The derivation provides a good illustration of the use of the index notation. Consider the ith component of the first formula. Using the epsilon symbol to express each cross product, we obtain

$$\begin{aligned}
[\mathbf{a} \times (\mathbf{b} \times \mathbf{c})]_i &= \epsilon_{ijk} a_j (\mathbf{b} \times \mathbf{c})_k \\
&= \epsilon_{ijk} a_j \epsilon_{klm} b_l c_m \\
&= \epsilon_{kij} \epsilon_{klm} a_j b_l c_m \\
&= (\delta_{il} \delta_{jm} - \delta_{im} \delta_{jl}) a_j b_l c_m \\
&= b_i (a_j c_j) - c_i (a_j b_j).
\end{aligned} \tag{5.34}$$

In the third line, we rearranged terms and cyclically permuted the indices of ϵ_{ijk}; in the fourth line, we used eq. (5.33); and in the fifth we identified combinations like $\delta_{jm} a_j c_m$ with $a_j c_j$. The final result is nothing but the ith component of the right-hand side of eq. (5.15).

Exercise 5.5 Prove eq. (5.16) using the ϵ symbol directly, i.e., without reducing the problem to proving eq. (5.15).

Exercise 5.6 Prove eq. (5.18) using the index notation.

Exercise 5.7 Show that

$$\epsilon_{ijk}\epsilon_{ijm} = 2!\delta_{km},$$ (5.35)

$$\epsilon_{ijk}\epsilon_{ijk} = 3!$$ (5.36)

Exercise 5.8 Let A be a 3×3 matrix, with elements A_{ij}. Show that

$$\det A = \frac{1}{3!}\epsilon_{ijk}\epsilon_{\bar{i}\bar{j}\bar{k}} A_{i\bar{i}} A_{j\bar{j}} A_{k\bar{k}}.$$ (5.37)

Solution: On the right-hand side, any term containing two A's from the same column is zero, since then either $i = j$, or $i = k$, or $j = k$. Likewise for two A's from the same row. Second, the signs of the nonzero terms are obviously correct. Lastly, the 3! corrects for overcounting—there are 3! nontrivial choices for each of the triples (i, j, k) and $(\bar{i}, \bar{j}, \bar{k})$, giving 36 terms in all, but each term is repeated 6 times, as the same permutation acting on both the triples leads to the same term in the sum. In fact, we can remove this redundancy and write the result in the simpler but less symmetrical forms

$$\det A = \epsilon_{ijk} A_{i1} A_{j2} A_{k3} = \epsilon_{ijk} A_{1i} A_{2j} A_{3k}.$$ (5.38)

6 Derivatives of vector fields

We now wish to consider vectors and scalar fields, i.e., vectors and scalars that depend on position $\mathbf{r}$ in ordinary three-dimensional space, and to consider spatial derivatives of these fields. To this end, let us recall how derivatives are defined when the independent variable is a single real variable. We denote this variable by t, which readers may think of as the time if they wish. For a scalar function $f(t)$, we have

$$\frac{df}{dt} = \lim_{a \to 0} \frac{f(t+a) - f(t)}{a}.$$ (6.1)

For a vector-valued function $\mathbf{v}(t)$, the derivative is defined analogously:

$$\frac{d\mathbf{v}}{dt} = \lim_{a \to 0} \frac{\mathbf{v}(t+a) - \mathbf{v}(t)}{a}.$$ (6.2)

It is clear from the definitions that df/dt and $d\mathbf{v}/dt$ are themselves scalar and vector functions.

Let us now turn to fields in three-space. The fields are now functions of three variables, x_1, x_2, and x_3, the components of $\mathbf{r}$, so we must deal with partial derivatives $\partial f/\partial x_i$ and $\partial \mathbf{v}/\partial x_i$, etc. What is noteworthy now is that certain sets and combinations of these derivatives are themselves scalar and vector fields. Since equations of physics must equate like objects (scalars to scalars, and vectors to vectors), these combinations are of some importance.

The first object to be studied is the *gradient of a scalar field*. This is the triple (read "grad f")

$$\nabla f = \left(\frac{\partial f}{\partial x_1}, \frac{\partial f}{\partial x_2}, \frac{\partial f}{\partial x_3} \right).$$ (6.3)

We assert that this is a vector field. To see this, let us consider f at two nearby points, $\mathbf{r}$ and $\mathbf{r}+\mathbf{a}$, and expand the difference in a Taylor series to first order in the components of $\mathbf{a}$ as $\mathbf{a} \to 0$:

$$f(\mathbf{r}+\mathbf{a}) - f(\mathbf{r}) \approx \sum_i a_i \frac{\partial f}{\partial x_i} + \cdots . \tag{6.4}$$

The left-hand side of this equation is the difference of two scalars, hence a scalar itself. Therefore, the right-hand side must also be a scalar. For the terms linear in the a_i, this implies that

$$\sum_i a_i \frac{\partial f}{\partial x_i} = a_i \frac{\partial f}{\partial x_i} = \text{scalar}. \tag{6.5}$$

(Note that we have also rewritten the sum using the summation convention.) Let us consider this equation in different coordinate frames which differ in how the coordinate axes are oriented. As the orientation of the axes is varied, the numerical values of the components a_i of $\mathbf{a}$ will change, as will the partial derivatives $\partial f/\partial x_i$. However, since the right-hand side is a scalar, the sum on the left will stay unchanged. This sum has exactly the same form as a dot product of two vectors, and so we conclude that the members of the triple (6.3) must change in exactly the same way as the components of a vector.

It follows that if we define an abstract gradient operator,

$$\nabla = \left(\frac{\partial}{\partial x_1}, \frac{\partial}{\partial x_2}, \frac{\partial}{\partial x_3} \right), \tag{6.6}$$

it will behave like a vector. Ignoring for the moment the fact that it must act on something in order to give a meaningful answer, we can argue that ∇f is a vector because it is like multiplying a vector (∇) with a scalar (f). The same idea can now be used to make other scalars and vectors. Thus,

$$\nabla \cdot \mathbf{v} = \sum_i \frac{\partial v_i}{\partial x_i} \equiv \frac{\partial v_i}{\partial x_i}, \tag{6.7}$$

read as the *divergence of* $\mathbf{v}$, is a scalar field because it has the structure of a dot product. Likewise, the object $\nabla \times \mathbf{v}$, read *curl of* $\mathbf{v}$, and defined like a cross product, i.e, as a triple with members given by

$$(\nabla \times \mathbf{v})_i = \epsilon_{ijk} \frac{\partial v_k}{\partial x_j}, \tag{6.8}$$

is a vector field.

The three objects—gradient, divergence, and curl—are the fundamental derivatives in vector field theory. Higher derivatives can be constructed by combining them. Before we consider these, however, it is useful to introduce two alternative notations for the gradient. In the first, we write

$$(\nabla f)_i = \partial_i f; \quad \partial_i = \frac{\partial}{\partial x_i}. \tag{6.9}$$

This shorthand can save quite a lot of writing. For instance, eq. (6.7) becomes

$$\nabla \cdot \mathbf{v} = \partial_i v_i, \tag{6.10}$$

and eq. (6.8) becomes

$$(\nabla \times \mathbf{v})_i = \epsilon_{ijk} \partial_j v_k. \tag{6.11}$$

The second way is to write $\nabla f = \partial f / \partial \mathbf{r}$. In fact, this notion can be made more general. We define the derivative of a scalar with respect to a vector to be a vector whose components are equal to the derivatives with respect to the components.

The simplest second derivative is obtained by taking the divergence of ∇f. This combination appears so often it is given its own symbol, denoted $\nabla^2 f$. We have

$$\nabla^2 f = \nabla \cdot (\nabla f) = \partial_i \partial_i f = \sum_i \frac{\partial^2 f}{\partial x_i^2}. \tag{6.12}$$

The operator ∇^2 is also known as the *Laplacian*, and is read *grad square* or *del square*. It has the important property of being a scalar. We have written it with an explicit sum at the end of eq. (6.12) because of its importance.

Likewise, we can differentiate $\nabla \cdot \mathbf{v}$ and $\nabla \times \mathbf{v}$ further. We have the obvious combinations

$$(a)\ \ \nabla(\nabla \cdot \mathbf{v}), \quad (b)\ \ \nabla \cdot (\nabla \times \mathbf{v}), \quad (c)\ \ \nabla \times (\nabla \times \mathbf{v}). \tag{6.13}$$

In addition, we have $\nabla \times (\nabla f)$. However, this and the combination $\nabla \cdot (\nabla \times \mathbf{v})$ are both identically zero. If we go back to the idea that ∇ is a vector without worrying about what it is acting on, the argument is essentially that behind $\mathbf{a} \times \mathbf{a} = 0$ for any vector $\mathbf{a}$. It is instructive to see how it comes out in the summation convention. Consider the first component of $\nabla \times \nabla f$. We have

$$\begin{aligned}
(\nabla \times (\nabla f))_1 &= \epsilon_{1jk} \partial_j (\nabla f)_k \\
&= \epsilon_{1jk} \partial_j \partial_k f \\
&= \epsilon_{123} \partial_2 \partial_3 f + \epsilon_{132} \partial_3 \partial_2 f \\
&= \frac{\partial^2 f}{\partial x_2 \partial x_3} - \frac{\partial^2 f}{\partial x_3 \partial x_2} \\
&= 0. \tag{6.14}
\end{aligned}$$

The third and fourth lines follow from the properties of the ϵ symbol, and the last line follows from the equality of mixed partials. The other components are similarly zero.

Exercise 6.1 Show that $\nabla \cdot (\nabla \times \mathbf{v}) = 0$.

In addition to the second derivatives $\nabla(\nabla \cdot \mathbf{v})$ and $\nabla \times (\nabla \times \mathbf{v})$, we can also define the Laplacian of a vector field, i.e.,

$$\nabla^2 \mathbf{v} = (\nabla \cdot \nabla)\mathbf{v}. \tag{6.15}$$

We simply do it component by component,[6] i.e.,

$$(\nabla^2 \mathbf{v})_i = \nabla^2 v_i. \tag{6.16}$$

Exercise 6.2 Show that

$$\nabla \times (\nabla \times \mathbf{v}) = \nabla(\nabla \cdot \mathbf{v}) - \nabla^2 \mathbf{v}. \tag{6.17}$$

Lastly, let us consider derivatives of products of two fields. The number of such products and derivatives is a handful, but still not too large. We list all the results.

(a) $\nabla(fg) = (\nabla f)g + f(\nabla g)$,

(b) $\nabla \cdot (f\mathbf{v}) = (\nabla f) \cdot \mathbf{v} + f \nabla \cdot \mathbf{v}$,

(c) $\nabla \times (f\mathbf{v}) = (\nabla f) \times \mathbf{v} + f \nabla \times \mathbf{v}$,

(d) $\nabla(\mathbf{v} \cdot \mathbf{w}) = (\mathbf{v} \cdot \nabla)\mathbf{w} + (\mathbf{w} \cdot \nabla)\mathbf{v} + \mathbf{v} \times (\nabla \times \mathbf{w}) + \mathbf{w} \times (\nabla \times \mathbf{v})$,

(e) $\nabla \cdot (\mathbf{v} \times \mathbf{w}) = \mathbf{w} \cdot (\nabla \times \mathbf{v}) - \mathbf{v} \cdot (\nabla \times \mathbf{w})$,

(f) $\nabla \times (\mathbf{v} \times \mathbf{w}) = \mathbf{v}(\nabla \cdot \mathbf{w}) - \mathbf{w}(\nabla \cdot \mathbf{v}) + (\mathbf{w} \cdot \nabla)\mathbf{v} - (\mathbf{v} \cdot \nabla)\mathbf{w}$.

$$\tag{6.18}$$

Readers who have not seen these results before should try and prove them for themselves before going on. They follow straightforwardly if one remembers the product rule of one-variable calculus and manipulates the ∇ operator as a vector, using the identities for the triple scalar and cross products. Result (a), e.g., is an almost immediate rewrite of the product rule. Result (b) follows from the equalities $\mathbf{a} \cdot (f\mathbf{b}) = f(\mathbf{a} \cdot \mathbf{b}) = (\mathbf{a} f) \cdot \mathbf{b}$ if one remembers that the ∇ acts on everything to the right. When all else fails, one can resort to the summation convention along with the Kronecker and epsilon symbols. To show (c), e.g., we proceed thus:

$$
\begin{aligned}
(\nabla \times (f\mathbf{v}))_i &= \epsilon_{ijk} \partial_j (f\mathbf{v})_k \\
&= \epsilon_{ijk} \partial_j (f v_k) \\
&= \epsilon_{ijk}((\partial_j f)v_k + f \partial_j v_k) \\
&= ((\nabla f) \times \mathbf{v})_i + f(\nabla \times \mathbf{v})_i.
\end{aligned}
\tag{6.19}
$$

In addition to the index notation, there is a clever method for such manipulations due to Feynman[7] that is worth knowing. Let us take the last result, (f). We write

$$\nabla \times (\mathbf{v} \times \mathbf{w}) = \nabla_{\mathbf{v}} \times (\mathbf{v} \times \mathbf{w}) + \nabla_{\mathbf{w}} \times (\mathbf{v} \times \mathbf{w}), \tag{6.20}$$

---

[6] We stress that this definition is made in terms of Cartesian coordinates, and anticipating the introduction of curvilinear or non-Cartesian coordinate systems, note that it does not work in these coordinate systems. In other words, one cannot equate the Laplacian of the components to the components of the Laplacian in curvilinear coordinates.

[7] Feynman, Leighton, and Sands (1964), vol. ll, sec. 27-3.

where the operators ∇_v and ∇_w act only on the fields $\mathbf{v}$ and $\mathbf{w}$, respectively. We now manipulate the separate parts of eq. (6.20) using our rules for vector algebra. Thus, for the first term, we use eq. (5.15), substituting ∇, $\mathbf{v}$, and $\mathbf{w}$ for $\mathbf{a}$, $\mathbf{b}$, and $\mathbf{c}$. This yields,

$$\nabla_v \times (\mathbf{v} \times \mathbf{w}) = (\nabla_v \cdot \mathbf{w})\mathbf{v} - (\nabla_v \cdot \mathbf{v})\mathbf{w}$$
$$= (\mathbf{w} \cdot \nabla_v)\mathbf{v} - \mathbf{w}(\nabla_v \cdot \mathbf{v})$$
$$= (\mathbf{w} \cdot \nabla)\mathbf{v} - \mathbf{w}(\nabla \cdot \mathbf{v}), \tag{6.21}$$

where in the second line we have used the fact that ∇_v does not act on $\mathbf{w}$, and in the third line we have removed the now unnecessary subscript on the gradient operator. The second part of eq. (6.20) is simplified similarly. Indeed, no further work is necessary. We write the inner cross product as $-(\mathbf{w} \times \mathbf{v})$, and use eq. (6.21), merely interchanging $\mathbf{v}$ and $\mathbf{w}$ everywhere. Putting together the two parts we obtain the desired formula.[8]

Exercise 6.3 Consider the operator

$$\mathbf{L}_{op} = -i(\mathbf{r} \times \nabla). \tag{6.22}$$

We write the Cartesian components of $\mathbf{L}_{op}$ as L^i_{op}, and $\mathbf{L}_{op} \cdot \mathbf{L}_{op} = L^2_{op}$. We also define

$$L^{\pm}_{op} = L^x_{op} \pm i L^y_{op}. \tag{6.23}$$

Show that

(a) $\mathbf{L}_{op} f(r) = 0,$ (f is a function of $r = |\mathbf{r}|$ only),

(b) $L^i_{op} L^j_{op} - L^j_{op} L^i_{op} = i\epsilon_{ijk} L^k_{op},$

(c) $\mathbf{L}^2_{op} L^i_{op} - L^i_{op} \mathbf{L}^2_{op} = 0,$

(d) $L^+_{op} L^-_{op} - L^-_{op} L^+_{op} = 2L^z_{op},$ (6.24)

(e) $L^{\pm}_{op} L^z_{op} - L^z_{op} L^{\pm}_{op} = \mp L^{\pm}_{op},$

(f) $L^{\pm}_{op} L^{\mp}_{op} = \mathbf{L}^2_{op} - (L^z_{op})^2 \pm L^z_{op},$

(g) $\mathbf{L}^2_{op} = -r^2\nabla^2 + \dfrac{\partial}{\partial r}\left(r^2 \dfrac{\partial}{\partial r}\right).$

Comments: The operator $\mathbf{L}_{op}$ may be familiar to the reader from quantum mechanics, where it has the physical meaning of angular momentum.[9] However, it is of great utility in electrodynamics and other vector field theories as well. We shall use the above results and those of exercise 8.11 when we introduce spherical harmonics, and in the multipole

---

[8] In using this method, one must also keep in mind that the gradient operators must act on the appropriate operand and not be left hanging. If, for instance, we had written the first term in eq. (5.15) as $\mathbf{b}(\mathbf{a} \cdot \mathbf{c})$, a literal substitution of the symbols would give for the first term in the first line of eq. (6.21) the expression $\mathbf{v}(\nabla_v \cdot \mathbf{w})$. As it stands, this is meaningless, and one must rearrange it into $(\mathbf{w} \cdot \nabla_v)\mathbf{v}$ to make sense of it.

[9] Note that $\mathbf{L}_{op}$ is dimensionless as defined. To give it the proper physical dimensions, one multiplies it by $\hbar$, Planck's constant.

expansion of electromagnetic fields. Results (b)–(e) give the *commutators*, the difference of the product of two operators in opposite order. Result (g) can also be written as

$$\nabla^2 = \frac{1}{r^2}\frac{\partial}{\partial r}\left(r^2\frac{\partial}{\partial r}\right) + \nabla^2_\Omega, \tag{6.25}$$

where

$$\nabla^2_\Omega = -r^{-2}\mathbf{L}^2_{\text{op}} \tag{6.26}$$

is known as the angular part of the Laplacian, as it gives zero when acting on a function of r alone.

7 Integration of vector fields

Integrals of scalar fields are exactly the same as multiple integrals of ordinary number valued functions. Thus, for a scalar field $f(\mathbf{r})$, we have the line, surface, and volume integrals:

$$\int_C f\,d\ell, \quad \int_S f\,ds, \quad \int_V f\,d^3x. \tag{7.1}$$

The subscripts on the integration sign indicate the curve, surface, or volume over which the integration is to be performed. On occasion, these subscripts are omitted. We shall consistently use the notations $d\ell$, ds, and d^3x for elements of length, surface, and volume, respectively. Sometimes, one writes d^2s instead of ds in the surface integral as an additional reminder to oneself about the dimensionality of the integral. The notations d^3r and dV for the volume element are also commonly encountered.

It is sufficient for us to think of integration in the sense of Riemann. The line integral is defined by adding together the values of the function (times infinitesimal length elements $\Delta\ell$) over the specified curve C in space. This curve may be open or closed. In the former case, we sometimes indicate the endpoints of the curve, $\mathbf{r}_1$ and $\mathbf{r}_2$, as limits of integration, and in the latter case, we sometimes put a circle around the integration symbol thus: $\oint$.

The surface integral is analogously defined by adding together the values of the function (times infinitesimal area elements Δs) over the specified surface S in space. This surface may be open, i.e., have as a boundary a closed curve, or be closed, in which case it has no boundary. In the latter case, one again uses the notation $\oint$ on occasion.

Lastly, the volume integral is defined by adding together the values of the function (times infinitesimal volume elements $\Delta x_1\Delta x_2\Delta x_3$) over the specified volume V. The region of integration may or may not extend to infinity. In the latter case, the volume V is finite and clearly has a boundary. As examples of the former, we could consider integrals over all space, over all space excluding a sphere of some radius at the origin, or over a right circular cylindrical region of finite radius, but extending infinitely along the axis. In all these cases, it is useful to think of the integration volume as the limit of a finite region, all or part of whose bounding surface is allowed to approach infinity. In particular, the case where the integral extends over all space arises quite often, and we commonly indicate this by not attaching any suffix to the integral sign.

In all the three types of integrals, note that the region of integration may consist of two or more disjoint pieces. Even though the integral is evaluated by adding together the separate integrals over each piece, it is sometimes convenient to regard the sum as a single integral. Secondly, the region of integration need not be simply connected. For example, we could have a surface integral over a plane annular region. In this case, its boundary would consist of both the inner and outer circles defining the annulus.

Integrals of vector fields are more interesting. First, it is obviously possible to extend eq. (7.1) to any component v_i of a vector field $\mathbf{v}(\mathbf{r})$:

$$\int_C v_i d\ell, \quad \int_S v_i ds, \quad \int_V v_i d^3x. \tag{7.2}$$

The integrals so defined, however, are not the objects that arise most naturally in physical theories. The most natural objects are scalars. Thus, the line integral is

$$\int_C \mathbf{v} \cdot d\ell, \tag{7.3}$$

which is defined by breaking up the curve of integration into infinitesimal segments. In each segment we introduce a vector $\Delta\ell$ of length $\Delta\ell$ and pointing along the tangent to the curve. The integral is then given by the limit as the number of segments tends to infinity:

$$\lim_{N\to\infty} \sum_{a=1}^{N} \mathbf{v}(\mathbf{r}_a) \cdot \Delta\ell_a. \tag{7.4}$$

Naturally, the various length elements $\Delta\ell_a$ must not reverse orientation abruptly, i.e., they must maintain the same sense as we move along the curve. If the curve is a straight line from $\mathbf{r}_1$ to $\mathbf{r}_2$, e.g., all $\Delta\ell_a$ must point from $\mathbf{r}_1$ to $\mathbf{r}_2$. For each line integral, therefore, we must specify an orientation.

When the curve C is closed, the line integral (7.3) is known as the *circulation* of $\mathbf{v}$ around the curve.

The surface integral is defined similarly. We associate with each element of the surface a vector $\Delta\mathbf{s}$ of magnitude equal to the area of the element, and orientation normal or perpendicular to the surface. The limit

$$\lim_{N\to\infty} \sum_{a=1}^{N} \mathbf{v}(\mathbf{r}_a) \cdot \Delta\mathbf{s}_a \tag{7.5}$$

then defines the surface integral, which we write as

$$\int_S \mathbf{v} \cdot d\mathbf{s} \quad \text{or} \quad \int_S \mathbf{v} \cdot \hat{\mathbf{n}} \, ds. \tag{7.6}$$

Again, there are two possible choices for the orientation $d\mathbf{s}$, and which one is meant must be specified. When the surface is closed, one generally chooses $d\mathbf{s}$ as the *outward* normal, i.e., pointing to infinity. The second form in eq. (7.6) stresses this aspect: the unit vector $\hat{\mathbf{n}}$ is the normal to the surface at a general point.

When the surface S is closed, the surface integral (7.6) is known as the *flux* of $\mathbf{v}$ through the surface.

Exercise 7.1 Evaluate the line integral (7.3) for $\mathbf{v} = \mathbf{r}$ over the following curves lying in the xy plane and centered on the origin: (a) a circle of radius R, (b) a square of side a. (**Answer:** (a) 0, (b) 0.)

Exercise 7.2 Evaluate the surface integral (7.6) for $\mathbf{v} = \mathbf{r}$ over (a) sphere of radius R centered on the origin, and (b) a cube of side a centered on the origin. (**Answer:** (a) $4\pi R^3$, (b) $3a^3$.)

8 The theorems of Stokes and Gauss

Next, let us discuss what happens when we integrate the various derivatives of a vector field. In the calculus of one variable, the integral of df/dt is the function f itself (up to an additive constant). The analogous result now is for a line integral over any curve running from a point $\mathbf{r}_1$ to another point $\mathbf{r}_2$:

$$\int_1^2 \nabla f \cdot d\boldsymbol{\ell} = f(\mathbf{r}_2) - f(\mathbf{r}_1). \tag{8.1}$$

Exercise 8.1 Prove eq. (8.1).

Equation (8.1) may be viewed as transforming the integral of a function over a zero-dimensional region (the right-hand side) into the integral of its derivative over a one-dimensional region. Note that the lower-dimensional region is the boundary of the higher dimensional region. Analogous results exist for higher dimensional (line and surface) integrals; indeed, they are of great importance in vector field theory.

The first result, known as *Stokes's theorem*, says that if C is a closed curve, then

$$\oint_{C = \partial S} \mathbf{v} \cdot d\boldsymbol{\ell} = \int_S \nabla \times \mathbf{v} \cdot d\mathbf{s}, \tag{8.2}$$

where S is *any* surface bounded by the curve C. The notation ∂S is used to denote the boundary of S. Note that there is an implicit convention for the signs of the integrals appearing in this theorem. This convention is that of the *right-hand rule*: if the fingers of the right-hand are curled so as to go around the loop C in the direction of the line integral, the normal to S must be chosen to point in the direction of the thumb.

The second result, known as *Gauss's theorem*, says that if S is any closed surface, then

$$\oint_{S = \partial V} \mathbf{v} \cdot d\mathbf{s} = \int_V \nabla \cdot \mathbf{v} d^3x, \tag{8.3}$$

where V is the volume enclosed by S. As in eq. (8.2), the notation $S = \partial V$ means that S is the boundary of V. The convention for the orientation of the surface normal is that it should point outward, i.e., away from the interior V.

Exercise 8.2 Obtain the results of exercises 7.1 and 7.2 using Stokes's and Gauss's theorems.

Exercise 8.3 Apply Stokes's theorem with two different surfaces S_1 and S_2 bounded by the closed curve C. According to the theorem, the two surface integrals of $\nabla \times \mathbf{v}$ are equal. Show this fact directly.

Solution: By reversing the sign of the surface normal in the integral over S_2, say, the difference between the two surface integrals is equivalent to an integral over a single closed surface S, composed of S_1 and S_2, and with a surface normal that is everywhere pointing outward or inward. Applying Gauss's theorem to this latter integral, and denoting the volume enclosed within S by V, we obtain

$$\int_{S_1} \nabla \times \mathbf{v} \cdot d\mathbf{s}_1 - \int_{S_2} \nabla \times \mathbf{v} \cdot d\mathbf{s}_2 = \int_{S=\partial V} \nabla \times \mathbf{v} \cdot d\mathbf{s}$$

$$= \int_V \nabla \cdot (\nabla \times \mathbf{v}) \, d^3x, \tag{8.4}$$

which vanishes, since the divergence of a curl is identically zero.

The theorems of Stokes and Gauss provide a vivid and physical way of thinking of the curl and divergence of a vector field. In fact, we could define the curl by using Stokes's theorem, since we could choose the surface S to be infinitesimally small. Similarly, the divergence could be defined in terms of Gauss's theorem.

The proofs of these two theorems are given in all books on vector calculus. "Proofs" that emphasize the physical ideas underlying the theorems are easily constructed as follows. For Stokes's theorem, one first establishes the result for an infinitesimal square loop (oriented along any of the three Cartesian coordinate planes) by Taylor expanding the vector field $\mathbf{v}$ about a point inside the loop. One then argues that any large loop can be approximated by adding together many small square loops. The line integrals along the interior legs all cancel in pairs, and the surface integrals over the small squares simply add. For Gauss's theorem, one proceeds similarly, by employing infinitesimal cubes.[10] Readers are encouraged to fill in the missing steps in this argument for themselves.

Exercise 8.4 (Green's identities) Given two scalar fields, $f_1(\mathbf{r})$ and $f_2(\mathbf{r})$, show that

$$\int_V (f_1 \nabla^2 f_2 + \nabla f_1 \cdot \nabla f_2) \, d^3x = \oint_{\partial V} f_1 \nabla f_2 \cdot d\mathbf{s}, \tag{8.5}$$

$$\int_V (f_1 \nabla^2 f_2 - f_2 \nabla^2 f_1) \, d^3x = \oint_{\partial V} (f_1 \nabla f_2 - f_2 \nabla f_1) \cdot d\mathbf{s}. \tag{8.6}$$

It should be noted that there are many other results that transform an integral over a d-dimensional region to one over its boundary, a $(d-1)$-dimensional region. For example, as a vector variant of Gauss's theorem, we have

$$\int_V (\nabla \times \mathbf{v}) d^3x = \oint_{S=\partial V} (\hat{\mathbf{n}} \times \mathbf{v}) d^2s. \tag{8.7}$$

---

[10] See, e.g., Feynman et al. (1964), vol. II, chap. 3.

Note that both integrands are now vectors. Superficially, the integrand on the left-hand side looks very different from a divergence. However, if we write

$$(\nabla \times \mathbf{v})_i = \epsilon_{ijk}\partial_j v_k = \partial_j(\epsilon_{jki}v_k), \tag{8.8}$$

for fixed i, then $\epsilon_{jki}v_k$ may be regarded as the j component of some vector $\mathbf{w}$, and the integrand *is* a divergence, $\nabla \cdot \mathbf{w}$. Gauss's theorem is now applicable and yields

$$\int_V (\nabla \times \mathbf{v})_i d^3x = \oint_S \epsilon_{jki}v_k n_j d^2s = \oint_S (\hat{\mathbf{n}} \times \mathbf{v})_i d^2s, \tag{8.9}$$

which completes the proof. One can remember all these variants via the symbolic replacement

$$\int_V \nabla d^3x \rightarrow \oint_{S=\partial V} \hat{\mathbf{n}} d^2s. \tag{8.10}$$

Likewise, Stokes's theorem corresponds to symbolic replacement rule

$$\int_S (d\mathbf{s} \times \nabla) \rightarrow \oint_{C=\partial S} d\boldsymbol{\ell}. \tag{8.11}$$

Exercise 8.5 Rewrite the following volume integrals as surface integrals:

$$\int_V \nabla f d^3x, \quad \int_V (\mathbf{a} \cdot \nabla)\mathbf{v} d^3x. \tag{8.12}$$

Here $\mathbf{a}$ is a constant vector, and f and $\mathbf{v}$ are scalar and vector fields.

Exercise 8.6 Rewrite $\oint_C f(\mathbf{r})d\boldsymbol{\ell}$ as a surface integral.

Exercise 8.7 (vector variant of Stokes's theorem) Show that

$$\int_S (d\mathbf{s} \times \nabla) \times \mathbf{v} = \oint_{C=\partial S} (d\boldsymbol{\ell} \times \mathbf{v}). \tag{8.13}$$

The theorems of Stokes and Gauss also enable us to derive two more results of wide applicability. We already know that if $\mathbf{v} = \nabla f$, then $\nabla \times \mathbf{v} = 0$, and that if $\mathbf{v} = \nabla \times \mathbf{w}$, then $\nabla \cdot \mathbf{v} = 0$. Are the converses also true? Under certain conditions[11] they are, as the following theorems state.

Theorem for curl-free fields: Let $\mathbf{v}$ be a vector field satisfying

$$\nabla \times \mathbf{v} = 0 \tag{8.14}$$

everywhere in some region V that is simply connected, i.e., such that any closed loop in V can be continuously shrunk to a point. Then $\mathbf{v}$ is the gradient of a scalar field in V. In other words, there exists some scalar field f such that $\mathbf{v} = \nabla f$.

---

[11] The conditions pertain to the topology of the region in which $\nabla \times \mathbf{v}$ or $\nabla \cdot \mathbf{v}$ vanishes. The reasons for imposing them are discussed in exercises 8.9 and 8.10 near the end of this section.

Theorem for divergence-free fields: Let $\mathbf{v}$ be a vector field satisfying

$$\nabla \cdot \mathbf{v} = 0 \tag{8.15}$$

everywhere in some region V, which is such that any closed surface lying in V can be continuously shrunk to a point. Then $\mathbf{v}$ is the curl of a vector field in V. In other words, there exists some vector field $\mathbf{w}$ such that $\mathbf{v} = \nabla \times \mathbf{w}$.

A curl-free field is also referred to as *irrotational*. A divergence-free field is also referred to as *solenoidal*.

To prove the first theorem, let us integrate $\mathbf{v}$ over two open curves C_1 and C_2 lying entirely in V, both running from a point $\mathbf{r}_1$ to another point $\mathbf{r}_2$. By going from $\mathbf{r}_1$ to $\mathbf{r}_2$ along C_1 and back to $\mathbf{r}_1$ along C_2 we obtain a single closed curve $C_1 - C_2$. Stokes's theorem now yields

$$\oint_{C_1 - C_2} \mathbf{v} \cdot d\boldsymbol{\ell} = \int_S \nabla \times \mathbf{v} \cdot d\mathbf{s} = 0, \tag{8.16}$$

since $\nabla \times \mathbf{v} = 0$ everywhere in V. In other words, the line integral of $\mathbf{v}$ from $\mathbf{r}_1$ to $\mathbf{r}_2$ is independent of the path of integration:

$$\int_C \mathbf{v} \cdot d\boldsymbol{\ell} \quad \text{is independent of } C. \tag{8.17}$$

This means, however, that the integral depends only on the positions of the points $\mathbf{r}_1$ and $\mathbf{r}_2$, or that we may write it as the difference $f(\mathbf{r}_2) - f(\mathbf{r}_1)$ of some scalar function f. By eq. (8.1), it then follows that

$$\int_C \mathbf{v} \cdot d\boldsymbol{\ell} = \int_C \nabla f \cdot d\boldsymbol{\ell}, \tag{8.18}$$

where the curve C is arbitrary as long as it lies inside V. By transposing the right-hand side to the left and writing the difference as a single integral, we conclude that $\mathbf{v} = \nabla f$.

It is apparent that the function $f(\mathbf{r})$ is determined only up to an additive constant, since the gradient of a constant is zero.

The proof of the second theorem is analogous. We consider a closed curve C, and two surfaces S_1 and S_2 that span this curve, i.e., both S_1 and S_2 are bounded by C. By taking the surface integral of $\mathbf{v}$ over the difference $S_1 - S_2$, transforming to a volume integral by Gauss's theorem, and using the hypothesis $\nabla \cdot \mathbf{v} = 0$, we conclude the two surface integrals are equal:

$$\int_{S_1} \mathbf{v} \cdot d\mathbf{s} = \int_{S_2} \mathbf{v} \cdot d\mathbf{s} \quad (\partial S_1 = \partial S_2). \tag{8.19}$$

In other words, for a given curve C, the surface integral is independent of the spanning surface. In this way we can associate a function with any closed loop C in V. This function has the additional (easily shown) property that if we divide a loop into two subloops C_1 and C_2 by drawing a curve between any two points on C, then the function for C is the sum of the functions for C_1 and C_2. The most general function with such a property is the line integral of a vector field $\mathbf{w}$. We have therefore shown that if $\nabla \cdot \mathbf{v} = 0$, then there exists a vector field $\mathbf{w}$ such that

$$\int_S \mathbf{v} \cdot d\mathbf{s} = \oint_{C=\partial S} \mathbf{w} \cdot d\boldsymbol{\ell}. \tag{8.20}$$

Writing the right-hand side as a surface integral by Stokes's theorem, we conclude that $\mathbf{v} = \nabla \times \mathbf{w}$.

As in the case of curl-free fields, the vector field $\mathbf{w}(\mathbf{r})$ is not uniquely determined, since we may add to it the gradient of any other scalar field $g(\mathbf{r})$.

Exercise 8.8 Show that if for a vector field $\mathbf{v}(\mathbf{r})$,

$$\oint_S \mathbf{v} \cdot d\mathbf{s} = 0 \tag{8.21}$$

for any closed surface S, then $\mathbf{v}$ is a curl.

Exercise 8.9 Consider the vector field

$$\mathbf{v} = \frac{y\hat{\mathbf{x}} - x\hat{\mathbf{y}}}{x^2 + y^2}. \tag{8.22}$$

Show that $\nabla \cdot \mathbf{v} = 0$, and that $\mathbf{v} = \nabla \times \mathbf{w}$, with $\mathbf{w} = -\ln \rho \hat{\mathbf{z}}$, where $\rho = (x^2 + y^2)^{1/2}$.

Comments: In this example, both vector fields $\mathbf{v}$ and $\mathbf{w}$ are defined everywhere except on the line $x = y = 0$. Thus, we see that a solenoidal field can sometimes be represented as a curl even when all surfaces in the region under consideration are not shrinkable to a point, and that this condition is sufficient but not necessary. The next example is less trivial, and shows a case where this condition is not obeyed and the field cannot be written as a curl.

Exercise 8.10 Consider the vector field

$$\mathbf{v} = \frac{\mathbf{r}}{r^3}. \tag{8.23}$$

(a) Show that $\nabla \cdot \mathbf{v} = 0$ everywhere except at the origin ($r = 0$). In this case, however, if we try to write $\mathbf{v} = \nabla \times \mathbf{w}$, we find that $\mathbf{w}$ is "bad" at points besides the origin. Show, in particular, that $\mathbf{w}$ must be singular somewhere on any spherical surface of radius $R \neq 0$ centered on the origin.

(b) Show that the following choice of $\mathbf{w}$ "works":

$$\mathbf{w} = \frac{1 - \cos \theta}{r \sin \theta} \mathbf{e}_\varphi, \tag{8.24}$$

where $\mathbf{e}_\varphi$ is the unit vector along lines of increasing φ (circles of latitude) in spherical polar coordinates. Discuss how this result fits into the conclusions from part (a).

It is perhaps worth noting at this point that many of the considerations of the last few sections acquire powerful and elegant interpretations in the language of differential forms and exterior calculus. Thus, Stokes's and Gauss's theorems become instances of a single larger result. The formulas $\nabla \times (\nabla f) = 0$ and $\nabla \cdot (\nabla \times \mathbf{v}) = 0$ are special cases of a single result known as *Poincare's lemma*. We have proved the converse of these results, namely, that a curl-free field is a gradient, and that a divergence-free field is a curl, only when the relevant domain V is topologically simple. The correct formulation of the converse when the domain is not simple is governed by two theorems named after De Rham, and their discussion is best done in terms of differential forms. Many other topics in multivariable calculus, Riemannian geometry, and Hamiltonian mechanics also acquire simpler interpretations in this language.[12] For practical applications of electromagnetism, however, it seems best to stick with vectors and tensors as indexed quantities, which is what we shall do.

Exercise 8.11 Let $f(\hat{\mathbf{r}})$ and $g(\hat{\mathbf{r}})$ be arbitrary, continuous, complex-valued functions depending only on the orientation or unit vector $\hat{\mathbf{r}} = \mathbf{r}/r$, and let $\langle f, g \rangle$ denote the inner product

$$\langle f, g \rangle = \int f^*(\hat{\mathbf{r}})g(\hat{\mathbf{r}})d\Omega, \tag{8.25}$$

where $\int d\Omega$ denotes integration over the unit sphere. Show that

(a) $\langle f, L_{op}^i g \rangle = \langle L_{op}^i f, g \rangle$,

(b) $\langle f, \mathbf{L}_{op}^2 g \rangle = \langle \mathbf{L}_{op}^2 f, g \rangle$,

(c) $\langle f, L_{op}^\pm g \rangle = \langle L_{op}^\mp f, g \rangle$, $\qquad\qquad$ (8.26)

(d) $\langle f, \mathbf{L}_{op}^2 f \rangle \geq \langle f, (L_{op}^z)^2 f \rangle$.

9 Fourier transforms, delta functions, and distributions

Fourier transform methods often provide the quickest solution to many problems in electrodynamics. In addition, the Fourier representation of electromagnetic fields has a direct physical interpretation in terms of plane waves. Our aim in this section is to discuss the principal and most frequently used results from the basic theory. For the most part, these results are stated without proof.[13]

Let $f(x)$ be a function of a real variable x. We shall suppose that as $x \to \pm\infty$, $f(x)$ tends to zero, and that it does so faster than any power of x. In other words, $|x|^n f(x) \to 0$ as $|x| \to \infty$ for any $n > 0$. We then define the Fourier transform of $f(x)$ as

$$f_k = \int_{-\infty}^{\infty} f(x)e^{-ikx}dx. \tag{9.1}$$

The basic theorem, known as the *inversion theorem*, states that

$$f(x) = \int_{-\infty}^{\infty} f_k e^{ikx} \frac{dk}{2\pi}. \tag{9.2}$$

Because of the symmetry between eqs. (9.1) and (9.2), $f(x)$ and f_k are said to form a Fourier transform pair. There are many other conventions for defining these pairs. The

---

[12] A lucid introduction to these topics is given by Flanders (1989).
[13] For an excellent account of the subject, see the monograph by Lighthill (1958).

signs of the exponents in the phase factors may be reversed, the factor of $1/2\pi$ may be put in eq. (9.1) instead of eq. (9.2), or split into a factor of $1/\sqrt{2\pi}$ in each equation, or removed altogether by writing the phase factors as $\pm 2\pi i k x$. We shall generally use x to denote a spatial variable, in which case it is useful to think of eq. (9.2) as representing $f(x)$ as a sum of plane waves e^{ikx}; k has the dimensions and meaning of a wave vector, and f_k is said to be the kth Fourier component of $f(x)$. Because of the de Broglie relation from quantum mechanics, one often speaks of k as momentum, and k space as momentum space, even when the context is purely classical and involves no quantum mechanics.

These relations are immediately extended to a function of all three space variables $\mathbf{r} = (x, y, z)$, and/or of time t. In the first case, the Fourier transform variable is the wave vector $\mathbf{k} = (k_x, k_y, k_z)$, and in the second the frequency ω. We then have the following Fourier transform pairs:

$$f_{\mathbf{k}} = \int f(\mathbf{r}) e^{-i\mathbf{k}\cdot\mathbf{r}} d^3x, \quad f(\mathbf{r}) = \int f_{\mathbf{k}} e^{i\mathbf{k}\cdot\mathbf{r}} \frac{d^3k}{(2\pi)^3}; \tag{9.3}$$

$$f_{\omega} = \int_{-\infty}^{\infty} f(t) e^{i\omega t} dt, \quad f(t) = \int_{-\infty}^{\infty} f_{\omega} e^{-i\omega t} \frac{d\omega}{2\pi}; \tag{9.4}$$

$$f_{\mathbf{k},\omega} = \int f(\mathbf{r}, t) e^{-i(\mathbf{k}\cdot\mathbf{r}-\omega t)} d^3x \, dt, \quad f(\mathbf{r}, t) = \int f_{\mathbf{k},\omega} e^{i(\mathbf{k}\cdot\mathbf{r}-\omega t)} \frac{d^3k}{(2\pi)^3} \frac{d\omega}{2\pi}. \tag{9.5}$$

Our conventions in writing down Fourier transforms are as follows. First, we always put our factors of 2π in the k space or frequency space integral. The signs of the phases are a little more difficult to remember. The way that the author uses is to think of the Fourier representation for the real-space or time-dependent function $f(\mathbf{r}, t)$. The phase factor in this is $e^{i(\mathbf{k}\cdot\mathbf{r}-\omega t)}$, which is the canonical factor for a plane wave traveling in the $+\mathbf{k}$ direction. The phase factors in the inverse formulas then follow automatically.

Further, note that even though we have written down our relations for a scalar field, f could stand for any one component of a vector field, or for the vector itself, component by component. In that case, instead of eq. (9.3), e.g., we would write

$$\mathbf{v}_{\mathbf{k}} = \int \mathbf{v}(\mathbf{r}) e^{-i\mathbf{k}\cdot\mathbf{r}} d^3x, \quad \mathbf{v}(\mathbf{r}) = \int \mathbf{v}_{\mathbf{k}} e^{i\mathbf{k}\cdot\mathbf{r}} \frac{d^3k}{(2\pi)^3}. \tag{9.6}$$

To avoid cluttering the subsequent formulas unnecessarily, we shall stick to scalar-valued fields. The extension to vector fields is immediate.

Exercise 9.1 Show that (a) if $f(\mathbf{r})$ is real, then $f_{-\mathbf{k}} = f_{\mathbf{k}}^*$; (b) if $f(\mathbf{r})$ is inversion symmetric, then so is $f_{\mathbf{k}}$, i.e., that $f_{-\mathbf{k}} = f_{\mathbf{k}}$ if $f(-\mathbf{r}) = f(\mathbf{r})$.

Exercise 9.2 Show that for a spherically symmetric function $f(r)$ that depends only on $r = |\mathbf{r}|$,

$$f_{\mathbf{k}} = \int f(r) \frac{\sin kr}{kr} d^3x = 4\pi \int_0^{\infty} f(r) \frac{\sin kr}{kr} r^2 \, dr. \tag{9.7}$$

Comments: This result is best remembered as the rule that the factor $e^{\pm i\mathbf{k}\cdot\mathbf{r}}$ is replaced by $\sin kr/kr$, both of which tend to 1 as k tends to 0. And, since $f_{\mathbf{k}}$ is also spherically symmetric, we may write f_k instead of $f_{\mathbf{k}}$ (provided we remember that we are dealing with a function of three and not just one variable).

Secondly, we give the analog of eq. (9.7) in two dimensions for reference. For a function $f(x, y)$ that depends only on $r = (x^2 + y^2)^{1/2}$, we have

$$f_k = 2\pi \int_0^\infty f(r) J_0(kr) r \, dr, \tag{9.8}$$

where J_0 is the Bessel function of order 0.[14]

Exercise 9.3 If the Fourier transform of $f(\mathbf{r})$ is $f_{\mathbf{k}}$, show that the Fourier transform of ∇f is $i\mathbf{k} f_{\mathbf{k}}$, i.e.,

$$\int \nabla f(\mathbf{r}) e^{-i\mathbf{k}\cdot\mathbf{r}} d^3 x = i\mathbf{k} f_{\mathbf{k}}. \tag{9.9}$$

Solution: Consider the x component of the gradient. We have

$$(\partial_x f(\mathbf{r})) e^{-i\mathbf{k}\cdot\mathbf{r}} = \partial_x (f(\mathbf{r}) e^{-i\mathbf{k}\cdot\mathbf{r}}) + i k_x f(\mathbf{r}) e^{-i\mathbf{k}\cdot\mathbf{r}}. \tag{9.10}$$

If we integrate this relation over all space, the first term on the right can be integrated with respect to x immediately, and it vanishes, since $f(\mathbf{r}) \to 0$ as $x \to \pm\infty$. The second term gives the Fourier transform of $f(\mathbf{r})$, so

$$\int (\partial_x f(\mathbf{r})) e^{-i\mathbf{k}\cdot\mathbf{r}} d^3 x = i k_x f_{\mathbf{k}}. \tag{9.11}$$

This is the x component of eq. (9.9). The other components are handled similarly, which proves the result.

Alternatively, we differentiate under the integral sign in the second member of the pair (9.3). This gives

$$\nabla f(\mathbf{r}) = \int i\mathbf{k} f_{\mathbf{k}} e^{i\mathbf{k}\cdot\mathbf{r}} \frac{d^3 k}{(2\pi)^3}. \tag{9.12}$$

To justify this operation, the integral for $f(\mathbf{r})$ over $\mathbf{k}$ must be uniformly convergent with respect to $\mathbf{r}$. In most physical situations, such is the case, and demonstrating uniform convergence is an unnecessary nicety.

Equation (9.9) is one of the most useful results in Fourier analysis. It reduces differentiation to multiplication, turning some of the commonest differential equations into algebraic equations. We sometimes say that the ∇ "brings down" a $\mathbf{k}$. As another use of this mnemonic, note that the Fourier transform of $\nabla \cdot \mathbf{v}$ is $i\mathbf{k} \cdot \mathbf{v}_{\mathbf{k}}$.

An important concept that greatly simplifies work with Fourier transforms is that of the *Dirac delta function*. To motivate this, let us pose the following question. Does there

---

[14] Bessel functions are discussed in appendix B.

exist a function $\delta(x - x')$ such that

$$f(x) = \int_{-\infty}^{\infty} f(x')\delta(x - x')dx' \tag{9.13}$$

for *any* well-behaved function $f(x)$? The equation can be viewed as demanding that $f(x)$ be equal to its average over some region of x values with respect to the weighting function $\delta(x - x')$. The region of space over which the average is taken may be very narrow if δ is a sharply peaked function, but it is clear that if the result is to hold for *all* $f(x)$, δ must in fact be infinitely narrow, i.e., that $\delta(x - x') = 0$ if $x \neq x'$. But then, the entire contribution to the integral must come from a single point $x' = x$, which seems absurd.

Exercise 9.4 Why, in eq. (9.13), do we seek δ as a function of the difference $x - x'$, rather than as a general function of both x and x'? Show that it makes no difference, i.e., that if we wish to satisfy

$$f(x) = \int f(x')\delta(x, x')dx' \tag{9.14}$$

for all $f(x)$, then $\delta(x, x')$ can depend only on the combination $x - x'$ and, in fact, only on $|x - x'|$.

Exercise 9.5 Show that if eq. (9.13) holds, then

$$\int_{-\infty}^{\infty} \delta(x - x')dx' = 1. \tag{9.15}$$

It is clear that no ordinary function[15] $\delta(x)$ can satisfy the above requirements. They can be satisfied, however, if we think of $\delta(x)$ as a sequence of functions $g_n(x)$ such that

$$\int_{-\infty}^{\infty} g_n(x)dx = 1 \text{ for all } n, \tag{9.16}$$

$$\lim_{n \to \infty} g_n(x) = 0, \quad x \neq 0. \tag{9.17}$$

What is really meant by eq. (9.13) is then

$$f(x) = \lim_{n \to \infty} \int_{-\infty}^{\infty} f(x')g_n(x - x')dx'. \tag{9.18}$$

Loosely speaking, $\delta(x)$ is the limit of a sequence of functions all with unit area that become infinitely narrow. We may visualize it as an infinitely sharp spike at $x = 0$ with unit weight under it. An example is the sequence of Lorentzians

$$g_n(x) = \frac{1}{n\pi} \frac{1}{x^2 + n^{-2}}. \tag{9.19}$$

We could also take the Gaussians

$$g_n(x) = \frac{n}{\sqrt{2\pi}} e^{-n^2 x^2 / 2}. \tag{9.20}$$

---

[15] From now on we will often write δ as a function with a single argument x or x'. When it is necessary to write the argument as $x - x'$, we will use the result from exercise 9.4 that $\delta(x - x') = \delta(x' - x)$.

For mathematical nicety, it is generally convenient to choose the functions g_n to be infinitely differentiable. Obviously, there are many sequences that we can choose, and $\delta(x)$ is associated with the entire class of sequences obeying eqs. (9.16) and (9.17). The limit in eq. (9.18) does not depend on which sequence is chosen.

The function $\delta(x)$ is the Dirac delta function. Properly speaking, however, it is not a function, but a *generalized function* or *distribution*, a quantity that makes sense only when one integrates over it. This point is worth keeping in mind. Many formulas in physics are most conveniently written in terms of "naked" delta functions when one wishes to represent a quantity—charge, probability of observation, etc.—that is concentrated in a very, very narrow range of some variable. Any measurable quantity, however, will necessarily involve an integral over the delta function.

Since $\delta(x)$ is concentrated entirely at $x = 0$, one need not integrate it over all space, and any interval that contains the origin is equally good. In fact, assuming $b > a$,

$$\int_a^b f(x)\delta(x)dx = \begin{cases} f(0), & \text{if } 0 \in (a, b); \\ 0, & \text{otherwise.} \end{cases} \tag{9.21}$$

The Fourier transform of the delta function and the inverse relation are particularly important. Consider the transform pair

$$\int_{-\infty}^{\infty} \frac{n}{\sqrt{2\pi}} e^{-n^2 x^2/2} e^{-ikx} dx = e^{-k^2/2n^2}; \quad \int_{-\infty}^{\infty} e^{-k^2/2n^2} e^{ikx} \frac{dk}{2\pi} = \frac{n}{\sqrt{2\pi}} e^{-n^2 x^2/2}. \tag{9.22}$$

Taking the limit as $n \to \infty$, we may write

$$\int_{-\infty}^{\infty} \delta(x) e^{-ikx} dx = 1; \quad \int_{-\infty}^{\infty} e^{ikx} \frac{dk}{2\pi} = \delta(x). \tag{9.23}$$

Exercise 9.6 Verify eq. (9.22).

Exercise 9.7 Verify eq. (9.23) by using the Lorentzian sequence (9.19).

Exercise 9.8 Show that

$$\int_{-\infty}^{\infty} e^{\pm ikx} dx = 2\pi\delta(k). \tag{9.24}$$

There are many other generalized functions of use in physics. An obvious example is the derivative of a delta function, $\delta'(x)$. This is defined as the sequence of derivatives $g_n'(x)$ of any of the sequences equivalent to $\delta(x)$. The definition of generalized functions in terms of sequences allows one to perform most operations of analysis in the usual way, by carrying out the operations on the sequences.[16] In this way, many operations that would otherwise be improper become sensible. As an example, let us substitute the formula (9.1)

---

[16] Thus, in addition to being Fourier transformed, generalized functions may be multiplied by an ordinary function, added together, or differentiated, either with respect to the argument or with respect to a parameter. One may not, however, multiply two generalized functions of the same variable.

for f_k in eq. (9.2) for $f(x)$. This gives

$$f(x) = \iint f(x')e^{ik(x-x')}dx'\frac{dk}{2\pi}. \tag{9.25}$$

Comparing eq. (9.25) with eq. (9.13), we are led to identify the k integral of the exponential factor with $\delta(x - x')$:

$$\int e^{ik(x-x')}\frac{dk}{2\pi} = \delta(x - x'). \tag{9.26}$$

This is, of course, equivalent to the second of eq. (9.23); we simply write $x - x'$ for x. The point is that interchanging the orders of integration in the double integral above would normally have to be justified with a tedious argument, which generalized functions permit us to sidestep. They permit us to talk about Fourier transforms of a larger class of functions than we first considered, and the restrictions on how rapidly these functions must vanish at infinity can be lifted.

Exercise 9.9 Show, by constructing an appropriate sequence of functions, that

$$\int_{-\infty}^{\infty} xe^{-ikx} = 2\pi i\delta'(k). \tag{9.27}$$

Exercise 9.10 Show that

$$\int_{-\infty}^{\infty} f(x)\delta'(x)dx = -f'(0). \tag{9.28}$$

Solution: Let $g_n(x)$ be a sequence for $\delta(x)$. Integrate by parts the expression

$$\int_{-\infty}^{\infty} f(x)g_n'(x)dx, \tag{9.29}$$

and let $n \xrightarrow{\cdot} \infty$. Alternatively, we simply integrate by parts in eq. (9.28) and argue that the integrated term vanishes because $\delta(\pm\infty) = 0$.

Exercise 9.11 Show that if the Fourier transforms of $f(x)$ and $g(x)$ are f_k and g_k, then

$$\int_{-\infty}^{\infty} f(x)g(x)dx = \int_{-\infty}^{\infty} f_k g_{-k}\frac{dk}{2\pi}; \quad \int_{-\infty}^{\infty} f_k g_k\frac{dk}{2\pi} = \int_{-\infty}^{\infty} f(x)g(-x)dx. \tag{9.30}$$

This is known as *Parseval's theorem*. The cases $f(x) = g(x)$ and $f(x) = g^*(x)$ are particularly important in physics. The reader should write down the analogous results for functions of **r**.

Exercise 9.12 Prove the *convolution theorem*:

$$\int_{-\infty}^{\infty} f(x)g(x)e^{-ikx}dx = \int_{-\infty}^{\infty} f_{k'}g_{k-k'}\frac{dk'}{2\pi}, \tag{9.31}$$

$$\int_{-\infty}^{\infty} f_k g_k e^{ikx}\frac{dk}{2\pi} = \int_{-\infty}^{\infty} f(x')g(x - x')dx'. \tag{9.32}$$

The integrals on the right are known as *convolutions* and are sometimes denoted by a star product symbol:

$$f(x) * g(x) \equiv \int_{-\infty}^{\infty} f(x')g(x-x')dx', \tag{9.33}$$

$$f_k * g_k \equiv \int_{-\infty}^{\infty} f_{k'}g_{k-k'}\frac{dk'}{2\pi}. \tag{9.34}$$

The theorem can be remembered in words as "The FT of a product is the convolution of the FTs."

In dealing with fields, or functions of $\mathbf{r} = (x, y, z)$, one is led to define delta functions of vector arguments, such as

$$\delta(\mathbf{r} - \mathbf{r}') \equiv \delta(x - x')\delta(y - y')\delta(z - z'), \tag{9.35}$$

and similarly $\delta(\mathbf{k} - \mathbf{k}')$. The utility of this notation is evident from results such as

$$\int e^{i\mathbf{k}\cdot(\mathbf{r}-\mathbf{r}')}\frac{d^3k}{2\pi} = \delta(\mathbf{r} - \mathbf{r}'), \tag{9.36}$$

which is easily seen to be true by factorizing the integral into three separate integrals over k_x, k_y, and k_z. Sometimes one writes $\delta^{(3)}(\mathbf{r} - \mathbf{r}')$ to emphasize the dimensionality of the vector argument. One can now go ahead and introduce functions like $\nabla_r\delta(\mathbf{r} - \mathbf{r}')$, where the subscript on the gradient indicates that it is with respect to the variable $\mathbf{r}$.

Exercise 9.13 Show that the Fourier transform of $1/r$ is $4\pi/k^2$. (*Hint*: Introduce a convergence factor $e^{-\mu r}$ into the integral, and let $\mu \to 0$ at the end. If you find this disturbing, think about how you would solve the problem "Show that the FT of 1 is $2\pi\delta(k)$.")

Let us now consider the behavior of delta functions under changes of variable. Naively, one might guess that $\delta(x)$, $\delta(2x)$, and $\delta(x^2)$ are all equivalent, since they are all concentrated at $x = 0$. This is not so. First, let us consider $\delta(ax)$, where a is any nonzero constant. Then,

$$\delta(ax) = \frac{1}{|a|}\delta(x). \tag{9.37}$$

For, with any $f(x)$, the substitution $y = |a|x$ gives

$$\int_{-\infty}^{\infty} f(x)\delta(ax)dx = \frac{1}{|a|}\int_{-\infty}^{\infty} f(y/|a|)\,\delta(y)dy = \frac{1}{|a|}f(0). \tag{9.38}$$

The only subtlety is why we have $|a|$ and not a. If $a < 0$, the substitution $y = ax$ causes the upper and lower limits of the y integral to be interchanged, and the final answer is $-f(0)/a$, which equals $f(0)/|a|$ once again.

Next, let us consider $\delta[g(x)]$, where $g(x)$ is some function. In this case,

$$\delta[g(x)] = \sum_i \frac{1}{|g'(x_i)|}\delta(x - x_i), \tag{9.39}$$

where the x_i are zeros of $g(x)$, and the sum is over all of them. The result is easily proved by integrating the product $f(x)\delta[g(x)]$. The argument is very similar to that for eq. (9.37), and the details are left to the reader.

Exercise 9.14 Show that

$$g(x)\,\delta(x - x') = g(x')\,\delta(x - x').$$ (9.40)

As an application of these ideas, let us consider a very thin spherical shell of some insulating material, which is uniformly charged. The radius of the shell is a, and the total charge is Q. Since the shell is very thin, it is natural to write the charge distribution $\rho(\mathbf{r})$ using delta functions. How should it be done? In spherical polar coordinates, ρ is clearly proportional to $Q\delta(r - a)$. The constant of proportionality is obtained by noting that by spherical symmetry, ρ depends only on the radial distance r, so that

$$\int \rho(\mathbf{r})d^3x = \int_0^\infty \rho(r)4\pi r^2 dr.$$ (9.41)

The integral must equal Q. To obtain this we must choose

$$\rho(\mathbf{r}) = \frac{Q}{4\pi r^2}\delta(r - a) = \frac{Q}{4\pi a^2}\delta(r - a).$$ (9.42)

In physical applications, it is often useful to think of $\delta(X)$ as having dimensions of $1/X$, whatever X may be. The results (9.37) and (9.38) are clearly in agreement with this point, as is the basic formula for the integral of $\delta(x)$.

Exercise 9.15 Show that if $g(x)$ is a function with a zero at $x = x_i$,

$$\delta'[g(x)] = \operatorname{sgn} g'\left[\frac{1}{g'^2}\delta'(x - x_i) + \frac{g''}{g'^3}\delta(x - x_i)\right],$$ (9.43)

$$\delta''[g(x)] = \operatorname{sgn} g'\left[\frac{1}{g'^3}\delta''(x - x_i) + 3\frac{g''}{g'^4}\delta'(x - x_i) + \left(3\frac{g''^2}{g'^5} - \frac{g'''}{g'^4}\right)\delta(x - x_i)\right].$$ (9.44)

Here, the successive derivatives g', g'', and g''' are all evaluated at x_i, and if g has more than one zero, one must add similar contributions from each zero. We shall use eq. (9.44) in chapter 25 when we study synchrotron radiation.

In the discussion above, we have made use of the concept of the *principal value*, or *Cauchy principal value*, without mentioning it. Indeed, this concept is integral to the entire structure of Fourier transforms. Consider, for example, the formula

$$\int_{-\infty}^\infty \frac{e^{-ikx}}{x^2 + \epsilon^2}dx = \frac{\pi}{\epsilon}e^{-|k|\epsilon}.$$ (9.45)

The integral on the left is improper. The least restrictive way to interpret it would be to replace the limits of integration by $-L'$ and L, and let L' and L tend to infinity separately. This double limit may exist in some cases, but it often does not. We therefore interpret the integral as

$$\lim_{L\to\infty} \int_{-L}^{L} \frac{e^{-ikx}}{x^2 + \epsilon^2} dx, \tag{9.46}$$

where we integrate over the *symmetric* interval $(-L, L)$, and only then take the limit. This is the definition of a principal value integral, and it is sometimes indicated by putting the letter P before the integral. Readers will recognize that this is the definition implicitly assumed when one evaluates the integral by using the standard semicircular contour with a large arc that tends to infinity.

Sometimes, one also has to define a principal value when the limits are finite. Such is the case for the integral

$$\int_{-3}^{5} \frac{dx}{x}. \tag{9.47}$$

To deal with the singularity at $x = 0$, one has to excise a symmetric interval $(-\epsilon, \epsilon)$, and let $\epsilon \to 0$. Thus, the principal value definition of this integral is

$$P\int_{-3}^{5} \frac{dx}{x} = \lim_{\epsilon\to 0}\left[\int_{-3}^{-\epsilon} + \int_{\epsilon}^{5}\right] = \lim_{\epsilon\to 0}\left[\ln\frac{\epsilon}{3} + \ln\frac{5}{\epsilon}\right] = \ln\frac{5}{3}. \tag{9.48}$$

An asymmetric choice of limits would obviously not have led to a finite and unique answer.

The following exercise gives an example where both types of limits have to be taken.

Exercise 9.16 Find the Fourier transform of $1/x$, by performing the integration by contours.

Answer: $-i\pi\,\mathrm{sgn}\,k$.

Exercise 9.17 Find the Fourier transform of $1/x$ as follows. Consider the function $x/(x^2 + \epsilon^2)$. The FTs of x and $1/(x^2 + \epsilon^2)$ have been given or found above. Convolve them and take the limit as $\epsilon \to 0$.

10 Rotational transformations of vectors and tensors

In our discussion so far, we have relied heavily on the geometrical meaning of vectors and vector operations, especially that a vector is something that points somewhere in space. We now expand on these ideas. The key point is that whether a physical quantity is a scalar, vector, or tensor depends on how the physical property in question is seen by two observers who chose to orient their reference frames differently. In particular, a vector field should not be regarded merely as a collection of three numbers at every point in space, since this point of view ignores the vectorial content of the quantity in question, i.e., that upon a rotation from one reference frame to another, the Cartesian components of the quantity will combine with each other in a specific way so as to give its projections

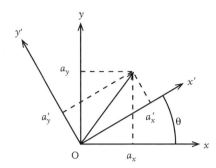

FIGURE 2.1. Components of a vector in two frames, one rotated with respect to the other.

along the new axes. With this in mind, let us now discuss the rotational properties of vectors and tensors in more detail.

In ordinary three-dimensional Euclidean space, the Pythagorean rule (5.3) for the length of a vector $\mathbf{a}$ yields an answer that is independent of how the coordinate axes are oriented. We say that the length, a, is a *scalar*. If we choose two sets of axes, the second rotated with respect to the first by an angle θ about z, as in fig. 2.1, e.g., then

$$a'_x = a_x \cos\theta + a_y \sin\theta,$$

$$a'_y = -a_x \sin\theta + a_y \cos\theta, \qquad\qquad (10.1)$$

$$a'_z = a_z.$$

Now it is easy to verify that

$$\sqrt{(a'_x)^2 + (a'_y)^2 + (a'_z)^2} = \sqrt{a_x^2 + a_y^2 + a_z^2}. \qquad\qquad (10.2)$$

With a little bit of work, one can also verify that eq. (10.2) holds for any rotation, not just those about the z axis.

The dot product is also a scalar. It is easily verified that its value in the two coordinate systems is the same, i.e.,

$$a'_x b'_x + a'_y b'_y + a'_z b'_z = a_x b_x + a_y b_y + a_z b_z. \qquad\qquad (10.3)$$

(The components b'_i are defined in complete analogy with eq. (10.1); we simply replace the letter a by b everywhere.)

The above relations have been written for a position vector, but they also hold for any other vector. Indeed, that is one way of *defining* a general vector. Thus, the vector equation $\mathbf{F} = m\mathbf{a}$ means that if the components of $\mathbf{a}$ in the two frames are related in a certain way, the components of $\mathbf{F}$ in the two frames are related in exactly the same way.

Let us suppose that to go from the unprimed to the primed frame, we must rotate by an angle θ about an axis $\hat{\mathbf{n}}$. It is obvious that any rotation can be specified in this way. Let the unit vectors along the unprimed and primed frame coordinate axes be denoted by $\mathbf{e}_i$

and $\mathbf{e}'_i$, respectively. We denote the abstract operation of rotation by $\mathcal{R}(\hat{\mathbf{n}}, \theta)$, and write

$$\mathbf{e}'_i = \mathcal{R}\mathbf{e}_i, \tag{10.4}$$

omitting the arguments of $\mathcal{R}$ for brevity. Let us also write

$$\mathbf{e}_i \cdot \mathbf{e}'_j = \mathbf{e}_i \cdot \mathcal{R}\mathbf{e}_j = R_{ij}. \tag{10.5}$$

The nine numbers R_{ij} (or $R_{ij}(\hat{\mathbf{n}}, \theta)$) form the so-called rotation matrix. Note the order of the indices carefully. We can write the relationship among the components of a fixed vector $\mathbf{a}$ in the two frames in terms of this matrix. For,

$$a'_i = \mathbf{a} \cdot \mathbf{e}'_i = \mathbf{a} \cdot (\mathcal{R}\mathbf{e}_i). \tag{10.6}$$

But since the $\mathbf{e}_i$ are complete, $\mathbf{a} = (\mathbf{a} \cdot \mathbf{e}_j)\mathbf{e}_j$ (using the summation convention), so

$$a'_i = (\mathbf{a} \cdot \mathbf{e}_j)(\mathbf{e}_j \cdot \mathcal{R}\mathbf{e}_i) = a_j R_{ji} = R_{ji}a_j. \tag{10.7}$$

The reader should note the order of indices in eqs. (10.5) and (10.7) carefully. One should also note that we have been thinking of rotating the unit vectors bodily while keeping the vector $\mathbf{a}$ fixed in space. The quantities a'_i and a_i are just the components of this same vector along different axes. This is known as the passive point of view. There is also the active point of view, in which one rotates the vector and relates components of the original and the rotated vector with respect to a fixed set of axes. Both points of view are useful, and the reader must always ascertain which is being considered. It would perhaps be better to put the prime on the subscript and write $a_{i'}$, but that is inconvenient.

Exercise 10.1 Show that

$$\mathbf{e}'_i = R_{ij}(\hat{\mathbf{n}}, \theta)\mathbf{e}_j. \tag{10.8}$$

Now show that the opposite rule applies to an active rotation, that is, if the vector a is taken into a' under a rotation by an angle θ about the axis $\hat{\mathbf{n}}$, then

$$a'_i = R_{ji}(\hat{\mathbf{n}}, \theta)a_j. \tag{10.9}$$

Note, however, the different meanings of the indices. In eq. (10.8), the indices on e and e' are essentially counting labels for the different vectors, while in eq. (10.9), they denote the components of two diferent vectors.

Let us now obtain the basic properties of rotation matrices. Since $a'_i a'_i = a_i a_i$,

$$R_{ji}a_j R_{ki}a_k = \delta_{jk}a_j a_k, \tag{10.10}$$

for all vectors $\mathbf{a}$, and we have

$$R_{ji}R_{ki} = \delta_{jk}. \tag{10.11}$$

In matrix notation, this means

$$R^{\mathrm{T}}R = RR^{\mathrm{T}} = 1, \quad R^{-1} = R^{\mathrm{T}}, \tag{10.12}$$

where R^T is the transpose of R. Any matrix whose inverse is equal to its transpose is said to be orthogonal. It is easy to show that the product of two orthogonal matrices is also orthogonal. Further, by taking the determinant of both sides of eq. (10.12) we obtain

$$(\det R)^2 = 1. \tag{10.13}$$

Hence, $\det R = \pm 1$. It is plain that the matrix for no rotation is the identity matrix, which has determinant $+1$, so by continuity, an infinitesimal rotation also has determinant 1. Since any finite rotation can be built up by composing a large number of smaller ones, it follows that the corresponding matrices all have determinant $+1$. The case $\det R = -1$ arises because the demand of length invariance is also satisfied by the operation of inversion, $a_i' = -a_i$, which clearly has determinant -1, as does any combination of pure rotations followed by an inversion. Rotations with $\det R = 1$ are said to be *proper*, while those with $\det R = -1$ are said to be *improper*.

Exercise 10.2 Show that any eigenvalue of a rotation matrix must be unimodular, and find the solutions to

$$\mathcal{R}(\hat{\mathbf{n}}, \theta)\mathbf{v} = \lambda \mathbf{v}, \tag{10.14}$$

i.e., the eigenvalues and eigenvectors of the proper rotation $\mathcal{R}(\hat{\mathbf{n}}, \theta)$. (*Hint:* The vector $\mathbf{v}$ need not be real.)

Exercise 10.3 Let $\mathcal{R}(\hat{\mathbf{n}}, d\theta)$ be an infinitesimal rotation. Show that

$$\mathbf{r}' = \mathcal{R}(\hat{\mathbf{n}}, d\theta)\mathbf{r} = \mathbf{r} + d\theta\, \hat{\mathbf{n}} \times \mathbf{r} + O(d\theta)^2. \tag{10.15}$$

Exercise 10.4 The previous exercise shows that an infinitesimal rotation can be associated with a vector $d\boldsymbol{\theta} = \hat{\mathbf{n}}\, d\theta$. Consider two infinitesimal rotations, $\mathcal{R}(\hat{\mathbf{n}}_i, d\theta_i) \equiv \mathcal{R}_i$. Show that

$$\mathcal{R}_1 \mathcal{R}_2 \leftrightarrow d\boldsymbol{\theta}_1 + d\boldsymbol{\theta}_2, \tag{10.16}$$

$$(\mathcal{R}_1 \mathcal{R}_2 - \mathcal{R}_2 \mathcal{R}_1)\mathbf{r} = (d\boldsymbol{\theta}_1 \times d\boldsymbol{\theta}_2) \times \mathbf{r} + \cdots. \tag{10.17}$$

The first equation states that infinitesimal rotations add like vectors. The second shows that this is not true beyond linear order, and that rotations about different axes do not commute.

Exercise 10.5 Consider two successive rotations $\mathcal{R}_A$ and $\mathcal{R}_B$. Let $\mathbf{e}_i' = \mathcal{R}_A \mathbf{e}_i$, and $\mathbf{e}_i'' = \mathcal{R}_B \mathbf{e}_i'$. We can write $\mathbf{e}_i'' = \mathcal{R}_C \mathbf{e}_i$, where $\mathcal{R}_C = \mathcal{R}_B \mathcal{R}_A$. Show that the corresponding rotation matrices multiply in the same order, i.e., $(R_C)_{ij} = (R_B)_{ik}(R_A)_{kj}$.

Solution: Straightforward. The only thing to understand is how R_B must be defined. The correct way is $(R_B)_{ik} = \mathbf{e}_i \cdot \mathcal{R}_B \mathbf{e}_k$. It is instructive to find $(R_B')_{ik} \equiv \mathbf{e}_i' \cdot \mathcal{R}_B \mathbf{e}_k'$ in terms of R_A and R_B.

In addition to vectors, there exist physical quantities whose rotational properties are more complex. For example, in an isotropic metal, Ohm's law may be written as

$$\mathbf{j} = \sigma \mathbf{E}, \tag{10.18}$$

where $\mathbf{j}$ and $\mathbf{E}$ are the current density and electric field at any point inside the metal, and σ is the conductivity, a material property. Many materials, however, are not isotropic, and so, while the relationship between $\mathbf{E}$ and $\mathbf{j}$ is still linear—doubling $\mathbf{E}$ leads to a doubling of $\mathbf{j}$, and superposition of two different $\mathbf{E}$'s yields the sum of the corresponding $\mathbf{j}$'s in response—it is not isotropic. That is, an electric field of the same magnitude applied along different directions will produce currents of different magnitudes. Linearity implies that eq. (10.18) must be generalized to

$$j_i = \sigma_{ij} E_j, \tag{10.19}$$

where we are once again using the summation convention. The set of nine numbers $\{\sigma_{ij}\}$ forms a *tensor of rank two*. To see how the measurements of observers in differently oriented frames are related to one another, we use the fact that $\mathbf{j}$ and $\mathbf{E}$ are vectors. Thus, $j_i' = R_{ji} j_j$, and $E_j = R_{jk} E_k'$ (observe index order), so

$$j_i' = R_{ji} j_j = R_{ji} \sigma_{jk} E_k = R_{ji} \sigma_{jk} R_{km} E_m'. \tag{10.20}$$

However, the primed-frame observer would write $j_i' = \sigma_{im}' E_m'$, so

$$\sigma_{im}' = R_{ji} R_{km} \sigma_{jk}. \tag{10.21}$$

This is exactly how the nine products $a_i b_m$ would transform if $\mathbf{a}$ and $\mathbf{b}$ were vectors:

$$a_i' b_m' = (R_{ji} a_j)(R_{km} b_k) = R_{ji} R_{km}(a_j b_k). \tag{10.22}$$

Any object that transforms in this way under rotations is said to be a tensor of rank two.

We can construct tensors of higher rank in exactly the same way. For a rank-three tensor, T_{ijk}, e.g., the transformation law is

$$T_{mnp}' = R_{im} R_{jn} R_{kp} T_{ijk}. \tag{10.23}$$

The generalization to higher rank is obvious. For each index, we simply include another rotation matrix in the transformation.

To give a few more examples of tensors, let us consider a solid subject to some stress. As a result, a small volume element that was centered at the point $\mathbf{r}$ is now displaced to $\mathbf{r} + \mathbf{u}(\mathbf{r})$. The state of the strained solid is naturally described by the displacement field $\mathbf{u}(\mathbf{r})$. If $\mathbf{u}$ were the same for all $\mathbf{r}$, however, the solid would just be uniformly displaced without any internal strain. Thus, physical properties depend on the gradients of $\mathbf{u}$ via the *strain tensor*[17]

$$u_{ij} = \frac{1}{2}\left(\frac{\partial u_i}{\partial x_j} + \frac{\partial u_j}{\partial x_i}\right). \tag{10.24}$$

---

[17] The antisymmetric combination $\partial u_i/\partial x_j - \partial u_j/\partial x_i$ does not appear, since this would be nonzero under a uniform rotation, which also does not produce any strain.

In certain materials, LiNbO$_3$, e.g., the presence of a strain leads to an internal electric field, given by

$$E_i = \gamma_{ijk} u_{jk}. \tag{10.25}$$

The object γ_{ijk} is the *piezoelectric tensor*, and it has rank three.

Exercise 10.6 Show that u_{ij} and γ_{ijk} are indeed tensors of the ranks stated.

It should be noted carefully that, in general, a tensor of rank two is not the same as a matrix (although see exercise 10.11), even though we sometimes speak of the matrix of such a tensor, and of its trace. This is simply because the algebra involved is the same. The physical meaning of the two entities is quite different. The distinction is also evident from the existence of higher-rank tensors.

Lastly, let us mention that just as one can produce higher-rank tensors by multiplying those of lower rank, one can do the converse by a process called *contraction*. Let $T_{ijkl\ldots}$ be a tensor of rank n. The objects

$$A_{kl\ldots} = T_{iikl\ldots}, \quad B_{jl\ldots} = T_{ijil\ldots} \tag{10.26}$$

are tensors of rank $n - 2$. We say that we have contracted on the first two indices in the first case, and on the first and third indices in the second case. The results will, in general, be different.

Exercise 10.7 Let T_{ij} be a second-rank tensor. Show that its trace, $\text{Tr}\, T \equiv T_{ii}$, is a scalar.

Exercise 10.8 Let D be a second-rank tensor whose components in some frame are $D_{ij} = \delta_{ij}$. Show that its components in any other frame are the same. This shows that δ_{ij} may be regarded as a tensor, known as the *unit tensor*.

Exercise 10.9 Let E be a third-rank tensor whose components in some frame are $E_{ijk} = \epsilon_{ijk}$. Show that its components in any other frame are the same. This shows that ϵ_{ijk} may be regarded as a tensor, known as the completely antisymmetric tensor of the third rank in three dimensions.

One can prove that δ_{ij} and ϵ_{ijk} are the only fundamental tensors invariant under all rotations, and all other tensors with this invariant property are products of these, e.g., $\delta_{ij}\delta_{mn}$.

Exercise 10.10 Consider an isotropic solid under stress. In exact analogy to the potential energy $kx^2/2$ of a Hookean spring, the energy density of the solid is written as

$$\mathcal{E} = \tfrac{1}{2} C_{ijkl} u_{ij} u_{kl}, \tag{10.27}$$

where u_{ij} is the strain tensor, and C_{ijkl} is known as the *elastic tensor*. Because the solid is isotropic, it has no preferred direction. Use this fact to show that C_{ijkl} can be written as a linear combination of just two rotationally invariant tensors. The coefficients of proportionality are material properties related to the velocities of longitudinal and transverse sound waves in the material.

Exercise 10.11 (subtle) Is the rotation matrix R_{ij} a rank-two tensor?

11 Orthogonal curvilinear coordinates

It is often convenient to do vector calculus in non-Cartesian coordinate systems. The two commonest systems are spherical polar coordinates and cylindrical coordinates.[18] Since the new coordinates can be written as functions of the Cartesian coordinates (or vice versa), the relationships between field derivatives in different coordinate systems may be obtained by purely algebraic manipulations involving the chain rule. This is almost certainly the wrong way to view these relationships, not to mention that it is also the more cumbersome way to derive them. It is simpler to draw upon the geometrical meaning of the various operations, as we shall do.

We suppose that every point in space can be uniquely addressed by giving three coordinates (u_1, u_2, u_3), which are smooth functions of $\mathbf{r} = (x, y, z)$: $u_i(\mathbf{r})$. We will consider only *orthogonal* coordinate systems, which means that the surfaces of constant u_i intersect those of constant u_j ($j \neq i$) at right angles everywhere. In that case, the unit vectors $\mathbf{e}_i$ normal to the coordinate surfaces $u_i =$ constant are identical with unit vectors tangent to the coordinate curves, and there is no need to distinguish them. It should be carefully noted that these vectors are now *not necessarily the same* at different points in space, i.e., $\{\mathbf{e}_i(\mathbf{r})\} \neq \{\mathbf{e}_i(\mathbf{r}')\}$ for $\mathbf{r} \neq \mathbf{r}'$.

The essence of any coordinate system is contained in the formula for the length interval between two infinitesimally close points. If the system is orthogonal, Pythagoras's theorem forces the interval to have the form

$$ds^2 = (h_1 du_1)^2 + (h_2 du_2)^2 + (h_3 du_3)^2, \tag{11.1}$$

where the h_i are nonnegative *scale factors*, varying smoothly with $\mathbf{r}$ or u_i. We illustrate this in fig. 2.2a, using two dimensions for clarity.

For example, for the spherical polar coordinates $(u_1, u_2, u_3) = (r, \theta, \varphi)$, related to Cartesian coordinates in the usual way: $x = r \sin \theta \cos \varphi$, $y = r \sin \theta \sin \varphi$, and $z = r \cos \theta$, we have

$$ds^2 = dr^2 + r^2 d\theta^2 + r^2 \sin^2 \theta \, d\varphi^2. \tag{11.2}$$

Hence,

$$h_r = 1, \quad h_\theta = r, \quad h_\varphi = r \sin \theta. \tag{11.3}$$

---

[18] Other systems, such as parabolic or ellipsoidal coordinates, are of use in more limited circumstances. Detailed discussion of these systems may be found in Morse and Feshbach (1953); see tables following chap. 5.

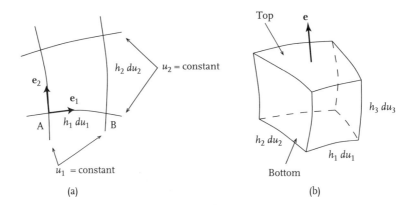

FIGURE 2.2. Curvilinear coordinates. (a) Two-dimensional illustration. (b) Integration regions for derivation of expressions for divergence and curl.

For cylindrical coordinates $(u_1, u_2, u_3) = (r, \varphi, z)$, related to Cartesian coordinates by $x = r \cos \varphi$, $y = r \sin \varphi$, and $z = z$,

$$ds^2 = dr^2 + r^2 d\varphi^2 + dz^2. \tag{11.4}$$

Hence,

$$h_r = 1, \quad h_\varphi = r, \quad h_z = 1. \tag{11.5}$$

All vector field derivatives can be written immediately once the scale factors are known. We first quote all the results, and derive them later:

$$\nabla f = \frac{1}{h_1} \frac{\partial f}{\partial u_1} \mathbf{e}_1 + \cdots , \tag{11.6}$$

$$\nabla \cdot \mathbf{A} = \frac{1}{h_1 h_2 h_3} \left[\frac{\partial}{\partial u_1} (h_2 h_3 A_1) + \cdots \right], \tag{11.7}$$

$$\nabla \times \mathbf{A} = \frac{1}{h_2 h_3} \left[\frac{\partial}{\partial u_2} (h_3 A_3) - \frac{\partial}{\partial u_3} (h_2 A_2) \right] \mathbf{e}_1 + \cdots , \tag{11.8}$$

$$\nabla^2 f = \frac{1}{h_1 h_2 h_3} \left[\frac{\partial}{\partial u_1} \left(\frac{h_2 h_3}{h_1} \frac{\partial f}{\partial u_1} \right) + \cdots \right]. \tag{11.9}$$

Here, $A_i = \mathbf{A} \cdot \mathbf{e}_i$, and the ellipses indicate terms that can be written down by cyclic permutation of the indices 1, 2, and 3. These formulas apply to any orthogonal coordinate system. Specific formulas for spherical polar and cylindrical coordinates can be obtained by using eqs. (11.3) and (11.5), respectively.

The gradient: To derive eq. (11.6), consider a function $f(\mathbf{r})$ at two neighboring points A and B, which differ only in the u_1 coordinate, as shown in fig. 2.2. It is then obvious that

$$df \equiv f_B - f_A = \frac{\partial f}{\partial u_1} du_1. \tag{11.10}$$

At the same time, $\mathbf{r}_B - \mathbf{r}_A = h_1 du_1 \mathbf{e}_1$, so

$$df = \nabla f \cdot \mathbf{e}_1 h_1 du_1. \tag{11.11}$$

Equating these two expressions, we obtain $\nabla f \cdot \mathbf{e}_1 = h_1^{-1}(\partial f/\partial u_1)$. The other components follow in the same way, and eq. (11.6) follows in turn.

The divergence: Let us consider the integral of $\nabla \cdot \mathbf{A}(\mathbf{r})$ over an infinitesimal "cube" as shown in fig. 2.2. Since the edges of this volume are $h_1 du_1$, $h_2 du_2$, and $h_3 du_3$, the volume element is $h_1 h_2 h_3\, du_1 du_2 du_3$, and

$$\int_{\text{cube}} \nabla \cdot \mathbf{A}\, d^3x = \nabla \cdot \mathbf{A}\, h_1 h_2 h_3\, du_1 du_2 du_3. \tag{11.12}$$

We now use Gauss's theorem to evaluate the integral:

$$\int_{\text{cube}} \nabla \cdot \mathbf{A}\, d^3x = \oint_{\partial(\text{cube})} \mathbf{A} \cdot d\mathbf{s}. \tag{11.13}$$

Let us consider the contribution to this surface integral from the surfaces marked "top" and "bottom" in the figure. The area of each face may be written as $h_1 h_2\, du_1 du_2$, but we must remember that the h_i now vary with position. Thus, this contribution is

$$[\mathbf{A} \cdot \mathbf{e}_3 h_1 h_2]_{\text{bottom}}^{\text{top}}\, du_1 du_2 = \frac{\partial}{\partial u_3}(h_1 h_2 A_3)(du_1 du_2 du_3). \tag{11.14}$$

By comparing this result with eq. (11.12), we obtain the contribution to $\nabla \cdot \mathbf{A}$ due to variation with u_3. The contributions from the faces normal to $\pm \mathbf{e}_1$ and $\pm \mathbf{e}_2$ may be written down by permuting the indices, and the result (11.7) follows.

The curl: The method is analogous to that for the divergence. We consider the surface integral of $\nabla \times \mathbf{A}$ over a small "square," such as the top face of the volume in fig. 2.2, and employ Stokes's theorem. The direct evaluation gives

$$\int_{\text{square}} (\nabla \times \mathbf{A}) \cdot d\mathbf{s} = (\nabla \times \mathbf{A}) \cdot \mathbf{e}_3\, h_1 h_2\, du_1 du_2, \tag{11.15}$$

while Stokes's theorem gives

$$\oint_{\partial(\text{square})} \mathbf{A} \cdot d\boldsymbol{\ell} = \left[-\frac{\partial}{\partial u_2}(A_1 h_1) + \frac{\partial}{\partial u_1}(A_2 h_2) \right] du_1 du_2, \tag{11.16}$$

the partial derivative with respect to u_1, e.g., arising from differences in $h_2 A_2$ between the edges of the square parallel to $\mathbf{e}_1$. Comparison of these two expressions gives the $\mathbf{e}_3$ component of $\nabla \times \mathbf{A}$. The other components are found similarly, and eq. (11.8) follows.

The Laplacian: This follows from eqs. (11.6) and (11.7) since $\nabla^2 f = \nabla \cdot (\nabla f)$.

We now give explicit formulas for these derivatives in cylindrical and spherical polar coordinates for reference. The formulas for the Laplacian and the gradient are worth remembering. The formulas for the divergence and the curl, on the other hand, are involved, and the reader will see that we will use them very infrequently.

Spherical polar coordinates:

$$\nabla f = \frac{\partial f}{\partial r}\mathbf{e}_r + \frac{1}{r}\frac{\partial f}{\partial \theta}\mathbf{e}_\theta + \frac{1}{r \sin \theta}\frac{\partial f}{\partial \varphi}\mathbf{e}_\varphi, \tag{11.17}$$

$$\nabla^2 f = \frac{1}{r^2}\frac{\partial}{\partial r}\left(r^2\frac{\partial f}{\partial r}\right) + \frac{1}{r^2}\frac{\partial^2 f}{\partial \theta^2} + \frac{1}{r^2 \sin^2 \theta}\frac{\partial^2 f}{\partial \varphi^2}, \tag{11.18}$$

$$\nabla \cdot \mathbf{A} = \frac{1}{r^2}\frac{\partial}{\partial r}\left(r^2 A_r\right) + \frac{1}{r \sin \theta}\frac{\partial}{\partial \theta}\left(\sin \theta A_\theta\right) + \frac{1}{r \sin \theta}\frac{\partial A_\varphi}{\partial \varphi}, \tag{11.19}$$

$$\nabla \times \mathbf{A} = \frac{1}{r \sin \theta}\left[\frac{\partial}{\partial \theta}(\sin \theta A_\varphi) - \frac{\partial A_\theta}{\partial \varphi}\right]\mathbf{e}_r + \left[\frac{1}{r \sin \theta}\frac{\partial A_r}{\partial \varphi} - \frac{1}{r}\frac{\partial}{\partial r}(r A_\varphi)\right]\mathbf{e}_\theta$$

$$+ \frac{1}{r}\left[\frac{\partial}{\partial r}(r A_\theta) - \frac{\partial A_r}{\partial \theta}\right]\mathbf{e}_\varphi. \tag{11.20}$$

Cylindrical polar coordinates:

$$\nabla f = \frac{\partial f}{\partial r}\mathbf{e}_r + \frac{1}{r}\frac{\partial f}{\partial \varphi}\mathbf{e}_\varphi + \frac{\partial f}{\partial z}\mathbf{e}_z, \tag{11.21}$$

$$\nabla^2 f = \frac{1}{r}\frac{\partial}{\partial r}\left(r\frac{\partial f}{\partial r}\right) + \frac{1}{r^2}\frac{\partial^2 f}{\partial \varphi^2} + \frac{\partial^2 f}{\partial z^2}, \tag{11.22}$$

$$\nabla \cdot \mathbf{A} = \frac{1}{r}\frac{\partial}{\partial r}\left(r A_r\right) + \frac{1}{r}\frac{\partial A_\varphi}{\partial \varphi} + \frac{\partial A_z}{\partial z}, \tag{11.23}$$

$$\nabla \times \mathbf{A} = \left[\frac{1}{r}\frac{\partial A_z}{\partial \varphi} - \frac{\partial A_\varphi}{\partial z}\right]\mathbf{e}_r + \left[\frac{\partial A_r}{\partial z} - \frac{\partial A_z}{\partial r}\right]\mathbf{e}_\varphi + \frac{1}{r}\left[\frac{\partial}{\partial r}(r A_\varphi) - \frac{\partial A_r}{\partial \varphi}\right]\mathbf{e}_z. \tag{11.24}$$

3 | Electrostatics in vacuum

With this chapter we begin our study of electrodynamics proper. We consider the simplest possible situation, static electric charges, and how they interact, the electric field and potential they produce, and the accompanying work and energy concepts.

A special situation arises when these static charges reside on conducting bodies. This situation is of great practical importance and is often studied along with the general problem of electrostatics. We expect that the reader has seen at least an elementary treatment of electrostatics in the presence of conductors, and so we defer it to chapter 14, after we have covered the key concepts of vacuum electrodynamics.

12 Coulomb's law

It is an empirically established fact that two point charges q_1 and q_2 that are held fixed at positions $\mathbf{r}_1$ and $\mathbf{r}_2$, respectively, exert equal and opposite forces on each other.[1] The force on charge 2 is given by

$$\mathbf{F}_{\text{on } 2} = \frac{q_1 q_2}{r^2} \hat{\mathbf{r}}, \tag{12.1}$$

$$\mathbf{F}_{\text{on } 2} = \frac{q_1 q_2}{4\pi\epsilon_0 r^2} \hat{\mathbf{r}}, \quad \text{(SI)}$$

where $\mathbf{r} = \mathbf{r}_2 - \mathbf{r}_1$, $r = |\mathbf{r}|$, and $\hat{\mathbf{r}} = \mathbf{r}/r$ is the unit vector from $\mathbf{r}_1$ to $\mathbf{r}_2$.

---

[1] Readers may wonder how such an assertion can be made. After all, one cannot really study point charges in the macroscopic realm where classical concepts are applicable. Thus, our statement is really an indirect deduction made from studying the forces between extended charged bodies, limits on the mass of the photon, and attempts to measure a potential difference at points in the interior of a hollow, closed, conducting shell. How these lattermost experiments imply an inverse-square force law is touched on briefly in section 88. Further, when the distances between particles are microscopic—on the scale of a few angstroms or less—the entire question must be posed in quantum mechanical terms, nonrelativistic at low speeds, and relativistic at speeds close to those of light. When this is done, one finds that the description of how charges interact continues to be the logical extension of Coulomb's law. For further discussion of these points, see A. S. Goldhaber and M. Nieto, *Rev. Mod. Phys.* **43**, 277 (1971), and R. E. Crandall, *Am. J. Phys.* **51**, 698 (1983).

Equation (12.1) is Coulomb's law. In the Gaussian system, the unit of charge is determined once units for force and distance are chosen. This unit is called the electrostatic unit (esu) or statcoulomb; it is such that two charges of 1 esu each repel each other with a force of 1 dyne at a separation of 1 cm. In the SI system, the unit of charge is called the coulomb (C), and the constant ϵ_0 is such that

$$\frac{1}{4\pi\epsilon_0} = 9 \times 10^9 \, \mathrm{N\,m^2/C^2}. \tag{12.2}$$

Further,

$$1\,\mathrm{C} = 3 \times 10^9 \, \mathrm{esu}. \tag{12.3}$$

The charge of an electron is -4.8×10^{-10} esu or -1.6×10^{-19} C.

The electric field at a point due to a set of charges is defined as the ratio of the force on a test charge at the point to the value of the test charge:

$$\mathbf{E} = \frac{\mathbf{F}}{q}. \quad \text{(Gaussian and SI)} \tag{12.4}$$

If the test charge is an extended body, then one must also invoke a limiting procedure in which the test charge is assumed to become vanishingly small. Since we have written the force law for point particles to begin with, this limit is automatic. Thus, the field at a point $\mathbf{r}$ due to a charge q at the origin is given by

$$\mathbf{E}(\mathbf{r}) = \frac{q}{r^2}\hat{\mathbf{r}} = q\frac{\mathbf{r}}{r^3}. \tag{12.5}$$

$$\mathbf{E}(\mathbf{r}) = \frac{q}{4\pi\epsilon_0 r^2}\hat{\mathbf{r}} = \frac{q}{4\pi\epsilon_0}\frac{\mathbf{r}}{r^3}. \quad \text{(SI)}$$

Next, let us consider a collection of point charges,[2] q_a, located at the points $\mathbf{r}_a$ (known as *source points*), $a = 1, 2, \ldots, N$. By the superposition principle, the electric field at a point $\mathbf{r}$ (known as the *field point*) is given by

$$\mathbf{E}(\mathbf{r}) = \sum_a q_a \frac{\mathbf{r} - \mathbf{r}_a}{|\mathbf{r} - \mathbf{r}_a|^3}. \tag{12.6}$$

In macroscopic problems, one usually deals with a very large number of charges in any small region of space. It is then more convenient to describe the source via a charge distribution, $\rho(\mathbf{r}')$, such that the charge in a tiny volume element ΔV centered at $\mathbf{r}'$ is $\rho(\mathbf{r}')\Delta V$. The sum in eq. (12.6) then passes into an integral:

$$\mathbf{E}(\mathbf{r}) = \int \rho(\mathbf{r}')\frac{\mathbf{r} - \mathbf{r}'}{|\mathbf{r} - \mathbf{r}'|^3}d^3x'. \tag{12.7}$$

$$\mathbf{E}(\mathbf{r}) = \frac{1}{4\pi\epsilon_0} \int \rho(\mathbf{r}')\frac{\mathbf{r} - \mathbf{r}'}{|\mathbf{r} - \mathbf{r}'|^3}d^3x'. \quad \text{(SI)}$$

---

[2] We shall use letters a, b, c, etc., to label collections of charges, currents, conductors etc., and letters in the middle of the alphabet, i, j, k, etc., to label Cartesian components of vectors.

Exercise 12.1 Lines of **E** (or any other vector field) are defined as space curves such that at any point on the line, **E** is tangent to it. Show that a line of **E** obeys the differential equations

$$\frac{dx}{E_x} = \frac{dy}{E_y} = \frac{dz}{E_z} \tag{12.8}$$

or, equivalently,

$$\frac{d\mathbf{r}}{ds} = \frac{\mathbf{E}}{E}, \tag{12.9}$$

where s is the arc length along the line. Yet one more way of stating these conditions is that in any orthogonal curvilinear coordinate system, the ratios $h_i du_i / E_i$ are all equal. We shall utilize lines of **B** in chapter 11 when we study the motion of charges in external magnetic fields.

13 The electrostatic potential

The electric field produced by a point charge can be written in the alternative form

$$\mathbf{E}(\mathbf{r}) = -\nabla\left(\frac{q}{r}\right). \tag{13.1}$$

(This is easily verified, either in Cartesian components or by noting that the gradient of a function only of r is $\hat{\mathbf{r}}(df/dr)$.) Since the gradient is a linear operation, the electric field from a collection of charges can also be written as the gradient of a scalar field. We therefore define the *electrostatic* or *scalar potential* $\phi(\mathbf{r})$ through the relation

$$\mathbf{E}(\mathbf{r}) = -\nabla\phi(\mathbf{r}). \quad \text{(Gaussian and SI)} \tag{13.2}$$

For a point charge,

$$\phi(\mathbf{r}) = \frac{q}{r}. \tag{13.3}$$

$$\phi(\mathbf{r}) = \frac{q}{4\pi\epsilon_0 r}. \quad \text{(SI)}$$

Note that according to its definition, ϕ is determined only up to an additive constant. In eq. (13.3) this constant is chosen so that $\phi \to 0$ as $r \to \infty$. This is often a convenient choice.

The potential from a collection of charges or from a distribution is immediately written down by superposition:

$$\phi(\mathbf{r}) = \sum_a \frac{q_a}{|\mathbf{r} - \mathbf{r}_a|}; \tag{13.4}$$

$$\phi(\mathbf{r}) = \int \frac{\rho(\mathbf{r}')}{|\mathbf{r} - \mathbf{r}'|} d^3x'. \tag{13.5}$$

$$\phi(\mathbf{r}) = \frac{1}{4\pi\epsilon_0} \int \frac{\rho(\mathbf{r}')}{|\mathbf{r} - \mathbf{r}'|} d^3x'. \quad \text{(SI)}$$

The advantage of introducing the potential is obvious. It is much easier to deal with a scalar than a vector field.

Exercise 13.1 Find the potential and electric field due to a uniformly charged thin spherical shell of radius R with total charge Q.

Solution: We first find the potential. The requisite integrals are the same as those for the gravitational potential due to a spherical shell, and may be found in most elementary texts on the subject.

$$\phi(\mathbf{r}) = \frac{Q}{r}, \quad \mathbf{E}(\mathbf{r}) = \frac{Q}{r^2}\hat{\mathbf{r}}, \quad r > R,$$

$$\phi(\mathbf{r}) = \frac{Q}{R}, \quad \mathbf{E}(\mathbf{r}) = 0, \quad r < R. \tag{13.6}$$

Exercise 13.2 Same as exercise 13.1, for a uniform spherical charge distribution of radius R, and total charge Q.

Answer:

$$\phi(\mathbf{r}) = \frac{Q}{r}, \qquad \mathbf{E}(\mathbf{r}) = \frac{Q}{r^2}\hat{\mathbf{r}}, \quad r > R,$$

$$\phi(\mathbf{r}) = \frac{Q}{2R^3}\left(3R^2 - r^2\right), \quad \mathbf{E}(\mathbf{r}) = \frac{Qr}{R^3}\hat{\mathbf{r}}, \quad r < R. \tag{13.7}$$

Exercise 13.3 An amusing application of the previous exercise is to J. J. Thomson's model of the atom (1904), in which the positive charge, Q, of the atom was assumed to be uniformly spread over a sphere of radius R, and the electrons were assumed to be point negative charges embedded in this sphere. Show that the electrons undergo simple harmonic motion in this model, and find the frequency of this motion. How does the value compare with the spectral frequencies found in hydrogen?

Answer: $\omega = (Qe/mR^3)^{1/2}$ (Gaussian system), where m is the electron mass. For hydrogen, taking $Q = e$ and $R = a_0 = 0.529$ Å, we find $\omega/2\pi = 6.6 \times 10^{15}$ Hz, double the limit of the Lyman series. One century plus later, the interesting point is not that this number is incorrect, but rather that it is *correct*, i.e., of the right magnitude. Basically, there is only one combination of the quantities Qe, m, and R with dimensions of frequency. Knowing Q, e, m, and the spectral frequencies, Thomson and his contemporaries could use the argument in reverse to reach sensible conclusions about the size of atoms.

14 Electrostatic energy

In addition to enabling easy calculation of fields, the potential has an important physical meaning. Let us consider a test charge q in an external electrostatic field, by which we mean that the sources of this field are not affected by the presence or motion of the test charge. Since the force on the test charge at any point $\mathbf{r}$ is $q\mathbf{E}(\mathbf{r})$, the mechanical work done on the charge in moving it from a point $\mathbf{r}_1$ to another $\mathbf{r}_2$ is given by

$$W = -q \int_1^2 \mathbf{E}(\mathbf{r}) \cdot d\boldsymbol{\ell}$$

$$= q \int_1^2 \nabla\phi \cdot d\boldsymbol{\ell}$$

$$= q[\phi(\mathbf{r}_2) - \phi(\mathbf{r}_1)], \quad \text{(Gaussian and SI)} \tag{14.1}$$

where we have used eq. (8.1) for the integral of a gradient. Clearly, $q\phi$ can be regarded as the potential energy of the test charge.

Note that the work done is independent of the path taken from $\mathbf{r}_1$ to $\mathbf{r}_2$, and that it is zero if $\mathbf{r}_2 = \mathbf{r}_1$. We express this fact by saying that the electrostatic field is conservative.

We can use these results to obtain the potential energy of an assembly of point charges. Let us start with two charges. By imagining these charges to be brought in from infinity to their final position, and calculating the work done in the process, we conclude that the potential energy is

$$U = \frac{q_a q_b}{r_{ab}}, \tag{14.2}$$

where $r_{ab} = |\mathbf{r}_a - \mathbf{r}_b|$. For more than two charges, we can calculate the energy by superposition, by considering the work done pairwise. Thus,

$$U = \sum_{\text{pairs}} \frac{q_a q_b}{r_{ab}}. \tag{14.3}$$

$$U = \sum_{\text{pairs}} \frac{1}{4\pi\epsilon_0} \frac{q_a q_b}{r_{ab}}. \quad \text{(SI)}$$

But $\sum_{b, b \neq a} q_b / r_{ab}$ is the potential at $\mathbf{r}_a$ due to all the charges except q_a. Thus, we can also write

$$U = \frac{1}{2} \sum_a q_a \phi(\mathbf{r}_a). \quad \text{(Gaussian and SI)} \tag{14.4}$$

These two formulas can immediately be adapted to a continuous charge distribution $\rho(\mathbf{r})$ by replacing sums by integrals. The case of a spherically symmetric charge distribution $\rho(\mathbf{r}) = \rho(r)$ is a particularly simple one, yet important. The potential due to such a distribution can be written as

$$\phi(r) = \frac{1}{r} \int_0^r 4\pi r^2 \rho(r) \, dr + \int_r^\infty 4\pi r \rho(r) \, dr. \tag{14.5}$$

The first term is the potential due to the charge inside a sphere of radius r, while the second is from the charge outside it (see exercise 13.1). If we now use the formula $U = \frac{1}{2} \int \phi\rho$, we get

$$U = \frac{1}{2}(4\pi)^2 \int_0^\infty dr_1 r_1 \rho(r_1) \int_0^{r_1} dr_2 r_2^2 \rho(r_2) + \frac{1}{2}(4\pi)^2 \int_0^\infty dr_1 r_1^2 \rho(r_1) \int_{r_1}^\infty dr_2 r_2 \rho(r_2). \tag{14.6}$$

But the two terms on the right are equal. For if we interchange the indices 1 and 2 in the first term, each term is the integral of $r_1^2 r_2 \rho(r_1)\rho(r_2)$, over the two-dimensional region of r_1-r_2 space $0 < r_1 < r_2 < \infty$. Hence, we can also write

$$U = (4\pi)^2 \int_0^\infty dr_1 r_1 \rho(r_1) \int_0^{r_1} dr_2 r_2^2 \rho(r_2) \tag{14.7}$$

$$= (4\pi)^2 \int_0^\infty dr_1 r_1^2 \rho(r_1) \int_{r_1}^\infty dr_2 r_2 \rho(r_2). \tag{14.8}$$

Depending on the problem, one or the other of these forms may be more convenient.

Exercise 14.1 Consider the electron as a sphere of radius r_e. Find the electrostatic potential energy when the total charge $-e$ is distributed uniformly (a) over the volume of the sphere; (b) over the surface of the sphere. [**Answer:** (a) $3e^2/5r_e$, (b) $e^2/2r_e$.]

Exercise 14.2 In an approximate model of the helium atom, each electron is assumed to be distributed around the nucleus with a charge distribution

$$\rho(\mathbf{r}) = -\frac{8e}{\pi a_0^3} \exp(-4r/a_0), \tag{14.9}$$

where a_0 is the Bohr radius. Find the electrostatic energy of repulsion between the two electrons.

Solution: The energy in question may be written as

$$U = \iint \frac{\rho(\mathbf{r}_1)\rho(\mathbf{r}_2)}{|\mathbf{r}_1 - \mathbf{r}_2|} d^3 x_1 d^3 x_2. \tag{14.10}$$

This is twice the expression for the energy of a single distribution, since the two electrons are distinct. The integral can be evaluated using either of the forms (14.7) or (14.8). However, there is yet another method that is often useful. We showed in exercise 9.13 that the Fourier transform of $1/r$ is $4\pi/k^2$. If we use this result and the convolution theorem, we obtain

$$U = \int \frac{d^3 k}{(2\pi)^3} \frac{4\pi}{k^2} \rho_{\mathbf{k}}\rho_{-\mathbf{k}}, \tag{14.11}$$

where $\rho_{\mathbf{k}}$ is the Fourier transform of $\rho(\mathbf{r})$. In the present problem,

$$\rho_{\mathbf{k}} = -\frac{256e}{(k^2 a_0^2 + 16)^2}. \tag{14.12}$$

The integral (14.11) is now simple, and the final answer is

$$U = \frac{5}{4} \frac{e^2}{a_0}. \tag{14.13}$$

The following exercises and discussion (modeled after that in the *Feynman Lectures*), show some more instances in physics where the electrostatic energy is important.

Exercise 14.3 (the electrostatic energy of ionic crystals) In the NaCl structure, each Na^+ ion is surrounded by six near-neighbor Cl^- ions at a distance a_{nn} along the $\pm x$, $\pm y$, and $\pm z$ axes, and vice versa. The value of a_{nn} is 0.281 nm. Calculate the electrostatic energy of this crystal and compare it with the measured value of 183 kcal/mol (43.7 kJ/mol) for the energy required to dissociate it into well-separated ions.[3]

Solution (partial): Let the energy per pair of Na^+ and Cl^- ions be U. Since the energy of any pair of charges is *half* of $q_a q_b / |\mathbf{r}_a - \mathbf{r}_b|$, and since the lattice as seen by Na^+ and Cl^- ions is identical, U is *twice* the energy of a single Na^+ ion. Let us denote the ion positions by $(n_x, n_y, n_z) a_{nn} \equiv \mathbf{n} a_{nn}$, where the n's are integers, and place a Na^+ ion at the origin. Then the charge at site $\mathbf{n}$ is $(-1)^{n_x + n_y + n_z} e$, and we may write

$$U = \frac{e^2}{a_{nn}} \sum_{\mathbf{n} \neq 0} (-1)^{n_x + n_y + n_z} \frac{1}{|\mathbf{n}|}. \tag{14.14}$$

The problem is thus to evaluate the sum (14.14). A head-on approach, in which we successively add the contributions from nearest neighbors, second neighbors, third neighbors, and so on, is not efficient, as the convergence is very slow. (Try it.) In fact, the sum is only conditionally convergent and can be made to give any value by rearrangement of the terms. Physically, this reflects the fact that as we build the crystal out from the origin it is possible to have an excess surface charge and to distribute this charge in such a way that at every stage of the building up, it leads to an arbitrary excess potential at the origin, the arbitrariness of which gets ever larger as the crystal gets larger. Thus, the physically sensible way to evaluate U is to arrange the terms so that each one corresponds to the contribution from an electrical neutral grouping of charges. Again, this can be done in many ways. One that requires little bookkeeping and converges rapidly is as follows.

We divide the charges into lines or rods parallel to the z axis. The contribution to U from the rod through the origin is clearly

$$U_0 = -\frac{2e^2}{a_{nn}} \left(1 - \frac{1}{2} + \frac{1}{3} - \frac{1}{4} + - \cdots \right), \tag{14.15}$$

$$= -\frac{e^2}{a_{nn}} (2 \ln 2).$$

(Note that the sum U_0 is also only conditionally convergent.) Up to an overall sign, the contribution from a rod through $(n_x, n_y, 0) a_{nn}$ is e^2 / a_{nn} times $S(\sqrt{n_x^2 + n_y^2})$, where

$$S(x) = \sum_{n=-\infty}^{\infty} \frac{(-1)^n}{(n^2 + x^2)^{1/2}}. \tag{14.16}$$

Evaluate and tabulate $S(x)$ for x's equal to the first few inter-rod distances, multiply each $S(x)$ by the number of such neighbors, and a sign if necessary, and in this way find U.

---

[3] This problem is discussed in all standard texts on solid-state physics. The chief mathematical difficulty lies in the evaluation of sums such as that in eq. (14.14), known as *Madelung* sums. This is most systematically done through a procedure known as *Ewald summation*. See, e.g., Kittel (1975), appendix A.

With $U = -(e^2/a_{nn})A$, up to five significant figures (which can be determined by going up to 8th neighbors in the xy plane), you should find $A = 1.7476$. The value of U is thus -206 kcal/mol. The electrostatic attraction is somewhat offset by the Pauli repulsion between nearby ions, which is why the true binding energy is slightly less.

Exercise 14.4 Repeat the preceding exercise for CsCl. In CsCl, each Cs^+ ion sits at the center of a cube at the vertices of which sit Cl^- ions, and vice versa. The nearest-neighbor Cs^+ to Cl^- distance is $a_{nn} = 3.57$ Å. (The indexing of the ions is now more involved than for NaCl. One approach is as follows. Write the energy per ion pair, U, as in eq. (14.14), with a Cs^+ ion the origin, and denote one of the vectors from this central ion to a near neighbor Cl^- by $\mathbf{a}_{nn}$. Organize the ions into rods parallel to $\mathbf{a}_{nn}$, and consider the lattice planes perpendicular to them. The plane through the origin contains only Cs^+ ions. The planes above and below this plane by a distance $a_{nn}/3$ contain only Cl^- ions. The rods now fall into three classes: those containing ions in the central plane, and those containing Cl^- ions in the planes immediately above and below it. It is best to continue labeling the ions in these three planes with Cartesian coordinates. Further symmetries become evident as one proceeds. You should find that $U = -A(e^2/a_{nn})$, with $A = 1.7627$. This is the Madelung constant for the CsCl structure. Note how close it is to that for the NaCl structure.)

Exercise 14.5 Quantify, approximately, the arbitrariness in the potential at the origin of a finite crystal of NaCl of linear size R, and thus verify that this arbitrariness diverges as $R \to \infty$. The crystal surface should not be excessively convoluted, i.e., its area should be of order R^2.

Electrostatic energy in nuclei: The strong force between nucleons (protons and neutrons) is known to be independent of charge to high accuracy. Although it is the dominant contributor to the binding energy of a nucleus, the electrostatic contribution is not insubstantial. One way to see its presence is to consider pairs of *mirror nuclei*, in which the number of protons in one nucleus equals the number of neutrons in the other, and vice versa. Examples are ^{7}Li (3 protons, 4 neutrons) and ^{7}Be (4 protons, 3 neutrons), or ^{11}B (5 protons, 6 neutrons) and ^{11}C (6 protons, 5 neutrons). The charge independence of the strong force is evidenced by a close similarity in the low-lying energy spectrum of such pairs—energy differences between the levels, and the sequence of quantum numbers.[4] Thus, the difference in the binding energy of two mirror nuclei is expected to be largely due to electrostatic effects. Let us see how true this is for the ^{11}B–^{11}C system.

To do this, we first describe how the binding energy is measured. If we imagine a nucleus of Z protons and N neutrons to be separated into its constituent nucleons, just as we imagined the NaCl crystal separated into ions, then the mass–energy equivalence gives the binding energy as

$$B = -[M(Z, N) - ZM_p - NM_n]c^2, \tag{14.17}$$

---

[4] See, e.g., fig. 8.7 of Feynman et al. (1964), vol. II, for the ^{11}C–^{11}B system, or Williams (1991), fig. 9.10, for the ^{7}Be–^{7}Li system.

where $M(Z, N)$, M_p, and M_n are the masses of the composite nucleus, one proton, and one neutron, respectively. The mass $M(Z, N)$ can be estimated as the mass of the neutral atom (which can be very accurately measured by mass spectroscopy) minus the mass Zm_e of Z electrons, since the binding energy of the electrons is much less than B—of order keV versus MeV. The mass of the neutral ^{11}B atom is 1.984 MeV less than that of ^{11}C. Adding to this $(M_p + m_e - M_n)c^2 = 0.7823$ MeV, we conclude that the electrostatic energy of ^{11}C exceeds that of ^{11}B by 2.16 MeV.

Next, we note that nuclei may be modeled as having a fairly sharply defined radius, R, given by

$$R = (Z + N)^{1/3} r_0, \tag{14.18}$$

where $r_0 = 1.2 \times 10^{-13}$ cm. For ^{11}C and ^{11}B, this gives $R = 2.67 \times 10^{-13}$ cm.

Let us now imagine the Z protons as being equally probably distributed in a sphere of radius R. We leave it to the reader to show that the Coulomb energy of such a system equals

$$U_C = \frac{3}{5} Z(Z-1) \frac{e^2}{R}. \tag{14.19}$$

With the value of R as given above, this yields $\Delta U_C = 6e^2/R = 3.23$ MeV. The agreement with the measured number is reasonable. To discuss the discrepancy any further would take us too far into the realm of nuclear physics.

15 Differential form of Coulomb's law

Up to now, we have written Coulomb's law and the field due to a charge distribution as a global expression giving its value at all points in space. Since we wish eventually to consider time-dependent phenomena, it is desirable also to have a local form of this law. Such a form is obtained by calculating the derivatives of **E**.

The two obvious derivatives to find are the curl and the divergence. From the fact that $\mathbf{E} = -\nabla\phi$, we have immediately,

$$\nabla \times \mathbf{E} = 0. \quad \text{(Gaussian and SI)} \tag{15.1}$$

To find the divergence, let us begin with the field of a point charge. We use the form (12.5) and the facts that $\nabla \cdot \mathbf{r} = 3$, $\nabla r^{-3} = -3r^{-4}\hat{\mathbf{r}}$:

$$
\begin{aligned}
\nabla \cdot \mathbf{E} &= q \left(\frac{\nabla \cdot \mathbf{r}}{r^3} + \mathbf{r} \cdot \nabla \frac{1}{r^3} \right) \\
&= q \left(\frac{3}{r^3} - 3 \frac{\mathbf{r} \cdot \hat{\mathbf{r}}}{r^4} \right) \\
&= 0, \quad (\mathbf{r} \neq 0).
\end{aligned}
\tag{15.2}
$$

The caveat $\mathbf{r} \neq 0$ is needed because neither $(1/r^3)$ nor its gradient is defined there. So, we cannot assert that $\nabla \cdot \mathbf{E}$ vanishes everywhere. To study this further, let us integrate $\nabla \cdot \mathbf{E}$

over a spherical volume of radius R, where R can be arbitrarily small or large, and use Gauss's theorem. This yields

$$
\begin{aligned}
\int_{|\mathbf{r}|\leq R} \nabla \cdot \mathbf{E}\, d^3x &= \oint_{r=R} \mathbf{E}\cdot d\mathbf{s} \\
&= \frac{q}{R^2}(4\pi R^2) \\
&= 4\pi q,
\end{aligned}
\tag{15.3}
$$

since $\mathbf{E}\cdot\hat{\mathbf{r}} = q/R^2$ everywhere on the surface of the sphere. Equations (15.2) and (15.3) can be consistent only if

$$
\nabla \cdot \mathbf{E} = 4\pi q\, \delta(\mathbf{r}). \quad \text{(point charge)}
\tag{15.4}
$$

For a system of point charges, by superposition, we obtain

$$
\nabla \cdot \mathbf{E} = 4\pi \sum_a q_a \delta(\mathbf{r} - \mathbf{r}_a).
\tag{15.5}
$$

The sum can also be written as the charge distribution $\rho(\mathbf{r})$, so along with the equation for $\nabla \times \mathbf{E}$, we have the local equations that we sought:

$$
\nabla \cdot \mathbf{E} = 4\pi \rho.
\tag{15.6}
$$

$$
\nabla \cdot \mathbf{E} = \rho/\epsilon_0. \quad \text{(SI)}
$$

This is sometimes known as *Gauss's law*, especially when it is written in the equivalent integral form

$$
\oint_{\partial V} \mathbf{E}\cdot d\mathbf{s} = 4\pi \int_V \rho\, d^3x.
\tag{15.7}
$$

Together, eqs. (15.1) and (15.6) are equivalent to Coulomb's law.

Equations (15.6) and (15.1) can be combined into a single equation for the potential by substituting $\mathbf{E} = -\nabla\phi$ in Gauss's law:

$$
\nabla^2\phi = -4\pi\rho.
\tag{15.8}
$$

$$
\nabla^2\phi = -\rho/\epsilon_0. \quad \text{(SI)}
$$

This is known as *Poisson's equation*. In regions of space where the charge density vanishes, it reduces to *Laplace's equation*,

$$
\nabla^2\phi = 0.
\tag{15.9}
$$

We note here that eqs. (15.6) and (15.1) are obtained from Maxwell's equations when there are no currents ($\mathbf{j} = 0$) and the charges are static. It is then consistent to take $\mathbf{B} = 0$ and $\mathbf{E}$ independent of time, which leaves us with just the two equations for $\mathbf{E}$.

16 Uniqueness theorem of electrostatics

In stating that eqs. (15.6) and (15.1) are equivalent to Coulomb's law, we have implicitly asserted that they determine the field uniquely everywhere. This is a general consequence of a basic theorem of vector calculus that a field whose curl and divergence are known everywhere is uniquely determined, provided the sources vanish at infinity[5] and the field is also required to vanish at infinity, at least as rapidly as $1/r^2$. Because of the importance of this theorem, we shall spend some time discussing it.

Let us therefore consider a vector field $\mathbf{v}(\mathbf{r})$ whose curl and divergence are given:

$$\nabla \cdot \mathbf{v} = f(\mathbf{r}), \tag{16.1}$$

$$\nabla \times \mathbf{v} = \mathbf{w}(\mathbf{r}). \tag{16.2}$$

We will first prove uniqueness, and then find the solution. Suppose these equations have two solutions, $\mathbf{v}_1$ and $\mathbf{v}_2$. The difference, $\mathbf{v}'$, obeys

$$\nabla \cdot \mathbf{v}' = \nabla \times \mathbf{v}' = 0. \tag{16.3}$$

We may therefore write $\mathbf{v}' = \nabla \psi$, with $\nabla^2 \psi = 0$. By Gauss's theorem— or Green's first identity, eq. (8.5)—we have

$$\int_V (\psi \nabla^2 \psi + (\nabla \psi)^2) d^3x = \oint_{S=\partial V} \psi \nabla \psi \cdot d\mathbf{s}. \tag{16.4}$$

Let us take the surface S to approach infinity. By the conditions of the theorem, $\nabla \psi = \mathbf{v}'$ vanishes as $1/r^2$ or faster, so ψ vanishes as $1/r$ or faster, and $\psi \nabla \psi$ vanishes at least as fast as $1/r^3$. The area of S, on the other hand, grows as r^2, so the surface integral vanishes. The first term on the left-hand side is identically zero, so we get

$$\int (\nabla \psi)^2 d^3x = \int (\mathbf{v}')^2 d^3x = 0. \tag{16.5}$$

Since the integrand cannot be negative at any point, the integral can vanish only if $\mathbf{v}' = 0$ everywhere, i.e., $\mathbf{v}_1 = \mathbf{v}_2$.

The solution itself can be found in several ways. We will use Fourier transforms, both for practice and to obtain results that will be of use later. We begin by noting that in examining the potential for a point charge, we have shown that

$$\nabla^2 \left(\frac{1}{r} \right) = -4\pi \delta(\mathbf{r}). \tag{16.6}$$

More generally,

$$\nabla^2 \frac{1}{|\mathbf{r} - \mathbf{r}'|} = -4\pi \delta(\mathbf{r} - \mathbf{r}'). \tag{16.7}$$

---

[5] In fact, the sources must obey the somewhat stronger condition that their integrals over all space are finite, which follows from eqs. (16.20) and (16.21) derived below. In physical applications, these conditions are always met.

Now, taking Fourier transforms of both sides of eq. (16.6), and denoting this operation by the notation F.T.( ), we get

$$k^2 \, \text{F.T.} \left(\frac{1}{r} \right) = 4\pi,$$

$$\text{F.T.} \left(\frac{1}{r} \right) = \frac{4\pi}{k^2}. \tag{16.8}$$

We found this result in exercise 9.13 also. From it, we also obtain, by taking gradients, that

$$\text{F.T.} \left(\frac{\mathbf{r}}{r^3} \right) = -4\pi i \frac{\mathbf{k}}{k^2}, \tag{16.9}$$

and, by interchanging the roles of $\mathbf{r}$ and $\mathbf{k}$, that

$$\text{F.T.} \left(\frac{1}{r^2} \right) = \frac{2\pi^2}{k}. \tag{16.10}$$

Next, taking Fourier transforms of eqs. (16.1) and (16.2), we obtain

$$\mathbf{k} \cdot \mathbf{v_k} = -i f_k, \tag{16.11}$$

$$\mathbf{k} \times \mathbf{v_k} = -i \mathbf{w_k}. \tag{16.12}$$

Taking the cross product of the second equation with $\mathbf{k}$, and using the first for $\mathbf{k} \cdot \mathbf{v_k}$, we can solve for $\mathbf{v_k}$:

$$\mathbf{v_k} = i \frac{\mathbf{k} \times \mathbf{w_k} - \mathbf{k} f_k}{k^2}. \tag{16.13}$$

It remains to find the inverse Fourier transform. We do this in two steps. First, the factor of $i\mathbf{k}$ is equivalent to a ∇ operation in $\mathbf{r}$ space. Thus, denoting inverse transforms also by F.T.( ),

$$\text{F.T.}(\mathbf{v_k}) = \nabla \times \text{F.T.} \left(\frac{\mathbf{w_k}}{k^2} \right) - \nabla \left[\text{F.T.} \left(\frac{f_k}{k^2} \right) \right]. \tag{16.14}$$

In the second step, we find the inverse transforms of f_k/k^2 and $\mathbf{w_k}/k^2$, by making use of the convolution theorem and the result (16.8) derived above. Thus,

$$\text{F.T.} \left(\frac{f_k}{k^2} \right) = [\text{F.T.} \, f_k] * \left[\text{F.T.} \frac{1}{k^2} \right] = f(\mathbf{r}) * \frac{1}{4\pi r} = \frac{1}{4\pi} \int \frac{f(\mathbf{r'})}{|\mathbf{r} - \mathbf{r'}|} d^3 x'. \tag{16.15}$$

Likewise,

$$\text{F.T.} \left(\frac{\mathbf{w_k}}{k^2} \right) = \frac{1}{4\pi} \int \frac{\mathbf{w}(\mathbf{r'})}{|\mathbf{r} - \mathbf{r'}|} d^3 x'. \tag{16.16}$$

Putting it all together, we obtain the solution to eqs. (16.1) and (16.2):

$$\mathbf{v}(\mathbf{r}) = -\nabla g(\mathbf{r}) + \nabla \times \mathbf{u}(\mathbf{r}), \tag{16.17}$$

where

$$g(\mathbf{r}) = \frac{1}{4\pi} \int \frac{f(\mathbf{r'})}{|\mathbf{r}-\mathbf{r'}|} d^3x', \tag{16.18}$$

$$\mathbf{u}(\mathbf{r}) = \frac{1}{4\pi} \int \frac{\mathbf{w}(\mathbf{r'})}{|\mathbf{r}-\mathbf{r'}|} d^3x'. \tag{16.19}$$

Note that if the source densities f and $\mathbf{w}$ are nonvanishing only in a bounded region of space, then g and $\mathbf{u}$ both diminish as $1/r$ as $r \to \infty$. For, taking the origin inside the region where the sources are nonvanishing, and choosing r to be much greater than the dimensions of this region, we may approximate $|\mathbf{r} - \mathbf{r'}| \approx r$ in the integrals, obtaining

$$g(\mathbf{r}) \approx \frac{1}{4\pi r} \int f(\mathbf{r'}) d^3x', \tag{16.20}$$

$$\mathbf{u}(\mathbf{r}) \approx \frac{1}{4\pi r} \int \mathbf{w}(\mathbf{r'}) d^3x'. \tag{16.21}$$

The field $\mathbf{v}$ thus vanishes as $1/r^2$, in accordance with the conditions imposed.

It is worth examining further why, in this theorem, we demand that $\mathbf{v}$ vanish at infinity. We have already seen that it helps establish uniqueness, but exactly what kinds of improprieties are thereby eliminated is not yet clear. In one dimension, a differential equation such as

$$\frac{df}{dx} = g(x) \tag{16.22}$$

is immediately solved by integration, but only up to an additive constant. This constant must be fixed by imposing an initial or boundary condition, i.e., a condition that gives $f(x)$ at one point, which may be at plus or minus infinity. In the three-dimensional case, the degree of indeterminacy is much greater. In addition to a constant field, there are many other functions that may be added to $\mathbf{v}$ without changing $\nabla \cdot \mathbf{v}$ or $\nabla \times \mathbf{v}$. Going back to the proof of uniqueness, it is clear that we may add to $\mathbf{v}$ any function of the form $\nabla \psi$ that obeys Laplace's equation,

$$\nabla^2 \psi = 0. \tag{16.23}$$

This has solutions such as z, $x^2 - y^2$, and $z(2z^2 - 3(x^2 + y^2))$. The corresponding fields do not die off at infinity and may even diverge. In electromagnetism, such fields are either unphysical or they represent "external" fields imposed on the system of interest and are not created by the sources themselves.

In the case of electrostatics, the curl of the field $\mathbf{E}$ vanishes everywhere. Thus, the analog of the field $\mathbf{u}$ is zero, and the solution (16.18) to the more general problem reduces to eq. (13.5) with $g = \phi$ and $f = 4\pi\rho$.

Exercise 16.1 Verify by direct substitution that eqs. (16.17)–(16.19) solve eqs. (16.1) and (16.2).

Solution: We have

$$\nabla \cdot \mathbf{v} = -\nabla^2 g = -\frac{1}{4\pi} \nabla^2 \int \frac{f(\mathbf{r}')}{|\mathbf{r} - \mathbf{r}'|} d^3 x'$$

$$= \int f(\mathbf{r}')\delta(\mathbf{r} - \mathbf{r}') d^3 x', \tag{16.24}$$

where we have taken the Laplacian inside the integral (since it operates on the point $\mathbf{r}$, and the variable of integration is $\mathbf{r}'$) and used eq. (16.7). The integral over the delta function is trivial and yields eq. (16.1) immediately.

For the curl part, we get

$$\nabla \times \mathbf{v} = \nabla(\nabla \cdot \mathbf{u}) - \nabla^2 \mathbf{u}$$

$$= \frac{1}{4\pi} \left[\int (\mathbf{w} \cdot \nabla)\nabla \left(\frac{1}{|\mathbf{r} - \mathbf{r}'|} \right) d^3 x' - \int \mathbf{w} \nabla^2 \left(\frac{1}{|\mathbf{r} - \mathbf{r}'|} \right) d^3 x' \right]. \tag{16.25}$$

The second term leads to $\mathbf{w}(\mathbf{r})$, and so we need to show that the first term vanishes. We abbreviate $|\mathbf{r} - \mathbf{r}'| = R$ and use the summation convention. We have, first of all, that $\partial_i f(R) = -\partial_i' f(R)$ for any function f, where ∂_i' is a differential with respect to x_i'. Thus, the ith component of the integral is

$$\int w_j \partial_j' \partial_i' \frac{1}{R} d^3 x' = \int w_j \partial_j' \frac{R_i}{R^3} d^3 x'$$

$$= \int \partial_j' \left(w_j \frac{R_i}{R^3} \right) d^3 x', \tag{16.26}$$

where the last line follows from $\nabla \cdot \mathbf{w} = 0$. But the integral is now a total divergence. Transforming it to a surface integral at infinity by Gauss's theorem, we conclude that it vanishes because $\mathbf{w}$ vanishes at infinity.

Exercise 16.2 The mean electrostatic potential of a hydrogen atom in its ground state is given by

$$\phi(r) = \frac{e}{r} \left(1 + \frac{r}{a_0} \right) e^{-2r/a_0}, \tag{16.27}$$

where a_0 is the Bohr radius. Find (a) the corresponding electric field; (b) the charge density, making sure it integrates to zero.

17 Solving Poisson's equation: a few examples

For any charge distribution, the potential is given by eq. (13.5). Often, the direct evaluation of the integral is infeasible, and it is simpler to solve Poisson's differential equation. We forgo a complete discussion of this problem, as we shall see many more related problems in later chapters. Instead, as a first example, we consider the potential produced by a periodic array of charged wires. We take the wires to lie in the xz plane, parallel to the z axis, and with x coordinates $0, \pm a, \pm 2a$, etc. Taking the charge per unit length to be λ, the charge density is therefore given by

$$\rho(\mathbf{r}) = \lambda \delta(y) \sum_{n=-\infty}^{\infty} \delta(x - na). \tag{17.1}$$

Since this function is periodic in x, so is the potential $\phi(\mathbf{r})$, and both functions may be expanded as a Fourier series in x. For ϕ, let us write

$$\phi(\mathbf{r}) = \sum_{m=-\infty}^{\infty} \phi_m(y) e^{ik_m x}, \tag{17.2}$$

where $k_m = 2\pi m/a$. Note that the Fourier coefficients ϕ_m are functions of y, but not of z, since ρ is independent of z. Substituting eq. (17.2) into Poisson's equation and using the standard result

$$\sum_{n=-\infty}^{\infty} \delta(x - na) = \frac{1}{a} \sum_{m=-\infty}^{\infty} e^{ik_m x}, \tag{17.3}$$

we obtain

$$\frac{d^2 \phi_m}{dy^2} - k_m^2 \phi_m = -\frac{4\pi\lambda}{a} \delta(y). \tag{17.4}$$

The solution to this inhomogeneous differential equation is most easily obtained by taking Fourier transforms. Multiplying by e^{-iky} and integrating over all y, we obtain

$$\tilde{\phi}_m(k) = \frac{4\pi\lambda}{a} \frac{1}{k^2 + k_m^2}, \tag{17.5}$$

where

$$\tilde{\phi}_m(k) = \int_{-\infty}^{\infty} \phi_m(y) e^{-iky} dy. \tag{17.6}$$

To find $\phi_m(y)$, we must find the inverse transform. In doing this, it clearly suffices to consider $y > 0$ and $m \geq 0$. A simple contour integration gives

$$\phi_m(y) = \frac{4\pi\lambda}{a} \int_{-\infty}^{\infty} \frac{dk}{2\pi} \frac{e^{iky}}{k^2 + k_m^2} \tag{17.7}$$

$$= \frac{\lambda}{m} e^{-k_m y} \quad (m > 0, \ y > 0). \tag{17.8}$$

Thus, we see that the higher Fourier components die off very rapidly as we move farther away from the plane of the wires. As $y \to \infty$, only the $m = 0$ component, $\phi_0(y)$, which we have not yet found, survives. To find this, we use the fact that at large y, the electric field must be that of a sheet of surface charge density λ/a. Using Gauss's law, we find that $E_y = 2\pi\lambda/a$. Since $E_y = -\partial\phi/\partial y$, we get $\phi_0 = -2\pi\lambda y/a$. Along with eq. (17.8), this finally yields the real-space potential as

$$\phi(\mathbf{r}) = -\frac{2\pi\lambda}{a} y + 2\lambda \sum_{m>0} \frac{1}{m} e^{-k_m y} \cos(k_m x). \tag{17.9}$$

To reiterate, the potential can be found in terms of Fourier components in x at each value of y. The higher components die off very rapidly with increasing y, and for $y \geq 6a/2\pi \simeq a$, say, only the uniform part survives.

Although at this point, the problem is essentially solved, it is amusing to note that the sum in eq. (17.9) can be written in simple closed form. If we write the summand as $\mathrm{Re}(z^m/m)$, where $z = \exp[-2\pi(y - ix)/a]$, and note that

$$\sum_{m=1}^{\infty} \frac{z^m}{m} = -\ln(1 - z), \tag{17.10}$$

we obtain, after some simple algebra,

$$\phi(\mathbf{r}) = -\lambda \ln\left[2\cosh\left(\frac{2\pi y}{a}\right) - 2\cos\left(\frac{2\pi x}{a}\right)\right]. \tag{17.11}$$

At large y, of course, we recover our earlier conclusions. This formula enables quantitative calculations for $y = O(a)$, and also gives the field at very small y. It is not difficult to show that as $y \to 0$ and $x \to 0$,

$$\phi(\mathbf{r}) = -2\lambda \ln\left(\frac{2\pi}{a}\sqrt{x^2 + y^2}\right) + \frac{\pi^2}{3}\lambda\frac{x^2 - y^2}{a^2} + \cdots. \tag{17.12}$$

Exercise 17.1 Find the potential produced by a one-dimensional array of point charges $\pm q$ spaced a distance b from each other and alternating in sign.

Solution: This problem can be solved in analogy with the method used for the line charge array. The charge density is now given by

$$\rho(\mathbf{r}) = q \sum_{n=-\infty}^{\infty} (-1)^n \delta(z - nb)\delta(x)\delta(y). \tag{17.13}$$

Since

$$\sum_{n=-\infty}^{\infty} (-1)^n \delta(z - nb) = \frac{1}{b} \sum_{m\ \text{odd}} e^{ik_m z}, \tag{17.14}$$

where now $k_m = \pi m/b$, only wave vectors with z components of the form $(2j + 1)\pi/b$ will appear in $\phi(\mathbf{r})$. We solve the problem by Fourier transforms in the xy plane, and the analog of eq. (17.7) is

$$\phi_m(r_\perp) = \frac{4\pi q}{b} \int \frac{e^{i(k_x x + k_y y)}}{k^2 + k_m^2} \frac{d^2k}{(2\pi)^2}, \tag{17.15}$$

$$= \frac{4\pi q}{b} \frac{1}{2\pi} K_0(|k_m| r_\perp). \tag{17.16}$$

Here, $r_\perp = (x^2 + y^2)^{1/2}$, $k^2 = (k_x^2 + k_y^2)$, and K_0 is the modified Bessel function of second kind of order 0. The second line is a basic and frequently occurring two-dimensional inverse Fourier transform result.[6]

---

[6] The principal properties of Bessel functions are covered in appendix B. Equation (17.16) follows from eq. (B.48).

Thus, the analog of eq. (17.9) is

$$\phi(\mathbf{r}) = \frac{4q}{b} \sum_{m=1,3,\dots} \cos(k_m z) K_0(k_m r_\perp). \tag{17.17}$$

There does not seem to be a simple closed form for this sum. However, we can again see the rapid decay of the higher z-direction Fourier components with increasing $r_\perp$, as $K_0(w) \approx (\pi/2w)^{1/2} e^{-w}$ for large w. It is a challenging exercise in classical analysis to show that eq. (17.17) agrees with the potential due to a point charge in the limit $z \to 0$, $r_\perp \to 0$.

We can understand why the method of computing the Madelung energy of NaCl in section 14 converged rapidly on the basis of this exercise. The sum $S(x)$ in eq. (14.16) is exactly (b/q) times $\phi(x, 0, 0)$ in eq. (17.17), and so decays exponentially with x. Readers may enjoy seeing this decay by plotting $S(x)$ versus x.

Exercise 17.2 For readers not willing to accept that eq. (17.15) leads to so-and-so's function with such-and-such properties on someone else's authority, we offer this exercise. Write

$$\frac{1}{k^2 + k_m^2} = \int_0^\infty e^{-s(k^2 + k_m^2)} ds \tag{17.18}$$

for the factor of $1/(k^2 + k_m^2)$ in the integrand of eq. (17.15). This expresses $\phi_m(r_\perp)$ as a triple integral. But the integral over k_x and k_y is now simple. Do this to show that

$$\phi_m(r_\perp) = \frac{2q}{b} \int_0^\infty \frac{1}{2s} \exp\left[-\left((k_m r_\perp)^2 s + \frac{1}{4s}\right)\right] ds. \tag{17.19}$$

The integral is another way of writing $K_0(k_m r_\perp)$. Evaluate it by the saddle-point method (valid for large $k_m r_\perp$), and obtain the asymptotic behavior cited above.

Exercise 17.3 Consider an infinitely long, cylindrical, insulating, plastic rod, of radius a, carrying a uniform charge density ρ. Find the potential inside and outside the rod by solving Poisson's equation. Differentiate to find the electric field, and compare with the answer from Gauss's law. We will use the result when finding the inductance of the two-wire transmission line (exercise 31.5).

18 Energy in the electric field

Since the transmission of energy constitutes a signal, it cannot take place at a speed faster than that of light. This makes it both meaningful and important to ask where the electrostatic energy of a system of charges *resides*. We will see that this energy must be associated with the electric field itself and that there is a density for it that depends only on the *local* value of the field. To this end, let us return to eq. (14.4) for the electrostatic energy of a collection of point charges. If we write this in terms of a continuous distribution, we obtain

$$U = \frac{1}{2} \int \rho(\mathbf{r}) \phi(\mathbf{r}) d^3 x. \quad \text{(Gaussian and SI)} \tag{18.1}$$

We further transform this result by using Poisson's equation to eliminate ρ:

$$U = -\frac{1}{8\pi} \int \phi \nabla^2 \phi. \tag{18.2}$$

Integrating by parts,[7] we obtain

$$U = -\frac{1}{8\pi} \oint_S \phi \nabla \phi \cdot d\mathbf{s} + \frac{1}{8\pi} \int (\nabla \phi)^2 d^3x. \tag{18.3}$$

The first term vanishes because the surface is at infinity and hence has an area that grows as r^2 while $\phi \nabla \phi$ vanishes as $1/r^3$ by the same argument as that following eq. (16.4). The second term may be written in the terms of the electric field, so that

$$U = \frac{1}{8\pi} \int \mathbf{E}^2(\mathbf{r}) d^3x. \tag{18.4}$$

$$U = \frac{\epsilon_0}{2} \int \mathbf{E}^2(\mathbf{r}) d^3x. \quad \text{(SI)}$$

Equation (18.4) expresses the energy directly in terms of the electric field and suggests that we interpret $\mathbf{E}^2/8\pi$ as a density for the energy stored in the electric field.

Exercise 18.1 Repeat exercise 14.1 by using eq. (18.4).

There is an important difference between the discrete form (14.4) and the continuum forms (18.1) and (18.4). Equation (18.1) is perfectly all right if we treat the charge as smeared into a jelly, but we run into trouble if we apply it to a genuine point charge. In other words, if we try and transform it back into eq. (14.4), we have the problem that for the charge at $\mathbf{r}_a$, we must decide whether to include the potential due to that charge or not. The problem is clearly seen in eq. (18.4). Near the charge at $\mathbf{r}_a$, $\mathbf{E}^2$ diverges as $1/(\mathbf{r} - \mathbf{r}_a)^4$, and the integral is infinite. Or, to put it yet another way, eqs. (18.1) and (18.4) include the infinite *self-energy* of the point charges, while eq. (14.4) does not.

Since an infinite self-energy is physically absurd, we can argue that the interaction of a point particle with itself is meaningless, that this energy should be excluded from the accounting, and that eq. (14.4) is the only correct expression. This point of view would be perfectly valid if only static phenomena were to be considered. Unfortunately, the problem reappears when we consider moving charges, in particular, charges undergoing accaleration. It is found that an accelerating charge emits radiation and thus loses energy. The resulting slowdown of the charge may be described in terms of a "radiative damping" force acting on the particle. But now one is back to thinking in terms of the charge acting on itself. In the simplest model, the damping force is proportional to the time derivative of the acceleration. This model has the unsatisfactory feature that a charge moving through an external field can undergo infinite "self-acceleration." This difficulty can be cured by making the damping force nonlocal in time, but the theory is then not causal.

---

[7] This is another way of invoking Gauss's theorem. The integrand may be written as $\nabla \cdot (\phi \nabla \phi) - (\nabla \phi)^2$, and the divergence term transformed into a surface integral.

We are forced to conclude that the theory of classical electrodynamics contains internal contradictions. This situation is in contrast to the classical theory of mechanics, which is completely consistent in itself, even though it must be supplanted by quantum mechanics in light of experimental facts. The contradictions in classical electrodynamics are removed only in a fully relativistic quantum field theory, in the sense that any measurable quantity can be calculated. Even then, however, there is no theory of a point electron. Nevertheless, the classical theory has a very large domain of validity where it describes physical phenomena with extremely good accuracy. We discuss this point further in sections 39 and 67, but it is useful to make some order-of-magnitude estimates of the distances and time scales at which trouble arises. Let us try and eliminate the divergence of the self-energy by supposing the electron to have a nonzero size r_e. Its electrostatic energy is then e^2/r_e times a geometrical factor of order unity, with an exact value that depends on the details of how the charge is distributed (see exercise 14.1). The only other energy scale in the problem is the rest energy, mc^2. By equating these two energies, we find that

$$r_e = \frac{e^2}{mc^2}. \tag{18.5}$$

This length, equal to 2.82×10^{-13} cm, is known as the *classical electron radius*.[8] The time required for light to traverse this distance is

$$\tau_e = \frac{e^2}{mc^3}, \tag{18.6}$$

which equals about 10^{-23} sec. Unless we deal with moving charges whose motion changes significantly over length scales of the order of r_e or on time scales of the order of τ_e, we need not worry about the absurdities arising from an infinite self-energy and the related difficulties. In fact, it should be noted that quantum effects become important much earlier—at length scales of the order of $\hbar/mc$, known as the *Compton wavelength* of the electron, and at times of order $\hbar/mc^2$. These are larger than r_e and τ_e by the factor $c\hbar/e^2 \approx 137$.

19 The multipole expansion

Let us now consider a charge distribution occupying a finite region of space of size L. At distances far from the charges, i.e., much greater than L, the potential and the field may be expanded in inverse powers of this distance. This expansion is of considerable use as the field can be well approximated in terms of a few parameters characterizing the distribution.

In the following, we consider only a system of point charges. The analogous formulas for a continuous distribution are written immediately. Taking the origin to lie somewhere

---

[8] For comparison, the Bohr radius, which gives the mean electron–proton separation in hydrogen, is 0.529×10^{-8} cm.

within the region of space occupied by the charges, we obtain

$$\phi(\mathbf{r}) = \sum_a \frac{q_a}{|\mathbf{r} - \mathbf{r}_a|}. \tag{19.1}$$

If $r \gg r_a$, we may expand the denominator as follows:

$$\frac{1}{|\mathbf{r} - \mathbf{r}_a|} = \frac{1}{r} + \mathbf{r}_a \cdot \left[\nabla_a \frac{1}{|\mathbf{r} - \mathbf{r}_a|} \right]_{\mathbf{r}_a=0} + \frac{1}{2!} x_{a,i} x_{a,j} \left[\nabla_{a,i} \nabla_{a,j} \frac{1}{|\mathbf{r} - \mathbf{r}_a|} \right]_{\mathbf{r}_a=0}$$

$$+ \frac{1}{3!} x_{a,i} x_{a,j} x_{a,k} \left[\nabla_{a,i} \nabla_{a,j} \nabla_{a,k} \frac{1}{|\mathbf{r} - \mathbf{r}_a|} \right]_{\mathbf{r}_a=0} + \cdots, \tag{19.2}$$

where ∇_a denotes the gradient with respect to $\mathbf{r}_a$. We now note that since ∇_a operates only on functions of the difference $\mathbf{r} - \mathbf{r}_a$, we may replace it by $-\nabla$, after which we may set $\mathbf{r}_a$ to zero inside all the square brackets. Thus, we have

$$\frac{1}{|\mathbf{r} - \mathbf{r}_a|} = \frac{1}{r} - (\mathbf{r}_a \cdot \nabla) \frac{1}{r} + \frac{1}{2!} (\mathbf{r}_a \cdot \nabla)^2 \frac{1}{r} - \frac{1}{3!} (\mathbf{r}_a \cdot \nabla)^3 \frac{1}{r} + \cdots. \tag{19.3}$$

Substituting in eq. (19.1) and using the summation convention as needed, we obtain

$$\phi(\mathbf{r}) = \frac{\sum q_a}{r} - \sum q_a \mathbf{r}_a \cdot \nabla \frac{1}{r} + \frac{1}{2} \sum_a q_a x_{a,i} x_{a,j} \partial_i \partial_j \frac{1}{r} + \cdots. \tag{19.4}$$

We shall denote the first term in this expansion by $\phi^{(0)}$, and the succeeding ones by $\phi^{(1)}$, $\phi^{(2)}$, etc. Note that $\phi^{(n)} \sim 1/r^{n+1}$ as $r \to \infty$.

Let us discuss the first few terms in this expansion individually. The leading, or *monopole* term, is just the potential produced by the net charge of the system, acting as if it were concentrated into a point at the origin.

The next term is the *dipole* term. The vector

$$\mathbf{d} = \sum_a q_a \mathbf{r}_a \tag{19.5}$$

is known as the *dipole moment* of the system. In terms of this quantity, the dipole potential is given by

$$\phi^{(1)} = -\mathbf{d} \cdot \nabla \frac{1}{r} = \frac{\mathbf{d} \cdot \mathbf{r}}{r^3}. \tag{19.6}$$

The corresponding electric field is given by

$$\mathbf{E} = -\nabla \frac{\mathbf{d} \cdot \mathbf{r}}{r^3}. \tag{19.7}$$

We leave it as an exercise for the reader to show that this equals

$$\mathbf{E} = \frac{3(\hat{\mathbf{n}} \cdot \mathbf{d})\hat{\mathbf{n}} - \mathbf{d}}{r^3}, \tag{19.8}$$

where $\hat{\mathbf{n}} = \mathbf{r}/r$ is the unit vector along $\mathbf{r}$. This field is sketched in fig. 3.1. Note that it dies at infinity as $1/r^3$, i.e., with one higher power of r^{-1} than the field due to a point charge. The case where the charges are balanced, i.e., the total charge is zero, is especially

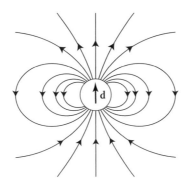

FIGURE 3.1. The electric field of a point dipole.

important. In this case, the dipole moment is independent of the choice of origin. If we shift to a different origin, so that $\mathbf{r} \to \mathbf{r} + \mathbf{u}$, $\mathbf{r}_a \to \mathbf{r}_a + \mathbf{u}$, where $\mathbf{u}$ is some fixed vector, then

$$\mathbf{d} \to \sum_a q_a \mathbf{r}_a + \mathbf{u} \sum_a q_a = \sum_a q_a \mathbf{r}_a, \tag{19.9}$$

since $\sum_a q_a = 0$.

The third, or *quadrupole*, term in eq. (19.4), is given by

$$\phi^{(2)} = \frac{1}{2} \sum q_a x_{a,i} x_{a,j} \partial_i \partial_j \frac{1}{r}. \tag{19.10}$$

Since $\nabla^2 (1/r) = 0$, we may write this equally as

$$\phi^{(2)} = \frac{1}{2} \sum q_a \left(x_{a,i} x_{a,j} - \frac{1}{3} r_a^2 \delta_{ij} \right) \partial_i \partial_j \frac{1}{r},$$

$$= \frac{1}{6} D_{ij} \partial_i \partial_j \frac{1}{r}, \tag{19.11}$$

where

$$D_{ij} = \sum_a q_a (3 x_{a,i} x_{a,j} - r_a^2 \delta_{ij}) \tag{19.12}$$

is known as the *quadrupole moment* of the system. This is a symmetric tensor of second rank. Such a tensor usually has six independent components (D_{11}, D_{22}, D_{33}, D_{12}, D_{13}, and D_{23}), but in the case of D_{ij}, the *trace*, or sum of diagonal elements, vanishes:

$$D_{ii} = 0. \tag{19.13}$$

So, only five components of D_{ij} are independent, and the potential due to a quadrupole is determined by five parameters. Since

$$\partial_i \partial_j \frac{1}{r} = \frac{3 x_i x_j - \delta_{ij} r^2}{r^5}, \tag{19.14}$$

and since $D_{ij}\delta_{ij} = 0$, this potential can be written as

$$\phi^{(2)}(\mathbf{r}) = \frac{D_{ij}n_i n_j}{2r^3}. \tag{19.15}$$

The electric field due to a quadrupole is obtained by taking the gradient of this expression.

Exercise 19.1 Show that for a system of charges with zero total charge and zero dipole moment, the quadrupole moment is independent of the choice of origin.

Exercise 19.2 The term $\phi^{(3)}$ is known as the *octupole* term. Define an octupole moment tensor, and find how many independent components it possesses.

The higher moments are of use only in specialized circumstances. We shall consider them largely in order to establish some formal results and to develop some mathematical results of wider utility. The basic result is what we shall call the spherical harmonic expansion of the inverse separation:

$$\frac{1}{|\mathbf{r} - \mathbf{r}'|} = \sum_{\ell=0}^{\infty} \sum_{m=-\ell}^{\ell} \frac{4\pi}{2\ell+1} \frac{r_<^\ell}{r_>^{\ell+1}} Y_{\ell m}^*(\hat{\mathbf{r}}') Y_{\ell m}(\hat{\mathbf{r}}). \tag{19.16}$$

In this formula, $r_<$ and $r_>$ are, respectively, the lesser and the greater of r and r', and the functions $Y_{\ell m}$ are the so-called spherical harmonics. By taking its complex conjugate, we also see that we can write it equally well with the complex conjugate on either one of the $Y_{\ell m}$'s.

A closely related expansion is obtained by thinking of $|\mathbf{r} - \mathbf{r}'|$ as a function of $\cos\vartheta = \hat{\mathbf{r}} \cdot \hat{\mathbf{r}}'$ in addition to r and r'. We then have

$$\frac{1}{|\mathbf{r} - \mathbf{r}'|} = \sum_{\ell=0}^{\infty} \frac{r_<^\ell}{r_>^{\ell+1}} P_\ell(\cos\vartheta), \tag{19.17}$$

where P_ℓ is a polynomial of degree ℓ in its argument, known as the *Legendre polynomial*.

We discuss the spherical harmonics and their relation to the Legendre polynomials at much greater length in appendix A. For our present purposes, we can regard them as being defined via the generating function

$$\frac{1}{\ell!}\left(z - \frac{\lambda}{2}\xi + \frac{1}{2\lambda}\eta\right)^\ell = r^\ell \sqrt{\frac{4\pi}{2\ell+1}} \sum_{m=-\ell}^{\ell} \frac{\lambda^m Y_{\ell m}(\hat{\mathbf{r}})}{\sqrt{(\ell+m)!\,(\ell-m)!}}, \tag{19.18}$$

where

$$\xi = x + iy, \quad \eta = x - iy. \tag{19.19}$$

The first few spherical harmonics are listed in table 3.1. Their most important property from the practical viewpoint is that they are orthonormal (this defines our choice of normalization), i.e.,

$$\int Y_{\ell m}(\hat{\mathbf{r}}) Y_{\ell' m'}^*(\hat{\mathbf{r}}) d\Omega = \delta_{\ell\ell'}\delta_{mm'}. \tag{19.20}$$

Here, the integration is over all orientations $\hat{\mathbf{r}}$, and $d\Omega$ denotes an element of solid angle.

Table 3.1. The first few spherical harmonics

ℓ	m	$r^\ell Y_{\ell m}$	Normalization factor
0	0	1	$(1/4\pi)^{1/2}$
1	1	$(x+iy)$	$-(3/8\pi)^{1/2}$
1	0	z	$(3/4\pi)^{1/2}$
2	2	$(x+iy)^2$	$(15/32\pi)^{1/2}$
2	1	$(x+iy)z$	$-(15/8\pi)^{1/2}$
2	0	$(3z^2-r^2)$	$(5/16\pi)^{1/2}$
3	3	$(x+iy)^3$	$-(35/64\pi)^{1/2}$
3	2	$(x+iy)^2z$	$(105/32\pi)^{1/2}$
3	1	$(5z^2-r^2)(x+iy)$	$-(21/64\pi)^{1/2}$
3	0	$(5z^2-3r^2)z$	$(7/64\pi)^{1/2}$

Notes: 1. The polynomial is to be multiplied by the normalization factor. For example, $r\,Y_{10} = (3/4\pi)^{1/2}z$.
2. $Y_{\ell,-m} = (-1)^m Y_{\ell m}^*$.

Exercise 19.3 Use the generating function (19.18) to obtain the Y_{3m}'s.

Substituting eq. (19.16) in eq. (19.1) with $\mathbf{r}_a$ for $\mathbf{r}'$ (and, naturally, $r_a < r$), we obtain a series in increasing powers of r_a/r for each term in the sum over a. The term in $1/r^{\ell+1}$ gives the potential due to the 2^ℓ-pole:

$$\phi^{(\ell)}(\mathbf{r}) = \frac{1}{r^{\ell+1}} \sqrt{\frac{4\pi}{2\ell+1}} \sum_{m=-\ell}^{\ell} q_{\ell m} Y_{\ell m}(\hat{\mathbf{r}}), \tag{19.21}$$

where the $2\ell+1$ quantities

$$q_{\ell m} = \sqrt{\frac{4\pi}{2\ell+1}} \sum_a q_a r_a^\ell Y_{\ell m}^*(\hat{\mathbf{r}}_a) \tag{19.22}$$

collectively define the 2^ℓ-pole moment of the system.[9] Note that even though $Y_{\ell m}$ is complex, there are only $2\ell+1$ real parameters involved, because

$$Y_{\ell,m}^* = (-1)^m Y_{\ell,-m}. \tag{19.23}$$

For the same reason, the expression (19.21) is real.

---

[9] Some authors prefer to put the entire factor of $4\pi/(2\ell+1)$ in eq. (19.21) instead of dividing it symmetrically between eqs. (19.21) and (19.22), as we have done.

For the special case $\ell = 1$, the dipole moment vector $\mathbf{d}$ is related to the q_{1m}'s as follows:

$$q_{10} = d_z, \qquad q_{1,\pm 1} = \mp \frac{1}{\sqrt{2}} (d_x \mp i d_y). \tag{19.24}$$

Similarly, for $\ell = 2$, the Cartesian tensor components D_{ij} are related to the q_{2m}'s as follows:

$$q_{20} = D_{zz},$$

$$q_{2,\pm 1} = \mp \frac{1}{\sqrt{6}} (D_{xz} \mp i D_{yz}), \tag{19.25}$$

$$q_{2,\pm 2} = \frac{1}{2\sqrt{6}} (D_{xx} - D_{yy} \mp 2i D_{xy}).$$

The quantities q_{1m} and q_{2m} are known as the *spherical tensor components* of the dipole and quadrupole tensors. In Cartesian components, the 2^ℓ-pole tensor (defined by a natural extension of the way we used to define a quadrupole tensor) is completely symmetric and vanishes if contracted on any two of its indices. Such Cartesian tensors are known as *irreducible*, and they can always be represented by $2\ell + 1$ spherical tensor components. Thus, a vector is always equivalent to a spherical tensor with $\ell = 1$, a rank-2 symmetric traceless tensor to a spherical tensor with $\ell = 2$, and so on. The general relation between spherical and Cartesian components is given by eqs. (19.24) and (19.25). Readers who have solved exercise 19.2 will appreciate this point immediately.

Exercise 19.4 For a spherically symmetric charge distribution, obtain eq. (14.8) by starting with the continuous analog of the formula $U = \frac{1}{2} \sum_{a,b} q_a q_b / r_{ab}$, and using eq. (19.16).

Exercise 19.5 (amusement) The methane molecule (CH_4) has tetrahedral symmetry. What is the lowest-order multipole in its far field? Repeat for the icosahedrally symmetric molecule buckminsterfullerene (C_{60}).

Answer: $\ell = 4$ (CH_4), $\ell = 6$ (C_{60}).

As another application of eq. (19.16), we derive the *mean value theorem* for electrostatics (also known as *Earnshaw's theorem*), which states that in any electric field, the average of the potential over the surface of any sphere lying entirely in a charge-free region of space is equal to the potential at the center of the sphere. To prove this theorem, let us choose a coordinate system centered at the sphere (see fig. 3.2). Let the sphere have a radius r, and let us consider a point charge q at a point $\mathbf{R}$, with $R > r$. The average of the potential due to this charge over the surface of the sphere is given by

$$\langle \phi \rangle = \frac{1}{4\pi r^2} \int \frac{q}{|\mathbf{r} - \mathbf{R}|} r^2 d\Omega. \tag{19.26}$$

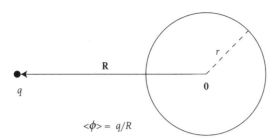

FIGURE 3.2. Basic geometry for proving the mean value theorem.

Although the integration can be done directly, let us use eq. (19.16) for $|\mathbf{r} - \mathbf{R}|^{-1}$. This gives

$$\langle \phi \rangle = \sum_{\ell,m} \frac{q}{2\ell+1} \int \frac{r^\ell}{R^{\ell+1}} Y_{\ell m}(\hat{\mathbf{r}}) Y_{\ell m}^*(\hat{\mathbf{R}}) d\Omega. \tag{19.27}$$

In the integral over orientations of $\hat{\mathbf{r}}$, only the $\ell = m = 0$ term survives. Since $Y_{00} = (4\pi)^{-1/2}$, we get

$$\langle \phi \rangle = \frac{q}{R} = \phi(\mathbf{0}). \tag{19.28}$$

This proves the theorem for this special configuration. But, since any potential obeying the conditions of the theorem can be viewed as arising from a collection of charges all outside the sphere, the theorem is proved in general.

Physically, the mean value theorem implies that there can be no maxima in the electrostatic field, because if there were, the average of the potential over the surface of a tiny sphere surrounding the maximum would be less than the value at the center in contravention of the theorem. Likewise, there can be no minima. It follows that no configuration of charges can be in equilibrium under purely electrostatic forces.

One sometimes encounters the following weaker argument against the existence of a minimum in a region where Laplace's equation holds. Suppose the contrary. At the supposed minimum, if we take x_1, x_2, and x_3 as Cartesian coordinates, all first derivatives $\partial\phi/\partial x_i$ vanish, and all second derivatives $\partial^2\phi/\partial x_i^2$ are strictly positive. But then $\nabla^2\phi > 0$, in violation of Laplace's equation. The following exercises show that this argument is not watertight.

Exercise 19.6 Let two identical point charges $+q$ be located at $(0, 0, \pm a)$. Find the potential near the origin in powers of x, y, and z, going up to terms of second order.

Answer:

$$\phi = 2\frac{q}{a} - \frac{q}{a^3}(x^2 + y^2 - 2z^2) + \cdots. \tag{19.29}$$

The origin is a saddle point.

Exercise 19.7 Let six identical point dipoles **d**, all pointing along $\hat{z}$, be located at the vertices of a regular octahedron: $(\pm a, 0, 0)$, $(0, \pm a, 0)$, $(0, 0, \pm a)$. Find the potential near the origin in powers of x, y, and z, going up to terms of third order.

Answer:

$$\phi = -\frac{7d}{a^5}(5z^3 - 3zr^2) + \cdots . \tag{19.30}$$

Now, all second derivatives $\partial^2\phi/\partial x_i^2$ vanish at the origin, and one would have to examine higher derivatives to decide if the origin was a local minimum or maximum. Of course, we know that it is not by the mean value theorem. One can also see that by noting that ϕ is odd under inversion about the origin, i.e., $\mathbf{r} \to -\mathbf{r}$.

20 Charge distributions in external fields

The concept of multipole moments is also useful in dealing with a charge distribution or system of point charges in a fixed external electric field created by sources that are far away from the charges of interest and are assumed not to be influenced by them. We may write the potential energy of the system as

$$U = \sum_a q_a \phi(\mathbf{r}_a), \tag{20.1}$$

where $\phi(\mathbf{r})$ is the potential of the external field. If this field varies slowly over the region of space occupied by the charges of interest, then by choosing the origin somewhere in this region, we may expand $\phi(\mathbf{r})$ in powers of $\mathbf{r}$ multiplied by gradients of ϕ:

$$\phi(\mathbf{r}) = \phi(0) + \mathbf{r} \cdot \nabla\phi + \tfrac{1}{2}x_i x_j \partial_i \partial_j \phi + \cdots , \tag{20.2}$$

where all the derivatives of ϕ are evaluated at the origin.

The first, or monopole, term in this expansion yields

$$U_{\mathrm{mono}} = \phi(0) \sum_a q_a . \tag{20.3}$$

The next, or dipole, term is of particular importance:

$$U_{\mathrm{dip}} = \sum_a q_a \mathbf{r}_a \cdot (\nabla\phi)_0 = -\mathbf{d} \cdot \mathbf{E}. \tag{20.4}$$

Here, **E** is the electric field at the origin chosen. If, as is frequently the case, the system of charges is neutral as a whole, this is the leading term in the energy. If the external field is uniform, this term terminates the expansion. It is often useful to consider the dipole moment to have no size at all, i.e., as located at a point. In that case, we can regard the electric field in eq. (20.4) as being evaluated at the location of the dipole. Since this energy depends on the angle between **d** and **E**, it follows that the dipole experiences a torque,

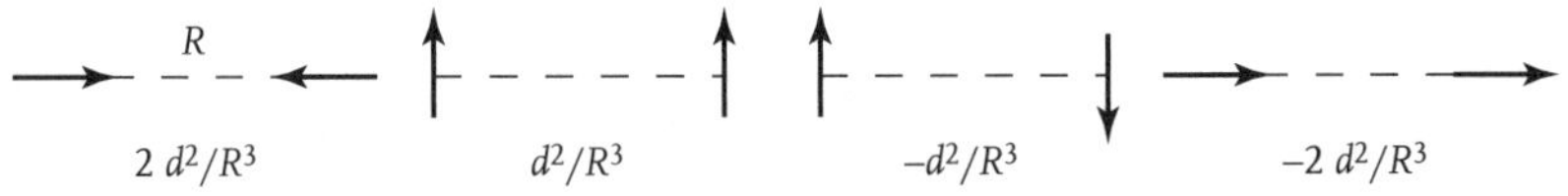

FIGURE 3.3. Interaction energy of two identical dipoles in various relative orientations.

which can be evaluated directly when the field is uniform:

$$N = \sum r_a \times (q_a E) = d \times E. \tag{20.5}$$

If the field is nonuniform, there is also a force on the dipole, given by

$$F = \sum_a q_a E(r_a) \approx \sum_a q_a (r_a \cdot \nabla) \, E|_0 = (d \cdot \nabla)E. \tag{20.6}$$

Together, eqs. (20.4)–(20.6) imply that a dipole in an external field lowers its energy by orienting itself parallel to the field and moving to regions of larger field magnitude.

Exercise 20.1 Show that the quadrupole contribution to the energy is given by

$$U_{\text{quad}} = \tfrac{1}{6} D_{ij} \partial_i \partial_j \phi. \tag{20.7}$$

Let us now consider two clusters or distributions of charge separated by a distance much greater than their own dimensions. The energy of interaction between these two distributions can clearly be expanded in terms of the moments of each distribution. The monopole–monopole term is clearly $Q_1 Q_2 / R$, where Q_1 and Q_2 are total charges of the two distributions, and R is the vector from distribution 1 to distribution 2. The monopole–dipole and dipole–monopole terms are

$$-Q_1 \frac{d_2 \cdot R}{R^3}, \qquad Q_2 \frac{d_1 \cdot R}{R^3}, \tag{20.8}$$

and the dipole–dipole term is

$$U_{\text{dd}} = \frac{(d_1 \cdot d_2) \, R^2 - 3(d_1 \cdot R)(d_2 \cdot R)}{R^5}. \tag{20.9}$$

For fixed R, this energy is minimized if both dipoles are parallel to each other and to R (see fig. 3.3). It is maximized if $d_1 \| R$, and $d_2 \| -R$, i.e., if the two dipoles are oriented head-to-head (or tail-to-tail). If the first dipole d_1 is held fixed parallel to $\hat{z}$ at the origin, the energy depends on the angle between d_2 and R. If $R \| \pm \hat{z}$ it is minimized for $d_2 \| \hat{z}$, and if R is in the xy plane, it is minimized for $d_2 \| -\hat{z}$.

4 | Magnetostatics in vacuum

Operationally, a magnetic field **B** may be defined via the Lorentz force law.[1] Whatever a magnetic field is, it is something to which a moving charge responds as per this law. By a vague principle of reciprocity, one might expect a moving charge to be a source of magnetic field. This turns out to be correct, but a static magnetic field requires not one moving charge, but a steady current of charges. This makes the mathematical treatment of even the simplest magnetostatic fields quite different from that of electrostatics.

In addition, however, magnetic fields are also created by point magnetic dipoles.[2] Two such dipoles interact in the same way as two electric dipoles. This provides a major similarity with electrostatics, and we shall adopt this as our point of entry for the study of magnetostatics. But magnetic dipoles are profoundly different from electric dipoles in that they produce a divergenceless field. The resulting subtleties are studied in section 26.

The existence of two types of sources of magnetic field obligates us to study the relationship between them. We shall use the Lorentz force law for this purpose. We shall also see how magnetic fields may be described by potentials.

The study of magnetic work and energy is deferred to the next chapter.

21 Sources of magnetic field

That like poles of two magnets repel and unlike poles attract has been known since antiquity. In 1785, Coulomb established by careful experiments that the force between

---

[1] There is, alas, no uniform terminology for **B**. It is common to find it called the *magnetic induction* and to use *magnetic field* for another field, **H**, encountered in the study of magnetic matter. The problem with this usage is that the pithy and valuable qualifier "field" is left out in speaking of **B**—nobody talks of the "magnetic induction field"—and **B** is the more fundamental entity anyway. We shall therefore adopt an older engineering convention and call **B** the *magnetic* field, and **H** the *magnetizing* field. Excessive rationalization of this choice is silly; what is needed is an easily verbalized and clearly distinct set of terms, and this our choice achieves. A reason, weak though it is, is given in section 107, when we study hysteresis in ferromagnets.

[2] There are no free magnetic charges, or monopoles. At least, none are known as of this writing.

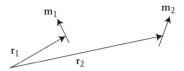

FIGURE 4.1. Two interacting dipoles.

two very long and thin magnetic rods can be described in terms of poles residing at points at the ends of the rods, and that this force has exactly the same form as that between two point *electric* charges. Since, however, the poles of a magnet are always equal and opposite, and a single pole or two unbalanced poles can never be created by breakage or any other treatment of the magnet, a mental synthesis of a magnet must be sought not in terms of isolated magnetic charges, but in terms of infinitesimally separated pairs of equal and opposite charges, i.e., *dipoles*. By the process of abstraction, we are naturally led to consider point dipoles.[3]

A second reason for considering point dipoles is that we now know that point particles, such as the electron and its siblings, the muon and the taon, possess magnetic dipole moments that arise from the spin of these particles, a quantum mechanical property with no classical counterpart. These moments are an important constituent of the permanent magnetism of materials such as iron and lodestone or magnetite (Fe_3O_4).[4]

It is simplest to begin by considering the potential energy of interaction between two magnetic dipoles, with moments $\mathbf{m}_1$ and $\mathbf{m}_2$, located at points $\mathbf{r}_1$ and $\mathbf{r}_2$ (see fig. 4.1). It follows from the similarity of the force between magnetic poles and electric charges that the expression for the potential energy is identical to that for electric dipoles:[5]

$$U(\mathbf{r}_{21}) = \frac{(\mathbf{m}_1 \cdot \mathbf{m}_2)r_{21}^2 - 3(\mathbf{m}_1 \cdot \mathbf{r}_{21})(\mathbf{m}_2 \cdot \mathbf{r}_{21})}{r_{21}^5}, \tag{21.1}$$

$$U(\mathbf{r}_{21}) = \frac{\mu_0}{4\pi}\frac{(\mathbf{m}_1 \cdot \mathbf{m}_2)r_{21}^2 - 3(\mathbf{m}_1 \cdot \mathbf{r}_{21})(\mathbf{m}_2 \cdot \mathbf{r}_{21})}{r_{21}^5}, \quad \text{(SI)}$$

where $\mathbf{r}_{21} = \mathbf{r}_2 - \mathbf{r}_1$. As in the electrostatic case, we would like to interpret this in terms of a field. One way to do that is to note that by differentiating $U(\mathbf{r}_{21})$ with respect to $\mathbf{r}_2$ and $\hat{\mathbf{m}}_2$, we can obtain the force and torque on dipole number 2. This suggests the following procedure. We take a test magnetic dipole moment $\mathbf{m}$ at a point. The torque on this moment,

$$\mathbf{N} = \mathbf{m} \times \mathbf{B}, \quad \text{(Gaussian and SI)} \tag{21.2}$$

---

[3] We are following here the same route as Maxwell (1954); see vol. II, chap. 1.

[4] If one is willing to regard the proton and neutron as elementary particles, then they also possess point dipole moments. This is certainly an excellent approximation for the macroscale, and even for much of atomic physics, although the magnitude of these moments is much less than that of the electron, and they are consequently less important for macroscopic physics. We refrain from talking of the "quark magnetic moments" because it is difficult and not particularly fruitful to give operational meaning to this term.

[5] We delay discussion of the units of various quantities until the end of section 24, but we give equations in both the Gaussian and SI systems from the outset.

defines the magnetic field **B** at that point. By choosing differently oriented moments, we can measure all components of **B**. If the field is nonuniform, the dipole also experiences a force, given by[6]

$$\mathbf{F} = \nabla(\mathbf{m} \cdot \mathbf{B}). \quad \text{(Gaussian and SI)} \tag{21.3}$$

Both these expressions are contained in the statement that a point magnetic dipole in an external magnetic field has a potential energy given by

$$U = -\mathbf{m} \cdot \mathbf{B}(\mathbf{r}), \quad \text{(Gaussian and SI)} \tag{21.4}$$

where **r** is the location of the dipole.

It now follows from eq. (21.1) that we must ascribe to a magnetic dipole **m** at the origin a magnetic field

$$\mathbf{B}(\mathbf{r}) = \frac{3(\hat{\mathbf{n}} \cdot \mathbf{m})\hat{\mathbf{n}} - \mathbf{m}}{r^3} \quad (\mathbf{r} \neq 0), \tag{21.5}$$

$$\mathbf{B}(\mathbf{r}) = \frac{\mu_0}{4\pi} \frac{3(\hat{\mathbf{n}} \cdot \mathbf{m})\hat{\mathbf{n}} - \mathbf{m}}{r^3} \quad (\mathbf{r} \neq 0), \quad \text{(SI)}$$

where $\hat{\mathbf{n}} = \mathbf{r}/r$.[7] Just as for the electric field, the superposition principle applies to **B** as well. That is, the field due to a collection of magnetic dipoles is given by adding together the fields due to each of the dipoles separately. In fact, we have implicitly used superposition in our arguments already.

However, we have also defined the magnetic field in terms of the Lorentz force on a moving charged particle:

$$\mathbf{F} = q\frac{\mathbf{v}}{c} \times \mathbf{B}. \tag{21.6}$$

$$\mathbf{F} = q\mathbf{v} \times \mathbf{B}. \quad \text{(SI)}$$

If we accept, for the moment, the equivalence of the two definitions, a current loop placed in the field of a point dipole must experience a torque and a force. To determine these, we first develop a simple model of a current loop. We visualize it as a physical loop of wire in which a large number of electrons are moving so that the net charge crossing any cross section of the wire is a constant in time. The current in the loop is defined as

$$\text{Current} = \text{Charge flowing through any cross section per unit time.} \tag{21.7}$$

---

[6] This expression differs from $(\mathbf{m} \cdot \nabla)\mathbf{B}$, the analog of eq. (20.6) for an electric dipole. As we shall see, this is because a magnetic dipole must be viewed as a current loop and not a pair of equal and opposite monopoles. Mathematically, $\nabla(\mathbf{m} \cdot \mathbf{B}) - (\mathbf{m} \cdot \nabla)\mathbf{B} = \mathbf{m} \times (\nabla \times \mathbf{B})$, which vanishes when $\nabla \times \mathbf{B} = 0$. Since many magnetostatic fields do have vanishing curl in large regions of space, it is easy to overlook such subtle differences in derivations that rely on simple situations.

[7] We shall see in section 26 that we must add a term proportional to $\mathbf{m}\delta(\mathbf{r})$ to eq. (21.5) to obtain an expression valid at $\mathbf{r} = 0$. Such a term cannot have any consequences as long we are not interested in the field at the origin. Thus, for the most part, we can forget about the caveat $\mathbf{r} \neq 0$ in eq. (21.5).

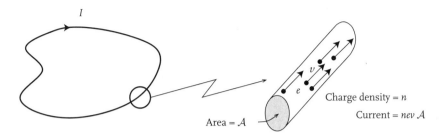

FIGURE 4.2. A mathematical current loop, and a more physical view of a current-carrying wire.

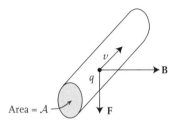

FIGURE 4.3. The force on a current-carrying wire in a magnetic field.

In fig. 4.2, we show a wire of uniform cross-sectional area $\mathcal{A}$, with a uniform electron number density, n, all moving with velocity v.[8] The current is then $I = nqv\mathcal{A}$, or $-nev\mathcal{A}$ if $q = -e$.

It follows from the definition that in a wire carrying a steady current, the charge density due to flowing charges at any point in the wire remains constant in time. Usually, there is an equal and opposite density of fixed charges (positive ions) such that the net charge density is zero.[9] A physical wire must, of course, have some thickness, but it is convenient to first consider the mathematical idealization in which a nonzero current flows through a wire of zero thickness. We also ignore the phenomenon of heat dissipation that occurs in actual material wires.

We now consider a small element $d\ell_2$ of the wire. This element contains $n\mathcal{A}d\ell_2$ charges, and therefore experiences a force (see fig. 4.3)

$$d\mathbf{F} = -(n\mathcal{A}d\ell_2/c)q\mathbf{B} \times \mathbf{v} = -(I/c)\mathbf{B} \times d\boldsymbol{\ell}_2. \tag{21.8}$$

The force and torque on a complete loop can then be obtained by adding up the forces and torques on its various elements. Let us find the torque an infinitesimal square loop

---

[8] It should be stated at once that this is a grossly incorrect view of how current flows in a real metal. In particular, v is at best described as a mean *drift velocity*, the average over a large number of electrons, or of one electron over a duration in which it suffers many collisions; it is *not* the actual velocity of an electron. Nevertheless, this picture suffices for understanding macroscopic EM fields.

[9] Provided relativistic effects are ignored. Otherwise, if the wire is neutral when no current is flowing, then when it is carrying a current, the negative charge density exceeds the positive by a factor $(1 - v^2/c^2)^{-1/2}$. The difference, of order (v^2/c^2), can be neglected if $v \ll c$, as is almost always the case.

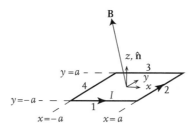

FIGURE 4.4. An infinitesimal square loop in a magnetic field.

of side $2a$. We choose a coordinate system with origin at the center of this loop, and with the z axis normal to the loop in such a way that it points along the thumb if the fingers of the right hand curl around in the direction of current flow (fig. 4.4). (We shall refer to this as the right-hand rule for normals to current loops.) The torque on leg 1 is $-a\hat{\mathbf{y}} \times [(-I\mathbf{B}/c) \times 2a\hat{\mathbf{x}}] = -2(I/c)a^2 B_y\hat{\mathbf{x}}$, where the magnetic field is evaluated at the center of the loop. The torque on the other legs is similarly calculated, and the total torque is

$$\mathbf{N}_{\text{on loop}} = \frac{4Ia^2}{c}(B_x\hat{\mathbf{y}} - B_y\hat{\mathbf{x}})$$

$$= \mathbf{m}_{\text{loop}} \times \mathbf{B}, \tag{21.9}$$

where, with $\mathcal{A}$ being the area of the loop, and $\hat{\mathbf{n}}$ the normal to it in the right-hand sense,

$$\mathbf{m}_{\text{loop}} = \frac{I\mathcal{A}}{c}\hat{\mathbf{n}}. \tag{21.10}$$

$$\mathbf{m}_{\text{loop}} = I\mathcal{A}\hat{\mathbf{n}}. \quad \text{(SI)}$$

That is, the loop feels exactly the same torque as a magnetic dipole of strength and direction $\mathbf{m}_{\text{loop}}$ would. Since eqs. (21.9) and (21.10) are written as vector equations without explicit reference to the coordinate system, they hold generally.

If the field is not uniform, the forces on legs 1 and 3 will not cancel exactly, nor will those on legs 2 and 4. Expanding in powers of a, it is easy to show that the total force is

$$\mathbf{F}_{\text{on loop}} = \frac{4Ia^2}{c}\left[\frac{\partial B_z}{\partial x}\hat{\mathbf{x}} + \frac{\partial B_z}{\partial y}\hat{\mathbf{y}} - \left(\frac{\partial B_x}{\partial x} + \frac{\partial B_y}{\partial y}\right)\hat{\mathbf{z}}\right], \tag{21.11}$$

where the field gradients are again evaluated at the origin. If the field is produced by a dipole or collection of dipoles, then exactly as in the electric case, its divergence vanishes at the site of the loop, and so $\partial B_x/\partial x + \partial B_y/\partial y = -\partial B_z/\partial z$. Hence,

$$\mathbf{F}_{\text{on loop}} = \frac{4Ia^2}{c}\left[\frac{\partial B_z}{\partial x}\hat{\mathbf{x}} + \frac{\partial B_z}{\partial y}\hat{\mathbf{y}} + \frac{\partial B_z}{\partial z}\hat{\mathbf{z}}\right]$$

$$= \frac{4Ia^2}{c}\nabla B_z$$

$$= \nabla(\mathbf{m}_{\text{loop}} \cdot \mathbf{B}), \quad \text{(Gaussian and SI)} \tag{21.12}$$

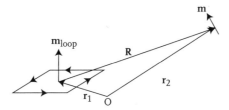

FIGURE 4.5. The interaction between an infinitesimal current loop and a point magnetic dipole.

usinq eq. (21.10) again. Thus, the force on the loop is also exactly that on an equivalent magnetic dipole $\mathbf{m}_{\text{loop}}$.

That the square loop responds to a magnetic field in exactly the same way as a magnetic dipole makes one suspect that it should also produce a dipole magnetic field. We can show that this is true by considering the interaction between the loop and a point dipole moment, $\mathbf{m}$. It is now convenient to shift the origin so that the loop is located at $\mathbf{r}_1$ and the point dipole at $\mathbf{r}_2$, and to define $\mathbf{R} = \mathbf{r}_2 - \mathbf{r}_1$ (fig. 4.5). The force and torque on the loop are given by eqs. (21.12) and (21.9), where $\mathbf{B}$ is the field due to the dipole. By Newton's third law, the dipole experiences an equal and opposite force:

$$\mathbf{F}_{\text{on dipole}} = -\nabla_1(\mathbf{m}_{\text{loop}} \cdot \mathbf{B}) = -(\mathbf{m}_{\text{loop}} \cdot \nabla_1)\mathbf{B}, \tag{21.13}$$

where the last replacement is possible because $\nabla_1 \times \mathbf{B}$ vanishes for the point dipole field. Note that $\mathbf{B}$ and its gradient are to be evaluated at the point $\mathbf{r}_1$, i.e., at the loop.

Similarly, the torque on the dipole is equal and opposite to that on the loop:

$$\mathbf{N}_{\text{on dipole}} = -\mathbf{m}_{\text{loop}} \times \mathbf{B}(\mathbf{r}_1). \tag{21.14}$$

This formula, however, gives the torque about $\mathbf{r}_1$. To obtain the torque on the dipole about the location of the dipole itself, i.e., the point $\mathbf{r}_2$, we must subtract $\mathbf{R} \times \mathbf{F}_{\text{on dipole}}$. Denoting this torque by $\mathbf{N}_0$, we have

$$\mathbf{N}_0 = -\mathbf{m}_{\text{loop}} \times \mathbf{B}(\mathbf{r}_1) + \mathbf{R} \times (\mathbf{m}_{\text{loop}} \cdot \nabla_1)\mathbf{B}. \tag{21.15}$$

But

$$\mathbf{R} \times (\mathbf{m}_{\text{loop}} \cdot \nabla_1)\mathbf{B} = (\mathbf{m}_{\text{loop}} \cdot \nabla_1)(\mathbf{R} \times \mathbf{B}) - [(\mathbf{m}_{\text{loop}} \cdot \nabla_1)\mathbf{R}] \times \mathbf{B}(\mathbf{r}_1)$$

$$= (\mathbf{m}_{\text{loop}} \cdot \nabla_1)(\mathbf{R} \times \mathbf{B}) + \mathbf{m}_{\text{loop}} \times \mathbf{B}(\mathbf{r}_1), \tag{21.16}$$

where the last line follows on noting $\nabla_{1,i} R_j = -\delta_{ij}$. Hence,

$$\mathbf{N}_0 = (\mathbf{m}_{\text{loop}} \cdot \nabla_1)(\mathbf{R} \times \mathbf{B}). \tag{21.17}$$

But

$$\mathbf{R} \times \mathbf{B} = \mathbf{R} \times \frac{3(\widehat{\mathbf{R}} \cdot \mathbf{m})\widehat{\mathbf{R}} - \mathbf{m}}{R^3} = \frac{\mathbf{m} \times \mathbf{R}}{R^3}, \tag{21.18}$$

so

$$\mathbf{N}_0 = \mathbf{m} \times (-\mathbf{m}_{\text{loop}} \cdot \nabla_2) \frac{\mathbf{R}}{R^3}, \tag{21.19}$$

where we have replaced ∇_1 by $-\nabla_2$. The last gradient is now very similar to that evaluated in eq. (19.7) for the gradient of an electrostatic dipole potential. We have

$$(-\mathbf{m}_{\text{loop}} \cdot \nabla_2) \frac{\mathbf{R}}{R^3} = \frac{3(\hat{\mathbf{R}} \cdot \mathbf{m}_{\text{loop}})\hat{\mathbf{R}} - \mathbf{m}_{\text{loop}}}{R^3} \equiv \mathbf{B}(\mathbf{r}_2). \tag{21.20}$$

We have defined this expression as $\mathbf{B}(\mathbf{r}_2)$, because it is exactly the magnetic field at $\mathbf{r}_2$ due to a point dipole $\mathbf{m}_{\text{loop}}$ at $\mathbf{r}_1$. The torque on the point dipole is

$$\mathbf{N}_0 = \mathbf{m} \times \mathbf{B}(\mathbf{r}_2), \tag{21.21}$$

which confirms the physical meaning of this assignment.

We have thus shown that a point magnetic dipole and an infinitesimal square current loop both behave in exactly the same way as sources of magnetic fields and also in their response to external magnetic fields. Our argument assumes that the $\mathbf{B}$ used to describe the dipole–dipole interaction is the same as that appearing in the Lorentz force law. Of course, as first found by Oersted in 1820, we know that current-carrying wires deflect magnetic needles. This observation may be regarded as supplying an empirical proof of the equivalence of the two definitions of $\mathbf{B}$. In the next few sections, we shall find the field of large current loops, the force between two loops, etc. These results have been extensively confirmed by experiment, so there is little doubt that the equivalence is exact.[10]

Exercise 21.1 Two identical point dipoles are located at fixed centers a distance a apart but are free to rotate about these centers. A uniform magnetic field $\mathbf{B}$ is applied perpendicular to the line joining the dipoles. Show that the configuration where the dipoles point along the applied field is stable only if B exceeds some critical value B_c, and find B_c.

Exercise 21.2 Find the equation for the lines of $\mathbf{B}$ for a dipole field in spherical polar coordinates, and find B along a line.

Solution: The equations for the lines of a vector field were given in exercise 12.1. If we take the dipole to point along $\hat{\mathbf{z}}$, these lines lie in planes of constant φ, and the equation for a line may be written as $dr/d\theta = r B_r / B_\theta$ (see fig. 4.6). Substituting $B_r = 2m\cos\theta/r^3$, $B_\theta = m\sin\theta/r^3$, and integrating, we get

$$r = \left(\frac{m}{B_{\text{eq}}}\right)^{1/3} \sin^2\theta \tag{21.22}$$

---

[10] We dwell on this point especially because we shall later show that the magnetism of matter can be represented either in terms of a dense collection of magnetic dipoles or in terms of atomic-sized current loops. These are, in fact, known as *Amperean currents*. However, we also know that the dipole moments of the electron, muon, etc., cannot at a fundamental level be regarded as classical spinning charge distributions, or any collection of microscopic current loops.

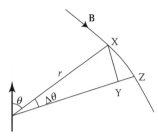

FIGURE 4.6. Equation for the lines of **B** for a dipole in spherical polar coordinates. $B_r / B_\theta \approx YZ/XY \approx \Delta r/r\,\Delta\theta$. See also exercise 12.1.

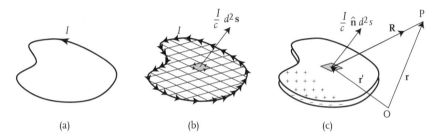

FIGURE 4.7. Equivalence between a current loop and a dipole sheet. The loop in (a) produces the same magnetic field as the mesh of loops in part (b), each of which carries a current I. Each tiny loop of area d^2s and normal $\hat{\mathbf{n}}$ has a magnetic moment $(I/c)\hat{\mathbf{n}}d^2s$, and hence this mesh produces the same field as a sheet with dipole moment per unit area equal to (I/c). This sheet is shown in (c) as two infinitesimally separated layers of positive and negative charges.

where B_{eq} is the value of B where the line crosses the equatorial plane. Along a line,

$$B = B_{eq}\frac{(1+3\cos^2\theta)^{1/2}}{\sin^6\theta}. \tag{21.23}$$

We shall use this result in appendix G when we discuss charged particle motion in the earth's magnetic field.

22 The law of Biot and Savart

Let us now find the magnetic field produced by an arbitrary current loop C of finite extent. We can fill this loop with a mesh of tiny square loops, each carrying a current I in the same sense as the big loop (fig. 4.7). The currents along the common legs of adjacent loops cancel each other, and we are left with the current on the outer edges. The field due to the big loop is given by superposing the tiny loop fields, each of which is that of a dipole. The tiny loops together form a surface S that spans the loop C. A small element d^2s of this surface carries a dipole moment $I\,d^2s/c$ directed along the normal $\hat{\mathbf{n}}$ in the sense of the right-hand rule for loops. The problem is thus reduced to finding the field due to a *dipole sheet*. Since we can fill the loop C with infinitely many different surfaces S, the final answer should depend only on C and not on the exact choice of S.

Since the forms of the electric and magnetic dipole fields are identical, the field desired can be found by solving the equivalent problem in electrostatics. We take the electric dipole moment per unit area of the sheet to be $I\hat{\mathbf{n}}/c$, find the electrostatic potential due to this sheet, and take the gradient to find the magnetic field. The potential at point P (fig. 4.7) is given by

$$\phi(\mathbf{r}) = \frac{I}{c} \int_S \frac{\mathbf{R}\cdot\hat{\mathbf{n}}'}{R^3} ds', \tag{22.1}$$

where $\mathbf{R} = \mathbf{r} - \mathbf{r}'$, and the prime in ds' shows that the integration is with respect to the variable $\mathbf{r}'$. To obtain $\mathbf{B}$, we take the gradient under the integration sign, and replace ∇ by $-\nabla'$, since the integrand is a function of $\mathbf{r} - \mathbf{r}'$:

$$\mathbf{B}(\mathbf{r}) = \frac{I}{c} \int_S \nabla' \frac{d\mathbf{s}\cdot\mathbf{R}}{R^3}. \tag{22.2}$$

We now note that because $\nabla' \cdot (\mathbf{R}/R^3) = 0$,

$$\nabla' \frac{d\mathbf{s}'\cdot\mathbf{R}}{R^3} = \nabla' \frac{d\mathbf{s}'\cdot\mathbf{R}}{R^3} - d\mathbf{s}'\nabla' \cdot \frac{\mathbf{R}}{R^3}$$

$$= (d\mathbf{s}' \times \nabla') \times \frac{\mathbf{R}}{R^3}. \tag{22.3}$$

The integral can now be transformed into a line integral by the vector variant of Stokes's theorem, eq. (8.13), to yield

$$\mathbf{B}(\mathbf{r}) = \frac{I}{c} \oint_C \frac{d\boldsymbol{\ell}' \times \mathbf{R}}{R^3}, \quad \mathbf{R} = \mathbf{r} - \mathbf{r}'. \tag{22.4}$$

$$\mathbf{B}(\mathbf{r}) = \frac{\mu_0 I}{4\pi} \oint_C \frac{d\boldsymbol{\ell}' \times \mathbf{R}}{R^3}, \quad \mathbf{R} = \mathbf{r} - \mathbf{r}'. \quad \text{(SI)}$$

This equation gives the magnetic field due to a general current loop. It is known as the *law of Biot and Savart*. It is equivalent to eq. (12.7), which gave us the electric field due to any charge distribution. It should be noted that eq. (22.4) makes sense only for a complete current loop, and one cannot regard the integrand as the field due to an element of the loop that is then superposed over all the elements.

Let us interrupt the formal development to consider a few applications of the Biot-Savart law and the equivalence between a current loop and a dipole sheet. The simplest example is of an infinite straight wire carrying a current I (fig. 4.8). The result that emerges is an approximation to the field very close to the wire in a general loop. Let us take the wire to pass through the origin, with current along $\hat{\mathbf{z}}$. By symmetry it is obvious that $\mathbf{B}$ can depend only on the perpendicular distance to the field point. If we choose the field point in the xz plane, the vector $d\boldsymbol{\ell} \times \mathbf{R}$ in eq. (22.4) points along $\hat{\mathbf{y}}$. It follows that the lines of $\mathbf{B}$ form circles that go around the wire in the right-hand sense. The

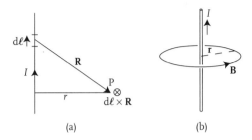

FIGURE 4.8. The magnetic field of a straight current-carrying wire.

magnitude of **B** is given by

$$B = \frac{I}{c} \int_{-\infty}^{\infty} \frac{r}{(r^2 + z^2)^{3/2}} dz = \frac{2I}{cr}.$$ (22.5)

$$= \frac{\mu_0}{2\pi} \frac{I}{r}. \quad \text{(SI)}$$

Like the electric field due to an infinite line charge, this field also dies away like $1/r$.

As the second example, we consider an infinite *sheet* of current. Let the sheet lie in the xy plane, with a current along $\hat{\mathbf{x}}$. We define the surface current $\mathbf{j}_s = j_s\hat{\mathbf{x}}$, where j_s is the amount of charge crossing a unit length segment in unit time. A narrow strip parallel to $\hat{\mathbf{x}}$ of width Δy can then be thought of as a wire carrying a current $j_s\Delta y$. The field due to a wire is given by eq. (22.5). By adding up the fields due to all the wires, we get the total magnetic field, **B**. Note that by symmetry, **B** must point along $\pm\hat{\mathbf{y}}$ everywhere. We thus get

$$B_y = \frac{2j_s}{c} \int_{-\infty}^{\infty} \frac{(-z)}{(y^2 + z^2)} dy = -\frac{2\pi j_s}{c} \text{sgn}(z),$$ (22.6)

where sgn(z) is the sign of z, i.e., $+1$ for $z > 0$, and -1 for $z < 0$. (In SI, $B_y = -(\mu_0 j_s/2)\text{sgn}(z)$.) As in the case of the infinite sheet of electric charge, the field does not die off with $|z|$. Note also the discontinuity in the magnetic field: $\Delta\mathbf{B} = \mathbf{B}(z = 0+) - \mathbf{B}(z = 0-) = (4\pi/c)\mathbf{j}_s \times \hat{\mathbf{z}}$.

Exercise 22.1 Find the magnetic field along the axis of a circular current loop.

Answer: If the loop is taken in the xy plane, of radius a with a current I flowing anticlockwise, then the field on the axis is along $\hat{\mathbf{z}}$, with $B_z = 2\pi I a^2/c(a^2 + z^2)^{3/2}$ [$\mu_0 I a^2/2(a^2 + z^2)^{3/2}$ in SI].

Exercise 22.2 Repeat the previous exercise for a square current loop.

Exercise 22.3 Consider two parallel wires a distance a apart, carrying equal currents I in opposite directions. Find the magnetic field far from the wires, at distances much greater than a. Do this by replacing the wires by a dipole ribbon of width a, which can then be approximated by a line of dipoles to get the far field. Check your answer by directly superposing the fields of the separate wires. Sketch the field.

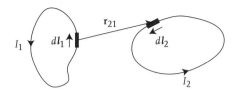

FIGURE 4.9. Force between two current loops.

Exercise 22.4 Find the magnetic field in the plane of a circular current loop, as a function of the distance from the center. The Biot-Savart line integral is best done numerically, even though an analytic expression for it exists in terms of elliptic integrals. Plot your results. What is the area in which the field is uniform to 5%? to 10%?

Exercise 22.5 Repeat the previous exercise for a square current loop. The integral can be done in simple closed form this time. Plot the field along a line from the center to the midpoint of a side, and to a corner.

The Biot-Savart law immediately enables us to find the forces between two current loops, carrying currents I_1 and I_2 (see fig. 4.9). We find the field at loop 2 due to loop 1 and use eq. (21.8) to calculate the force on an element of loop 2. Adding together all length elements gives the net force on loop 2,

$$\mathbf{F}_{\text{on }2} = \frac{I_1 I_2}{c^2} \oint_{C_1} \oint_{C_2} \frac{d\boldsymbol{\ell}_2 \times (d\boldsymbol{\ell}_1 \times \mathbf{r}_{21})}{r_{21}^3}, \tag{22.7}$$

with $\mathbf{r}_{21} = \mathbf{r}_2 - \mathbf{r}_1$ being the vector from the element $d\boldsymbol{\ell}_1$ to $d\boldsymbol{\ell}_2$. Expanding out the triple vector product, we obtain

$$\mathbf{F}_{\text{on }2} = \frac{I_1 I_2}{c^2} \oint_{C_1} \oint_{C_2} \left(\frac{(d\boldsymbol{\ell}_2 \cdot \mathbf{r}_{21}) d\boldsymbol{\ell}_1}{r_{21}^3} - \frac{(d\boldsymbol{\ell}_2 \cdot d\boldsymbol{\ell}_1) \mathbf{r}_{21}}{r_{21}^3} \right). \tag{22.8}$$

The first term can be written as $-d\boldsymbol{\ell}_1 (d\boldsymbol{\ell}_2 \cdot \nabla_2) r_{21}^{-1}$. When integrated over loop 2, it yields zero. Hence,

$$\mathbf{F}_{\text{on }2} = -\frac{I_1 I_2}{c^2} \oint_{C_1} \oint_{C_2} \frac{(d\boldsymbol{\ell}_2 \cdot d\boldsymbol{\ell}_1) \mathbf{r}_{21}}{r_{21}^3}. \tag{22.9}$$

$$\mathbf{F}_{\text{on }2} = -\frac{\mu_0}{4\pi} I_1 I_2 \oint_{C_1} \oint_{C_2} \frac{(d\boldsymbol{\ell}_2 \cdot d\boldsymbol{\ell}_1) \mathbf{r}_{21}}{r_{21}^3}. \quad \text{(SI)}$$

By interchanging the indices 1 and 2, and noting that $\mathbf{r}_{12} = -\mathbf{r}_{21}$, we see that

$$\mathbf{F}_{\text{on }1} = -\mathbf{F}_{\text{on }2}. \tag{22.10}$$

In other words, the forces on the loops are equal and opposite, and obey Newton's third law.

It is useful at this point to discuss more closely the meaning of the term *magnetostatics* and what we mean by a steady current. After all, the charges in a current-carrying wire are not static. Indeed, if the loop is curved, as it must be to be closed, the individual charges carrying the current undergo acceleration. If we remember that accelerating charges produce radiation, the question arises as to how the magnetic field can be constant in time. The answer is that the radiative effects cancel out when we consider a large number of charges that flow in a continuous stream. The stipulation of large numbers is important. If the discreteness of the individual charges is evident, it is meaningless to talk of a steady current. Many authors prefer the term *magnetism of steady currents* to magnetostatics for this reason.

The same issue underlies our application of Newton's third law to dipoles and infinitesimal current loops, and its emergence for large loops, eq. (22.10). The forces exerted on each other by two moving point charges do not obey this law, even when retardation effects arising from the finite speed of light can be neglected. This can be seen in the first term in eq. (22.8), which vanishes only when integrated.

23 Differential equations of magnetostatics; Ampere's law

From the Biot-Savart law, plus the expression for the field due to a magnetic dipole, we can obtain the field due to any distribution of sources of magnetic field by using the principle of superposition. However, as in the electrostatic case, it is highly profitable to obtain differential equations for $\mathbf{B}$. The natural objects to seek are $\nabla \cdot \mathbf{B}$ and $\nabla \times \mathbf{B}$. This must be done for a point dipole and a current loop.

For a point magnetic dipole at the origin, we know from the corresponding electrostatic problem that $\nabla \cdot \mathbf{B}$ and $\nabla \times \mathbf{B}$ both vanish for $\mathbf{r} \neq 0$. But, since $\mathbf{B} \to 0$ as $r \to \infty$, we also know from the uniqueness theorem of section 15 that both $\nabla \cdot \mathbf{B}$ and $\nabla \times \mathbf{B}$ cannot vanish *everywhere*; at least one of them must have a nonzero value at the origin. This value will clearly take the form of a singular distribution or generalized function, such as the delta function or one of its derivatives, but, whatever it is, we need to find it in order to superpose a collection of point dipoles.

The above problem turns out to be somewhat subtle, and we shall consider it in section 26. For the rest of this section we shall consider only current loops.

Direct differentiation of the Biot-Savart law for the field due to a current loop gives

$$\nabla \cdot \mathbf{B} = -\frac{I}{c} \oint_C d\boldsymbol{\ell}' \cdot \left(\nabla \times \frac{\mathbf{R}}{R^3} \right). \tag{23.1}$$

But $\nabla \times (\mathbf{R}/R^3) = -\nabla \times \nabla R^{-1} = 0$. Hence,

$$\nabla \cdot \mathbf{B} = 0. \tag{23.2}$$

Pictorially, this states that lines of $\mathbf{B}$ have no beginning or end; they go around in closed loops.

To find $\nabla \times \mathbf{B}$, it is illustrative to use an indirect approach rather than operate on the Biot-Savart integral head-on. We return to the equivalence between a current loop C and

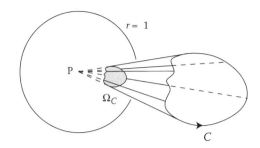

FIGURE 4.10. How to find the solid angle subtended by a loop C at a point P. Consider a sphere of unit radius around P, and draw a bundle of rays from P to points on C. The solid angle Ω_C is the area of the portion of the unit sphere intercepted by these rays.

a dipole sheet S spanning C and note that the integrand in eq. (22.1) for the magnetic scalar potential is precisely the solid angle $d\Omega$ subtended by the surface element d^2s at P. Hence,

$$\phi(\mathbf{r}) = \frac{I}{c}\Omega_C(\mathbf{r}), \tag{23.3}$$

where Ω_C is the total solid angle subtended by the loop C at P (see fig. 4.10). Let us consider a current loop that can be spanned by a flat surface. Then, just above the sheet at the point P (fig. 4.11), the solid angle Ω subtended by C is 2π. It follows that just below the sheet at P', the solid angle must be -2π. We can see this by thinking of the layer of dipoles as a *double layer* of charges, one positive and the other negative, infinitesimally displaced from each other. In fig. 4.11, the positive sheet lies above the negative one, in accord with the dipole direction drawn. We now consider the work done in taking a unit positive test charge from P to P' along the circuit marked 1 in fig. 4.11. We first take the charge from P to a point R infinitely far away above the sheet. The solid angle at R is 0, so $\phi(R) - \phi(P) = -2\pi I/c$. This makes complete sense: the potential is lower at R than at P because a unit positive test charge is repelled by the positive sheet more strongly than it is attracted by the negative sheet. Next we transport the charge from R to R', keeping it very far from the loop at all times. No work is done in this process, i.e., $\phi(R') = \phi(R)$. Lastly, we take the test charge from R' to P'. The charge now feels a net attraction because it is closer to the negative sheet, and so $\phi(P') - \phi(R') = -2\pi I/c$. For the complete loop, we have

$$\phi(P) - \phi(P') = 4\pi I/c. \tag{23.4}$$

It is not difficult to see now that the result is valid for a loop of any shape, not just one that is planar. If we remember that the points P and P' are infinitesimally separated, the difference $\phi(P) - \phi(P')$ is the work done on the test charge in taking it around the closed loop marked 1 in fig. 4.11. The work done is, in turn, the line integral of the electric field. Hence, we have shown that

$$\oint_{\text{loop } 1} \mathbf{E} \cdot d\boldsymbol{\ell} = \frac{4\pi}{c} I. \tag{23.5}$$

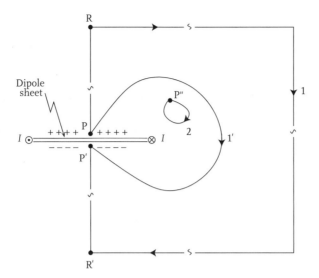

FIGURE 4.11. Proof of Ampere's law. The figure shows a cross section through a loop carrying a current I, and the equivalent dipole sheet, which is represented as a double layer of charges. The work done in transporting a unit (magnetic) charge from point P infinitesimally above the dipole sheet to a point P′ infinitesimally below it, around the circuits 1 or 1′, can be calculated in two ways: as I/c times the difference in the solid angle subtended by the current loop at these two points or as the line integral of **B** around the circuit. Equating these two answers leads to Ampere's law, eq. (23.7).

The same result is obtained if we deform loop 1 into 1′, since the electric field is conservative. For a loop such as 2, on the other hand, if we start at point P″, the solid angle first decreases and then increases, returning to its original value at the end. Thus,

$$\oint_{\text{loop 2}} \mathbf{E} \cdot d\boldsymbol{\ell} = 0. \tag{23.6}$$

The electric field due to the dipole sheet is identical to the magnetic field due to the current loop. Hence, we can replace **E** by **B** in the above equations and write

$$\oint_C \mathbf{B} \cdot d\boldsymbol{\ell} = \frac{4\pi}{c} I \quad (I = \text{current through } C). \tag{23.7}$$

$$\oint_C \mathbf{B} \cdot d\boldsymbol{\ell} = \mu_0 I \quad (I = \text{current through } C). \quad \text{(SI)}$$

This is *Ampere's law*. It is equivalent to Gauss's law for the flux of **E** through a closed surface.

Before leaving this discussion, let us comment on the fact that the potential ϕ is discontinuous on crossing the dipole layer, even though its gradient (**B**) is continuous. If one were considering a true electric dipole sheet, this discontinuity would have real physical significance. In the magnetic case, however, the dipole sheet is merely a device for calculation, and physical answers should not depend on where it is chosen to lie. This

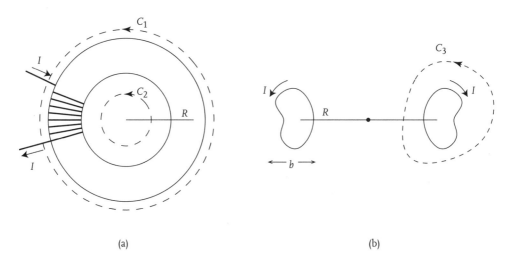

FIGURE 4.12. A toroidal solenoid in top view (a) and in a cross-sectional side view (b). In (a), only a partial current winding is shown. C_1, C_2, and C_3 are contours for applying Ampere's law.

immediately implies that there can be no discontinuity in **B** itself. As for ϕ, the key point is that a jump must exist somewhere. Its location can be changed by choosing S differently, but its magnitude cannot be changed. Since the line integral of **B** is precisely equal to the size of this jump, this quantity is also independent of where S is located.

It is instructive to apply eq. (23.7) to the infinite, straight, current-carrying wire. By symmetry, lines of **B** must form circles around the wire. The line integral of **B** around a circle at a distance r from the wire is $2\pi r B$. The current passing through this circular contour is I. Hence, $B = 2I/cr$ (or $\mu_0 I/2\pi r$ in SI), exactly as found before. It is also instructive to apply Ampere's law to the infinite sheet of current and to two parallel current sheets carrying equal and opposite currents. The next two exercises show two more common situations where the law is useful.

Exercise 23.1 A *solenoid* is a device for producing reasonably uniform fields. A wire is tightly wound in a spiral on a tube or former made of material through which lines of magnetic field pass essentially as in vacuum. The number of turns per unit length is generally very large. The cross section of the tube may be of any shape (square, e.g.), but is most commonly circular.

Find the magnetic field due to a tightly wound toroidal solenoid with N turns carrying a current I (fig. 4.12). Take the cross section of the torus to be arbitrary. Assume, however, that $2\pi R/N \ll b$, where b is the linear dimension of this cross section, and R is the mean large radius of the torus.

Solution: The condition $2\pi R/N \ll b$ allows us to assume that **B** is cylindrically symmetric. By considering the line integral of **B** along circles such as C_1 and C_2 we see that the azimuthal component of **B** vanishes outside the solenoid. Inside the solenoid, however,

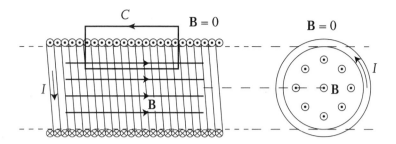

FIGURE 4.13. The magnetic field of an infinite ideal solenoid. Application of Ampere's law to the rectangular contour C shows that the field vanishes outside the solenoid and is uniform and parallel to the axis of the solenoid inside it.

$B_\varphi = 2NI/cr_\perp$ ($\mu_0 NI/2\pi r_\perp$ in SI). By considering a path such as C_3, we see that since a net current I must pass through the surface bounded by it, there is also a small nonazimuthal field $\mathbf{B}'$ equivalent to that of a single loop of wire carrying a current I; $|\mathbf{B}'|/B_\varphi \sim R/Nb \ll 1$.

Exercise 23.2 Find the magnetic field due to an ideal infinitely long solenoid of n turns per unit length, carrying a current I (fig. 4.13).

Solution: By considering such a solenoid as a limiting case of a toroidal one, we see that $\mathbf{B}$ must point along the axis of the solenoid ($\hat{z}$) and that it vanishes outside the solenoid. The field inside follows by applying the integral form of Ampere's law to the rectangular contour C. We get $B = (4\pi/c)nI$ ($\mu_0 nI$ in SI). On the axis this agrees with what we found using the Biot-Savart law. Note also the relation between the discontinuity in B at the surface of the solenoid and the current sheet at that surface.

The integral form of Ampere's law, eq. (23.7), leads directly to an expression for $\nabla \times \mathbf{B}$. To obtain this, let us return to the picture of a current loop in terms of a real physical wire, and instead of looking at a complete cross section that cuts through the entire width of the wire, let us consider a smaller area element lying within the wire and ask how much charge crosses this in unit time. If we denote the area element by $d\mathbf{s}$ and take the charge carriers to be electrons, then it is apparent that this number is

$$-ne\mathbf{v} \cdot d\mathbf{s}. \tag{23.8}$$

This leads us to define the *volume current density* as

$$\mathbf{j} = -ne\mathbf{v}. \tag{23.9}$$

In words, $\mathbf{j}$ is the charge crossing a unit area per unit time and has a direction parallel to the charge flow. More generally, if we have a collection of charges q_a, each located at a point $\mathbf{r}_a$ at a given instant of time and moving with velocity $\mathbf{v}_a$, the current density is given by

$$\mathbf{j}(\mathbf{r}) = \sum_a q_a \mathbf{v}_a \delta(\mathbf{r} - \mathbf{r}_a). \tag{23.10}$$

It is easily seen that the integral of this expression over any finite area gives the charge crossing that area in unit time.

We now note that Ampere's law in the form (23.7) applies to a thick wire. This follows by superposition if we think of the wire as built of many infinitesimally thin filaments. The same argument shows that the loop C need not enclose a thick wire completely. Rather, it can cut through it. All these cases are covered if we write

$$\text{Current through } C = \int_{S'} \mathbf{j} \cdot d\mathbf{s}, \tag{23.11}$$

where the surface integral is over any surface S' spanning the curve C. Since, by Stokes's theorem, we can write

$$\oint_C \mathbf{B} \cdot d\boldsymbol{\ell} = \int_{S'} (\nabla \times \mathbf{B}) \cdot d\mathbf{s}, \tag{23.12}$$

and since the surface S' is arbitrary, we have

$$\nabla \times \mathbf{B} = \frac{4\pi}{c} \mathbf{j}. \tag{23.13}$$

$$\nabla \times \mathbf{B} = \mu_0 \mathbf{j}. \quad \text{(SI)}$$

This is the differential form of Ampere's law. Along with eq. (23.2), $\nabla \cdot \mathbf{B} = 0$, it forms the basis for all of magnetostatics.[11]

At this point it is important to note that the current density cannot be arbitrarily specified. The law of charge conservation imposes a significant constraint. Let us consider an arbitrary volume V of space, bounded by a surface S. The total outward flux of charge through S is given by $\int_S \mathbf{j} \cdot d\mathbf{s}$. By charge conservation, this must be the negative rate of change of the total charge inside V. That is,

$$\int_S \mathbf{j} \cdot d\mathbf{s} = -\frac{d}{dt} \int_V \rho \, d^3x, \tag{23.14}$$

where ρ is the charge density as before. Transforming the left-hand side by Gauss's theorem, and noting that V is arbitrary, we obtain

$$\nabla \cdot \mathbf{j} + \frac{\partial \rho}{\partial t} = 0. \tag{23.15}$$

This is known as the *equation of continuity*. It expresses the *local* conservation of charge. Not only is the total charge in the universe unchanging with time (global conservation), it can change in a local region only by flowing out of it. Charge cannot disappear in one spot and appear in a distant spot instantaneously.

For magnetostatics, where charge does not build up or get depleted at any point, $\partial\rho/\partial t$ is zero. Hence,

$$\nabla \cdot \mathbf{j} = 0. \quad \text{(magnetostatics)} \tag{23.16}$$

---

[11] Pending the demonstration in section 26 that this makes sense for point dipoles.

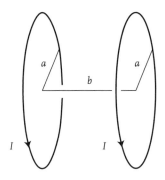

FIGURE 4.14. A Helmholtz coil pair.

Happily, this is exactly what is implied by Ampere's law, since $\nabla \cdot (\nabla \times \mathbf{B}) = 0$. Were $\nabla \cdot \mathbf{j}$ nonzero, the problem would be ill-posed.

Secondly, it should be noted that although the expression (23.10) is generally valid, even for time-dependent situations, the source term in Ampere's law for magnetostatics must be a steady current. Equation (23.10) can yield a steady current only in the limit of a large number of charges, and we once again see the necessity of this stipulation.

In many situations, the magnetic field has cylindrical symmetry. In such cases, if the field along the axis can be found (by the Biot-Savart law, e.g.), the differential equations $\nabla \cdot \mathbf{B} = \nabla \times \mathbf{B} = 0$ enable us to find it at points a little bit off the axis. This idea is explored in the next few exercises.

Exercise 23.3 (Helmholtz coils) If two identical circular current loops of radius a, carrying equal currents I, are arranged parallel and coaxial to one another, and separated by a distance b along the axis, one obtains a pair of *Helmholtz coils* (fig. 4.14). Such coils are often used in the laboratory to provide a quasi-uniform field. In so-called vector magnets, a combination of two or three such pairs of coils at right angles to each other (or two coils and a solenoid) with independent current feeds provides a way of changing all three components of the field in an open space that is easily accessible to probes, laser beams, gas lines, etc.

Taking the z axis as the axis of the coils, and the origin to lie at the midpoint between them, show that on the axis, dB_z/dz and $d^2 B_z/dz^2$ both vanish near the origin if $b = a$. Near the origin, $\nabla \cdot \mathbf{B} = \nabla \times \mathbf{B} = 0$. Using these relations, find $\mathbf{B}$ for x, y, and z all much smaller than a, when $b = a$.

Answer: $\mathbf{B} \approx B_0 \hat{z} + (144/125a^4) B_0 \mathbf{b}$, where $B_0 = 16\pi I/5\sqrt{5}ac$ ($4\mu_0 I/5\sqrt{5}a$ in SI), and

$$b_z = -z^4 + 3(x^2 + y^2)z^2 - \tfrac{3}{8}(x^2 + y^2)^2,$$

$$b_x = 2xz^3 - \tfrac{3}{2}x(x^2 + y^2)z, \tag{23.17}$$

$$b_y = 2yz^3 - \tfrac{3}{2}y(x^2 + y^2)z. \tag{23.18}$$

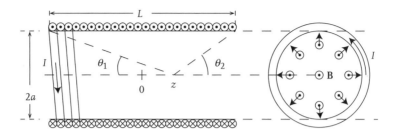

FIGURE 4.15. A finite circular solenoid. The field inside is no longer uniform and parallel to the axis but flares out as shown on the right, in the cross-sectional view, at a distance z along the solenoid.

The reader may enjoy plotting this to see just how uniform the field is. Along the axis, for example, the exact expression shows that B_z drops by only 5% as we move from the midpoint between the coils to the center of one of them.

Exercise 23.4 (anti-Helmholtz coils) Suppose the current is reversed in one coil of a Helmholtz pair. Such a device is used when one wishes to produce a field whose magnitude is not large but which has a large and uniform gradient. Find the field near the center of an anti-Helmholtz coil pair, and the value of b/a that maximizes the field gradient.

Exercise 23.5 (the finite solenoid) Consider a finite circular solenoid of radius a and length L, with n turns per unit length, and carrying a current I (see fig. 4.15). Find **B** (a) at a general point on the axis; (b) in the region near the center of the solenoid, assuming $L \gg a$; and (c) in the region near the axis at one end of the solenoid, assuming $L \gg a$.

Answer: Using cylindrical coordinates $r_\perp$, φ, and z, we have

(a) $\mathbf{B} = (2\pi n I/c)(\cos\theta_1 + \cos\theta_2)\hat{\mathbf{z}}$, where θ_1 and θ_2 are defined in the figure.

(b) With z measured from the center of the solenoid,

$$B_z = \frac{2\pi n I}{c}\left(\frac{2L}{\sqrt{L^2+4a^2}} + \frac{24a^2}{L^4}(r_\perp^2 - 2z^2) + \cdots\right),$$

$$B_\perp = \frac{96\pi n I}{c}\frac{a^2}{L^4}r_\perp z.$$

The azimuthal component B_φ is zero.

(c) With z measured from the edge of the solenoid where the field exits,

$$B_z = \frac{2\pi n I}{c}\left(\frac{L}{\sqrt{L^2+a^2}} - \frac{z}{a} + \cdots\right),$$

$$B_\perp = \frac{\pi n I}{c}\frac{r_\perp}{a}.$$

The azimuthal component B_φ is again zero.

Exercise 23.6 For a flat, circular loop carrying a current I, use the result for the magnetic field along the axis of the loop to find the field a small distance $r_\perp$ away from the axis in powers of $r_\perp$ going up to $O(r_\perp^4)$.

Exercise 23.7 Try and repeat the previous exercise for a square current loop. Show that the on-axis field determines the off-axis field up to terms of $O(r_\perp^2)$ only and that the terms of $O(r_\perp^4)$ must be found in some other way.

24 The vector potential

Let us examine further the issue of representing the magnetic field as the derivative of some potential function. We have already seen in our derivation of Ampere's law that a scalar potential has undesirable properties in that it must be discontinuous across some surface spanning every current loop.[12] (More generally, we could regard the scalar potential as a multiply-valued function in the same sense as $\log z$ is of the complex variable z, and the surfaces of discontinuity as analogous to the branch cuts used to define a given branch of $\log z$.) However, since

$$\nabla \cdot \mathbf{B} = 0 \quad \text{(Gaussian and SI)}, \tag{24.1}$$

we may write $\mathbf{B}$ as the curl of another *vector* field:

$$\mathbf{B} = \nabla \times \mathbf{A}. \quad \text{(Gaussian and SI)} \tag{24.2}$$

This equation defines the *vector potential*, $\mathbf{A}$. Such a representation is always possible for $\mathbf{B}$ fields due to a current distribution. We will ask in section 26 whether it also holds for point dipoles.

To find the vector potential due to a current distribution, let us first rewrite the Biot-Savart law (22.4) in terms of volume currents. Following the argument from eq. (23.8) to (23.11), we may write (replace $I d\boldsymbol{\ell}$ by $\mathbf{j}\, d^3x$)

$$\mathbf{B}(\mathbf{r}) = \frac{1}{c} \int \frac{\mathbf{j}(\mathbf{r}') \times \mathbf{R}}{R^3}\, d^3x', \tag{24.3}$$

with $\mathbf{R} = \mathbf{r} - \mathbf{r}'$. But $\mathbf{R}/R^3 = -\nabla(|\mathbf{r} - \mathbf{r}'|^{-1})$, and taking the gradient outside the integration, we obtain

$$\mathbf{B}(\mathbf{r}) = \frac{1}{c} \nabla \times \int \frac{\mathbf{j}(\mathbf{r}')}{|\mathbf{r} - \mathbf{r}'|}\, d^3x'. \tag{24.4}$$

Comparing with eq. (24.2), we see that

$$\mathbf{A}(\mathbf{r}) = \frac{1}{c} \int \frac{\mathbf{j}(\mathbf{r}')}{|\mathbf{r} - \mathbf{r}'|}\, d^3x'. \tag{24.5}$$

$$\mathbf{A}(\mathbf{r}) = \frac{\mu_0}{4\pi} \int \frac{\mathbf{j}(\mathbf{r}')}{|\mathbf{r} - \mathbf{r}'|}\, d^3x'. \quad \text{(SI)}$$

---

[12] Nevertheless, the use of a scalar potential is expedient in many cases.

This is the magnetic analog of eq. (13.5), Poisson's integral solution for the electric potential of a charge distribution.

We now give a few examples of the use of the vector potential to calculate magnetic fields for some simple circuits. As becomes rapidly evident, the advantages are not as great as for the electrostatic potential. To find the **B** field due to a circular loop, e.g., one must find **A** at points where there is no symmetry. This is so even if we want **B** only on the axis, because it is necessary to differentiate **A**. In such cases, it may be simpler to use the Biot-Savart law directly.

The examples that we consider will all have cylindrical symmetry. Since the use of cylindrical coordinates is then often advantageous, we give, for reference, the formulas for **A** and **B**. If **A** is independent of the azimuthal angle φ, then

$$
\begin{aligned}
B_\perp &= -\frac{\partial A_\varphi}{\partial z}, \\
B_\varphi &= \frac{\partial A_\perp}{\partial z} - \frac{\partial A_z}{\partial r_\perp}, \\
B_z &= \frac{1}{r_\perp} \frac{\partial}{\partial r_\perp}(r_\perp A_\varphi).
\end{aligned}
\tag{24.6}
$$

Infinite straight wire: With the wire running along the z axis, and carrying a current I, we have $\mathbf{j} = I\delta(x)\delta(y)\hat{\mathbf{z}}$. Then

$$
A_z = \frac{2I}{c} \int_0^\infty \frac{dz}{\sqrt{z^2 + r_\perp^2}}.
\tag{24.7}
$$

The other components of **A** are zero. The integral as written is divergent. To regulate it, we replace the upper limit by L, where $L \gg r_\perp$. Then it equals $\sinh^{-1}(L/r_\perp) \approx \ln(2L/r_\perp)$. This yields $A_z \approx -(2I/c)\ln(2L/r_\perp)$. The L dependence drops out upon taking the curl, and we recover our previous result,

$$
\mathbf{B} = \frac{2I}{c} \frac{1}{r_\perp} \mathbf{e}_\varphi.
\tag{24.8}
$$

Infinite circular solenoid: As before, we take the solenoid to have radius a, n turns per unit length, and carry a current I. Then,

$$
\mathbf{j} = nI\delta(r_\perp - a)\mathbf{e}_\varphi.
\tag{24.9}
$$

When substituted in eq. (24.5), this yields a double integral over z' and φ'. The only nonvanishing component of **A** is A_φ. By symmetry, this is independent of z and φ, so we set $z = 0$ and $\varphi = 0$ at the outset. As in the case of the straight wire, it is convenient to regulate the integral by taking the solenoid to have a finite length L, and let $L \to \infty$ at the end. With these maneuvers, we obtain

$$
A_\varphi = \frac{2nIa}{c} \int_0^{2\pi} d\varphi \int_0^{L/2} dz \frac{\cos\varphi}{(z^2 + r_\perp^2 + a^2 - 2ar_\perp\cos\varphi)^{1/2}}.
\tag{24.10}
$$

The z integral yields, as $L \to \infty$,

$$
\left[\ln(L) - \ln\zeta + \frac{\zeta^2}{L^2} + \cdots\right]\cos\varphi,
\tag{24.11}
$$

with $\zeta^2 = r_\perp^2 + a^2 - 2ar_\perp \cos\varphi$. The term in $\ln(L)$ vanishes when the φ integration is performed. The remaining expression is well behaved, as $L \to \infty$, and the limit can be taken at this stage. We thus find

$$A_\varphi = -\frac{nIa}{c} \int_0^{2\pi} \cos\varphi \ln(r_\perp^2 + a^2 - 2ar_\perp \cos\varphi) \, d\varphi. \tag{24.12}$$

An integration by parts transforms this to

$$A_\varphi = \frac{2nIa^2 r_\perp}{c} \int_0^{2\pi} \frac{\sin^2\varphi}{r_\perp^2 + a^2 - 2ar_\perp \cos\varphi} \, d\varphi, \tag{24.13}$$

which can be evaluated by contours or the substitution $u = \tan(\varphi/2)$. The result is

$$A_\varphi = \frac{2\pi nI}{c} \times \begin{cases} r_\perp, & r_\perp < a, \\ (a^2/r_\perp), & r_\perp > a. \end{cases} \tag{24.14}$$

Taking the curl, we obtain $\mathbf{B} \| \hat{\mathbf{z}}$, with

$$B_z = \begin{cases} (4\pi nI/c), & r_\perp < a, \\ 0, & r_\perp > a. \end{cases} \tag{24.15}$$

This result agrees with the answer we obtained using Ampere's law, and gives a more convincing demonstration of the vanishing of $\mathbf{B}$ outside the solenoid. Curiously, we see that $\mathbf{A}$ *does not vanish* outside the solenoid. We will comment on this point in the next section.

Exercise 24.1 Find the vector potential and magnetic field of an infinite straight wire with a circular cross section of radius a, carrying a current I distributed uniformly over its cross section. We will use the result for $\mathbf{A}$ when finding the inductance of a two-wire transmission line (exercise 31.5).

Exercise 24.2 Find the leading $1/L$ corrections to the magnetic field for a finite solenoid of length L in a plane passing through the center of the solenoid perpendicular to the solenoid axis.

Exercise 24.3 Consider a hollow spherical shell of radius a and uniform charge density σ, spinning about an axis of symmetry with angular velocity ω. Find the magnetic field everywhere inside and outside this shell. (The key to efficiently solving this problem is to *not* use cylindrical or spherical polar coordinates.)

Solution: For a collection of charged moving particles, $\mathbf{j} = \sum_a q_a \mathbf{v}_a \delta(\mathbf{r} - \mathbf{r}_a)$, so $\mathbf{A} = \sum_a q_a \mathbf{v}_a/c|\mathbf{r} - \mathbf{r}_a|$. As the ath "particle," let us take a small element on the sphere at the point $\mathbf{r}' = a\hat{\mathbf{r}}'$, occupying a solid angle $d\Omega'$. The charge on this particle is $\sigma a^2 d\Omega'$, and its velocity is $\omega a \hat{\mathbf{z}} \times \hat{\mathbf{r}}'$. Summing over all particles, i.e., integrating over all surface elements, we get

$$A(r) = \frac{\omega\sigma a^3}{c} \int \frac{\hat{z} \times \hat{r}'}{|r - r'|} d\Omega'. \tag{24.16}$$

Using the spherical harmonic expansion for $|r - r'|^{-1}$, we get

$$A(r) = \frac{\omega\sigma a^3}{c} \sum_{\ell m} \frac{4\pi}{2\ell + 1} \frac{r_<^\ell}{r_>^{\ell+1}} \int (\hat{z} \times \hat{r}') \, Y_{\ell m}(\hat{r}) Y_{\ell m}^*(\hat{r}') \, d\Omega'. \tag{24.17}$$

The essential observation now is that $\hat{z} \times \hat{r}'$ is a linear combination of $Y_{\ell m}$'s with $\ell = 1$ and that the sum $\sum_m Y_{\ell m}(\hat{r}) Y_{\ell m}^*(\hat{r}')$ acts as a projection operator onto the spherical harmonics of order ℓ. In other words,

$$\int (\hat{z} \times \hat{r}') \sum_m Y_{\ell m}(\hat{r}) Y_{\ell m}^*(\hat{r}') d\Omega' = (\hat{z} \times \hat{r})\delta_{\ell,1}. \tag{24.18}$$

Hence,

$$A(r) = \frac{\omega\sigma a^3}{c} \frac{4\pi}{3} \frac{r_<}{r_>^2} \hat{z} \times \hat{r}. \tag{24.19}$$

In other words,

$$A(r) = \begin{cases} \dfrac{m \times r}{r^3}, & r > a, \\[3mm] \dfrac{m \times r}{a^3}, & r < a, \end{cases} \tag{24.20}$$

where the quantity

$$m = \frac{4\pi}{3} \frac{\omega\sigma a^4}{c} \hat{z} \tag{24.21}$$

can be identified as the total magnetic moment of the spinning shell by looking at the form of A for $r > a$. Taking the curl, we obtain

$$B = \begin{cases} \dfrac{3(m \cdot \hat{r})\hat{r} - m}{r^3}, & r > a, \\[3mm] \dfrac{2m}{a^3}, & r < a. \end{cases} \tag{24.22}$$

It is amusing that the field outside the shell is purely dipolar right up to its surface and is uniform inside it.

This is a good place to discuss the conversion between SI and Gaussian units for various quantities introduced in this chapter. We have already discussed the conversion for I and B in chapter 1. The conversion for magnetic moment follows by noting that $-m \cdot B$ is the potential energy in both systems and recalling the conversion factor for B. We thus get

$$m_{SI} = \sqrt{\frac{4\pi}{\mu_0}} m_{Gau}. \tag{24.23}$$

The conversion factors for vector potential and magnetic flux are identical to that for **B**:

$$\mathbf{A}_{SI} = \sqrt{\frac{\mu_0}{4\pi}} \mathbf{A}_{Gau}, \quad \Phi_{SI} = \sqrt{\frac{\mu_0}{4\pi}} \Phi_{Gau}. \tag{24.24}$$

Next, let us discuss the units. In the Gaussian system, the unit of current is the *statampere*, which equals one statcoulomb per second. We have mentioned earlier that in the Gaussian system, **B** has the same dimensions as **E**. However, the unit for B is called the *gauss*. Although it is numerically the same as a statvolt/cm, the different name helps in distinguishing which quantity one is talking about. The unit for magnetic flux does not have a special name.

The SI unit of current is the *ampere*, which is equal to one coulomb per second. Consequently,

$$1 \text{ ampere} \equiv 3 \times 10^9 \text{ statamperes}. \tag{24.25}$$

The ampere is connected with mechanical units via eq. (22.9). The dimensional constant μ_0, known as the *permittivity of free space*, has dimensions of force/(current)^2, and a value

$$\mu_0 \equiv 4\pi \times 10^{-7} \text{ N/A}^2. \tag{24.26}$$

As stated earlier, the combination $\mu_0 \epsilon_0$ is exactly equal to $1/c^2$. We shall see why this is so when we discuss electromagnetic waves. The SI unit of **B** is the *tesla*, and that of flux is the *weber*. The conversions to Gaussian units are

$$1 \text{ tesla} \equiv 10^4 \text{ gauss} \tag{24.27}$$

$$1 \text{ weber} \equiv 10^8 \text{ gauss} \cdot \text{cm}^2. \tag{24.28}$$

The Gaussian and SI units of magnetic moment are the electromagnetic unit (emu) and amp·m^2. The conversion is

$$1 \text{ A} \cdot \text{m}^2 \equiv 10^3 \text{ emu} \tag{24.29}$$

25 Gauge invariance

Just as we may add a constant to the electrostatic potential without changing **E**, we may also change the vector potential without changing **B**. The indeterminacy in **A** is much greater than that in ϕ, however. If

$$\mathbf{A}' = \mathbf{A} + \nabla \psi, \tag{25.1}$$

where ψ is any scalar field, then **A**′ is also a valid vector potential, since

$$\nabla \times \mathbf{A}' = \nabla \times \mathbf{A} + \nabla \times \nabla \psi = \nabla \times \mathbf{A} = \mathbf{B}. \tag{25.2}$$

Thus, two vector potentials differing by a total gradient are physically equivalent, and eq. (24.5) is only one of many solutions that could have been written.

In the electrostatic case, we often found it convenient to demand that ϕ vanish at infinity in order to fix it completely. How should we fix $\mathbf{A}$? Recall that a vector field is uniquely determined if its curl and divergence are given. We already know $\nabla \times \mathbf{A}'$. Taking the divergence of eq. (25.1), we get

$$\nabla \cdot \mathbf{A}' = \nabla \cdot \mathbf{A} + \nabla^2 \psi. \tag{25.3}$$

Hence, by choosing ψ appropriately, we can change the divergence of the vector potential to anything we please. For magnetostatics, it is convenient to choose

$$\nabla \cdot \mathbf{A} = 0. \tag{25.4}$$

This condition leads directly to eq. (24.5). For, with $\mathbf{B} = \nabla \times \mathbf{A}$,

$$\nabla \times \mathbf{B} = \nabla \times (\nabla \times \mathbf{A}) = -\nabla^2 \mathbf{A} + \nabla(\nabla \cdot \mathbf{A}), \tag{25.5}$$

and with $\nabla \cdot \mathbf{A} = 0$, Ampere's law becomes

$$\nabla^2 \mathbf{A} = -\frac{4\pi}{c}\mathbf{j}. \tag{25.6}$$

$$\nabla^2 \mathbf{A} = -\mu_0 \mathbf{j}. \quad \text{(SI)}$$

Each Cartesian component of this equation is a Poisson equation, which we can solve as in the electrostatic case. The solution is precisely eq. (24.5).[13]

The indeterminacy in $\mathbf{A}$ is referred to as a *gauge degree of freedom*, and the removal of this indeterminacy is known as *fixing the gauge*. Depending on the type of problem at hand, different gauge choices are convenient. Equation (25.4), e.g., is known as the *Coulomb gauge* condition. Since forces and torques on charged particles and current loops depend only on the fields $\mathbf{E}$ and $\mathbf{B}$, all measurable quantities must be independent of the gauge choice. This is the principle of *gauge invariance*. Although this invariance is not a symmetry of nature in the same sense as rotational invariance, or relativistic frame invariance, it is critical to the quantum theory of electromagnetism. For classical electromagnetism, a description in terms of potentials is ultimately a convenience (albeit a rather great one), but it introduces a certain redundancy. Witness the arbitrary constant in ϕ.[14] Gauge invariance is the statement that physical quantities must be independent of these redundant degrees of freedom.

Among gauge-invariant quantities, the line integral of $\mathbf{A}$ around any closed loop C is especially notable. For,

$$\oint_{C=\partial S} \mathbf{A} \cdot d\boldsymbol{\ell} = \int_S (\nabla \times \mathbf{A}) \cdot d\mathbf{s} = \int_S \mathbf{B} \cdot d\mathbf{s} = \Phi, \tag{25.7}$$

---

[13] We must add the proviso that $\mathbf{A}$ vanish at infinity, or, more generally, that it behave in some other specified way on a distant boundary. The same proviso must be added to eq. (25.4), because we could add to $\mathbf{A}$ a function like $x\hat{\mathbf{x}} - y\hat{\mathbf{y}}$ (which is $\frac{1}{2}\nabla(x^2 - y^2)$) without changing $\nabla \cdot \mathbf{A}$.

[14] And, as we shall show in sections 33 and 79, the so-called longitudinal part of $\mathbf{A}$. See especially exercise 33.2.

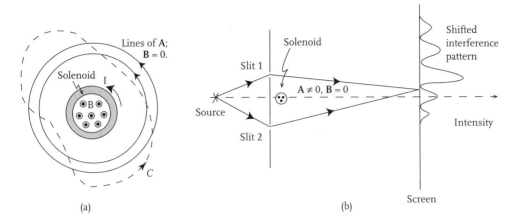

FIGURE 4.16. The Aharanov-Bohm effect. (a) Even though **B** vanishes outside the solenoid, the vector potential **A** does not. (b) The effect of the vector potential is seen in the interference phenomenon. The source produces charged particles, which go around the solenoid precisely in those regions where **B** = 0, **A** ≠ 0. The presence of the solenoid shifts the interference pattern.

where Φ is the magnetic flux threading the loop. This is a physically measurable object and must be independent of gauge. This result explains why $\mathbf{A} \neq 0$ outside an infinite solenoid, even while $\mathbf{B} = 0$ (see eqs. (24.14) and (24.15)). Equation (25.7) implies that the circulation of **A** around a contour C as shown in fig. 4.16 is given by

$$\oint_C \mathbf{A} \cdot d\boldsymbol{\ell} = \int_S \mathbf{B} \cdot d\mathbf{s} = \pi a^2 B_0, \tag{25.8}$$

where $B_0 = 4\pi n I / c$ is the magnitude of the field inside the solenoid. Hence, the line integral cannot vanish, and we conclude that **A** cannot be made zero everywhere outside the solenoid just by changing the gauge. The reader can easily verify that our solution for **A**, eq. (24.14), gives the correct circulation around C. The $1/r_\perp$ decrease is precisely what is needed to ensure this.

Exercise 25.1 Find the potential energy of a large loop carrying a current I in an *external* magnetic field $\mathbf{B}(\mathbf{r})$.

Answer: $U = -I\Phi/c$, where Φ is the external flux threading the loop.

We said above that the use of ϕ and **A** is a convenience in the classical theory and that the fundamental fields are **E** and **B**. A clear illustration that this is not so when quantum mechanics enters the picture is found in the *Aharanov-Bohm effect*. Suppose that we perform a two-slit interference experiment with electrons (or any other charged particles) and place a very thin solenoid behind the slits in a region through which there is essentially no amplitude for the electrons to pass (fig. 4.16). Since there is no magnetic field in the region outside the solenoid, classical thinking would say that the solenoid can have no influence on the interference. What is actually observed is that the interference pattern is shifted as if there were an additional phase difference between the quantum mechanical amplitudes for the two particle trajectories,

$$\Delta\phi = \frac{e}{\hbar c}\oint_{(1-2)} \mathbf{A}\cdot d\boldsymbol{\ell} = \frac{e\Phi}{\hbar c}. \tag{25.9}$$

By (1–2) we mean the closed loop formed by following path 1 from the source to the screen through slit number 1 and then back to the source along path 2 through slit 2. The rightmost member of this equation expresses the circulation of $\mathbf{A}$ in terms of Φ, the flux through the solenoid. This phase difference increases by 2π if the flux increases by $\Phi_0 = hc/e = 4.135 \times 10^{-7}$ G cm^2. This number is known as the *flux quantum*. What is remarkable about eq. (25.9) is that it shows that the local field felt by the electrons is not $\mathbf{B}$ but $\mathbf{A}$. In quantum mechanics, it is the potentials that are more fundamental.[15]

The "reality" of $\mathbf{A}$ and the Aharanov-Bohm idea are particularly confirmed in many experiments involving superconductors. Discussion of these is well outside our scope, but we mention that because superconductivity involves a pairing of electrons, the *superconducting flux quantum* is $hc/2e = 2 \times 10^{-7}$ G cm^2.[16]

26 $\nabla \cdot \mathbf{B}$ and $\nabla \times \mathbf{B}$ for a point dipole

Let us at last turn to the problem of finding the divergence and curl of the point-dipole field carefully. As mentioned in section 23, these will be given by singular generalized function(s) at the origin, and the issue is to find the correct form.

For reference, let us repeat the form of the dipole magnetic field:

$$\mathbf{B}(\mathbf{r}) = \frac{3(\hat{\mathbf{n}}\cdot\mathbf{m})\hat{\mathbf{n}} - \mathbf{m}}{r^3} \quad (\mathbf{r} \neq 0). \tag{26.1}$$

There are two obvious choices for representing this $\mathbf{B}$. We could, by analogy with electrostatics, try the gradient of a scalar potential ϕ. The surface of discontinuity in ϕ is now shrunk to a point, and since this point (the origin) is a singularity of $\mathbf{B}$ anyway, perhaps the discontinuity will not matter. Taking $\phi = \mathbf{m}\cdot\mathbf{r}/r^3$, we have

$$\mathbf{B}_s = -\nabla\frac{\mathbf{m}\cdot\mathbf{r}}{r^3} = (\mathbf{m}\cdot\nabla)\nabla\frac{1}{r}, \tag{26.2}$$

where the subscript s indicates the use of a scalar potential. Let us now note that

$$\nabla_i\nabla_j\frac{1}{r} = \frac{3x_i x_j - r^2\delta_{ij}}{r^5} - \frac{4\pi}{3}\delta_{ij}\delta(\mathbf{r}). \tag{26.3}$$

The key point here is the presence of the second term. To see its necessity, let us take the trace of this equation. The left-hand side becomes $\nabla^2(1/r)$, which we know to be $-4\pi\delta(\mathbf{r})$. The first term on the right-hand side is traceless. The second term is the only one that

---

[15] A lucid discussion of the Aharanov-Bohm effect is given by Baym (1969), chap. 3. Baym's discussion is based on the Feynman approach to quantum mechanics in terms of summation over paths. A simpler approach may be found in Feynman et al. (1964), vol. II, sec. 15-5.

[16] Flux quantization in superconductors was proposed by F. London in 1950 and observed almost simultaneously by Deaver and Fairbank, and by Doll and Nabauer, in 1961. Aharanov and Bohm proposed their experiment in 1959, and their predictions were verified by Chambers in 1960.

vanishes when $\mathbf{r} \neq 0$, has the correct rotational tensor properties, and satisfies the trace requirement.

It follows that

$$\mathbf{B}_s = \mathbf{B}_{\text{far}} - \frac{4\pi}{3} \mathbf{m}\delta(\mathbf{r}), \tag{26.4}$$

where by $\mathbf{B}_{\text{far}}$ we mean the field for $\mathbf{r} \neq 0$. With the choice $\phi = \mathbf{m} \cdot \mathbf{r}/r^3$, some careful algebra has revealed another term in $\mathbf{B}$ at the origin.

The second option is to try the curl of a vector potential, $\mathbf{A}$. To find a suitable $\mathbf{A}$, let us take $\mathbf{m} \| \hat{\mathbf{z}}$ and evaluate the flux of $\mathbf{B}$ through a spherical cap of radius r, extending from the north pole to a polar angle θ_0. This is given by

$$\int_0^{2\pi} \int_0^{\theta_0} \frac{2m \cos\theta}{r^3} r^2 \sin\theta \, d\theta \, d\varphi = \frac{2m\pi}{r} \sin^2 \theta_0. \tag{26.5}$$

Equating this to the circulation of $\mathbf{A}$ around the edge of the cap, i.e., $2\pi r \sin\theta_0 A_\varphi$, we see that a vector potential with only an azimuthal component will fit the bill. In vector form, we have

$$\mathbf{A} = \frac{\mathbf{m} \times \mathbf{r}}{r^3}. \tag{26.6}$$

Note that this is in the Coulomb gauge, i.e., $\nabla \cdot \mathbf{A} = 0$ everywhere, *including the origin*. This generates a $\mathbf{B}$ field (which we distinguish with a subscript v to show the use of a vector potential):

$$\mathbf{B}_v = \nabla \times \frac{\mathbf{m} \times \mathbf{r}}{r^3} = -(\mathbf{m} \cdot \nabla)\frac{\mathbf{r}}{r^3} + \mathbf{m}\left(\nabla \cdot \frac{\mathbf{r}}{r^3}\right). \tag{26.7}$$

The first term is exactly $\mathbf{B}_s$ in eq. (26.4). The second term is $\mathbf{m}(4\pi\delta(\mathbf{r}))$. Hence,

$$\mathbf{B}_v = \mathbf{B}_{\text{far}} + \frac{8\pi}{3} \mathbf{m}\delta(\mathbf{r}). \tag{26.8}$$

Our vector potential has led to a different term in $\mathbf{B}$ at the origin.

Are the delta-function terms in eqs. (26.4) and (26.8) meaningful, and, if so, is the correct coefficient $-(4\pi/3)$, $(8\pi/3)$, or something in between? In fact, the correct answer is eq. (26.8). There is a powerful theoretical reason for this. As we shall see when we discuss relativity, the electric and magnetic potentials ϕ and $\mathbf{A}$ together form a so-called four-vector, and these quantities mix with one another when we transform between inertial frames moving relative to one another. Relativistic invariance demands that the theory be constructed only out of four-vectors, four-scalars, etc., and this demand could not be met if part of the magnetic field were the gradient of a scalar potential without further unwelcome modifications of the electric field.

Quite marvelously, there is also an experimental way to directly confirm the correctness of eq. (26.8). We can probe the field right on top of a point dipole, by studying the hyperfine splitting of hydrogen. The proton has a magnetic moment $\mathbf{m}_p$ and so produces a magnetic field

$$\mathbf{B}_p = \mathbf{B}_{\text{far}} + k\mathbf{m}_p\delta(\mathbf{r}), \tag{26.9}$$

where the coefficient k of the second term is left arbitrary to avoid prejudice. The electron, with its magnetic moment $\mathbf{m}_e$, interacts with this field, giving the so-called hyperfine term in the Hamiltonian,

$$\mathcal{H}_{\text{hf}} = -\mathbf{m}_e \cdot \mathbf{B}_p. \tag{26.10}$$

In the ground state of hydrogen, the electron's spatial distribution, given by the square $|\psi_e(\mathbf{r})|^2$ of the wave function, is spherically symmetric. Hence, the contribution from the term $\mathbf{B}_{\text{far}}$ is zero when averaged over the electron distribution. Only the delta-function term survives, and $\delta(\mathbf{r})$ is replaced by $|\psi_e(0)|^2$, the probability density for the electron to be at the nucleus.[17] Experiment shows unambiguously that the energy depends on the orientations of the two moments as

$$E_{\text{hf}} = -\frac{8\pi}{3} \mathbf{m}_p \cdot \mathbf{m}_e |\psi_e(0)|^2. \tag{26.11}$$

In other words, the vector-potential-derived form (26.8) is the correct one. In particular, the configuration in which the moments are parallel to each other is lowest in energy. It lies below the antiparallel one by an energy that corresponds to microwaves of wavelength 21.4 cm. This *21-centimeter line* is famous for being one of the most accurately measured quantities in physics and is the standard signature of interstellar hydrogen.[18]

We can now work out $\nabla \cdot \mathbf{B}$ and $\nabla \times \mathbf{B}$ for a point dipole. Omitting the suffix v in eq. (26.8), manipulations that are familiar by now yield

$$\nabla \cdot \mathbf{B} = 0, \tag{26.12}$$

$$\nabla \times \mathbf{B} = \nabla \times 4\pi \mathbf{m} \delta(\mathbf{r}). \tag{26.13}$$

These equations invite two comments. First, the divergence of $\mathbf{B}$ is *always* zero. Second, the source term in the equation for $\nabla \times \mathbf{B}$ has exactly the form that would be given by a current distribution

$$\mathbf{j}_{\text{dip}} = c\nabla \times \mathbf{m}\delta(\mathbf{r}). \tag{26.14}$$

$$\mathbf{j}_{\text{dip}} = \frac{1}{\mu_0} \nabla \times \mathbf{m}\delta(\mathbf{r}). \quad \text{(SI)}$$

With this equivalence, we can now firmly restate that, along with $\nabla \cdot \mathbf{B} = 0$, Ampere's law (23.13) forms the basis of all magnetostatics. Note that $\nabla \cdot \mathbf{j}_{\text{dip}} = 0$, as it should for a steady current. Hence, from now on, when we talk of the $\mathbf{B}$ field produced by a current distribution, the case of a point dipole will be included via eq. (26.14), and we will not always distinguish between these two types of sources.

---

[17] See Baym (1969), chap. 23.

[18] The above argument can be sharpened even further by considering the hyperfine splittings of positronium (e^+e^-) and muonium (μ^+e^-), since—unlike the proton—e^+, e^-, and μ^+ are all true point particles. The hyperfine splitting in positronium is of the same order as the fine structure but is easily separable in muonium. Nevertheless, in both cases, the coefficient ($8\pi/3$) is firmly indicated. Indeed, the muonium hyperfine splitting is part of the present-day battery of high-precision tests of quantum electrodynamics. For further discussion of these matters, see Sakurai (1967), sec. 4-6, and Peskin and Schroeder (1995), sec. 6.3.

It should now be apparent that the field of an *electric* dipole contains a delta function contribution analogous to eq. (26.4). The reason is that the interior fields of magnetic and electric dipoles—the field inside a finite-sized current loop or that in the space between two equal and opposite charges—inside a finite-sized dipole, so to speak, are very different. These differences continue to survive in the limit of point dipoles and are reflected in the different delta-function contributions.

Exercise 26.1 Find the magnetic field for the charged spinning spherical shell of exercise 24.3 by using the equivalence between a current loop and a dipole sheet, and the magnetic scalar potential.

Solution: If we think of the spinning shell as a collection of current loops along circles of latitude and replace each loop by a dipole sheet, we see that the shell may be replaced by a sphere filled with a uniform density of magnetic dipole moments equal to $\mathbf{M} = (3/4\pi a^3)\mathbf{m} = (\omega\sigma a/c)\hat{\mathbf{z}}$. The scalar potential due to this sphere is given by

$$\phi = \int_{r<a} \frac{\mathbf{M} \cdot (\mathbf{r} - \mathbf{r}')}{|\mathbf{r} - \mathbf{r}'|^3} d^3x' = -\mathbf{M} \cdot \nabla \int_{r<a} \frac{1}{|\mathbf{r} - \mathbf{r}'|} d^3x'. \tag{26.15}$$

The integral is just the electrostatic potential due to a solid sphere with uniform unit charge density. Using the result for this from exercise 13.2, we obtain

$$\phi(\mathbf{r}) = \mathbf{m} \cdot \mathbf{r} \times \begin{cases} (1/r^3), \ r > a, \\ \\ (1/a^3), \ r < a. \end{cases} \tag{26.16}$$

Let us denote the field $-\nabla\phi$ by $\mathbf{B}_s$. Then $\mathbf{B}_s$ is the true dipolar magnetic field for $r > a$. For $r < a$, however, $\mathbf{B}_s = -\mathbf{m}/a^3$, which does not agree with our previous calculation. The reason is the incorrect delta-function piece in the field of a point dipole given by a magnetic scalar potential. Let the field from a vector potential (which we know will be correct) be denoted by $\mathbf{B}_v$. If we think of the dipole-filled sphere as a set of point dipoles $\mathbf{m}_a$ at points $\mathbf{r}_a$, then eqs. (26.4) and (26.8) show that the contributions from the ath dipole are related by

$$\mathbf{B}_{v,a}(\mathbf{r}) = \mathbf{B}_{s,a}(\mathbf{r}) + 4\pi\mathbf{m}_a\delta(\mathbf{r} - \mathbf{r}_a). \tag{26.17}$$

Summing over the all the dipoles in the sphere, for $r < a$, we obtain

$$\mathbf{B}_v = \mathbf{B}_s + 4\pi\mathbf{M} = \frac{2\mathbf{m}}{a^3}, \tag{26.18}$$

which is correct.[19]

---

[19] We shall see in chapter 16 that the magnetic field due to a uniformly magnetized sphere is identical to that of the surface current distribution of this problem; the equivalence is more than mathematical. The field that we have called $\mathbf{B}_s$ here will turn out to be the same as $\mathbf{H}$.

27 Magnetic multipoles

Let us now find the magnetic field due to a current distribution at large distances from the region where the currents are located. As in the electrostatic case, the result can be written in terms of multipoles, obtained by taking moments of the current distribution.

The leading moment is the magnetic dipole. For a current loop, we have already seen that the field at a general point is equivalent to that of a dipole sheet. It is evident that the far field will be that of a dipole with a moment $\mathbf{m}$ obtained by adding together all the infinitesimal moments in the sheet. In other words, if the dimensions of the current loop are L, then for $r \gg L$ we get

$$\mathbf{A(r)} \approx \frac{\mathbf{m} \times \mathbf{r}}{r^3},$$

$$\mathbf{B(r)} \approx \frac{3(\hat{\mathbf{n}} \cdot \mathbf{m})\hat{\mathbf{n}} - \mathbf{m}}{r^3},$$

where

$$\mathbf{m} = \frac{I}{c} \int_S \hat{\mathbf{n}} \, d^2s. \tag{27.1}$$

We can transform this into a quantity that manifestly depends only on the loop by using the vector variant of Stokes's theorem, eq. (8.13).[20] We can write

$$\hat{\mathbf{n}} \, d^2s = -\tfrac{1}{2}(d\mathbf{s} \times \nabla) \times \mathbf{r}. \tag{27.2}$$

Hence,

$$\mathbf{m} = \frac{I}{2c} \oint_C \mathbf{r} \times d\boldsymbol{\ell}. \tag{27.3}$$

$$\mathbf{m} = \frac{I}{2} \oint_C \mathbf{r} \times d\boldsymbol{\ell}. \quad \text{(SI)}$$

The extension to volume currents is immediate:

$$\mathbf{m} = \frac{1}{2c} \int \mathbf{r} \times \mathbf{j(r)} \, d^3x. \tag{27.4}$$

$$\mathbf{m} = \frac{1}{2} \int \mathbf{r} \times \mathbf{j(r)} \, d^3x. \quad \text{(SI)}$$

The higher magnetic multipoles appear only rarely, but formulas for them can be derived in exact analogy with the electrostatic case. The vector potential for the 2^ℓ-pole

---

[20] For a planar loop, the surface integral gives $\mathcal{A}\hat{\mathbf{n}}$, where $\mathcal{A}$ is the area of the loop. Hence, $\mathbf{m} = I\mathcal{A}\hat{\mathbf{n}}/c$, as for an infinitesimal loop; c.f. eq. (21.10).

is given by

$$\mathbf{A}^{(\ell)}(\mathbf{r}) = \frac{1}{r^{\ell+1}} \sqrt{\frac{4\pi}{2\ell+1}} \sum_{m=-\ell}^{\ell} \mathbf{M}_{\ell m} Y_{\ell m}(\hat{\mathbf{r}}), \tag{27.5}$$

$$\mathbf{M}_{\ell m} = \frac{1}{c} \sqrt{\frac{4\pi}{2\ell+1}} \int \mathbf{j}(\mathbf{r}) r^{\ell} Y_{\ell m}^{*}(\hat{\mathbf{r}}) d^{3}x. \tag{27.6}$$

Exercise 27.1 Find the vector potential and magnetic field of the anti-Helmholtz coils of exercise 23.4 at great distances from the coils. Express the field in both Cartesian and spherical polar coordinates. Sketch the resulting field. (The formal multipole expansion is less useful here than direct differentiation of the vector potential of one coil.)

Exercise 27.2 Express the far field of the previous exercise in terms of a *scalar* potential, ϕ_{far}. Find a similar potential ϕ_{near} for the field near the origin found in exercise 23.4. Compare the spherical harmonic structure of ϕ_{near} and ϕ_{far} and explain any similarity or difference.

5 | Induced electromagnetic fields

So far, we have considered fields that are created either by static charges or by steady currents. In this chapter we consider electromagnetic fields that are induced by the time dependence of the fields themselves. A varying magnetic field is always accompanied by an electric field, and vice versa. These phenomena are described by the Faraday and Ampere-Maxwell laws. When we have these laws, a major milestone will have been reached, since we will then have the basic laws of electromagnetism in full.

We also consider the energy in a *static* magnetic field in this chapter. Readers may wonder why this discussion is placed with that of *time-dependent* fields. The reason is that only *changes* in energy are meaningfully considered, and a change in a field configuration necessarily requires that we consider time dependence. However, it will turn out that the induced fields make a contribution only to the magnetic energy, and our earlier expression for the electrostatic energy will remain unchanged.

Lastly, we will see how the scalar and vector potentials need to be modified to include the time dependence.

28 Induction

In 1831 Faraday discovered that if a current-carrying coil (the primary) was moved back and forth near a static loop of wire (the secondary), a current was induced in the secondary. The same effect was observed if instead of the primary coil, a magnet was used. If instead of moving the primary, the current in it was turned off or on by means of a switch, a transitory pulse of current was seen to arise in the secondary immediately after the switch was operated.

The common feature in the above experiments is that the magnetic field near the secondary coil is changing, which was eventually interpreted by Faraday as indicating that a changing magnetic field produces an electric field. This is because the current in the secondary shows that there is a force acting on the electrons in the wire, which can arise only if there is an electric field. (Recall that a static magnetic field by itself cannot produce a current in a nearby loop. Energy would be dissipated in the loop, and we have seen that a steady magnetic field can deliver no work to a charge.) This effect is known as *induction*, and the quantitative law for it, which is named after Faraday, is

$$\nabla \times \mathbf{E} = -\frac{1}{c}\frac{\partial \mathbf{B}}{\partial t}. \tag{28.1}$$

$$\nabla \times \mathbf{E} = -\frac{\partial \mathbf{B}}{\partial t}. \quad \text{(SI)}$$

Faraday's law generalizes the $\nabla \times \mathbf{E} = 0$ law of electrostatics to time-dependent situations. In such situations, the electric field is not irrotational. Note, however, that Gauss's law continues to hold unchanged.

If we integrate eq. (28.1) over a surface S in space, and use Stokes's theorem, we obtain

$$\oint_{C=\partial S} \mathbf{E} \cdot d\boldsymbol{\ell} = -\frac{1}{c}\int_S \frac{\partial \mathbf{B}}{\partial t} \cdot d\mathbf{s}. \tag{28.2}$$

The line integral of $\mathbf{E}$ around a circuit is one instance of an *electromotive force*, or emf for short.[1] We shall denote the emf by the symbol $\mathcal{E}$. Since the surface integral of $\mathbf{B}$ is the magnetic flux Φ passing through that surface, we can also write

$$\mathcal{E} = -\frac{1}{c}\frac{\partial \Phi}{\partial t}. \tag{28.3}$$

$$\mathcal{E} = -\frac{\partial \Phi}{\partial t}. \quad \text{(SI)}$$

In words, the emf around a circuit is the negative rate of change of the magnetic flux passing through that circuit (divided by the speed of light in the Gaussian system). This is, in fact, the form in which Faraday first gave his law, and so stated, it is known as the *flux rule*. The direction of the induced emf can be deduced from eq. (28.3) but is often stated in terms of *Lenz's rule*. Let us note that the induced current in a loop creates its own magnetic field. Lenz's rule states that the direction of the induced emf around a circuit is such that the induced flux opposes the impressed change in flux. The rule is easily verified by considering a circular loop in the xy plane with a steadily increasing $\mathbf{B}$ field in the z direction.

Let us now vary our experiment somewhat. We keep the primary (current-carrying) coil fixed and move the secondary coil instead. By the principle of relativity, we still expect to find an emf in the secondary. Indeed, just this was observed by Faraday. But the magnetic field is now time independent, so eq. (28.1) does not apply, and the induced emf must be governed by some other law. This law is the Lorentz force law. To see this, we consider yet another variant of the experiment, in which the secondary circuit is not just moved rigidly

---

[1] More generally, a source of emf is any physical agency that can drive a current around a circuit, and the emf is the work delivered by this agency for a unit charge to go once around the circuit. In the present case, this agency is whatever causes the time dependence of the magnetic field, either through motion of the magnet or through motion of the coil, and since $\mathbf{E}$ is the force per unit charge, the line integral of $\mathbf{E}$ is the emf. We discuss other sources of emf in chapter 17.

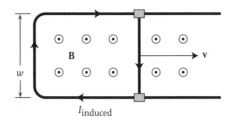

FIGURE 5.1. An adjustable loop made by attaching a wire with sliding contacts to another U-shaped wire. If a magnetic field is applied normal to the loop and the adjustable arm pulled to the right, a current will be induced in the closed loop in the direction shown.

but changes its shape. We take a rectangular loop of wire with a movable arm (fig. 5.1), placed in a steady and uniform magnetic field perpendicular to the loop. If the movable arm is moved with a speed v, the electrons in that arm must move with it, and each of them will experience a Lorentz force $-ev B/c$, or $v B/c$ per unit charge. The total emf is found by taking the line integral of the force per unit charge around the loop. Since the force acts only over the length of the movable arm, we get

$$\mathcal{E} = -wv B/c, \tag{28.4}$$

where the sign indicates that the emf is clockwise.

We now note that this emf is also given by Faraday's flux rule. For the flux through the loop at any time is $w L B$, and since $d L/dt = v$,

$$-\frac{1}{c}\frac{d\Phi}{dt} = -\frac{wv B}{c}, \tag{28.5}$$

equal to $\mathcal{E}$. This must mean that the Lorentz force law and Faraday's law in the form (28.1) are relativistically connected, but it is still remarkable that Faraday should have found a single rule that sums up all phenomena as seen by different observers.

Exercise 28.1 Consider a rectangular loop of wire, with arms along the x and y axes. Let the loop be rigidly moved in the x direction in a static magnetic field with a z component that varies only with x. Find the emf and show that it given by the flux rule.

Exercise 28.2 Consider an arbitrarily shaped moving loop in an arbitrary static magnetic field, and also allow the loop to deform as it moves. Find the rate of change of the flux through the loop, and verify the flux rule.

Answer:

$$\frac{d\Phi}{dt} = -\oint_C (\mathbf{v} \times \mathbf{B}) \cdot d\boldsymbol{\ell}, \tag{28.6}$$

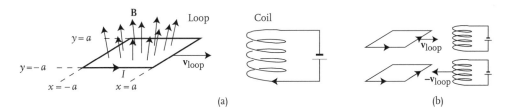

FIGURE 5.2. (a) An infinitesimal current loop being pulled through an inhomogeneous magnetic field, created by a coil. Electrical work must be done to keep the current in the loop fixed. (b) The electrical work done to keep the currents in the coil fixed can be found by going to a frame in which the loop is stationary and the coil moves with an opposite velocity.

where $\mathbf{v}$ is the velocity of an element $d\boldsymbol{\ell}$ of the loop. The integral, divided by c is just the emf around the loop.

All this talk of wire loops moving through magnetic fields naturally suggests that we also consider the motion of extended conductors through magnetic fields. This will be done in section 131.

Our discussion is focused on basic principles of electromagnetism rather than its applications. Faraday's law underlies electrical motors, generators, transformers, and the unfathomable technological revolution wrought by these devices. Since it would be remiss to completely pass over these developments, we discuss the ac generator very briefly in section 118.

29 Energy in the magnetic field—Feynman's argument

We now wish to examine the energy of steady currents more carefully. Our approach will parallel what we did for electrostatics. We imagine assembling the system of currents by bringing in a suitable set of elementary current loops from infinity and ask how much work is required for this assembly. That work will be the energy of the system, and we will calculate it by slightly modifying a very pretty argument due to Feynman.[2]

It is apparent that we need only consider the problem of bringing two infinitesimal circuits in from infinity. The general system can be built up from this. Second, the energy of an infinitesimal loop by itself need not be considered. That would be like the self-energy of a point charge, which is physically meaningless. Third, purely for purposes of distinguishing them, we will call one of the circuits a "loop" and the other one a "coil." Further, we will take the loop to be square, and oriented as shown in fig. 5.2.

As a guess, we might expect that the energy is given by eq. (21.4), $-\mathbf{m} \cdot \mathbf{B}$, where $\mathbf{m}$ is the magnetic moment of the loop, and $\mathbf{B}$ is the field at the loop due to the coil. This is incorrect. Suppose the loop is brought toward the coil, which remains stationary. As the loop is moved, the magnetic field from the coil grows stronger, and the flux through the

---

[2] Feynman et al. (1964), vol. II, secs. 15-1–15-3. See also Eyges (1972), sec. 11.3.

loop changes. By Faraday's law, this induces an emf which tends to oppose the currents in the loop. Work must then be done on the battery driving the current in the loop in order to keep the current that it delivers fixed. Otherwise, we will end up with a situation where the currents in the circuits decrease as they are brought together, and that is not the configuration whose energy we want. Let us call the work required to counteract the induced emf the electrical work, W_e. This work is left out of the accounting in eq. (21.4). Let us now calculate W_e for the loop. Suppose it moves with a velocity $v_{loop}\hat{x}$. In a time Δt, it moves a distance $v_{loop}\Delta t$, and the flux through it changes by

$$\Delta\Phi_{loop} = 4a^2 \frac{\partial B_z}{\partial x} v_{loop}\Delta t. \tag{29.1}$$

The emf induced in the loop is given by

$$\mathcal{E} = -\frac{1}{c}\frac{\Delta\Phi}{\Delta t} = -\frac{4a^2}{c}\frac{\partial B_z}{\partial x} v_{loop}. \tag{29.2}$$

A quantity of charge equal to $I\Delta t$ must be pushed around the loop against this emf, requiring an amount of work

$$\Delta W_{e,loop} = -(I\Delta t)\mathcal{E} = m\frac{\partial B_z}{\partial x} v_{loop}\Delta t. \tag{29.3}$$

Next, let us note that because the field at the loop is inhomogeneous, it exerts a force whose x component is $m\partial B_z/\partial x$ (see eq. (21.11)). To counteract this force, we must do an amount of mechanical work

$$\Delta W_m = -m\frac{\partial B_z}{\partial x} v_{loop}\Delta t. \tag{29.4}$$

Not surprisingly, this is just the change in the mechanical energy $-\mathbf{m}\cdot\mathbf{B}$.

The key point now is that ΔW_m is exactly equal and opposite to $\Delta W_{e,loop}$, i.e.,

$$\Delta W_{e,loop} + \Delta W_m = 0. \tag{29.5}$$

There are three points to note about eq. (29.5). First, the individual contributions depend only on the product $v_{loop}\Delta t$, so they depend only on the initial and final positions of the loop, not on how fast it is moved. By moving the loop extremely slowly, we can make the energy radiated as small as we please and thus put to rest any worries that radiation has not been included. Second, it is not a coincidence that the total work is zero. The force on the electrons, both along the wire and along the direction of motion of the loop, arises from the $\mathbf{v}\times\mathbf{B}$ term in the Lorentz force. As a general result, this force is always perpendicular to $\mathbf{v}$, and so it can never do any work on a charged particle. (See also exercise 29.1.) Third, eq. (29.5) clearly holds for any shaped loop, not just a square one.

It is obvious that as the loop moves, the flux through the coil changes, so it too experiences an opposing emf. To calculate the electrical work that must be done to keep the currents in the coil fixed, let us go to a frame in which the loop is stationary and the coil moves toward the loop with velocity $-v_{loop}\hat{x}$. By exactly the same argument as before,

we obtain

$$\Delta W_{e,\text{coil}} + \Delta W_m = 0. \tag{29.6}$$

The mechanical work in this equation is exactly the same as that in the previous one, since the force on the coil is equal and opposite to that on the loop.

Adding the two equations together, we obtain

$$\Delta W_{e,\text{loop}} + \Delta W_{e,\text{coil}} + 2\Delta W_m = 0. \tag{29.7}$$

We can write this as

$$\Delta W_{\text{tot}} = \Delta W_{e,\text{loop}} + \Delta W_{e,\text{coil}} + \Delta W_m = -\Delta W_m. \tag{29.8}$$

This is the basic result we have been seeking. Since the mechanical energy is $-\mathbf{m} \cdot \mathbf{B}$, the energy of the two infinitesimal circuits is

$$U_{\text{tot}} = \mathbf{m} \cdot \mathbf{B}. \tag{29.9}$$

Exercise 29.1 Calculate the electrical work that must be supplied to the moving loop by considering the Lorentz force in the direction of the current along the various legs of the loop. Let the loop be made of wire of cross-sectional area $\mathcal{A}$, and let it have n electrons per unit volume, all moving with a speed v_c around the wire.

Let us now recast eq. (29.9) in a way that will give useful expressions for the energy of a general system of currents. Recalling that the magnetic moment is proportional to the current times the area for a planar loop, for our infinitesimal loop pair, we have

$$U_{\text{tot}} = \frac{I_1}{c} \int_{S_1} \mathbf{B}_2(\mathbf{r}_1) \cdot d\mathbf{s}_1. \tag{29.10}$$

The integral is taken over a surface spanning any *one* of the two loops (number 1, the way we have indicated by the subscripts), and $\mathbf{B}_2$ is the field due to the other loop. But, since $\mathbf{B}_2 = \nabla \times \mathbf{A}_2$, we can rewrite this by Stokes's theorem as

$$U_{\text{tot}} = \frac{I_1}{c} \oint_{C_1} \mathbf{A}_2(\mathbf{r}_1) \cdot d\boldsymbol{\ell}_1. \tag{29.11}$$

For a general system of macroscopic loops, we first regard each loop as composed of tiny loops. We then write an expression like eq. (29.11) for each *pair* of tiny loops. It is easier to write one term for every loop and halve the total to correct for double counting. At the same time, it is better to express the answer in terms of volume currents. In this way we obtain

$$U_{\text{tot}} = \frac{1}{2c} \int \mathbf{j} \cdot \mathbf{A} \, d^3x. \tag{29.12}$$

$$U_{\text{tot}} = \frac{1}{2} \int \mathbf{j} \cdot \mathbf{A} \, d^3x. \quad \text{(SI)}$$

The integral is over all space. This result is the analog of eqs. (14.4) and (18.1), and it holds for a general steady current distribution.

To write eq. (29.12) in terms of the **B** field, we use Ampere's law. Thus,

$$U_{\text{tot}} = \frac{1}{8\pi} \int (\nabla \times \mathbf{B}) \cdot \mathbf{A} \, d^3x, \tag{29.13}$$

where the curl acts on **B** only. But,

$$(\nabla \times \mathbf{B}) \cdot \mathbf{A} = \nabla \cdot (\mathbf{B} \times \mathbf{A}) + \mathbf{B} \cdot (\nabla \times \mathbf{A}). \tag{29.14}$$

The total divergence is transformed into a vanishing surface integral at infinity, and, invoking $\nabla \times \mathbf{A} = \mathbf{B}$ once more, we finally get

$$U_{\text{tot}} = \frac{1}{8\pi} \int \mathbf{B}^2(\mathbf{r}) \, d^3x. \tag{29.15}$$

$$U_{\text{tot}} = \frac{1}{2\mu_0} \int \mathbf{B}^2(\mathbf{r}) \, d^3x. \quad \text{(SI)}$$

This is the magnetic analog of eq. (18.4). It comes with exactly similar caveats about divergent self-energies, which must be excluded when we deal with point dipoles or idealized mathematical current-carrying wires of zero thickness.

30 Energy in the magnetic field—standard argument

The approach of the previous section is extremely useful in illuminating the difference between the energy of a system of currents and the energy of a loop in an *external* magnetic field. There is also a standard argument for calculating the former energy, which we now present because it has the advantage of brevity.

As before, we imagine that a current distribution $\mathbf{j}(\mathbf{r})$ has been created by slowly building it up from nothing. The objective is to calculate the work required to change it by an incremental amount $\Delta\mathbf{j}$ in a duration Δt. By adding up all the work increments, we will obtain the energy of the current distribution.

The change $\Delta\mathbf{j}$ leads to a change $\Delta\mathbf{B}$ in the magnetic field $\mathbf{B}(\mathbf{r}, t)$, which by Faraday's law results in an electric field **E**. This electric field exerts forces on the moving charges constituting the currents. By Lenz's rule, these forces must necessarily be such as to oppose the change $\Delta\mathbf{j}$. The work required to overcome this opposition is

$$\Delta W = -\Delta t \int \mathbf{j} \cdot \mathbf{E} \, d^3x. \tag{30.1}$$

Since $\mathbf{j} = c\nabla \times \mathbf{B}/4\pi$, we have

$$\Delta W = -\frac{c\,\Delta t}{4\pi} \int \mathbf{E} \cdot \nabla \times \mathbf{B} \, d^3x = -\frac{c\,\Delta t}{4\pi} \int \mathbf{B} \cdot \nabla \times \mathbf{E} \, d^3x, \tag{30.2}$$

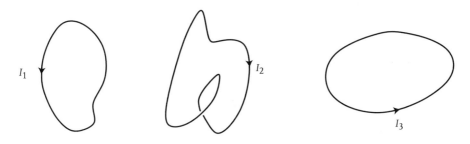

FIGURE 5.3. A system of interacting current loops.

where the second equality follows by integration by parts. But $c\nabla \times \mathbf{E} = -\partial\mathbf{B}/\partial t \approx -\Delta\mathbf{B}/\Delta t$, so

$$\Delta W = \frac{1}{4\pi}\int \mathbf{B}\cdot\Delta\mathbf{B}\,d^3x = \Delta\int \frac{B^2}{8\pi}d^3x. \tag{30.3}$$

Adding up the work increments, we once again arrive at eq. (29.15).

31 Inductance

Let us consider a system of current-carrying loops (fig. 5.3). Since the energy is quadratic in $\mathbf{B}$, and $\mathbf{B}$ is separately linear in each of the loop currents, it follows that the energy should be expressible as a quadratic function of the currents themselves. We now find this expression. Let the loops be labeled by a subscript a. From eq. (29.12) the total energy can be written as

$$U_{\text{tot}} = \sum_a U_{aa} + \sum_{a>b} U_{ab}, \tag{31.1}$$

where

$$U_{aa} = \frac{1}{2c}\int \mathbf{j}_a\cdot\mathbf{A}_a d^3x_a = \frac{1}{2c^2}\iint \frac{\mathbf{j}_a(\mathbf{r}_a)\cdot\mathbf{j}_a(\mathbf{r}'_a)}{|\mathbf{r}_a - \mathbf{r}'_a|}d^3x_a d^3x'_a, \tag{31.2}$$

$$U_{ab} = \frac{1}{c}\int \mathbf{j}_a\cdot\mathbf{A}_b d^3x_a = \frac{1}{c^2}\iint \frac{\mathbf{j}_a(\mathbf{r}_a)\cdot\mathbf{j}_b(\mathbf{r}_b)}{|\mathbf{r}_a - \mathbf{r}_b|}d^3x_a d^3x_b. \tag{31.3}$$

The subscripts a and b on $\mathbf{j}$ indicate over which current distribution the integration is to be performed. The second of these equations shows that $U_{ab} = U_{ba}$.

The above formulas are applicable to any current distributions, including currents in extended conductors. For the practically important case of currents carried by thin wires, we can get approximate formulas by replacing $\mathbf{j}\,d^3x$ by $I\,d\boldsymbol{\ell}$, and volume integrals by loop integrals. The energy can be written as a symmetrical double sum

$$U_{\text{tot}} = \frac{1}{2}\sum_{a,b} L_{ab} I_a I_b, \tag{31.4}$$

where

$$L_{ab} \simeq \frac{1}{c^2} \oint \oint \frac{d\boldsymbol{\ell}_a \cdot d\boldsymbol{\ell}_b}{|\mathbf{r}_a - \mathbf{r}_b|}. \tag{31.5}$$

$$L_{ab} \simeq \frac{\mu_0}{4\pi} \oint \oint \frac{d\boldsymbol{\ell}_a \cdot d\boldsymbol{\ell}_b}{|\mathbf{r}_a - \mathbf{r}_b|}. \quad \text{(SI)}$$

The coefficient $L_{ab} = L_{ba}$ for $a \neq b$ is known as the *mutual inductance* of loops a and b, while L_{aa} is the self-inductance of loop a. The quadratic form (31.4) is clearly positive definite, which implies, among other similar formulas, that

$$L_{aa} > 0, \quad L_{aa} L_{bb} > L_{ab}^2. \tag{31.6}$$

Exercise 31.1 Show that the total energy and the inductance coefficients can be written as

$$U_{\text{tot}} = \frac{1}{2c} \sum_a I_a \Phi_a, \quad L_{ab} = \frac{\Phi_{a,b}}{c\, I_b}, \tag{31.7}$$

$$U_{\text{tot}} = \frac{1}{2} \sum_a I_a \Phi_a, \quad L_{ab} = \frac{\Phi_{a,b}}{I_b}, \quad \text{(SI)}$$

where Φ_a is the total flux through loop a, and $\Phi_{a,b}$ is that part of this flux that is due to the current in loop b. (Note that $\Phi_{a,b} \neq \Phi_{b,a}$.)

We shall calculate the inductances (self and mutual) of some simple setups in the rest of this section. It is useful, however, to first discuss the units and conversion factors for inductance. The conversion factor follows by noting that $L I^2/2$ is the energy in both systems and recalling the conversion for I:

$$L_{\text{SI}} = \frac{\mu_0 c^2}{4\pi} L_{\text{Gau}} = \frac{1}{4\pi\epsilon_0} L_{\text{Gau}}. \tag{31.8}$$

The SI unit of inductance is the *henry* (H), and

$$1 \text{ henry} = \frac{1}{9} \times 10^{-11} \text{ Gaussian unit.} \tag{31.9}$$

One henry is a largish inductance, but not excessively so. The Gaussian unit for inductance does not have a special name. We are using the electrostatic unit for inductance. Some authors prefer to use the electromagnetic unit, which is c^2 times our unit and can be given in centimeters, as it has dimensions of length.

Equation (31.5) shows that the inductances of a system of current loops depend only on the geometry. In reality, there is also a weaker dependence on the exact distribution of current over the cross section of the wires. This equation leads to an immediate method of calculating mutual inductances, although the integrals must in general be done numerically. For the self-inductance of a loop, on the other hand, it leads to a meaningless answer if applied literally, since for any fixed $\mathbf{r}_a$, the integral over $\mathbf{r}_b$ is logarithmically divergent due to points arbitrarily close to $\mathbf{r}_a$. It is clear that when the nonzero dimensions of any physical wire are considered, this divergence must get cut off at distances of the order of a, the radius of the wire, since the true expression for the energy is finite. We can

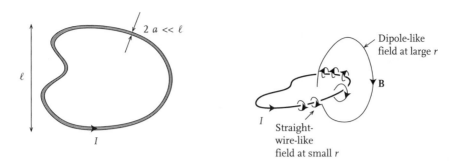

FIGURE 5.4. How to find the self-inductance of a wire loop that does not wind close to or back on itself. The field near the wire can be approximated as that of an infinite, straight, current-carrying wire, and that far from the wire can be approximated as that of a dipole.

obtain an estimate for the self-inductance of a loop of circumference ℓ made from a wire of radius $a \ll \ell$ (see fig. 5.4) by equating the expression $L I^2/2$ for the total energy to the energy stored in the magnetic field and finding the latter energy as follows.[3] At distances $r \gg \ell$ from the loop, the magnetic field is given by $B \sim m/r^3$, where $m \sim I\ell^2/c$. Ignoring all multiplicative factors of order unity, the field energy in this region of space is therefore

$$\int_{r>\ell} \frac{B^2}{8\pi} d^3x \sim \frac{m^2}{\ell^3} \sim \frac{I^2\ell}{c^2}. \tag{31.10}$$

Next, let us consider points close to the wire, but outside it. At such points, we may neglect the local curvature of the wire; the B field is then essentially that of an infinite wire. Denoting the distance from the axis of the wire by $r_\perp$, $B \simeq 2I/cr_\perp$. The field energy from this region is

$$\int_a^{\xi\ell} \frac{B^2}{8\pi} (2\pi\ell r_\perp \, dr_\perp) \simeq \frac{I^2\ell}{c^2} \int_a^{\xi\ell} \frac{dr_\perp}{r_\perp} \simeq \frac{I^2\ell}{c^2} \ln\left(\frac{\xi\ell}{a}\right). \tag{31.11}$$

Note that we have cut off the integral at the upper end at a length $\xi\ell$, where ξ is a dimensionless number of order unity. The precise value of this number is not important at the present level of approximation.

Lastly, let us find the field energy inside the wire. This depends on the exact current distribution. If we assume the current to flow entirely on the surface of the wire, for example, then $B = 0$ inside it, and the energy from this region is zero. If we assume the current to be uniformly spread over the wire, on the other hand, then $B = 2Ir_\perp/ca^2$, and a calculation similar to eq. (31.11) gives the energy from this region as

$$\frac{I^2\ell}{c^2} \int_0^a \frac{r_\perp^2}{a^4} r_\perp \, dr_\perp = \frac{I^2\ell}{4c^2}. \tag{31.12}$$

[3] We are assuming that the linear dimensions of the loop are of order ℓ. In other words, the wire is not wound back and forth or into so tight a spiral as to fill out a two-dimensional region. When this is the case, as it is for a good practical solenoid, the calculation must be done differently. See below.

Of the three contributions to the energy, that from the midrange, eq. (31.11), is the largest when $\ell \gg a$. Indeed, the other two contributions can be incorporated into the undetermined numerical constant ξ. Thus, we have the following *logarithmically accurate* formula for the self-inductance:

$$L \simeq \frac{2\ell}{c^2} \ln\left(\frac{\xi\ell}{a}\right). \tag{31.13}$$

For $\ell = 20$ cm and $a = 1$ mm, this yields $L \simeq 200$ nH.

The following examples and exercises explore the self-inductance for some practically important circuits.

Ideal solenoid: Consider a tightly wound, long solenoid, of length h, number of turns per unit length n, and radius R. Ignoring end effects, we know that the field inside the solenoid is given by $B = 4\pi n I/c$. Integrating the energy density $B^2/8\pi$ over the volume of the solenoid and equating this to $LI^2/2$, we obtain

$$L = \left(\frac{2\pi n R}{c}\right)^2 h = \frac{2\pi n R\ell}{c^2}, \tag{31.14}$$

$$L = \frac{\mu_0 n R\ell}{2}, \quad \text{(SI)}$$

where $\ell = 2\pi n Rh$ is the total length of wire. This expression is much greater than eq. (31.13) for an open loop of the same length; the reason is the large mutual inductance of neighboring turns.

Exercise 31.2 For the solenoid just described, estimate the energy in the region $r \gg h$, and thus show that the correction to the inductance due to the finiteness of the length is of relative order $(h/R)^2$.

Exercise 31.3 (toroidal solenoid) Find the inductance of a tightly wound toroidal solenoid with (a) a rectangular cross section and (b) a circular cross section.

Exercise 31.4 (coaxial cable) Consider a coaxial cable made of a central conductor of radius a and an outer conductor (which carries the return current) in the form of a thin sheath of radius $b > a$. Find the self-inductance per unit length, assuming that the current in the central conductor is distributed uniformly over (a) its cross section, and (b) its surface. How can you think of the circuit in (b) as a single loop of wire?

Answer: Using $U_{\text{tot}} = LI^2/2$, we get (a) $c^2 L = 2\ln(b/a) + \frac{1}{2}$, and (b) $c^2 L = 2\ln(b/a)$. Think of a tightly wound toroidal solenoid of rectangular cross section with inner radius a, outer radius b, and height of rectangle h. The cable is obtained by letting $h \to \infty$.

Exercise 31.5 (two-wire transmission line) Consider a transmission line made of two infinite parallel wires, both of radius a and center-to-center separation b. The current flows up one wire and back down the other. Find the self-inductance of this line per unit length, assuming that the current is distributed uniformly over (a) the cross section of each wire and (b) the surface of each wire.

Solution: (a) As the problem is two dimensional, we will let $\mathbf{r}$ stand for the vector in the xy plane. Let the center of wire 1 be at $\mathbf{r} = 0$ and of wire 2 at $\mathbf{r} = b\hat{\mathbf{x}}$, and let the vector potentials due to the two wires be $\mathbf{A}_1$ and $\mathbf{A}_2$, with $\mathbf{A}_2(\mathbf{r}) = -\mathbf{A}_1(\mathbf{r} - b\hat{\mathbf{x}})$. From eq. (29.12), the energy per unit length is seen to be

$$u = \frac{1}{c} \int \mathbf{j}_1 \cdot (\mathbf{A}_1 + \mathbf{A}_2) \, d^2x, \tag{31.15}$$

where $\mathbf{j}_1$ is the current density in wire 1. Let us call the contribution from the two terms u_1 and u_2. For $\mathbf{A}_1$, we use (exercise 24.1)

$$\mathbf{A}_1(\mathbf{r}) = \begin{cases} -\dfrac{I}{ca^2} r^2 \hat{\mathbf{z}}, & r \le a, \\[2ex] -\dfrac{I}{c}(1 + 2\ln(r/a))\hat{\mathbf{z}}, & r \ge a. \end{cases} \tag{31.16}$$

Then $u_1 = -I^2/2c^2$, and $u_2 = (I^2/c^2) - (I/ac)^2\phi(b\hat{\mathbf{x}})$, where

$$\phi(\mathbf{r}) = -\int_{r' < a} 2\ln\left|\frac{\mathbf{r} - \mathbf{r}'}{a}\right| d^2x' \tag{31.17}$$

is the electrostatic potential of a wire of radius a with uniform unit charge density. By solving Poisson's equation for this potential directly, we get (exercise 17.3)

$$\phi(\mathbf{r}) = \begin{cases} \pi(a^2 - r^2), & r \le a, \\[1ex] -2\pi a^2 \ln(r/a), & r \ge a. \end{cases} \tag{31.18}$$

(The constants of integration are fixed by demanding continuity at $r = a$ and matching to either $\phi(0)$ or the asymptotic form of $\phi(\mathbf{r})$ for $r \to \infty$, which can be found directly from the integral. Note that ϕ is essentially the same function as A_1, up to an additive constant and a scale factor.) Thus, $u = I^2(1 + 4\pi\ln(b/a))/2c^2$, and the inductance per unit length is $L = (1 + 4\pi\ln(b/a))/c^2$.

For part (b), we note that the **B** field is the same as in part (a) outside the wires and zero inside. Thus, we subtract from the energy in part (a) the integral of $B^2/8\pi$ over the interior of the wires. Equating this new energy to $L I^2/2$ yields $L = (4\pi/c^2)\ln(b/a)$.

32 The Ampere-Maxwell law

Faraday's discovery showed that a changing magnetic field produces an electric field. Some thirty years later, Maxwell deduced that a changing electric field must produce a magnetic field. Today, therefore, we would not say that either type of time-varying field *produces* the other in a causative sense, only that when you have one type of time-varying field, the other must be present. The true causative agents are the time-varying charges

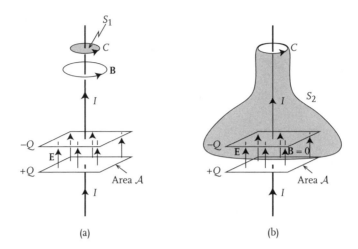

FIGURE 5.5. Application of Ampere's law to a slowly charging capacitor. In (a), the loop C is shown spanned by the disk S_1. Since the magnetic field around C is essentially that of an infinitely long wire, and the current puncturing S_1 is I, Ampere's law is obeyed, and all is well. In (b), the same loop C is shown spanned by the flask-shaped surface S_2, whose base passes between the capacitor plates. The circulation of $\mathbf{B}$ around C is the same as in (a), but the current puncturing S_2 is zero. All is not well. See text for the cure.

and currents, and when these change so as to produce a time-dependent $\mathbf{E}$ or $\mathbf{B}$ field, they also produce a (generally time-dependent) field of the other kind.

The equation that must be modified in light of Maxwell's discovery is that which gives Ampere's law for static magnetic fields,

$$\nabla \times \mathbf{B} = \frac{4\pi}{c}\mathbf{j}. \tag{32.1}$$

We can appreciate why this is so by considering a parallel plate capacitor that is being charged. The wires leading to the plates are long and straight, and the charging battery is far away (fig. 5.5). We imagine that the current I in the wire is very small, so the charge Q on the capacitor plates is increasing very slowly. Then the magnetic field near the wire, far from the capacitor, cannot be very different from the static value. To find it, let us apply Ampere's law in integral form to the circular loop C, spanned by the flat disk S_1:

$$\oint_{C=\partial S_1} \mathbf{B} \cdot d\boldsymbol{\ell} = \frac{4\pi}{c}\int \mathbf{j} \cdot d\mathbf{s} = \frac{4\pi}{c}I. \tag{32.2}$$

Letting the radius of C be R, the left-hand side is $2\pi R B$, yielding $B = 2I/Rc$, as is expected for a long, straight wire.

If, however, instead of S_1, we take the surface spanning C in the shape of a flask, S_2, as shown in fig. 5.5b, then a contradiction arises. The base of this flask lies between the capacitor plates, so no current flows through this part of S_2. No current flows through the sides of S_2 either, so

$$\int_{S_2} \mathbf{j} \cdot d\mathbf{s} = 0. \tag{32.3}$$

We are led to conclude, incorrectly, that $B = 0$ on the loop C. Thus, our starting point, eq. (32.1), must be wrong, and there must be another term in the expression for $\nabla \times \mathbf{B}$, which is such that its surface integral over S_2 is $4\pi I/c$. We note that $\mathbf{E}$ is nonzero between the capacitor plates. If we imagine the plates as very closely spaced, we may ignore the fringing effects and take $\mathbf{E}$ to be uniform and nonzero only between the plates. Taking the area of the plates as $\mathcal{A}$, we have $E = 4\pi Q/\mathcal{A}$, so that[4]

$$\int_{S_2} \mathbf{E} \cdot d\mathbf{s} \simeq 4\pi Q. \tag{32.4}$$

This is not good enough; what we need is $4\pi I/c$. Since $dQ/dt = I$, it is not too hard to see that $(1/c)\partial\mathbf{E}/\partial t$ will do the job. This is, indeed, the missing term, and the correct law is

$$\nabla \times \mathbf{B} = \frac{4\pi}{c}\mathbf{j} + \frac{1}{c}\frac{\partial\mathbf{E}}{\partial t}. \tag{32.5}$$

$$\nabla \times \mathbf{B} = \mu_0\mathbf{j} + \mu_0\epsilon_0\frac{\partial\mathbf{E}}{\partial t}. \quad \text{(SI)}$$

This is sometimes known as the *Ampere-Maxwell law* or *equation*.

The term $(1/c)\partial\mathbf{E}/\partial t$ in eq. (32.5) is often called the *displacement current*. This terminology, due to Maxwell, is historical and rooted in the mechanical picture for electromagnetism that was felt necessary in his time. Nowadays, we prefer to think using the abstract notion of fields, so the meaning of the words "displacement current" is obscure to us. Still, the terminology remains.

Maxwell's modification of Ampere's law has one more important consequence. It makes the equations consistent with charge conservation. If we take the divergence of eq. (32.5), we get

$$0 = \frac{4\pi}{c}\nabla \cdot \mathbf{j} + \frac{1}{c}\frac{\partial}{\partial t}\nabla \cdot \mathbf{E}. \tag{32.6}$$

Since Gauss's law allows us to substitute $4\pi\rho$ for $\nabla \cdot \mathbf{E}$, this implies

$$\nabla \cdot \mathbf{j} + \frac{\partial\rho}{\partial t} = 0, \tag{32.7}$$

which is the equation of continuity, i.e., the law for charge conservation. Without the displacement current, we would obtain $\nabla \cdot \mathbf{j} = 0$, which is not true in general.

Let us quickly dispose of the question whether our expression for the electrostatic energy, eq. (18.4), needs to be modified in light of induced magnetic fields. The answer is

---

[4] Note that if C is oriented as drawn, and we are to apply Stokes's theorem, then the normal to S_2 must be taken to point "into" the flask.

no. Imagine, as before, that two charges are brought together from infinity. As the charges move, the electric fields change, which leads to induced magnetic fields. However, a magnetic field does no work on a charge, and hence does not affect the energy.

With eq. (32.5), we now have all the laws of electromagnetism. We write these out all together so as to have them in one place.

Law	Equation (Gaussian)	Equation (SI)
Gauss's law	$\nabla \cdot \mathbf{E} = 4\pi\rho$	$\nabla \cdot \mathbf{E} = \dfrac{\rho}{\epsilon_0}$
Faraday's law	$\nabla \times \mathbf{E} = -\dfrac{1}{c}\dfrac{\partial \mathbf{B}}{\partial t}$	$\nabla \times \mathbf{E} = -\dfrac{\partial \mathbf{B}}{\partial t}$
Ampere-Maxwell law	$\nabla \times \mathbf{B} = \dfrac{4\pi}{c}\mathbf{j} + \dfrac{1}{c}\dfrac{\partial \mathbf{E}}{\partial t}$	$\nabla \times \mathbf{B} = \mu_0\mathbf{j} + \mu_0\epsilon_0\dfrac{\partial \mathbf{E}}{\partial t}$
No magnetic monopoles	$\nabla \cdot \mathbf{B} = 0$	$\nabla \cdot \mathbf{B} = 0$
Lorentz force law	$\mathbf{F} = q\left(\mathbf{E} + \dfrac{1}{c}\mathbf{v} \times \mathbf{B}\right)$	$\mathbf{F} = q\left(\mathbf{E} + \mathbf{v} \times \mathbf{B}\right)$

$$(32.8)$$

It should be stressed that although we have given compelling reasons for the presence of the displacement current, in no sense have we derived it. Indeed, it would be a delusion to think that we can ever do anything but guess the laws of physics. The truth of any law is confirmed by observation of the consequences of that law. In the case of eq. (32.5), or, rather, of the full set (32.9) of equations of electromagnetism, the consequences are immense and immensely varied, and amply confirmed by experiment. We shall learn about them in more detail in the following chapters.

33 Potentials for time-dependent fields

As we shall see in later chapters, potentials continue to be powerful conceptual and problem-solving tools for time-dependent problems. We therefore examine how they must be modified for time-dependent fields.

The equation $\nabla \cdot \mathbf{B} = 0$ holds as before, so we can continue to write

$$\mathbf{B} = \nabla \times \mathbf{A}. \quad \text{(Gaussian and SI)} \tag{33.1}$$

Using this result we can write Faraday's law as

$$\nabla \times \left(\mathbf{E} + \frac{1}{c}\frac{\partial \mathbf{A}}{\partial t}\right) = 0. \tag{33.2}$$

The vector in parentheses can thus be written as the gradient of a scalar field, which we denote as $-\phi$ so as to get the right limit in the static case. Hence, the electric field can be written as

$$\mathbf{E} = -\nabla\phi - \frac{1}{c}\frac{\partial\mathbf{A}}{\partial t}. \tag{33.3}$$

$$\mathbf{E} = -\nabla\phi - \frac{\partial\mathbf{A}}{\partial t}. \quad \text{(SI)}$$

In terms of the potentials, the field equations with source terms become

$$-\nabla^2\phi - \frac{1}{c}\frac{\partial}{\partial t}\nabla\cdot\mathbf{A} = 4\pi\rho, \tag{33.4}$$

$$-\nabla^2\mathbf{A} + \nabla(\nabla\cdot\mathbf{A}) + \frac{1}{c}\frac{\partial}{\partial t}\nabla\phi + \frac{1}{c^2}\frac{\partial^2\mathbf{A}}{\partial t^2} = \frac{4\pi}{c}\mathbf{j}. \tag{33.5}$$

To fruitfully solve these equations, we can make use of the gauge degree of freedom to impose additional restrictions on the potentials. As in static situations, this freedom allows certain transformations of the potentials that leave the fields unchanged. First, $\mathbf{B}$ is unaltered if we add the gradient of a scalar function, $\nabla\psi$, to $\mathbf{A}$. It then follows that we must subtract $(1/c)\partial\psi/\partial t$ from ϕ in order that $\mathbf{E}$ be unaltered. Thus, the most general gauge transformation is

$$\mathbf{A} \to \mathbf{A}' = \mathbf{A} + \nabla\psi, \tag{33.6}$$

$$\phi \to \phi' = \phi - \frac{1}{c}\frac{\partial\psi}{\partial t}, \tag{33.7}$$

where ψ is a scalar field.

In the *Lorenz gauge*,[5] we impose the subsidiary condition (which we will show below to be allowed)

$$\nabla\cdot\mathbf{A} + \frac{1}{c}\frac{\partial\phi}{\partial t} = 0. \tag{33.8}$$

We use this condition to eliminate $\mathbf{A}$ in eq. (33.4) in favor of ϕ and to eliminate the middle two terms on the left of eq. (33.5). The resulting equations are

$$\nabla^2\phi - \frac{1}{c^2}\frac{\partial^2\phi}{\partial t^2} = -4\pi\rho, \tag{33.9}$$

$$\nabla^2\mathbf{A} - \frac{1}{c^2}\frac{\partial^2\mathbf{A}}{\partial t^2} = -\frac{4\pi}{c}\mathbf{j}. \tag{33.10}$$

These are both inhomogeneous wave equations. With the Lorenz condition, (33.8), they are completely equivalent to the four first-order equations (32.9). We discuss how to solve them in section 54.

---

[5] Named after L. V. Lorenz, not to be confused with H. A. Lorentz. This confusion is widespread and very unfortunate. Writing in 1867, only six years after Maxwell, and independently of him, Lorenz gave the correct retarded solutions for the potentials, eqs. (54.16) and (54.17), and suggested the equivalence of light and EM waves. Lorenz's equations unequivocally imply the existence of the displacement current, but it is unclear if he saw the significance of this connection. For a fuller account of Lorenz's work, and the history of this miscitation, see J. D. Jackson and L. B. Okun, *Rev. Mod. Phys.* **73**, 663 (2001).

To show that the Lorenz condition can always be imposed, suppose that it is not obeyed to begin with. We then make a gauge transformation as per eqs. (33.6) and (33.7). The transformed potentials obey

$$\nabla \cdot \mathbf{A}' + \frac{1}{c}\frac{\partial \phi'}{\partial t} = \nabla \cdot \mathbf{A} + \frac{1}{c}\frac{\partial \phi}{\partial t} + \nabla^2 \psi - \frac{1}{c^2}\frac{\partial^2 \psi}{\partial t^2} \tag{33.11}$$

and will therefore be in the Lorenz gauge if the function ψ obeys

$$\nabla^2 \psi - \frac{1}{c^2}\frac{\partial^2 \psi}{\partial t^2} = -\left(\nabla \cdot \mathbf{A} + \frac{1}{c}\frac{\partial \phi}{\partial t}\right), \tag{33.12}$$

which is yet another inhomogeneous wave equation! If the reader is willing to temporarily accept that we can solve such equations, then we have shown that the Lorenz gauge is always a possible gauge.

The Lorenz gauge is well suited to formal considerations and to situations where relativity is important. In particular, the Lorenz condition holds in a boosted frame if it holds in one frame.

A second gauge, which is useful for radiation problems, is the *Coulomb* gauge, also known as the *transverse*, or *radiation* gauge. In this, the subsidiary condition is

$$\nabla \cdot \mathbf{A} = 0. \tag{33.13}$$

The equations for the potentials are thereby simplified to

$$\nabla^2 \phi = -4\pi\rho, \tag{33.14}$$

$$\nabla^2 \mathbf{A} - \frac{1}{c^2}\frac{\partial^2 \mathbf{A}}{\partial t^2} = -\frac{4\pi}{c}\mathbf{j} + \frac{1}{c}\frac{\partial}{\partial t}\nabla\phi. \tag{33.15}$$

The equation for $\mathbf{A}$ is again a wave equation, but that for ϕ is a Poisson equation. We know that the solution to this equation is given by

$$\phi(\mathbf{r}, t) = \int \frac{\rho(\mathbf{r}', t)}{|\mathbf{r} - \mathbf{r}'|}\, d^3x'. \tag{33.16}$$

The utility of the Coulomb gauge becomes clear only after one understands how to decompose fields into "transverse" and "longitudinal" parts (see the following exercise). This decomposition also resolves the apparent acausality of this gauge, as evidenced by eq. (33.16). In this equation, the potential at any point $\mathbf{r}$ is given by the instantaneous value of the charge density at distant points, whereas we know that electromagnetic influences cannot travel faster than light. It turns out, however, that when we differentiate the potentials, the fields $\mathbf{E}$ and $\mathbf{B}$ are completely causal.[6] We will use the Coulomb gauge in section 79 when we develop the action formalism for interacting charges and fields.

---

[6] For a fuller discussion of these points, see O. J. Brill and J. Goodman, *Am. J. Phys.* **35**, 832 (1967).

Exercise 33.1 (a) Any vector field $V(r)$ can be decomposed into the sum of a transverse or divergenceless part V^T and a longitudinal or curl-free part V^L. Show that

$$V^T(\mathbf{r}) = \nabla \times \left[\nabla \times \frac{1}{4\pi} \int \frac{V(\mathbf{r}')}{|\mathbf{r} - \mathbf{r}'|} d^3x' \right], \tag{33.17}$$

$$V^L(\mathbf{r}) = -\frac{1}{4\pi} \nabla \left[\nabla \cdot \int \frac{V(\mathbf{r}')}{|\mathbf{r} - \mathbf{r}'|} d^3x' \right]. \tag{33.18}$$

These forms show at once that $\nabla \cdot V^T = \nabla \times V^L = 0$.

(b) Show that the D'Alembertian of A in the Coulomb gauge (the left-hand side of eq. (33.15)) equals $-(4\pi/c)\mathbf{j}^T$.

Solution: (a) Let the spatial Fourier transform of V be denoted by V_k. In k space, the conditions $\nabla \cdot V^T = \nabla \times V^L = 0$ become $\mathbf{k} \cdot V_k^T = \mathbf{k} \times V_k^L = 0$. Thus, V_k^T and V_k^L are, respectively, the parts of V_k perpendicular and parallel to $\mathbf{k}$, which are given by

$$V_k^T = -\mathbf{k} \times \left(\mathbf{k} \times \frac{V_k}{k^2} \right), \quad V_k^L = \mathbf{k} \left(\mathbf{k} \cdot \frac{V_k}{k^2} \right). \tag{33.19}$$

If we recall that by the convolution theorem, V_k/k^2 is the FT of $\int d^3x' [V(\mathbf{r}')/4\pi|\mathbf{r} - \mathbf{r}'|]$, the results in real space follow by taking inverse transforms.

(b) Let us denote the right-hand side of eq. (33.15) by $-4\pi w(\mathbf{r})/c$. We have

$$w = \mathbf{j} - \frac{1}{4\pi} \frac{\partial}{\partial t} \nabla \phi. \tag{33.20}$$

Taking spatial Fourier transforms, we obtain $w_k = \mathbf{j}_k - i\mathbf{k}\dot{\phi}_k/4\pi$. Equation (33.16) yields $\phi_k = 4\pi\rho_k/k^2$, and the equation of continuity implies $\dot{\rho}_k = -i\mathbf{k} \cdot \mathbf{j}_k$. Hence, $w_k = \mathbf{j}_k - \mathbf{k}(\mathbf{k} \cdot \mathbf{j}_k/k^2)$. But this is precisely the FT of $\mathbf{j}^T$. This proves the desired result.

Comment: The theorem that guarantees the decomposition of a vector field into its transverse and longitudinal parts is sometimes known as the *Helmholtz theorem* or the *fundamental theorem of vector calculus*. Note that this decomposition is not local in position space, i.e., V^L and V^T at a point $\mathbf{r}$ depend on V at points $\mathbf{r}' \neq \mathbf{r}$. This decomposition also enables one to better understand the uniqueness theorem discussed in section 16, namely, that any field whose curl and divergence are known is uniquely determined. The curl determines the transverse part of the field, and the divergence determines the longitudinal part.

Exercise 33.2 Show that the transverse part of A is gauge invariant.

6 Symmetries and conservation laws

In this chapter we examine the symmetries of the laws of electromagnetism, and the conservation laws for energy, momentum, and angular momentum. We also touch on invariance of the laws under boosts, i.e., changes in the velocity of an inertial reference frame. This invariance is, of course, what is more commonly called relativity and is a basic requirement of any fundamental physical law. However, the laws of electromagnetism obey the modern relativity of Einstein, Poincare, Lorentz, et al., not of Galileo. We shall study modern relativity in greater detail in chapter 23. Here, we consider only frames moving at low relative speeds, i.e., at speeds much less than that of light.

We have already seen in section 32 that charge conservation is contained in Maxwell's equations, and we shall relate it to gauge invariance in section 80.

34 Discrete symmetries of the laws of electromagnetism

Since the laws of electromagnetism can be written in vector form, it follows at once that these laws obey rotational invariance, i.e., that they are the same for two observers who choose to orient their reference frames differently. Similarly, the laws are invariant under a translation of either space or time coordinates, i.e., they are the same for two observers who choose different origins for space or time. These invariances or symmetries are *continuous*, in that the amounts by which the frames are rotated or translated may be varied continuously, and they are basic requirements of any fundamental physical law. Each symmetry leads to a conservation law: time translation to energy conservation, spatial translation to momentum conservation, and rotation to angular momentum conservation.[1]

In addition to the above symmetries, we find that (barring weak interaction effects) the laws of physics are also invariant under the discrete symmetries of space–time, namely,

---

[1] The connection between continuous symmetries and conservation laws is formally brought out by Noether's theorem, for which see Goldstein, Poole, and Safko (1992), sec. 13.7.

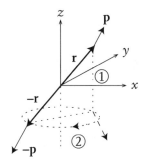

FIGURE 6.1. Equivalence of inversion, or parity, with rotoreflection. Suppose a particle is at point **r** moving with momentum **p**. A reflection in the xy plane (step 1), followed by a 180° rotation about the z axis (step 2) changes the position to $-\mathbf{r}$ and the momentum to $-\mathbf{p}$.

mirror reflection or *parity*, and *time* (or *motion*) *reversal*. In this section, we shall see how these symmetries are realized for electromagnetism.

Let us first consider how mechanical quantities such as a particle's position, momentum, angular momentum, etc., behave under these symmetries, starting with parity. The first point to note is that reflection in a mirror plane is equivalent to inversion of all coordinate and momentum components when it is combined with rotational symmetry, which we have already established. Consider a particle at **r** moving with a momentum **p**, as in fig. 6.1. We now perform a reflection in the xy plane, and follow it up with a 180° rotation about the z axis. Under these operations, the coordinates and momenta transform as

$$(x, y, z) \rightarrow (x, y, -z) \rightarrow (-x, -y, -z), \tag{34.1}$$

$$(p_x, p_y, p_z) \rightarrow (p_x, p_y, -p_z) \rightarrow (-p_x, -p_y, -p_z). \tag{34.2}$$

In short, $\mathbf{r} \rightarrow -\mathbf{r}$, $\mathbf{p} \rightarrow -\mathbf{p}$. It is easier to deal with inversion of all coordinates, rather than only one, since one does not have to specify the orientation of the mirror plane. Of course, by rotational invariance, if a law of physics is invariant under reflection in one particularly oriented mirror plane, it is invariant under reflection in any mirror plane.

Vectors such as **r** and **p**, which change sign under parity, are known as *true* or *polar* vectors, or just "vectors" without any qualifier. Consider, now, the angular momentum. This behaves as

$$\mathbf{L} = \mathbf{r} \times \mathbf{p} \rightarrow (-\mathbf{r}) \times (-\mathbf{p}) = \mathbf{L}. \tag{34.3}$$

Quantities such as this, which behave like vectors under rotations but do not change sign under parity, are known as *pseudo* or *axial* vectors. It should be noted that this behavior is entirely due to our definition of the cross product. We could, if we wanted, work entirely with the second-rank tensor

$$M_{ij} = \epsilon_{ijk} L_k = x_i p_j - x_j p_i, \tag{34.4}$$

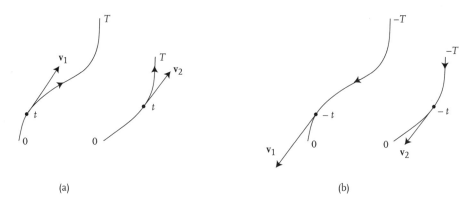

FIGURE 6.2. Motion reversal. Part (a) shows the trajectories of two particles under some time-reversal symmetric interaction, and part (b) shows the trajectories that would be followed if we reversed the velocities of both particles at some instant.

and then there would be no need for *pseudo* anything. The visual and notational benefits of the cross product are so great, however, that most physicists are unwilling to give it up.

Let us consider, finally, how an equation such as Newton's second law is parity invariant. We write it for a particle moving in a scalar potential U:

$$\frac{d\mathbf{p}}{dt} = -\nabla U. \tag{34.5}$$

Under parity, $\mathbf{p} \to -\mathbf{p}$, and $\nabla \to -\nabla$, but U, being a scalar, is unchanged. Thus, the equation as a whole is transformed to

$$\frac{d(-\mathbf{p})}{dt} = -(-\nabla U), \tag{34.6}$$

which is the same as before. One sometimes says that the law is *form invariant*.

Next, let us consider time reversal. (The alternative term, motion reversal, is preferred by some people.) Symmetry under this operation is most easily understood in terms of the following commonly given explanation. Imagine that we could make a motion picture of a distant solar system. Time-reversal symmetry means that if someone accidently ran this picture backward, he or she would not be able to discover that fact. In other words, the motion of the planets in the reversed movie would be completely consistent with the laws of mechanics. Another way of understanding time reversal is as follows. Let us consider two particles moving under the influence of some interaction such that their trajectories are as depicted in fig. 6.2. Starting with coordinates $\mathbf{r}_a$ $(a = 1, 2)$ and momenta $\mathbf{p}_a$ at time $t = 0$, we evolve to $\mathbf{r}'_a$ and $\mathbf{p}'_a$ at time T, i.e.,

$$(\mathbf{r}_a, \mathbf{p}_a)_{t=0} \to (\mathbf{r}'_a, \mathbf{p}'_a)_{t=T}. \tag{34.7}$$

Time-reversal symmetry means that if we start the particles at $\mathbf{r}'_a$ with *reversed* momenta $-\mathbf{p}'_a$, the particles will retrace their trajectory, i.e., they will reach positions $\mathbf{r}_a$ with

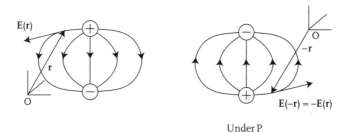

Under P

FIGURE 6.3. Effect of parity on electric field. On the left is shown a configuration of charges and the **E** field it produces. On the right is shown the same configuration under an inversion about the origin, O. The field produced by the inverted configuration is such that the electric field at an inverted point is the negative of the field at the original point: $\mathbf{E}(-\mathbf{r}) = -\mathbf{E}(\mathbf{r})$.

momenta $-\mathbf{p}_a$ a time T later. In symbols,

$$(\mathbf{r}'_a, -\mathbf{p}'_a)_{t=t_0} \rightarrow (\mathbf{r}_a, -\mathbf{p}_a)_{t=t_0+T}. \tag{34.8}$$

Note that we have written the starting time in the reversed motion as t_0, which we can choose to be an arbitrary number by time translation invariance. It is generally convenient to choose this constant so that the positions of the particles in the direct and reversed motions coincide at $t = 0$. Time reversal can then be formally written as a transformation under which

$$\mathbf{r} \rightarrow \mathbf{r}, \quad t \rightarrow -t, \quad \mathbf{p} \rightarrow -\mathbf{p}. \tag{34.9}$$

We could, in fact, have left out the rule for transforming $\mathbf{p}$ and derived it. Let us do this at low speeds. Then $\mathbf{p} = m d\mathbf{r}/dt$, which gets transformed to $m d\mathbf{r}/d(-t) = -m d\mathbf{r}/dt = -\mathbf{p}$. Readers can show the same result for themselves at speeds close to that of light.

Let us see how Newton's law, eq. (34.5), fares under time reversal. Since the potential is unchanged, we get

$$\frac{d(-\mathbf{p})}{d(-t)} = -\nabla U. \tag{34.10}$$

The two minus signs on the left-hand side cancel, leaving us with the same equation as before. Newton's law is time-reversal invariant.

We are now ready to see how $\mathbf{E}$ and $\mathbf{B}$ behave under P and T. (This is standard shorthand: one uses the letters P and T in lieu of the full words parity and time reversal, and quantities are said to be "even under P" or "odd under T" instead of "even under parity" or "odd under time reversal," etc.) Consider a static charge distribution $\rho(\mathbf{r})$ and the field $\mathbf{E}(\mathbf{r})$ it produces. The inverted distribution (see fig. 6.3) $\rho'(\mathbf{r}) = \rho(-\mathbf{r})$ clearly produces an oppositely directed field: $\mathbf{E}'(\mathbf{r}) = -\mathbf{E}(-\mathbf{r})$. T reversal, on the other hand, does nothing to the charge distribution, so $\mathbf{E}$ is unchanged. To see the behavior of $\mathbf{B}$, let us consider a steady current in a circular loop (fig. 6.4). Under parity, both the position and direction of current flow of every current element are reversed, so the current loop as a whole is

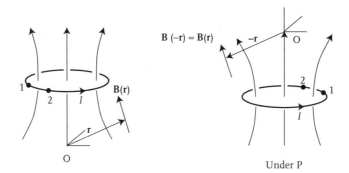

FIGURE 6.4. Effect of parity on magnetic field. On the left is shown a current loop and the **B** field it produces. On the right is shown the same current loop under an inversion about the origin, O. The direction of current flow in the inverted loop may be seen by noting that it flows from the point 1 to point 2 along the short arc connecting these points and the position of these points in the inverted loop. The field produced by the inverted loop is such that the magnetic field at an inverted point is the same as the field at the original point: $\mathbf{B}(-\mathbf{r}) = \mathbf{B}(\mathbf{r})$.

unaltered. We conclude that $\mathbf{B} \to \mathbf{B}$. Under time reversal, on the other hand, the current flow direction is reversed, so $\mathbf{B} \to -\mathbf{B}$.

The reader will have noted that to derive the P and T transformations of **E** and **B**, we assumed that the electric charge was invariant under both operations. Although it would be completely consistent to assume the opposite (and multiply the transformation laws for **E** and **B** by minus signs), the natural assumption is that charge is a scalar.

We show the P and T transformation properties of **E** and **B**, along with those of several other mechanical and electromagnetic quantities, in table 6.1. Note in particular that **B** behaves exactly like the angular momentum **L** under both P and T. The physical significance of some of the quantities will be discussed later.

Let us now see if the laws of electromagnetism obey P and T. Consider the Lorentz force law. Denoting the transformed fields by $\mathbf{E}'$ and $\mathbf{B}'$, the P-transformed law is

$$-\frac{d\mathbf{p}}{dt} = q\left(\mathbf{E}' - \frac{\mathbf{v}}{c} \times \mathbf{B}'\right). \tag{34.11}$$

But since $\mathbf{E}' = -\mathbf{E}$, $\mathbf{B}' = \mathbf{B}$, we recover the same law as before. In the same way, under T, the law is transformed to

$$\frac{d\mathbf{p}}{dt} = q\left(\mathbf{E}' - \frac{\mathbf{v}}{c} \times \mathbf{B}'\right). \tag{34.12}$$

But now $\mathbf{E}' = \mathbf{E}$, $\mathbf{B}' = -\mathbf{B}$, so we again recover the same law as before.

Likewise, if we consider Gauss's law, we will see that both sides of the equation are even under P and even under T, so that this law also obeys P and T. The other equations in eq. (32.8) can be examined in the same way and will all be found to be P and T invariant.

Table 6.1. Behavior of various mechanical and electromagnetic quantities under parity and time reversal

Quantity	Behavior under P	Behavior under T
$\mathbf{r}$	$-\mathbf{r}$	$\mathbf{r}$
$\mathbf{p}$	$-\mathbf{p}$	$-\mathbf{p}$
$\mathbf{L} = \mathbf{r} \times \mathbf{p}$	$\mathbf{L}$	$-\mathbf{L}$
$\mathcal{E} = \frac{1}{2}mv^2$ (e.g.)	$\mathcal{E}$	$\mathcal{E}$
ρ	ρ	ρ
$\mathbf{j}$	$-\mathbf{j}$	$-\mathbf{j}$
$\mathbf{E}$	$-\mathbf{E}$	$\mathbf{E}$
$\mathbf{B}$	$\mathbf{B}$	$-\mathbf{B}$
ϕ ($\mathbf{E} = -\nabla\phi$)	ϕ	ϕ
$\mathbf{A}$ ($\mathbf{B} = \nabla \times \mathbf{A}$)	$-\mathbf{A}$	$-\mathbf{A}$
$\mathbf{S} = \mathbf{E} \times \mathbf{B}$	$-\mathbf{S}$	$-\mathbf{S}$
$\dfrac{1}{8\pi}(E^2 + B^2)$	same	same
$\mathbf{E} \cdot \mathbf{B}$	$-\mathbf{E} \cdot \mathbf{B}$	$-\mathbf{E} \cdot \mathbf{B}$

35 Energy flow and the Poynting vector

Let us now examine how the concept of electromagnetic field energy is modified when we consider time-dependent fields. It is apparent that we must allow this energy density to change with time, which raises the issue of how it is conserved. We shall discover that this conservation is *local*, i.e., that energy flows from one place to another by passing through the intervening space, so there is a current associated with this. This is the Poynting vector.

Let us begin by asking how the expression for energy density that we have obtained by static considerations, $(E^2 + B^2)/8\pi$, changes with time. Using eq. (32.8), we have

$$\frac{1}{2c}\frac{\partial}{\partial t}(E^2 + B^2) = \frac{1}{c}\left(\mathbf{E} \cdot \frac{\partial \mathbf{E}}{\partial t} + \mathbf{B} \cdot \frac{\partial \mathbf{B}}{\partial t}\right)$$

$$= \mathbf{E} \cdot \left(\nabla \times \mathbf{B} - \frac{4\pi}{c}\mathbf{j}\right) - \mathbf{B} \cdot (\nabla \times \mathbf{E})$$

$$= -\frac{4\pi}{c}\mathbf{j} \cdot \mathbf{E} - \nabla \cdot (\mathbf{E} \times \mathbf{B}). \tag{35.1}$$

The current may always be regarded as a collection of charges in motion:

$$\mathbf{j} = \sum_a q_a \mathbf{v}_a \delta(\mathbf{r} - \mathbf{r}_a). \tag{35.2}$$

Substituting this in eq. (35.1) and integrating over a fixed volume V large enough to contain all the particles, we get

$$\frac{d}{dt} \int_V (\mathbf{E}^2 + \mathbf{B}^2) d^3x = -8\pi \sum_a q_a \mathbf{v}_a \cdot \mathbf{E}(\mathbf{r}_a) - 2c \int_{\partial V} (\mathbf{E} \times \mathbf{B}) \cdot d\mathbf{s}. \tag{35.3}$$

Now consider the $\mathbf{v} \cdot \mathbf{E}$ term for a single particle. Making use of the Lorentz force law, we have

$$q\mathbf{v} \cdot \mathbf{E} = \mathbf{v} \cdot \left(\frac{d\mathbf{p}}{dt} - q\frac{\mathbf{v}}{c} \times \mathbf{B} \right) = \mathbf{v} \cdot \frac{d\mathbf{p}}{dt}. \tag{35.4}$$

At low velocities, $\mathbf{v} = \mathbf{p}/m$, and the right-hand side is the time derivative of $\mathbf{p}^2/2m$, the kinetic energy. This continues to be true at high velocities. To show this, we must use the relativistically correct relations between energy ($\mathcal{E}$), momentum, and velocity,

$$\mathbf{p} = \frac{m\mathbf{v}}{\sqrt{1 - v^2/c^2}}, \tag{35.5}$$

$$\mathbf{v} = \frac{\mathbf{p}c}{\sqrt{p^2 + m^2c^2}}, \tag{35.6}$$

$$\mathcal{E} = (p^2c^2 + m^2c^4)^{1/2}. \tag{35.7}$$

With these relations,

$$\mathbf{v} \cdot \frac{d\mathbf{p}}{dt} = \frac{\mathbf{p}c}{\sqrt{p^2 + m^2c^2}} \cdot \frac{d\mathbf{p}}{dt}$$

$$= c\frac{d}{dt}(p^2 + m^2c^2)^{1/2}$$

$$= \frac{d\mathcal{E}}{dt}. \tag{35.8}$$

For our collection of particles, we get

$$\sum_a q_a \mathbf{v}_a \cdot \mathbf{E}(\mathbf{r}_a) = \frac{d}{dt}\mathcal{E}_{\text{tot,kin}}, \tag{35.9}$$

where $\mathcal{E}_{\text{tot, kin}}$ is the total kinetic energy of the particles. Equation (35.1) can therefore be written as

$$\frac{d}{dt}\left[\int_V \frac{\mathbf{E}^2 + \mathbf{B}^2}{8\pi} d^3x + \mathcal{E}_{\text{tot,kin}} \right] = -\oint_{\partial V} \mathbf{S} \cdot d\mathbf{s}, \tag{35.10}$$

$$\frac{d}{dt}\left[\int_V \frac{1}{2}\left(\epsilon_0 \mathbf{E}^2 + \mu_0^{-1}\mathbf{B}^2 \right) d^3x + \mathcal{E}_{\text{tot,kin}} \right] = -\oint_{\partial V} \mathbf{S} \cdot d\mathbf{s}, \quad \text{(SI)}$$

where

$$S = \frac{c}{4\pi}(E \times B),$$ (35.11)

$$S = \frac{1}{\mu_0}(E \times B). \quad \text{(SI)}$$

Equation (35.10) has exactly the form required of a conservation law. The quantity in square brackets is the total energy inside the volume V, mechanical plus electromagnetic. Its rate of change is the integral of a certain vector S over the surface bounding V. It is natural to interpret S, which is known as the *Poynting vector*, as the energy flux density, i.e., its normal component to a surface as the amount of energy crossing a unit area of that surface in unit time.

The most important aspect of eq. (35.10) is that it is local, i.e., that the energy in a given region of space can only change by flowing into or out of the adjoining space. The contrary possibility would be for energy to disappear from one region of space and appear simultaneously in another. That would represent a form of action at a distance, which we know to be incorrect. In the locality of eq. (35.10), we see yet another hint that the laws of electromagnetism as they were known in the late 1860s obeyed relativity from the outset.

There are two other comments to be made about our derivation of eq. (35.10). First, we have shown, as a by-product, that the quantity

$$\int j \cdot E \, d^3x$$ (35.12)

is the rate at which mechanical work is received by a system of charges and currents in an external field. Note that the magnetic field makes no contribution. This is because the force that it exerts on a particle is perpendicular to its velocity. This fact is also consistent with our earlier assertion that a steady current distribution (which does not produce any E field) radiates no energy. It produces only a B field, which cannot increase the energy of any test charges in its vicinity.

The second comment is that it is not obvious that the expressions for the energy density and the energy flux vector that we have found are the only possible ones. It turns out to be so only when one includes the considerations of relativity. The most simple-minded modification, adding the curl of some other vector to S without modifying the energy density, can also be ruled out because of incompatibility with angular momentum conservation.

The formula for the Poynting vector leads to certain seemingly paradoxical conclusions when we examine it further. For example, let us consider a particle such as an electron, which has both an electric charge, q, and magnetic dipole moment, m, and let the particle be at rest. The particle produces both an electric and a magnetic field, and the Poynting vector is easily found to be

$$S = \frac{c}{4\pi} \frac{q m \times r}{r^6},$$ (35.13)

which is nonzero. The particle is at rest and obviously cannot radiate any energy. And, indeed, $S \cdot \hat{r} = 0$, showing that no energy flows in the radial direction. Nevertheless, it is somewhat odd that energy is swirling around the particle, going around and around endlessly! It may become somewhat less bizarre to the reader when we consider angular momentum conservation (section 37).

Exercise 35.1 Write the law for energy conservation for a system of charged particles interacting purely electromagnetically in local form.

Exercise 35.2 Consider a parallel plate capacitor with circular plates of radius R and spacing d ($d \ll R$) that is fed by a sinusoidal current $I(t) = I_0 \cos \omega t$. Assume that $\omega \ll c/d$, so that radiation effects are negligible. **(a)** Find the electromagnetic fields in the region within the capacitor plates (far away from the edges so fringing effects may also be neglected) as follows. First, find $\mathbf{E}(\mathbf{r}, t)$ due to the instantaneous charge on the plates. Then use this $\mathbf{E}(\mathbf{r}, t)$ as the source term in the Ampere-Maxwell law to find $\mathbf{B}(\mathbf{r}, t)$. Third, solve the Faraday law with this $\mathbf{B}(\mathbf{r}, t)$ to find the corrections to the electric field found initially. Iterate this process a few times to obtain successive corrections to $\mathbf{E}(\mathbf{r}, t)$ and $\mathbf{B}(\mathbf{r}, t)$. What is the small parameter in the series so generated? **(b)** Find the energy density and the Poynting vector, going up to the second-order corrections in both $\mathbf{E}(\mathbf{r}, t)$ and $\mathbf{B}(\mathbf{r}, t)$. Is the direction of the Poynting vector physically reasonable? **(c)** Discuss energy conservation by considering the changes in the energy in a cylindrical volume of height d and radius $r < R$, and coaxial with the plates.

36 Momentum conservation

Relativity also impels the conclusion that the electromagnetic field carries momentum. A proper discussion of this point will have to wait till section 163, but we give the result now in view of its importance. It is a general rule that the momentum *density*, $\mathbf{g}$, is $1/c^2$ times the energy flux. We can verify this for a simple example using the formulas (35.5)–(35.7) above. Suppose we have a beam of particles all carrying the same momentum p traveling along the x direction, such that n of them cross a unit area of the yz plane per unit time. Let each particle have energy $\mathcal{E}$ and velocity v. In unit time, all the particles in a cylinder of unit base area and height v along the x direction will cross the yz plane (see fig. 6.5). Since there are n particles in this cylinder, and the volume of the cylinder is v, the momentum density is np/v. From eqs. (35.5)–(35.7), we see that $p/v = \mathcal{E}/c^2$. Thus, the momentum density is $n\mathcal{E}/c^2$, which is precisely $1/c^2$ times the energy flux $n\mathcal{E}$.

Applied to electromagnetic fields, this rule says that $\mathbf{g} = \mathbf{S}/c^2$, i.e.,

$$\mathbf{g} = \frac{1}{4\pi c} \mathbf{E} \times \mathbf{B}. \tag{36.1}$$

$$\mathbf{g} = \frac{1}{\mu_0 c^2} \mathbf{E} \times \mathbf{B}. \quad \text{(SI)}$$

Let us now study the conservation of momentum in light of the electromagnetic field momentum. We proceed exactly as we did for the energy and find the rate of change

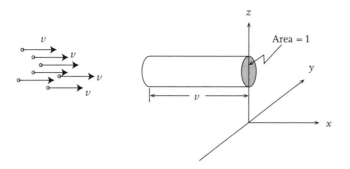

FIGURE 6.5. A swarm of particles, all moving along the x axis, with velocity **v**. In unit time, all particles in the cylinder will cross through the shaded area in the yz plane.

of **g**:

$$\frac{\partial g_i}{\partial t} = \frac{1}{4\pi c}\epsilon_{ijk}\left(\frac{\partial E_j}{\partial t}B_k + E_j\frac{\partial B_k}{\partial t}\right). \tag{36.2}$$

We now use the Faraday and Ampere-Maxwell laws to substitute for $\partial E_j/\partial t$ and $\partial B_j/\partial t$. After standard manipulations of the epsilon symbols, this yields

$$\frac{\partial g_i}{\partial t} = \frac{1}{4\pi}[(E_j\partial_j E_i - E_j\partial_i E_j) + (E \to B)] - \frac{1}{c}(\mathbf{j}\times\mathbf{B})_i. \tag{36.3}$$

We now rewrite the very first term, $E_j\partial_j E_i$ as $\partial_j(E_j E_i) - (\nabla\cdot\mathbf{E})E_i$, and use Gauss's law to write $\nabla\cdot\mathbf{E} = 4\pi\rho$. The same maneuver on the corresponding B term gives just $\partial_j(B_j B_i)$, since $\nabla\cdot\mathbf{B} = 0$. The second term, $-E_j\partial_i E_j$, may be written as $-\partial_i(E^2/2)$. Likewise for the corresponding B term. In this way, we get

$$\frac{\partial g_i}{\partial t} = -\frac{\partial}{\partial x_j}T_{ij} - \left(\rho E_i + \frac{1}{c}(\mathbf{j}\times\mathbf{B})_i\right), \tag{36.4}$$

where T_{ij} is the tensor,

$$T_{ij} = \frac{1}{4\pi}\left(\frac{1}{2}(E^2 + B^2)\delta_{ij} - E_i E_j - B_i B_j\right). \tag{36.5}$$

$$T_{ij} = \frac{1}{2}(\epsilon_0 E^2 + \mu_0^{-1}B^2)\delta_{ij} - \epsilon_0 E_i E_j - \mu_0^{-1}B_i B_j. \quad \text{(SI)}$$

It is apparent that this tensor is symmetric, i.e.,

$$T_{ij} = T_{ji}. \tag{36.6}$$

The physical content of eq. (36.4) can be seen by considering a collection of charged particles as before and integrating over a volume V containing them. The first term on the right can be transformed into a surface integral, and the second term becomes

$$-\sum_a\left(q_a\mathbf{E}(\mathbf{r}_a) + q_a\frac{\mathbf{v}_a}{c}\times\mathbf{B}(\mathbf{r}_a)\right)_i = -\sum_a\frac{dp_{a,i}}{dt} = -\frac{dP_{\text{mech},i}}{dt}, \tag{36.7}$$

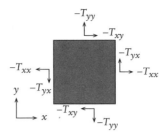

FIGURE 6.6. Forces on a small object and their relation to the stress tensor components.

where $\mathbf{P}_{\text{mech}}$ is the total mechanical momentum. Hence, our conservation law reads

$$\frac{d}{dt}\left[\int_V \mathbf{g}\,d^3x + \mathbf{P}_{\text{mech}}\right]_i = -\oint_{\partial V} T_{ij}\,d^2s_j. \tag{36.8}$$

The left-hand side is the rate of change of the ith component of the total momentum, mechanical plus electromagnetic, inside the volume V. Hence, the right-hand side must represent the amount of this momentum component flowing *into* V. More precisely, T_{ij} is the j flux of i momentum. Secondly, in mechanics, the rate of change of momentum of an object equals the force acting on that object. If we are to preserve this interpretation, $-T_{ij}$ must be interpreted as the ith component of force per unit area on a surface normal to $\mathbf{e}_j$. Note that the force acts on everything inside V. In fig. 6.6 we show the forces on a small region to clarify this point further. The meaning we have assigned to T_{ij} exactly parallels that of the stress tensor in the theory of elasticity. For this reason, T_{ij} is known as the *Maxwell stress tensor*. Note, however, that we must regard the electromagnetic stresses as existing in space, ready to act on a material surface if one should be present. This concept is perhaps not so strange if one recalls that this is more or less how the $\mathbf{E}$ and $\mathbf{B}$ fields were conceived initially.

[At this point, we should alert readers that there is a dispute between pointy-endians and rounded-endians on the sign of the stress tensor. Many authors (Jackson, for example) define it with a sign opposite of ours, while others (Landau and Lifshitz, for example) do it our way. People who find all the minus signs in fig. 6.6 ugly and would therefore like the x component of the force per unit area on a surface normal to $\hat{\mathbf{y}}$ to be T_{xy} instead of $-T_{xy}$, side with Jackson. Our definition seeks to preserve the parallel with energy conservation. The rate of increase of either quantity (energy or momentum) in a volume necessarily equals the flux of that quantity *into* the volume. Our T_{ij} is the momentum flux, not its negative. Further, we shall see in section 163 that in relativity the stress tensor generalizes to something called the stress–energy tensor. If we wish the time–time component of this tensor to be the energy density (and not the negative of the energy density), then the space–space components, which form just the stress tensor, must be defined with our sign choice.]

We can gain some more understanding of the stress tensor if we consider purely static fields. In that case, the force on an object is equal to the total momentum flowing *into* a volume surrounding that object, i.e., the right-hand side of eq. (36.8). Consider, for

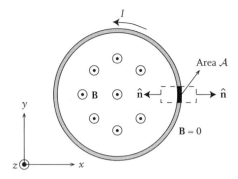

FIGURE 6.7. Stresses in a magnetic field. The figure shows an ideal solenoid in cross section. The force on a small piece of the frame, of area $\mathcal{A}$, is found by integrating the stress tensor over a surface (shown by the dashed line) that encloses this piece.

example, a long, thin solenoid made by winding the wire on a hollow cylindrical frame. The magnetic field is uniform and along the axis of the solenoid, which we take as $\hat{\mathbf{z}}$. The only nonvanishing components of the stress tensor inside the solenoid are

$$T_{xx} = T_{yy} = \frac{B^2}{8\pi}, \quad T_{zz} = -\frac{B^2}{8\pi}. \tag{36.9}$$

Let us now consider the net force on a small piece of the frame, of area $\mathcal{A}$, along the x axis. This is obtained by integrating $-T_{ij}$ over a surface surrounding the piece, as shown in fig. 6.7. T_{ij} vanishes outside the solenoid, the integrals from the surfaces parallel to the x axis cancel in pairs, and only the inner surface parallel to the frame contributes. On this surface, the outward normal $\hat{\mathbf{n}} = -\hat{\mathbf{x}}$, and, hence, the force has only an x component, given by

$$F_x = -\int_S T_{xx} n_x ds = \frac{B^2}{8\pi} \mathcal{A}. \tag{36.10}$$

Dividing by $\mathcal{A}$, we see that the frame experiences a pressure $B^2/8\pi$ tending to push it outward. This pressure is sometimes called magnetic pressure, and it can often be considerable. For a field of 2×10^5 G (20 T), e.g., this pressure is 1.6×10^9 dyn/cm^2, or about 1600 atmospheres!

Exercise 36.1 Verify eq. (36.3).

Exercise 36.2 Find the force on the plates of an ideal parallel plate capacitor.

Exercise 36.3 Consider a surface in the xy plane, and let $\mathbf{E}$ vanish below the surface ($z < 0$), and let it be in the yz plane above it ($z > 0$), at an angle θ to the z axis. (See fig. 6.8.) Find the force on a unit area of the surface.

Answer: $\mathbf{f} = (E^2/8\pi)(\cos 2\theta \hat{\mathbf{z}} + \sin 2\theta \hat{\mathbf{y}})$.

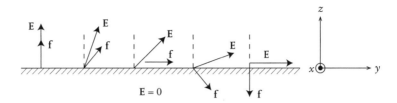

FIGURE 6.8. Force on a surface due to an electric field on one side of it.

Exercise 36.4 We shall see in section 109 that if a type I superconductor of any shape is placed in a low enough magnetic field, $\mathbf{B} = 0$ inside the superconductor (the flux expulsion or *Meissner effect*), while outside it, $\mathbf{B}$ is everywhere tangential to the surface. Find the force on any surface element of the superconductor (direction and magnitude).

Exercise 36.5 Find the force on two equal and opposite charges a distance $2a$ apart by integrating the stress tensor over the plane halfway between the charges. Repeat for two equal and like charges.

Exercise 36.6 Using the stress tensor, find the force per unit length on two parallel wires, both carrying equal currents in the same direction and separated by a distance $2a$. Repeat for equal currents in opposite directions.

37 Angular momentum conservation*

If $\mathbf{g}$ is the momentum density of the electromagnetic field, it is natural to regard

$$\mathcal{L} = \mathbf{r} \times \mathbf{g} \tag{37.1}$$

as the angular momentum density. Its conservation can be discussed in close parallel with that of momentum. Using eq. (36.4), we obtain

$$\frac{\partial \mathcal{L}_i}{\partial t} = -\epsilon_{ijk}\left[x_j \frac{\partial}{\partial x_m} T_{km} + x_j \left(\rho E_k + \frac{1}{c}(\mathbf{j} \times \mathbf{B})_k \right) \right]. \tag{37.2}$$

If we note that $\epsilon_{ijk} T_{jk} = 0$, we can write the first term as

$$\epsilon_{ijk}\left[\frac{\partial}{\partial x_m}(x_j T_{km}) - \delta_{jm} T_{km} \right] = \frac{\partial}{\partial x_m}(\epsilon_{ijk} x_j T_{km}). \tag{37.3}$$

This now has the form of a divergence,[2] so when it is integrated over a volume, we will be able to transform it to a surface integral of the tensor

$$M_{im} = \epsilon_{ijk} x_j T_{km}. \tag{37.4}$$

---

[2] This argument is sometimes used in reverse to deduce the symmetry of the stress tensor. One takes the position that angular momentum *has* to be conserved because of the isotropy of space and *concludes* that the rate of change of the angular momentum density must be a divergence. For this to be so, $\epsilon_{ijk} T_{jk}$ must vanish, i.e., T_{ij} must be symmetric.

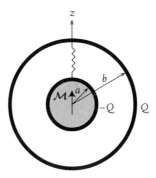

FIGURE 6.9. The Pugh and Pugh angular momentum making machine. The inner sphere is magnetic with a total moment $\mathcal{M}$ and coated with a thin metal film. It is concentric with an outer spherical metal shell. Initially the spheres are given charges $\pm Q$ as shown. At $t = 0$, the spheres are connected by a wire of high resistance running along the z axis. The resulting current flow exerts torques on both spheres, making them spin. Where does the angular momentum of the spheres come from?

The remaining term in eq. (37.2) describes the rate of change of the mechanical angular momentum. This fact comes out if we once again consider a collection of charges and integrate over a volume large enough to contain them all. For one charge, the mechanical term becomes

$$-\epsilon_{ijk}x_j\frac{dp_k}{dt} = -\left(\mathbf{r}\times\frac{d\mathbf{p}}{dt}\right)_i = -\frac{d}{dt}(\mathbf{r}\times\mathbf{p})_i, \tag{37.5}$$

since $\mathbf{v}\times\mathbf{p} = 0$. Adding together the contributions from all particles, we obtain the ith component of $-d\mathbf{L}_{\text{mech}}/dt$, where $\mathbf{L}_{\text{mech}}$ is the total mechanical angular momentum.

Collecting together all these manipulations, we thus arrive at the conservation law

$$\frac{d}{dt}\left[\int_V \mathcal{L}_i d^3x + L_{\text{mech},i}\right] = -\oint_{\partial V} M_{ij}\, d^2s_j. \tag{37.6}$$

This is interpreted in the same way as the law for conservation of momentum. The tensor M_{ij} has the meaning of flux of angular momentum.

Exercise 37.1 Suppose that we added to $\mathbf{S}$ the curl of some vector $\mathbf{V}$. Since the momentum density is $\mathbf{S}/c^2$, the expression for $\mathbf{g}$ would have to be modified, and so would that for $\mathcal{L}$. Show that momentum conservation would still hold, but angular momentum conservation would not.

Exercise 37.2 A very pretty thought experiment that shows the reality of electromagnetic angular momentum is due to Pugh and Pugh.[3] A solid spherical magnet with total magnetic moment $\mathcal{M}$ and radius a is coated with a thin metallic film and is concentric with a thin metal shell of radius b (fig. 6.9). Both spheres are mounted on spindles with

[3] E. W. Pugh and G. E. Pugh, *Am. J. Phys.* **35**, 153 (1967).

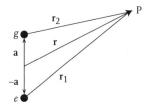

FIGURE 6.10. A mixed dipole.

negligible friction and are free to rotate independently about the z axis. The outer sphere is given a charge Q, and the inner a charge $-Q$, forming a spherical capacitor. At $t = 0$, the two spheres are connected by a wire of very high resistance running along the z axis. When the current starts to flow in the outer sphere, say, it does so in the presence of a magnetic field created by the inner magnet. A torque $\mathbf{N}(t)$ is thus exerted on the outer sphere, causing it to start spinning. Likewise for the inner sphere. Assuming that the current is so small that the charge distribution on each sphere may be taken to be completely uniform at all times, find the total angular momentum of the contraption when the spheres are fully discharged. Show that this equals the angular momentum in the EM field at $t = 0$.

Solution: Let the charge on the outer sphere at time t be $q(t)$. Since it is uniformly distributed, the amount that is south of the circle of colatitude θ is $(1 + \cos\theta)q(t)/2$, and the current flowing through this circle is $-(1 + \cos\theta)\dot{q}/2$. Thus, the torque on the circular ribbon lying between colatitudes θ and $\theta + d\theta$ is

$$-\frac{1 + \cos\theta}{2}\dot{q}\,\frac{2\mathcal{M}\cos\theta}{cb^3}b^2 \sin\theta\, d\theta\, \hat{\mathbf{z}}. \tag{37.7}$$

Integrating over all ribbons and over time, we see that the final angular momentum of the outer sphere is $\mathbf{L}_{\text{out}} = (2Q\mathcal{M}/3bc)\hat{\mathbf{z}}$. Likewise, for the inner sphere $\mathbf{L}_{\text{in}} = -(2Q\mathcal{M}/3ac)\hat{\mathbf{z}}$, and the total final mechanical angular momentum is

$$\mathbf{L}_{\text{final}} = -\frac{2}{3}\frac{Q\mathcal{M}}{c}\left(\frac{1}{a} - \frac{1}{b}\right)\hat{\mathbf{z}}. \tag{37.8}$$

It is easy to find the initial electromagnetic angular momentum and see that it is the same.

Another interesting conclusion is reached by supposing that magnetic monopoles exist and considering a "mixed dipole," consisting of an electric charge e and magnetic charge g. Let us take the charges to be at the points $\pm\mathbf{a}$ with respect to the origin (fig. 6.10), so that

$$\mathbf{E} = \frac{e}{r_1^3}\mathbf{r}_1, \quad \mathbf{B} = \frac{g}{r_2^3}\mathbf{r}_2. \tag{37.9}$$

The angular momentum density in the field is then

$$\mathcal{L} = \frac{1}{4\pi c}\mathbf{r} \times (\mathbf{E} \times \mathbf{B}) = \frac{eg}{2\pi c r_1^3 r_2^3}[r^2\mathbf{a} - \mathbf{r}(\mathbf{r} \cdot \mathbf{a})]. \tag{37.10}$$

The total angular momentum is obtained by integrating this expression over all space. If we take $\mathbf{a} \| \hat{\mathbf{z}}$, it is obvious by symmetry that only the z component of the total angular momentum is nonzero. This is given by

$$L_z^{tot} = \frac{aeg}{2\pi c} \int d^3x \frac{x^2 + y^2}{|\mathbf{r} + a\hat{\mathbf{z}}|^3 |\mathbf{r} - a\hat{\mathbf{z}}|^3}. \tag{37.11}$$

To evaluate the integral, we employ Parseval's theorem. We note that the integrand can be written as the product $f_+(\mathbf{r})f_-(\mathbf{r})$, where

$$f_\pm(\mathbf{r}) = \frac{x \pm iy}{[x^2 + y^2 + (z \pm a)^2]^{3/2}}. \tag{37.12}$$

Let the Fourier transforms of $f_\pm(\mathbf{r})$ be denoted by $\tilde{f}_\pm(\mathbf{k})$. Since

$$f_\pm(\mathbf{r}) = -(\partial_x \pm i\partial_y)\frac{1}{[x^2 + y^2 + (z \pm a)^2]^{1/2}}, \tag{37.13}$$

we have

$$\tilde{f}_\pm(\mathbf{k}) = -i(k_x \pm ik_y) \times \text{F.T.}\left[\frac{1}{|\mathbf{r} \pm a\hat{\mathbf{z}}|}\right]$$

$$= -i(k_x \pm ik_y)\frac{4\pi}{k^2}e^{\mp i\mathbf{k} \cdot a\hat{\mathbf{z}}}. \tag{37.14}$$

Parseval's theorem now yields

$$\int d^3x f_-(\mathbf{r})f_+(\mathbf{r}) = \int \frac{d^3k}{(2\pi)^3}\tilde{f}_-(\mathbf{k})\tilde{f}_+(-\mathbf{k}) \tag{37.15}$$

$$= 4\int_{-\infty}^{\infty}\int_0^{\infty}\frac{k_\perp^2 e^{2iak_z}}{(k_z^2 + k_\perp^2)^2}k_\perp dk_\perp dk_z, \tag{37.16}$$

where we have used eq. (37.14) and employed cylindrical polar coordinates in k space in the last line. Next, let us do the k_z integral. A simple contour integration yields

$$\int_{-\infty}^{\infty}\frac{e^{2iak_z}}{(k_z^2 + k_\perp^2)^2}dk_z = \frac{\pi}{2k_\perp^3}(2k_\perp a + 1)e^{-2k_\perp a}. \tag{37.17}$$

Hence,

$$\int d^3x f_-(\mathbf{r})f_+(\mathbf{r}) = 2\pi\int_0^{\infty}e^{-2k_\perp a}(2k_\perp a + 1)dk_\perp = \frac{2\pi}{a}, \tag{37.18}$$

and the total z angular momentum is given by

$$L_z^{tot} = \frac{eg}{c}. \tag{37.19}$$

The result is independent of the separation $2a$.

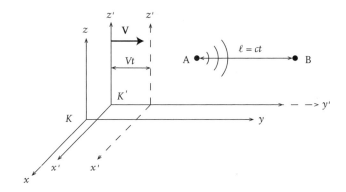

FIGURE 6.11. Speed of light in Galilean relativity. The frame K' is moving at velocity $\mathbf{V}$ relative to K. A light signal is sent from A to B, two points fixed in the K frame, a distance ℓ apart. As per K or K', the light transit time is $t = \ell/c$. In this time, K' has advanced a distance Vt, so K' measures the distance traveled by the light as $\ell - Vt$ and assigns the light a speed of $(\ell - Vt)/t = c - V$.

If we now use the quantum mechanical result that angular momentum is quantized in units of $\hbar/2$, we conclude that

$$\frac{eg}{c} = n\frac{\hbar}{2}, \tag{37.20}$$

where n is a positive or a negative integer. This conclusion was reached by Dirac in 1931 in a different way. If a magnetic monopole were to exist, Dirac's result would show that the electric charge of any particle would have to be an integer multiple of $c\hbar/2g$. In particular, we would have a reason for why the charges of the electron and proton are equal in magnitude, a fact that is otherwise quite mysterious. The beauty of this idea has led to searches for magnetic monopoles ever since, but so far none have been found.

38 Relativity at low speeds

As we have said many times, electromagnetism does not obey Galilean relativity. Among other things, this relativity would imply an observer-dependent speed of light. Suppose there are two observers K and K', as shown in fig. 6.11, with K' moving at a velocity V relative to K along the y direction. Let a signal be sent from point A to B, and let K measure the speed of this signal to be c. Galilean relativity would imply that K' would measure the speed of this signal to be $c - V$. Likewise, if a signal sent from B to A had a speed c per K, it would have a speed $c + V$ per K'. We know that this value is false for a light signal.

If $V \ll c$, however, the differences in the speeds may be ignored, and Galilean invariance is approximately true. It is therefore useful to investigate its implications further, in particular the relation between the electric and magnetic fields measured by observers in relative motion.

To obtain the transformation law for $\mathbf{E}$, suppose that in frame K, a charge q is at some point P, moving with velocity $\mathbf{v}$. The Lorentz force on the particle is

$$\mathbf{F} = q\left(\mathbf{E} + \frac{\mathbf{v}}{c} \times \mathbf{B}\right), \tag{38.1}$$

where $\mathbf{E}$ and $\mathbf{B}$ are the fields at P. Let K' be the comoving frame, i.e., moving at velocity $\mathbf{v}$ relative to K. In K', the particle is at rest, and so the force on it is

$$\mathbf{F}' = q\mathbf{E}'. \tag{38.2}$$

Since $\mathbf{F}' = \mathbf{F}$ in Galilean relativity,

$$\mathbf{E}' = \mathbf{E} + \frac{\mathbf{v}}{c} \times \mathbf{B}. \tag{38.3}$$

$$\mathbf{E}' = \mathbf{E} + \mathbf{v} \times \mathbf{B}. \quad \text{(SI)}$$

To obtain the transformation law for $\mathbf{B}$, we will use the Ampere-Maxwell law in a charge- and current-free region. In the frame K, this reads

$$\nabla \times \mathbf{B} - \frac{1}{c}\frac{\partial \mathbf{E}}{\partial t} = 0. \tag{38.4}$$

The law must have the same form in frame K'. To exploit this fact, we first note that the coordinates of a point in space and the times of events as measured by the two observers are related by

$$\mathbf{r}' = \mathbf{r} - \mathbf{v}t, \quad t' = t. \tag{38.5}$$

Hence,

$$\frac{\partial}{\partial t} = \frac{\partial}{\partial t'} - \mathbf{v} \cdot \nabla', \quad \nabla = \nabla'. \tag{38.6}$$

Equation (38.4) can now be rewritten as

$$0 = \nabla' \times \mathbf{B} - \frac{1}{c}\left(\frac{\partial}{\partial t'} - \mathbf{v} \cdot \nabla'\right)\mathbf{E}$$

$$= \nabla' \times \mathbf{B} - \frac{1}{c}\frac{\partial \mathbf{E}'}{\partial t'} - \frac{\mathbf{v}}{c} \cdot \nabla'\mathbf{E}', \tag{38.7}$$

where we have replaced $\mathbf{E}$ by $\mathbf{E}'$, since the corrections are of order $1/c^2$. Because $\nabla' \cdot \mathbf{E}' = 0$, we can write the last term as $\nabla' \times (\mathbf{v} \times \mathbf{E}')/c$ and again replace $\mathbf{E}'$ by $\mathbf{E}$ ignoring corrections of order $1/c^2$. Thus,

$$\nabla' \times \left(\mathbf{B} - \frac{\mathbf{v}}{c} \times \mathbf{E}\right) - \frac{\partial \mathbf{E}'}{\partial t'} = 0. \tag{38.8}$$

If this is to be the Ampere-Maxwell law in K', we must have

$$\mathbf{B}' = \mathbf{B} - \frac{\mathbf{v}}{c} \times \mathbf{E}. \tag{38.9}$$

$$\mathbf{B}' = \mathbf{B} - \frac{\mathbf{v}}{c^2} \times \mathbf{E}. \quad \text{(SI)}$$

39 Electromagnetic mass*

We saw in section 18 that the energy stored in the electrostatic field of a point electron diverges. Let us now explore the consequences of trying to regulate this divergence. We take the electron as a sphere of radius a, with some distribution of charge, whose details may or may not matter. The electrostatic energy, which we now denote by $\mathcal{E}_{em}$, is

$$\mathcal{E}_{em} = \frac{e^2}{2a}\zeta, \tag{39.1}$$

where ζ is a number of order unity. Let us now imagine that this electron is moving uniformly at a very low velocity $\mathbf{v}$. By the results of the previous section, the electric field is not modified to first order in $\mathbf{v}$, and the magnetic field is given by

$$\mathbf{B} = \frac{1}{c}\mathbf{v} \times \mathbf{E}. \tag{39.2}$$

Hence, to first order in $\mathbf{v}$, the energy in the field is not modified. However, it now has a net momentum. The momentum density is given by

$$\mathbf{g} = \frac{1}{4\pi c}(\mathbf{E} \times \mathbf{B}) = \frac{1}{4\pi c^2}[\mathbf{v}E^2 - \mathbf{E}(\mathbf{E} \cdot \mathbf{v})]. \tag{39.3}$$

We integrate this over all space to obtain the total momentum $\mathbf{p}_{em}$. The integral of the second term is $1/3$ of the first term by spherical symmetry, so

$$\mathbf{p}_{em} = \frac{1}{6\pi c^2}\int E^2 \mathbf{v}\, d^3x = \frac{4}{3}\frac{\mathcal{E}_{em}\mathbf{v}}{c^2}. \tag{39.4}$$

By the relativistic equivalence of mass and energy, the quantity $\mathcal{E}_{em}/c^2$ should be viewed as an electromagnetic mass m_{em}. It was at one time hoped that this electromagnetic field energy surrounding any charged particle was the origin of mass! There are two problems with this idea, at least in its simplest form. First, we find the electron radius to be of order e^2/mc^2, the classical electron radius, $\sim 3 \times 10^{-13}$ cm, well above all experimental upper bounds. Second, the coefficient between the momentum and velocity is $4/3$ this mass, and not the mass. This discrepancy is particularly bad. Let us suppose that the electron is set moving from rest by some external agency. We have shown that momentum is conserved only if the field momentum is included. Therefore, the external agency must provide this momentum, which we could measure by measuring its recoil. We could also measure the velocity of the electron, and the ratio would give us its mass. This ratio is turning out to have an extra factor of $4/3$.

One solution to this problem was given by Poincare. He observed that a charge distribution would fly apart because of Coulomb repulsion unless it was held together by other forces. When these forces, known as *Poincare stresses*, are included, we obtain a negative nonelectromagnetic contribution to the mass. When added to the electromagnetic one, we presumably obtain the experimentally measured mass. Further, it turns out that both the energy and momentum are modified in such a way that the mass–energy relation and the momentum–velocity relation lead to the same answer for the mass. However, there is not the slightest evidence for internal stresses in the electron! In sum, there is no satisfactory theory for mass at present.

This problem foreshadows the logical inconsistencies in classical electrodynamics. One could live with the self-energy for electrons at rest and in uniform motion, because this is independent of the velocity, and even if it is infinite, one could argue that this is not observable because it never changes. However, the problem reappears ever more acutely when we consider accelerated motion, as we shall discuss in section 67. Nevertheless, with careful understanding of its limitations, we obtain a theory with a very wide range of applicability.

7 | Electromagnetic waves

We mentioned in chapter 1 that the electromagnetic field equations possess nontrivial solutions even when no sources are present and that these solutions describe electromagnetic waves, or light. In this chapter, we study the properties and propagation of these waves. The simplest solutions are *plane waves*, in which the fields depend only on one spatial coordinate and time. We also discuss the polarization of light, which relates to the vectorial nature of the EM fields. Nonplane waves are introduced via the geometrical optics approximation, which establishes the connection with the elementary description of light in terms of rays that propagate in straight lines. To illustrate these ideas, we study the fields in a laser beam. Other topics include the representation of the free EM field as a collection of oscillators, and the angular momentum in the free EM field.

40 The wave equation for E and B

In vacuum, and in the absence of sources, the equations of electromagnetism are (Gaussian system)

$$\nabla \cdot \mathbf{E} = 0,$$

$$\nabla \cdot \mathbf{B} = 0,$$

$$\nabla \times \mathbf{E} = -\frac{1}{c}\frac{\partial \mathbf{B}}{\partial t},$$

$$\nabla \times \mathbf{B} = \frac{1}{c}\frac{\partial \mathbf{E}}{\partial t}.$$

Taking the curl of the third equation, on the left we get

$$\nabla \times (\nabla \times \mathbf{E}) = \nabla(\nabla \cdot \mathbf{E}) - \nabla^2 \mathbf{E} = -\nabla^2 \mathbf{E} \tag{40.1}$$

(using $\nabla \cdot \mathbf{E} = 0$), while on the right we get

$$-\frac{1}{c}\frac{\partial}{\partial t}\nabla \times \mathbf{B} = -\frac{1}{c^2}\frac{\partial^2 \mathbf{E}}{\partial t^2},$$ (40.2)

making use of the fourth field equation (Faraday's equation). Thus, we have

$$\nabla^2 \mathbf{E} - \frac{1}{c^2}\frac{\partial^2 \mathbf{E}}{\partial t^2} = 0.$$ (40.3)

In the same way, taking the curl of the Faraday equation leads to

$$\nabla^2 \mathbf{B} - \frac{1}{c^2}\frac{\partial^2 \mathbf{B}}{\partial t^2} = 0.$$ (40.4)

If the same operations are performed on the equations in the SI system, we obtain

$$\nabla^2 \mathbf{E} - \mu_0 \epsilon_0 \frac{\partial^2 \mathbf{E}}{\partial t^2} = 0,$$ (40.5)

$$\nabla^2 \mathbf{B} - \mu_0 \epsilon_0 \frac{\partial^2 \mathbf{B}}{\partial t^2} = 0.$$ (40.6)

Comparing these with the Gaussian system equations, we conclude that the constants μ_0 and ϵ_0 must be related to the speed of light by

$$\mu_0 \epsilon_0 = \frac{1}{c^2}.$$ (40.7)

Equations (40.3) and (40.4) are both *vector wave equations*. The reader may recall (or verify directly) that the wave equation for a scalar field in one dimension,

$$\frac{\partial^2 f}{\partial x^2} - \frac{1}{c^2}\frac{\partial^2 f}{\partial t^2} = 0,$$ (40.8)

possesses as a general solution

$$f(x, t) = f_1(x - ct) + f_2(x + ct),$$ (40.9)

where f_1 and f_2 are arbitrary functions. The first term, $f_1(x - ct)$, represents a disturbance or wave form traveling to the right at speed c. To see this, consider it at a point x_1 at time t_1, and at another point $x_2 = x_1 + a$ at time t_2. The argument of f_1 will be unchanged if $t_2 - t_1 = a/c$. In the same way we see that the term $f_2(x + ct)$ represents a disturbance moving to the left.

It is clear that similar nonzero solutions will exist for the $\mathbf{E}$ and $\mathbf{B}$ fields obeying the vector wave equations. However, the behavior now is much richer, first, because the equations are three dimensional, and second, because $\mathbf{E}$ and $\mathbf{B}$ are not independent but coupled together by one of the first-order field equations, e.g., Faraday's equation. We can obtain an uncoupled description in terms of the potentials. Let us work in the Coulomb gauge,

$$\nabla \cdot \mathbf{A} = 0.$$ (40.10)

But now the equation $\nabla \cdot \mathbf{E} = 0$ becomes $\nabla^2 \phi = 0$, which has as a solution

$$\phi = 0.$$ (40.11)

Thus, $\mathbf{B} = \nabla \times \mathbf{A}$, $\mathbf{E} = -(1/c)\partial\mathbf{A}/\partial t$. Substituting these expressions into Faraday's equation and using $\nabla \cdot \mathbf{A} = 0$ once more, we obtain

$$\nabla^2 \mathbf{A} - \frac{1}{c^2}\frac{\partial^2 \mathbf{A}}{\partial t^2} = 0. \tag{40.12}$$

The general solution for $\mathbf{E}$ and $\mathbf{B}$ is then obtained by applying the operations of $\partial/\partial t$ and curl to the general solution for $\mathbf{A}$. Of course, by applying these operations to eq. (40.12) directly, we obtain the wave equations for the fields.

41 Plane electromagnetic waves

The simplest type of electromagnetic wave is a plane wave, in which the fields vary only in one direction. Let us take this direction to be z. We shall see that this is also the direction of propagation of the wave.

For plane waves, the equations for $\mathbf{E}$ and $\mathbf{B}$ can be solved directly without great effort, and we need not solve for $\mathbf{A}$ first. Let us start by examining E_z. The wave equation for this component becomes

$$\frac{\partial^2 E_z}{\partial z^2} - \frac{1}{c^2}\frac{\partial^2 E_z}{\partial t^2} = 0, \tag{41.1}$$

while the condition $\nabla \cdot \mathbf{E} = 0$ implies that

$$\frac{\partial E_z}{\partial z} = 0. \tag{41.2}$$

Hence, E_z is independent of z, and the wave equation reduces to $\partial^2 E_z/\partial t^2 = 0$. Integrating, we obtain $E_z = at + b$, where a and b are constants. The first term is an unphysical solution, as it implies an electric field growing without bound with time. Hence, we must set $a = 0$, leaving $E_z = b$. This, however, is a trivial uniform field that can be superimposed on the wave solution that we will find below. We ignore this henceforth, i.e., we set

$$E_z = 0. \tag{41.3}$$

In the same way, we can also set $B_z = 0$.

The other two components of $\mathbf{E}$ also satisfy a one-dimensional wave equation. From this point on, let us consider only a right-moving solution for these components. From the Ampere-Maxwell equation, we then see that $\mathbf{B}$ must also be a right-moving wave. That is, we may take $\mathbf{E} = \mathbf{E}(z - ct)$ and $\mathbf{B} = \mathbf{B}(z - ct)$, where the vectors are both in the xy plane. The Faraday and Ampere-Maxwell equations imply that

$$\hat{\mathbf{z}} \times \mathbf{E}' = \mathbf{B}', \quad \hat{\mathbf{z}} \times \mathbf{B}' = -\mathbf{E}', \tag{41.4}$$

$$\hat{\mathbf{z}} \times \mathbf{E}' = c\mathbf{B}', \quad \hat{\mathbf{z}} \times \mathbf{B}' = -\mu_0\epsilon_0 c\mathbf{E}', \quad \text{(SI)}$$

where primes denote differentiation with respect to $z - ct$. Integration of these equations yields $\mathbf{B} = \hat{\mathbf{z}} \times \mathbf{E} + \text{const}$. The constant can again be set to zero, as it is a superposition of a trivial uniform and static solution. In this way we conclude that in a plane wave traveling along a general direction, $\hat{n}$,

$$\mathbf{E} \perp \hat{n}, \quad \mathbf{B} \perp \hat{n}, \quad \mathbf{B}(\mathbf{r}, t) = \hat{n} \times \mathbf{E}(\mathbf{r}, t). \tag{41.5}$$

$$\mathbf{E} \perp \hat{n}, \quad \mathbf{B} \perp \hat{n}, \quad \mathbf{B}(\mathbf{r}, t) = \frac{1}{c}\hat{n} \times \mathbf{E}(\mathbf{r}, t). \quad \text{(SI)}$$

Both $\mathbf{E}$ and $\mathbf{B}$ are transverse to the direction of propagation and are mutually orthogonal at all times. Further, $\mathbf{E}$ and $\mathbf{B}$ (or $c\mathbf{B}$ in SI) have equal magnitudes at the same space–time point.

Exercise 41.1 Find the electromagnetic fields in a right-moving plane wave by solving for the vector potential first.

Answer: In the Gaussian system, $\mathbf{A} = \mathbf{A}(z - ct)$, $\mathbf{A} \perp \hat{n}$, $\mathbf{E} = \mathbf{A}'(z - ct)$, $\mathbf{B} = \hat{n} \times \mathbf{A}'(z - ct)$.

Let us now compute the energy density, energy flux, and momentum density associated with a plane wave. The energy density is

$$u = \frac{1}{8\pi}(E^2 + B^2) = \frac{1}{4\pi}E^2. \tag{41.6}$$

$$u = \tfrac{1}{2}(\epsilon_0 E^2 + \mu_0^{-1} B^2) = \epsilon_0 E^2. \quad \text{(SI)}$$

The energy flux is given by the Poynting vector,

$$\mathbf{S} = \frac{c}{4\pi}\mathbf{E} \times \mathbf{B} = \frac{c}{4\pi}E^2\hat{n}. \tag{41.7}$$

$$\mathbf{S} = \frac{1}{\mu_0}\mathbf{E} \times \mathbf{B} = \epsilon_0 c E^2\hat{n}. \quad \text{(SI)}$$

The magnitude of the energy flux is also known as the *intensity* of the light.

The momentum density $\mathbf{g}$ is given by $\mathbf{S}/c^2$. Using the above results for $\mathbf{S}$ and u, we obtain

$$\mathbf{g} = \frac{u}{c}\hat{n}. \quad \text{(Gaussian and SI)} \tag{41.8}$$

Equation (41.8) relates the momentum density of an electromagnetic plane wave to its energy density (which is not the same thing as relating the momentum density to the energy *flux*, which can be done generally). This relationship is the same as that for a zero-mass particle in special relativity:

$$\mathcal{E} = (p^2 c^2 + m^2 c^4)^{1/2} \to pc \quad \text{(as } m \to 0). \tag{41.9}$$

Further, we note that such a particle can move only at the speed c. For, if we look at the formulas for energy and momentum for a massive particle

$$\mathcal{E} = \frac{mc^2}{\sqrt{1 - v^2/c^2}}, \quad p = \frac{mv}{\sqrt{1 - v^2/c^2}}, \tag{41.10}$$

we see that if we put $m = 0$, we get the trivial result $\mathcal{E} = p = 0$, unless we simultaneously have $v = c$. In that case p and $\mathcal{E}$ are not determined by the speed but are related by $\mathcal{E} = pc$. The result (41.8) implies that in an electromagnetic plane wave, we may think of the region of the field as filled by zero-mass particles, all traveling along the direction $\hat{n}$ at the speed of light. Such an interpretation arises naturally when we consider the quantum theory of light, and the particles are known as *photons*.

Let us also find the Maxwell stress tensor (36.5) for a plane wave. Take the x axis parallel to $\mathbf{E}$. Then $\mathbf{B} \parallel \hat{y}$, and

$$T_{xx} = \frac{1}{4\pi} \left(\frac{1}{2}(E^2 + B^2) - E_x^2 \right), \tag{41.11}$$

which vanishes by virtue of eq. (41.5). Likewise, $T_{yy} = 0$. Next, T_{xy} is proportional to $(E_x E_y + B_x B_y)$, which also vanishes. The components T_{xz} and T_{yz} are clearly zero too. The only nonvanishing component is

$$T_{zz} = \frac{1}{4\pi} E^2 = u. \tag{41.12}$$

Thus, if we put a perfectly absorbing surface normal to the direction of propagation of light in its path, it will experience a force per unit area, or pressure, equal to the energy density in the light. This pressure, known as *light pressure*, is experimentally detectable.

We can also ask for the angular momentum carried by a plane electromagnetic wave. See section 47.

Exercise 41.2 (blackbody radiation) Consider a cavity whose walls are maintained at temperature T. The atoms in the cavity walls will emit and reabsorb light until the cavity is filled with radiation in equilibrium with the walls. This is known as *blackbody radiation*, and its spectral distribution, i.e., the energy per unit volume in a frequency range $d\omega$ about ω, is given by Planck's formula,

$$du = \frac{\hbar \omega^3}{\pi^2 c^3} \frac{1}{e^{\hbar \omega / k_B T} - 1} d\omega. \tag{41.13}$$

(a) Show that the energy density is given by the *Stefan-Boltzmann law*, $u = \sigma_{SB} T^4$, where $\sigma_{SB} = \pi^2 k_B^4 / 15 \hbar^3 c^3$. (Use $\int_0^\infty x^3 (e^x - 1)^{-1} dx = \pi^4/15$.) (b) Find $I(\omega, \hat{n})$, the energy flux of radiation traveling along $\hat{n}$ per unit frequency per unit solid angle. (c) Find the Maxwell stress tensor and the pressure on the walls of the cavity. (d) Suppose a small metal disk of area A is suspended inside the cavity, and one side is painted black (perfectly absorbing) and the other side silver (perfectly reflecting). Find the net force on the disk.

42 Monochromatic plane waves and polarization

In the previous section, we did not specify the form of the functions $\mathbf{E}(z - ct)$ or $\mathbf{B}(z - ct)$. A particularly important case is where this function is periodic in time with the simple form $\sin(\omega t)$ or $\cos(\omega t)$. Such plane waves are called monochromatic. Every other plane wave can be resolved in a Fourier series or Fourier integral of monochromatic waves.

The time dependence $\sin(\omega t)$ means that the fields must be a linear combination of $\sin(kz - \omega t)$ and $\cos(kz - \omega t)$, where $k = \omega/c$. It is mathematically easier to deal with complex exponentials and so we write, for the **E** field, e.g.,

$$\mathbf{E}(\mathbf{r}, t) = \boldsymbol{\mathcal{E}}_0 e^{i(\mathbf{k}\cdot\mathbf{r}-\omega t)}, \tag{42.1}$$

where

$$\mathbf{k} = \frac{\omega}{c}\hat{\mathbf{n}}, \quad \boldsymbol{\mathcal{E}}_0 \perp \hat{\mathbf{n}}. \tag{42.2}$$

The **B** field is then given by eq. (41.5):

$$\mathbf{B}(\mathbf{r}, t) = \boldsymbol{\mathcal{B}}_0 e^{i(\mathbf{k}\cdot\mathbf{r}-\omega t)}; \quad \boldsymbol{\mathcal{B}}_0 = \hat{\mathbf{k}} \times \boldsymbol{\mathcal{E}}_0. \tag{42.3}$$

Note that $\boldsymbol{\mathcal{E}}_0$ and $\boldsymbol{\mathcal{B}}_0$ are complex vectors, i.e., their components are complex numbers. It is to be understood that we must take the real part of an expression such as eq. (42.1) to obtain physically meaningful numbers. Note also that $\hat{\mathbf{k}} = \hat{\mathbf{n}}$.

We wish to consider the time averages of the energy density and the Poynting vector at a fixed point in space. The relevant expressions are in eqs. (41.6) and (41.7), both of which are bilinear in the fields. In appendix C we discuss how to obtain averages of such bilinear expressions starting directly with complex exponentials. Using these methods, we get

$$\bar{u} = \frac{1}{8\pi}\boldsymbol{\mathcal{E}}_0 \cdot \boldsymbol{\mathcal{E}}_0^*, \tag{42.4}$$

$$\bar{\mathbf{S}} = \frac{c}{8\pi}\boldsymbol{\mathcal{E}}_0 \cdot \boldsymbol{\mathcal{E}}_0^* \hat{\mathbf{n}}. \tag{42.5}$$

To proceed further, let us write the complex vector $\boldsymbol{\mathcal{E}}_0$ in the form

$$\boldsymbol{\mathcal{E}}_0 = (\mathbf{E}_{0r} + i\mathbf{E}_{0i})e^{-i\alpha}, \tag{42.6}$$

where $\mathbf{E}_{0r}$ and $\mathbf{E}_{0i}$ are real vectors, and α is real. This can always be done, since it amounts to multiplying $\boldsymbol{\mathcal{E}}_0$ by $e^{i\alpha}$. The freedom to choose α is now exploited to make $\mathbf{E}_{0r}$ and $\mathbf{E}_{0i}$ orthogonal to each other. From eq. (42.6) we first obtain (writing $\boldsymbol{\mathcal{E}}_{0r}$ and $\boldsymbol{\mathcal{E}}_{0i}$ for the real and imaginary parts of $\boldsymbol{\mathcal{E}}_0$)

$$\mathbf{E}_{0r} = \boldsymbol{\mathcal{E}}_{0r}\cos\alpha - \boldsymbol{\mathcal{E}}_{0i}\sin\alpha,$$

$$\mathbf{E}_{0i} = \boldsymbol{\mathcal{E}}_{0r}\sin\alpha + \boldsymbol{\mathcal{E}}_{0i}\cos\alpha. \tag{42.7}$$

Hence,

$$\mathbf{E}_{0r} \cdot \mathbf{E}_{0i} = \tfrac{1}{2}(\mathcal{E}_{0r}^2 - \mathcal{E}_{0i}^2)\sin 2\alpha + \boldsymbol{\mathcal{E}}_{0r} \cdot \boldsymbol{\mathcal{E}}_{0i}\cos 2\alpha, \tag{42.8}$$

which will vanish if

$$\tan 2\alpha = -\frac{2\boldsymbol{\mathcal{E}}_{0r} \cdot \boldsymbol{\mathcal{E}}_{0i}}{\mathcal{E}_{0r}^2 - \mathcal{E}_{0i}^2}. \tag{42.9}$$

The real vectors $\mathbf{E}_{0r}$ and $\mathbf{E}_{0i}$ may be regarded as the real and imaginary parts of a single complex vector

$$\mathbf{E}_0 = \mathbf{E}_{0r} + i\mathbf{E}_{0i}. \tag{42.10}$$

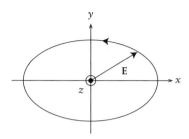

FIGURE 7.1. Elliptically polarized light. The figure illustrates *right* elliptical polarization.

Since $\mathcal{E}_0 = E_0 e^{-i\alpha}$, the electric field can be written

$$\mathbf{E}(\mathbf{r}, t) = \mathbf{E}_0 e^{i(\mathbf{k}\cdot\mathbf{r}-\omega t-\alpha)}. \tag{42.11}$$

Let us choose $\mathbf{E}_{0r}$ to be along $\hat{\mathbf{x}}$. Then we must take $\mathbf{E}_{0i}$ to be along $\hat{\mathbf{y}}$, and we may write E_{0x} and E_{0y} for the magnitudes of these vectors. The electric field then becomes

$$\mathbf{E}(\mathbf{r}, t) = (E_{0x}\hat{\mathbf{x}} + iE_{0y}\hat{\mathbf{y}})e^{i(\mathbf{k}\cdot\mathbf{r}-\omega t-\alpha)}. \tag{42.12}$$

Separating this equation into its x and y components and dropping the imaginary parts, we obtain, finally, the most general form for the real electric field in a harmonic or monochromatic plane wave:

$$E_x(\mathbf{r}, t) = E_{0x}\cos(\mathbf{k}\cdot\mathbf{r} - \omega t - \alpha), \tag{42.13}$$

$$E_y(\mathbf{r}, t) = -E_{0y}\sin(\mathbf{k}\cdot\mathbf{r} - \omega t - \alpha). \tag{42.14}$$

The magnetic field components are given by $B_x(\mathbf{r}, t) = -E_y(\mathbf{r}, t)$, $B_y(\mathbf{r}, t) = E_x(\mathbf{r}, t)$.

The equations (42.13) and (42.14) imply that

$$\frac{E_x^2(\mathbf{r}, t)}{E_{0x}^2} + \frac{E_y^2(\mathbf{r}, t)}{E_{0y}^2} = 1, \tag{42.15}$$

which is the equation of an ellipse. Hence, if we sit at a fixed space point $\mathbf{r}$, the tip of the electric field vector describes an ellipse in the xy plane, with semi-axes E_{0x} and E_{0y} (fig. 7.1). The light is correspondingly said to be *elliptically polarized*. The sense in which the ellipse is described is determined by the sign of the product $E_{0x}E_{0y}$.

For a fixed choice of x and y axes, one must also give an angle specifying the orientation of the ellipse, i.e., of the vectors $\mathbf{E}_{0r}$ and $\mathbf{E}_{0i}$ with respect to $\hat{\mathbf{x}}$ and $\hat{\mathbf{y}}$. This angle, plus another quantity, such as $E_{0r}E_{0i}/(E_{0r}^2 + E_{0i}^2)$ (which is equivalent to the signed eccentricity of the ellipse), make up two parameters that are required to specify the most general polarization state of a plane wave. We shall encounter other ways of describing the polarization in sections 45 and 53. Elliptical polarization is the most general possibility. There are two special cases that deserve to be mentioned separately. First, if $E_{0y} = E_{0x}$, we have *right circular polarization*, and if $E_{0y} = -E_{0x}$, we have *left circular polarization*. These two cases are frequently abbreviated as RCP and LCP. We summarize the relevant formulas:

$$\text{RCP}: \quad E_{0x} = +E_{0y}, \quad \mathbf{E}(\mathbf{r}, t) = E_{0x}(\hat{\mathbf{x}} + i\hat{\mathbf{y}})e^{i(\mathbf{k}\cdot\mathbf{r} - \omega t - \alpha)},$$

$$\text{LCP}: \quad E_{0x} = -E_{0y}, \quad \mathbf{E}(\mathbf{r}, t) = E_{0x}(\hat{\mathbf{x}} - i\hat{\mathbf{y}})e^{i(\mathbf{k}\cdot\mathbf{r} - \omega t - \alpha)}.$$

It should be noted that the choice of x and y axes is now arbitrary. It should also be noted that our convention for right and left circular polarization is exactly the opposite of an older convention, one still used in optics. With our usage, $\mathbf{E}$ moves in the sense of a right-hand screw for RCP light, and the angular momentum of the light is along $+\hat{\mathbf{z}}$, while for LCP, $\mathbf{E}$ moves as a left-hand screw and the angular momentum is parallel to $-\hat{\mathbf{z}}$.[1] The older convention is based on the sense of rotation of $\mathbf{E}$ as seen by an observer viewing the light coming head-on toward him or her.

The second case is when $E_{0y} = 0$ or $E_{0x} = 0$. Now the electric field at a fixed point $\mathbf{r}$ moves back and forth along a straight line through the origin in the xy plane. The light is said to be *linearly polarized* (or plane polarized) along the direction of $\mathbf{E}$, i.e., along $\hat{\mathbf{x}}$ if $E_{0y} = 0$ or along $\hat{\mathbf{y}}$ if $E_{0x} = 0$. Obviously, any other direction in the xy plane is also possible. Also, it should be noted that although $\mathbf{B}$ also moves along a line in the xy plane (at 90° to $\mathbf{E}$), the direction of polarization always refers to $\mathbf{E}$.

The most general elliptically polarized light can be represented as a superposition of two linearly polarized light waves along $\hat{\mathbf{x}}$ and $\hat{\mathbf{y}}$. For this reason, the latter are referred to as a basis for polarization states. Similarly, it is also possible to use RCP and LCP as a polarization basis. These concepts are especially of value in counting the number of independent degrees of freedom in the electromagnetic field.

Exercise 42.1 A light beam traveling along the z axis is elliptically polarized. One semi-axis of the ellipse can always be taken to lie in the first quadrant of the xy plane. Let the angle this semi-axis makes with $\hat{\mathbf{x}}$ be β ($0 \leq \beta \leq \pi/2$), and let the electric field amplitude along that axis be E_1. Let the amplitude along the other semi-axis of the ellipse be E_2. Write the field in the form

$$\mathbf{E} \sim \text{Re}(\sin \Psi e^{i\Delta}\hat{\mathbf{x}} - \cos \Psi \hat{\mathbf{y}})e^{-i\omega t}, \tag{42.16}$$

and determine the parameters Ψ and Δ in terms of β and the ratio E_2/E_1. This form is standard in ellipsometric studies (see section 137).

Answer: It is simpler to write the answers in terms of both E_1 and E_2, although only the ratio matters:

$$\tan \Psi = \left[\frac{E_1^2 \cos^2 \beta + E_2^2 \sin^2 \beta}{E_2^2 \cos^2 \beta + E_1^2 \sin^2 \beta} \right]^{1/2}, \tag{42.17}$$

$$\cos \Delta = \frac{(E_1^2 - E_2^2) \sin 2\beta}{[(E_1^2 - E_2^2)^2 \sin^2 2\beta + 4E_1^2 E_2^2]^{1/2}}. \tag{42.18}$$

---

[1] The angular momentum associated with each polarization is discussed further in section 47.

All square roots are taken with positive sign. It suffices to take $0 \leq \Psi \leq \pi/2$, and $-\pi \leq \Delta \leq \pi$. Further, $\Delta < 0$ for right and $\Delta > 0$ for left elliptical polarization.

43 Nonplane monochromatic waves; geometrical optics*

For the scalar wave equation in three dimensions

$$\nabla^2 f - \frac{1}{c^2}\frac{\partial^2 f}{\partial t^2} = 0, \tag{43.1}$$

nonplane solutions are easily found. For example, we have the spherical wave solution[2]

$$f(\mathbf{r}, t) = f_0 \frac{e^{i(kr-\omega t)}}{r}. \tag{43.2}$$

For electromagnetic waves, the first-order Maxwell equations couple together the different field components, making the task of writing down such solutions more difficult. However, the features that are physically important can be understood on the basis of an approximate analysis. Consider, e.g., the field

$$\mathbf{E}(\mathbf{r}, t) = \hat{\mathbf{n}} \times (\hat{\mathbf{n}} \times \mathbf{u}_0) \frac{e^{i(kr-\omega t)}}{r}, \tag{43.3}$$

$$\mathbf{B}(\mathbf{r}, t) = -\hat{\mathbf{n}} \times \mathbf{u}_0 \frac{e^{i(kr-\omega t)}}{r}. \tag{43.4}$$

Here, $\mathbf{u}_0$ is a constant vector, and $\hat{\mathbf{n}} = \mathbf{r}/r$. Now, $\nabla \cdot \mathbf{B} = 0$, while

$$\nabla \times \mathbf{B} = -\hat{\mathbf{n}} \times (\hat{\mathbf{n}} \times \mathbf{u}_0) \left[ik\frac{e^{i(kr-\omega t)}}{r} - \frac{e^{i(kr-\omega t)}}{r^2} \right]. \tag{43.5}$$

If we consider the field only at distances large compared with the wavelength, i.e., when $kr \gg 1$, the second term in square brackets may be neglected in comparison with the first. We then note that the first term is precisely $c^{-1}\partial\mathbf{E}/\partial t$, so that the Maxwell equation for $\nabla \times \mathbf{B}$ is satisfied. Similarly, $\nabla \cdot \mathbf{E}$ and $\nabla \times \mathbf{E} + c^{-1}\partial\mathbf{B}/\partial t$ fail to vanish by terms of order $(1/r)$ relative to $\mathbf{E}$ and $\mathbf{B}$. Thus, all the Maxwell equations are satisfied in the same approximation. The fields (43.3) and (43.4) are essentially those radiated by an oscillating electric dipole, as we shall see in chapter 10.

We now observe that in the field (43.3) and (43.4), at all points in space, $\mathbf{E}(\mathbf{r}, t)$, $\mathbf{B}(\mathbf{r}, t)$, and $\hat{\mathbf{n}}$ form an orthogonal triad, and $|\mathbf{E}| = |\mathbf{B}|$. These are all properties common to a plane electromagnetic wave. Lastly, in a region of space far from the origin, of size L such that $r \gg L \gg \lambda$, the direction of propagation and the amplitude of the wave are essentially uniform, and the wave behaves in all respects as a plane wave. This property, of behaving like a plane wave in a sufficiently small region of space that is at the same time large on the scale of the wavelength, holds for a very large number of electromagnetic waves.

---

[2] If we consider monochromatic waves, i.e., take $\omega = kc$, and suppress the $e^{-i\omega t}$ time dependence, then it is easy to show that the spherical wave $g(\mathbf{r}) = e^{ikr}/r$ is the wave produced by a point source. Formally, it is the Green function for the scalar Helmholtz equation: $(\nabla^2 + k^2)g(\mathbf{r}) = -4\pi\delta(\mathbf{r})$.

For visible light, in particular, $\lambda \simeq 0.5~\mu$m, this condition is met for almost all common situations, such as natural sunlight, or illumination by incandescent or fluorescent lamps. The propagation of light under these conditions is the subject of *geometrical optics*.

It is clear that in this approximation, the small parameter is the wavelength of the radiation itself. Over the region of space of interest, therefore, the wave goes through a very large number of cycles, and its phase varies by a very large amount compared with unity. This idea is the key to developing a systematic theory. We consider only monochromatic waves, and write

$$\mathbf{E} = \mathbf{E}_0(\mathbf{r})e^{ik\psi(\mathbf{r})}, \quad \mathbf{B} = \mathbf{B}_0(\mathbf{r})e^{ik\psi(\mathbf{r})}. \tag{43.6}$$

We have omitted the factor $e^{-i\omega t}$ and shall continue to do so, since this is just carried passively throughout. Second, $k = \omega/c$, as in the example studied above. Third, $\mathbf{E}_0(\mathbf{r})$ and $\mathbf{B}_0(\mathbf{r})$ are complex vectors, and ψ is real. The point of writing the fields this way is that the function $\psi(\mathbf{r})$, which is known as the *eikonal*, is expected to behave approximately as $\hat{\mathbf{n}} \cdot \mathbf{r}$, where $\hat{\mathbf{n}}$ is the local direction of propagation. Thus, it should vary on a length scale much greater than $1/k$. The same holds for the amplitudes $\mathbf{E}_0$ and $\mathbf{B}_0$. We therefore substitute the forms (43.6) in the Maxwell equations and keep only terms that are formally of lowest order in $1/k$. For example, in

$$\nabla \cdot \mathbf{B} = (\nabla \cdot \mathbf{B}_0 + ik\mathbf{B}_0 \cdot \nabla\psi)e^{ik\psi(\mathbf{r})}, \tag{43.7}$$

we neglect the first term. Carrying out this approximation in all the Maxwell equations, we obtain

$$\nabla\psi \times \mathbf{E}_0 - \mathbf{B}_0 = 0, \tag{43.8}$$

$$\nabla\psi \times \mathbf{B}_0 + \mathbf{E}_0 = 0, \tag{43.9}$$

$$\nabla\psi \cdot \mathbf{E}_0 = 0, \tag{43.10}$$

$$\nabla\psi \cdot \mathbf{B}_0 = 0. \tag{43.11}$$

Equations (43.8)–(43.11) imply that $\mathbf{E}_0$, $\mathbf{B}_0$, and $\nabla\psi$ are mutually perpendicular. In addition, if we substitute $\mathbf{B}_0$ from the first equation in the second and use the last two equations, we obtain $(\nabla\psi)^2\mathbf{E}_0 = \mathbf{E}_0$. Similarly, $(\nabla\psi)^2\mathbf{B}_0 = \mathbf{B}_0$. Since $\mathbf{E}_0$ and $\mathbf{B}_0$ do not vanish everywhere, we conclude that

$$(\nabla\psi)^2 = 1. \tag{43.12}$$

This is the basic equation of geometrical optics; it is known as the *eikonal equation*. It implies that $\nabla\psi$ is a unit vector, exactly as we should expect on the basis of our qualitative argument for the local behavior of ψ. We denote

$$\nabla\psi = \hat{\mathbf{n}}, \tag{43.13}$$

anticipating the fact that $\hat{\mathbf{n}}$ has the same meaning as it did for plane waves, namely, the direction of energy flow. To show this, we find the time-averaged energy density and

Poynting vector in the wave. The energy density is

$$\bar{u} = \frac{1}{16\pi}(\mathbf{E}_0 \cdot \mathbf{E}_0^* + \mathbf{B}_0 \cdot \mathbf{B}_0^*) = \frac{1}{8\pi}(\mathbf{E}_0 \cdot \mathbf{E}_0^*). \tag{43.14}$$

The last result follows from the equality of the electric and magnetic energies. To see this, dot eq. (43.8) into $\mathbf{B}_0^*$, and the complex conjugate of eq. (43.9) into $\mathbf{E}_0$, and compare. In the same way, $\mathbf{E}_0 \times \mathbf{B}_0^* = \mathbf{E}_0^* \times \mathbf{B}_0$. Hence,

$$\begin{aligned}
\bar{\mathbf{S}} &= \frac{c}{8\pi}(\mathbf{E}_0^* \times \mathbf{B}_0) \\
&= \frac{c}{8\pi}(\mathbf{E}_0^* \times (\nabla\psi \times \mathbf{E}_0)) \\
&= \frac{c}{8\pi}(\mathbf{E}_0^* \cdot \mathbf{E}_0)\nabla\psi.
\end{aligned} \tag{43.15}$$

Hence,

$$\bar{\mathbf{S}} = c\bar{u}\nabla\psi, \quad \bar{u} = \frac{1}{c}\bar{\mathbf{S}} \cdot \nabla\psi. \tag{43.16}$$

Exactly as asserted above, the energy flows along $\nabla\psi$, i.e., perpendicular to the surfaces $\psi = $ const, and the energy density and its flux are locally related exactly as for a plane wave. The surfaces of constant ψ are known as *wave fronts*, and the lines of $\bar{\mathbf{S}}$ are known as *rays*. Finding these lines is equivalent to solving the eikonal equation, since we could integrate along the appropriate ray to find ψ at any point, given any initial wave front. We turn, therefore, to this problem.

Let the arc length along a ray be denoted by s, and let the ray have the equation $\mathbf{r}(s)$. Then,

$$\frac{d\mathbf{r}}{ds} = \hat{\mathbf{n}} = \nabla\psi[\mathbf{r}(s)]. \tag{43.17}$$

Hence,

$$\frac{d^2\mathbf{r}}{ds^2} = \left(\frac{d\mathbf{r}}{ds} \cdot \frac{d}{d\mathbf{r}}\right)(\nabla\psi) = (\nabla\psi \cdot \nabla)\nabla\psi. \tag{43.18}$$

We now note the vector identity

$$\nabla(\mathbf{u} \cdot \mathbf{u}) = 2(\mathbf{u} \cdot \nabla)\mathbf{u} + 2\mathbf{u} \times (\nabla \times \mathbf{u}). \tag{43.19}$$

If we use $\nabla\psi$ for $\mathbf{u}$, the second term vanishes. Hence,

$$\frac{d^2\mathbf{r}}{ds^2} = \frac{1}{2}\nabla[(\nabla\psi)^2] = 0, \tag{43.20}$$

upon using the eikonal equation. It follows that $d\mathbf{r}/ds = $ const, i.e., that the rays are straight lines. This fact is, of course, *the* property of light that comes to mind when one sees sharp shadows, and it must have been pondered by the earliest philosophers.

We digress briefly at this point to note that as part of the subject of electromagnetism, geometrical optics is only a small chapter. However, it is a rich and vast subject in its own right. We have introduced it with a view to studying electromagnetic waves in vacuum in the short-wavelength approximation. The term as it is more generally used includes the propagation of light in transparent media. In the simplest description, such a medium is characterized by a refractive index n, and the eikonal equation becomes $(\nabla \psi)^2 = n^2$. In the general case, n may be a function of position. In a homogeneous medium (constant n), we again see that light rays are straight lines. At an interface between two such media, the propagation directions are related by the law of refraction. These few facts, along with the law of reflection, underlie almost the entire study of optical instruments, in particular the study of image formation and aberrations. Further, the phenomena of polarization, interference, and diffraction can be understood by relatively small extensions of the geometrical theory, which often go under the name of *wave optics*. Thus, interference from split sources or from thin films or wedges can be understood by attaching a phase to every point of the light ray, as if the light were made of particles carrying an internal clock that ticks away the time elapsed. This phase is, of course, the eikonal in the eikonal approximation. Diffraction—the existence of a nonzero intensity in the geometrical shadow—can be understood by making one additional assumption, known as *Huygens's principle*, which states that each point of a wave front acts as a source of secondary spherical wavelets, and that the envelope of all such wavelets an infinitesimal time later gives the new position of the wave front. How this comes about is discussed in elementary physics texts. This principle is merely a rule for solving the eikonal equation. It acquires additional content if the light encounters obstacles or screens with apertures. In such cases, solving the full electromagnetic equations is very difficult. Huygens's principle, along with the assumption of a phase attached to every ray, provides a simplified and physical way of approaching the problem. We shall discuss diffraction in the next chapter.[3]

Returning to our main discussion, if the next order in $1/k$ is retained in the Maxwell equations, one obtains the so-called transport equations, which describe the amplitudes $\mathbf{E}_0$ and $\mathbf{B}_0$ to be attached to every point of a ray, in other words, how the intensity and polarization of a wave change as it propagates along. We will not derive these equations, but we can understand the intensity variation from energy considerations. Consider a small section of a wave front, which is concave toward the direction of propagation. Any such small section has two principal sections of curvature, which are orthogonal to each other (denoted $A_1 B_1$ and $A_2 B_2$ in fig. 7.2), and corresponding radii of curvature R_1 and R_2.[4] Thus, the rays emanating from every point of the infinitesimal section $A_1 B_1$ will pass through the point C_1 a distance R_1 away, while those from $A_2 B_2$ will pass through C_2, a distance R_2 away. The area of the infinitesimal section shown is $R_1 R_2 \theta_1 \theta_2$, where the θ_i

---

[3] A comprehensive discussion of the geometrical theory, interference, and diffraction is given in the treatise by Born and Wolf (1975).

[4] Take the point in question to be the origin of a coordinate system, the plane tangent to the surface at this point to be the xy plane, and suppose the surface is concave toward positive z. The most general equation for the infinitesimal surface element is $z = \alpha x^2 + 2\beta xy + \gamma y^2$. This is a symmetric positive definite form (else z would be negative for some x, y). Hence, its eigenvectors—the principal sections of curvature—are orthogonal to each other. The eigenvalues are $1/2 R_1$ and $1/2 R_2$, where R_1 and R_2 are the principal radii of curvature. The product $1/R_1 R_2$ is known as the *Gaussian curvature* of the surface.

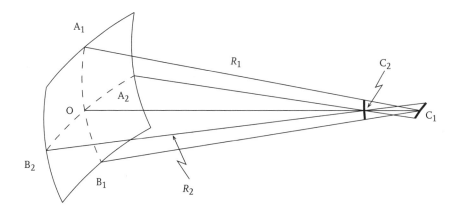

FIGURE 7.2. Propagation of wave fronts and formation of caustics.

are the angular openings of the arcs $A_i O B_i$. These angles remain fixed as the wave front propagates, and since the energy flowing through the section considered must remain constant, it follows that as we proceed along the ray emanating from O, the intensity, or energy flux density, varies as

$$I = \frac{\text{const}}{R_1 R_2}.$$

(43.21)

Note that the constant as well as the values of R_1 and R_2 will, in general, vary as we move from point to point on the wave front.

From fig. 7.2, we see that the tiny section of wave front drawn collapses to a line at C_1 and C_2. At C_1, $R_1 = 0$, and at C_2, $R_2 = 0$, so the formula (43.21) implies that intensity diverges. The locus of the centers of curvature of all the points on a wave front forms a surface known as a *caustic* or "burning" curve. In general, every wave front has two caustics. In the special case $R_1 = R_2$, the wave front propagates to a point, known as a *focus*.

The divergence of the intensity at caustics and foci is clearly unphysical and exposes a breakdown of the geometrical optics approximation. In reality, when the wave nature of light is considered, the divergence is eliminated. Nevertheless, there is a large and sharply localized maximum in the intensity in the neighborhood of the caustics or foci. The study of these singularities is therefore of some interest. Two commonly encountered caustics are the rainbow and the number 3–shaped curve seen inside a teacup when light from a compact source falls on it. We discuss their geometry in appendix D, along with a simple wave optics treatment of the intensity profile across a caustic.

We conclude this section by noting that there are deep associations between geometrical optics with other branches of physics. First, there is a close analogy between it and the Hamiltonian formulation of classical mechanics. This analogy was discovered by Hamilton himself. Thus, the eikonal function is analogous to the action, and the eikonal equation is the Hamilton-Jacobi equation. If the solution of these equations is formulated in variational terms, we obtain Fermat's principle of least time in optics, and

Maupertuis's principle of least action in mechanics.[5] Second, the passage from the full electrodynamic equations to geometrical optics is closely parallel to that from quantum mechanics to classical mechanics. Indeed, this passage in reverse played a crucial role in the development of quantum mechanics.

44 Electromagnetic fields in a laser beam*

In this section we discuss the electromagnetic fields in a common type of laser beam.[6] The example illustrates some of the ideas of geometrical optics, since the intensity varies on a length scale much greater than the wavelength, yet the analysis can be carried out using the Maxwell equations directly.

We consider a beam traveling in the z direction and take the frequency of the light to be ω. We write

$$\mathbf{E}(\mathbf{r}, t) = \mathbf{e}(\mathbf{r})e^{i(kz-\omega t)}, \quad \mathbf{B}(\mathbf{r}, t) = \mathbf{b}(\mathbf{r})e^{i(kz-\omega t)}, \tag{44.1}$$

where $k = \omega/c$ as usual. We further suppose the beam to be polarized mainly along $\hat{\mathbf{x}}$, i.e., $\mathbf{e} \parallel \hat{\mathbf{x}}$ to good approximation. We then expect $\mathbf{b} \parallel \hat{\mathbf{y}}$ to the same approximation. We will seek solutions that are of finite transverse extent, as is appropriate for a beam of finite width, and see how the width, polarization, and wave fronts change as the beam propagates along.

Let us write

$$e_x(\mathbf{r}) = \chi(\mathbf{r}) \tag{44.2}$$

and assume that χ varies little on a length scale $1/k$. In particular, $\partial\chi/\partial z \ll k\chi$, $\partial^2\chi/\partial z^2 \ll k\partial\chi/\partial z$. With these approximations, the wave equation for E_x becomes

$$\frac{\partial^2\chi}{\partial x^2} + \frac{\partial^2\chi}{\partial y^2} + 2ik\frac{\partial\chi}{\partial z} = 0. \tag{44.3}$$

We have kept the term in $\partial\chi/\partial z$ because otherwise we would get Laplace's equation in two dimensions, the only solution of which that is regular both at $x = y = 0$ and as $(x^2 + y^2)^{1/2} \to \infty$ is $\chi = $ constant, which describes an infinite plane wave. Equation (44.3) shows that the characteristic length scale of variation along z is much greater than that along x and y. Since we are looking for a solution that is localized in x and y, we try an ansatz of the form

$$\chi(\mathbf{r}) = h(z)e^{-(x^2+y^2)/2g(z)}. \tag{44.4}$$

---

[5] Since the terms "action" and "principle of least action" are used for more than one thing, let us explain precisely what we mean to avoid misunderstanding. By action, we mean the integral $\int pdq$ along any specified path, and not the integral $\int Ldt$ of the Lagrangian. It may also be referred to as the action at fixed energy. Maupertuis's principle states that for a given total energy, the true trajectory connecting any two points q and q' is that which minimizes the action.

[6] Our discussion follows W. L. Erikson and S. Singh, *Phys. Rev. E* **49**, 5778 (1994), who give a more general analysis and also experimentally confirm the predicted structure of the cross-polarization component intensity—see eq. (44.28) and fig. 7.4.

This yields

$$\frac{\partial^2 \chi}{\partial x^2} + \frac{\partial^2 \chi}{\partial y^2} = h(z)\left[\frac{x^2 + y^2}{g^2(z)} - \frac{2}{g(z)}\right] e^{-(\,)}, \tag{44.5}$$

$$\frac{\partial \chi}{\partial z} = \left[h'(z) + h\frac{(x^2 + y^2)}{2g^2}g'(z)\right] e^{-(\,)}, \tag{44.6}$$

where the primes denote d/dz. Equation (44.3) will be satisfied if

$$h' = -\frac{i}{k}\frac{h}{g}, \quad \frac{hg'}{g^2} = \frac{i}{k}\frac{h}{g^2}. \tag{44.7}$$

These equations have the solution

$$g(z) = \frac{i}{k}(z - iz_0), \quad h(z) = \frac{\alpha_0}{z - iz_0}, \tag{44.8}$$

where z_0 and α_0 are arbitrary constants. It suffices to take z_0 to be real, since the imaginary part merely shifts the zero of z.

We have thus found a solution for χ of the form

$$\chi(\mathbf{r}) = \frac{\alpha_0}{z - iz_0} \exp\left(i\frac{k}{2}\frac{x^2 + y^2}{(z - iz_0)}\right). \tag{44.9}$$

A more revealing way to write the answer is

$$\chi(\mathbf{r}) = \frac{\alpha_0}{z - iz_0} \exp\left[-\frac{x^2 + y^2}{2w^2(z)} + i\frac{k}{2}\left(\frac{x^2 + y^2}{R(z)}\right)\right], \tag{44.10}$$

where

$$w(z) = \left(\frac{z^2 + z_0^2}{kz_0}\right)^{1/2}, \quad R(z) = \frac{z^2 + z_0^2}{z}. \tag{44.11}$$

The quantity $w(z)$ is the transverse width of the beam as it propagates along. If we evaluate the integral of $|E_x|^2$ in the xy plane, we find that 99% of the intensity is contained in a disk of radius $2.15w$. This width is minimum at $z = 0$, where it equals

$$w_0 = (z_0/k)^{1/2}. \tag{44.12}$$

This point is known as the *waist* of the beam. The width grows linearly with z for $z \gg z_0$. Equivalently, we can say that the beam has an angular half-width $1/kw_0$ or $\lambda/2\pi w_0$. This is similar to the result that a beam of light that is made to pass through a narrow circular aperture of diameter D diverges with an angular width $1.22\lambda/D$.

The quantity $R(z)$ can be understood by examining the wave fronts. If, as in the previous section, we write $\mathbf{E} = \mathbf{E}_0(\mathbf{r})e^{i(k\psi(\mathbf{r}) - \omega t)}$, these are the surfaces of constant ψ, i.e.,

$$\psi \equiv z + \frac{x^2 + y^2}{2R(z)} = \text{const.} \tag{44.13}$$

Thus, $R(z)$ is the radius of curvature of the wave front passing through the point $(0, 0, z)$ on the axis of the beam. For $z \gg z_0$, $R(z) \approx z$, so the wave fronts are essentially spherical with the center at 0 (see fig. 7.3).

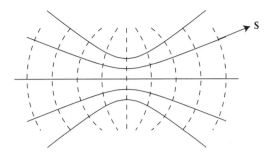

FIGURE 7.3. Wave fronts (dashed lines) and rays for the laser beam of a Gaussian intensity profile.

In the previous section we saw that the energy flow is along $\nabla\psi$, perpendicular to the wave fronts, and that the vectors $\mathbf{E}_0$ and $\mathbf{B}_0$ are orthogonal to $\nabla\psi$. Since the wave fronts are spherical for points off the axis of the beam, both $\mathbf{e}$ and $\mathbf{b}$ should acquire a longitudinal or z component. This is in fact so. In addition, there are also small *cross-polarization* components, i.e., nonzero e_y and b_x. We can find all these components by solving the Maxwell equations iteratively to higher orders in $1/k$. With eq. (44.1), the equations read

$$\nabla \cdot \mathbf{e} + ike_z = 0, \tag{44.14}$$

$$\nabla \cdot \mathbf{b} + ikb_z = 0, \tag{44.15}$$

$$\nabla \times \mathbf{e} + ik\hat{z} \times \mathbf{e} - ik\mathbf{b} = 0, \tag{44.16}$$

$$\nabla \times \mathbf{b} + ik\hat{z} \times \mathbf{b} + ik\mathbf{e} = 0. \tag{44.17}$$

We start with eq. (44.2) as our zeroth-order solution. Equation (44.16) then implies

$$b_y \simeq e_x = \chi, \tag{44.18}$$

as suspected earlier. We will see that the other two transverse components, e_y and b_x, are of order $1/k^2$. If we accept this temporarily, then eqs. (44.14) and (44.15) imply that

$$e_z = \frac{i}{k}\frac{\partial\chi}{\partial x} = -\frac{x}{z - iz_0}\chi, \tag{44.19}$$

$$b_z = \frac{i}{k}\frac{\partial\chi}{\partial y} = -\frac{y}{z - iz_0}\chi, \tag{44.20}$$

where we have also used the explicit results for $\chi(\mathbf{r})$. These formulas suffice to find the various components of the Poynting vector. We obtain

$$S_z = \frac{c}{8\pi}\mathrm{Re}(e_x b_y^* - e_y b_x^*) = \frac{c}{8\pi}|\chi|^2, \tag{44.21}$$

$$S_x = \frac{c}{8\pi}\mathrm{Re}(e_y b_z^* - e_z b_y^*) = \frac{c}{8\pi}\frac{x}{R(z)}|\chi|^2. \tag{44.22}$$

In the same way, $S_y = (cy/8\pi R)|\chi|^2$, so

$$\mathbf{S} = \frac{c}{8\pi}|\chi|^2 \left(\hat{\mathbf{z}} + \frac{x\hat{\mathbf{x}} + y\hat{\mathbf{y}}}{R(z)}\right). \tag{44.23}$$

This shows that the energy flow is indeed perpendicular to the wave fronts. We can also find the rays. A ray that passes through the waist at a distance r_0 has the equation

$$r_{\perp} = r_0 \left(1 + (z/z_0)^2\right)^{1/2}. \tag{44.24}$$

So far the analysis has followed expectations. A surprise awaits us, however, in the form of the cross-polarization components, e_y and b_x. Substituting the results for e_x, b_y, e_z, and b_z into eqs. (44.16) and (44.17), we obtain equations for these components:

$$b_x = -e_y + \frac{1}{k^2}\frac{\partial^2 \chi}{\partial x \partial y},$$

$$e_y = -b_x + \frac{1}{k^2}\frac{\partial^2 \chi}{\partial x \partial y}.$$

Obviously, one solution would be to have $b_x = -e_y$, with both quantities large and obeying the same equation as e_x, i.e., (44.3). This would amount to superposing a second wave polarized at right angles to the first. We are not interested in this. Hence, we set the combination $b_x - e_y$ equal to zero and obtain

$$e_y = b_x = \frac{1}{2k^2}\frac{\partial^2 \chi}{\partial x \partial y}. \tag{44.25}$$

With the explicit solution (44.9), we get

$$e_y(\mathbf{r}) = -\frac{xy}{2(z - iz_0)^2}\chi(\mathbf{r}). \tag{44.26}$$

By inserting suitably oriented polarizers into the beam, we can measure the components e_x and e_y or, more precisely, their intensities. If we define scaled distances $X = x/w(z)$, $Y = y/w(z)$, then the intensity or energy flux *per unit scaled area* in each component remains fixed as the beam propagates, and is given by

$$I_x(X, Y) = I_0 e^{-(X^2 + Y^2)}, \tag{44.27}$$

$$I_y(X, Y) = \frac{I_0}{4(kw_0)^4} X^2 Y^2 e^{-(X^2 + Y^2)}, \tag{44.28}$$

where I_0 is a constant. Thus, while a contour plot of I_x should have a single maximum at $X = Y = 0$, that of I_y should show a four-lobed structure, with maxima at $(X, Y) = (\pm 1, \pm 1)$. The experiment of Erikson and Singh confirms this structure beautifully. Their data are shown in fig. 7.4.

45 Partially polarized (quasimonochromatic) light*

Real light is never perfectly monochromatic and must be considered as a superposition of plane waves in some frequency range, $\Delta\omega$, centered about a particular frequency ω_0. Of special interest is the case of almost monochromatic light, wherein $\Delta\omega \ll \omega_0$. At a fixed

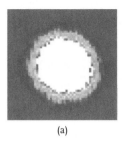 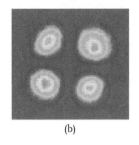 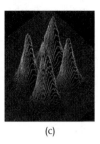

(a)   (b)   (c)

FIGURE 7.4. Intensity contours for the (a) x- and (b), (c) y-polarization components of the Gaussian laser beam as measured by Erikson and Singh. The cross-polarization component [parts (b) and (c)] has a clear four-lobed structure. Images kindly provided by Professor Surendra Singh.

point in space, we can then write

$$\mathbf{E}(t) = \mathbf{E}_0(t)e^{-i\omega_0 t}, \tag{45.1}$$

where the amplitude $\mathbf{E}_0(t)$ varies slowly on the time scale $2\pi/\omega_0$.

Many physically measurable properties of this kind of light are determined by considering bilinear functions of $\mathbf{E}$ and $\mathbf{B}$, averaged over many cycles $2\pi/\omega_0$. The polarization, in particular, is described by the tensor

$$\rho_{ij} = \overline{E_{0i} E_{0j}^*} / \overline{E_{0k} E_{0k}^*}, \tag{45.2}$$

where the overline denotes a time average. It should be noted that the indices i, j, k run over x and y only (assuming the light is propagating along the z direction) and that the summation convention is in effect. Also, the tensor ρ has been defined to have unit trace

$$\text{Tr}\,\rho = \rho_{ii} = 1, \tag{45.3}$$

and to have a Hermitean matrix,

$$\rho_{ji}^* = \rho_{ij}. \tag{45.4}$$

It is useful to see what the tensor ρ is like for fully polarized, or monochromatic light. In this case $\mathbf{E}_0 = \text{const}$. Then, the matrix of ρ is given by

$$\rho = \text{const} \begin{pmatrix} E_{0x} E_{0x}^* & E_{0x} E_{0y}^* \\ E_{0y} E_{0x}^* & E_{0y} E_{0y}^* \end{pmatrix}. \tag{45.5}$$

It is easy to see that in this case,

$$\det \rho = 0. \tag{45.6}$$

This result is true for all specific types of full polarization. For linear polarization along $\hat{x}$, $\hat{y}$, or $\hat{x} + \hat{y}$, e.g.,

$$\rho = \begin{pmatrix} 1 & 0 \\ 0 & 0 \end{pmatrix}, \quad \begin{pmatrix} 0 & 0 \\ 0 & 1 \end{pmatrix}, \quad \text{or} \quad \begin{pmatrix} 1/2 & 1/2 \\ 1/2 & 1/2 \end{pmatrix}. \tag{45.7}$$

Likewise for right and left circular polarization,

$$\rho = \frac{1}{2} \begin{pmatrix} 1 & \mp i \\ \pm i & 1 \end{pmatrix}. \tag{45.8}$$

In each case, one can see that $\text{Tr}\,\rho = 1$ and $\det\rho = 0$.

The condition $\det\rho = 0$ is necessary and sufficient for complete polarization. In the general case, one has

$$\det\rho \propto \overline{E_{0x} E_{0x}^*}\; \overline{E_{0y} E_{0y}^*} - \overline{E_{0x} E_{0y}^*}\; \overline{E_{0y} E_{0x}^*} \geq 0, \tag{45.9}$$

where the final inequality follows from a Schwarz inequality. (Consider $\overline{|E_{0x} + \alpha E_{0y}|^2} \geq 0$, and minimize the left-hand side with respect to α.) The case of equality can be obtained only when E_{0x} and E_{0y} are proportional to each other (with a fixed constant of proportionality) at all times, i.e., for purely monochromatic light.

In the general case, the matrix of the polarization tensor can be written as

$$\rho = \frac{1}{2} \begin{pmatrix} 1 & 0 \\ 0 & 1 \end{pmatrix} + \frac{1}{2} \begin{pmatrix} a_3 & a_1 - i a_2 \\ a_1 + i a_2 & -a_3 \end{pmatrix}, \tag{45.10}$$

where the a_i are all real quantities known as the *Stokes parameters*. Readers familiar with the Pauli matrices will recognize eq. (45.10) as the general expression of a 2×2 Hermitean matrix as a linear combination of the unit matrix and the Pauli matrices. The condition $\det\rho \geq 0$ implies that

$$\frac{1}{4}(1 - a_3^2 - a_1^2 - a_2^2) \geq 0, \tag{45.11}$$

i.e., that

$$P \equiv (a_1^2 + a_2^2 + a_3^2)^{1/2} \leq 1. \tag{45.12}$$

(It follows, as a necessary condition, that each Stokes parameter must be less than 1 in absolute value.) The quantity P is known as the *degree of polarization*.

The physical meaning of the Stokes parameters can be seen by examining the special cases of fully polarized light discussed above. The quantity $(1 + a_3)/2$ is thus the fraction of intensity in a partially polarized beam of light that would pass through an $\hat{x}$ polarizer, $(1 + a_1)/2$ is the fraction that would pass through a polarizer designed to let through light that is linearly polarized along $\hat{x} + \hat{y}$, and $(1 + a_2)/2$ is the fraction that would pass through an RCP pass filter. These meanings also suggest immediate operational means of measuring the Stokes parameters. It is obvious that a redefinition of the x and y axes will change a_1 and a_3, but not a_2. Hence, the quantities $a_\ell = \sqrt{a_1^2 + a_3^2}$ and $a_c = a_2$ may be referred to as the degrees of linear and circular polarization, respectively. They are the only inherent polarization characteristics of the light, and we shall show in section 166 that they are invariant under Lorentz transformations.

Lastly, the important case of completely unpolarized light is obtained when all $a_i = 0$. In this case, a filter that blocks light of *any one definite polarization* (linear, circular, or elliptical) will let half the light go through.

For perfectly polarized light, the three Stokes parameters (a_1, a_2, a_3) may be regarded as the components of a vector lying on the surface of a sphere, because of the condition $a_1^2 + a_2^2 + a_3^2 = 1$. This sphere is known as the *Poincare sphere*, and each point on it represents a different polarization state. Once again, we see that to specify the most general such state, two parameters are needed.

Exercise 45.1 Find how the Stokes parameters transform under a rotation about the axis of the beam.

Exercise 45.2 Show that the Stokes parameters are invariant under time reversal.

Exercise 45.3 A quasimonochromatic light beam traveling along $\hat{z}$ passes through an ideal compensator, a device that introduces a delay, δ, into the phase of the x component of the electric field, relative to the y component, without altering the amplitude of either component. It then passes through a polarizer that allows only linear polarization at an angle θ to the x axis. Let the intensity of the final outcoming beam be $I(\delta, \theta)$, and let the absolute maximum and minimum of this intensity under variations in both δ and θ be I_{max} and I_{min}. Find $I(\delta, \theta)$, and show that the intensity modulation, defined as

$$\frac{I_{max} - I_{min}}{I_{max} + I_{min}}, \tag{45.13}$$

is given by P, the degree of polarization.

46 Oscillator representation of electromagnetic waves

Another way to represent electromagnetic waves that is extremely useful is as a collection of harmonic oscillators. This representation is used in many situations, e.g., in statistical mechanics, when one wishes to understand the properties of radiation in equilibrium at a certain temperature (blackbody radiation), and in quantum mechanics, in describing the interaction of light with matter. We shall use it chapter 12 when we discuss the action formulation of electromagnetism.

The mathematical problem is to represent the most general solution of the electromagnetic equations without source terms. As stated, this problem is incompletely posed, and one must specify the volume in which these solutions are sought, along with the boundary conditions on the surface of this volume. The simplest procedure is to choose the volume in the shape of a rectangular parallelepiped and to impose *periodic boundary conditions* on the fields. That is, if the sides of the parallelepiped are taken to be L_x, L_y, and L_z, we demand that

$$f(x, y, z) = f(x + L_x, y, z) = f(x, y + L_y, z) = f(x, y, z + L_z), \tag{46.1}$$

where f is any component of any of the fields. These conditions are preferable to others that may seem more physical at first sight, e.g., $\mathbf{E} = 0$ on the boundaries. In the limit of an infinite volume, $V = L_x L_y L_z$, additive quantities such as energy and momentum per unit volume are independent of the exact shape and the boundary conditions.

We have already seen that the most general solution for $\mathbf{E}$ and $\mathbf{B}$ can be obtained by working in the Coulomb gauge, solving the wave equation for $\mathbf{A}$, and taking the time derivative and the curl. If we write $\mathbf{A}(\mathbf{r}, t)$ as a spatial Fourier series,

$$\mathbf{A}(\mathbf{r}, t) = \sum_{\mathbf{k}} \mathbf{A}_{\mathbf{k}}(t) e^{i\mathbf{k}\cdot\mathbf{r}}, \tag{46.2}$$

and impose the periodic boundary conditions (46.1) on $\mathbf{A}$ also, the allowed values of $\mathbf{k}$ are found to obey

$$e^{ik_x L_x} = e^{ik_y L_y} = e^{ik_z L_z} = 1, \tag{46.3}$$

yielding

$$k_x = \frac{2\pi}{L_x} n_x, \quad k_y = \frac{2\pi}{L_y} n_y, \quad k_z = \frac{2\pi}{L_z} n_z, \tag{46.4}$$

where n_x, n_y, and n_z are integers. Secondly, since $\mathbf{A}$ is real, we must have

$$\mathbf{A}_{-\mathbf{k}}(t) = \mathbf{A}_{\mathbf{k}}^*(t). \tag{46.5}$$

If we feed the expansion (46.2) into the wave equation, the coefficient $\mathbf{A}_{\mathbf{k}}(t)$ is seen to obey a harmonic oscillator equation with a general solution of the form

$$\mathbf{A}_{\mathbf{k}}(t) = \mathbf{a}_{\mathbf{k}0} e^{-i\omega_k t} + \mathbf{a}'_{\mathbf{k}0} e^{i\omega_k t}, \tag{46.6}$$

where

$$\omega_{\mathbf{k}} = ck. \tag{46.7}$$

The vector-valued coefficients, $\mathbf{a}_{\mathbf{k}0}$ and $\mathbf{a}'_{\mathbf{k}0}$, are arbitrary except that the reality requirement (46.5) implies that $\mathbf{a}'_{\mathbf{k}0} = \mathbf{a}^*_{-\mathbf{k}0}$. Thus, writing

$$\mathbf{a}_{\pm\mathbf{k}}(t) = \mathbf{a}_{\pm\mathbf{k}0} e^{-i\omega_k t}, \tag{46.8}$$

we get

$$\mathbf{A}_{\mathbf{k}}(t) = \mathbf{a}_{\mathbf{k}}(t) + \mathbf{a}^*_{-\mathbf{k}}(t), \tag{46.9}$$

$$\mathbf{A}(\mathbf{r}, t) = 2\,\mathrm{Re} \sum_{\mathbf{k}} \mathbf{a}_{\mathbf{k}0}\, e^{i(\mathbf{k}\cdot\mathbf{r} - \omega_k t)}. \tag{46.10}$$

To get eq. (46.10), we used eq. (46.8), and converted a sum on $\mathbf{k}$ to one over $-\mathbf{k}$. In what follows, the time dependence of $\mathbf{a}_\mathbf{k}(t)$ and $\mathbf{A}_\mathbf{k}(t)$ will not always be made explicit.

There is another restriction on the expansion (46.10). The transversality condition $\nabla \cdot \mathbf{A} = 0$ necessitates that

$$\mathbf{k} \cdot \mathbf{a}_\mathbf{k} = 0. \tag{46.11}$$

This can be met by taking $\mathbf{a}_\mathbf{k}$ as a linear combination of two arbitrarily chosen basis vectors $\mathbf{e}_{\mathbf{k},s}$ ($s = 1, 2$), which form an orthonormal triad along with $\hat{\mathbf{k}}$ for every $\mathbf{k}$. In other words, we write

$$\mathbf{a}_\mathbf{k} = a_{\mathbf{k},1} \mathbf{e}_{\mathbf{k},1} + a_{\mathbf{k},2} \mathbf{e}_{\mathbf{k},2}, \tag{46.12}$$

where

$$\mathbf{e}_{\mathbf{k},s} \cdot \mathbf{k} = 0, \quad \mathbf{e}_{\mathbf{k},1} \cdot \mathbf{e}_{\mathbf{k},2} = 0, \quad \mathbf{e}_{\mathbf{k},1} \times \mathbf{e}_{\mathbf{k},2} = \hat{\mathbf{k}}. \tag{46.13}$$

These basis vectors clearly describe two independent linear polarizations. Other choices for the polarization basis are examined in exercise 46.1.

We now Fourier expand the $\mathbf{E}$ and $\mathbf{B}$ fields in parallel with eq. (46.2). The coefficients are related by

$$\mathbf{E}_\mathbf{k} = -\frac{1}{c}\dot{\mathbf{A}}_\mathbf{k} = ik(\mathbf{a}_\mathbf{k} - \mathbf{a}^*_{-\mathbf{k}}), \tag{46.14}$$

$$\mathbf{B}_\mathbf{k} = i\mathbf{k} \times \mathbf{A}_\mathbf{k} = i\mathbf{k} \times (\mathbf{a}_\mathbf{k} + \mathbf{a}^*_{-\mathbf{k}}). \tag{46.15}$$

Note that these Fourier coefficients also obey the reality conditions $\mathbf{E}_{-\mathbf{k}} = \mathbf{E}^*_\mathbf{k}$, $\mathbf{B}_{-\mathbf{k}} = \mathbf{B}^*_\mathbf{k}$. Substituting these expansions in the expression for the energy density, u, integrating over the parallelepiped to obtain the total energy U, and making use of the facts that, first,

$$\int e^{i(\mathbf{k}\cdot\mathbf{r} - \mathbf{k}'\cdot\mathbf{r})}d^3x = V\delta_{\mathbf{k}\mathbf{k}'}, \tag{46.16}$$

and that, second, $\mathbf{a}_\mathbf{k} \perp \mathbf{k}$, we obtain

$$U = \frac{1}{8\pi} \int (\mathbf{E}^2 + \mathbf{B}^2)d^3x$$

$$= \frac{V}{8\pi} \sum_\mathbf{k} (\mathbf{E}_\mathbf{k} \cdot \mathbf{E}_{-\mathbf{k}} + \mathbf{B}_\mathbf{k} \cdot \mathbf{B}_{-\mathbf{k}})$$

$$= \frac{V}{8\pi} \sum_\mathbf{k} [k^2 |\mathbf{a}_\mathbf{k} - \mathbf{a}^*_{-\mathbf{k}}|^2 + k^2 |\mathbf{a}_\mathbf{k} + \mathbf{a}^*_{-\mathbf{k}}|^2]. \tag{46.17}$$

In the summand, we now get two terms, $\mathbf{a}_\mathbf{k} \cdot \mathbf{a}^*_\mathbf{k}$ and $\mathbf{a}_{-\mathbf{k}} \cdot \mathbf{a}^*_{-\mathbf{k}}$. By changing the summation over the second term to one over $-\mathbf{k}$ instead of $\mathbf{k}$, we see that both sums are the same. Hence,

$$U = \frac{V}{2\pi} \sum_\mathbf{k} k^2 \mathbf{a}_\mathbf{k} \cdot \mathbf{a}^*_\mathbf{k} = \frac{V}{2\pi} \sum_{\mathbf{k},s} k^2 a_{\mathbf{k},s} a^*_{\mathbf{k},s}. \tag{46.18}$$

The total momentum is similarly found to be

$$\mathbf{G} = \frac{V}{2\pi c} \sum_{\mathbf{k},s} k^2 \hat{\mathbf{k}}\, a_{\mathbf{k},s} a_{\mathbf{k},s}^*. \tag{46.19}$$

The above expansion can be regarded as a decomposition of the electromagnetic field into independent, or noninteracting, plane wave modes, two for each wave vector $\mathbf{k}$. The energy is a sum over the energy in each mode, and likewise the momentum. Further, the momentum in a mode is $\hat{\mathbf{k}}/c$ times the energy in that mode, exactly as eq. (41.8) would imply. In lieu of the fields, the dynamical variables in this description are the amplitudes $a_{\mathbf{k},s}(t)$. Equation (46.8) implies that they obey the equation of motion for a simple harmonic oscillator of frequency $\omega_{\mathbf{k}}$. We can see this more clearly if we define real amplitudes $x_{\mathbf{k},s}$ and $p_{\mathbf{k},s}$ according to

$$x_{\mathbf{k},s} = \sqrt{\frac{V}{4\pi c^2}}(a_{\mathbf{k},s} + a_{\mathbf{k},s}^*), \quad p_{\mathbf{k},s} = -i\omega_{\mathbf{k}}\sqrt{\frac{V}{4\pi c^2}}(a_{\mathbf{k},s} - a_{\mathbf{k},s}^*). \tag{46.20}$$

Then,

$$\dot{x}_{\mathbf{k},s} = p_{\mathbf{k},s}, \quad \dot{p}_{\mathbf{k},s} = -\omega_{\mathbf{k}}^2 x_{\mathbf{k},s}, \tag{46.21}$$

which are precisely the equations of motion for a simple harmonic oscillator coordinate $x_{\mathbf{k},s}$ and its canonically conjugate momentum $p_{\mathbf{k},s}$. In terms of the x's and p's, the energy U equals

$$\mathcal{H}_f = \frac{1}{2} \sum_{\mathbf{k},s} (p_{\mathbf{k},s}^2 + \omega_{\mathbf{k}}^2 x_{\mathbf{k},s}^2). \tag{46.22}$$

We have denoted this expression by the symbol $\mathcal{H}_f$ to indicate that it also serves as the Hamiltonian for the electromagnetic field, since the Hamilton equations $\dot{x}_{\mathbf{k},s} = \partial \mathcal{H}_f / \partial p_{\mathbf{k},s}$, $\dot{p}_{\mathbf{k},s} = -\partial \mathcal{H}_f / \partial x_{\mathbf{k},s}$ are precisely the equations of motion (46.21).

Exercise 46.1 Show that the most general polarization basis is obtained by replacing the vectors $\mathbf{e}_{\mathbf{k},s}$ in eq. (46.12) by complex vectors $\boldsymbol{\epsilon}_{\mathbf{k},s}$, where

$$\boldsymbol{\epsilon}_{\mathbf{k},s} = U_{ss'}\mathbf{e}_{\mathbf{k},s'}, \tag{46.23}$$

where U is a 2×2 unitary matrix. Develop the rules for orthonormality and the cross product in terms of $\boldsymbol{\epsilon}_{\mathbf{k},s}$ and apply them to the RCP-LCP basis.

47 Angular momentum of the free electromagnetic field*

Let us now turn to the question of the angular momentum in the source-free electromagnetic field, i.e., the field consisting entirely of electromagnetic waves, in particular, the issue of the separation of spin and orbital angular momentum. Readers should beware that this issue is rife with subtleties, misuse of terminology, and the potential for misunderstanding. Further, insofar as spin is essentially a quantum mechanical concept, some exposure to quantum mechanics is necessary.[7]

In the transverse gauge ($\nabla \cdot \mathbf{A} = 0$), the angular momentum in the EM field can be written as[8]

$$\mathbf{L} = \frac{1}{4\pi c} \int \left[E_i (\mathbf{r} \times \nabla) A_i + (\mathbf{E} \times \mathbf{A}) \right] d^3x. \qquad (47.1)$$

The first term depends explicitly on the choice of origin, since it contains the moment arm $\mathbf{r}$, while the second does not. For this reason these two terms are said to constitute the orbital and spin angular momenta, respectively. However, this separation is merely formal and a matter of convenience, and there is no true physical meaning to it. The reason is that the gauge condition couples the different vector components of the potential $\mathbf{A}$ (the spin part) to the way in which these components depend on the coordinates (the orbital part). A second way to see this is to note that the spin of a particle is related to the internal rotational symmetry properties of the particle in its rest frame. For light, however, there is no such frame. In particular, the often-heard statement that "the photon is a spin 1 particle" means no more and no less than that the EM field must be represented in terms of a potential $\mathbf{A}$ that is a tensor of rank 1, i.e., a vector. There are two ways in which a vector field changes under a bodily rotation. First, there is a change in the dependence on the coordinates. Second, a given Cartesian component at a point will be a linear combination of the components at the rotational preimage of that point. For a given $\mathbf{A}$, the second part of the change is fully determined by the fact that $\mathbf{A}$ is a vector and is independent of the actual coordinate dependence. However, the division of labor among these two changes can be altered by a change of gauge.[9]

Thus, for light, only the total angular momentum can be defined unambiguously. When quantized, the magnitude of this angular momentum can take on only integer values, except 0. The problem of determining the mathematical form of the fields that have definite angular momentum and parity is interesting and forms the subject of the so-called spherical waves and the *vector spherical harmonics*. We shall not enter into this question, except to say that such fields cannot at the same time have definite linear momentum, since the operators for linear and angular momentum do not commute.

It follows that plane electromagnetic waves do not have definite magnitude of angular momentum. However, it is possible for the component of the angular momentum along the direction of propagation to be simultaneously definite. This component is related to the polarization of the light, and the definite angular momentum states are nothing but the RCP and LCP states. One sometimes says that these states have "spin" or "helicity" ± 1, although our caveats about the meaning of the spin of the EM field must be kept

---

[7] A very clear discussion is given by Berestetskii, Lifshitz, and Pitaevskii (1964), secs. 2–8. However, this discussion is sophisticated and demands close reading.

[8] Readers may derive this result for themselves, or see, e.g., Gottfried (1966), p. 512.

[9] If, in eq. (47.1), we replace the vector potential $\mathbf{A}$ by its transverse part $\mathbf{A}^T$, the two terms *are* separately gauge invariant, but neither term obeys the commutators or Poisson brackets of an angular momentum. See S. J. van Enk and G. Nienhuis, *J. Mod. Opt.* **41**, 963 (1994).

in mind while doing this.[10] Secondly, the assignment of these values to the angular momentum is valid only in the sense of how the fields transform under a rotation of the coordinate axes about the direction of propagation. In the more physical sense of the integral of the angular momentum density, this quantity is ill-defined for a plane wave of truly infinite transverse extent, and we must make the transverse extent finite.[11]

In view of the previous remarks, it is instructive to calculate the angular momentum in the laser beam of section 44. The width of the beam is finite, so the calculation is simple. It is necessary to generalize the analysis somewhat, allowing both x and y components of $\mathbf{E}$ to be large. This can be done by superposing two solutions of the type found in section 44 with the same frequency and w_0 and z_0 parameters. We denote the solutions with large E_x and E_y by χ_x and χ_y, respectively, and their amplitude factors by α_x and α_y. The χ_y solution can be obtained by rotating the χ_x solution by $90°$ about $\hat{\mathbf{z}}$.

We start by finding the momentum density $\mathbf{g}(\mathbf{r})$. It is not difficult to show that

$$g_z = \frac{1}{8\pi c}(|\chi_x|^2 + |\chi_y|^2),$$

$$g_x = \frac{1}{8\pi c}\left[\frac{x}{R(z)}(|\chi_x|^2 + |\chi_y|^2) - \frac{2yz_0}{z^2 + z_0^2}\text{Im}(\chi_x^*\chi_y)\right], \tag{47.2}$$

$$g_y = \frac{1}{8\pi c}\left[\frac{y}{R(z)}(|\chi_x|^2 + |\chi_y|^2) - \frac{2xz_0}{z^2 + z_0^2}\text{Im}(\chi_y^*\chi_x)\right].$$

The angular momentum density is given by $\mathbf{r} \times \mathbf{g}$. The x component is given by

$$\mathcal{L}_x = yg_z - zg_y. \tag{47.3}$$

If we integrate this over the cross section of the beam, i.e., all x and y, it is clear that we will get zero. The same is true for $\mathcal{L}_y$. The only component that has a nonzero integral is $\mathcal{L}_z$. We find

$$\mathcal{L}_z = \frac{1}{8\pi c}\frac{2z_0}{z^2 + z_0^2}(x^2 + y^2)\text{Im}(\chi_x^*\chi_y). \tag{47.4}$$

Integrating this over all x, y, and a length ℓ in the z direction, we obtain

$$L_{\text{tot},z} = \frac{2\ell}{8ck}\frac{1}{(kw_0)^2}\text{Im}(\alpha_x^*\alpha_y). \tag{47.5}$$

The key factor is the last one. Defining

$$\alpha_{\text{RCP}} = \frac{\alpha_x - i\alpha_y}{\sqrt{2}}, \quad \alpha_{\text{LCP}} = \frac{\alpha_x + i\alpha_y}{\sqrt{2}}, \tag{47.6}$$

we can rewrite

$$L_{\text{tot},z} = \frac{\ell}{8ck}\frac{1}{(kw_0)^2}(|\alpha_{\text{RCP}}|^2 - |\alpha_{\text{LCP}}|^2). \tag{47.7}$$

---

[10] There is no state with helicity 0. This is again due to the transversality of the field.

[11] However, one can extract the answers from the formulas for an infinite plane wave with careful attention to this aspect of the problem. See, e.g., Baym (1969), pp. 13–16.

The quantities α_{RCP} and α_{LCP} are the amplitudes of the right- and left-circularly polarized components of the laser beam. As asserted in section 42, we see that the RCP component has positive angular momentum along the direction of propagation, while the LCP component has a negative value.

To relate our calculation to the quantum mechanical interpretation, let us look at the total momentum and energy in the same volume, i.e., a length ℓ along the light beam. The former is found by integrating the momentum density, eq. (47.2), and the latter by integrating the energy density $(E^2 + B^2)/8\pi$. Noting that

$$|\alpha_x|^2 + |\alpha_y|^2 = |\alpha_{RCP}|^2 + |\alpha_{LCP}|^2, \tag{47.8}$$

we obtain

$$\mathbf{g}_{tot} = \frac{\ell}{8c} \frac{1}{(kw_0)^2} (|\alpha_{RCP}|^2 + |\alpha_{LCP}|^2) \hat{\mathbf{z}}, \tag{47.9}$$

$$U_{tot} = \frac{\ell}{8} \frac{1}{(kw_0)^2} (|\alpha_{RCP}|^2 + |\alpha_{LCP}|^2). \tag{47.10}$$

As expected, $U_{tot} = cg_{tot}$. This follows naturally if we regard the beam as a stream of photons, each of which carries a momentum $\hbar\mathbf{k}$ and an energy $\hbar\omega = \hbar ck$. In the volume we are considering, there are a large number of photons. To get the right values for U_{tot} and g_{tot}, the number of RCP photons must be

$$N_{RCP} = \frac{1}{\hbar\omega} \frac{\ell}{8(kw_0)^2} |\alpha_{RCP}|^2, \tag{47.11}$$

and similarly for LCP photons. In terms of these numbers,

$$L_{tot,z} = (N_{RCP} - N_{LCP})\hbar. \tag{47.12}$$

This immediately suggests that individual RCP and LCP photons carry z angular momentum $+\hbar$ and $-\hbar$, respectively. This is indeed correct; experiments show that an atom that absorbs an RCP or LCP photon gains angular momentum $\pm\hbar$ along the photon's direction of propagation.[12]

Exercise 47.1 Show that the signs in eq. (47.6) are correct, i.e., that in an RCP beam, $\alpha_{LCP} = 0$, $\alpha_{RCP} \neq 0$, and vice versa.

Exercise 47.2 Verify eqs. (47.2)–(47.10).

---

[12] We should add two caveats here for readers not fully conversant with quantum mechanics. First, all the quantities we have found classically, the angular momentum, the energy, etc., must be regarded as the means of measured values of these quantities in many identical runs of the experiment. When the light intensity is large, the fluctuations about these means from run to run are so small that, for all practical purposes, they are nonexistent, and the quantities may be regarded as having definite values. This is the limit in which our analysis holds. Second, what about a linearly polarized beam of light, one in which $\alpha_y = 0$, say? The energy and the momentum expressions can still be interpreted in terms of a certain number of *linearly polarized photons*, but that for the angular momentum is bizarre. Now, $L_{tot,z} = 0$. Does it mean that linearly polarized photons have zero angular momentum, and what would happen if *one* of these photons were absorbed by an atom? This question gets to the heart of quantum mechanics, and we refer the reader to Baym's excellent discussion, cited earlier, for the answer.

8 | Interference phenomena

We turn in this chapter to optical phenomena where the intensity of the light field changes rapidly, either in space or in time, due to the superposition of two or more waves with well-defined phase relationships. Historically, these phenomena, particularly the observation that orthogonal polarizations did not interfere, played a crucial role in elucidating the nature of light as a transverse wave. Their study is a vast and continuing area of research, and our treatment is necessarily cursory. However, we also discuss some twentieth-century developments in this area, such as the description of partially polarized light in terms of correlation functions, the Hanbury-Brown and Twiss effect, and the Pancharatnam phase.

48 Interference and diffraction

We saw in section 44 that a beam that is focused down to a spot of diameter w_0 diverges with an angular half-width $\Delta\theta \simeq \lambda/2\pi w_0$. This is a general consequence of the finiteness of the wavelength. Suppose that at a distance z along the beam, it has a finite extent Δx and Δy in the transverse directions. The field can be represented as

$$\mathbf{E}(x, y, z) = e^{ik_0 z} \int \frac{d^3 k}{(2\pi)^3} \mathbf{E_k} e^{i\mathbf{k}\cdot\mathbf{r}}, \tag{48.1}$$

where k_0 is the mean wave vector. Because of the finite transverse extent, the Fourier amplitudes $\mathbf{E_k}$ will be spread out over widths Δk_x, Δk_y, where

$$\Delta k_x \gtrsim \frac{1}{\Delta x}, \quad \Delta k_y \gtrsim \frac{1}{\Delta y}. \tag{48.2}$$

Taking the minimum spread in real space for Δx and Δy, we get $\Delta k_{x,y} \sim w_0^{-1}$. Thus, the angular width is

$$\Delta\theta \sim \frac{\Delta k_{x,y}}{k_0} \sim \frac{\lambda}{2\pi w_0}. \tag{48.3}$$

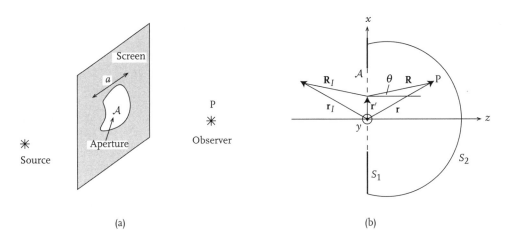

FIGURE 8.1. (a) Basic diffraction setup. (b) Definitions of various vectors and integration surfaces used in deriving the integral equation for the diffracted field. The surface S_1 consists of the entire xy plane, screen plus aperture.

Phenomena such as the above, which may be regarded as corrections to geometrical optics because of the nonzero wavelength of light, go under the names of interference and diffraction. These terms are sometimes more narrowly used to describe effects that arise when light falls on obstacles whose dimensions are much greater than the wavelength of the light.[1] The prototypical phenomenon is the existence of a nonzero intensity in the geometrical shadow. Suppose that a plane EM wave falls on an aperture $\mathcal{A}$ in a flat screen, which we take to lie in the xy plane (see fig. 8.1). The linear dimensions of the aperture are denoted by a. What is the field beyond the aperture, in the region $z > 0$? A heuristic answer is given by Huygens's principle. This states that each point of a wave front acts as a source of secondary spherical wavelets. Thus, if $\mathbf{r}'$ is a point in the aperture, and the incident field there is $\psi_{\mathrm{inc}}(\mathbf{r}')$, it produces a spherical wave, whose contribution to the field at a point $\mathbf{r}$ is proportional to $e^{ik|\mathbf{r}-\mathbf{r}'|}/|\mathbf{r}-\mathbf{r}'|$. The total field at $\mathbf{r}$ is given by adding up the contribution from all source points:

$$\psi(\mathbf{r}) \sim \int_{\mathcal{A}} \psi_{\mathrm{inc}}(\mathbf{r}') \frac{e^{ik|\mathbf{r}-\mathbf{r}'|}}{|\mathbf{r}-\mathbf{r}'|} d^2 s'. \tag{48.4}$$

The integral extends over the aperture only. The field ψ can stand for any one of the components of $\mathbf{E}$ or $\mathbf{B}$, but this is not strictly true, because as we have seen in our discussion of the laser beam, propagation at oblique directions mixes polarizations. Thus, eq. (48.4) must be understood as approximate.

Let us now try and put eq. (48.4) on a firmer footing. For simplicity, we first ignore the vector nature of the EM field and consider only scalar wave fields. That is, in the

---

[1] When the size of the body is much smaller than the wavelength, one speaks of the *scattering* of light. When the light interacts with a macroscopic medium, the terminology switches to reflection, refraction, and transmission. It should be clear that there are no sharp boundaries between these phenomena, and they share many conceptual underpinnings. We study scattering in chapter 22 and the propagation of electromagnetic waves in material media in chapter 20.

diffraction region $z \geq 0$, ψ obeys the equation

$$(\nabla^2 + k^2)\psi(\mathbf{r}) = 0. \tag{48.5}$$

We also introduce a Green function $G(\mathbf{r}, \mathbf{r}')$ obeying

$$(\nabla^2 + k^2)G(\mathbf{r}, \mathbf{r}') = -4\pi\delta(\mathbf{r} - \mathbf{r}') \tag{48.6}$$

for all $\mathbf{r}$ and $\mathbf{r}'$ with $z \geq 0$, $z' \geq 0$. The boundary conditions on this function are left unspecified for now, but it is obvious that whatever they are, we must have $G \sim e^{ikr}/r$ for $r \to \infty$ in the diffraction region.

We now use Green's theorem to construct an integral equation for $\psi(\mathbf{r})$. To this end, we multiply the equation for G by ψ, that for ψ by G, and subtract. This yields

$$-4\pi\psi(\mathbf{r})\delta(\mathbf{r} - \mathbf{r}') = \psi(\nabla^2 + k^2)G - G(\nabla^2 + k^2)\psi$$

$$= \nabla \cdot [\psi\nabla G - G\nabla\psi]. \tag{48.7}$$

Next, we integrate over all $\mathbf{r}$ with $z \geq 0$, and use Gauss's theorem. At the same time, we switch the $\mathbf{r}$ and $\mathbf{r}'$ labels. We thus obtain

$$4\pi\psi(\mathbf{r}) = \int_{S_1 + S_2} \left[\psi(\mathbf{r}')\nabla' G(\mathbf{r}', \mathbf{r}) - G(\mathbf{r}', \mathbf{r})\nabla'\psi(\mathbf{r}') \right] \cdot \hat{\mathbf{n}}'_{\text{in}} d^2 s'. \tag{48.8}$$

Note that $\hat{\mathbf{n}}'_{\text{in}}$ is the inward normal. This is the desired integral equation, first obtained by Kirchhoff. The surface integral extends over S_1, the plane of the screen, and S_2, a hemispherical surface at infinity. We now argue that the integral over the latter vanishes. This is because both $\psi(\mathbf{r}')$ and $G(\mathbf{r}', \mathbf{r})$ behave as $e^{ikr'}/r'$. Thus, the gradient terms of order $1/r'^2$ cancel directly, while those of order $1/r'^3$ are harmless, because the area of S_2 varies as r'^2 only.

The remaining surface integral over S_1 provides a solution for $\psi(r)$ if we know the field and/or its normal derivative in the plane of the screen. It is reasonable to assume that ψ equals the incident field ψ_{inc} inside the aperture and that it vanishes elsewhere in the screen plane. That is,

$$\psi(\mathbf{r}') = \begin{cases} \psi_{\text{inc}}, & \mathbf{r}' \in \mathcal{A}, \\ 0, & \mathbf{r}' \in S_1 - \mathcal{A}. \end{cases} \tag{48.9}$$

We can further eliminate the $\nabla'\psi$ term in eq. (48.8) by choosing the Green function to obey

$$G(\mathbf{r}', \mathbf{r}) = 0, \quad \mathbf{r}' \in \mathcal{A}. \tag{48.10}$$

This boundary condition can be met by choosing G to be the field of a point source located at $\mathbf{r} = (x, y, z)$ in the diffraction region, minus that of an image source at $\mathbf{r}_I = (x, y, -z)$:

$$G(\mathbf{r}', \mathbf{r}) = \frac{e^{ik|\mathbf{r}' - \mathbf{r}|}}{|\mathbf{r}' - \mathbf{r}|} - \frac{e^{ik|\mathbf{r}' - \mathbf{r}_I|}}{|\mathbf{r}' - \mathbf{r}_I|}. \tag{48.11}$$

With this choice for $\psi(\mathbf{r}')$ and G, the integral equation (48.8) turns into a formula for ψ:

$$\psi(\mathbf{r}) = \frac{1}{4\pi} \int_{\mathcal{A}} \psi_{\text{inc}}(\mathbf{r}') \left[\hat{\mathbf{z}} \cdot \nabla' G(\mathbf{r}', \mathbf{r}) \right] d^2 s'. \tag{48.12}$$

It remains to evaluate $\hat{\mathbf{z}} \cdot \nabla' G$. For this, we note that for $\mathbf{r}'$ on $\mathcal{A}$,

$$\hat{\mathbf{z}} \cdot \nabla' |\mathbf{r}' - \mathbf{r}| = -\frac{\hat{\mathbf{z}} \cdot \mathbf{r}}{|\mathbf{r}' - \mathbf{r}|} = -\frac{z}{R} = -\cos\theta, \tag{48.13}$$

where $R = |\mathbf{r}' - \mathbf{r}|$, and θ is shown in fig. 8.1. In the same way, $\hat{\mathbf{z}} \cdot \nabla' |\mathbf{r}' - \mathbf{r}_I| = -z_I / R_I = \cos\theta$. If $\mathbf{r}$ is far from the screen, then only the phase terms in G need to be differentiated, and we get

$$\psi(\mathbf{r}) = -\frac{ik}{2\pi} \int_{\mathcal{A}} \psi_{\text{inc}}(\mathbf{r}') \frac{e^{ik|\mathbf{r}-\mathbf{r}'|}}{|\mathbf{r} - \mathbf{r}'|} \cos\theta \, d^2 r'. \tag{48.14}$$

Apart from the explicit constant of proportionality, and the so-called obliquity factor of $\cos\theta$, eq. (48.14) is identical with eq. (48.4). However, this $\cos\theta$ factor is not terribly significant, since the dominant contribution to the integral comes from a very small region of the aperture close to the geometrical ray. We shall see below that in many cases this region is of linear size $(R\lambda)^{1/2}$, from which it follows that $\cos\theta$ is essentially a constant, which may be taken out of the integral.

It would now seem to be a simple matter to extend this argument to the vector wave equation. However, while it is straightforward to obtain a vector equivalent of the integral equation (48.8), specifying the $\mathbf{E}$ and $\mathbf{B}$ fields on the screen in a consistent manner is not. These boundary conditions depend on the material properties of the screen, which are usually too complex to incorporate with any degree of rigor. One usually ends up assuming that the fields in the aperture are the same as the incident fields, and zero elsewhere on the screen, which renders the use of the vector integral equation over the scalar one a minor refinement. A consistent treatment *is* possible for the case of a thin, flat, perfectly conducting metal screen with an aperture, and the diffracted fields can be viewed as being produced by the induced currents in the screen. This case is relevant to microwave diffraction. The calculations are rather involved, however, and must usually be performed numerically.[2]

From our perspective, the scalar theory is perfectly adequate. We are interested mainly in the fact of diffraction, not its fine details. The key physical point is the phase attached to a wave, showing up in the factor $e^{ik|\mathbf{r}-\mathbf{r}'|}$ in the integrand in eq. (48.8). The theory works extremely well for EM waves in the optical range, and apertures much bigger than the wavelength.[3]

---

[2] The problem of diffraction from a conducting knife edge can be solved exactly. See Schwinger et al. (1998), chap. 48.

[3] This is an appropriate place to remark that there is also no essential difference between the phenomena of interference and diffraction. The term *interference* is used when there are a small number of wave forms that need to be added together. These wave forms can be secondary wavelets from point-like apertures in a screen, or two beams of light that have been created by splitting a single beam, or the waves from two independent sources, etc. The term *diffraction* applies to the interference between the infinitely many wavelets arising from an extended aperture.

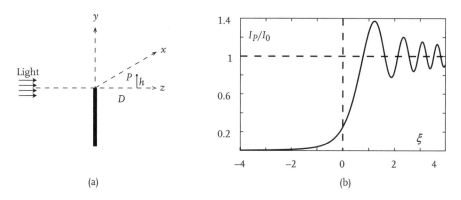

FIGURE 8.2. Fresnel diffraction from a straight edge. (a) Setup. (b) Intensity profile. The variable ξ is $(k/\pi D)^{1/2}h$.

49 Fresnel diffraction

We now apply our theory to a few standard examples. The mathematical problem is to evaluate the integral in eq. (48.4). Unless $r \gg a$, the size of the aperture, the factor $|\mathbf{r} - \mathbf{r}'|^{-1}$ will be important and the field will essentially be that of the incident wave. Interference between the wavelets from different secondary source points in the aperture will start to happen only when $r \gg a$. The most rapidly varying factor in the integrand is then $e^{ik|\mathbf{r}-\mathbf{r}'|}$ (times a similar factor in ψ_{inc} if the source is close to the aperture). Since $r \gg r'$, we have

$$k|\mathbf{r} - \mathbf{r}'| = kr - k\frac{\mathbf{r} \cdot \mathbf{r}'}{r} + \frac{k}{2r^3}(\mathbf{r} \times \mathbf{r}')^2 + \cdots . \tag{49.1}$$

Thus, if $r \sim D$ and $r' \sim a$, the successive terms are of order kD, ka, and ka^2/D. The first term yields an overall phase factor and may be ignored. The second term is always important. However, even if $a/D \ll 1$, we may still have $ka^2/D \gg 1$, and then the third term must also be kept. One is then in the *Fresnel diffraction* regime. If D is large enough, though, the third term can be ignored, and one is in the regime of *Fraunhofer diffraction* or *diffraction at infinity*.

Fresnel diffraction from a straight edge: Consider the setup shown in fig. 8.2. We take the screen to occupy the $y < 0$ part of the xy plane and consider a normally incident plane wave. Then $\psi_{\text{inc}} = \psi_0$ everywhere. The observation point P is taken to be at a distance D along the z axis, and a height h along y. Any offset along the x axis is irrelevant because of translation invariance. The distance from a point (x, y) in the aperture plane to P is, approximately,

$$D + \frac{1}{2D}\left(x^2 + (y - h)^2\right) + \cdots . \tag{49.2}$$

The inverse distance factor in eq. (48.14) may be taken as just $1/D$, and the obliquity factor, $\cos\theta$, as 1, so that

$$\psi_P \approx -\frac{ik}{2\pi}\frac{e^{ikD}}{D}\psi_0 \int_0^\infty \int_{-\infty}^\infty e^{ik[x^2+(y-h)^2]/2D}dx\,dy. \tag{49.3}$$

The x integral is a Gaussian and yields $(2\pi D/ik)^{1/2}$. For the y integral, we substitute $y-h=(\pi D/k)^{1/2}u$ and thus get

$$\psi_P \approx -\sqrt{\frac{\pi}{2}}e^{ikD}\psi_0 \int_{-\xi}^\infty e^{i\pi u^2/2}du, \tag{49.4}$$

where

$$\xi = \left(\frac{k}{\pi D}\right)^{1/2}h. \tag{49.5}$$

Denoting the incident intensity as I_0, the intensity I_P at P is thus given by

$$I_P = \frac{I_0}{2}\left|\int_{-\xi}^\infty e^{i\pi u^2/2}du\right|^2. \tag{49.6}$$

We now break the integral into one from 0 to ∞ and another from $-\xi$ to 0. The first integral equals $(1+i)/2$, and the second can be written as $C(\xi)+iS(\xi)$, where C and S are the Fresnel integrals[4]

$$C(\xi) = \int_0^\xi \cos\left(\frac{\pi}{2}u^2\right)du, \quad S(\xi) = \int_0^\xi \sin\left(\frac{\pi}{2}u^2\right)du. \tag{49.7}$$

Thus,

$$\frac{I_P}{I_0} = \frac{1}{8}\left|(1+i)+2\bigl(C(\xi)+iS(\xi)\bigr)\right|^2. \tag{49.8}$$

A plot of I_P/I_0 is shown in fig. 8.2b. The ratio rises rapidly from 0 to a maximum of about 1.4 at $\xi \simeq 1$. It then oscillates about 1 with diminishing amplitude, and maxima and minima (fringes) that get more and more closely spaced as $\xi \to \infty$. We can understand this behavior qualitatively, along with the length scale $(D\lambda)^{1/2}$ for the fringe separation, as follows. At large heights, h, it is apparent that the presence of the screen hardly diminshes the intensity. For small h, the most important question is the variation in the phase with which the wavelets from points close to the origin of our coordinate system arrive at the observation point. The path length from the origin to point P is $D+h^2/D$. This distance (minus the constant D) is of order λ when $h \sim (D\lambda)^{1/2}$. This is the height scale over which variations in phase are important, and there is the possibility of getting significant interference.

Exercise 49.1 The essential behavior of the intensity in straight-edge Fresnel diffraction can be obtained by directly examining the integral in eq. (49.6) in the limits $\xi \to -\infty$, $\xi \to \infty$,

---

[4] There are many definitions of the Fresnel integrals. We follow the convention of Abramowitz and Stegun (1972).

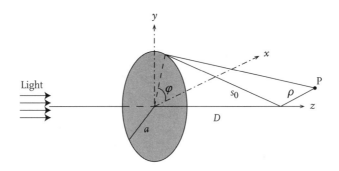

FIGURE 8.3. Diffraction from a circular disk.

and $\xi = 0$. Show that

$$\frac{I_P}{I_0} \approx \begin{cases} \dfrac{1}{2\pi^2\xi^2}, & \xi \to -\infty, \\[2ex] 1 + \dfrac{1}{\pi\xi}\left(\sin\dfrac{\pi}{2}\xi^2 - \cos\dfrac{\pi}{2}\xi^2\right), & \xi \to \infty, \\[2ex] \dfrac{1}{4}, & \xi = 0. \end{cases} \tag{49.9}$$

Exercise 49.2 Consider diffraction from a straight edge, when, in the coordinate system of fig. 8.2, the source is placed at the point $(0, 0, -D')$ and the observer is at $(0, h, D)$. Show that the intensity is still given by eqs. (49.6) and (49.9), provided ξ is now defined as

$$\xi = (kD'/\pi D(D' + D))^{1/2}h,$$

and I_0 is the intensity at points where $\xi \gg 1$.

Diffraction from an opaque circular disk: Let light from a distant source be normally incident on an opaque circular disk of radius a (see fig. 8.3). We seek the diffracted intensity behind the disk, close to the axis. Let us first find the intensity on the axis, at a distance D from the disk. To do this, we define

$$s(r') = |\mathbf{r} - \mathbf{r}'| = (D^2 + r'^2)^{1/2}. \tag{49.10}$$

In terms of this variable, the obliquity factor $\cos\theta$ equals D/s. Then, with ψ_0 being the field in the xy plane, and

$$s_0 = (D^2 + a^2)^{1/2}, \tag{49.11}$$

we have

$$\psi_P = -\frac{ik}{2\pi}\psi_0 \int_a^\infty \frac{De^{ik|\mathbf{r}-\mathbf{r}'|}}{|\mathbf{r}-\mathbf{r}'|^2} 2\pi r' dr'$$

$$= -ikD\psi_0 \int_{s_0}^\infty \frac{e^{iks}}{s} ds. \tag{49.12}$$

We now integrate repeatedly by parts to get a series in powers of $(1/ks_0)$. Keeping only the first term and squaring the magnitude, we get

$$I_P = I_0 \frac{D^2}{D^2+a^2}. \tag{49.13}$$

It is now simple to extend the calculation to off-axis points. Let the distance from the axis be denoted by ρ. It is clear that the dominant contribution to the field comes from points just on the edge of the disk, and so we merely have to replace e^{iks_0} in our previous answer by the average of the phase factor over this edge. For source points on the edge, we have

$$s = |\mathbf{r}-\mathbf{r}'| \approx s_0 - \frac{a\rho}{s_0}\cos\varphi + \cdots. \tag{49.14}$$

Hence, with J_0 being the Bessel function of order zero,

$$\psi_P \approx \psi_0 \frac{D}{s_0}\frac{1}{2\pi}\int_0^{2\pi} e^{iks_0} e^{-i(ka\rho/s_0)\cos\varphi} d\varphi = \frac{D}{s_0}J_0(ka\rho/s_0)e^{iks_0}\psi_0 \tag{49.15}$$

and

$$I_P = I_0 \frac{D^2}{s_0^2}J_0^2\left(\frac{ka\rho}{s_0}\right). \tag{49.16}$$

Thus, if a screen is placed at a distance D from the disk, we will see a series of concentric light and dark rings, with a bright spot at the center. The diameter of this spot is $4.8s_0/ka$, as found from the first zero of J_0.

At first, the presence of a bright spot precisely on the axis of the disk, where geometrical optics predicts zero intensity, is startling.[5] However, its physical origin is easy to understand. Right on the axis, light from all points on the edge of the disk arrives in phase and adds constructively. Light from more distant points can be out of phase, but it is more attenuated in comparison and cannot cancel the contribution from the edge points completely. The radial length scale for fringes can be understood as follows. From a point at a radial distance ρ, the minimum and maximum path lengths to points on the edge of the disk differ by $\sim 2a\rho/s_0$. When this difference equals λ, wavelets from these two points on the disk edge are in phase, and wavelets from points a quarter of the way

[5] It is obligatory to give the history of this spot, variously named after Poisson, Arago, and Fresnel. In 1818, Fresnel presented a paper to the French Academy of Sciences propounding the wave theory of light. Poisson, who was an advocate of the corpuscular theory of light, saw that the wave theory implied the existence of a bright spot on the axis of the disk and argued that the absurdity of this result disproved the wave theory. He was soon hoist by his own petard, however, when Arago did the experiment and found the spot! It is reported by some writers that Poisson did not wish the spot to be named after him, but so it goes.

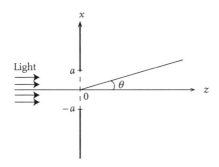

FIGURE 8.4. Fraunhofer diffraction from a thin slit.

around the edge are out of phase. Thus, the intensity minimum can be expected when $\rho \simeq \lambda s_0/2a$, i.e., the diameter of the bright spot is approximately $\lambda s_0/a$, close to the more accurate value $0.76\lambda s_0/a$.

50 Fraunhofer diffraction

When the observation point is very far away, the phase of a wavelet, $k|\mathbf{r} - \mathbf{r}'|$, can be approximated by the first two terms in eq. (49.1), i.e., by $kr - k\mathbf{r} \cdot \mathbf{r}'/r$. Defining

$$\mathbf{k} = k\mathbf{r}/r, \tag{50.1}$$

the diffracted field can be written as

$$\psi_P \approx -\frac{ik}{2\pi}\frac{e^{ikr}}{r}\int_{\mathcal{A}} \psi_{\text{inc}}(\mathbf{r}')e^{-i\mathbf{k}\cdot\mathbf{r}'}d^2 s'. \tag{50.2}$$

Since $\mathbf{r}'$ lies in the xy plane, the vector $\mathbf{k}$ can be replaced by its projection onto this plane. Thus, the Fraunhofer diffraction pattern is essentially given by the two-dimensional Fourier transform of the incident wave field in the aperture. If we write the observation direction as

$$\hat{\mathbf{r}} \approx \hat{\mathbf{z}} + \theta_x\hat{\mathbf{x}} + \theta_y\hat{\mathbf{y}}, \tag{50.3}$$

where the angles θ_x, θ_y are very small, and define

$$\mathbf{q} = (q_x, q_y) = k(\theta_x, \theta_y), \tag{50.4}$$

then

$$\psi_P \approx -\frac{ik}{2\pi}\frac{e^{ikr}}{r}\tilde{\psi}_{\mathbf{q}}, \tag{50.5}$$

where

$$\tilde{\psi}_{\mathbf{q}} = \int_{\mathcal{A}} \psi_{\text{inc}}(\mathbf{r}')e^{-i\mathbf{q}\cdot\mathbf{r}'}d^2 s'. \tag{50.6}$$

Taking $|\psi_P|^2$ gives us the intensity. This is the basic formula of Fraunhoffer diffraction.

Diffraction from a thin slit: We take the slit to be along the y axis and to have a width $2a$. The incident light is normal to the slit. The geometry is shown in fig. 8.4. The intensity is independent of y by symmtery. Ignoring the normalization, the diffracted field at angle θ is given by

$$\psi_P \sim \int_{-a}^{a} e^{-ik\theta x} dx \sim 2\frac{\sin(ka\theta)}{k\theta}. \tag{50.7}$$

Squaring this, we obtain the diffracted intensity. If we denote the intensity per unit length of the slit in an angular range $d\theta$ by dI, then

$$dI = \frac{I_{\text{inc}}}{\pi ak} \frac{\sin^2 ka\theta}{\theta^2} d\theta. \tag{50.8}$$

The normalization is fixed by demanding that the integral over all θ (from $-\infty$ to ∞) yield I_{inc}, the incident intensity.

The diffraction pattern appears as a series of bright and dark lines (fringes) parallel to the slits. The maxima are located at $\theta = 0$, $(n+\frac{1}{2})\pi/ka$, and the minima (zeros of intensity) are at $\theta = n\pi/ka$ $(n \neq 0)$, where n is an integer. The central maximum is over 20 times more intense than the next one.

Exercise 50.1 Consider diffraction from a thin slit in the Fresnel limit, and interpret the pattern in terms of two Fresnel diffraction patterns from straight edges. Examine how the pattern goes over into the Fraunhofer pattern as the distance from the slits to the observation point gets large.

51 Partially coherent light

Many light sources, especially blackbodies, and other thermal sources, such as flames, incandescent lamps, the sun, and other stars, produce stochastic and stationary radiation. In such sources, a fraction of the atoms are in some excited state from which they decay, but the fraction is maintained at a steady value by heating and thereby continual repopulation of the excited state. The electric field seen by an observer can therefore be regarded as a sum of many independent wave trains. It is physically obvious that its mean properties, ignoring the rapid fluctuations, are constant over time. It is also obvious that such light cannot be very coherent. By contrast, laser light is generally highly coherent. Heuristically, the lack of coherence arises because the electric field is not given by a definite function at all points in space and time but has some element of randomness or stochasticity. In this section, we make these concepts a little more precise.[6]

To simplify the discussion, we will describe the radiation by a scalar field $f(\mathbf{r}, t)$ instead of the vectors $\mathbf{E}(\mathbf{r}, t)$ and $\mathbf{B}(\mathbf{r}, t)$. Little is lost by doing so, especially if the polarization of the light is not being studied. The quantity f could denote one of the components of $\mathbf{E}$, for example. The fluctuations in the other components are assumed not to behave differently,

---

[6] An exceptionally detailed discussion of coherence is given by Mandel and Wolf (1995); see especially chaps. 4–7. The examples that we use for illustration are taken from this source.

and the behavior of **B** is also similar, because of the first-order differential equations connecting **E** and **B**. The stochastic properties of the radiation are then described by time averages of the type

$$G(\mathbf{r}_1, \mathbf{r}_2; \tau) \equiv \langle f(\mathbf{r}_1, t) f(\mathbf{r}_2, t + \tau) \rangle \equiv \lim_{T \to \infty} \frac{1}{T} \int_{t_0}^{t_0+T} f(\mathbf{r}_1, t) f(\mathbf{r}_2, t + \tau) dt \tag{51.1}$$

and similar integrals of products of more than two f's. We shall assume that $\langle f(\mathbf{r}, t) \rangle = 0$ and that the field is stationary, i.e., that all averages are independent of the choice of starting time t_0. The particular average G is known as the *autocorrelation function*.[7]

On physical grounds, one expects the time average of f^2 at a given point **r** to be the power or intensity of the light at that point, and the square of the temporal Fourier transform of f to be the spectral distribution of this power in some sense. The exact realtionship is given by the *Wiener-Khinchine theorem*, which we discuss in appendix F. According to this theorem, the *power spectrum* of f is given by

$$W(\mathbf{r}, \omega) = \frac{1}{\pi} \int_{-\infty}^{\infty} G(\mathbf{r}, \mathbf{r}; \tau) e^{i\omega\tau} d\tau, \quad (\omega > 0), \tag{51.2}$$

in terms of which the total power is given by

$$P = \int_0^{\infty} W(\omega) \, d\omega. \tag{51.3}$$

Suppose that f is a superposition of independent wave trains of average duration Δt. It follows that the autocorrelation function will also die off over a time scale Δt. By the usual relationship between a function and its Fourier transform, it follows that the spectral density will have a bandwidth $\Delta\omega$, where

$$\Delta\omega \simeq \frac{1}{\Delta t}. \tag{51.4}$$

The Wiener-Khinchine theorem also generalizes to a set of functions $f_i(t)$, which collectively form a stationary stochastic process, by which we mean that correlation functions of different f's also do not depend on the origin of the time interval chosen for averaging. Denoting the correlation of f_i and f_j by $G_{ij}(\tau)$, we obtain the cross-spectral density

$$W_{ij}(\omega) = \frac{1}{\pi} \int_{-\infty}^{\infty} G_{ij}(\tau) e^{i\omega\tau} d\tau. \tag{51.5}$$

This does not have the interpretation of a power distribution, but it is still true that it is given by the product of the FTs of f_i and f_j, with suitable scaling factors that can be obtained by following the argument of appendix F. It is plain that the same idea applies to the correlation $G(\mathbf{r}_1, \mathbf{r}_2; t)$ with different spatial arguments.

The finite bandwidth of the light has immediate consequences when we study interference phenomena. To see this, let us consider a Michelson interferometer as shown in fig. 8.5a. We take the light to be quasimonochromatic, so $\Delta\omega \ll \omega_0$. Let the path

---

[7] It is clear that if we were using the full description in terms of $\mathbf{E}(\mathbf{r}, t)$ and $\mathbf{B}(\mathbf{r}, t)$, we would need to deal with a set of tensorial correlation functions instead of G. Much of the formalism generalizes rather obviously.

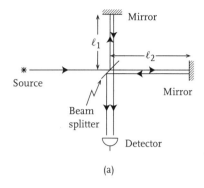
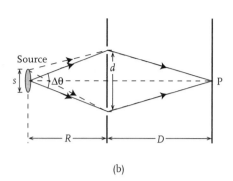

(a)   (b)

FIGURE 8.5. Temporal coherence of light can be revealed by means of a Michelson interference experiment (a), and spatial coherence by a Young interference experiment (b). In part (b), the distances D and R are much greater than s and d.

length of the two beams differ by $\Delta\ell = 2(\ell_1 - \ell_2)$. The corresponding time delay is $\Delta\ell/c$. The beam splitter essentially splits each constituent wave train in the light into two equal parts. It follows that if the delay $\Delta\ell/c$ is substantially greater than the autocorrelation time Δt discussed above, the waves that will be superposed in the viewing plane will be independent, with a random phase difference, and the interference fringes will be washed out. For this reason, the time Δt is said to be the *coherence time* of the light, and the length $c\Delta t$ the *coherence length*.

It is apparent that the coherence time characterizes the correlations between the radiation field at the same spatial point at unequal times. We can also ask for the correlations at different positions. These are revealed by a Young interference setup as in fig. 8.5b. Suppose the source has transverse dimensions s and that light from different points in the source is uncorrelated as it is emitted by independent atoms. Each point in the source will then yield an interference pattern on the screen. If these different patterns are out of step, the net visibility of the interference fringes will be reduced. Suppose the source is symmetrically located with respect to the slits as in the figure, and consider the intensity at the center of the screen, P. The path lengths of the rays from the center of the source to the slits to P are equal, so the interference pattern due to this point on the source is at a maximum. The rays from the edge of the source, on the other hand, differ in path length by $\Delta\ell = s\Delta\theta/2$. If the pattern due to the light from the edge of the source has a minimum at P, we may regard the overall interference as washed out. The condition for seeing the interference is, therefore, $\Delta\ell \le \lambda_0/2$, or in terms of d, the distance between the slits,

$$d \le \lambda_0 R/s, \tag{51.6}$$

since $\Delta\theta = d/R$. The length d is known as the *transverse coherence length* at a distance R from the source. The radiation at two points (in the plane of the slits) closer to each other than d may be regarded as coherent, while that at points farther apart is incoherent. Sometimes, one speaks of the coherence area,

$$\mathcal{A}_{\text{coh}} \simeq \frac{\lambda_0^2 R^2}{\mathcal{A}_s} \simeq \frac{\lambda_0^2}{\Omega_s}, \tag{51.7}$$

where $\mathcal{A}_s$ is the area of the source. In the second form, we have used $\Omega_s = \mathcal{A}_s / R^2$, the solid angle subtended by the source at the region where the coherence is being investigated.

Let us now give the coherence area for a few sources by way of example. In each case, $\lambda_0 = 5000\,\text{Å}$. For a source with $\mathcal{A}_s = 1\,\text{mm}^2$, $\mathcal{A}_{\text{coh}} \simeq 1\,\text{mm}^2$ at a distance of 2 m. For solar light at the earth, $\Omega_s = 6.81 \times 10^{-5}$ sr, $\mathcal{A}_{\text{coh}} \sim (0.06\,\text{mm})^2$. For the star Betelgeuse in Orion, $\Omega_s = 4.15 \times 10^{-14}$ sr, and $\mathcal{A}_{\text{coh}} \simeq 6\,\text{m}^2$.

The product of $\mathcal{A}_{\text{coh}}$ and $\Delta\ell$, the longitudinal coherence length, is sometimes known as the *coherence volume*, V_{coh}. Let us take the three sources described above and filter the light so that a wavelength range $\Delta\lambda \simeq 10^{-7}\lambda_0$ is transmitted. The longitudinal coherence length is then $\sim 10^7 \lambda_0 = 5$ m in each case, and the coherence volumes are 5 cm^3, 18 mm^3, and 30 m^3, respectively.

The notions of coherence length, area, etc., apply more generally, however. Take, for example, a garden-variety He–Ne laser, with $\lambda_0 = 6000\,\text{Å}$ and a beam cross section of $1\,\text{mm}^2$. Over a few seconds, the frequency can be stabilized enough to yield a bandwidth $\Delta\omega \simeq 2\pi \times 10^6$ sec^{-1}. The longitudinal coherence length over this duration is then $\simeq 300$ m. If the coherence area is taken as the cross section of the beam, $V_{\text{coh}} \simeq 300\,\text{cm}^3$.

The example of Betelgeuse is particularly interesting. The light, though very weak, is highly coherent. This is a consequence of the great distance from the source. The variations in the distance due to the spatial extent of the source can then be neglected, and one is essentially looking at a point source, and every wavelet from the source spreads out so much by the time it reaches the observer that one is looking at a single wave front over the entire region of observation. This high degree of coherence is the basis of Michelson's stellar interferometer, discussed in the following exercise.

Exercise 51.1 (Michelson's stellar interferometer) Light from a star of diameter D at a distance R falls upon the two mirrors M$_1$ and M$_2$ as shown in fig. 8.6 and combined by the optical system shown. The distance d between M$_1$ and M$_2$ is variable. Assuming that each point on the disk of the star is an equally strong independent source of light, find the intensity seen at the interferometer as a function of d.

Solution: Let the mirrors be located along the x axis and the direction to the star along the z axis. Then, for a point with coordinates (x, y) in the plane of the star's disk, the path length difference is

$$R_1 - R_2 \approx xd / R. \tag{51.8}$$

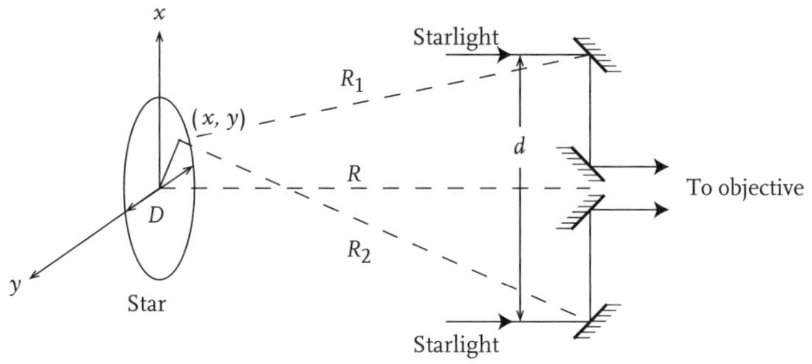

FIGURE 8.6. The Michelson stellar interferometer.

Using polar coordinates in the disk of the star, the total intensity observed obeys

$$I \propto \frac{4}{\pi D^2} \int_0^{2\pi} \int_0^{\frac{1}{2}D} \left[1 + \cos\left(\frac{kdr}{R} \cos\theta \right) \right] r \, dr \, d\theta$$

$$= 1 + \frac{2}{v} J_1(v),$$ (51.9)

where J_1 is the Bessel function of order 1, and with α being the angular diameter of the star,

$$v = kd \, D/2 \, R = kd\alpha/2.$$ (51.10)

The intensity drops from a maximum at $v = 0$ to a minimum of about 45% of the maximum at $v \simeq 5$. Michelson was able to go up to a mirror separation d of about 6 m and in this way determined the angular diameters of Betelgeuse and Sirius A as 0.06 and 0.006 seconds of arc, respectively.

52 The Hanbury-Brown and Twiss effect; intensity interferometry[*]

Michelson stellar interferometry has been used in the radio wavelength part of the spectrum since 1950 or so. In a modern version, known as *very long baseline interferometry* (VLBI), the radio signal amplitudes obtained at *different* radio telescopes are added together. By using separations of order 10^4 km, the product kd can be made 100 to 1000 times that feasible in the optical region, and the angular resolution can be made as fine as 10^{-9}. To achieve this, however, the signals at the two (or more) telescopes must be measured with extreme timing precisions. This was not possible around 1950. It was realized by Hanbury-Brown and Twiss that if the *amplitudes* at two points were correlated, the *intensities* would also be correlated. Since measuring the intensities places much weaker demands on the timing accuracy, Hanbury-Brown and Twiss were able to measure the sizes of several radio sources. They also demonstrated the viability

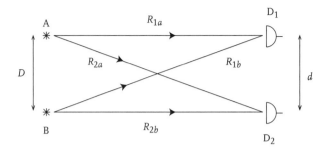

FIGURE 8.7. Schematic setup for intensity interferometry.

of intensity interferometry in the optical and obtained an angular diameter of 0.0068 seconds for Sirius A. The basic idea behind intensity interferometry can be understood by considering two detectors, D_1 and D_2, which receive signals from two sources, A and B. The distances involved are shown in fig. 8.7. We denote the average distance from the sources to the detectors by R and assume that $R \gg D \gg d$. The electric field at detector 1 can then be written as

$$E_1 = \frac{1}{R}\left(a(t'_1)e^{ik R_{1a}} + b(t'_1)e^{ik R_{1b}}\right). \tag{52.1}$$

Several aspects of this expression need explanation: a and b are the complex amplitudes of the signals from sources A and B, respectively; the argument t'_1 is meant to indicate the retarded times when these signals originated (which need not be the same for the two sources, but we use the same symbol for economy); and R_{1a} is the distance from source A to detector 1, etc. The field is attenuated by the factor $1/R$, as appropriate for a spherical wave. The most important point is that the amplitude $a(t'_1)$ is itself stochastic, as it may be the sum of a very large number of independent wave trains. Likewise for $b(t'_1)$. Measured quantities must therefore be obtained by taking averages. We will assume that because of cancellation of a large number of random phases,

$$\langle a(t'_1)\rangle = \langle b(t'_1)\rangle = 0; \quad \langle a^2(t'_1)\rangle = \langle b^2(t'_1)\rangle = 0. \tag{52.2}$$

Further, averages such as $\langle ab\rangle$ and $\langle a^*b\rangle$ are zero because the two sources are uncorrelated. Only averages of quantities such as $|a|^2$ and $|b|^2$ that involve one a and one a^* or one b and one b^* are nonzero.

The unaveraged intensity at detector 1 is obtained by taking the absolute square of E_1:

$$I_1 = \frac{1}{2R^2}\left(|a(t'_1)|^2 + |b(t'_1)|^2 + 2\,\mathrm{Re}\, a^*(t'_1)b(t'_1)e^{ik(R_{1b}-R_{1a})}\right). \tag{52.3}$$

In the same way,

$$I_2 = \frac{1}{2R^2}\left(|a(t'_2)|^2 + |b(t'_2)|^2 + 2\,\mathrm{Re}\, a^*(t'_2)b(t'_2)e^{ik(R_{2b}-R_{2a})}\right). \tag{52.4}$$

If we average these expressions, we get

$$\langle I_1 \rangle = \langle I_2 \rangle = \frac{1}{2R^2}\left(\langle |a|^2\rangle + \langle |b|^2\rangle\right). \tag{52.5}$$

We have assumed that the signals are stationary and therefore we have omitted the time arguments.

If we multiply I_1 and I_2 and then average, however, we obtain additional terms, as the averages $\langle a(t_1')a^*(t_2')\rangle$ need not be zero:

$$\langle I_1 I_2 \rangle = \langle I_1\rangle\langle I_2\rangle + \frac{1}{4R^4}\left(\langle a(t_1')a^*(t_2')\rangle\langle b(t_1')b^*(t_2')\rangle e^{ik(R_{1b}-R_{1a}-R_{2b}+R_{2a})} + \text{c.c.}\right). \tag{52.6}$$

We write the additional terms in terms of the excess correlation

$$g_2(d) = \frac{\langle I_1 I_2\rangle}{\langle I_1\rangle\langle I_2\rangle} - 1. \tag{52.7}$$

To simplify our expression for $g_2(d)$, we first note that

$$R_{1a} = |\mathbf{r}_a - \mathbf{r}_1| \approx r_a - \hat{\mathbf{r}}_a \cdot \mathbf{r}_1, \tag{52.8}$$

where $\mathbf{r}_a$ and $\mathbf{r}_1$ are the vectors to source A and detector 1 from an origin near the detectors, and $\hat{\mathbf{r}}_a = \mathbf{r}_a/r_a$. The other distances R_{1b}, etc., are similiarly approximated, and we get

$$R_{1b} - R_{1a} - R_{2b} + R_{2a} = -(\hat{\mathbf{r}}_a - \hat{\mathbf{r}}_b)\cdot(\mathbf{r}_1 - \mathbf{r}_2)$$

$$= \pm d\alpha, \tag{52.9}$$

where α is the angular separation of the two sources. Secondly, we assume for simplicity that $\langle a(t_1')a^*(t_2')\rangle$ is equal to its complex conjugate and do not show the time arguments explicitly. Thus,

$$g_2(d) = 2\frac{\langle aa^*\rangle\langle bb^*\rangle}{\left(\langle |a|^2\rangle + \langle |b|^2\rangle\right)^2}\cos(kd\alpha). \tag{52.10}$$

Our analysis shows that the intensity correlation also contains an interferometric term. By varying the distance d and seeing the change in the correlation, we can measure the angular separation α between two sources, or the diameter of a single extended source. We can also see that as the times t_1' and t_2' grow very far apart, the correlation functions $\langle a(t_1')a^*(t_2')\rangle$ and $\langle b(t_1')b^*(t_2')\rangle$ will have additional decay because of limited temporal coherence.

In fact, one does not need two spatially separated sources to see intensity interference in the first place. Hanbury-Brown and Twiss also measured the correlation $\langle I(t_1)I(t_2)\rangle$ for a single beam of light and found that there was an excess that gradually went to zero as the time delay $|t_1 - t_2|$ grew larger. Indeed, this may have been the greatest of their insights. One can understand the effect by writing the complex electric field in the beam as

$$E(x, t) = (E_1 + iE_2)e^{i(kx-\omega t)}, \tag{52.11}$$

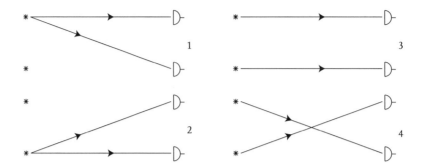

FIGURE 8.8. Different ways in which two photons can reach the detectors D_1 and D_2. Alternatives 3 and 4 are indistinguishable, even in principle.

where E_1 and E_2 are identically distributed, independent, real Gaussian random variables with variance I_0. This description is appropriate to a thermal source of light. Then,

$$\langle I(t) \rangle = \frac{1}{2} \langle E_1^2 + E_2^2 \rangle = I_0. \tag{52.12}$$

On the other hand, since $\langle E_1^4 \rangle = 3 I_0^2$,

$$\langle I^2(t) \rangle = \frac{1}{4} \langle (E_1^2 + E_2^2)^2 \rangle = 2 I_0^2. \tag{52.13}$$

Thus, at zero time delay, the excess correlation g_2 is 1. For obvious reasons, this effect is known as *photon bunching*.

The work of Hanbury-Brown and Twiss had a profound impact on the field of quantum optics and led to concepts such as photon counting and photon statistics. Basically, since it was clear by the 1950s that all interference phenomena are ultimately quantum mechanical, it became critical to discover how intensity interference fit into the quantum framework. Quantum mechanically, ordinary *amplitude* interference, such as in the two-slit experiment of Young, is understood by saying that a photon can travel from the source to the detector in two possible ways, via slit 1 or via slit 2, and the quantum mechanical amplitudes for these two indistinguishable alternatives add or interfere. Intensity interference must also be understood in terms of two alternatives: photons a and b are detected in detectors 1 and 2, but one cannot tell which photon ends up in which detector. In more detail, for the setup of fig. 8.7, one photon can reach each of the detectors 1 and 2 in four possible ways, as shown in fig. 8.8. Alternatives 1 and 2, in which the photons come from one source only, are distinguishable from each other, and from 3 and 4, as a matter of principle. Alternatives 3 and 4, on the other hand, are indistinguishable from each other, and so one must add their quantum mechanical amplitudes before squaring to find the probability of the event. It is this addition before squaring that leads to intensity interference.

In the photon bunching experiment also, one has two alternatives. Let us say that photons a and b are emitted by the source at times t_a and t_b and detected at times t_1 and t_2. But, since the time of emission of a photon is uncertain, one cannot say whether photon a is detected at t_1 and b at t_2, or vice versa. The amplitudes for these two alternatives must be added before squaring, leading to interference. One can also see why the interference dies away as $|t_2 - t_1|$ gets large. For then, because of the finite frequency bandwidth of the light, one *can* determine the time of emission of the photons to sufficient accuracy to say which photon is detected when, and the alternatives become distinguishable, or noninterfering.

Our classical description of intensity interference is nevertheless valid, because photons are bosonic particles. Thus, it is possible to have states in which the mean value of the electric field operator is large and behaves as a classical physical observable, which can be added together, or superposed. We thus see that the bosonic character of photons underlies the classical superposition principle of electrodynamics, and this is all that we used in our analyses. Fermionic particles would behave very differently: alternatives 3 and 4 in fig. 8.8 would have to be added together with a relative minus sign, since they differ from each other by an exchange of particles. Instead of bunching, we would see electron *antibunching*. Electron beams are inherently nonclassical. Most photon beams, on the other hand, admit the possibility of a classical description.[8]

53 The Pancharatnam phase*

The vector nature of the electromagnetic field has not played any essential part in the discussion in this chapter up to now, except in the comment that orthogonal polarizations do not interfere. It is possible, however, to change the state of polarization of a light beam by means of devices such as birefringent crystals,[9] and the question then arises of the phase to be assigned to a polarization change. Suppose, for example, that in a Young-type two-slit interference experiment, we change the polarization of one of the beams in such a way that its initial and final polarization state are the same. How will this beam interfere with the other one?[10] This question was posed and answered by Pancharatnam in 1956.[11]

To describe Pancharatnam's phase, it is first necessary to extend the Poincare sphere description of perfectly polarized light. Consider the two-component complex column vector,

$$|\psi\rangle = \frac{1}{N} \begin{pmatrix} E_{0x} \\ E_{0y} \end{pmatrix}, \quad N = \left(|E_{0x}|^2 + |E_{0y}|^2\right)^{1/2}, \tag{53.1}$$

and its conjugate row vector,

$$\langle\psi| = \frac{1}{N} \begin{pmatrix} E_{0x}^* & E_{0y}^* \end{pmatrix}. \tag{53.2}$$

---

[8] Indeed, it is quite difficult to produce a manifestly nonclassical light beam. This was first done in a beautiful demonstration of *photon antibunching* by Leonard Mandel and his students: H. J. Kimble, M. Dagenais, and L. Mandel, *Phys. Rev. Lett.* **39**, 691 (1977).

[9] For a discussion of how such crystals work, see Landau and Lifshitz (1960), secs. 77 and 78.

[10] Naturally, we assume that all other changes in phase, due to reflection off mirrors, passage through a refractive medium, etc., are accounted for.

[11] S. Pancharatnam, *Proc. Ind. Acad. Sci. A*, **44**, 247 (1956).

This vector is normalized in the sense that

$$\langle \psi | \psi \rangle \equiv \frac{1}{N^2} \begin{pmatrix} E_{0x}^* & E_{0y}^* \end{pmatrix} \begin{pmatrix} E_{0x} \\ E_{0y} \end{pmatrix} = 1. \tag{53.3}$$

We now note that the outer product of $\langle \psi |$ and $| \psi \rangle$ is the matrix of the polarization tensor introduced in section 45:

$$\rho = \frac{1}{N^2} \begin{pmatrix} E_{0x} E_{0x}^* & E_{0x} E_{0y}^* \\ E_{0y} E_{0x}^* & E_{0y} E_{0y}^* \end{pmatrix} = \frac{1}{N} \begin{pmatrix} E_{0x} \\ E_{0y} \end{pmatrix} \frac{1}{N} \begin{pmatrix} E_{0x}^* & E_{0y}^* \end{pmatrix} = | \psi \rangle \langle \psi |. \tag{53.4}$$

The normalization now implies that $| \psi \rangle$ is the eigenvalue of ρ with unit eigenvalue,

$$\rho | \psi \rangle = | \psi \rangle. \tag{53.5}$$

If we now introduce the Poincare matrix M via the Stokes parameters as

$$M = 2\rho - 1 = \begin{pmatrix} a_3 & a_1 - i a_2 \\ a_1 + i a_2 & -a_3 \end{pmatrix}, \tag{53.6}$$

it follows that $| \psi \rangle$ is also the unit eigenvalue eigenvector of M:

$$M | \psi \rangle = | \psi \rangle. \tag{53.7}$$

Note that both M and $| \psi \rangle$ require two parameters for their specification. For M this is obvious, and the parameters could be taken as the spherical polar angles α and β (the co-latitude and the azimuth) on the Poincare sphere, or the equivalent unit vector $\hat{n}$. For ψ, four real numbers are needed to give E_{0x} and E_{0y}, but these are reduced to two because of normalization and the indeterminacy of the overall phase.

Since eq. (53.7) does not fix the phase of $| \psi \rangle$, Pancharatnam postulated that the intrinsic relative phase of two polarization states $| \psi_1 \rangle$ and $| \psi_2 \rangle$ corresponding to Poincare sphere directions $\hat{n}_1$ and $\hat{n}_2$ is determined by the condition that the scalar product $\langle \psi_2 | \psi_1 \rangle$ be real and positive. He then showed that if one considers three points on the Poincare sphere, $\hat{n}_1$, $\hat{n}_2$, and $\hat{n}_3$, and constructs corresponding states such that $| \psi_2 \rangle$ is in phase with $| \psi_1 \rangle$, and $| \psi_3 \rangle$ is in phase with $| \psi_2 \rangle$, then $| \psi_1 \rangle$ will not be in phase with $| \psi_3 \rangle$. Instead, if one constructs another eigenstate $| \psi_1' \rangle$ for the point $\hat{n}_1$ to be in phase with $| \psi_3 \rangle$, then

$$\langle \psi_1 | \psi_1' \rangle = e^{-i\Omega_{123}/2}, \tag{53.8}$$

where Ω_{123} is the solid angle formed by connecting the points 1, 2, and 3 with the shorter geodesic arcs between each pair of points. (See fig. 8.9a.) The solid angle is regarded as positive if the geodesic triangle is traversed in the anticlockwise sense. Thus, if the polarization of one of the light beams in an interference experiment is taken in a closed circuit around this geodesic triangle, it will interfere with the other beam with a relative phase $\Omega_{123}/2$. Pancharatnam also went on to confirm this result experimentally. Pancharatnam's rule is similar to that for parallel transport on a curved surface, in that just as the latter tells us how to compare two vectors at different points on the surface,

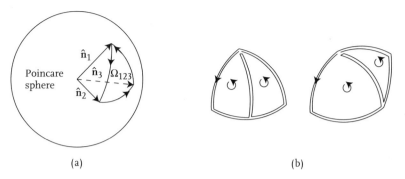

FIGURE 8.9. (a) The Pancharatnam phase for a polarization circuit on the Poincare sphere in the form of a geodesic triangle is $-1/2$ times the solid angle enclosed by that triangle. (b) Composition of circuits in terms of geodesic triangles.

the former tells us how to compare two different polarization states. Further, just as a vector that is parallel transported around a closed circuit on a curved surface need not return to its original direction, a light beam that is taken on a closed circuit in polarization space need not have the same phase as that at the start. Both phenomena are examples of *anholonomy.*

A direct proof of Pancharatnam's result is algebraically tedious. We will therefore first show it for two special circuits. We will then argue that any circuit can be built up as a sequence of infinitesimal geodesic triangle circuits and prove that the total phase acquired in that circuit is given by half the negative of the solid angle enclosed by it.

To this end, let us write

$$
M_i = \begin{pmatrix} \cos \alpha_i & e^{-i\beta_i} \sin \alpha_i \\ e^{i\beta_i} \sin \alpha_i & -\cos \alpha_i \end{pmatrix}, \quad |\bar{\psi}_i\rangle = \begin{pmatrix} \cos \frac{1}{2}\alpha_i \\ \sin \frac{1}{2}\alpha_i \, e^{i\beta_i} \end{pmatrix}.
\tag{53.9}
$$

The bar in $|\bar{\psi}_i\rangle$ indicates that it is an eigenstate of M_i with the preset phase convention that the first component is real and positive. Let $|\psi_1\rangle = |\bar{\psi}_1\rangle$, and $|\psi_2\rangle = e^{i\gamma_{21}}|\bar{\psi}_2\rangle$ be in phase as per Pancharatnam's rule. Then γ_{21} is fixed by the requirement that

$$
\langle \psi_2 | \psi_1 \rangle = e^{-i\gamma_{21}} \left[\cos \frac{1}{2}\alpha_1 \cos \frac{1}{2}\alpha_2 + \sin \frac{1}{2}\alpha_1 \sin \frac{1}{2}\alpha_2 e^{-i(\beta_2 - \beta_1)} \right]
\tag{53.10}
$$

be real and positive.

The explicit formula for γ_{21} in terms of arctangents, etc., is messy and unilluminating.

For a general geodesic triangle circuit, we introduce two more phases γ_{32} and γ_{13} according to eq. (53.10) above, so that the net phase after the circuit is

$$
\gamma = \gamma_{21} + \gamma_{32} + \gamma_{13}.
\tag{53.11}
$$

Our goal is to show that $\gamma = -\Omega_{123}/2$.

Let us now consider the first of our special circuits. Let point 1 be at the north pole, and 2 and 3 be on the equator. The solid angle of the geodesic triangle 123 (a semilune) is then given by

$$\Omega_{123} = \beta_{32} \equiv \beta_3 - \beta_2. \tag{53.12}$$

To find γ_{21}, we put $\alpha_1 = 0$ and $\alpha_2 = \pi/2$ in eq. (53.10). This gives $\langle \psi_2 | \psi_1 \rangle = e^{-i\gamma_{21}}/\sqrt{2}$, so $\gamma_{21} = 0$. In the same way, $\gamma_{13} = 0$. To find γ_{32}, we put $\alpha_2 = \alpha_3 = \pi/2$. Then,

$$\langle \psi_3 | \psi_2 \rangle = e^{-\gamma_{32}} \left[\frac{1}{2}(1 + e^{-i\beta_{32}}) \right] = \cos \tfrac{1}{2}\beta_{32}\, e^{-i(\gamma_{32} + \frac{1}{2}\beta_{32})}. \tag{53.13}$$

Hence, choosing the azimuths so that $-\pi \le \beta_{32} \le \pi$, which is always possible, we get

$$\gamma_{32} = -\frac{1}{2}\beta_{32}, \quad \left(\alpha_1 = \alpha_2 = \frac{1}{2}\pi \right). \tag{53.14}$$

Thus, the net phase is $-\frac{1}{2}\beta_{32}$, and Pancharatnam's result is verified.

For the second special circuit, let 1, 2, and 3 be equally spaced directions on the equator, so that $\beta_{21} = \beta_{32} = \beta_{13} = 2\pi/3$. By eq. (53.14), each of the γ_{ij}'s is $-(\frac{1}{2})(2\pi/3) = -\pi/3$, so the net phase is $-\pi$, and since the solid angle Ω_{123} is 2π, Pancharatnam's result is confirmed once again.

Let us finally work out the phase for an arbitrary circuit $\mathcal{C}$ on the Poincare sphere. We observe that (see fig. 8.9b) (1) any geodesic triangular circuit can be composed of two smaller geodesic triangles, (2) any geodesic polygonal circuit can be composed as a sum of geodesic triangles. In both cases, the phase for the larger circuit is obtained by adding up the phases for the subcircuits. We now approximate the circuit $\mathcal{C}$ by a sequence of infinitesimal geodesic arcs connecting $\hat{\mathbf{n}}_1$, $\hat{\mathbf{n}}_2$, ..., $\hat{\mathbf{n}}_N$, $\hat{\mathbf{n}}'_1 \equiv \hat{\mathbf{n}}_1$ on $\mathcal{C}$. It follows that if Pancharatnam's result is true for a geodesic triangle, then the phase $\gamma_{\mathcal{C}}$ for $\mathcal{C}$ is $-\Omega_{\mathcal{C}}/2$, where $\Omega_{\mathcal{C}}$ is the solid angle enclosed by $\mathcal{C}$. However, we can find the net phase for $\mathcal{C}$ by adding up the infinitesimal phase changes $\gamma_{i+1,i}$ for each of the arcs connecting $\hat{\mathbf{n}}_i$ and $\hat{\mathbf{n}}_{i+1}$. Writing $\alpha_{i+1} = \alpha_i + \delta\alpha$, $\beta_{i+1} = \beta_i + \delta\beta$, eq. (53.10) gives

$$\langle \psi_{i+1} | \psi_i \rangle = e^{-i\gamma_{i+1,i}} \left[1 - \frac{i}{2}(1 - \cos \alpha_i)\delta\beta \right]. \tag{53.15}$$

Hence,

$$\gamma_{i+1,i} = -\frac{1}{2}(1 - \cos \alpha_i)\delta\beta \tag{53.16}$$

and

$$\gamma_{\mathcal{C}} = -\frac{1}{2}\oint_{\mathcal{C}}(1 - \cos\alpha)d\beta = -\frac{1}{2}\Omega_{\mathcal{C}}, \tag{53.17}$$

where the last result follows by Stokes's theorem.

We conclude by noting that Pancharatnam's phase is the first instance of anholonomy in physics and of many similar phases that have now come to be known as *geometric phases*. It bears a deep resemblance to the Aharanov-Bohm phase and to its subsequent generalization by Berry for the phase acquired in adiabatic evolution of a general quantum system.[12]

---

[12] See M. V. Berry, *J. Mod. Opt.* **34**, 1401 (1987).

Exercise 53.1 The second circuit considered above (just after eq. (53.14)) corresponds to a complete great circle on the Poincare sphere. The net phase change for the special great circle perpendicular to the a_2 axis can be found by an elementary argument. This circuit corresponds to starting at the north pole, with a state of linear polarization, and rotating the electric field by an angle π about the direction of the beam. The polarization tensor and Poincare matrix of the beam return to their starting values at the end of the circuit, but the fields themselves are multiplied with a minus sign. Show that application of the rule (53.10) leads to a net phase change of $-2\pi/2$, in accord with the elementary argument. (*Hint:* Take $\hat{n}_1$ along the north pole, $\alpha_2 = \alpha_3 = \pi/2 + \epsilon$, where ϵ is infinitesimal, and $\beta_2 = 0$, $\beta_3 = \pi$.)

9 | The electromagnetic field of moving charges

In this chapter we will obtain the EM field of a moving charge. We first obtain a solution for the potentials due to an arbitrary time-dependent charge and current distribution that is analogous to the solution of Poisson's equation in electrostatics. We then find the fields of a steadily moving charge. Next, we use the general solution for the potentials to find the fields of an arbitrarily moving charge. This will lead us to one of the most important phenomena in electromagnetism, namely, the fact that accelerating charges radiate energy in the form of electromagnetic fields. The critical fact will turn out to be that there is a part to these fields that dies only inversely with distance, as opposed to static fields that die inversely with the square of the distance. This slower dying part will be identified with the radiated fields.[1] We conclude the chapter with a qualitative discussion of radiation from nonrelativistic charges.

The results of this chapter will be used in the next one, where we consider the radiation from a collection of moving charges, and in chapter 25, where we consider radiation from a single relativistic charge.

54 Green's function for the wave equation

Recall that in the Lorenz gauge,

$$\left(\nabla^2 - \frac{1}{c^2}\frac{\partial^2}{\partial t^2}\right)\phi = -4\pi\rho(\mathbf{r}, t), \tag{54.1}$$

$$\left(\nabla^2 - \frac{1}{c^2}\frac{\partial^2}{\partial t^2}\right)\mathbf{A} = -\frac{4\pi}{c}\mathbf{j}(\mathbf{r}, t). \tag{54.2}$$

---

[1] Section 5.7 and appendix B of Purcell (1985) contain an ingeniously simple argument, made almost entirely by drawing pictures of field lines, that shows the transversality, $1/r$ nature, and power of the radiated field. This argument was first given by Thomson (1904), chap. III, but in our view Purcell has improved on Thomson's exposition. Readers who have never seen this argument will jump with joy once they do.

Both these equations are inhomogeneous wave equations. It is therefore useful to try and solve such equations generally, just as we solved the inhomogeneous Laplace equation (i.e., Poisson equation) in the form of an integral.

Let us therefore consider the mathematical problem when the source is an impulse, i.e., a delta function, in both space and time:

$$\left(\nabla^2 - \frac{1}{c^2}\frac{\partial^2}{\partial t^2}\right)G(\mathbf{r}, t; \mathbf{r}', t') = -4\pi\delta(\mathbf{r} - \mathbf{r}')\delta(t - t'). \tag{54.3}$$

G is the Green function for the wave equation. It is directly the solution for ϕ for a unit source that flashes on at $\mathbf{r}'$ at some time t' and then flashes off immediately, and it also solves analogous problems for components of $\mathbf{A}$. Clearly, G depends only on the differences

$$\mathbf{R} = \mathbf{r} - \mathbf{r}', \quad \tau = t - t'. \tag{54.4}$$

When we wish to emphasize this fact, we will write $G(\mathbf{R}, \tau)$ instead of showing all four arguments.

Next, let us Fourier transform G with respect to the time,

$$G(\mathbf{R}, \tau) = \int_{-\infty}^{\infty} \frac{d\omega}{2\pi} e^{-i\omega\tau}\tilde{G}(\mathbf{R}, \omega), \tag{54.5}$$

and define

$$k = \omega/c. \tag{54.6}$$

Substituting eq. (54.5) in the wave equation and using the Fourier representation for $\delta(\tau)$, we get

$$(\nabla^2 + k^2)\tilde{G}(\mathbf{R}, \omega) = -4\pi\delta(\mathbf{R}). \tag{54.7}$$

We now note that as $\omega \to 0$, this is the equation for the electrostatic potential due to a point charge. Hence,

$$\tilde{G}(\mathbf{R}, 0) = \frac{1}{R}. \tag{54.8}$$

Secondly, for all other ω too, $\tilde{G}$ can be a function of only $R = |\mathbf{R}|$. Thus, eq. (54.7) becomes

$$\frac{1}{R}\left[\frac{d^2}{dR^2}(R\tilde{G}) + k^2(R\tilde{G})\right] = -4\pi\delta(\mathbf{R}). \tag{54.9}$$

Now, for both $R = 0$ and $R \neq 0$, $R\delta(\mathbf{R}) = 0$.[2] We can therefore solve for $R\tilde{G}$ immediately:

$$R\tilde{G}(\mathbf{R}, \omega) = Ae^{ikR} + Be^{-ikR}. \tag{54.10}$$

Here, A and B are general constants. But to be consistent with eq. (54.8), we must have $A + B = 1$. The most general solution to eq. (54.7) is therefore

$$\tilde{G}(\mathbf{R}, \omega) = A\frac{e^{ikR}}{R} + (1 - A)\frac{e^{-ikR}}{R}. \tag{54.11}$$

---

[2] To see this, note that the integral of $R\delta(\mathbf{R})$ over an arbitrarily small sphere surrounding the origin is zero.

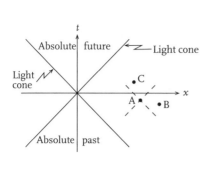
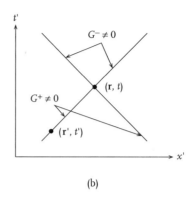

(a)   (b)

FIGURE 9.1. Space–time diagrams showing (a) the light cones and the future and past regions, and (b) the cones on which $G^\pm$ do not vanish. Note that in (b), the event $(\mathbf{r}', t')$ should be thought of as represented by a point that can be anywhere in the diagram, and $(\mathbf{r}, t)$ is a fixed event about which the light cones are drawn.

The two limiting cases of this solution, where $A = 1$ or $A = 0$, are especially important. We write these as

$$\tilde{G}^\pm(\mathbf{R}, \omega) = \frac{e^{\pm ikR}}{R}, \quad (k = \omega/c). \tag{54.12}$$

Inverting the Fourier transform on ω, we get

$$G^\pm(\mathbf{R}, \tau) = \frac{1}{R} \int_{-\infty}^{\infty} \frac{d\omega}{2\pi} e^{-i\omega(\tau \mp R/c)} = \frac{1}{R} \delta(\tau \mp R/c). \tag{54.13}$$

It is useful to represent the solutions (54.13) on a *space–time diagram* (fig. 9.1). Since a four-dimensional space can be visualized only in the mind's eye, on paper we draw one space dimension, x, and time, t. Points in this space–time diagram are known as *events*. A particle moving at a uniform speed v and passing through $x = 0$ at $t = 0$ is represented by a straight line through the origin, of slope $\pm v$ with respect to the t axis. This slope cannot exceed c in magnitude, so it follows that a particle starting at the origin can never reach points outside the region that is bounded by the lines $x = \pm ct$ and includes the positive t axis. This region is known as the *forward light cone*. Likewise, the region bounded by these same lines including the negative t axis is called the *backward light cone*. Only a particle starting inside the backward light cone can reach the origin. We can also construct similar cones around any other point in space–time; these are then said to be the light cones of that particular point. If, as in the figure, an event B is outside the light cones of event A, A and B are said to be *space-like* separated, and no particle, no light, no signal, no information can get from A to B or vice versa. If, by contrast, an event is inside the light cones of another event, the two events are said to be *time-like* separated (e.g., A and C in the figure). Events that are space-like separated appear to be simultaneous in an inertial reference frame that is moving at a sufficiently high speed (always less than c, of course). Time-like separated events, on the other hand, cannot appear simultaneous in any inertial frame. For this reason, the forward light cone of a point is also said to be its *absolute future*, while the backward light cone is said to be its *absolute past*.

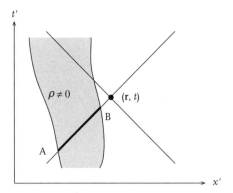

FIGURE 9.2. Graphical representation of the integrand in the inhomogeneous part of the potential due to a time-dependent charge distribution. See text for explanation.

It follows that a signal can propagate from a point only to points in its forward light cone, while it can reach that point only from points in its backward light cone. The function G^+ can be nonzero only if $\tau > 0$, i.e., $t > t'$, and only if $R = c\tau$. Thus, if we consider the point $(\mathbf{r}, t)$ as in fig. 9.1, G^+ is nonzero only on the backward light cone of that point, and G^- is nonzero only on the forward light cone. The solution G^- therefore violates causality. It represents a situation where the signal reaches points from a source that lies in the future, i.e., has not yet been turned on. While it may be an acceptable solution to eq. (54.3), it is not compatible with physically meaningful boundary conditions and must be discarded. Only the solution G^+ is physically meaningful. One also calls G^+ and G^- the *retarded* and *advanced* Green functions:

$$G^+(\mathbf{R}, \tau) \equiv G_{\text{ret}}(\mathbf{R}, \tau), \quad G^-(\mathbf{R}, \tau) \equiv G_{\text{adv}}(\mathbf{R}, \tau). \tag{54.14}$$

With the Green function in hand, we can write the solution to eq. (54.1), the inhomogeneous wave equation for ϕ, immediately:

$$\phi(\mathbf{r}, t) = \phi_0(\mathbf{r}, t) + \int\!\!\int d^3r' \, dt' \frac{\rho(\mathbf{r}', t')}{|\mathbf{r} - \mathbf{r}'|} \delta(t' - [t - R/c]). \tag{54.15}$$

Here, $\mathbf{R} = \mathbf{r} - \mathbf{r}'$, and ϕ_0 is any solution to the homogeneous wave equation. How is this solution to be determined? One way would be to fix initial conditions on $\phi(\mathbf{r}, t)$, analogous to what is commonly done in mechanical problems. In electromagnetism, though, this is not the usual situation. Usually, we are given conditions far from the system at all times; e.g., we may know that radiation is incident upon the system. In that case, ϕ_0 represents the potential of the incident radiation, while the integral term describes the radiation emitted by the system.

It is useful to represent the inhomogeneous term in eq. (54.15) graphically. We show the field point $(\mathbf{r}, t)$ along with its light cones and depict the region where $\rho(\mathbf{r}', t')$ is nonzero by a shaded band (fig. 9.2). The delta function in the integrand is nonvanishing only on the backward light cone of $(\mathbf{r}, t)$, and so only those regions of nonzero ρ lying on the backward light cone can contribute to the potential. These points lie on a three-dimensional surface (depicted by the segment AB in the figure).

It is obviously possible to integrate over t' in eq. (54.15), leaving the answer in the form of an integral over $\mathbf{r}'$ alone. There are two commonly encountered ways of writing the result (we drop the ϕ_0 from now on):

$$\phi(\mathbf{r}, t) = \int d^3 r' \frac{\rho(\mathbf{r}', t - R/c)}{R} = \int d^3 r' \frac{[\rho(\mathbf{r}', t')]_{\text{ret}}}{R}, \quad (R = \mathbf{r} - \mathbf{r}'). \tag{54.16}$$

The "ret" suffix stands for "retarded" and when applied to any quantity is meant to remind us that the quantity is to be evaluated only at space–time points satisfying the retardation condition.

Lastly, we can write the vector potential in complete parallel:

$$\mathbf{A}(\mathbf{r}, t) = \frac{1}{c} \int d^3 r' \frac{\mathbf{j}(\mathbf{r}', t - R/c)}{R} = \frac{1}{c} \int d^3 r' \frac{[\mathbf{j}(\mathbf{r}', t')]_{\text{ret}}}{R}, \quad (R = \mathbf{r} - \mathbf{r}'). \tag{54.17}$$

Equations (54.16) and (54.17) are the solutions to the Lorenz gauge wave equations for ϕ and $\mathbf{A}$, eqs. (54.1) and (54.2).

55 Fields of a uniformly moving charge

Let us now consider a charge in uniform straight-line motion, with velocity $v\hat{\mathbf{x}}$. Then, the charge and current density are given by

$$\rho(\mathbf{r}, t) = q\delta(\mathbf{r} - vt\hat{\mathbf{x}}), \quad \mathbf{j}(\mathbf{r}, t) = q v \hat{\mathbf{x}} \delta(\mathbf{r} - vt\hat{\mathbf{x}}). \tag{55.1}$$

The field due to such a moving charge can be found in various ways. Perhaps the simplest is to solve for the potentials directly. We do this in the Lorenz gauge using Fourier transform methods. It pays to introduce the abbreviations

$$x_t = x - vt, \quad \gamma = (1 - v^2/c^2)^{-1/2}. \tag{55.2}$$

Let us first consider the equation for the potential. It is obvious that the solution can depend on x and t only in the combination $x - vt$. Hence, we can write

$$\phi(\mathbf{r}, t) = \int \frac{d^3 k}{(2\pi)^3} \phi_{\mathbf{k}} e^{i\mathbf{k}\cdot\mathbf{r}} e^{-ik_x vt}. \tag{55.3}$$

Feeding this form into eq. (54.1), the wave equation for ϕ, and using the Fourier representation for $\rho(\mathbf{r}, t)$, we get

$$\left(k^2 - \frac{v^2}{c^2}k_x^2\right)\phi_{\mathbf{k}} = 4\pi q. \tag{55.4}$$

It follows that

$$\phi(\mathbf{r}, t) = \int \frac{d^3 k}{(2\pi)^3} \frac{4\pi q}{\gamma^{-2}k_x^2 + k_y^2 + k_z^2} e^{i(k_x x_t + k_y y + k_z z)}. \tag{55.5}$$

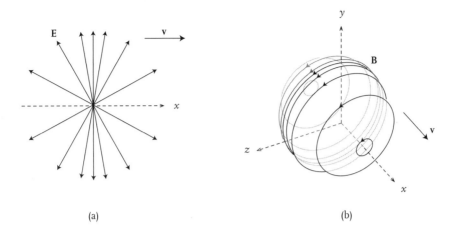

FIGURE 9.3. The fields of a uniformly moving charge.

Let us now define $k'_x = \gamma^{-1}k_x$, $k'_{y,z} = k_{y,z}$. The integral can then be written as

$$\phi(\mathbf{r}, t) = \gamma \int \frac{d^3 k'}{(2\pi)^3} \frac{4\pi q}{k'^2} e^{i(\gamma k'_x x_t + k'_y y + k'_z z)}$$
$$= \gamma \frac{q}{[\gamma^2 (x - vt)^2 + y^2 + z^2]^{1/2}}, \tag{55.6}$$

where the last line follows from the FT of $1/k'^2$.

The vector potential is now easy to determine. It is apparent that $A_y = A_z = 0$ and that the equation for A_x is essentially the same as that for ϕ. Hence,

$$\mathbf{A}(\mathbf{r}, t) = \frac{\mathbf{v}}{c}\phi = \gamma \frac{q}{c[\gamma^2 (x - vt)^2 + y^2 + z^2]^{1/2}}\mathbf{v}. \tag{55.7}$$

We are now ready to find the fields. Because ϕ and $\mathbf{A}$ depend on t only through $x - vt$, it follows that $\partial/\partial t$ can be replaced by $-(\mathbf{v} \cdot \nabla)$. Hence,

$$\mathbf{E} = -\nabla\phi - \frac{1}{c}\frac{\partial \mathbf{A}}{\partial t} = -\nabla\phi + \frac{1}{c^2}\mathbf{v}(\mathbf{v} \cdot \nabla\phi). \tag{55.8}$$

Explicitly, we have

$$\mathbf{E}(\mathbf{r}, t) = q\left(1 - \frac{v^2}{c^2}\right)\frac{\mathbf{R}}{\left[(x - vt)^2 + \left(1 - \frac{v^2}{c^2}\right)(y^2 + z^2)\right]^{3/2}}, \tag{55.9}$$

where $\mathbf{R} = \mathbf{r} - \mathbf{r}_0(t)$, with $\mathbf{r}_0(t) = (vt, 0, 0)$ being the position of the charge at time t. The magnetic field is even simpler:

$$\mathbf{B} = \nabla \times \mathbf{A} = -\frac{\mathbf{v}}{c} \times \nabla\phi = \frac{1}{c}\mathbf{v} \times \mathbf{E}. \tag{55.10}$$

We draw these fields in fig. 9.3. There are several features worthy of note. First, the fields at time t can be written purely in terms of the instantaneous position of the charge.[3]

---

[3] Although this seems to imply action at a distance, that is not true. The future position of a uniformly moving particle can be predicted perfectly. An observer far away from the particle can in effect do this by looking at how the fields at his location have been changing—this is what the Maxwell's equations do! We shall see that this feature will fail to hold when we consider accelerated charges.

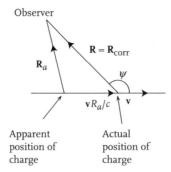

FIGURE 9.4. Actual (instantaneous at time t) and apparent position of a uniformly moving charge. This figure should also be compared with fig. 9.6 after the reader has read section 57.

Secondly, the electric field points radially outward from the charge, but it is not spherically symmetric. When $\mathbf{R} \perp \mathbf{v}$, $\mathbf{E}$ is $(1 - v^2/c^2)^{-1/2}$ times the field $q\mathbf{R}/R^3$ of a static charge. When $\mathbf{R}\|\mathbf{v}$, $\mathbf{E}$ is $(1 - v^2/c^2)$ times the static charge value. Thus, the field is increased over its static value at right angles to the direction of motion but reduced in the direction of motion. If we draw snapshots of $\mathbf{E}$ and $\mathbf{B}$, the lines of $\mathbf{E}$ all emanate from the charge, but they are squished like a pancake along the direction of motion. The lines of $\mathbf{B}$ circulate around this direction (just like they would for a steady current); $|\mathbf{B}|$ is biggest at right angles to the motion and reduced fore and aft. Thirdly, because $\mathbf{E}\|\mathbf{R}$, and $\mathbf{S} \propto \mathbf{E} \times \mathbf{B}$, $\mathbf{S} \perp \mathbf{R}$. Hence, $\mathbf{S} \cdot \widehat{\mathbf{R}} = 0$, i.e., there is no radially outward flow of energy. A charge in uniform motion does not radiate. This is as it should be, since an observer moving at the same velocity as the charge would see the charge at rest. This observer would obviously see no radiation, and since his reference frame is inertial, there can be no radiation in the lab frame as well.

Exercise 55.1 Show that the field (55.9) can also be written as

$$\mathbf{E} = \frac{q\mathbf{R}}{R^3} \frac{\left(1 - (v^2/c^2)\right)}{\left(1 - (v^2/c^2)\sin^2 \psi\right)^{3/2}}, \tag{55.11}$$

where $\mathbf{R} = \mathbf{r} - \mathbf{r}_0(t)$, and ψ is the angle between $\mathbf{v}$ and $\mathbf{R}$ (see fig. 9.4).

Exercise 55.2 Obtain eq. (55.6) by solving the wave equation by making suitable changes of variables, without using Fourier transforms.

Exercise 55.3 Find the EM field of a current-carrying wire. What density of neutralizing background charge is needed?

Exercise 55.4 Find the force between two electrons abreast of each other in a beam in which all electrons move at speed v.

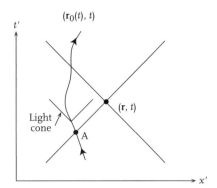

FIGURE 9.5. Space–time diagram showing the world line of a charged particle moving along the path $\mathbf{r}_0(t)$. This line must lie inside the forward and backward light cones drawn about *any* point on the line. Only the point A on this world line contributes to the potentials and the fields at the point $(\mathbf{r}, t)$.

56 Potentials of an arbitrarily moving charge—the Lienard-Wiechert solutions

Let us apply the general formulas for the potentials due to an arbitrary current and charge distribution to a single moving charge, whose path $\mathbf{r}_0(t)$, we assume to be given. Note that here we do not care how this motion is produced, although it must not be physically meaningless. Thus, in the space–time diagram (fig. 9.5), its trajectory, or *world line*, must lie inside the forward light cone at all times. If we take the particle charge to be q, the charge and current density it produces may be written as

$$\rho(\mathbf{r}, t) = q\delta[\mathbf{r} - \mathbf{r}_0(t)], \tag{56.1}$$

$$\mathbf{j}(\mathbf{r}, t) = q\dot{\mathbf{r}}_0\delta[\mathbf{r} - \mathbf{r}_0(t)]. \tag{56.2}$$

When we feed these formulas into the integral solutions for the potentials, eqs. (54.16) and (54.17), the only point $\mathbf{r}'$ that contributes to the integrals is that which satisfies

$$t_r + \frac{1}{c}|\mathbf{r}' - \mathbf{r}_0(t_r)| = t. \tag{56.3}$$

The point $\mathbf{r}_0(t_r)$ is the *retarded position* of the charge, and t_r is the *retarded time*, the time at which it had this position. Feynman presents a very nice way of thinking about this equation. Imagine, he says, that the charge is very far away, and attached to it is a little source of light. As the light from this source speeds on its way to us, the observer, the charge moves, so that by the time the light reaches us, it is no longer where it was when the light started on its journey. We therefore see the charge as it was earlier, at a position $\mathbf{r}_0(t_r)$. For that reason, we also call $\mathbf{r}_0(t_r)$ the *apparent position* of the charge (see fig. 9.6). We write this as $\mathbf{r}_a(t)$, regarding it as a function of the observation time t:

$$\mathbf{r}_a(t) = \mathbf{r}_0(t_r). \tag{56.4}$$

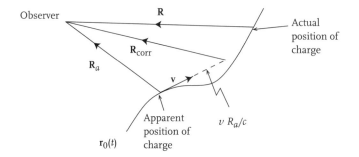

FIGURE 9.6. Actual (instantaneous at time t) and apparent position of a charge moving along an arbitrary path $\mathbf{r}_0(t)$. We also show the extrapolated or "corrected" position that the charge would have at time t if it were to continue to move uniformly with the velocity it had at time t_r, where $(t - t_r) = R_a/c$ is the time light takes to reach the observer from the apparent position.

Exercise 56.1 Find $\mathbf{r}_a(t)$ for a particle moving in a circle.

Equation (56.3) is an implicit relationship between t and t_r, and this implicitness presents the chief difficulty in finding the potentials and fields in tractable forms. It is physically obvious that the fields measured by a distant observer can depend only on characteristics of the motion at the retarded time. In order not to interrupt the subsequent argument, we discuss the kinematics of the observed motion now. First, for a fixed observer, we have the retarded velocity and acceleration,

$$\mathbf{v}_r = \frac{d}{dt}\mathbf{r}_0(t)\bigg|_{t=t_r}, \quad \mathbf{a}_r = \frac{d^2}{dt^2}\mathbf{r}_0(t)\bigg|_{t=t_r}. \tag{56.5}$$

However, we can also describe the motion entirely in terms of (the negative of) the apparent position of the source relative to the observer,[4]

$$\mathbf{R}_a(t) = \mathbf{r} - \mathbf{r}_0(t_r) = \mathbf{r} - \mathbf{r}_a(t). \tag{56.6}$$

Since we can reconstruct the actual motion if we know $\mathbf{R}_a(t)$, we should also be able to express the fields in terms of the *apparent* velocity and acceleration, obtained by differentiating $\mathbf{R}_a$ with respect to the observer's time:

$$\mathbf{v}_a = -\frac{d}{dt}\mathbf{R}_a(t), \quad \mathbf{a}_a = -\frac{d^2}{dt^2}\mathbf{R}_a(t). \tag{56.7}$$

It is also convenient to introduce the ratios

$$\boldsymbol{\beta}_r = \mathbf{v}_r/c, \quad \boldsymbol{\beta}_a = \mathbf{v}_a/c. \tag{56.8}$$

Next, let us find the relation between $\mathbf{v}_r$ and $\mathbf{v}_a$. For this, let us imagine that the source emits a very short pulse of radiation from time t_r to $t_r + \Delta t_r$ (see fig. 9.7). The pulse

---

[4] We could also define a retarded position relative to the observer, $-\mathbf{R}_r$, but it is obvious that $\mathbf{R}_r = \mathbf{R}_a$.

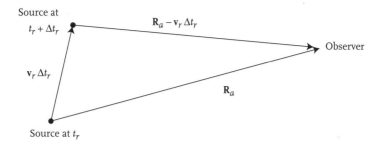

FIGURE 9.7. Position of a moving source at the start and end of a pulse of radiation, and a distant observer. Since the distance traveled by the radiation at the start and end of the pulse is different, the time of travel will be different, and the observer will see a contracted pulse if the source is getting closer to her.

reaches the observer at time

$$t_r + \frac{1}{c} R_a. \tag{56.9}$$

At the end of the pulse, the source is at a position $-\mathbf{R}_a + \mathbf{v}_r \Delta t_r$, so the pulse terminates at the observation time

$$t_r + \Delta t_r + \frac{1}{c} |\mathbf{R}_a - \mathbf{v}_r \Delta t_r| \approx t_r + \Delta t_r + \frac{1}{c}(R_a - \mathbf{v}_r \cdot \widehat{\mathbf{R}}_a \Delta t_r). \tag{56.10}$$

Therefore, the observed pulse duration is $(1 - \boldsymbol{\beta}_r \cdot \widehat{\mathbf{R}}_a)\Delta t_r$. That is,

$$\Delta t = \frac{1}{\zeta} \Delta t_r, \tag{56.11}$$

with

$$\zeta = \frac{1}{1 - \boldsymbol{\beta}_r \cdot \widehat{\mathbf{R}}_a}. \tag{56.12}$$

If the source is moving toward the observer, $\Delta t < \Delta t_r$. Hence, we shall refer to the quantity $1/\zeta$ as the *pulse compression factor* even if it need not always be less than 1. To avoid misunderstanding, we note that ζ is not related to relativistic time dilation. Rather, it is closely related to the Doppler shift.

   Equation (56.11) may be written equivalently as

$$\frac{d}{dt} = \zeta \frac{d}{dt_r}, \quad \frac{d}{dt_r} = \frac{1}{\zeta} \frac{d}{dt}. \tag{56.13}$$

It then follows that

$$\mathbf{v}_a = -\frac{d\mathbf{R}_a}{dt} = \zeta \mathbf{v}_r. \tag{56.14}$$

It also pays to derive two more formulas, for dR_a/dt, and dR_a/dt_r. We have

$$\frac{dR_a}{dt} = \widehat{\mathbf{R}}_a \cdot \frac{d\mathbf{R}_a}{dt} = -\widehat{\mathbf{R}}_a \cdot \mathbf{v}_a = -\zeta \widehat{\mathbf{R}}_a \cdot \mathbf{v}_r, \tag{56.15}$$

using eq. (56.14) in the last step. The right-hand side can be written entirely in terms of ζ, and a simple rearrangement gives

$$\zeta = 1 - \frac{1}{c}\frac{dR_a}{dt}. \tag{56.16}$$

The derivative with respect to t_r now follows by using eq. (56.13). It is easily shown that

$$\zeta^{-1} = 1 + \frac{1}{c}\frac{dR_a}{dt_r}. \tag{56.17}$$

One can relate $\mathbf{a}_a$ and $\mathbf{a}_r$ in the same way, although we shall not need to do so.

It is now straightforward to find the potentials for a moving charge. We substitute eq. (56.1) into the solution for ϕ in the form (54.15). This yields

$$\phi(\mathbf{r}, t) = q \iint d^3r' \, dt' \frac{1}{|\mathbf{r}-\mathbf{r}'|} \delta(\mathbf{r}' - \mathbf{r}_0(t')) \, \delta\big[t' - t + \tfrac{1}{c}|\mathbf{r} - \mathbf{r}_0(t')|\big]. \tag{56.18}$$

The second delta function is satisfied only at $t' = t_r$. If we recall that for a function $f(x)$ with a zero at $x = x_0$, $\delta[f(x)] = \delta(x - x_0)/|f'(x_0)|$, we find

$$\delta[t' - \cdots] = \delta(t' - t_r) \times \left[1 + \frac{1}{2|\mathbf{r} - \mathbf{r}_0(t_r)|}\left(-2\frac{\mathbf{v}_r}{c}\cdot\mathbf{R}_a\right)\right]^{-1} = \zeta\delta(t' - t_r). \tag{56.19}$$

This result is completely understandable in light of our previous discussion. It renders the t' integral trivial and converts the argument of the remaining delta function to $\mathbf{r}' - \mathbf{r}_0(t_r)$. The $\mathbf{r}'$ integral is then also easy, and we get

$$\phi(\mathbf{r}, t) = q\frac{\zeta}{R_a} = \frac{q}{R_a - \boldsymbol{\beta}_r \cdot \mathbf{R}_a}. \tag{56.20}$$

In exactly the same way, we have

$$\mathbf{A}(\mathbf{r}, t) = \frac{q}{c} \iint d^3r' \, dt' \frac{\mathbf{v}_0(t')}{|\mathbf{r}-\mathbf{r}'|} \delta(\mathbf{r}' - \mathbf{r}_0(t')) \, \delta\big[t' - t + \tfrac{1}{c}|\mathbf{r} - \mathbf{r}_0(t')|\big], \tag{56.21}$$

which simplifies to

$$\mathbf{A}(\mathbf{r}, t) = q\frac{\zeta\boldsymbol{\beta}_r}{R_a} = \frac{q\boldsymbol{\beta}_r}{R_a - \boldsymbol{\beta}_r \cdot \mathbf{R}_a}. \tag{56.22}$$

Equations (56.20) and (56.22) are known as the *Lienard-Wiechert potentials*.

57 Electromagnetic fields of an arbitrarily moving charge

The fields produced by an arbitrarily moving charge can be presented either in terms of the apparent position, velocity, and acceleration or in terms of retarded quantities. It pays to find both forms, as they provide different insights. Historically, the retarded forms were found first, but we shall proceed in reverse order.

Let us find $\mathbf{E}$ first. It is easiest to start from eqs. (56.18) and (56.21) directly, but with $\mathbf{r}'$ instead of $\mathbf{r}_0(t)$ in the time delta function. There are three pieces to $\mathbf{E}$, which we call $\mathbf{E}_1$,

E_2, and E_3. The first two arise from $\nabla\phi$, and the last from $\partial A/\partial t$. To save writing, we set $q = 1$ and restore it at the end. With $\mathbf{R} = \mathbf{r} - \mathbf{r}'$, we have

$$E_1 = \iint d^3r' \, dt' \frac{\widehat{\mathbf{R}}}{R^2} \delta(\mathbf{r}' - \mathbf{r}_0(t')) \, \delta\left[t - t' - \tfrac{1}{c}R\right] = \zeta \frac{\widehat{\mathbf{R}}_a}{R_a^2}. \tag{57.1}$$

Next,

$$E_2 = \frac{1}{c} \iint d^3r' \, dt' \frac{\widehat{\mathbf{R}}}{R} \delta(\mathbf{r}' - \mathbf{r}_0(t')) \, \delta'\left[t - t' - \tfrac{1}{c}R\right],$$

$$= \frac{1}{c}\frac{d}{dt} \iint d^3r' \, dt' \frac{\widehat{\mathbf{R}}}{R} \delta(\mathbf{r}' - \mathbf{r}_0(t')) \, \delta\left[t - t' - \tfrac{1}{c}R\right]. \tag{57.2}$$

Last,

$$E_3 = -\frac{1}{c^2}\frac{d}{dt} \iint d^3r' \, dt' \frac{\mathbf{v}_0(t')}{R} \delta(\mathbf{r}' - \mathbf{r}_0(t')) \, \delta\left[t - t' - \tfrac{1}{c}R\right]. \tag{57.3}$$

Writing E_{23} for $E_2 + E_3$, we have

$$E_{23} = \frac{1}{c}\frac{d}{dt} \iint d^3r' \, dt' \frac{\widehat{\mathbf{R}} - \boldsymbol{\beta}(t')}{R} \delta(\mathbf{r}' - \mathbf{r}_0(t')) \, \delta\left[t - t' - \tfrac{1}{c}R\right]$$

$$= \frac{1}{c}\frac{d}{dt}\left[\frac{\zeta}{R_a}(\widehat{\mathbf{R}}_a - \boldsymbol{\beta}_r)\right]. \tag{57.4}$$

We now rearrange the total field $E_1 + E_{23}$ by the order of the time derivative. In doing this, we use eq. (56.16) for ζ. It is clear that we will then get three terms, $\mathbf{E}^{(0)}$, $\mathbf{E}^{(1)}$, and $\mathbf{E}^{(2)}$, where the superscript denotes the order of the time derivative. The term $\mathbf{E}^{(0)}$ arises entirely from E_1 and is given by

$$\mathbf{E}^{(0)} = \frac{\widehat{\mathbf{R}}_a}{R_a^2}. \tag{57.5}$$

The term $\mathbf{E}^{(1)}$ arises from E_1 and the first term in E_{23}:

$$\mathbf{E}^{(1)} = -\frac{1}{c}\frac{\widehat{\mathbf{R}}_a}{R_a^2}\frac{dR_a}{dt} + \frac{1}{c}\frac{d}{dt}\frac{\widehat{\mathbf{R}}_a}{R_a}. \tag{57.6}$$

In the second term, we write $\widehat{\mathbf{R}}_a/R_a$ as the product of two factors, $\widehat{\mathbf{R}}_a/R_a^2$ and R_a, and differentiate by the product rule. This gives

$$\mathbf{E}^{(1)} = \frac{R_a}{c}\frac{d}{dt}\frac{\widehat{\mathbf{R}}_a}{R_a^2}. \tag{57.7}$$

Lastly, $\mathbf{E}^{(2)}$ arises from both terms in E_{23}. We have

$$\mathbf{E}^{(2)} = \frac{1}{c^2}\frac{d}{dt}\left[-\frac{\widehat{\mathbf{R}}_a}{R_a}\frac{dR_a}{dt} + \frac{1}{R_a}\frac{d\mathbf{R}_a}{dt}\right]. \tag{57.8}$$

Writing $\mathbf{R}_a = R_a \widehat{\mathbf{R}}_a$ in the second term and using the product rule to differentiate yields

$$\mathbf{E}^{(2)} = \frac{1}{c^2}\frac{d}{dt}\frac{d\widehat{\mathbf{R}}_a}{dt}. \tag{57.9}$$

Adding together the various orders and restoring q, we get

$$\mathbf{E} = q\frac{\widehat{\mathbf{R}}_a}{R_a^2} + q\frac{R_a}{c}\frac{d}{dt}\frac{\widehat{\mathbf{R}}_a}{R_a^2} + q\frac{1}{c^2}\frac{d^2\widehat{\mathbf{R}}_a}{dt^2}. \tag{57.10}$$

The $\mathbf{B}$ field is obtained by applying the curl to eq. (56.21). We denote by $\mathbf{B}_1$ and $\mathbf{B}_2$, respectively, the terms where the curl acts on the $1/R$ and the delta function in the integrand:

$$\mathbf{B}_1 = \iint d^3r'\,dt'\frac{\mathbf{v}_0(t')\times\widehat{\mathbf{R}}}{c\,R^2}\delta(\mathbf{r}'-\mathbf{r}_0(t'))\,\delta\big[t-t'-\tfrac{1}{c}R\big],$$

$$= \zeta\frac{\mathbf{v}_r\times\widehat{\mathbf{R}}_a}{c\,R_a^2} = \frac{\mathbf{v}_a\times\widehat{\mathbf{R}}_a}{c\,R_a^2}; \tag{57.11}$$

$$\mathbf{B}_2 = \iint d^3r'\,dt'\frac{\mathbf{v}_0(t')\times\widehat{\mathbf{R}}}{c^2\,R}\delta(\mathbf{r}'-\mathbf{r}_0(t'))\,\delta'\big[t-t'-\tfrac{1}{c}R\big],$$

$$= \frac{1}{c^2}\frac{d}{dt}\zeta\frac{\mathbf{v}_r\times\widehat{\mathbf{R}}_a}{R_a} = \frac{d}{dt}\frac{\mathbf{v}_a\times\widehat{\mathbf{R}}_a}{c^2\,R_a}. \tag{57.12}$$

Thus, again restoring q,

$$\mathbf{B} = q\frac{\mathbf{v}_a\times\widehat{\mathbf{R}}_a}{c\,R_a^2} + q\frac{d}{dt}\frac{\mathbf{v}_a\times\widehat{\mathbf{R}}_a}{c^2\,R_a}. \tag{57.13}$$

Equation (57.10) is due to Feynman, and eq. (57.13) is due to Heaviside. Collectively, they are known as the *Heaviside-Feynman formulas* for the fields.

One important consequence of eqs. (57.10) and (57.13) is that

$$\mathbf{B} = \widehat{\mathbf{R}}_a\times\mathbf{E}. \tag{57.14}$$

Thus, $\mathbf{B}\perp\mathbf{E}$, but $|\mathbf{B}|\neq|\mathbf{E}|$. To see that eq. (57.14) is true, note that $\widehat{\mathbf{R}}_a\times\mathbf{E}^{(0)}$ vanishes. The first term in eq. (57.13) is easily seen to be $\widehat{\mathbf{R}}_a\times\mathbf{E}^{(1)}$. The second term is seen to be $\widehat{\mathbf{R}}_a\times\mathbf{E}^{(2)}$ if we note that

$$\mathbf{v}_a\times\widehat{\mathbf{R}}_a = \widehat{\mathbf{R}}_a\times\left(R_a\frac{d}{dt}\widehat{\mathbf{R}}_a + \widehat{\mathbf{R}}_a\frac{d}{dt}R_a\right) = R_a\widehat{\mathbf{R}}_a\times\frac{d}{dt}\widehat{\mathbf{R}}_a. \tag{57.15}$$

Equation (57.14) then follows.

The first term in eq. (57.10) gives the Coulomb field produced by a charge at the retarded or apparent position. The second term takes the time derivative of this field and multiplies it with the delay. Thus, this term offsets the effect of the delay to some extent. Together, the first two terms produce a field very close to the instantaneous Coulomb field for a slow-moving charge. This is especially true close to the charge, where these terms dominate. The last term in eq. (57.10) is the most interesting one, and it contains the radiation field. We shall show in the next section that it decays as just $1/R$ for an accelerating charge, and so it is the dominant field at very large distances.

The Heaviside-Feynman formulas are compact and elegant, but they do not explicitly separate the $1/R^2$ and $1/R$ parts of the field. This is achieved by the older, Lienard-Wiechert forms, in terms of retarded quantities. To find $\mathbf{E}$ in this form, we write the time derivative in eq. (57.4) in terms of t_r using eq. (56.13):

$$\mathbf{E}_{23} = \frac{1}{c} \zeta \frac{d}{dt_r} \left[\frac{\zeta}{R_a} (\widehat{\mathbf{R}}_a - \boldsymbol{\beta}_r) \right]. \tag{57.16}$$

We now observe that $\zeta / R_a = 1/(R_a - \boldsymbol{\beta}_r \cdot \mathbf{R}_a)$. Thus,

$$\frac{d}{dt_r} \frac{\zeta}{R_a} = -\frac{\zeta^2}{R_a^2} \left(\frac{dR_a}{dt_r} - \frac{\mathbf{a}_r}{c} \cdot \mathbf{R}_a + \frac{v_r^2}{c} \right), \tag{57.17}$$

$$\frac{d}{dt_r} (\widehat{\mathbf{R}}_a - \boldsymbol{\beta}_r) = - \left(\frac{\mathbf{v}_r}{R_a} + \frac{\mathbf{R}_a}{R_a^2} \frac{dR_a}{dt_r} - \frac{\mathbf{a}_r}{c} \right). \tag{57.18}$$

The procedure for finding $\mathbf{E}$ is now as follows. We use the two preceding rules to express the time derivative in eq. (57.16). We then eliminate dR_a/dt_r everywhere in favor of ζ using eq. (56.17). Adding the result for $\mathbf{E}_1$, we obtain a lengthy expression for $\mathbf{E}$. However, no further unobvious manipulations are needed, and a page of straightforward algebra yields the final answer,

$$\mathbf{E} = \frac{q(1 - \beta_r^2)}{(R_a - \boldsymbol{\beta}_r \cdot \mathbf{R}_a)^3} (\mathbf{R}_a - \boldsymbol{\beta}_r R_a) + \frac{q}{c^2 (R_a - \boldsymbol{\beta}_r \cdot \mathbf{R}_a)^3} (\dot{\mathbf{v}}_r \times (\mathbf{R}_a - \boldsymbol{\beta}_r R_a)) \times \mathbf{R}_a. \tag{57.19}$$

The $\mathbf{B}$ field can then be found by eq. (57.14).

The form (57.19) naturally separates into two parts: the second term depends on the acceleration at the retarded time, $\dot{\mathbf{v}}_r$, while the first does not. The first term is all that survives for a uniformly moving charge. It decays as $1/R^2$ as $R \to \infty$ and is a sort of retarded Coulomb field. Since it decays as $1/R^2$, so does the corresponding $\mathbf{B}$ field, and the contribution to the Poynting vector decays as $1/R^4$. The integral of this part of the energy flux over a distant surface of radius R therefore decays as $1/R^2$, which vanishes as $R \to \infty$. This part of the field is also known as the *nonradiative* or the *induction* part.

The second term in eq. (57.19), on the other hand, decays only as $1/R$. The corresponding contribution to $\mathbf{B}$ also varies as $1/R$, and to $\mathbf{S}$ varies as $1/R^2$, so the total energy flux through a surface at distance R remains constant as $R \to \infty$. The presence of this term means that energy is being radiated away, and it is therefore the radiation field. We study this in more depth below and in the next chapter.

Another feature of eq. (57.19) whose implications we shall see only in chapter 25 is worth mentioning here. The powers of $(R_a - \boldsymbol{\beta}_r \cdot \mathbf{R}_a)$ in the denominators arise from the pulse compression factor and can become very small in the direction of the particle's motion if its speed is close to that of light. At such speeds, the fields and the radiated power are concentrated in the direction of motion. This effect is exploited in synchrotrons.

For completeness, let us note that the surface integral of the part of the Poynting vector due to the radiative part of $\mathbf{E}$ and the nonradiative part of $\mathbf{B}$ (or vice versa) vanishes as $1/R$ as $R \to \infty$. Thus, very far from the moving charge, only the radiative parts of $\mathbf{E}$ and $\mathbf{B}$ need be kept in order to understand energy flow. The cross terms are necessary, however, for understanding the radiation of angular momentum.

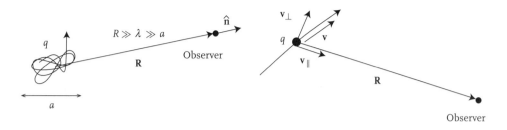

FIGURE 9.8. A charge q moving in a region of size $a \ll R$. On the right we show the components of the velocity of the charge parallel and perpendicular to the line to the observer.

58 Radiation from accelerated charges: qualitative discussion

The formulas (57.19) and (57.10) enable us to draw some important conclusions regarding radiation from moving charges in the special case where the motion of the charge is nonrelativistic and confined to a region whose size, a, is small compared with the distance to the observer (fig. 9.8). Let us frame the discussion using eq. (57.10). Since we are considering the radiation, we need focus only on the third term. We first note that $d\widehat{\mathbf{R}}_a/dt$ is perpendicular to $\widehat{\mathbf{R}}_a$, by virtue of $\widehat{\mathbf{R}}_a$ being a unit vector. Next, we note that if the acceleration of the particle were directed toward the observer, the unit vector $\widehat{\mathbf{R}}_a$ would not change at all. Thus, the largest part of $(d^2\widehat{\mathbf{R}}_a/dt^2)$ comes from the component of the acceleration $\ddot{\mathbf{R}}_a$ perpendicular to $\mathbf{R}_a$. For any vector $\mathbf{w}$ let us define the components parallel and perpendicular to $\mathbf{R}_a$:

$$\mathbf{w}_\parallel = (\mathbf{w} \cdot \widehat{\mathbf{R}}_a)\widehat{\mathbf{R}}_a, \quad \mathbf{w}_\perp = -(\mathbf{w} \times \widehat{\mathbf{R}}_a) \times \widehat{\mathbf{R}}_a. \tag{58.1}$$

Then, ignoring terms of $O(1/R_a^2)$, we have

$$\frac{\partial^2 \widehat{\mathbf{R}}_a}{\partial t^2} \approx \frac{\ddot{\mathbf{R}}_{a,\perp}}{R_a} = -\frac{1}{R_a}(\ddot{\mathbf{r}}_0(t_r))_\perp \tag{58.2}$$

and

$$\mathbf{E} \simeq -\frac{q}{R_a c^2}\ddot{\mathbf{r}}_{0\perp}\left(t - \frac{R_a}{c}\right). \tag{58.3}$$

Next, because the distance R_a is large compared with the range of motion, we can neglect the difference between R_a and R, the instantaneous distance. Similarly, we can neglect the variation in the delay. Thus, we can replace R_a by R everywhere, obtaining

$$\mathbf{E} \simeq -\frac{q}{Rc^2}\ddot{\mathbf{r}}_{0\perp}\left(t - \frac{R}{c}\right). \tag{58.4}$$

The same conclusion is reached by starting from eq. (57.19) if we recall that R is very large and that $v \ll c$. Only the second term in this equation survives, and we can drop all corrections to $\mathbf{R}_a$ of relative order v/c.

Equation (58.4) is a marvelously simple result. The electric field of the radiation is proportional to two factors. The first is $1/R$; the field dies off as the inverse of the distance,

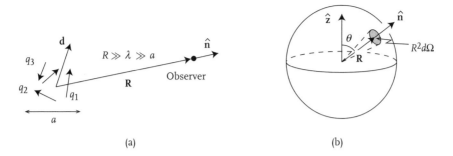

FIGURE 9.9. (a) A collection of moving charges q_1, q_2, etc., confined to a region of size $a \ll R$, with instantaneous dipole moment $\mathbf{d}$. (b) Geometrical relationship between $\mathbf{R}$, $\hat{\mathbf{n}}$, etc., showing the definition of solid angle.

as opposed to the inverse square for static fields. The second factor is the negative of the acceleration of the apparent position of the particle as seen by the observer, perpendicular to the line of sight to the apparent position. A charge accelerating in a straight line does not radiate along that line. We will study this point in greater detail in the next chapter.

We can apply eq. (58.4) to a collection of charges, all moving at nonrelativistic speeds, and all limited to a region of size a (fig. 9.9). Let $\mathbf{R}$ be the location of the field or observation point with respect to an origin somewhere in the region of the charges. Let us denote the unit vector along $\mathbf{R}$ by $\hat{\mathbf{n}}$ and assume that $a \ll R$. The delay can then be taken to be the same for all the charges, equal to R/c. In eq. (58.4), the passage from one to many charges is made by making the replacement

$$q\ddot{\mathbf{r}}_{0\perp} \to \sum_a q_a \ddot{\mathbf{r}}_{a\perp} = \ddot{\mathbf{d}}_{\perp}, \tag{58.5}$$

where $\mathbf{d}$ is the dipole moment of the collection. Hence, the fields are given by

$$\mathbf{E} = \frac{1}{c^2 R} \hat{\mathbf{n}} \times \left(\hat{\mathbf{n}} \times \ddot{\mathbf{d}}_{\text{ret}} \right), \tag{58.6}$$

$$\mathbf{B} = \hat{\mathbf{n}} \times \mathbf{E}. \tag{58.7}$$

In fact, as we shall see in the next chapter, these formulas require the stronger condition $R \gg ca/v$ in order to be valid. The quantity $(ca/v) \equiv \lambda$ is the characteristic wavelength of the radiation emitted. The region $R \gg \lambda$ is known as the *radiation zone* or the *far zone*.

The Poynting vector of the radiation is given by

$$\mathbf{S} = \frac{c}{4\pi} \mathbf{E} \times \mathbf{B} = \frac{c}{4\pi} E^2 \hat{\mathbf{n}}. \tag{58.8}$$

The area subtended by a cone of solid angle $d\Omega$ on a sphere of radius R is $R^2 d\Omega$, so the power emitted in this cone is

$$dP = \frac{c}{4\pi} \frac{1}{c^4 R^2} \left| \left(\hat{\mathbf{n}} \times \ddot{\mathbf{d}}_{\text{ret}} \right) \right|^2 R^2 d\Omega. \tag{58.9}$$

The angular distribution of radiation is traditionally written as a formula for the ratio $dP/d\Omega$.[5] In the present case,

$$\frac{dP}{d\Omega} = \frac{1}{4\pi c^3} \left| (\hat{\mathbf{n}} \times \ddot{\mathbf{d}}_{\mathrm{ret}}) \right|^2. \tag{58.10}$$

When we integrate this over all $\hat{\mathbf{n}}$, we obtain the total power radiated:

$$P = \frac{2}{3c^3} \ddot{\mathbf{d}}_{\mathrm{ret}}^2. \tag{58.11}$$

When we apply eq. (58.11) to a single charge, we obtain

$$P = \frac{2q^2}{3c^3} \dot{\mathbf{v}}_{\mathrm{ret}}^2. \tag{58.12}$$

This is known as *Larmor's formula*.

Exercise 58.1 Consider an electron moving in a circular orbit of radius r. Describe the polarization of the radiation and the total power radiated, assuming that the motion is nonrelativistic.

Solution: Let the orbit be anticlockwise, in the xy plane, and at a frequency ω. The speed is then $v = \omega r$; by assumption $v \ll c$. The radiation is at a frequency ω. It is left-circularly polarized along the $+\hat{\mathbf{z}}$ axis, right-circularly polarized along $-\hat{\mathbf{z}}$, and linearly polarized along $\hat{\mathbf{z}} \times \hat{\mathbf{n}}$ in the xy plane. Since the orbit is circular, $\ddot{\mathbf{r}} = -\omega^2 \mathbf{r}$, and $\ddot{\mathbf{d}} = e\omega^2 \mathbf{r}$. Thus, the total power radiated is

$$P = \frac{2e^2 \omega^4 r^2}{3c^3}. \tag{58.13}$$

Since energy is being lost to radiation, the orbit cannot be perfectly circular. If we denote the kinetic energy $m\omega^2 r^2/2$ by $\mathcal{E}$ and equate $\dot{\mathcal{E}}$ to $-P$, we obtain $d\mathcal{E}/dt = -\mathcal{E}/\tau$, with $\tau = 3mc^3/4e^2\omega^2$. Thus, $\omega\tau \sim 10^{23}\,\mathrm{sec}^{-1}/\omega$; unless $\omega \geq 10^{22}\,\mathrm{sec}^{-1}$ (which is a very high frequency for most situations where classical physics is applicable), $\omega\tau \gg 1$, implying that energy is lost very slowly, and the electron spirals into the origin slowly, with the decrease in the radius in one orbit being very small in comparison with the radius itself.

Note that we have not specified how the circular orbit is maintained. Thus, the analysis applies to an electron in a magnetic field (cyclotron motion) or in a (classical) atom.

---

[5] Note that $dP/d\Omega$ is not a derivative of some function of $\hat{\mathbf{n}}$, and the only reason for writing things this way is to remind ourselves that this is an angular power distribution, which when integrated over all angles will yield the total power.

10 | Radiation from localized sources

In this chapter we continue our study of radiation and consider the problem in which the current and charge distributions are localized, specified functions of **r** and t. We first develop formalism for representing the EM fields in the frequency domain. We then show that far from the source, the radiation field at a given frequency, ω, is very simply related to the Fourier transform of the current distribution at that frequency and a wave vector **k**, whose direction is that from the source to the observer and whose magnitude is that of light at a frequency ω, i.e., $k = \omega/c$. This result enables us to derive formulas for the power spectrum of several types of sources. We then specialize to the case where the source is nonrelativistic, i.e., the charges are moving much more slowly than light. In this case, the Fourier transform of the current distribution can be developed in a multipole expansion analogous to that in electro- and magnetostatics, so the EM field is given by successively higher time derivatives of the various multipoles. We also study the field close to a radiating system, and the angular momentum radiated by it. Finally, we examine the issue of radiation reaction, i.e., the question of how the loss of energy by an accelerated charge is to be reflected in the equations of motion for that charge.

Relativistic sources are studied in chapter 25.

59 General frequency-domain formulas for fields

In this section, we rewrite the integral solutions for the fields from a general time-dependent charge and current distribution that we found in section 54 in the frequency domain. Such an approach is extremely profitable. If the source is turned on and then turned off after some time, it can be Fourier analyzed in time. The same is true if the motion is periodic or quasiperiodic. In both cases, Fourier analysis provides a natural way of discussing the problem.

Let us first write the Lorenz gauge solution for the potentials as

$$\phi(\mathbf{r}, t) = \int d^3x' \frac{\rho(\mathbf{r}', t_r)}{|\mathbf{r} - \mathbf{r}'|}, \tag{59.1}$$

$$\mathbf{A}(\mathbf{r}, t) = \frac{1}{c} \int d^3x' \frac{\mathbf{j}(\mathbf{r}', t_r)}{|\mathbf{r} - \mathbf{r}'|}, \tag{59.2}$$

where we have introduced the retarded or emission time,

$$t_r = t - \frac{1}{c}|\mathbf{r} - \mathbf{r}'|, \tag{59.3}$$

which depends on both the source and field points. Next, let us write the FT with respect to time as

$$\rho(\mathbf{r}, t) = \int \frac{d\omega}{2\pi} e^{-i\omega t} \rho_\omega(\mathbf{r}), \tag{59.4}$$

$$\mathbf{j}(\mathbf{r}, t) = \int \frac{d\omega}{2\pi} e^{-i\omega t} \mathbf{j}_\omega(\mathbf{r}), \tag{59.5}$$

where for brevity we show the frequency argument as a subscript rather than in parentheses. Similarly, the FT with respect to both space and time is written as

$$\rho(\mathbf{r}, t) = \int \int \frac{d^3q\, d\omega}{(2\pi)^4} e^{i(\mathbf{q}\cdot\mathbf{r} - \omega t)} \rho_{\mathbf{q}\omega}, \tag{59.6}$$

etc. We will use $\mathbf{q}$ for the general momentum–space vector, reserving $\mathbf{k}$ for a particular value. The reader should not confuse $\mathbf{q}$ with the charge.

In terms of ρ_ω, and expanding t_r in full, eq. (59.1) reads (with $R = |\mathbf{r} - \mathbf{r}'|$)

$$\phi(\mathbf{r}, t) = \int d^3x' \int \frac{d\omega}{2\pi} \frac{\rho_\omega(\mathbf{r}')}{|\mathbf{r} - \mathbf{r}'|} e^{-i\omega(t - R/c)}$$

$$= \int \frac{d\omega}{2\pi} e^{-i\omega t} \left[\int d^3x' \frac{\rho_\omega(\mathbf{r}')}{|\mathbf{r} - \mathbf{r}'|} e^{ikR} \right], \quad (k = \omega/c). \tag{59.7}$$

If we define the Fourier transforms of the potentials and fields in parallel with eqs. (59.4) and (59.5), then the expression in square brackets is $\phi_\omega(\mathbf{r})$. A similar expression applies to $\mathbf{A}_\omega(\mathbf{r})$. Thus,

$$\begin{pmatrix} \phi_\omega(\mathbf{r}) \\ \mathbf{A}_\omega(\mathbf{r}) \end{pmatrix} = \int d^3x' \frac{e^{ikR}}{R} \begin{pmatrix} \rho_\omega(\mathbf{r}') \\ c^{-1}\mathbf{j}_\omega(\mathbf{r}') \end{pmatrix}, \quad (k = \omega/c). \tag{59.8}$$

If we note that the kernel in eq. (59.8) is nothing but the Green function $\tilde{G}(\mathbf{R}, \omega)$ for the wave equation in the frequency domain, we realize that the result could also have been derived directly by taking the time Fourier transform of the wave equation for ρ, say, and using the Green function $\tilde{G}$ of the operator $\nabla^2 + k^2$. The derivation given above has the benefit of showing where the retardation effects are contained. They lie in the e^{ikR} phase factor in the integrand. This phase factor will lead to the most remarkable feature of radiation fields, namely, that if R is the distance from the source, they die off as $1/R$, as opposed to $1/R^2$ for static fields. The phenomenon of retardation is essential for this fact, as we also saw when we considered the fields due to an arbitrarily moving point charge in the previous chapter. For the purposes of this chapter, retardation effects are easier to handle in the frequency domain than in the time domain.

To obtain the fields $\mathbf{E}_\omega$ and $\mathbf{B}_\omega$ in the frequency domain, we must differentiate the potentials. The relevant formulas are

$$\mathbf{E}_\omega = -\nabla\phi_\omega + ik\mathbf{A}_\omega, \tag{59.9}$$

$$\mathbf{B}_\omega = \nabla \times \mathbf{A}_\omega. \tag{59.10}$$

However, in the region outside the source, $\mathbf{j} = 0$, so the Ampere-Maxwell law implies

$$-ik\mathbf{E}_\omega = \nabla \times \mathbf{B}_\omega. \tag{59.11}$$

Likewise, Faraday's law reads

$$ik\mathbf{B}_\omega = \nabla \times \mathbf{E}_\omega. \tag{59.12}$$

Hence, it is enough to know only one of $\mathbf{E}_\omega$ or $\mathbf{B}_\omega$ outside the source in order to find the other. The physical reason for this is that in the source region ρ and $\mathbf{j}$ are connected via the continuity relation, which in frequency space reads

$$-i\omega\rho_\omega + \nabla \cdot \mathbf{j}_\omega = 0. \tag{59.13}$$

60 Far-zone fields

Let us apply the above formulas to the case where the sources are localized to a region of size a, and the characteristic frequency associated with the motion of the charges or currents is ω_0. We will see that the fields have a fundamentally different character depending on whether R is large or small compared to the length $\lambda_0 \equiv 2\pi c/\omega_0$. This length is, of course, the wavelength of light whose frequency is ω_0. In this section we find the fields in the *far zone, radiation zone,* or *wave zone,* defined by the conditions $a \ll R$, $\lambda_0 \ll R$. There is no restriction on the ratio a/λ_0. If v is the characteristic velocity of the charges, then $a \sim v/\omega_0$, and $a/\lambda_0 \sim v/c$. Hence, the results of this section hold for relativistic sources too.

It is convenient in this section to use $\mathbf{R}$ for the field point rather than $\mathbf{r}$. If $R \gg a$, then for $\mathbf{r}'$ in the source region, fig. 10.1 shows that

$$|\mathbf{R} - \mathbf{r}'| = R - \mathbf{r}' \cdot \hat{\mathbf{n}} + O(r'^2/R), \tag{60.1}$$

where $\hat{\mathbf{n}} = \mathbf{R}/R$. This corresponds to approximating t_r as

$$t_r \approx t - \frac{1}{c}(R - \mathbf{r}' \cdot \hat{\mathbf{n}}). \tag{60.2}$$

Using eq. (60.1) in eq. (59.8), and remembering that $\mathbf{R}$ is now the field point, we get

$$\begin{pmatrix} \phi_\omega(\mathbf{R}) \\ \mathbf{A}_\omega(\mathbf{R}) \end{pmatrix} \approx \int d^3x' \frac{e^{ik(R-\mathbf{r}'\cdot\hat{\mathbf{n}})}}{R} \begin{pmatrix} \rho_\omega(\mathbf{r}') \\ c^{-1}\mathbf{j}_\omega(\mathbf{r}') \end{pmatrix} = \frac{e^{ikR}}{R} \begin{pmatrix} \rho_{\mathbf{k}\omega} \\ c^{-1}\mathbf{j}_{\mathbf{k}\omega} \end{pmatrix}, \tag{60.3}$$

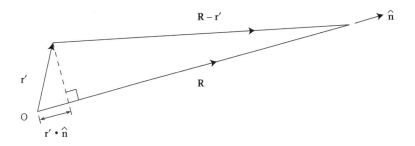

FIGURE 10.1. Diagram to show eq. (60.1).

where

$$\mathbf{k} = \frac{\omega}{c}\hat{\mathbf{n}}. \tag{60.4}$$

Note that we neglected the $\mathbf{r}' \cdot \hat{\mathbf{n}}$ correction to R in the $|\mathbf{R} - \mathbf{r}'|^{-1}$ factor, since it would give rise to a term of order $1/R^2$, but that we kept it in the exponential phase factor. The reason is that even though $\mathbf{r}' \cdot \hat{\mathbf{n}}$ is small in comparison with R, the phase factor $\exp(-i\mathbf{k} \cdot \mathbf{r}')$ may still vary considerably with $\mathbf{r}'$.[1]

It should also be noted that whenever we write $\mathbf{j}_{\mathbf{k}\omega}$ or $\rho_{\mathbf{k}\omega}$ with a suffix $\mathbf{k}$ rather than $\mathbf{q}$, we will mean that the spatial Fourier transform is to be taken at the special momentum–space vector related to the frequency ω via eq. (60.4).

We now need to take gradients of ϕ_ω and $\mathbf{A}_\omega$. In doing so, it is necessary to operate only on the e^{ikR} factor, for operating on the factor $1/R$ produces a term of order $1/R^2$, while operating on the e^{ikR} factor produces a term of order $1/R$. There is also an $\mathbf{R}$ dependence in $\rho_{\mathbf{k}\omega}$ and $\mathbf{j}_{\mathbf{k}\omega}$ because $\mathbf{k}\|\hat{\mathbf{n}}$, and $\hat{\mathbf{n}}$ varies with $\mathbf{R}$. However, it is obvious that $\nabla\hat{\mathbf{n}} \sim 1/R$. Hence, these gradients may also be neglected. Thus,

$$\mathbf{E}_\omega = -\nabla\phi_\omega + ik\mathbf{A}_\omega$$
$$= ik\frac{e^{ikR}}{cR}(-c\rho_{\mathbf{k}\omega}\hat{\mathbf{n}} + \mathbf{j}_{\mathbf{k}\omega}), \tag{60.5}$$
$$\mathbf{B}_\omega = \nabla \times \mathbf{A}_\omega$$
$$= ik\frac{e^{ikR}}{cR}(\hat{\mathbf{n}} \times \mathbf{j}_{\mathbf{k}\omega}). \tag{60.6}$$

We now use the fact that the continuity equation implies

$$\omega\rho_{\mathbf{k}\omega} = k(\hat{\mathbf{n}} \cdot \mathbf{j}_{\mathbf{k}\omega}). \tag{60.7}$$

This enables us to write the fields entirely in terms of $\mathbf{j}_{\mathbf{k}\omega}$. Introducing the components parallel and perpendicular to the line of sight from the observer to the source,

$$\mathbf{j}_{\mathbf{k}\omega}^{\|} = \hat{\mathbf{n}}(\hat{\mathbf{n}} \cdot \mathbf{j}_{\mathbf{k}\omega}), \tag{60.8}$$

---

[1] The condition for this variation to be small at the characteristic frequency ω_0 is $\omega_0 a \ll c$, or $v \ll c$. We therefore expect further simplification of the results when the source is nonrelativistic, which we shall see in section 62.

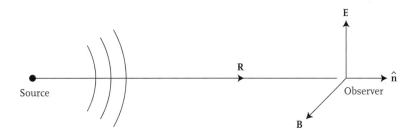

FIGURE 10.2. Radiation-zone fields. **E**, **B**, and the direction of propagation make an orthogonal triad, and $|\mathbf{E}| = |\mathbf{B}|$.

$$\mathbf{j}_{\mathbf{k}\omega}^{\perp} = \mathbf{j}_{\mathbf{k}\omega} - \mathbf{j}_{\mathbf{k}\omega}^{\parallel} = -\hat{\mathbf{n}} \times (\hat{\mathbf{n}} \times \mathbf{j}_{\mathbf{k}\omega}), \tag{60.9}$$

the fields can be written compactly as

$$\mathbf{E}_{\omega} = \frac{ike^{ikR}}{cR} \mathbf{j}_{\mathbf{k}\omega}^{\perp}, \tag{60.10}$$

$$\mathbf{B}_{\omega} = \frac{ik}{c}(\hat{\mathbf{n}} \times \mathbf{j}_{\mathbf{k}\omega}^{\perp})\frac{e^{ikR}}{R} = \hat{\mathbf{n}} \times \mathbf{E}_{\omega}. \tag{60.11}$$

We again stress that these formulas are not limited to nonrelativistic sources. Equations (60.10) and (60.11) are the far-zone fields. They have a number of important properties.[2] First, they decay inversely with distance. This behavior can be traced to the action of the gradients on the e^{ikR} factor, i.e., to the effects of retardation. Second, $\mathbf{E}_{\omega}$, $\mathbf{B}_{\omega}$, and $\hat{\mathbf{n}}$ form a right triad, and $|\mathbf{B}_{\omega}| = |\mathbf{E}_{\omega}|$ (fig. 10.2). Thus, locally, the EM field looks like a plane wave. Third, they are determined only by $\mathbf{j}_{\mathbf{k}\omega}^{\perp}$, the part of $\mathbf{j}_{\mathbf{k}\omega}$ perpendicular or transverse to **k**. Since $\mathbf{q} \cdot \mathbf{j}_{\mathbf{q}\omega}^{\perp} = 0$, $\nabla \cdot \mathbf{j}_{\omega}^{\perp} = 0$. So, $\mathbf{j}_{\mathbf{k}\omega}^{\perp}$ is the Fourier transform of the *transverse* or *divergenceless* part of the current.[3] Since $\nabla \cdot \mathbf{j} = -\dot{\rho}$, this means that a source such as a slowly discharging capacitor in which the only currents are due to a time-dependent charge density does not radiate EM waves.

It also pays to write the radiation zone fields (60.10) and (60.11) in the time domain. In full, eq. (60.10) reads

$$\mathbf{E}_{\omega}(\mathbf{R}) = \frac{i}{c^2 R} \int d^3x' \omega e^{i\omega(R - \mathbf{r}' \cdot \hat{\mathbf{n}})/c} \mathbf{j}_{\omega}^{\perp}(\mathbf{r}'). \tag{60.12}$$

Now for any function $f(t)$ with FT f_{ω},

$$\omega f_{\omega} = \text{F.T.}[i \partial f(t)/\partial t], \quad e^{i\omega\bar{t}} f_{\omega} = \text{F.T.}[f(t - \bar{t})]. \tag{60.13}$$

Using these rules to invert the Fourier transform of the integrand in eq. (60.12), we get

$$\mathbf{E}(\mathbf{R}, t) = -\frac{1}{c^2 R} \int d^3x' \frac{\partial}{\partial t} \mathbf{j}^{\perp}\left(\mathbf{r}', t - \frac{R}{c} + \frac{\mathbf{r}' \cdot \hat{\mathbf{n}}}{c}\right), \tag{60.14}$$

[2] The same properties were found in the previous chapter for the far-zone fields of a moving charge.
[3] As shown in exercise 33.1, **j**, like any other vector field, can be decomposed into longitudinal (curl-free) and transverse (divergence-free) parts: $\mathbf{j}(\mathbf{r}, t) = \mathbf{j}^T + \mathbf{j}^L$; $\nabla \cdot \mathbf{j}^T = 0$, $\nabla \times \mathbf{j}^L = 0$. Fourier transforms provide a compact way to see this. The FTs of $\mathbf{j}^L$ and $\mathbf{j}^T$ are the components of the FT of **j** parallel and perpendicular to **q**.

or, equally accurately,

$$E(\mathbf{R}, t) = -\frac{1}{c^2 R} \int d^3x' \frac{\partial}{\partial t} \mathbf{j}^\perp(\mathbf{r}', t_r), \tag{60.15}$$

where we use the exact expression for t_r. Two important features are evident in eq. (60.15). First, there is no radiation unless $\partial \mathbf{j}^\perp / \partial t$ is nonzero. Thus, a steady current does not radiate. Second, as seen in the previous chapter, only the part of the apparent acceleration of the charges that is perpendicular to the observer's line of sight contributes to the radiation.

Once the electric field is known, the magnetic field is obtained from eq. (60.11) very simply as

$$B(\mathbf{R}, t) = \hat{\mathbf{n}} \times E(\mathbf{R}, t). \tag{60.16}$$

Of course, it is also true that $E = -\hat{\mathbf{n}} \times B$.

Exercise 60.1 Derive eqs. (60.15) and (60.16) directly from the time-domain expressions for ϕ and A.

Exercise 60.2 Derive the radiation-zone fields by starting with the potentials in the Coulomb gauge.

Solution: In the Coulomb gauge, the scalar potential is given by the instantaneous charge distribution:

$$\phi(\mathbf{R}, t) = \int \frac{\rho(\mathbf{r}', t)}{|\mathbf{R} - \mathbf{r}'|} d^3x' \approx \frac{Q}{R} + \frac{\mathbf{d}(t) \cdot \mathbf{R}}{R^3} + O(R^{-3}). \tag{60.17}$$

Here, Q is the total charge, and $\mathbf{d}(t)$ the dipole moment of the source. Thus, $-\nabla\phi$ dies with R at least as fast as $1/R^2$ (the time-varying part as $1/R^3$) and so cannot be part of the radiation field. The formulas for these fields therefore simplify to $E = -\dot{A}/c$, $B = \nabla \times A$, for which only the vector potential is needed. This is given by the equation

$$\nabla^2 A - \frac{1}{c^2} \frac{\partial^2 A}{\partial t^2} = -\frac{4\pi}{c} \mathbf{j} + \frac{1}{c} \nabla \frac{\partial \phi}{\partial t}. \tag{60.18}$$

As shown in exercise 33.1, the right-hand side is $-4\pi \mathbf{j}^T / c$, where $\mathbf{j}^T$ is the transverse part of $\mathbf{j}$, which we are now writing as $\mathbf{j}^\perp$. Making this identification and taking the temporal Fourier transform, we get

$$\nabla^2 A_\omega + k^2 A_\omega = -\frac{4\pi}{c} \mathbf{j}_\omega^\perp. \tag{60.19}$$

Making use of the Green function solution for the wave equation, we get

$$A_\omega(\mathbf{R}) = \frac{1}{c} \int \frac{e^{ik|\mathbf{R}-\mathbf{r}'|}}{|\mathbf{R} - \mathbf{r}'|} \mathbf{j}_\omega^\perp(\mathbf{r}') d^3x'. \tag{60.20}$$

We expand this for large R in the now-familiar way:

$$\mathbf{A}_\omega(\mathbf{R}) = \frac{e^{ikR}}{cR} \int e^{-i\mathbf{k}\cdot\mathbf{r}'} \mathbf{j}_\omega^\perp(\mathbf{r}') d^3x' = \frac{e^{ikR}}{cR} \mathbf{j}_{\mathbf{k}\omega}^\perp.$$

(60.21)

The fields follow at once:

$$\mathbf{E}_\omega(\mathbf{R}) = \frac{ike^{ikR}}{cR} \mathbf{j}_{\mathbf{k}\omega}^\perp, \qquad \mathbf{B}_\omega(\mathbf{R}) = \frac{ike^{ikR}}{cR} (\hat{\mathbf{n}} \times \mathbf{j}_{\mathbf{k}\omega}^\perp).$$

(60.22)

61 Power radiated

Let us now consider the power radiated by the source. We first find the energy flux (Poynting vector) in the far zone. Since $\mathbf{B} = \hat{\mathbf{n}} \times \mathbf{E}$,

$$\mathbf{S} = \frac{c}{4\pi} E^2 \hat{\mathbf{n}} = \frac{1}{4\pi c^3 R^2} \left[\frac{\partial}{\partial t} \int d^3x' \mathbf{j}^\perp(\mathbf{r}', t_r) \right]^2 \hat{\mathbf{n}}.$$

(61.1)

Since this drops with distance as $1/R^2$, the total energy flowing into a solid-angle element $d\Omega$ around some direction is independent of R (except in the delay factor—the signal gets to a more distant point later). Hence, we can rewrite this formula in terms of the angular power distribution

$$\frac{dP}{d\Omega} = \frac{1}{4\pi c^3} \left[\frac{\partial}{\partial t} \int d^3x' \mathbf{j}^\perp(\mathbf{r}', t_r) \right]^2.$$

(61.2)

The equivalent SI system formula is obtained by inserting an extra factor of $1/4\pi\epsilon_0$ on the right. It is conventional in formulas for the power to write this factor as $cZ_0/4\pi$, where Z_0, a quantity with the dimensions of resistance and known as the *impedance of the vacuum*, is given by

$$Z_0 = (\mu_0/\epsilon_0)^{1/2} = 376.7 \text{ ohms.}$$

(61.3)

Hence,

$$\frac{dP}{d\Omega} = \frac{Z_0}{16\pi^2 c^2} \left[\frac{\partial}{\partial t} \int d^3x' \mathbf{j}^\perp(\mathbf{r}', t_r) \right]^2. \quad \text{(SI)}$$

(61.4)

Integrating eq. (61.2) over all $\hat{\mathbf{n}}$, we obtain the total power radiated:

$$P = \int \frac{dP}{d\Omega} d\Omega.$$

(61.5)

Equations (61.2) and (61.5) give the instantaneous power for a general source. Usually, the source has special characteristics—it is either monochromatic or periodic, etc. In each of these cases, we are interested in the integral or average of the power over some time period rather than its instantaneous value. We discuss four cases of special importance.

Burst source: Suppose the source is on only for a short period of time. For example, we could be considering a collision between two charged particles, which accelerate only within the short collision duration. In this case, we are interested in the *total energy* radiated rather than its rate. We obtain this by integrating the Poynting vector over all times and over a large sphere of radius R:

$$\mathcal{E} = \int R^2 d\Omega \int_{-\infty}^{\infty} dt \frac{c}{4\pi} E^2(\mathbf{R}, t), \tag{61.6}$$

$$= \int R^2 d\Omega \int_{-\infty}^{\infty} \frac{d\omega}{2\pi} \frac{c}{4\pi} |\mathbf{E}_\omega(\mathbf{R})|^2, \tag{61.7}$$

where the second line follows by Parseval's theorem. If we now use eq. (60.10) and the fact that $\mathbf{E}_{-\omega} = \mathbf{E}_\omega^*$, we obtain[4]

$$\mathcal{E} = \int d\Omega \int_0^{\infty} d\omega \frac{d^2\mathcal{E}}{d\omega d\Omega}, \tag{61.8}$$

$$\frac{d^2\mathcal{E}}{d\omega d\Omega} = \frac{\omega^2}{4\pi^2 c^3} |\mathbf{j}_{\mathbf{k}\omega}^\perp|^2. \tag{61.9}$$

The quantity $d^2\mathcal{E}/d\omega d\Omega$ tells us how the radiated energy is distributed both spectrally (in ω) and angularly (in $\hat{\mathbf{n}}$). By integrating over $\hat{\mathbf{n}}$, we obtain $d\mathcal{E}/d\omega$, the distribution in frequency without regard to the direction of emission, and by integrating over frequency we obtain the angular distribution $d\mathcal{E}/d\Omega$ without regard to frequency.[5] Note that the spectral distribution is defined only for positive frequencies.[6] Also note that we could equally well have written $|\hat{\mathbf{n}} \times \mathbf{j}_{\mathbf{k}\omega}^\perp|^2$ instead of $|\mathbf{j}_{\mathbf{k}\omega}^\perp|^2$ in this formula. We follow Jackson's convention in this regard and always give the angular distribution of the power as the absolute square of a vector parallel to $\mathbf{E}$. In this way, we simultaneously indicate the polarization of the radiation. When one is not interested in this information, however, a second form of the answer is often useful:

$$\frac{d^2\mathcal{E}}{d\omega d\Omega} = \frac{\omega^2}{4\pi^2 c^3} (|\mathbf{j}_{\mathbf{k}\omega}|^2 - c^2 |\rho_{\mathbf{k}\omega}|^2). \tag{61.10}$$

This is an immediate consequence of the continuity equation ($\mathbf{j}_{\mathbf{k}\omega} \cdot \hat{\mathbf{n}} = c\rho_{\mathbf{k}\omega}$).

Stochastic source: Second, let us consider the case of a source that produces a stochastic and stationary radiation field. We mentioned such sources in section 51 in connection with partially coherent light. An example of such a source is an incandescent bulb or a candle. When we say that a candle flame is blue or yellow, we are essentially talking of its power spectrum. How can we make this characterization quantitative? A heuristic way is

---

[4] For brevity, we give the power and energy radiated only in the Gaussian system in this section. SI formulas for these quantities are obtained by multiplying with $cZ_0/4\pi$, and are tabulated along with Gaussian formulas in eq. (61.26).

[5] As in the case of $dP/d\Omega$, the notations $d^2\mathcal{E}/d\omega d\Omega$, $d\mathcal{E}/d\omega$, etc., do not indicate the derivatives of a function with respect to ω or $\hat{\mathbf{n}}$. Rather, they indicate distributions that may be integrated over the appropriate variable(s).

[6] Some authors work with a *two-sided* frequency distribution that is half ours and is intended to be integrated over both positive and negative frequencies. Unfortunately, authors do not always identify whether they are working with one-sided or two-sided distributions, and the reader must be alert for consequent normalization errors. Similarly, one sometimes encounters a spectral distribution that gives the power in a range $d\omega/2\pi$ instead of $d\omega$.

to take some kind of average of the expression (61.9):

$$\frac{d^2\mathcal{E}}{d\omega d\Omega} \text{ "=" } \frac{\omega^2}{4\pi^2 c^3}\langle|\mathbf{j}_{\mathbf{k}\omega}^\perp|^2\rangle, \tag{61.11}$$

where the angular brackets denote this as yet ill-defined average. Next, we write out the FTs over time explicitly and get

$$\langle|\mathbf{j}_{\mathbf{k}\omega}^\perp|^2\rangle = \left\langle \int_{-\infty}^{\infty} \mathbf{j}_{\mathbf{k}}^\perp(t_1)e^{i\omega t_1}dt_1 \cdot \int_{-\infty}^{\infty} \mathbf{j}_{\mathbf{k}}^{\perp*}(t_2)e^{-i\omega t_2}dt_2 \right\rangle. \tag{61.12}$$

Writing $t_1 = t + \frac{1}{2}\tau$, $t_2 = t - \frac{1}{2}\tau$, and substituting in eq. (61.11), we get

$$\frac{d^2\mathcal{E}}{d\omega d\Omega} \text{ "=" } \frac{\omega^2}{4\pi^2 c^3} \int_{-\infty}^{\infty} dt \int_{-\infty}^{\infty} d\tau \langle \mathbf{j}_{\mathbf{k}}^\perp(t+\tfrac{1}{2}\tau) \cdot \mathbf{j}_{\mathbf{k}}^{\perp*}(t-\tfrac{1}{2}\tau)\rangle e^{i\omega\tau}$$

$$\text{"=" } \int_{-\infty}^{\infty} dt \left[\frac{\omega^2}{4\pi^2 c^3} \int_{-\infty}^{\infty} d\tau \langle \mathbf{j}_{\mathbf{k}}^\perp(t+\tfrac{1}{2}\tau) \cdot \mathbf{j}_{\mathbf{k}}^{\perp*}(t-\tfrac{1}{2}\tau)\rangle e^{i\omega\tau}\right]. \tag{61.13}$$

If the current is made up of many individual uncorrelated pieces, then its value at a given instant will be uncorrelated with the value after a certain time. In other words, the quantity in angular brackets will die off as the time interval τ gets large, and the integral over τ can be regarded as a quasi-instantaneous property. The outer integral over t can then be interpreted as the power spectrum at time t:

$$\frac{d^2 P}{d\omega d\Omega}(t) = \frac{\omega^2}{4\pi^2 c^3} \int_{-\infty}^{\infty} d\tau \langle \mathbf{j}_{\mathbf{k}}^\perp(t+\tfrac{1}{2}\tau) \cdot \mathbf{j}_{\mathbf{k}}^{\perp*}(t-\tfrac{1}{2}\tau)\rangle e^{i\omega\tau}. \tag{61.14}$$

This result is in fact correct, as shown by the honest equals sign sans quotes. When the current is strictly stationary, it follows from the *Wiener-Khinchin theorem*, discussed in appendix F (see eq. (F.6)). The average that appears is known as the *transverse current–current autocorrelation function*,

$$G_{jj}^\perp(\tau) = \langle \mathbf{j}_{\mathbf{k}}^\perp(t+\tfrac{1}{2}\tau) \cdot \mathbf{j}_{\mathbf{k}}^{\perp*}(t-\tfrac{1}{2}\tau)\rangle, \tag{61.15}$$

$$\equiv \lim_{T\to\infty} \frac{1}{T} \int_{t_0}^{t_0+T} \mathbf{j}_{\mathbf{k}}^\perp(t+\tfrac{1}{2}\tau) \cdot \mathbf{j}_{\mathbf{k}}^{\perp*}(t-\tfrac{1}{2}\tau)\, dt. \tag{61.16}$$

We have given the formal definition of $G_{jj}^\perp$ in eq. (61.16). This equation also explains that the averaging is a time average. In practice, the averaging time T must be large compared to the time scales on which the fields vary. A strictly stationary current is one for which the average is independent of the starting time t_0. In this case, the correlation function is independent of t, and the Wiener-Khinchin theorem states that the power spectrum is given by its Fourier transform, up to additional kinematic factors, as given in eq. (61.14). However, eq. (61.14) also makes sense if the conditions at the source vary slowly on the time scale of the dominant frequencies in the spectrum itself. For example, we could turn up the current to a mercury vapor lamp by turning a knob by hand. The time scale for doing this is on the order of seconds, much longer than 10^{-15} sec, the inverse of the characteristic frequency of the light from the lamp. It is then meaningful to talk about how the power spectrum changes as the knob is turned.

If we had started from eq. (61.10) instead of eq. (61.9), we would have obtained

$$\frac{d^2 P}{d\omega d\Omega}(t) = \frac{\omega^2}{4\pi^2 c^3} \int_{-\infty}^{\infty} d\tau \left[\langle \mathbf{j_k}(t + \tfrac{1}{2}\tau) \cdot \mathbf{j_k^*}(t - \tfrac{1}{2}\tau) \rangle - c^2 \langle \rho_k(t + \tfrac{1}{2}\tau) \rho_k^*(t - \tfrac{1}{2}\tau) \rangle \right] e^{i\omega\tau}.$$

(61.17)

This formula is sometimes more convenient than eq. (61.14). It should be noted though that the equivalence of the two forms follows only after an integration by parts, and the integrands themselves are not equal.

Monochromatic source: As discussed in appendix F, the Fourier transform of the correlation function gives the power spectral density not only for random signals but also for signals that may contain a periodic or pure harmonic component. Hence, the formula (61.14) also applies to such sources. Suppose the conditions at the source vary at a single frequency, i.e.,

$$\mathbf{j}(\mathbf{r}, t) = \text{Re}[\mathbf{j_0}(\mathbf{r})e^{-i\omega_0 t}],$$

(61.18)

and likewise for ρ. Note that we indicate the amplitude of the current by attaching a suffix 0. It will be clear from the context that this does not indicate the Fourier transform at zero frequency, so there is no scope for confusion. The spatial Fourier transform gives

$$\mathbf{j_k}(t) = \mathbf{j_{k0}} \cos \omega_0 t,$$

(61.19)

and the correlation function in eq. (61.14) equals

$$\langle \mathbf{j_k^\perp}(t + \tfrac{1}{2}\tau) \cdot \mathbf{j_k^{\perp*}}(t - \tfrac{1}{2}\tau) \rangle = \frac{1}{2} |\mathbf{j_{k0}^\perp}|^2 \cos \omega_0 \tau,$$

(61.20)

where

$$\mathbf{j_{k0}^\perp} = \int \mathbf{j_0^\perp}(\mathbf{r}')e^{-i(\omega/c)\hat{\mathbf{n}}\cdot\mathbf{r}'} d^3x'.$$

(61.21)

The power spectrum is therefore given by

$$\frac{d^2 P}{d\omega d\Omega} = \frac{\omega^2}{8\pi c^3} |\mathbf{j_{k0}^\perp}|^2 \delta(\omega - \omega_0).$$

(61.22)

It should be noted that the absolute square $|\mathbf{j_{k0}^\perp}|^2$ includes both a dot product and complex conjugation.

The formula (61.22) also applies to a quasimonochromatic source, since a strictly monochromatic source is an idealization. In fact, we could replace $\mathbf{j_{k0}}$ by $\mathbf{j_{k0}}(t)$, where the time dependence is very slow on the $2\pi/\omega$ time scale. This is how we would think of the time-dependent power put out by an amplitude-modulated radio station.

Periodic source: Lastly, we consider a source that is periodic with period T. In this case, we expand the current in a Fourier series of the form

$$\mathbf{j}(\mathbf{r}, t) = \mathbf{j_0}(\mathbf{r}) + \sum_{n \geq 1} \mathbf{j_n}(\mathbf{r}) \cos(\omega_n t + \delta_n),$$

(61.23)

where $\omega_n = (2\pi/T)n$, and (for $n \geq 1$),

$$j_n(\mathbf{r}) = 2 \left| \frac{1}{T} \int_0^T j(\mathbf{r}, t) e^{i\omega_n t} dt \right|. \tag{61.24}$$

The power, averaged over several periods, is given by

$$\frac{d^2 P}{d\omega d\Omega} = \frac{1}{8\pi c^3} \sum_{n \geq 1} \omega_n^2 |j_{\mathbf{k}n}^{\perp}|^2 \delta(\omega - \omega_n), \tag{61.25}$$

where $j_{\mathbf{k}n}$ is the spatial Fourier transform of $j_n(\mathbf{r})$ at the wave vector $\mathbf{k} = \hat{\mathbf{n}}\omega_n/c$.

The differing factors of 2 and π in the formulas for the various cases are vexing but hard to avoid. We tabulate them here for ready reference.

Source type	Distribution	Formula (Gaussian)	Formula (SI)
Burst	$\dfrac{d^2\mathcal{E}}{d\omega d\Omega}$	$\dfrac{\omega^2}{4\pi^2 c^3}|j_{\mathbf{k}\omega}^{\perp}|^2$	$Z_0\dfrac{\omega^2}{16\pi^3 c^2}|j_{\mathbf{k}\omega}^{\perp}|^2$
Stochastic	$\dfrac{d^2 P}{d\omega d\Omega}$	$\dfrac{\omega^2}{4\pi^2 c^3}\displaystyle\int_{-\infty}^{\infty} d\tau\, G_{jj}^{\perp}(\tau) e^{i\omega\tau}$	$Z_0\dfrac{\omega^2}{16\pi^3 c^2}\displaystyle\int_{-\infty}^{\infty} d\tau\, G_{jj}^{\perp}(\tau) e^{i\omega\tau}$
Monochromatic	$\dfrac{d^2 P}{d\omega d\Omega}$	$\dfrac{\omega^2}{8\pi c^3}|j_{\mathbf{k}0}^{\perp}|^2 \delta(\omega - \omega_0)$	$Z_0\dfrac{\omega^2}{32\pi^2 c^2}|j_{\mathbf{k}0}^{\perp}|^2 \delta(\omega - \omega_0)$
Periodic	$\dfrac{d^2 P}{d\omega d\Omega}$	$\dfrac{1}{8\pi c^3}\displaystyle\sum_{n \geq 1}\omega_n^2|j_{\mathbf{k}n}^{\perp}|^2\delta(\omega - \omega_n)$	$Z_0\dfrac{1}{32\pi^2 c^2}\displaystyle\sum_{n \geq 1}\omega_n^2|j_{\mathbf{k}n}^{\perp}|^2\delta(\omega - \omega_n)$

$$\tag{61.26}$$

Once again, we note that these results are not limited to nonrelativistic sources. Also, the alternative forms (61.10) and (61.17) should be remembered.

62 The long-wavelength electric dipole approximation

Let us now consider the important case of a nonrelativistic source, for which $v \ll c$. Equivalently, we may say that the source dimensions are much less than the wavelength of the radiation, i.e., $a \ll \lambda_0$. In this case, the factor $e^{i\mathbf{k}\cdot\mathbf{r}}$ in the integrand for the spatial Fourier transform $j_{\mathbf{k}\omega}$ may be expanded in powers of $\mathbf{k}\cdot\mathbf{r}$. The leading approximation is

$$j_{\mathbf{k}\omega}^{(0)} = \int j_\omega(\mathbf{r}) d^3 x, \tag{62.1}$$

which is just the volume integral of j_ω. To analyze this expression, let us switch to thinking in terms of a collection of charges. Then,

$$j(\mathbf{r}, t) = \sum_a q_a \dot{\mathbf{r}}_a \delta(\mathbf{r} - \mathbf{r}_a(t)),$$

$$\int j_\omega d^3 x = \int_{-\infty}^{\infty} dt\, e^{i\omega t} \sum_a q_a \dot{\mathbf{r}}_a(t) = \int_{-\infty}^{\infty} dt\, e^{i\omega t} \dot{\mathbf{d}}(t),$$

where $\mathbf{d}(t)$ is the time-dependent electric dipole moment of the source. Hence,

$$\mathbf{j}_{\mathbf{k}\omega}^{(0)} = (\dot{\mathbf{d}})_\omega = -i\omega\mathbf{d}_\omega, \tag{62.2}$$

and, by eq. (60.10),

$$\mathbf{E}_\omega^{(0)} = k^2 \mathbf{d}_\omega^\perp \frac{e^{ikR}}{R}. \tag{62.3}$$

This is precisely the result obtained in the previous chapter (section 58). To see this, recall that $k = \omega/c$, and Fourier transform back to the time domain. We get

$$\mathbf{E}^{(0)}(\mathbf{R}, t) = -\frac{1}{c^2 R}\ddot{\mathbf{d}}_{\text{ret}}^\perp. \tag{62.4}$$

The delay involved in computing the retardation is taken as R/c for every point in the source. The long-wavelength electric dipole approximation amounts to the complete neglect of the variation in this delay across the source.[7]

The energy radiated in this approximation is given by

$$\frac{d^2\mathcal{E}}{d\omega d\Omega} = \frac{ck^4}{4\pi^2}|\hat{\mathbf{n}} \times (\hat{\mathbf{n}} \times \mathbf{d}_\omega)|^2, \tag{62.5}$$

$$\frac{d\mathcal{E}}{d\omega} = \frac{2}{3\pi}ck^4|\mathbf{d}_\omega|^2. \tag{62.6}$$

The angular distribution of the radiation depends on the exact form of $\mathbf{d}_\omega$. As an example, suppose the real and imaginary parts of $\mathbf{d}_\omega$ are parallel. Taking them to be along $\hat{\mathbf{z}}$, $d^2\mathcal{E}/d\omega d\Omega \propto (1 - \cos^2\theta)$, where θ is the usual polar angle in spherical polar coordinates. There is a null in the radiated power along the direction in which the dipole oscillates ($\theta = 0$). We shall see this case in the short dipole antenna in section 64 below. As a second example, let us reconsider the electron in a circular orbit from exercise 58.1.

---

[7] There is, unfortunately, a serious source of confusion in talking of multipole moments of radiation fields. There are *two* moment expansions, and the terminology does not always differentiate between them. The expansion we develop in this and the next section is based on the small parameter ka and entails the same moments that arise in the discussion of *static* electric and magnetic fields. The second expansion is in terms of eigenfunctions of the angular momentum and parity that we mentioned while discussing the angular momentum in the free EM field (section 47) and is based on the identity (compare with eq. (19.16))

$$\frac{e^{ik|\mathbf{r}-\mathbf{r}'|}}{|\mathbf{r}-\mathbf{r}'|} = 4\pi ik \sum_{\ell,m} j_\ell(kr_<)h_\ell^{(1)}(kr_>)Y_{\ell m}(\hat{\mathbf{r}})Y_{\ell m}^*(\hat{\mathbf{r}}'),$$

where j_ℓ is the spherical Bessel function, and $h_\ell^{(1)}$ is the spherical Hankel function of the first kind. If this identity is fed into eq. (59.8), the potentials associated with a given angular pattern will have coefficients that are integrals of the source charge and current weighted by a spherical Bessel function, e.g.,

$$\int \rho_\omega(\mathbf{r}')j_\ell(kr')Y_{\ell m}^*(\hat{\mathbf{r}}')\,d^3x'.$$

We refer to this as a *Bessel moment* of the source. The Bessel moment expansion does not have any small parameter. If $ka \ll 1$, however, we may approximate $j_\ell(kr') \sim (kr')^\ell$, and the Bessel moments reduce to the static moments. For relativistic sources, on the other hand, only the Bessel moment expansion makes sense. This expansion is discussed in Jackson (1999), secs. 9.6–9.12.

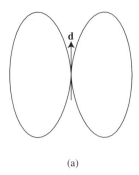 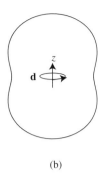

(a)   (b)

FIGURE 10.3. Polar plots of the angular power distribution for dipole radiation:
(a) $\mathbf{d} \parallel \hat{\mathbf{z}}$, corresponding to $\ell = 1$, $m = 0$; (b) $\mathbf{d} \propto (\hat{\mathbf{x}} + i\hat{\mathbf{y}})$, corresponding to $\ell = 1$,
$m = 1$.

Now $\ddot{\mathbf{d}} \propto (\hat{\mathbf{x}} \cos \omega t + \hat{\mathbf{y}} \sin \omega t)$, and $\mathbf{d}_\omega \propto \hat{\mathbf{x}} + i\hat{\mathbf{y}}$. We can think of this radiator as made of
two identical dipoles oscillating along $\hat{\mathbf{x}}$ and $\hat{\mathbf{y}}$, $90°$ out of phase. Hence, there is no null
direction, and $d^2 \mathcal{E}/d\omega d\Omega \propto (1 + \cos^2 \theta)$ (see fig. 10.3). We label the radiation patterns by
the indices ℓ and m if the relevant moment of the charge distribution behaves as $Y_{\ell m}$. It
should be noted that the power distribution itself is not given by $|Y_{\ell m}|^2$. We can also apply
the formulas of this section to a single nonrelativistic charge q. The time derivative of the
dipole moment is replaced by $q\,d\mathbf{r}/dt = q\mathbf{v}$. Hence,

$$\mathbf{E} = -\frac{q}{Rc^2} \dot{\mathbf{v}}_{\text{ret}}^\perp, \tag{62.7}$$

$$\frac{dP}{d\Omega} = \frac{q^2}{4\pi c^3} (\hat{\mathbf{n}} \times (\hat{\mathbf{n}} \times \dot{\mathbf{v}}_{\text{ret}}))^2, \tag{62.8}$$

$$P = \frac{2q^2}{3c^3} \dot{\mathbf{v}}_{\text{ret}}^2. \tag{62.9}$$

This is, of course, the same Larmor formula that we derived in section 58.

63 Higher multipoles*

Let us now examine the higher terms in the expansion of the $e^{-i\mathbf{k}\cdot\mathbf{r}}$ factor in the integral
for $\mathbf{j}_{\mathbf{k}\omega}$. We have

$$\mathbf{j}_{\mathbf{k}\omega} = \int \left[1 - i\mathbf{k}\cdot\mathbf{r} - \frac{1}{2}(\mathbf{k}\cdot\mathbf{r})^2 + \cdots\right] \mathbf{j}_\omega(\mathbf{r}) d^3x. \tag{63.1}$$

The leading term was analyzed in the preceding section. The next term,

$$\mathbf{j}_{\mathbf{k}\omega}^{(1)} = -\frac{i}{c} \int (\hat{\mathbf{n}}\cdot\mathbf{r}) \omega \mathbf{j}_\omega d^3x, \tag{63.2}$$

involves both the electric quadrupole and the magnetic dipole moments of the source. To
see this, we again write it for a system of moving charges:

$$\mathbf{j}_{\mathbf{k}\omega}^{(1)} = -\frac{i}{c} \int_{-\infty}^{\infty} dt\, e^{i\omega t} \omega \sum_a q_a \dot{\mathbf{r}}_a (\hat{\mathbf{n}}\cdot\mathbf{r}_a). \tag{63.3}$$

We now write $\omega e^{i\omega t}$ as $-i(\partial e^{i\omega t}/\partial t)$ and integrate by parts. This yields

$$\mathbf{j}_{k\omega}^{(1)} = \frac{1}{c} \int_{-\infty}^{\infty} dt\, e^{i\omega t} \frac{\partial}{\partial t} \sum_a q_a \mathbf{v}_a (\hat{\mathbf{n}} \cdot \mathbf{r}_a). \tag{63.4}$$

The sum over all charges can be expressed in terms of the moments of the source as follows:

$$q_a \mathbf{v}_a (\hat{\mathbf{n}} \cdot \mathbf{r}_a) = \frac{1}{2} \frac{\partial}{\partial t} [q_a \mathbf{r}_a (\hat{\mathbf{n}} \cdot \mathbf{r}_a)] + \frac{1}{2} q_a \mathbf{v}_a (\hat{\mathbf{n}} \cdot \mathbf{r}_a) - \frac{1}{2} q_a \mathbf{r}_a (\hat{\mathbf{n}} \cdot \mathbf{v}_a). \tag{63.5}$$

When we sum on a, the last two terms can be combined into the magnetic dipole moment

$$\mathbf{m} = \frac{1}{2c} \sum_a q_a (\mathbf{r}_a \times \mathbf{v}_a), \tag{63.6}$$

while the first term can be written in terms of the electric quadrupole moment:

$$D_{ij} = \sum_a q_a (3 x_{ai} x_{aj} - r_a^2 \delta_{ij}). \tag{63.7}$$

We define a vector $\mathbf{D}$, with components given by $D_i = D_{ij} \hat{n}_j$. Clearly,

$$\mathbf{D} = \sum_a q_a [3 \mathbf{r}_a (\hat{\mathbf{n}} \cdot \mathbf{r}_a) - r_a^2 \hat{\mathbf{n}}]. \tag{63.8}$$

Hence,

$$\sum_a q_a \mathbf{v}_a (\hat{\mathbf{n}} \cdot \mathbf{r}_a) = \frac{1}{6} \frac{\partial}{\partial t} \left(\mathbf{D} + \sum_a q_a r_a^2 \hat{\mathbf{n}} \right) + c(\mathbf{m} \times \hat{\mathbf{n}}) \tag{63.9}$$

and

$$\mathbf{j}_{k\omega}^{(1)} = \int_{-\infty}^{\infty} dt\, e^{i\omega t} \left[(\dot{\mathbf{m}} \times \hat{\mathbf{n}}) + \frac{1}{6c} \dot{\mathbf{D}} + \frac{1}{6c} \left(\frac{\partial^2}{\partial t^2} \sum_a q_a r_a^2 \right) \hat{\mathbf{n}} \right]. \tag{63.10}$$

The very last term is parallel to $\hat{\mathbf{n}}$, so it does not contribute to the transverse part of $\mathbf{j}$ and may be dropped. The electric field is then given by using the general formula (60.10):

$$\mathbf{E}_{\omega}^{(1)} = \frac{i\omega e^{ikR}}{c^2 R} \left(-\frac{\omega^2}{6c} \mathbf{D}_{\omega} - i\omega (\mathbf{m}_{\omega} \times \hat{\mathbf{n}}) \right). \tag{63.11}$$

Let us rewrite eq. (63.11) in the time domain and include the electric dipole part. Also, we give the magnetic field at the same time ($\mathbf{B} = \hat{\mathbf{n}} \times \mathbf{E}$).

$$\mathbf{E}(\mathbf{r}, t) = \frac{1}{c^2 R} \left[(\ddot{\mathbf{d}} \times \hat{\mathbf{n}}) \times \hat{\mathbf{n}} - (\ddot{\mathbf{m}} \times \hat{\mathbf{n}}) + \frac{1}{6c} (\dddot{\mathbf{D}} \times \hat{\mathbf{n}}) \times \hat{\mathbf{n}} \right]_{\text{ret}} + \cdots, \tag{63.12}$$

$$\mathbf{B}(\mathbf{r}, t) = \frac{1}{c^2 R} \left[\ddot{\mathbf{d}} \times \hat{\mathbf{n}} + (\ddot{\mathbf{m}} \times \hat{\mathbf{n}}) \times \hat{\mathbf{n}} + \frac{1}{6c} (\dddot{\mathbf{D}} \times \hat{\mathbf{n}}) \right]_{\text{ret}} + \cdots. \tag{63.13}$$

The symmetry between the electric and magnetic dipole terms is worth noting.

We can continue the multipole expansion to higher order in the same way. There is a commonly used terminology to refer to the various terms. The electric dipole, quadrupole, and octupole terms are known as E1, E2, and E3, respectively, and so on. Similarly, the magnetic dipole is known as M1, the quadrupole as M2, and so on. Poles higher than E2 or M1 are relatively rare. Note also that irrespective of the order of the pole, the fields die off as $1/R$.

Let us now consider the power radiated. If the source contains more than one multipole, the angular distribution of the power will contain cross terms from different multipoles. The total power, however, is a sum of contributions from individual poles. Consider, e.g., the term involving both electric and magnetic dipoles. The total power is

$$P = c\,R^2\langle(\mathbf{E} \times \mathbf{B}) \cdot \hat{\mathbf{n}}\rangle, \tag{63.14}$$

where the angular brackets denote an average over all angles. The term in question involves the average

$$\langle[(\ddot{\mathbf{d}} \times \hat{\mathbf{n}}) \times \hat{\mathbf{n}}] \cdot (\ddot{\mathbf{m}} \times \hat{\mathbf{n}})\rangle = \langle(\ddot{\mathbf{d}} \times \hat{\mathbf{n}}) \cdot \ddot{\mathbf{m}}\rangle = (\ddot{\mathbf{m}} \times \ddot{\mathbf{d}}) \cdot \langle\hat{\mathbf{n}}\rangle. \tag{63.15}$$

But $\langle\hat{\mathbf{n}}\rangle$ is clearly zero. Hence, there is no E1–M1 cross term in the power. All the other terms can be evaluated in exactly the same way. In this calculation, one encounters the average of various tensors, such as $\hat{n}_i\hat{n}_j$, $\hat{n}_i\hat{n}_j\hat{n}_k$, etc. The average of any such tensor of odd rank is zero by parity. For the even-rank tensors, the answer can only involve products of the unit tensor δ_{ij}, because that is the only way to get a rotationally invariant result. Thus,

$$\langle\hat{n}_i\hat{n}_j\rangle = \frac{1}{3}\delta_{ij}, \tag{63.16}$$

where the constant of proportionality can be fixed by taking the trace of both sides. Similarly,

$$\langle\hat{n}_i\hat{n}_j\hat{n}_k\hat{n}_l\rangle = \frac{1}{15}(\delta_{ij}\delta_{kl} + \delta_{ik}\delta_{jl} + \delta_{il}\delta_{jk}). \tag{63.17}$$

The tensor on the right-hand side is the only one fully symmetric in i, j, k, and l, and the proportionality constant can be fixed by contracting on any two of the indices.

The final answer for the total power radiated is

$$P = \frac{2}{3c^3}\ddot{\mathbf{d}}^2 + \frac{2}{3c^3}\ddot{\mathbf{m}}^2 + \frac{1}{180c^5}\dddot{D}_{ij}\dddot{D}_{ij} + \cdots . \tag{63.18}$$

Note that the magnitudes of the M1 and E2 terms are smaller than the E1 term by a factor of $(v/c)^2$. Similarly, the M2 and E3 terms are smaller by yet another power of $(v/c)^2$, and so on. For a source at one frequency, the formula becomes

$$P = \frac{1}{3}ck^4|\mathbf{d}|^2 + \frac{1}{3}ck^4|\mathbf{m}|^2 + \frac{1}{360}ck^6 D_{ij}D_{ij}^* + \cdots , \tag{63.19}$$

where $\mathbf{d}$, $\mathbf{m}$, and D_{ij} are the complex amplitudes of oscillation of the respective moments.

We have already discussed the angular distribution of E1 radiation. The same pattern clearly apples to pure M1 radiation. What if both the electric and magnetic dipoles are zero for some source? This happens for certain nuclei and for some atomic and molecular transitions as well. The quadrupole moment must be found quantum mechanically for such systems, of course, but the rest of our discussion is still applicable. In particular, the angular pattern is unchanged. Keeping only the D_{ij} terms in eqs. (63.12) and (63.13) and

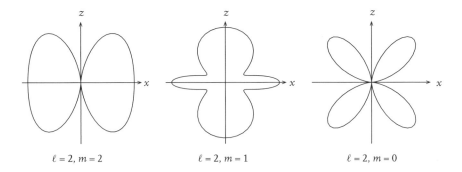

$\ell = 2, m = 2$ $\ell = 2, m = 1$ $\ell = 2, m = 0$

FIGURE 10.4. Polar plots of the angular power distribution for quadrupole radiation from pure $Y_{\ell m}$ sources. The reader should imagine rotating the plots about the z axis to visualize the distributions in three dimensions.

forming the Poynting vector, etc., we obtain

$$\frac{dP}{d\Omega} = \frac{1}{144c^5} \dddot{D}_{ij} \dddot{D}_{kl} (\hat{\mathbf{n}}_j \hat{\mathbf{n}}_k \delta_{il} - \hat{\mathbf{n}}_i \hat{\mathbf{n}}_j \hat{\mathbf{n}}_k \hat{\mathbf{n}}_l). \tag{63.20}$$

The result can be better stated if we use spherical tensor components instead of Cartesian ones for the quadrupole moment. The patterns for Y_{22}, Y_{21}, and Y_{20} sources are shown in fig. 10.4. Note that, as for dipole radiation (fig. 10.3), the angular power distribution itself is not given by $|Y_{\ell m}|^2$. Nor is it always four-lobed.

Exercise 63.1 Show that the E1–E2 and M1–E2 cross terms in the total power vanish.

Exercise 63.2 Find the angular distribution of power for a quadrupole source in which only one of the spherical tensor components q_{2m} is nonzero.

Answer: For given m, the vector $\mathbf{D}$ and the angular distribution are tabulated below. A proportionality factor is omitted in $\mathbf{D}$.

m	$\mathbf{D}$	$dP/d\Omega$	
2	$(n_x + in_y)(\hat{\mathbf{x}} + i\hat{\mathbf{y}})$	$\dfrac{5}{16\pi}(1 - \cos^4\theta)$	
1	$n_z(\hat{\mathbf{x}} + i\hat{\mathbf{y}}) + (n_x + in_y)\hat{\mathbf{z}}$	$\dfrac{5}{16\pi}(1 - 3\cos^2\theta + 4\cos^4\theta)$	(63.21)
0	$2n_z\hat{\mathbf{z}} - (n_x\hat{\mathbf{x}} + n_y\hat{\mathbf{y}})$	$\dfrac{15}{8\pi}\sin^2\theta\cos^2\theta$	

For $m \to -m$, $\mathbf{D} \to \mathbf{D}^*$, and $dP/d\Omega$ is unchanged.

Exercise 63.3 Consider two nonrelativistic particles, each of charge q, in a common circular orbit of radius a, 180° apart from one another. Find the power radiated and its angular distribution. Compare with the power radiated by a single charge $2q$ in the same orbit at the same speed (so the total time-averaged current around the circle is the same).

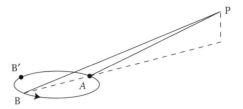

FIGURE 10.5. Radiation from two charges moving on a circle. The plane ABP is normal to the circle, and P is very far from the charges (AP ≫ AB). When one charge appears to be at A, the other appears to be at B′, not B.

Solution: If the angular velocity of each particle is ω, the radiation is at a frequency 2ω, since that is the periodicity of the source. The source has no electric dipole moment and a time-independent magnetic dipole moment. The lowest-order time-dependent moment is electric quadrupole, so the radiation is E2. Taking the plane of motion as the xy plane, the only nonzero components of D_{ij} are

$$D_{xx} = -D_{yy} = 3qa^2 \cos 2\omega t, \quad D_{xy} = D_{yx} = 3qa^2 \sin 2\omega t. \tag{63.22}$$

Thus, the radiation pattern is of the Y_{22} type, and the total power radiated is $P_2 = 32q^2a^4\omega^6/5c^5$. For the single charge, the power is $P_1 = 8q^2a^4\omega^4/3c^3$, so $P_2/P_1 = (12/5)(v/c)^2 \ll 1$. The physical reason is easy to see from fig. 10.5. At the point P, the only thing that keeps **E** from being exactly zero is the slight extra time the light takes to go from point B to P, as compared to that from A to P. So when P sees a charge at A, the apparent position of the second charge is not B, but B′. The difference in the orthogonal components of the acceleration means that the **E** fields from the two charges don't cancel exactly.

64 Antennas

Controlled production and detection of radiation is achieved via antennas, which are among the most ubiquitous electromagnetic devices today. Even a low-end cell phone has four to five antennas in it, and antenna design and theory is a vast and learned field, far beyond the author's expertise. We shall be able only to scratch its surface and go over a few basic ideas. Secondly, we shall discuss antennas only as transmitters; their role as receivers is governed by concepts discussed in chapter 19. Finally, we shall give the power in SI units in this section.

The *center-fed linear antenna* (fig. 10.6) is perhaps the simplest one. One takes two straight metal rods of equal length and connects them as shown to a high-frequency generator. The direction of current flow is shown at one instant in the figure. Half a cycle later, it is reversed. As a first approximation, let us take the current in the rods to be uniform:

$$I(z, t) = I_0 \cos \omega t. \tag{64.1}$$

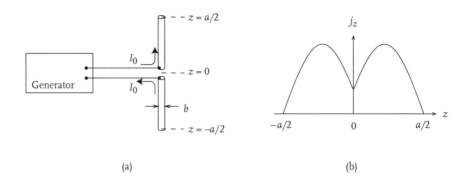

FIGURE 10.6. Center-fed linear antenna: (a) schematic of device; (b) instantaneous current distribution as per model (64.5).

Charge must then pile up and get depleted at the ends of the rod, leading to a time-dependent dipole moment, whose derivative is given by

$$\dot{\mathbf{d}} = \hat{\mathbf{z}} \int_{-a/2}^{a/2} I(z, t)\, dz = I_0 a \cos \omega t \hat{\mathbf{z}}. \tag{64.2}$$

This process can be aided by attaching large conductors at the ends of the rod, essentially forming a capacitor, as was done by Hertz in 1886 in his pioneering demonstration of radio waves. For the magnitude of the far-zone electric field, we get, using eq. (62.4),

$$E^{(0)}(\mathbf{R}, t) = \left| \frac{1}{c^2 R} \ddot{\mathbf{d}}^{\perp}_{\text{ret}} \right| = \frac{I_0 a \omega}{c^2 R} \sin \omega t_r \sin \theta. \tag{64.3}$$

Hence, writing the results in terms of wavelength λ rather than k or ω, we get

$$\frac{dP}{d\Omega} = \frac{1}{8} I_0^2 \left(\frac{a}{\lambda} \right)^2 Z_0 \sin^2 \theta; \quad P = \frac{\pi}{3} I_0^2 \left(\frac{a}{\lambda} \right)^2 Z_0. \quad \text{(SI)} \tag{64.4}$$

The angular distribution is that of a dipole oscillating in one direction, $\hat{\mathbf{z}}$.

The approximation of a uniform current is poor if the antenna is made of just two thin rods ($b \ll a$), without capacitive loading. Finding the actual distribution is, alas, rather difficult, since one must determine the electric field both inside and outside the rod, match their tangential components at the surface, and ensure that the field outside is appropriate to the time-dependent current distribution. It turns out that for very thin rods, the charge flow is approximately wavelike at the speed of light.[8] Since charge cannot flow into or out of the rods at the ends, the current must vanish there. Further, the current entering the upper rod must equal that leaving the lower rod. To meet all these conditions, we choose

$$\mathbf{j}(\mathbf{r}, t) = I_m \sin(\tfrac{1}{2}ka - k|z|) \cos \omega t\, \delta(x)\delta(y)\hat{\mathbf{z}}, \tag{64.5}$$

with $k = \omega/c$. Note that the maximum current, I_m, differs from the input ($z = 0$) current, $I_0 = I_m \sin(ka/2)$.

---

[8] See, e.g., Kraus (1988), chap. 9.

To find the power, we note that our source is monochromatic. Hence, eq. (61.22) applies, for which we need $\mathbf{j}_{k0}^{\perp}$. From eq. (64.5), we get

$$\mathbf{j}_{k0} = I_m \hat{\mathbf{z}} \int_{-a/2}^{a/2} \sin(\tfrac{1}{2}ka - k|z|)e^{-ikz\cos\theta}\,dz \tag{64.6}$$

$$= \frac{2I_m}{k\sin^2\theta}\left[\cos(\tfrac{1}{2}ka\cos\theta) - \cos(\tfrac{1}{2}ka)\right]\hat{\mathbf{z}}. \tag{64.7}$$

Hence,

$$|\mathbf{j}_{k0}^{\perp}| = |\hat{\mathbf{n}} \times (\mathbf{j}_{k0} \times \hat{\mathbf{n}})| = |\mathbf{j}_{k0}|\sin\theta, \tag{64.8}$$

and the power is given by

$$\frac{dP}{d\Omega} = \frac{I_m^2 Z_0}{8\pi^2}\left[\frac{\cos(\tfrac{1}{2}ka\cos\theta) - \cos(\tfrac{1}{2}ka)}{\sin\theta}\right]^2. \quad \text{(SI)} \tag{64.9}$$

$$P = \int_0^{\pi} \frac{dP}{d\Omega}\, 2\pi\sin\theta\,d\theta. \tag{64.10}$$

(The integral over θ is not elementary for general ka.)

In the limit of a very short antenna, also known as a *short* or *Hertzian dipole*, $ka \ll 1$, we expect the electric dipole approximation to hold, and $dP/d\Omega \sim \sin^2\theta$.[9] Now, $j(\mathbf{r}, t) \sim k(\tfrac{1}{2}a - |z|)$ to good approximation, so

$$\dot{\mathbf{d}} = \int \mathbf{j}(\mathbf{r}, t)d^3x = \frac{I_m}{4}ka^2\cos\omega t\,\hat{\mathbf{z}}. \tag{64.11}$$

To obtain the total power, we should square the coefficient of $\cos\omega t$, multiply by $\omega^2/4\pi c^3$, another factor of $1/2$ from the time average, and a factor of $8\pi/3$ from the integration over all angles:

$$P = \frac{cZ_0}{4\pi} \times \frac{I_m^2}{16}(ka^2)^2 \times \frac{\omega^2}{4\pi c^3} \times \frac{1}{2} \times \frac{8\pi}{3} = \frac{I_m^2}{192\pi}(ka)^4 Z_0. \quad \text{(SI)} \tag{64.12}$$

The same result is obtained by letting $ka \to 0$ in eq. (64.9) directly and integrating over angles.

From the viewpoint of the generator, the antenna consumes energy, which it then radiates. An ac current of amplitude I_0 flowing through a resistor R would lead to a power loss $I_0^2 R/2$. We therefore define the *radiation resistance*, R_{rad}, of the antenna, such that

$$P = \tfrac{1}{2}I_0^2 R_{\text{rad}}, \tag{64.13}$$

with I_0 being the feed or input current. For the short dipole, $I_0 \approx I_m ka/2$ from eq. (64.5), so

$$R_{\text{rad}} = Z_0 \frac{\pi}{6}\left(\frac{a}{\lambda}\right)^2. \quad \text{(SI)} \tag{64.14}$$

---

[9] In the spirit of the footnote in section 62, we mention again the potential confusion between the long-wavelength moments and the Bessel moments. Thus, fig. 9.7 of Jackson (1999), showing polar plots of the radiated power from a center-fed antenna, is based on the *Bessel* moments. Jackson's discussion in section 9.2, however, pertains to the *long-wavelength* electric dipole approximation. Caveat lector!

For $a/\lambda = 0.2$, we get $R_{\mathrm{rad}} \simeq 8\,\Omega$, which is rather small. A short dipole is not a very efficient radiator. For better efficiencies, one commonly uses half-wave ($a = \lambda/2$) or full-wave ($a = \lambda$) antennas.

Next, let us discuss the angular power distribution (64.9). As a starts to increase, this distribution stays of the general $\sin^2\theta$ shape, but gets more peaked in the equatorial plane. When $a > \lambda$, it develops additional null directions besides $\theta = 0$. New nulls arise every time a passes a multiple of λ, and the distribution becomes multilobed. This feature is common to all large antennas, and it arises because of interference between the radiation from different parts of the antenna.

Exercise 64.1 By numerical integration of eq. (64.9), find, for half- and full-wave antennas, (i) the radiation resistance; (ii) the gain, defined as the power in the direction where it is maximal, normalized by its average over all directions; (iii) the half-power beamwidth in the xz plane, defined as the angle between the directions where the power falls to half its maximum value.

Answer: For $a = \lambda/2$: $R_{\mathrm{rad}} = 73\,\Omega$, gain $= 1.64$, beamwidth $= 78°$. For $a = \lambda$: R_{rad} is ill-defined if we use eq. (64.5) and hard to compute realistically, for the true feed current depends sensitively on the width of the rods and the gap near $z = 0$. The gain is 1.80, and the beamwidth is $47°$. By comparison, for a short dipole, gain $= 1.5$, beamwidth $= 90°$.

Exercise 64.2 A circular current loop of radius a carries a current $I(t) = I_0 \cos \omega t$. Find the radiation pattern and the total power assuming $ka \ll 1$.

Solution: Let the loop lie in the xy plane, and let the origin be at the center of the loop. The power radiated is given by eq. (61.22), since the source conditions vary at a single frequency. The basic object required is $\mathbf{j}_{k0}$, the spatial FT of the current distribution. Recalling the rule $\mathbf{j}d^3x = I d\boldsymbol{\ell}$ for a filamentary current, we obtain

$$\mathbf{j}_{k0} = I_0 \oint e^{-i\mathbf{k}\cdot\mathbf{r}}d\boldsymbol{\ell}. \tag{64.15}$$

Because of symmetry, it is enough to consider the observer, and, therefore, the vector $\mathbf{k}$, to lie in the xz plane. Letting φ be the polar angle in the xy plane, we have

$$\mathbf{k}\cdot\mathbf{r} = k_x a \cos\varphi,$$

$$d\boldsymbol{\ell} = a(-\sin\varphi\,\hat{\mathbf{x}} + \cos\varphi\,\hat{\mathbf{y}})d\varphi.$$

Hence,

$$\mathbf{j}_{k0} = I_0 a \int_0^{2\pi} e^{-ik_x a \cos\varphi}(-\sin\varphi\,\hat{\mathbf{x}} + \cos\varphi\,\hat{\mathbf{y}})d\varphi. \tag{64.16}$$

The x component vanishes, as the integrand is odd under $\varphi \to 2\pi - \varphi$. For the y component, since $ka \ll 1$, we may expand the $e^{-ik_x a \cos\varphi}$ factor. Keeping only the leading nonzero term, we obtain

$$\mathbf{j}_{\mathbf{k}0} \approx -i\pi I_0 k_x a^2 \hat{\mathbf{y}} \quad (ka \ll 1). \tag{64.17}$$

This equals $\mathbf{j}_{\mathbf{k}0}^{\perp}$, as it is already perpendicular to $\mathbf{k}$. Writing $k_x = k \sin \theta$, substituting in eq. (61.22), and performing the trivial integration over frequency, we obtain

$$\frac{d P}{d\Omega} = \frac{I_0^2}{32} \left(\frac{2\pi a}{\lambda}\right)^4 \sin^2 \theta \, Z_0. \quad \text{(SI)} \tag{64.18}$$

The total power radiated is

$$P = \frac{\pi}{12} I_0^2 \left(\frac{2\pi a}{\lambda}\right)^4 Z_0. \quad \text{(SI)} \tag{64.19}$$

This radiation pattern is that of an $m = 0$ dipole. We can further say it is of the magnetic dipole (M1) type because the electric dipole moment of the source is clearly zero, and the oscillating current gives rise to an oscillating magnetic dipole moment of amplitude $\pi I_0 a^2/c$. By eq. (63.19) the total power is then $\pi^2 I_0^2 (ka)^4/3c$, in accord with eq. (64.19).

For general ka, we can express the angular power distribution in terms of the Bessel function J_1. Making use of the representation (B.5) in eq. (64.16), we obtain $\mathbf{j}_{\mathbf{k}0} = -2\pi i I_0 a J_1(k_x a)\hat{\mathbf{y}}$. Hence,

$$\frac{d P}{d\Omega} = \frac{\pi}{2c}[I_0 ka J_1(ka \sin \theta)]^2. \tag{64.20}$$

As for the linear antenna, this is multilobed for large ka.

Exercise 64.3 An astonishing variety of angular power patterns can be obtained by combining two or more antennas next to each other and driving them in specific phase relationships. Even for two antennas, the results are surprising. (See, e.g., fig. 11-11 of Kraus (1988).) Consider two half-wave center-fed linear antennas, both oriented vertically (along z), separated by a distance d, and driven by equal currents out of phase by δ. Plot the patterns in the xy plane for (i) $d = \lambda/2, \delta = 0$ (broadside array); (ii) $d = \lambda/2, \delta = \pi$ (end-fire array); (iii) $d = 3\lambda/8, \delta = \pi/2$; (iv) $d = \lambda/4, \delta = 3\pi/4$.

65 Near-zone fields

Let us now determine the fields in the *near zone*, i.e., when $R \ll \lambda_{\mathrm{av}}$, the characteristic wavelength of the radiation. We shall also assume the source is nonrelativistic, i.e., $a \ll \lambda_{\mathrm{av}}$. In this case, $k R \ll 1$ for the bulk of the frequencies in the Fourier decomposition, and in the general integral expressions for the frequency domain potentials, eq. (59.8), we may replace $e^{ik R}$ by 1. Then,

$$\phi_\omega \simeq \int d^3 x' \frac{\rho_\omega(\mathbf{r}')}{|\mathbf{r} - \mathbf{r}'|}; \quad \mathbf{A}_\omega \simeq \frac{1}{c} \int d^3 x' \frac{\mathbf{j}_\omega(\mathbf{r}')}{|\mathbf{r} - \mathbf{r}'|}. \tag{65.1}$$

It is useful to estimate the relative size of these two potentials. Since $\nabla \cdot \mathbf{j}_\omega \sim \mathbf{j}_\omega/a$, the continuity equation yields $\mathbf{j}_\omega \sim a\omega \rho_\omega$. Hence, $\mathbf{A}_\omega \sim (a/\lambda_{\mathrm{av}})\phi_\omega \ll \phi_\omega$.

The fields are obtained from the potentials via eqs. (59.9) and (59.10):

$$-\nabla\phi_\omega = \int d^3x' \frac{(\mathbf{r}-\mathbf{r}')}{|\mathbf{r}-\mathbf{r}'|^3}\rho_\omega(\mathbf{r}'). \tag{65.2}$$

The other part of $\mathbf{E}_\omega$, $ik\mathbf{A}_\omega$, on the other hand, is of magnitude

$$\int d^3x' \frac{k|\mathbf{j}_\omega(\mathbf{r}')|}{c|\mathbf{r}-\mathbf{r}'|} \sim \int d^3x' \frac{a}{\lambda_{av}^2}\frac{|\rho_\omega(\mathbf{r}')|}{R} \sim \frac{aR}{\lambda_{av}^2}|\nabla\phi_\omega| \ll |\nabla\phi_\omega|. \tag{65.3}$$

Thus, in the near zone $\mathbf{E}_\omega \simeq -\nabla\phi_\omega$ and is given by eq. (65.2). The magnetic field $\mathbf{B}_\omega$ equals $\nabla \times \mathbf{A}_\omega$ and is given by a very similar expression. If we resynthesize the Fourier components, we can write the results back in the time domain:

$$\mathbf{E}(\mathbf{r},t) = \int d^3x' \frac{(\mathbf{r}-\mathbf{r}')}{|\mathbf{r}-\mathbf{r}'|^3}\rho(\mathbf{r}',t), \tag{65.4}$$

$$\mathbf{B}(\mathbf{r},t) = \frac{1}{c}\int d^3x' \frac{(\mathbf{r}-\mathbf{r}')}{|\mathbf{r}-\mathbf{r}'|^3} \times \mathbf{j}(\mathbf{r}',t). \tag{65.5}$$

Not surprisingly, these look like the instantaneous fields due to a charge and current distribution. In the near zone, retardation effects may be neglected. Corrections to these expressions will involve time derivatives of the instantaneous higher electric and magnetic multipoles of the source. It should also be noted that we assumed only $R \ll \lambda_{av}$ and $a \ll \lambda_{av}$ in our derivation. The relative size of a and R is arbitrary. Thus, eqs. (65.4) and (65.5) may be used inside the source region as well.

We can, in fact, find the fields at somewhat larger distances, of the order of the wavelength, i.e., when $kR \sim 1$. We continue to assume $a \ll \lambda_{av}$. The analysis is then simplified by keeping only the electric dipole moment of the source. Under these conditions, in the formula for the scalar potential,

$$\phi_\omega(\mathbf{R}) = \int \frac{e^{ik|\mathbf{R}-\mathbf{r}'|}}{|\mathbf{R}-\mathbf{r}'|}\rho_\omega(\mathbf{r}')d^3x', \tag{65.6}$$

we may expand the kernel in powers of $\mathbf{r}'$. Up to terms of order $\mathbf{r}'$, we get

$$\phi_\omega(\mathbf{R}) = \int \frac{e^{ikR}}{R}\left(1 + \frac{\hat{\mathbf{n}}\cdot\mathbf{r}'}{R} - i\mathbf{k}\cdot\mathbf{r}' + \cdots\right)\rho_\omega(\mathbf{r}')d^3x'. \tag{65.7}$$

By considering a collection of moving charges, we can show that

$$\int d^3x' \rho_\omega(\mathbf{r}')\begin{pmatrix}1\\\mathbf{r}'\end{pmatrix} = \begin{pmatrix}0\\\mathbf{d}_\omega\end{pmatrix}. \tag{65.8}$$

Hence,

$$\phi_\omega(\mathbf{R}) = \frac{e^{ikR}}{R^2}(1 - ikR)\hat{\mathbf{n}}\cdot\mathbf{d}_\omega. \tag{65.9}$$

We can find the vector potential in the same way. The calculation is now simpler, because

$$\int d^3x' \mathbf{j}_\omega(\mathbf{r}') = -i\omega\mathbf{d}_\omega \tag{65.10}$$

already. Any additional powers of $\mathbf{r}'$ in the integrand would lead to higher moments. Hence,

$$
\mathbf{A}_\omega(\mathbf{R}) = \frac{1}{c} \int \frac{e^{ik|\mathbf{R}-\mathbf{r}'|}}{|\mathbf{R}-\mathbf{r}'|} \mathbf{j}_\omega(\mathbf{r}') d^3x'
$$

$$
\approx \frac{-ike^{ikR}}{R} \mathbf{d}_\omega. \tag{65.11}
$$

The fields follow from the potentials by taking gradients. We leave it as an exercise for the reader to show that

$$
\mathbf{E}_\omega = \frac{e^{ikR}}{R} k^2 \mathbf{d}_\omega^\perp + \frac{e^{ikR}}{R^3} (1 - ikR)(3\mathbf{d}_\omega^\parallel - \mathbf{d}_\omega), \tag{65.12}
$$

$$
\mathbf{B}_\omega = \frac{e^{ikR}}{R} k^2 (\hat{\mathbf{n}} \times \mathbf{d}_\omega^\perp) + ik \frac{e^{ikR}}{R^2} (\hat{\mathbf{n}} \times \mathbf{d}_\omega). \tag{65.13}
$$

The terms of order $1/R$ are in fact the far-zone fields in the E1 approximation. In the near zone, on the other hand, where $kR \ll 1$, we get

$$
\mathbf{E}_\omega \approx \frac{1}{R^3} (3\mathbf{d}_\omega^\parallel - \mathbf{d}_\omega). \tag{65.14}
$$

The magnetic field is zero to the same order; the leading term is of order $1/\lambda_0 R^2$. Note also that $\mathbf{B}_\omega$ is purely perpendicular to $\hat{\mathbf{n}}$.

Exercise 65.1 Find the near-zone fields for a magnetic dipole source.

Answer: Use the electric dipole case formulas, with $\mathbf{d}_\omega \to \mathbf{m}_\omega$, $\mathbf{E}_\omega \to \mathbf{B}_\omega$, $\mathbf{B}_\omega \to -\mathbf{E}_\omega$.

66 Angular momentum radiated*

It is also interesting to examine the angular momentum radiated by a current source. By the angular momentum conservation law (37.6), the rate of radiation is given by

$$
\frac{dL_i}{dt} = \int M_{ij} \hat{n}_j d^2s, \tag{66.1}
$$

where the integral is over the surface of a large sphere of radius R, $\hat{\mathbf{n}}$ is the outward normal to this sphere, and

$$
M_{ij} = \epsilon_{imn} T_{jn} x_m, \tag{66.2}
$$

$$
T_{jn} = - \left(E_j E_n + B_j B_n - \tfrac{1}{2}(E^2 + B^2)\delta_{jn} \right). \tag{66.3}
$$

Note the sign on the right-hand side of eq. (66.1). Now, $d\mathbf{L}/dt$ stands for the rate at which angular momentum is *radiated away*, which is the negative of the rate of change of angular momentum inside the sphere.

It is easy to see that the part of the stress tensor T_{jn} proportional to δ_{jn} contributes nought to the integral in eq. (66.1). The remaining terms can be rearranged to yield

$$\frac{d\mathbf{L}}{dt} = -\frac{R^3}{4\pi} \int \left[(\hat{\mathbf{n}} \times \mathbf{E})(\hat{\mathbf{n}} \cdot \mathbf{E}) + \mathbf{E} \leftrightarrow \mathbf{B} \right] d\Omega. \tag{66.4}$$

The essential point here is that because of the appearance of the longitudinal parts $\hat{\mathbf{n}} \cdot \mathbf{E}$ and $\hat{\mathbf{n}} \cdot \mathbf{B}$, it is not enough to consider only the radiation zone fields, and we must include the next corrections in powers of $1/R$. This is laborious to do in complete generality, and we limit ourselves to considering only electric dipole radiation. The fields are given to adequate accuracy by eqs. (65.12) and (65.13). To leading order $\hat{\mathbf{n}} \times \mathbf{E} \sim R^{-1}$ and $\hat{\mathbf{n}} \cdot \mathbf{E} \sim R^{-2}$, so the product varies as R^{-3}, exactly what is needed to cancel the R^3 factor in front. The magnetic field terms make no contribution, since $\hat{\mathbf{n}} \cdot \mathbf{B}$ vanishes. In the frequency domain,

$$\hat{\mathbf{n}} \times \mathbf{E}_\omega = \frac{\omega^2 e^{ikR}}{c^2 R} (\hat{\mathbf{n}} \times \mathbf{d}_\omega), \tag{66.5}$$

$$\hat{\mathbf{n}} \cdot \mathbf{E}_\omega = -2i \frac{\omega e^{ikR}}{c R^2} (\hat{\mathbf{n}} \cdot \mathbf{d}_\omega). \tag{66.6}$$

Translating to the time domain, we get

$$\hat{\mathbf{n}} \times \mathbf{E} = -\frac{1}{c^2 R} (\hat{\mathbf{n}} \times \ddot{\mathbf{d}}_{\text{ret}}), \tag{66.7}$$

$$\hat{\mathbf{n}} \cdot \mathbf{E} = \frac{1}{c R^2} (\hat{\mathbf{n}} \cdot \dot{\mathbf{d}}_{\text{ret}}). \tag{66.8}$$

Hence,

$$\frac{d\mathbf{L}}{dt} = \frac{2}{3c^3} (\dot{\mathbf{d}}_{\text{ret}} \times \ddot{\mathbf{d}}_{\text{ret}}). \tag{66.9}$$

This is the formula for the angular momentum radiated in the electric dipole approximation.

Let us apply eq. (66.9) to a charge q rotating in a circle of radius a at frequency ω. Then,

$$\mathbf{d}(t) = q a (\cos \omega t \, \hat{\mathbf{x}} + \sin \omega t \, \hat{\mathbf{y}}), \tag{66.10}$$

and

$$\frac{d\mathbf{L}}{dt} = \frac{2q^2 a^2 \omega^3}{3c^3} \hat{\mathbf{z}}. \tag{66.11}$$

As expected, only the $\hat{\mathbf{z}}$ component of the total angular momentum is nonzero. In the same approximation, the energy radiation rate is

$$\frac{d\mathcal{E}}{dt} = \frac{2}{3c^3} (\ddot{\mathbf{d}}_{\text{ret}})^2 = \frac{2q^2 a^2 \omega^4}{3c^3}. \tag{66.12}$$

The ratio $\dot{L}/\dot{\mathcal{E}}$ equals $1/\omega$, which has a suggestive quantum mechanical interpretation as the angular momentum per radiated photon ($\hbar$) divided by the energy per radiated photon ($\hbar\omega$).

67 Radiation reaction

The radiative energy lost by an accelerating charge tends to slow it down. This slowing down can be attributed to a backreaction of the radiation on the particle itself. In the equations of motion for the charge, therefore, we must include an additional force, $\mathbf{F}_{rad}$, describing this reaction. Let us try and find this force, known as the *radiation reaction* or *radiation damping*, for a nonrelativistic electron to begin with.

According to the Larmor formula, the power radiated is given by

$$P = \frac{2}{3}\frac{e^2}{c^3}\dot{\mathbf{v}}^2. \tag{67.1}$$

The radiation reaction must be such that, over time, the work done by it on the charge equals the negative of the energy lost to radiation. Hence,

$$\overline{\mathbf{F}_{rad}\cdot\mathbf{v}} = -\frac{2}{3}\frac{e^2}{c^3}\overline{\dot{\mathbf{v}}^2}, \tag{67.2}$$

where the bars denote an average over some characteristic time of the motion. Such an average is necessary, if for no other reason than that without it the formula cannot be satisfied. We now write

$$\dot{\mathbf{v}}^2 = \frac{d}{dt}(\mathbf{v}\cdot\dot{\mathbf{v}}) - \mathbf{v}\cdot\ddot{\mathbf{v}}. \tag{67.3}$$

The time average of the total time derivative term vanishes, so

$$\overline{\mathbf{F}_{rad}\cdot\mathbf{v}} = \frac{2}{3}\frac{e^2}{c^3}\overline{\mathbf{v}\cdot\ddot{\mathbf{v}}}. \tag{67.4}$$

We may, therefore, take the radiation reaction to be

$$\mathbf{F}_{rad} = \frac{2}{3}\frac{e^2}{c^3}\ddot{\mathbf{v}} = \frac{2}{3}m\tau_e\ddot{\mathbf{v}}. \tag{67.5}$$

In the second form, $\tau_e = e^2/mc^3$ is the characteristic electron time introduced in section 18; it is the classical electron radius divided by c.

The presence of the force (67.5) means that the equation of motion for an electron subject to an external force $\mathbf{F}_{ext}$ must be modified to

$$m\left(\dot{\mathbf{v}} - \frac{2}{3}\tau_e\ddot{\mathbf{v}}\right) = \mathbf{F}_{ext}. \tag{67.6}$$

But, we now have the problem that this equation has unphysical solutions. If $\mathbf{F}_{ext}$ vanishes, e.g., one solution is $\mathbf{v} = 0$, but another is $\dot{\mathbf{v}} \propto \exp(3t/2\tau_e)$. Such "runaway solutions" show that the description of the radiation reaction in terms of an instantaneous force cannot be truly correct. The root of this problem is that we are trying to evaluate the effect of the electron's field back on itself. Such an evaluation must be based on a model of an electron as a finite charge distribution, such as that considered in our discussion of electromagnetic mass in section 39. Let us summarize what we learned at that point. As the dimensions of the charge distribution approach zero, we discover that (i) the electromagnetic field energy $\mathcal{E}_0$ diverges in a way dependent on the assumed structure of the charge distribution, and (ii) the electromagnetic field momentum $\mathbf{p}_0$ equals $(4/3)(\mathcal{E}_0 \mathbf{v}/c^2)$, where $\mathbf{v}$ is the electron velocity, assumed small. It is tempting to interpret $\mathcal{E}_0/c^2$ as an "electromagnetic mass," but the fact that it diverges for a point electron, and the factor of 4/3 in the momentum–energy relation forces us to include a negative nonelectromagnetic mass, originating in the Poincare stresses. This negative mass, when added to the electromagnetic one, produces the experimentally observed mass. Even after surviving this logical Scylla and Charybdis, however, we are not home, for we discover that there remains a radiation reaction force that is precisely the heuristically found (67.5), which, as already noted, is unsatisfactory. The upshot is that there is no known satisfactory procedure in classical electrodynamics that overcomes these difficulties.[10] Even quantum electrodynamics provides only a partial resolution of these problems.

Nevertheless, eq. (67.6) can be used if we understand the limitations on it. Suppose the external force acts only for a time T during which the mean acceleration of the particle is $\dot{v}$. The particle acquires an energy $\sim m(\dot{v}T)^2$ due to this force. It radiates an amount of energy equal to $e^2 \dot{v}^2 T/c^3$ using the Larmor formula again. The condition for the radiation reaction to be small can be found by demanding that the energy lost be small relative to the total energy itself, i.e.,

$$\frac{e^2}{c^3}\dot{v}^2 T \ll m(\dot{v}T)^2, \tag{67.7}$$

which is equivalent to

$$T \gg \tau_e = \frac{e^2}{mc^3}. \tag{67.8}$$

Thus, any impulsive external force must be applied over times long compared to the classical time τ_e. Similarly, if the motion is confined to a dimension a and has a typical frequency ω, the mean energy of the particle is $m\omega^2 a^2$, and the energy lost in one period is of order $(e^2/c^3)(\omega^2 a)^2(1/\omega)$. Requiring the latter to be small compared to the former, we obtain the condition

$$\omega \tau_e \ll 1. \tag{67.9}$$

The same condition applies to the frequency of an electromagnetic field incident on the particle. If the acceleration is caused by a magnetic field, this condition requires the field to obey

$$B \ll m^2 c^4/e^3. \tag{67.10}$$

Thus, there is a very large regime of parameters in which our simple-minded treatment of radiation reaction is valid. The time $\tau_e \sim 10^{-23}$ sec, and the limit on B is enormous, $\sim 6 \times 10^{15}$ G. In reality, quantum effects set in at frequencies and magnetic fields smaller than these limits by a factor of $\hbar c/e^2 \simeq 137$, but even the new limits are exceedingly liberal. Thus, the classical treatment is an excellent way to study the trajectories and stability of particle beams in accelerators.

---

[10] For this reason, we shall not discuss the numerous attempts to obtain integrodifferential or difference equations that seek to describe radiation reaction without the problems of acausality or runaway solutions.

As an example, let us reconsider the classical model of an atom (exercise 58.1), including radiation reaction. Let the nuclear charge be Z, and let the orbital radius be a. In the absence of damping, we would have an orbital frequency $\omega_0 = (Ze^2/ma^3)^{1/2}$. With radiation damping, the equation of motion is

$$m\frac{d^2\mathbf{r}}{dt^2} = -\frac{Ze^2}{r^3}\mathbf{r} + \frac{2}{3}m\tau_e\frac{d^3\mathbf{r}}{dt^3}. \tag{67.11}$$

Feeding into this equation a solution close to the undamped one, in the form

$$\mathbf{r} = \text{Re } a(\hat{\mathbf{x}} + i\hat{\mathbf{y}})e^{-i\omega t}, \tag{67.12}$$

and writing the radius a in terms of the original frequency ω_0, we obtain

$$\omega^2 = \omega_0^2 - \frac{2}{3}i\omega^3\tau_e. \tag{67.13}$$

If we solve this equation assuming $\omega_0\tau_e \ll 1$, we find that ω acquires an imaginary part $-i\gamma/2$, as well as a small shift $\Delta\omega_0$ in the real part, given by

$$\gamma = \frac{2}{3}\omega_0^2\tau_e, \quad \Delta\omega_0 = -\frac{1}{3}\omega_0^3\tau_e^2. \tag{67.14}$$

It is interesting to find the spectral distribution of the energy radiated by this atom. We may regard the source as quasimonochromatic and use the FT of the current–current autocorrelation function (see eq. (61.14)) to find the power spectrum. Writing $\bar{\omega}_0 = \omega_0 + \Delta\omega_0$, we have

$$\mathbf{j}_k(t) = -e\mathbf{v}(t) = ea\bar{\omega}_0(\sin\bar{\omega}_0 t\,\hat{\mathbf{x}} - \cos\bar{\omega}_0 t\,\hat{\mathbf{y}})e^{-\gamma t/2}. \tag{67.15}$$

To compute the correlation function, we average for a time T that is much longer than the period $2\pi/\omega_0$, but much shorter than $1/\gamma$. This is permissible as $\omega_0 \gg \gamma$. Dropping terms of relative order $(1/\omega_0 T)$ and (γT), we obtain

$$G_{jj}^{\perp}(\tau) \approx \frac{1}{2}e^2a^2\omega_0^2\cos\bar{\omega}_0\tau\, e^{-\gamma|\tau|/2}(\hat{\mathbf{x}}_\perp\cdot\hat{\mathbf{x}}_\perp + \hat{\mathbf{y}}_\perp\cdot\hat{\mathbf{y}}_\perp), \tag{67.16}$$

where $\hat{\mathbf{x}}_\perp = -\hat{\mathbf{n}}\times(\hat{\mathbf{n}}\times\hat{\mathbf{x}})$, etc. The angular factor can also be written as $(1+n_z^2)$, and integrating it over all solid angles yields $16\pi/3$. Hence,

$$\frac{dP}{d\omega} = \frac{2e^2a^2\omega_0^2\omega^2}{3\pi c^3}\int_{-\infty}^{\infty}\cos\bar{\omega}_0\tau e^{-\gamma|\tau|/2}e^{i\omega\tau}d\tau$$

$$= \frac{2e^2a^2\omega_0^2\omega^2}{3\pi c^3}\text{Re}\left[\frac{1}{-i(\omega-\bar{\omega}_0)+\gamma/2} + \frac{1}{-i(\omega+\bar{\omega}_0)+\gamma/2}\right]. \tag{67.17}$$

Since we are only interested in positive frequencies, the second denominator in the square brackets can never be small, and the power is concentrated in the frequency range $\omega_0 \pm \gamma$. Thus, we may write

$$\frac{dP}{d\omega} \simeq \frac{2e^2a^2\omega_0^4}{3c^3} \times \frac{1}{\pi}\frac{\gamma/2}{(\omega-\omega_0-\Delta\omega_0)^2+(\gamma/2)^2}. \tag{67.18}$$

The damping broadens the lineshape from a delta function into a Lorentzian, of full width at half-maximum equal to γ. We have also shown the downward frequency shift explicitly. The linewidth and frequency shift are key features of real atomic spectra, although it should be mentioned that neither is correctly given by classical electrodynamics, especially the shift.

Finally, integrating over all frequencies, we get the total power radiated:

$$P = \int_0^\infty d\omega \frac{dP}{d\omega} = \frac{2e^2 a^2 \omega_0^4}{3c^3}, \tag{67.19}$$

in complete agreement with exercise 58.1.

11 | Motion of charges and moments in external fields

With this chapter we carry out the second part of the program indicated in chapter 1, namely, the motion of charges in specified **E** and **B** fields. Throughout, we ignore the energy radiated by these charges when their motion involves acceleration. In addition to charges, we also consider the motion of point magnetic moments, or spins, which are fundamental properties of many particles, atoms, ions, etc.

Since the Lorentz force law is relativistically correct, a large portion of the analysis in this chapter is relativistically correct. We will not see the relativistic transformation of the electromagnetic fields until chapter 24, however, so although we mention these transformations, we will not make any real use of them. We need only the relativistic relationship between the momentum, energy, and velocity of a particle, which we expect readers to have seen before. Explicit Lorentz transformations will not be invoked.

68 The Lorentz force law

A point charge q in an external electromagnetic field $\mathbf{E}(\mathbf{r}, t)$ and $\mathbf{B}(\mathbf{r}, t)$ moves according to the Lorentz force law,

$$\frac{d\mathbf{p}}{dt} = q\left(\mathbf{E} + \frac{1}{c}\mathbf{v} \times \mathbf{B}\right), \tag{68.1}$$

$$= q(\mathbf{E} + \mathbf{v} \times \mathbf{B}). \quad \text{(SI)}$$

Here, **p** is the momentum of the particle, and **v** its velocity. The fields are evaluated at the position of the particle.

Equation (68.1) is relativistically correct as written. For a nonrelativistic particle, i.e., one whose speed v is much less than the speed of light,

$$\mathbf{p} \approx m\mathbf{v} \quad (v \ll c), \tag{68.2}$$

where m is the mass of the particle.[1] If $v \sim c$, however, one must use the exact relationship between momentum and velocity,

$$\mathbf{p} = \frac{m\mathbf{v}}{\sqrt{1 - v^2/c^2}}. \tag{68.3}$$

Similarly, the relativistically correct formulas for the energy, $\mathcal{E}$, in terms of the velocity or the momentum are

$$\mathcal{E} = \sqrt{m^2 c^4 + p^2 c^2} \tag{68.4}$$

$$= \frac{mc^2}{\sqrt{1 - v^2/c^2}}. \tag{68.5}$$

Equations (68.3) and (68.5) also lead to the useful formula

$$\mathbf{v} = \frac{\mathbf{p} c^2}{\mathcal{E}}. \tag{68.6}$$

Note that $\mathcal{E}$ includes the rest energy, mc^2. At low speeds, we have

$$\mathcal{E} \approx mc^2 + \frac{1}{2} mv^2, \tag{68.7}$$

and one generally subtracts off the constant term mc^2 to arrive at the familiar expression $\frac{1}{2} mv^2$ for the kinetic energy.

It follows from eqs. (68.5) and (68.6) that

$$\frac{d\mathcal{E}}{dt} = \mathbf{v} \cdot \frac{d\mathbf{p}}{dt}. \tag{68.8}$$

For a charged particle in an EM field, therefore,

$$\frac{d\mathcal{E}}{dt} = q\mathbf{v} \cdot \mathbf{E}. \quad \text{(Gaussian and SI)} \tag{68.9}$$

It follows that only the electric field does work on a charged particle. The magnetic field does accelerate the particle, but because the force that it exerts, $q\mathbf{v} \times \mathbf{B}/c$, is perpendicular to $\mathbf{v}$, it does no work. This point is of great importance.

For completeness, we give the relativistically correct formula for the product of mass and acceleration:

$$m\dot{\mathbf{v}} = \frac{q}{\sqrt{1 - v^2/c^2}} \left(\mathbf{E} + \frac{1}{c} \mathbf{v} \times \mathbf{B} - \frac{1}{c^2} \mathbf{v}(\mathbf{v} \cdot \mathbf{E}) \right). \tag{68.10}$$

69 Motion in a static uniform electric field

Let us now suppose that only an electric field is present, i.e., $\mathbf{B} = 0$, and that $\mathbf{E}$ does not vary in space or with time. We take $\mathbf{E} \parallel \hat{\mathbf{z}}$. Then resolving eq. (69.1) into components, we

---

[1] Although the practice is now dying out, some texts still speak of a mass that depends on the velocity. This velocity-dependent mass, $m(v)$, equals $m(1 - v^2/c^2)^{-1/2}$ in our notation and is introduced in order to preserve the appearance of the Newtonian relation $\mathbf{p} = m\mathbf{v}$. We shall always regard the mass as a constant and never refer to the velocity-dependent object $m(v)$.

have

$$\dot{p}_x = 0, \quad \dot{p}_y = 0, \quad \dot{p}_z = q\,E. \tag{69.1}$$

If $v \ll c$, the problem is isomorphic to a mass in a uniform gravitational field. The z component of the velocity increases linearly in time, $v_z = v_{z0} + (q\,E/m)t$, the transverse component of velocity is constant, and if we take this transverse component to lie along $\hat{\mathbf{x}}$, the trajectory is a parabola in the xz plane: $z - z_0 = (q\,E/2mv_0^2)(x - x_0)^2$, where x_0 and z_0 are constants.

It is clear that if the particle continues accelerating, the nonrelativistic solution must become invalid after some time. However, this problem is simple enough to be solved fully relativistically. We take the motion to be in the xz plane as above, so that $p_y = 0$ at all time. Integrating eq. (69.1), we get

$$p_x = p_0, \quad p_z = q\,Et, \tag{69.2}$$

where p_0 is a constant, and we have chosen the origin of time so that $p_z = 0$ at $t = 0$. Then,

$$\mathcal{E} = \sqrt{\mathcal{E}_0^2 + (q\,Ect)^2}, \tag{69.3}$$

where $\mathcal{E}_0 = (m^2c^4 + p_0^2c^2)^{1/2}$. Next, from the formula $\mathbf{v} = \mathbf{p}c^2/\mathcal{E}$, we have

$$\frac{dx}{dt} = \frac{p_0c^2}{\sqrt{\mathcal{E}_0^2 + (q\,Ect)^2}}, \quad \frac{dz}{dt} = \frac{q\,Ec^2t}{\sqrt{\mathcal{E}_0^2 + (q\,Ect)^2}}. \tag{69.4}$$

Integrating these equations, and setting the constants of integration to zero, we obtain the complete solution

$$x = \frac{p_0c}{q\,E}\sinh^{-1}\left(\frac{q\,Ect}{\mathcal{E}_0}\right), \quad z = \frac{1}{q\,E}\sqrt{\mathcal{E}_0^2 + (q\,Ect)^2}. \tag{69.5}$$

Eliminating t from these equations, we obtain the equation for the trajectory,

$$z = \frac{\mathcal{E}_0}{q\,E}\cosh\left(\frac{q\,Ex}{p_0c}\right). \tag{69.6}$$

Exercise 69.1 (Rutherford scattering) The essence of this problem lies in solving for the trajectory of a charge in a static (albeit nonuniform) electric field. A particle of mass m and charge $Z'e$ scatters off a fixed heavy target of charge Ze. Let the particle have a velocity v_0 and impact parameter b when it is asymptotically far from the target. Treat the particle nonrelativistically throughout. (The impact parameter is the distance by which the particle would miss the target if it were not deflected by the scattering potential.)

(a) Show that in plane polar coordinates, the particle trajectory is of the form

$$r = \frac{\ell^2}{m\alpha(\epsilon\cos\varphi - 1)}, \tag{69.7}$$

where $\alpha = ZZ'e^2$, and ϵ and ℓ are constants. [*Hint*: Let $\ell = mr^2\dot{\varphi}$ be the angular momentum. Use the constancy of the angular momentum to express the energy $\mathcal{E}$—potential plus kinetic—as a function of r and $dr/d\varphi$. Energy conservation implies that $d\mathcal{E}/d\varphi = 0$; this equation is elementary when expressed in the variable $u = 1/r$. The solution should lead to eq. (69.7).]

(b) By evaluating the expression for $\mathcal{E}$ at $\varphi = 0$, show that

$$\epsilon = (1 + 2\mathcal{E}\ell^2/m\alpha^2)^{1/2}. \tag{69.8}$$

(c) Equation (69.7) describes a hyperbola with eccentricity ϵ and asymptotes along the directions $\varphi = \pm\varphi_0$ with $\cos\varphi_0 = 1/\epsilon$. Thus, the scattering angle $\theta = \pi - 2\varphi_0$. Show that $\ell = mv_0 b$, and find θ as a function of b and $\mathcal{E}$.

(d) Imagine a broad beam of particles all of energy $\mathcal{E}$ and a spread of impact parameters incident on the target. Incident particles that pass through the annulus of impact radii b and $b + db$ will suffer scattering through angles in the range $(\theta, \theta + d\theta)$, i.e., into a solid angle $d\Omega = 2\pi \sin\theta \, d\theta$. The differential scattering cross section $d\sigma/d\Omega$ is defined as $2\pi b |db/d\Omega|$. Show that

$$\frac{d\sigma}{d\Omega} = \frac{\alpha^2}{4m^2 v_0^4} \frac{1}{\sin^4(\theta/2)}. \tag{69.9}$$

This is the Rutherford scattering cross section.

70 Motion in a static uniform magnetic field

Let us now consider motion in a uniform and static magnetic field, which we take along the z axis. Let us first consider the nonrelativistic problem. In this case,

$$\dot{\mathbf{v}} = \frac{q}{mc}\mathbf{v} \times \mathbf{B}. \tag{70.1}$$

Hence, $\dot{v}_z = 0$, so $v_z = \text{const}$, and $z = v_z t + \text{const}$. Further, $\dot{\mathbf{v}}$ is independent of v_z, so the motion along the z axis does not affect the motion in the xy plane, and the net motion may be regarded as a superposition of these two motions.

For the motion in the xy plane, we note that (i) the acceleration is always perpendicular to the velocity, and (ii) the energy of the particle is a constant. This is exactly like a particle being swung in a circle at the end of a rope. If we denote the radius of the circular motion by r_c, its angular frequency by ω_c, and equate the centripetal force, $m\omega_c^2 r_c$, to the Lorentz force, $qvB/c = q\omega_c r_c B/c$, we obtain

$$\omega_c = \frac{qB}{mc}. \tag{70.2}$$

$$\omega_c = \frac{qB}{m}. \quad \text{(SI)}$$

This is known as the *cyclotron frequency*. Note that this frequency is independent of the cyclotron radius r_c and hence of its velocity or energy.

When we include the motion along the field, we see that the most general orbit is a helix, with its axis parallel to the magnetic field. The velocity along the axis is constant and arbitrary, and the angular frequency of the circular part of the motion is given by eq. (70.2). The motion is clockwise for positive charges and anticlockwise for negative charges. In each case, if we regard the moving charge as a current loop, the magnetic moment of this loop points opposite to **B**.

The relativistic problem is now easy to solve. We have

$$\dot{\mathbf{p}} = \frac{q}{c}\mathbf{v} \times \mathbf{B}, \tag{70.3}$$

but $\mathbf{p} \neq m\mathbf{v}$. Instead, we have $\mathbf{p} = \mathcal{E}\mathbf{v}/c^2$. Because a magnetic field can do no work, $\mathcal{E}$ is a constant, so $\dot{\mathbf{p}} = (\mathcal{E}/c^2)\dot{\mathbf{v}}$, and the equation of motion becomes

$$\dot{\mathbf{v}} = \frac{qc}{\mathcal{E}}\mathbf{v} \times \mathbf{B}. \tag{70.4}$$

This equation differs from eq. (70.1) only in the constant multiplying the $\mathbf{v} \times \mathbf{B}$ term. Once again, the motion along the z axis is uniform, and the motion in the xy plane is circular, with a frequency that can be read off by comparing eqs. (70.1) and (70.4):

$$\omega_c = \frac{qcB}{\mathcal{E}}. \tag{70.5}$$

Note that $v_\perp$, the speed in the xy plane, is a constant in this case too, and the trajectory is still a helix. However, the frequency of circular motion, ω_c, now depends on $\mathcal{E}$, and since $\mathcal{E}$ depends on v_z, the z axis and xy plane motions are no longer independent.

Another interesting result is obtained by taking the xy components of the equation $\mathbf{p} = \mathcal{E}\mathbf{v}/c^2$ and writing $v_\perp = \omega_c r_c$, where r_c is again the cyclotron radius. Using eq. (70.5), we obtain

$$p_\perp = \frac{qB}{c}r_c. \tag{70.6}$$

This result is true at all speeds. In many applications, the motion is purely transverse to the magnetic field. In such cases, a measurement of the radius of curvature of the particle trajectory in a known magnetic field yields its momentum and energy.

Exercise 70.1 Consider a nonrelativistic electron moving in an isotropic harmonic oscillator potential $V = \omega_0^2 r^2/2$ and a uniform magnetic field $\mathbf{B} = B\hat{z}$. Find the normal modes of motion. If $\omega_c = eB/mc \ll \omega_0$, this problem provides a toy model for the classical theory of the Zeeman effect, with the harmonic potential serving to confine the electron near the nucleus in lieu of the true Coulomb potential.

Exercise 70.2 If the magnetic field in a cyclotron is even slightly nonuniform, random perturbations of the particle's motion will cause it to drift out of the field after some time. One way to stabilize against such drift is to add guide fields that make the magnetic field depend on r, the distance perpendicular to the z axis: $\mathbf{B} = B(r)\hat{z}$. Assuming that the motion is nonrelativistic, and that the field is cylindrically symmetric, find the condition on dB/dr for stability.

Solution: Let us first assume that the motion is confined to the xy plane. Then, in polar coordinates, the nonrelativistic equations of motion are

$$2\dot{r}\dot{\varphi} + r\ddot{\varphi} = -\frac{q}{mc}B_z(r)\dot{r}, \tag{70.7}$$

$$\ddot{r} - r\dot{\varphi}^2 = \frac{q}{mc}B_z(r)r\dot{\varphi}. \tag{70.8}$$

To first approximation, let us put $r = r_0$, $\dot{\varphi} = \omega_0$. Then eq. (70.7) is nugatory, and eq. (70.8) gives

$$\omega_0 = -\frac{q}{mc}B_z(r_0). \tag{70.9}$$

This is just the cyclotron frequency. It is important, however, to keep track of the minus sign.

Let us now consider small perturbations of the motion,

$$r(t) = r_0 + \eta(t), \quad \dot{\varphi} = \omega_0 + \xi(t), \tag{70.10}$$

and linearize the equations of motion in η and ξ. This yields

$$\omega_0\dot{\eta} + r_0\dot{\xi} = 0, \tag{70.11}$$

$$\ddot{\eta} - r_0\omega_0\xi = -v\omega_0^2\eta, \tag{70.12}$$

where

$$v = \frac{r_0}{B_z(r_0)}\left(\frac{d B_z}{dr}\right)_{r_0}. \tag{70.13}$$

We differentiate the equation for $\ddot{\eta}$ and eliminate $\dot{\xi}$. With $u(t) \equiv \dot{\eta}$, we get

$$\ddot{u} + \omega_0^2(1 + v)u = 0. \tag{70.14}$$

The solution is oscillatory and bounded provided $v > -1$.

Next, let us analyze perturbations in z. We first note that because $\nabla \times \mathbf{B} = 0$, $\partial B_r/\partial z = \partial B_z/\partial r$, so the field must have some radial component. For small z and $r \approx r_0$, we may take

$$B_r \simeq z\left.\frac{\partial B_z}{\partial r}\right|_{r_0} \simeq \frac{z}{r_0}B_z(r_0)v. \tag{70.15}$$

The equation of motion for $z(t)$ thus reads

$$\ddot{z} = -\frac{q}{mc}r\dot{\varphi}B_r \approx v\omega_0^2 z, \tag{70.16}$$

where we have dropped terms like ηz and ξz on the right as they are of second order of smallness. Thus, $z(t)$ will be bounded provided $v < 0$.

Finally, let us discuss the effect of B_r on the xy motion. To eq. (70.7) we must add a term proportional to $\dot{z}B_r$. This, however, is of second order, so our linear stability analysis holds unchanged. Thus, by choosing $-1 < v < 0$, the motion can be stabilized in all directions.

71 Motion in crossed E and B fields; $E < B$

Let us now suppose that both **E** and **B** are present, but static and uniform. We will see (section 161) when we discuss the relativistic transformation of fields that depending on the fields in the laboratory frame, the problem can be transformed into one of three types in a suitable reference frame:

(a) If **E** $\perp$ **B** and $E < B$ in the lab frame, **E** can be made zero in the new frame.

(b) If **E** $\perp$ **B** and $E > B$ in the lab frame, **B** can be made zero in the new frame.

(c) If **E** $\cdot$ **B** $\neq 0$ in the lab frame, **E** and **B** can be made parallel in the new frame.

The problem is thus reduced to solving for one of three standard types of motion,[2] and backtransforming to the lab frame. By far the most interesting case is (a), and we shall devote most of our attention to that. Further, since we have not yet studied the relativistic transformation of fields, we will limit ourselves to low speeds. Luckily, this is enough to show all the essential features.

Let us choose our axes so that $\hat{\mathbf{z}} \parallel \mathbf{B}$ and $\hat{\mathbf{y}} \parallel \mathbf{E}$. Since $v \ll c$, we have $\mathbf{p} = m\mathbf{v}$, and separating the equation of motion into components, we get

$$m\dot{v}_x = \frac{qB}{c}v_y, \quad m\dot{v}_y = qE - \frac{qB}{c}v_x, \quad m\dot{v}_z = 0. \tag{71.1}$$

It follows that the particle moves uniformly in the z direction and that this motion is decoupled from the motion in the xy plane, i.e., the plane perpendicular to **B**. Hence, only the latter is interesting. Differentiating the equation for $\dot{v}_y$, and feeding in the equation for $\dot{v}_x$, we obtain

$$\ddot{v}_y = -\omega_c^2 v_y. \tag{71.2}$$

The solution is sinusoidal at the cyclotron frequency ω_c and may be taken in the form

$$v_y = -u \sin \omega_c t. \tag{71.3}$$

We have denoted the amplitude of v_y by u and chosen the origin of time so that $v_y = 0$ and $\dot{v}_y < 0$ at $t = 0$. Substituting this solution into the equation for $\dot{v}_y$ and introducing the *drift velocity*

$$v_d = cE/B, \tag{71.4}$$

$$v_d = E/B, \quad \text{(SI)}$$

we obtain

$$v_x = u \cos \omega_c t + v_d. \tag{71.5}$$

---

[2] There is also the exceptional case of **E** $\perp$ **B**, and $E = B$, for which see Landau and Lifshitz (1975), vol. 2, sec. 22, problem 2.

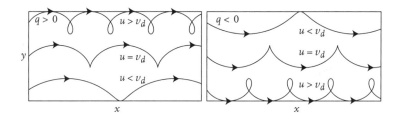

FIGURE 11.1. Particle trajectories in crossed **E** and **B** fields, with **E**$\|\hat{\mathbf{y}}$ and **B**$\|\hat{\mathbf{z}}$ with $E < B$. The quantity u is the maximum value of the y component of the particle's velocity.

Integrating the velocities, we obtain the trajectory:

$$x = \frac{u}{\omega_c} \sin \omega_c t + v_d t, \quad y = \frac{u}{\omega_c} \cos \omega_c t. \tag{71.6}$$

We have chosen **v** $\| \hat{\mathbf{x}}$ at $t = 0$. The general form of the trajectory (see fig. 11.1) depends on the ratio u/v_d. The quantity u is the maximum component of the velocity parallel to **E** and determines the average energy of the particle.

The motion we have found can be viewed as a cyclotron motion about a *guiding center*, which drifts uniformly with the velocity $v_d \hat{\mathbf{x}}$. More generally, the time-averaged velocity is

$$\langle \mathbf{v} \rangle = c \frac{\mathbf{E} \times \mathbf{B}}{B^2}, \tag{71.7}$$

which is also known as the *guiding center velocity*. This result shows that our initial assumption that $v \ll c$ can hold only if $v_d \ll c$, i.e., $E \ll B$. This is a necessary condition that must be obeyed by the fields for our analysis to be valid.

That the average velocity is at right angles to **E** is somewhat counterintuitive, for we would expect a steady force along **E**. We can understand it qualitatively by adding the **E** field as a perturbation to cyclotron motion in the **B** field (see fig. 11.2). Suppose the charge is positive and moving along $\hat{\mathbf{y}}$ as at point P. The Lorentz force causes the charge to bend in a circular arc. When it reaches point Q where **v** $\| \hat{\mathbf{x}}$, the electric force partly cancels the magnetic force, reducing the centripetal force. Thus, the radius of curvature of the orbit is larger than r_c at point Q. As the magnetic field continues to bend the particle, the velocity eventually points along $-\hat{\mathbf{x}}$. The point at which this happens, R, is displaced from the point Q along the x axis because of the reduced curvature at Q. At R, the electric and magnetic forces add, and the radius of curvature is less than r_c. Thus, the next point where **v** $\| \hat{\mathbf{x}}$, S, is further displaced from the point R along x. The net displacement QS after one period is along $\hat{\mathbf{x}}$, parallel to $\mathbf{E} \times \mathbf{B}$.

We leave it for the reader to construct the same argument for a negative charge and verify that, as implied by eq. (71.7), the drift is always along $+\hat{\mathbf{x}}$, independent of the sign of the charge.

The qualitative behavior of the motion is unchanged at relativistic speeds and still consists of a periodic motion about a uniformly moving guiding center. The trajectory is no longer a cycloid, and the frequency is not qB/mc. However, eq. (71.7) remains true for all speeds as long as $E < B$. In particular, if E is very small compared to B,

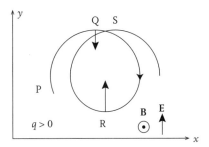

FIGURE 11.2. Why the drift is along $\mathbf{E} \times \mathbf{B}$. The centripetal force at R is greater than that at Q, leading to a tighter circular arc at R.

the drift velocity will be small compared to c, even if the cyclotron motion is quite relativistic.

Exercise 71.1 Obtain the results of this section by combining eq. (71.1) into a single differential equation for the quantity $v_x + iv_y$ and solving this equation. This is a common and powerful technique.

Exercise 71.2 The reader who is unconvinced by the argument following eq. (71.7) should numerically integrate the equations of motion, using, e.g., the simple fourth-order Runge-Kutta method described in many numerical methods textbooks. Convenient initial conditions are $x = y = 0$, $v_y = u$, and $v_x = gu$, and it is further convenient to take u as the unit of velocity. Plot the trajectories for several different choices of g. (This is how fig. 11.1 was obtained.)

Exercise 71.3 Find the trajectory of the center of curvature of a nonrelativistic particle in crossed **E** and **B** fields.

Solution: Let $\mathbf{r}(t)$ and $\mathbf{R}(t)$ be the positions of the particle and the center of curvature, respectively. Denoting by κ the inverse of the radius of curvature, and by $\hat{\mathbf{n}}$ the unit vector from $\mathbf{r}(t)$ to $\mathbf{R}(t)$, we have $\kappa\hat{\mathbf{n}} = v^{-1}d\hat{\mathbf{v}}/dt$. Some straightforward algebra yields

$$\mathbf{R}(t) = \mathbf{r}(t) - r\,\frac{1 + 2\zeta\cos\theta + \zeta^2}{1 + \zeta\cos\theta}[\sin\theta\hat{\mathbf{x}} + (\cos\theta + \zeta)\hat{\mathbf{y}}], \tag{71.8}$$

where $\zeta = v_d/u$, and $\theta = \omega_c t$. To lowest order in v_d/u,

$$\mathbf{R}(t) = \left(v_d t - \frac{v_d}{\omega_c}\sin\theta\cos\theta, \ -\frac{v_d}{\omega_c}(1 + \cos^2\theta)\right). \tag{71.9}$$

Thus, the frame in which the motion appears to be instantaneously circular is not inertial.

A special case of the $E \times B$ drift arises if the particle has a velocity precisely equal to the drift velocity at any instant. In this case, without restriction to low speeds, the electric and magnetic forces cancel exactly, and the particle continues to move uniformly with this velocity forever. This solution (put $r_c = 0$ in eq. (71.6)) is exploited in making velocity

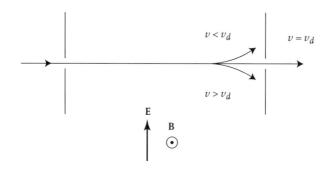

FIGURE 11.3. Velocity selection by crossed **E** and **B** fields.

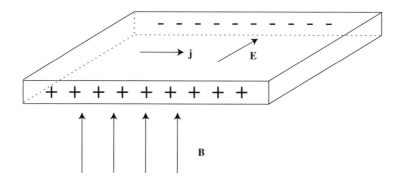

FIGURE 11.4. Schematic of Hall effect. The figure is drawn for positively charged current carriers. The reader should verify the signs of the surface charges and the direction of the electric field produced by them.

selectors. If the directions of **E** and **B** are chosen as shown in fig. 11.3 and a pair of slits is arranged along the x axis and a beam of charged particles traveling along $\hat{x}$ is incident on the first slit, then only those particles with a velocity close to v_d will get through the second slit.

This uniform drift also manifests itself in the *Hall effect* in solid-state physics. It is a remarkable fact, which we shall not try to explain here, that the flow of current in metals and semiconductors can be understood in terms of quasiparticles that are effectively noninteracting, ignoring the mess of strong interparticle Coulomb forces. The charge on a quasiparticle may be $+e$ or $-e$. (In some cases, charge carriers of both signs may be present.) Suppose we drive a current along the $+x$ direction in the presence of a **B** field along $\hat{z}$ (see fig. 11.4). This **B** field tends to deflect the charges along $\pm\hat{y}$, depending on their sign. The deflected charges accumulate on the edges of the sample until an electric field is set up in the y direction that counterbalances the $\mathbf{v} \times \mathbf{B}$ force. The sign of this electric field depends on the sign of the charges, and by measuring the field, we can discover this sign.

Let us briefly discuss the cases labeled (b) and (c) at the beginning of this section. If $\mathbf{E} \perp \mathbf{B}$, but $E > B$, there is no periodicity in the motion, as the B field is not big enough to bend the trajectory, and the particle is continually accelerated to higher energies. If $\mathbf{E} \not\perp \mathbf{B}$, the component of **E** along **B** will accelerate the particle. In the nonrelativistic case, the motion along **B** decouples, and the problem can be analyzed as above. In the relativistic case, the energy $\mathcal{E}$ will increase steadily, and the frequency of the periodic motion will decrease.

Exercise 71.4 In a *Penning trap*, an electron moves in a quadrupolar electric field $\mathbf{E} = K(x\hat{\mathbf{x}} + y\hat{\mathbf{y}} - 2z\hat{\mathbf{z}})$ and a uniform magnetic field $\mathbf{B} = B_0\hat{\mathbf{z}}$. Assuming that $v \ll c$, find and describe the motion of the electron, and find the condition on K for the motion to be completely bounded.

Solution: The equations of motion are completely linear, and the solution is therefore a sum of exponentials of the form $e^{\pm\mu t}$. The motion along $\hat{\mathbf{z}}$ separates from that in the xy plane and is bounded only if $K < 0$. It is then oscillatory at a frequency $\omega_z = (2|K|e/m)^{1/2}$. The motion in the xy plane is bounded if $|K| < m\omega_c^2/e$, with $\omega_c = eB_0/mc$ being the cyclotron frequency. If $|K| \ll m\omega_c^2/e$, this is a superposition of a cyclotron motion at frequency ω_c and a slower epicyclic motion at $\omega_+ = \omega_z^2/4\omega_c$—known as the *magnetron motion*.

72 Motion in a time-dependent magnetic field; the betatron

When a charge moves in a time-dependent magnetic field, in addition to the force due to the magnetic field, it also experiences a force due to the electric field that is induced as per Faraday's law. The resulting motion is very complicated in almost all situations imaginable, but one where the problem is tractable occurs in a device called the *betatron*, in which the induced electric field is used to accelerate electrons. The magnetic field in a betatron must have a specific space and time dependence, which we shall discover. In the interesting applications, electrons are accelerated to a few hundred keV, so the analysis that follows is fully relativistic. Suppose that a set of coils is used to create a magnetic field that is cylindrically symmetric and that points along the z axis everywhere in the xy plane (fig. 11.5). The induced electric field then points along the azimuthal direction $\hat{\mathbf{e}}_\varphi$ everywhere in this plane. (Henceforth, we will consider the fields only in this plane, since the motion of the particles is confined to it.) The circulation of **E** along a circle of radius r centered at the origin is

$$\oint \mathbf{E}(r) \cdot d\boldsymbol{\ell} = \int (\nabla \times \mathbf{E}) \cdot d\mathbf{s} = -\frac{1}{c} \int_S \frac{\partial \mathbf{B}}{\partial t} \cdot d\mathbf{s}, \tag{72.1}$$

where the surface integral extends over the disk of radius r. The left-hand side is just $2\pi r E_\varphi(r, t)$. So, defining the average magnetic field B_{avg} as

$$B_{\text{avg}}(r, t) = \frac{1}{\pi r^2} \int_S \mathbf{B} \cdot d\mathbf{s}, \tag{72.2}$$

we have

$$E_\varphi(r, t) = -\frac{r}{2c} \frac{d}{dt} B_{\text{avg}}. \tag{72.3}$$

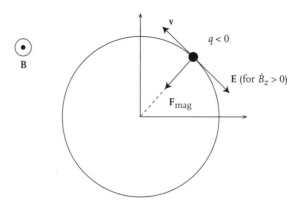

FIGURE 11.5. Particle in betatron orbit. $\mathbf{F}_{mag}$ is the force due to the B field.

Now suppose a particle of charge $q < 0$ is moving anticlockwise along this circle. This particle will be accelerated by the electric field. Since $\mathbf{v} \parallel \mathbf{p} \parallel \mathbf{E}$, the magnitude of the momentum increases according to

$$\frac{dp}{dt} = q\, E_\varphi = -\frac{qr}{2c}\frac{d}{dt}B_{avg}. \tag{72.4}$$

On the other hand,

$$\hat{\mathbf{e}}_r \cdot \frac{d\mathbf{p}}{dt} = \frac{qvB}{c}. \tag{72.5}$$

But, since the particle is moving in a circular orbit, the left-hand side of this equation is just $-\omega p = -vp/r$ by kinematics. Thus,[3]

$$p(t) = -\frac{qr}{c}B(r, t). \tag{72.6}$$

Differentiating this result with respect to t and writing B_{orb} for the field $B(r, t)$ at the orbital radius, we obtain

$$\frac{dp}{dt} = -\frac{qr}{c}\frac{d}{dt}B_{orb}. \tag{72.7}$$

Equating this to the "betatron acceleration" (72.4), we obtain

$$\frac{d}{dt}B_{orb} = \frac{1}{2}\frac{d}{dt}B_{avg}. \tag{72.8}$$

This is the condition that must be obeyed for the particle to move along a circle, and the coils used to produce the field must be designed accordingly.

In the actual application, the simplest way to vary B is by passing a sinusoidal ac current through one set of field coils. It is obvious from eq. (72.4) that the particle will gain energy over only that portion of the current cycle in which $q\,\dot{B}_{avg}$ is negative (increasing B_z if $q < 0$). Since the purpose of the device is to achieve high energies, electrons are injected into the field region and removed a quarter-cycle later, and the injection and removal are

---

[3] Note that this is just eq. (70.6), which we derived earlier for a static uniform $\mathbf{B}$ field.

carefully synchronized with the correct portion of the current cycle. The final momentum of the electron is given by eq. (72.6) with $B = B_{\text{max}}$, the peak magnetic field. Assuming $p \gg mc$, the final energy of the electron is then

$$\mathcal{E} \approx pc = |q| r\, B_{\text{max}}. \tag{72.9}$$

A classic betatron was built at the University of Illinois at Urbana-Champaign in 1949,[4] for which $r = 1.22\,$m, and B varied sinusoidally at $60\,$Hz with $B_{\text{max}} = 9.2\,$kG. The exit energy of electrons is $337\,$MeV according to our formula. The maximum energy actually obtained was $315\,$MeV. The difference is due to radiative energy loss which we have ignored (see section 169). Modern-day high-energy particles are produced in different types of machines—synchrotrons, linacs, etc., but betatrons still find many uses, in medicine, e.g. Their small size is a special advantage.

73 Motion in a quasiuniform static magnetic field—guiding center drift*

Let us now consider the motion of a charged particle in a magnetic field that is almost uniform everywhere. Such a situation arises in many instances in the interstellar medium, in the earth's magnetic field, and in magnetically confined plasmas. The motion, as was first discussed by H. Alfven in 1940, can be regarded as a rapid cyclotronic circular motion around a slowly drifting guiding center.[5] This description will hold provided

$$r_c \nabla B \ll B, \tag{73.1}$$

where r_c is the local cyclotron radius.

In order not to interrupt the physical argument, let us first discuss some mathematical aspects of a weakly inhomogeneous B field. In a source-free region, $\nabla \cdot \mathbf{B}$ and $\nabla \times \mathbf{B}$ both vanish, so the simplest nonvanishing gradient is ∇B, where $B = |\mathbf{B}|$. Suppose that in some local region, $\mathbf{B}$ is essentially along $\hat{\mathbf{z}}$, i.e., $B_x, B_y \ll B_z$. Then $B \simeq B_z + (B_x^2 + B_y^2)/2B_z$, and

$$\nabla B = \nabla B_z + \frac{1}{B_z}(B_x \nabla B_x + B_y \nabla B_y) \approx \nabla B_z. \tag{73.2}$$

So, in the simplest approximation, the expression for the drift velocity of the guiding center will contain only ∇B_z.

It is not permissible, however, to neglect everything but $B_z(\mathbf{r})$. Since $\nabla \times \mathbf{B} = 0$, $\partial B_x/\partial z = \partial B_z/\partial x$, e.g. Thus, the B_x and B_y components will become nonzero as we move along $\mathbf{B}$, giving rise to curvature of the field lines. There is a basic relation between this curvature and $\nabla_\perp B$. (We denote components of all vectors along and perpendicular to

---

[4] D. W. Kerst, G. D. Adams, H. W. Koch, and C. S. Robinson, *Phys. Rev.* **78**, 297 (1950).
[5] More comprehensive treatment of this subject may be found in several texts: Alfven and Falthammar (1963); Northrop (1963); Roederer (1970).

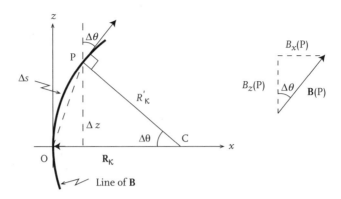

FIGURE 11.6. Proof of eq. (73.6). The quantities $\Delta\theta$, Δz, and Δs are all infinitesimal.

the local direction of **B** by the suffixes ∥ and ⊥, respectively.) To see this, let us choose a local coordinate system (see fig. 11.6), such that at the origin, $\mathbf{B} \parallel \hat{\mathbf{z}}$, and $\nabla_{\perp} B$ lies along $\hat{\mathbf{x}}$. That is,

$$\nabla_{\perp} B = \frac{\partial B_z}{\partial x}\hat{\mathbf{x}} = \frac{\partial B_x}{\partial z}\hat{\mathbf{x}}, \tag{73.3}$$

where the last step follows from $\nabla \times \mathbf{B} = 0$. But, referring to fig. 11.6, we get

$$\frac{\partial B_x}{\partial z} = \lim_{\Delta z \to 0} \frac{B_x(\mathrm{P}) - B_x(\mathrm{O})}{\Delta z} = \lim_{\Delta z \to 0} \frac{B_x(\mathrm{P})}{\Delta z}. \tag{73.4}$$

We now note that $B_x(\mathrm{P}) = B(\mathrm{P})\sin\Delta\theta$, and $\Delta z = R'_\kappa \sin\Delta\theta$, where R'_κ is the length of the segment PC. As $\Delta z \to 0$, $R'_\kappa = R_\kappa(1 + O(\Delta z))$, where R_κ is the radius of curvature. Hence,

$$\frac{\partial B_x}{\partial z} = \lim_{\Delta z \to 0} \frac{B(\mathrm{P})}{R'_\kappa} = \frac{B}{R_\kappa}, \tag{73.5}$$

where B is simply the magnitude of **B** at O. Hence, $\nabla_{\perp} B = (B/R_\kappa)\hat{\mathbf{x}}$. To write the relationship in general, let us denote the vector from the local center of curvature of the field line to the field point by $\mathbf{R}_\kappa$. Then,

$$\nabla_{\perp} B = -\frac{B}{R_\kappa}\widehat{\mathbf{R}}_\kappa = -\frac{B}{R_\kappa^2}\mathbf{R}_\kappa. \tag{73.6}$$

We are now ready to discuss the motion itself. Let us first ignore the curvature of the field lines and take $\mathbf{B} \parallel \hat{\mathbf{z}}$, $\nabla_{\perp} B \parallel \hat{\mathbf{x}}$, as above. Now, consider a charge $q > 0$ moving in a circular trajectory in the xy plane (fig. 11.7). We see from the figure that the field, and therefore the centripetal force, is less at point P than at Q. The average force over one orbit will be nonzero, along $-\hat{\mathbf{x}}$, and as a result the orbit will fail to close. Over time, the guiding center will drift. Which way does it drift? Along the average force, $-\hat{\mathbf{x}}$? No! The argument is much the same as that for the $E \times B$ drift in crossed **E** and **B** fields. The stronger centripetal force at Q means that at this point the local radius of curvature of the orbit will be smaller, so the charge will get pulled up in the y direction more quickly.

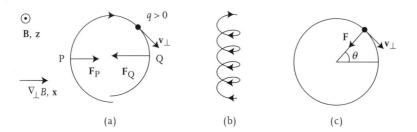

FIGURE 11.7. Drift in an inhomogeneous field. (a) Detailed view of orbit. (b) Drifting cyclotron trajectory. (c) Averaging the Lorentz force over one orbit.

The net drift is toward $+\hat{\mathbf{y}}$. If the charge were negative, the drift would be toward $-\hat{\mathbf{y}}$. The general direction of the drift is along $q\mathbf{B} \times \nabla B$. To find the drift velocity, let us average the Lorentz force over one cyclotron orbit. Denoting the magnetic field at a point on the orbit by $B_z(\theta)$, that at the guiding center by just B_z, and the cyclotron frequency by ω_c, we have

$$B_z(\theta) \approx B_z + \frac{v_\perp}{\omega_c}\frac{\partial B_z}{\partial x}\cos\theta. \tag{73.7}$$

Thus, the average force is

$$\langle \mathbf{F} \rangle = \int_0^{2\pi} \frac{q}{c} v_\perp B_z(\theta)(-\cos\theta\,\hat{\mathbf{x}} - \sin\theta\,\hat{\mathbf{y}})\frac{d\theta}{2\pi}$$

$$= -\frac{1}{2}\frac{q\,v_\perp^2}{c\,\omega_c}\frac{\partial B_z}{\partial x}\hat{\mathbf{x}}. \tag{73.8}$$

We now regard this as an effective $\mathbf{E}$ field $\langle\mathbf{F}\rangle/q$ and use the result for the $E \times B$ drift velocity. Using eq. (70.5) for ω_c and replacing $(\partial B_z/\partial x)\hat{\mathbf{x}}$ by $\nabla_\perp B$, we obtain the general formula for the perpendicular drift velocity of the guiding center:

$$\mathbf{v}_{d\perp} = \frac{\mathcal{E}v_\perp^2}{2qc\,B^2}\hat{\mathbf{B}} \times \nabla_\perp B \quad \text{(no field line curvature)}. \tag{73.9}$$

This formula is relativistically correct. The form at low speeds is obtained very simply by writing mc^2 for $\mathcal{E}$. Also, it gives the correct dependence of the direction of the drift on the sign of q. It is also obvious that motion along $\mathbf{B}$ does not affect the argument at all. This is also a suitable place to remark that the parallel component of the guiding center velocity is the same as that of the particle itself, i.e., $v_{G\parallel} = v_\parallel$.

Next, let us include the effect of field-line curvature. In this case, the guiding center follows the field line in a first approximation. It is somewhat difficult to obtain this result without backing into it, but we can give a self-consistent argument. If we assume that the guiding center follows the field line, then in a frame moving with the guiding center, the particle experiences a centrifugal force which is $m(v_\parallel^2/R_\kappa)\widehat{\mathbf{R}}_\kappa$ in the nonrelativistic case (see fig. 11.8). If we ascribe this to an effective electric field once again, we obtain a drift velocity

$$c\frac{1}{q}\frac{\mathcal{E}}{c^2}v_\parallel^2\frac{\mathbf{R}_\kappa}{R_\kappa^2} \times \frac{\mathbf{B}}{B^2}, \tag{73.10}$$

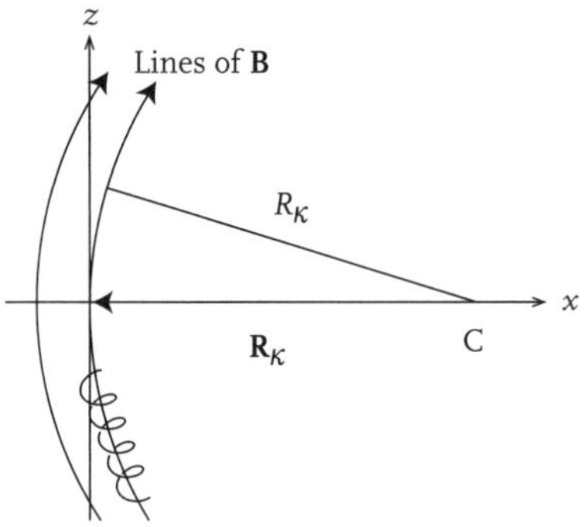

FIGURE 11.8. Effect of curved field lines. The drift is into the page for a positive charge.

where we have used the relativistically correct expression $\mathcal{E}/c^2$ instead of m. The fact that this drift velocity is not in the plane defined by $\mathbf{B}$ and $\nabla_\perp B$ (known as the *osculating plane*) means that there is no additional motion of the guiding center in this plane and makes the argument self-consistent.

Using eq. (73.6) for the curvature and adding eqs. (73.9) and (73.10), we obtain the total guiding center drift velocity

$$\mathbf{v}_{d\perp} = \frac{\mathcal{E}}{qc\,B^2}(v_\parallel^2 + \tfrac{1}{2}v_\perp^2)\hat{\mathbf{B}} \times \nabla_\perp B. \tag{73.11}$$

The two parts (73.9) and (73.10) are often referred to as the gradient drift and the curvature drift, but since they are always present together, there is no physical meaning to this separation. Also, note that we may replace $\nabla_\perp B$ by ∇B in eq. (73.11), since the cross product with $\mathbf{B}$ kills the parallel component.

Exercise 73.1 Up to a distance of about $8\,R_e$ ($R_e = 6.37 \times 10^8$ cm is the earth's radius), the geomagnetic field is close to that due to a dipole moment of magnitude $M_0 = 8.1 \times 10^{25}$ G cm^3 pointing to the south pole. For electrons whose motion lies entirely in the equatorial plane, find the time required to orbit the earth at a mean distance R. Take $R = 4\,R_e$ and find the numerical values for kinetic energies 1 MeV and 10 keV. All distances are measured from the center of the earth.

Solution: Let us first determine if the B field is sufficiently slowly varying for eq. (73.11) to be valid. The cyclotron radius is $r_c = pc\,R^3/e\,M_0$. This radius is $1.5 \times 10^{-3}\,R_e$ and $1.1 \times 10^{-4}\,R_e$ for electrons with energy 1 MeV and 10 keV, respectively. These are both much smaller than R_e, the length scale on which the field varies. Thus, eq. (73.11) is applicable. The actual application is elementary. The drift time for one complete orbit is

$$T_D = \frac{2}{3}\tau_c \left(\frac{R}{r_c}\right)^2, \tag{73.12}$$

where τ_c is the cyclotron orbital time. The drift time equals 16 min and 18 hr for 1-MeV and 10-keV electrons, respectively. By comparison, the cyclotron times are 0.22 msec and

74 μsec. Note that the drift is toward the east for electrons and to the west for positive charges.

74 Motion in a slowly varying magnetic field—the first adiabatic invariant*

The analysis of the previous section enables us to follow the motion of the charge in a quasi-uniform field *locally*, in that we know how the guiding center moves. Starting from an initial position, we can determine the position of the guiding center a few cyclotron orbits later. We can then find the cyclotron velocity at this new location, but it is not obvious how to determine the new values of $v_\parallel$ and $v_\perp$. This problem can be overcome by resorting to the powerful concept of adiabatic invariants of motion. These invariants give information about certain *global* characteristics of the motion, even when the motion cannot be followed in detail. In this section, we find the simplest such invariant for a charge in an inhomogeneous magnetic field.

Let us first consider a magnetic field of the form $B(t)\hat{\mathbf{z}}$, i.e., the field is time-dependent but remains uniform and along $\hat{\mathbf{z}}$ at all times. The motion along the field is then uniform and simple, and only motion in the xy plane need be considered. If B changes very little in one cyclotron orbit, then over this period the particle receives an amount of work

$$\Delta W = -q \oint \mathbf{E} \cdot d\boldsymbol{\ell} = \frac{q}{c} \int \frac{\partial \mathbf{B}}{\partial t} \cdot d\mathbf{s} = \frac{q}{c} \pi r_c^2 \dot{B}, \tag{74.1}$$

where $\mathbf{E}$ is the induced field, for which we have used Faraday's law. Thus, the particle's energy increases at a rate

$$\frac{d\mathcal{E}}{dt} \simeq \Delta W \frac{\omega_c}{2\pi} = \frac{q}{2c} r_c^2 \omega_c \dot{B}. \tag{74.2}$$

But, since $\mathcal{E}^2 = (m^2 c^4 + p^2 c^2)$,

$$\frac{d\mathcal{E}}{dt} = \frac{c^2}{2\mathcal{E}} \frac{dp^2}{dt}. \tag{74.3}$$

Now, because the force is always perpendicular to $\mathbf{B}$, $\dot{p}_\parallel = d(\mathbf{p} \cdot \hat{\mathbf{z}})/dt = \hat{\mathbf{z}} \cdot (d\mathbf{p}/dt) = 0$. Hence, $dp^2/dt = dp_\perp^2/dt$, which gives

$$\frac{dp_\perp^2}{dt} = \frac{2\mathcal{E}}{c^2} \frac{d\mathcal{E}}{dt} = \frac{q\mathcal{E}}{c^3} r_c^2 \omega_c \dot{B}. \tag{74.4}$$

Using eqs. (70.5) and (70.6) for ω_c and r_c, we can write this as

$$\frac{dp_\perp^2}{dt} = \frac{p_\perp^2}{B} \frac{dB}{dt}, \tag{74.5}$$

which is immediately integrated to yield

$$\frac{p_\perp^2}{B} = \text{constant}. \tag{74.6}$$

The combination $p_\perp^2 / B$ is called an *adiabatic invariant*, as it is not a true constant of motion. Rather, it is only approximately constant when the rate of change of the field is *small enough*. How small is small enough? This is a rather difficult question, but it is obvious that we must at the least have $\dot{B} \ll \omega_c B$. The question of just how well the quantity is conserved is even more difficult, and we must refer the reader to specialized texts in mechanics.[6] These texts also show that such adiabatic invariants are nothing but the action integrals $\oint P\,dQ$ taken over a complete orbit. Here, P and Q may be taken as any generalized momentum and coordinate. If we take $Q = \varphi$, the azimuthal angle, then $P = L_z$, the angular momentum along $\hat{\mathbf{z}}$. Since $L_z = r_c p_\perp = c p_\perp^2 / q B$ by eq. (70.6), we get $\oint P\,dQ = 2\pi q p_\perp^2 / c B$, which is basically the adiabatic invariant we have found.

The invariant (74.6) can be written in two other useful forms. By eq. (70.6), $p_\perp \propto B r_c$, so

$$\frac{p_\perp^2}{B} \propto B\pi r_c^2, \tag{74.7}$$

which is the flux through the orbit. If we use eq. (70.5) to replace B in this form, we obtain $(\mathcal{E}/qc)\omega_c r_c^2$. But $q\omega_c r_c^2/2c$ is the magnetic moment $\mathcal{M}$ of the equivalent current loop. Hence,

$$\frac{p_\perp^2}{B} \propto \frac{\mathcal{E}}{mc^2}\mathcal{M}. \tag{74.8}$$

This is the third form. It is useful in the nonrelativistic limit, for then $\mathcal{E}/mc^2 \approx 1$, and we can say that the invariant is the magnetic moment.

Let us now return to the problem of the previous section—a particle in a static but weakly nonuniform B field. How can one use the adiabatic invariant found above? One method is to transform to the guiding center frame. In this frame, the particle will see a magnetic field changing with time, and if $(v_G/\omega_c)\nabla B \ll B$, this change will be slow enough. Thus, the quantity $B'r_c'^2$ is invariant, where we use primes to denote quantities in the guiding center frame. Now, it turns out that since the guiding center moves essentially along $\mathbf{B}$, that $B' = B$, and $r_c' = r_c$, so the invariant may be taken as Br_c^2 or $p_\perp^2/B$ in the lab frame. This combination remains unchanged as the guiding center moves along the field lines. Since the strength and direction of $\mathbf{B}$ can vary quite a lot over large distances, this is a nontrivial result.

It is instructive to obtain this conclusion in another way, which is somewhat prosaic but does not require knowing how fields transform and does not invite worrisome questions about nonzero electric fields in the guiding center frame or from the fact that this frame is noninertial.

Let us choose a local coordinate system in the vicinity of the guiding center at any given time, and let $\mathbf{B} = B_0\hat{\mathbf{z}}$ at the origin. Since $\nabla \cdot \mathbf{B} = \nabla \times \mathbf{B} = 0$, we can write the field near the origin as

$$\mathbf{B}(\mathbf{r}) = B_0\hat{\mathbf{z}} + \frac{1}{2}\nabla\sum_{i=1}^{5}\alpha_i\phi_i, \tag{74.9}$$

---

[6] See, e.g., Landau and Lifhitz (1966), secs. 49–51; or for a mathematically more sophisticated treatment, see Arnold (1978), sec. 52.

where the ϕ_i are the five spherical harmonics of order 2, and the α_i are constants. Instead of the standard harmonics, it is convenient to take

$$(\phi_1, \ldots, \phi_5) = (x^2 - y^2, 2z^2 - x^2 - y^2, 2xy, 2yz, 2zx). \tag{74.10}$$

The α_i are then various gradients of **B** at $\mathbf{r} = 0$. For example,

$$\frac{\partial B}{\partial z} \approx \frac{\partial B_z}{\partial z} = 2\alpha_2. \tag{74.11}$$

We now note that

$$\frac{dp_\parallel}{dt} = \frac{q}{c}(v_x B_y - v_y B_x), \tag{74.12}$$

and, substituting the explicit forms, get

$$\frac{dp_\parallel}{dt} = \frac{q}{c}[v_x(-\alpha_1 y - \alpha_2 y + \alpha_3 x + \alpha_4 z) - v_y(\alpha_1 x - \alpha_2 x + \alpha_3 y + \alpha_5 z)]. \tag{74.13}$$

If we now take the trajectory as a helix,

$$(x, y, z) = (r_c \cos \omega_c(t - t_0), \; r_c \sin \omega_c(t - t_0), \; v_\parallel t),$$

$$(v_x, v_y, v_z) = (-v_\perp \sin \omega_c(t - t_0), \; v_\perp \cos \omega_c(t - t_0), \; v_\parallel), \tag{74.14}$$

and average $\dot{p}_\parallel$ over several orbits, only the products $v_x y$ and $v_y x$ survive the averaging, and we get

$$\frac{dp_\parallel}{dt} = -\frac{q}{c} v_\perp r_c \alpha_2 = -\frac{c^2 p_\perp^2}{2\mathcal{E} B} \frac{\partial B}{\partial z}. \tag{74.15}$$

In the last step, we have used eq. (74.11), written $v_\perp = \omega_c r_c$, and used eqs. (70.5) and (70.6).

We now note that since we have no electric field, the energy is conserved, so $p_\parallel^2 + p_\perp^2$ is conserved. Thus,

$$\frac{dp_\perp^2}{dt} = -\frac{dp_\parallel^2}{dt} = \frac{p_\perp^2}{B} \frac{p_\parallel c^2}{\mathcal{E}} \frac{\partial B}{\partial z}, \tag{74.16}$$

using eq. (74.15). But $p_\parallel c^2 / \mathcal{E} = v_\parallel$, and $v_\parallel(\partial B/\partial z) = d B_{gc}/dt$, where B_{gc} is the value of B at the guiding center. Thus,

$$\frac{dp_\perp^2}{dt} = \frac{p_\perp^2}{B_{gc}} \frac{d B_{gc}}{dt}, \tag{74.17}$$

which again implies that $p_\perp^2 / B_{gc}$ is a constant as the guiding center moves along the field. Along with the fact that $p_\perp^2 + p_\parallel^2$ is a constant of motion, this result allows us to determine the motion of the guiding center completely.

The invariance of $p_\perp^2 / B$ (or, equivalently, $v_\perp^2 / B$) has a very interesting consequence. Let the particle's velocity make an angle α with the local direction of **B**. If we use the arc length s along the field line as a coordinate, then $v_\parallel(s) = v_0 \cos \alpha(s)$, and $v_\perp = v_0 \sin \alpha(s)$.

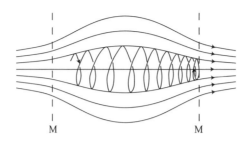

FIGURE 11.9. A magnetic bottle. The particle will be reflected at the mirror points M.

Since v_0 is constant,

$$\frac{\sin^2 \alpha(s)}{B(s)} = \text{const.} \tag{74.18}$$

Let α and B equal α_i and B_i at some convenient point on the field line along which the particle moves. Then,

$$v_\parallel = v_0 \left[1 - \sin^2 \alpha_i \frac{B(s)}{B_i} \right]^{1/2} \tag{74.19}$$

Let us suppose that the particle moves to regions where the field is stronger, i.e., $B(s) > B_i$. At a point where $B(s) = B_i / \sin^2 \alpha_i$, $v_\parallel$ vanishes, and the particle is moving purely transverse to the field. At this point the particle will be reflected back (fig. 11.9).

This phenomenon is used to design magnetic mirrors and bottles to confine plasmas. Consider a symmetric configuration such as that shown in fig. 11.9, and choose the reference point to be the midplane. Suppose we wish to confine the particles to the region between the mirrors marked M. Particles within a "loss cone" defined by $\sin^2 \alpha_i \leq B_i / B_M$ will escape, but the rest will be trapped.

The same kind of trapping also takes place in the earth's magnetic field. This field is stronger near the poles, so electrons and protons get trapped between certain latitudes around the equator. These trapped-particle regions are known as the *Van Allen radiation belts*. (The word "radiation" refers not to electromagnetic radiation but to the fact that the particles are highly energetic.) It is believed that particles can stay trapped in these belts for as long as a few years. We examine the special problem of motion of charged particles in the earth's field further in appendix G.

75 The classical gyromagnetic ratio and Larmor's theorem

We now turn to the behavior of magnetic moments in an external magnetic field. We begin by obtaining two results that pertain to current distributions consisting only of moving charges. In other words, currents equivalent to spin moments are excluded. Further, we must assume that the current is carried only by particles with the same charge-to-mass ratio, q/m.

The first result concerns the ratio of magnetic moment to angular momentum. We consider a current distribution $\mathbf{j}(\mathbf{r})$ of the type described above. This gives rise to a magnetic moment (which we again denote by $\mathcal{M}$ to avoid confusion with the mass),

$$\mathcal{M} = \frac{1}{2c} \int \mathbf{r} \times \mathbf{j} \, d^3x. \tag{75.1}$$

Since all the current is carried by charges with the same charge-to-mass ratio,

$$\mathbf{j} = \frac{q}{m} \rho_m \mathbf{v}, \tag{75.2}$$

where ρ_m is the mass density, and $\mathbf{v}$ the velocity of the differential element involved. Hence,

$$\mathcal{M} = \frac{q}{2mc} \int \mathbf{r} \times \rho_m \mathbf{v} \, d^3x. \tag{75.3}$$

But the integral is the total angular momentum, $\mathbf{L}$, so

$$\mathcal{M} = \frac{q}{2mc} \mathbf{L}. \tag{75.4}$$

In particular, this result holds for a spinning body of arbitrary shape in which the mass and the charge are uniformly distributed. For the spin angular momentum of the electron, $\mathbf{S}$, on the other hand, one finds

$$\mathcal{M} = g \frac{-e}{2m_e c} \mathbf{S}, \tag{75.5}$$

where $g \simeq 2$. This result underlies our frequent earlier assertions that electron spin is a nonclassical phenomenon. Since the spin angular momentum of the electron is $\hbar/2$, its magnetic moment has a magnitude approximately equal to the *Bohr magneton*,

$$\mu_B = \frac{e\hbar}{2mc} = 9.2741 \times 10^{-21} \text{ erg/G}. \tag{75.6}$$

$$\mu_B = \frac{e\hbar}{2m} = 9.2741 \times 10^{-24} \text{ J/T.} \quad \text{(SI)}$$

Note that 1 erg/G = 1 emu, and that 1 J/T = 1 A m^2.

Next, we consider a collection of charges q_a with masses m_a, with $q_a/m_a = q/m$ for all a. We assume that the charges interact via some potential and that the force on the ath charge in the absence of any applied $\mathbf{B}$ field is $\mathbf{F}_a$. We will also assume that the B field is weak and that the motion is nonrelativistic. The equation of motion for this charge is then

$$m_a \frac{d\mathbf{v}_a}{dt} = \mathbf{F}_a. \tag{75.7}$$

If a steady and uniform magnetic field $\mathbf{B}$ is applied, this is modified to

$$m_a \frac{d\mathbf{v}_a}{dt} = \mathbf{F}_a + \frac{q_a}{c} \mathbf{v}_a \times \mathbf{B}. \tag{75.8}$$

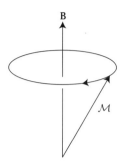

FIGURE 11.10. Larmor precession. The sense of precession shown is for $q > 0$.

Let us now view the motion from a reference frame rotating at an angular velocity ω. In this frame (which we denote by adding primes), the equation of motion is

$$m_a \frac{d\mathbf{v}'_a}{dt} = \mathbf{F}_a + \frac{q_a}{c}\mathbf{v}'_a \times \mathbf{B} - 2m_a(\boldsymbol{\omega} \times \mathbf{v}'_a) - m_a\boldsymbol{\omega} \times (\boldsymbol{\omega} \times \mathbf{r}'_a), \tag{75.9}$$

where the last two terms are the noninertial forces, Coriolis and centrifugal, respectively. If we now assume that the frequency ω is small, so that the centrifugal term can be neglected, then (since q_a/m_a is the same for all a), the Coriolis force will cancel the Lorentz force provided

$$\boldsymbol{\omega} = \boldsymbol{\omega}_L \equiv -\frac{q}{2mc}\mathbf{B}. \tag{75.10}$$

In other words, in the rotating frame, the motion will look exactly as if there were no magnetic field. This result is known as *Larmor's theorem*, and the frequency ω_L is known as the *Larmor frequency*. One sometimes restates the result by saying that the only effect of a weak magnetic field (weak so that ω_L is small and the centrifugal effects are negligible) is to cause the entire motion to precess around $\mathbf{B}$ at a frequency ω_L.

To further elucidate Larmor's theorem, let us consider the equation of motion for the net magnetic moment of the collection, $\mathcal{M}$. The torque, $\mathbf{N}$, on the moment is given by $\mathcal{M} \times \mathbf{B}$. But the torque is the rate of change of angular momentum, $\mathbf{L}$. Hence,

$$\frac{d\mathbf{L}}{dt} = \mathbf{N} = \mathcal{M} \times \mathbf{B}. \tag{75.11}$$

However, as per eq. (75.4), $\mathcal{M}$ is proportional to $\mathbf{L}$ with the gyromagnetic ratio $q/2mc$, so

$$\frac{d\mathcal{M}}{dt} = \mathcal{M} \times \frac{q\mathbf{B}}{2mc}. \tag{75.12}$$

This is in accord with the general theorem: the moment precesses about $\mathbf{B}$ at a frequency ω_L given by eq. (75.10) (see fig. 11.10). For electrons, with a negative charge $-e$, the precession is anticlockwise about $\mathbf{B}$.

The above results are mathematically correct so far as they go, but their physical relevance is limited by the existence of spin. The fact that a magnetic moment precesses in a $\mathbf{B}$ field is very general, however, and it is easy to modify the above treatment so as to make it generally applicable. For many systems, including protons, neutrons, electrons, nuclei, atoms, and ions, the magnetic moment and angular momentum continue to be proportional to each other, but with a nonclassical gyromagnetic ratio, γ. The total angular momentum is now denoted by $\mathbf{J}$ to indicate that it includes both orbital and spin contributions (for which one usually reserves the symbols $\mathbf{L}$ and $\mathbf{S}$), and eq. (75.4) is replaced by[7]

$$\mathcal{M} = \gamma \hbar \mathbf{J}. \tag{75.13}$$

It follows that the magnetic moment will precess with a frequency

$$\boldsymbol{\omega}_p = -\gamma \mathbf{B}. \tag{75.14}$$

In fact, in modern parlance, this phenomenon is also referred to as *Larmor precession*, extending the original meaning of the term. Further, one usually puts the gyromagnetic ratio in a form similar to eq. (75.4) and writes

$$\gamma = g \frac{q}{2mc}. \tag{75.15}$$

The constant g is known as the g-factor. For atoms and ions, one takes $q = -e$, $m = m_e$, and g usually lies between 1 and 2.[8] For protons, $m = m_p$, the proton mass, and $g = 5.586$. For neutrons, which have no net charge, but can still have a spin moment due to internal circulating currents, one continues to take $m = m_p$, and $g = -3.826$. The quantity $e\hbar/2m_p c$ is known as the *nuclear magneton*, in analogy with the Bohr magneton.

Exercise 75.1 A solid contains pairs of spins that are strongly coupled together. The spins in a pair, $\mathbf{S}_1$ and $\mathbf{S}_2$, are identical, have $g = 2$, and experience effective magnetic fields $\mathbf{B}_1 = \mathbf{B}_0 + \alpha \mathbf{S}_2$, $\mathbf{B}_2 = \mathbf{B}_0 + \alpha \mathbf{S}_1$, where $\mathbf{B}_0 \parallel \hat{\mathbf{z}}$ is an applied field, and $\alpha > 0$ is a constant known as the *exchange coupling*.

Suppose the spins are tipped by *small* angles away from the z axis. Linearize the equations of motion in the x and y components and find all normal mode frequencies along with the associated precession pattern.

Answer: There are two modes. In mode 1, the spins precess in phase, and $\omega_1 = \gamma B_0$. In mode 2, the spins precess $180°$ out of phase with each other, at a common frequency $\omega_2 = \gamma(B_0 + 2\alpha S_0)$, where S_0 is the magnitude of the spins.

Exercise 75.2 (action for spin precession) Let the orientation of the magnetic moment be described by spherical polar coordinates θ and φ. The action for a spin trajectory $(\theta(t), \varphi(t))$ may be taken as

$$S = \int_{t_i}^{t_f} [-J\hbar(1 - \cos\theta)\,\dot{\varphi} + \gamma\hbar\mathbf{J}\cdot\mathbf{B}]\,dt. \tag{75.16}$$

---

[7] Note that in this equation, we have written the angular momentum as $\hbar\mathbf{J}$, so that the quantity $\mathbf{J}$ is the *dimensionless* angular momentum. This is common practice, since atomic and subatomic angular momenta are small integer multiples of $\hbar/2$.

[8] If one is dealing with moments of electronic origin, the minus signs in eq. (75.14) and in the formula for γ become irksome, and one often writes $\boldsymbol{\omega}_p = \gamma\mathbf{B}$, with $\gamma = g\mu_B/\hbar$. It must be remembered, however, that in these cases, $\mathcal{M}$ is *antiparallel* to $\mathbf{J}$.

Show that this is true by showing that the Euler-Lagrange equations of motion are the same as those of Larmor precession.

76 Precession of moments in time-dependent magnetic fields*

If the field in which the moment is precessing is time dependent, the motion cannot be integrated in general. A few special types of motion can still be studied in broad terms.

Magnetic resonance: Suppose the magnetic field has a large static component and, perpendicular to this, a small ac component. That is,

$$\mathbf{B}(t) = B_0\hat{\mathbf{z}} + B_1\cos\omega t\,\hat{\mathbf{x}}, \tag{76.1}$$

with $B_1 \ll B_0$. Further, the ac field frequency is close to that of the frequency of Larmor precession in the steady field $B_0\hat{\mathbf{z}}$. Under these conditions, the moment will undergo a resonant absorption of energy from the ac field.

The simplest way to solve this problem is to transform to a rotating frame of reference. Suppose only the dc field $B_0\hat{\mathbf{z}}$ were present. The spin would precess anticlockwise (say) about $\hat{\mathbf{z}}$ at a frequency $\omega_0 = \gamma B_0$. In a corotating frame, the spin would appear to be stationary, i.e., not acted upon by a magnetic field at all. This implies that in a frame rotating at frequency $\boldsymbol{\omega} = \omega\hat{\mathbf{z}}$, where ω need not equal ω_0, the equation of motion will read

$$\frac{d\mathcal{M}}{dt} = \left(\frac{d\mathcal{M}}{dt}\right)_{\text{lab}} - \boldsymbol{\omega} \times \mathcal{M}, \tag{76.2}$$

where the suffix "lab" indicates the time derivative in the nonrotating, laboratory frame. This result is completely general and of wide utility.

Next, let us note that the ac field can be written as

$$B_1\cos\omega t\,\hat{\mathbf{x}} = \mathbf{B}_+(t) + \mathbf{B}_-(t); \quad \mathbf{B}_\pm(t) = \frac{1}{2}B_1(\cos\omega t\,\hat{\mathbf{x}} \pm \sin\omega t\,\hat{\mathbf{y}}). \tag{76.3}$$

The part $\mathbf{B}_+$ is a field rotating anticlockwise about the z axis, while $\mathbf{B}_-$ is rotating clockwise. If we transform to a frame rotating at ω, the first field will appear to be stationary, while the second will appear to be counterrotating at frequency 2ω. The equation of motion will read

$$\frac{d\mathcal{M}}{dt} = \mathbf{B}_{\text{eff}} \times \mathcal{M} + \frac{1}{2}B_1(\cos 2\omega t\hat{\mathbf{x}} - \sin 2\omega t\hat{\mathbf{y}}) \times \mathcal{M}, \tag{76.4}$$

where

$$\mathbf{B}_{\text{eff}} = \left(B_0 - \frac{1}{\gamma}\omega\right)\hat{\mathbf{z}} + \frac{1}{2}B_1\hat{\mathbf{x}} \tag{76.5}$$

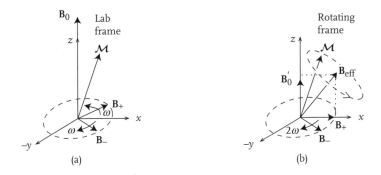

FIGURE 11.11. Magnetic fields in the (a) lab and (b) rotating frames. In (a) the ac field can be divided into anticlockwise and clockwise rotating parts $\mathbf{B}_{\pm}$. The magnitude of the field $\mathbf{B}_0$ is shown greatly reduced for clarity. In (b), the z component of the steady magnetic field is reduced to $B'_0 = B_0 - \omega/\gamma$, the component $\mathbf{B}_+$ appears stationary, and $\mathbf{B}_-$ appears to be counterrotating at a frequency 2ω.

is the static part of the effective field seen by the moment in the rotating frame (see fig. 11.11). We now argue that the static part of the field will cause a steady precession, while the precession about the rapidly rotating part of the field will essentially cancel out over many cycles and may therefore be ignored. The static field can tip the moment through relatively large angles. This phenomenon is rightly called a resonance because the potential energy of the moment in the lab frame, $-\mathcal{M} \cdot \mathbf{B}_0$, can be made to oscillate with a rather large amplitude. This *Rabi oscillation* occurs at a frequency

$$\omega_{\text{Rabi}} = \left[\frac{1}{4}(\gamma B_1)^2 + (\gamma B_0 - \omega)^2 \right]^{1/2}. \tag{76.6}$$

Exactly at resonance, i.e., when $\omega = \gamma B_0$, the steady field will point along $\hat{\mathbf{x}}$. If the moment initially points along $\hat{\mathbf{z}}$, this field can now tip the moment through as large an angle as we please. By applying the ac field as a pulse of the appropriate duration, the tipping angle can be made equal to $\pi/2$, π, etc. The corresponding pulses are known as $\pi/2$ pulses, π pulses, etc.

The above phenomenon is the basis of various magnetic resonance techniques, NMR (nuclear magnetic resonance) and ESR (electron spin resonance), in particular.[9] At the simplest level, the method can be used to measure γ. If the ac frequency ω is held fixed and B_0 is slowly swept, the resonant absorption of energy can be detected as a change in the impedance of the ac coil. In our treatment, there is no net absorption of energy, as the energy is periodically returned to the field. In reality, absorption always occurs, as there are always some damping mechanisms to dissipate the absorbed energy. Alternatively, if γ is known, a measurement of the precession frequency gives a precise measurement of the field acting on the moment. There are now many sophisticated pulse sequences and

---

[9] A proper study of these phenomena must be based on quantum mechanics. However, for a single magnetic moment in an external magnetic field, the equation of motion for the mean value of moment is the same as the classical one. The classical treatment, therefore, captures a great deal of the true behavior.

techniques, and NMR and ESR are vast areas of study, with applications in condensed matter physics, chemistry, and medical imaging of the human body. The list does not stop there. Magnetic resonance imaging is used in geology, archaeology, forensics, and practically any area of science one can think of.

Adiabatic following in a slowly varying field: If the field is slowly varying, by which we mean that $|\dot{\mathbf{B}}| \ll \omega_p |\mathbf{B}|$, we can find an adiabatic invariant and thus solve the problem. It is obvious that under the slow variation condition stated the moment will precess rapidly about the instantaneous direction of $\mathbf{B}$, completing many cycles before the direction of $\mathbf{B}$ changes by a few degrees. We will show that as the axis of precession, i.e., $\mathbf{B}$, moves about, the angle between the moment and this axis remains approximately constant.

The equation of motion now reads

$$\dot{\mathcal{M}} = \gamma \mathbf{B}(t) \times \mathcal{M}. \tag{76.7}$$

The energy of the moment is given by $U = -\mathcal{M} \cdot \mathbf{B}$. We differentiate this to obtain

$$\frac{dU}{dt} = -\frac{d\mathcal{M}}{dt} \cdot \mathbf{B} - \mathcal{M} \cdot \frac{d\mathbf{B}}{dt}$$

$$= -\mathcal{M} \cdot \frac{d\mathbf{B}}{dt}, \tag{76.8}$$

since $\dot{\mathcal{M}}$ is perpendicular to $\mathbf{B}$. Since the energy does not change at all for uniform precession, we can evaluate dU/dt by averaging it over one precession period $T = 2\pi/\omega_p$. In this way, we get

$$\frac{dU}{dt} = -\frac{1}{T} \int_0^T \mathcal{M} \cdot \frac{d\mathbf{B}}{dt} dt. \tag{76.9}$$

To evaluate the right-hand side, let us write $\mathbf{B}(t)$ for times close to $t = 0$ in the form

$$\mathbf{B}(t) = B_0 \hat{\mathbf{z}} + \mathbf{b}t, \tag{76.10}$$

where $|\mathbf{b}| \ll \omega_p B_0$. Denoting the angle between $\mathbf{B}$ and $\mathcal{M}$ by θ, the solution for $\mathcal{M}(t)$ is

$$\mathcal{M}(t) = M_0 \big[\cos\theta \, \hat{\mathbf{z}} + \sin\theta (\cos\omega_p t \, \hat{\mathbf{x}} + \sin\omega_p t \, \hat{\mathbf{y}}) \big]. \tag{76.11}$$

Then,

$$\frac{1}{T} \int_0^T \mathcal{M} \cdot \frac{d\mathbf{B}}{dt} dt = \frac{M_0}{T} \int_0^T \mathbf{b} \cdot \big[\cos\theta \, \hat{\mathbf{z}} + \sin\theta (\cos\omega_p t \, \hat{\mathbf{x}} + \sin\omega_p t \, \hat{\mathbf{y}}) \big] dt$$

$$= M_0 b_z \cos\theta. \tag{76.12}$$

But $b_z = dB/dt$, so

$$\frac{dU}{dt} = -M_0 \cos\theta \frac{dB}{dt}. \tag{76.13}$$

But, since $U = -M_0 B \cos\theta$, we also have

$$\frac{dU}{dt} = -M_0 \frac{dB}{dt} \cos\theta + M_0 B \sin\theta \frac{d\theta}{dt}. \tag{76.14}$$

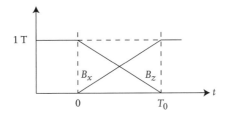

FIGURE 11.12. Time-dependent field for exercise 76.2.

Comparing with eq. (76.13), we obtain

$$\frac{d\theta}{dt} = 0. \tag{76.15}$$

In words, the angle between the moment and the field remains constant, which is what we set out to show.

The above result can be also be restated in terms of the area on the unit sphere swept out by the direction of the magnetic moment in one precession period. The area of the cap including the magnetic field direction is $2\pi(1 - \cos\theta)$. Since θ is an adiabatic constant, so is this area. The point of stating things this way is that this area is (up to a constant) the action (see eq. (75.16)), and, as stated earlier, the adiabatic invariant for a periodic motion is the action for that motion.

Exercise 76.1 Consider a magnetic moment in a magnetic field that varies as

$$\mathbf{B} = B_0(\cos\omega t\,\hat{\mathbf{x}} + \sin\omega t\,\hat{\mathbf{y}}), \tag{76.16}$$

where $\omega \ll \gamma B_0$. By transforming to a suitable rotating frame, solve the motion exactly and verify that the adiabatic treatment is correct.

Exercise 76.2 Numerically integrate the equation of motion for Larmor precession of an electron in the time-dependent magnetic field shown in fig. 11.12. Let the spin make an angle of $30°$ with the z axis at $t = 0$. Plot the angle between the spin and the **B** field as a function of time for various values of T_0 of the order of a nanosecond.

Solution: We plot the results from a Runge-Kutta integration in fig. 11.13, for $T_0 = 0.1$ and 1 ns. The Larmor precession time is of order γB_0, where $B_0 \sim 1$ T. For $T_0 = 1$ ns, $\gamma B_0 T_0 = 176$, so the angle between the field and the spin should then be nearly constant, as indeed it is. The variation is not very large even for $T_0 = 0.1$ ns. The additional structure in the graph for $T_0 = 1$ ns, i.e., the amplitude and period of the rapid wiggles, can also be understood using ideas from the previous exercise and is left as an exercise for the reader.

Response to a rapidly varying field: If the magnetic field varies rapidly, i.e., $|\dot{\mathbf{B}}| \gg \omega_p|\mathbf{B}|$, to first approximation, the moment simply does not move. The reason is that the instantaneous axis of precession changes direction before the moment can precess even

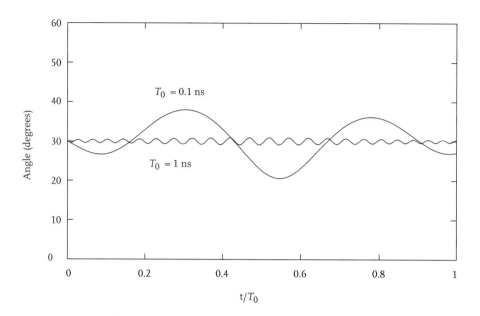

FIGURE 11.13. The angle between the spin and the field for the field of fig. 11.12.

once. Not knowing which way to turn, the hapless moment gets paralyzed. We illustrate this with a solvable example.

Suppose the magnetic field is $\mathbf{B} = B_0 \cos \omega t\, \hat{\mathbf{z}}$, with $\omega \gg \gamma B_0$. Now, $\dot{\mathcal{M}}_z = 0$, and

$$\dot{\mathcal{M}}_x = -\gamma B_z(t)\mathcal{M}_y, \quad \dot{\mathcal{M}}_y = \gamma B_z(t)\mathcal{M}_x. \tag{76.17}$$

Since the magnitude of $\mathcal{M}$ is fixed, it follows that if we write $\mathcal{M}_x = \mathcal{M}_\perp \cos\varphi$, $\mathcal{M}_y = \mathcal{M}_\perp \sin\varphi$, we need only to find the time dependence of φ. The equation of motion for φ is

$$\dot{\varphi} = \gamma B_z(t). \tag{76.18}$$

Taking $\varphi(0) = 0$, the integration yields

$$\varphi(t) = \frac{\gamma B_0}{\omega} \sin \omega t. \tag{76.19}$$

Since $\varphi(t) \ll 1$, it follows that to lowest order in $1/\omega$, the solution is

$$\mathcal{M}_z(t) = \mathcal{M}_z(0), \quad \mathcal{M}_x \approx \mathcal{M}_\perp, \quad \mathcal{M}_y \approx \frac{\gamma B_0}{\omega}\mathcal{M}_\perp \sin \omega t. \tag{76.20}$$

Exercise 76.3 Reconsider the problem in exercise 76.1 (the rotating magnetic field) when $\omega \gg \gamma B_0$, and write the answer in the lab frame. Show that for any initial condition, the moment makes excursions of order $1/\omega$ in leading order.

12 | Action formulation of electromagnetism

In this chapter, we will see that just like classical mechanics, classical electromagnetism can be formulated in terms of a least action principle, and Lagrangian and Hamiltonian methods can then be applied. We assume that readers have at least some familiarity with these concepts in the context of mechanics. The main advantage of this formulation is conceptual, in that it provides a unified way of thinking about the two apparently disparate subjects of mechanics and electromagnetism. This unification is of great value in developing a quantum theory of charges interacting with electromagnetic fields. However, it is also of value strictly within a classical framework. For example, the laws of energy and momentum conservation for the combined system of charges and fields emerge as consequences of a single general result, Noether's theorem.[1] Although we shall not say anything more about it, this point of view is also helpful in developing numerical methods of integrating Maxwell's equations. By discretizing the action in space and time suitably, one can ensure that the approximate solutions that result, being derived from an action principle, will automatically conserve certain physical quantities. Such so-called geometrical integrators are often superior to purely mathematical approaches, which view the Maxwell equations only as partial differential equations and do not guarantee any conservation laws.

77 Charged particle in given field

Let us begin by recasting the Lorentz force law in terms of a Lagrangian. For simplicity, we will first consider the nonrelativistic form of this law, i.e.,

$$m\frac{d\mathbf{v}}{dt} = q\left(\mathbf{E}(\mathbf{r}, t) + \frac{1}{c}\mathbf{v} \times \mathbf{B}(\mathbf{r}, t)\right). \tag{77.1}$$

---

[1] See, e.g., Jose and Saletan (1998), secs. 3.2.2, 9.2.

Recall that in the arguments of $\mathbf{E}(\mathbf{r}, t)$ and $\mathbf{B}(\mathbf{r}, t)$, $\mathbf{r}$ is the position of the particle. This is varying with time, so we should write $\mathbf{E}(\mathbf{r}(t), t)$, etc. for completeness, but this gets cumbersome.

Since the Lagrangian deals in energies, let us rewrite eq. (77.1) using the potentials ϕ and $\mathbf{A}$ instead of the fields. Expanding the triple cross product $\mathbf{v} \times (\nabla \times \mathbf{A})$ and remembering that the ∇ operator does not act on the velocity $\mathbf{v}$, we get

$$m\frac{d\mathbf{v}}{dt} = -q\nabla\phi - \frac{q}{c}\frac{\partial \mathbf{A}}{\partial t} - \frac{q}{c}(\mathbf{v}\cdot\nabla)\mathbf{A} + \frac{q}{c}\nabla(\mathbf{v}\cdot\mathbf{A}). \tag{77.2}$$

The first and last terms on the right can be written as a total gradient, and the second and third terms can be combined into the total time derivative $d\mathbf{A}/dt$, for

$$\frac{d}{dt}\mathbf{A}(\mathbf{r}(t), t) = \frac{\partial \mathbf{A}}{\partial t} + \frac{\partial \mathbf{A}}{\partial x_i}\frac{dx_i}{dt}$$

$$= \frac{\partial \mathbf{A}}{\partial t} + v_i\nabla_i\mathbf{A}. \tag{77.3}$$

Hence, the equation of motion can be written as

$$\frac{d}{dt}\left(m\mathbf{v} + \frac{q}{c}\mathbf{A}\right) = -q\nabla\left(\phi - \frac{1}{c}\mathbf{v}\cdot\mathbf{A}\right). \tag{77.4}$$

We wish to cast this into the form

$$\frac{d}{dt}\frac{\partial L}{\partial \mathbf{v}} = \frac{\partial L}{\partial \mathbf{r}}, \tag{77.5}$$

where L is the Lagrangian. If the terms involving $\mathbf{A}$ were absent, this recasting would be immediate. However, these terms can also be given this structure, if we note that

$$\frac{\partial}{\partial \mathbf{v}}\mathbf{v}\cdot\mathbf{A} = \mathbf{A}, \quad \frac{\partial}{\partial \mathbf{r}}\mathbf{v}\cdot\mathbf{A} = \nabla(\mathbf{v}\cdot\mathbf{A}). \tag{77.6}$$

Hence, the following expression will serve as a Lagrangian:

$$L(\mathbf{r}, \mathbf{v}, t) = \frac{1}{2}m\mathbf{v}^2 - q\phi(\mathbf{r}, t) + \frac{q}{c}\mathbf{v}\cdot\mathbf{A}(\mathbf{r}, t) \quad (v \ll c). \tag{77.7}$$

$$L(\mathbf{r}, \mathbf{v}, t) = \frac{1}{2}m\mathbf{v}^2 - q\phi(\mathbf{r}, t) + q\mathbf{v}\cdot\mathbf{A}(\mathbf{r}, t) \quad (v \ll c). \quad \text{(SI)}$$

With the Lagrangian in hand, it is easy to obtain the Hamiltonian, $\mathcal{H}$. This is given by

$$\mathcal{H} = \mathbf{p}\cdot\mathbf{v} - L, \tag{77.8}$$

where $\mathbf{p}$ is the momentum that is canonically conjugate to the coordinate:

$$\mathbf{p} = \frac{\partial L}{\partial \mathbf{v}}. \tag{77.9}$$

Applying this rule to eq. (77.7), we get

$$\mathbf{p} = m\mathbf{v} + \frac{q}{c}\mathbf{A}. \tag{77.10}$$

$$\mathbf{p} = m\mathbf{v} + q\mathbf{A}. \quad \text{(SI)}$$

This is not the usual product of mass and velocity, which we call the *kinetic* momentum:

$$\mathbf{p}_{\text{kin}} = m\mathbf{v}. \tag{77.11}$$

We will comment further on this difference below. For the present, we press ahead, noting simply that if we wish to reap the benefits of the Hamiltonian formalism, we must work with $\mathbf{p}$ as the basic momentum variable, not $\mathbf{p}_{\text{kin}}$.

The prescription (77.8) gives

$$\mathcal{H} = \left(m\mathbf{v} + \frac{q}{c}\mathbf{A}\right) \cdot \mathbf{v} - \left(\frac{1}{2}m v^2 + \frac{q}{c}\mathbf{v} \cdot \mathbf{A} - q\phi\right). \tag{77.12}$$

Writing this in terms of $\mathbf{p}$ via eq. (77.10), we get

$$\mathcal{H}(\mathbf{r}, \mathbf{p}, t) = \frac{1}{2m}\left(\mathbf{p} - \frac{q}{c}\mathbf{A}(\mathbf{r}, t)\right)^2 + q\phi(\mathbf{r}, t) \quad (v \ll c). \tag{77.13}$$

$$\mathcal{H}(\mathbf{r}, \mathbf{p}, t) = \frac{1}{2m}\left(\mathbf{p} - q\mathbf{A}(\mathbf{r}, t)\right)^2 + q\phi(\mathbf{r}, t) \quad (v \ll c). \quad \text{(SI)}$$

The equations of motion are Hamilton's equations:

$$\dot{\mathbf{r}} = \frac{\partial \mathcal{H}}{\partial \mathbf{p}} = \frac{1}{m}\left(\mathbf{p} - \frac{q}{c}\mathbf{A}\right), \tag{77.14}$$

$$\dot{\mathbf{p}} = -\frac{\partial \mathcal{H}}{\partial \mathbf{r}} = \frac{1}{m}\left(p_i - \frac{q}{c}A_i\right)\nabla\frac{q}{c}A_i - q\nabla\phi. \tag{77.15}$$

These two first-order differential equations are equivalent to the Lorentz force law. In particular, eq. (77.15) is the same as eq. (77.4).

The momentum $\mathbf{p}$ that appears in the Hamiltonian has a somewhat formal significance. For a charge moving in a given electromagnetic field, the kinetic momentum, $m\mathbf{v}$, is clearly unique. The canonical momentum, $\mathbf{p}$, on the other hand, is not, since the vector potential can be changed by a gauge transformation. Note, however, that the Hamiltonian itself is gauge invariant, since the first term is the kinetic energy (kinetic momentum squared divided by $2m$). In fact, with this realization, eq. (77.13) has a pleasing interpretation as the total energy, since the second term is the potential energy. The magnetic field does no work on the particle and so, although it can change the momentum, it cannot be part of the potential energy. This viewpoint immediately leads to another derivation of the work received by a charged particle in a time-independent electromagnetic field. Let us denote the first term in eq. (77.13) by $\mathcal{E}_{\text{kin}}$, the kinetic energy of the particle. Since the Hamiltonian is a constant of motion, we have

$$\frac{d}{dt}\mathcal{E}_{\text{kin}} = -q\frac{d}{dt}\phi(\mathbf{r}(t)) = -q\mathbf{v} \cdot \nabla\phi = q\mathbf{v} \cdot \mathbf{E}, \tag{77.16}$$

exactly as found before.

Relativistically correct expressions: The relativistically correct form of the Lorentz force law is

$$\frac{d}{dt}\frac{m\mathbf{v}}{\sqrt{1 - v^2/c^2}} = q\left(\mathbf{E}(\mathbf{r}, t) + \frac{1}{c}\mathbf{v} \times \mathbf{B}(\mathbf{r}, t)\right). \tag{77.17}$$

This differs from the approximate form (77.1) only in the formula for the kinetic momentum on the left-hand side. This suggests that in the Lagrangian (77.7), only the first term, the kinetic energy of a free particle, needs to be modified. This is indeed so, and it is easy to verify that the correct equation of motion follows from

$$L(\mathbf{r}, \mathbf{v}, t) = -mc^2 \left(1 - \frac{v^2}{c^2}\right)^{1/2} - q\phi(\mathbf{r}, t) + \frac{q}{c} \mathbf{v} \cdot \mathbf{A}(\mathbf{r}, t). \tag{77.18}$$

The kinetic momentum is given by differentiating the first term in L with respect to $\mathbf{v}$, i.e.,

$$\mathbf{p}_{kin} = \frac{m\mathbf{v}}{\sqrt{1 - v^2/c^2}}, \tag{77.19}$$

and the canonical momentum is given by differentiating all of L with respect to $\mathbf{v}$, i.e.,

$$\mathbf{p} = \frac{m\mathbf{v}}{\sqrt{1 - v^2/c^2}} + \frac{q}{c}\mathbf{A}$$

$$= \mathbf{p}_{kin} + \frac{q}{c}\mathbf{A}, \tag{77.20}$$

as before. The Hamiltonian now follows from the general recipe (77.8). We get

$$\mathcal{H} = \left[m^2 c^4 + c^2 \left(\mathbf{p} - \frac{q}{c}\mathbf{A}\right)^2 \right]^{1/2} + q\phi. \tag{77.21}$$

$$\mathcal{H} = \left[m^2 c^4 + c^2 (\mathbf{p} - q\mathbf{A})^2 \right]^{1/2} + q\phi. \quad \text{(SI)}$$

Exercise 77.1 Find the adiabatically invariant action integral of section 74 by evaluating the integral as $\oint \mathbf{p}_\perp \cdot d\mathbf{r}_\perp$, where $\perp$ signifies xy-plane components, and $\mathbf{p}$ is the canonical momentum.

78 The free field

Let us now consider the free electromagnetic field in the absence of any charges or currents. To treat this as a mechanical system we must first identify the coordinate-like variables. In general, whenever we wish to consider a field from this viewpoint, there are an infinite number of coordinates, corresponding to the value of the field at every space point. An alternative procedure is to expand the field in some set of modes and view the expansion coefficients as the dynamical coordinates. In the case of the EM field, it is particularly convenient to resort to the oscillator representation of EM waves developed in section 46, and to take the complex Fourier coefficients $\mathbf{A}_\mathbf{k}$ of the vector potential as the coordinates. These obey the equation of motion

$$\ddot{\mathbf{A}}_\mathbf{k} + \omega_\mathbf{k}^2 \mathbf{A}_\mathbf{k} = 0, \tag{78.1}$$

with $\omega_\mathbf{k} = ck$. We seek a Lagrangian L that will yield this equation as the Euler-Lagrange equation. This is almost trivial to do if we follow the analogy with a simple harmonic oscillator. The following will work as the contribution to L for each $\mathbf{k}$:

$$L_{\mathbf{k}} = \dot{\mathbf{A}}_{\mathbf{k}} \cdot \dot{\mathbf{A}}_{\mathbf{k}}^* - \omega_{\mathbf{k}}^2 \mathbf{A}_{\mathbf{k}} \cdot \mathbf{A}_{\mathbf{k}}^*. \tag{78.2}$$

For,

$$\frac{\partial L_{\mathbf{k}}}{\partial \dot{\mathbf{A}}_{\mathbf{k}}^*} = \dot{\mathbf{A}}_{\mathbf{k}}, \quad \frac{\partial L_{\mathbf{k}}}{\partial \mathbf{A}_{\mathbf{k}}^*} = -\omega_{\mathbf{k}}^2 \mathbf{A}_{\mathbf{k}}, \tag{78.3}$$

and the Euler-Lagrange equation

$$\frac{d}{dt}\frac{\partial L_{\mathbf{k}}}{\partial \dot{\mathbf{A}}_{\mathbf{k}}^*} - \frac{\partial L_{\mathbf{k}}}{\partial \mathbf{A}_{\mathbf{k}}^*} = 0 \tag{78.4}$$

reads

$$\ddot{\mathbf{A}}_{\mathbf{k}} + \omega_{\mathbf{k}}^2 \mathbf{A}_{\mathbf{k}} = 0, \tag{78.5}$$

exactly as desired.

To obtain the total Lagrangian, we must add up the individual $L_{\mathbf{k}}$'s. This yields, with $\omega_{\mathbf{k}} = ck$,

$$L = \frac{V}{8\pi c^2} \sum_{\mathbf{k}} (\dot{\mathbf{A}}_{\mathbf{k}} \cdot \dot{\mathbf{A}}_{\mathbf{k}}^* - c^2 k^2 \mathbf{A}_{\mathbf{k}} \cdot \mathbf{A}_{\mathbf{k}}^*), \tag{78.6}$$

where V is the volume of the box used to define the modes. We have also multiplied the sum by an overall constant $V/8\pi c^2$ for future convenience. Exactly as in section 46 we now rewrite the sum in terms of $\mathbf{E}$ and $\mathbf{B}$. Recalling that

$$\mathbf{E}_{\mathbf{k}} = -\frac{1}{c}\dot{\mathbf{A}}_{\mathbf{k}}, \quad \mathbf{B}_{\mathbf{k}} = i\mathbf{k} \times \mathbf{A}_{\mathbf{k}} \tag{78.7}$$

and that $\mathbf{k} \cdot \mathbf{A}_{\mathbf{k}} = 0$, we obtain

$$L = \frac{V}{8\pi} \sum_{\mathbf{k}} (\mathbf{E}_{\mathbf{k}} \cdot \mathbf{E}_{-\mathbf{k}} - \mathbf{B}_{\mathbf{k}} \cdot \mathbf{B}_{-\mathbf{k}}). \tag{78.8}$$

By Parseval's theorem, this can also be written in terms of the real space fields as

$$L = \int \frac{\mathbf{E}^2 - \mathbf{B}^2}{8\pi} d^3x. \tag{78.9}$$

$$L = \int \frac{1}{2}\left(\epsilon_0 \mathbf{E}^2 - \frac{\mathbf{B}^2}{\mu_0}\right) d^3x. \quad \text{(SI)}$$

It is significant that the potentials have dropped out of the final formula.

Let us comment at this point that in writing the sum (78.6), we could have introduced an arbitrary $\mathbf{k}$-dependent factor $K_{\mathbf{k}}$ in the summand without affecting the equations of motion. This would have the effect that, in terms of the real space fields, the Lagrangian would contain integrals of the form

$$\iint \mathbf{E}(\mathbf{r}) K(\mathbf{r} - \mathbf{r}') \mathbf{E}(\mathbf{r}') d^3x \, d^3x', \tag{78.10}$$

with some kernel K, in contrast to eq. (78.9), which involves only a local expression. If any kernel K were present, it would be a built-in property of space, for which there is no evidence.[2] Secondly, we would not obtain Maxwell's equations with source terms when we considered the interaction of the fields with charges and currents.

It is also interesting to obtain Maxwell's equations for a free field directly from the principle of least action starting from the Lagrangian (78.9). The action is

$$S = \int L \, dt = \int \frac{\mathbf{E}^2 - \mathbf{B}^2}{8\pi} d^3x \, dt. \tag{78.11}$$

We now regard $\mathbf{E}$ and $\mathbf{B}$ as functionals of the potentials, ϕ and $\mathbf{A}$, as per the standard rules

$$\mathbf{E} = -\frac{1}{c}\frac{\partial \mathbf{A}}{\partial t} - \nabla\phi, \quad \mathbf{B} = \nabla \times \mathbf{A}, \tag{78.12}$$

and vary each of them in turn. Suppose we vary ϕ by $\delta\phi$. Then

$$\delta\mathbf{E} = -\nabla(\delta\phi), \quad \delta\mathbf{B} = 0, \tag{78.13}$$

and

$$\delta S = -\frac{1}{4\pi}\int \mathbf{E} \cdot \nabla(\delta\phi) \, d^3x \, dt = \frac{1}{4\pi}\int (\nabla \cdot \mathbf{E})\delta\phi \, d^3x \, dt, \tag{78.14}$$

after an integration by parts. Setting $\delta S = 0$, we obtain

$$\nabla \cdot \mathbf{E} = 0. \tag{78.15}$$

Next, let us vary $\mathbf{A}$. Now, $\delta\mathbf{E} = -(1/c)\partial(\delta\mathbf{A})/\partial t$, and $\delta\mathbf{B} = \nabla \times \delta\mathbf{A}$. Hence,

$$\delta S = -\frac{1}{4\pi}\int \left[\mathbf{E} \cdot \frac{1}{c}\frac{\partial}{\partial t}\delta\mathbf{A} - \mathbf{B} \cdot \nabla \times \delta\mathbf{A} \right] d^3x \, dt$$

$$= \frac{1}{4\pi}\int \left[\frac{\partial\mathbf{E}}{\partial t}\frac{1}{c}\delta\mathbf{A} - (\nabla \times \mathbf{B}) \cdot \delta\mathbf{A} \right] d^3x \, dt. \tag{78.16}$$

Setting δS to zero yields the remaining Maxwell equation

$$\nabla \times \mathbf{B} - \frac{1}{c}\frac{\partial\mathbf{E}}{\partial t} = 0. \tag{78.17}$$

The source-free equations, $\nabla \cdot \mathbf{B} = 0$ and $\nabla \times \mathbf{E} + c^{-1}(\partial\mathbf{B}/\partial t) = 0$, are automatic consequences of the definitions of the fields in terms of the potentials. Alternatively, we could write the Maxwell equations entirely in terms of the potentials. These equations would, of course, be the wave equations for ϕ and $\mathbf{A}$.

Readers may wonder why, in varying the action (78.11), we did not impose the constraint that the potentials be in a certain gauge. For the answer, see the end of section 80.

The Hamiltonian for the free field is straightforward to obtain from the Lagrangian in terms of oscillator modes, eq. (78.6). The conjugate momentum to the variable $\mathbf{A_k}$ is

---

[2] An expression such as eq. (78.10) would also conflict with the demands of modern relativity, in that it would treat space and time unequally. This problem could be remedied, but the resulting theory would be nonlocal.

given by

$$\mathbf{P_k} = \frac{\partial L}{\partial \dot{\mathbf{A}}_\mathbf{k}} = 2 \times \frac{V}{8\pi c^2} \dot{\mathbf{A}}_\mathbf{k}^*. \tag{78.18}$$

(The factor of 2 arises because for a given $\mathbf{k} = \mathbf{k}_0$, say, the $\mathbf{k}_0$ and $-\mathbf{k}_0$ terms in the sum (78.6) both contribute to $\mathbf{P_k}$.) Hence,

$$\mathcal{H} = \sum_\mathbf{k} \mathbf{P_k} \cdot \dot{\mathbf{A}}_\mathbf{k} - L = \frac{V}{8\pi c^2} \sum_\mathbf{k} (\dot{\mathbf{A}}_\mathbf{k} \cdot \dot{\mathbf{A}}_\mathbf{k}^* + c^2 k^2 \mathbf{A_k} \cdot \mathbf{A_k^*}). \tag{78.19}$$

This is not written in terms of its natural variables, $\mathbf{P_k}$, but that is not of concern here. Let us simply note that in terms of the real space fields,

$$\mathcal{H} = \int \frac{\mathbf{E}^2 + \mathbf{B}^2}{8\pi} d^3x. \tag{78.20}$$

$$\mathcal{H} = \int \frac{1}{2} \left(\epsilon_0 \mathbf{E}^2 + \frac{\mathbf{B}^2}{\mu_0} \right) d^3x. \quad \text{(SI)}$$

This is, of course, the total energy of the EM field, and we chose the multiplicative prefactor in eq. (78.6) to get this right.

79 The interacting system of fields and charges

In deriving the charged particle Lagrangian, eqs. (77.7) and (77.18), the terms $q\phi$ and $q\mathbf{v} \cdot \mathbf{A}$ were viewed as describing the behavior of the charges in a given field. However, because the field and the charges are both dynamical entities on an equal footing, these terms must describe the general interaction between them. Thus, the full Lagrangian for the interacting system is given by

$$L = \sum_a \left[\frac{1}{2} m_a v_a^2 - q_a \phi(\mathbf{r}_a, t) + \frac{q_a}{c} \mathbf{v}_a \cdot \mathbf{A}(\mathbf{r}_a, t) \right] + \int \frac{\mathbf{E}^2 - \mathbf{B}^2}{8\pi} d^3x, \tag{79.1}$$

which may also be written as

$$L = \frac{1}{2} \sum_a m_a v_a^2 - \int \rho(\mathbf{r}, t)\phi(\mathbf{r}, t)\, d^3x + \int \mathbf{j}(\mathbf{r}, t) \cdot \mathbf{A}(\mathbf{r}, t)\, d^3x + \int \frac{\mathbf{E}^2 - \mathbf{B}^2}{8\pi} d^3x. \tag{79.2}$$

Here, and in the rest of this section, we write the kinetic energy of the charges in the nonrelativistic limit. The action is given by integrating the Lagrangian over time.

The appearance of the potentials in L means that our formulation contains redundant (gauge) degrees of freedom. The simplest way to eliminate these is to work in the Coulomb gauge. That is, we set

$$\phi(\mathbf{r}, t) = \sum_a \frac{q_a}{|\mathbf{r} - \mathbf{r}_a(t)|} \tag{79.3}$$

and choose $\mathbf{A}$ to satisfy $\nabla \cdot \mathbf{A} = 0$. The vector potential is once again Fourier analyzed,

$$\mathbf{A}(\mathbf{r}, t) = \sum_\mathbf{k} \mathbf{A_k}(t) e^{i\mathbf{k}\cdot\mathbf{r}}, \tag{79.4}$$

but the coefficients $A_k(t)$ no longer obey a free oscillator equation. Indeed, it is one of the tasks of the Lagrangian to furnish the correct equations of motion. To this end, let us try and write L entirely in terms of $\dot{\mathbf{r}}_a$ and $\dot{\mathbf{A}}_k$. We first note that the electric field naturally separates into transverse and longitudinal parts:

$$\mathbf{E}_L = -\nabla\phi, \quad \mathbf{E}_T = -\frac{1}{c}\frac{\partial \mathbf{A}}{\partial t}. \tag{79.5}$$

The integral of E^2 separates similarly (see exercise 79.1):

$$\int E^2\, d^3x = \int E_L^2\, d^3x + \int E_T^2\, d^3x. \tag{79.6}$$

Only the transverse part can be expressed in terms of the A_k, as can the entire integral of B^2. The manipulations are formally the same as for the free field, and we get

$$\frac{1}{8\pi}\int \left(E_T^2 - B^2\right) d^3x = \frac{V}{8\pi c^2}\sum_k (\dot{\mathbf{A}}_k \cdot \dot{\mathbf{A}}_k^* - c^2 k^2 \mathbf{A}_k \cdot \mathbf{A}_k^*). \tag{79.7}$$

The integral of E_L^2, on the other hand, can be written entirely in terms of the position of the charges using the explicit solution for $\phi(\mathbf{r})$, provided we discard the infinite self-energies. The same holds for the interaction term $-\sum_a q_a\phi(\mathbf{r}_a, t)$. Combining the two we get

$$\frac{1}{8\pi}\int E_L^2\, d^3x - \sum_a q_a\phi(\mathbf{r}_a, t) = -\frac{1}{2}\sum_{a,b}' \frac{q_a q_b}{|\mathbf{r}_a - \mathbf{r}_b|}. \tag{79.8}$$

Thus, the complete Lagrangian for the system of charges and fields is

$$L = \frac{1}{2}\sum_a m_a v_a^2 - \frac{1}{2}\sum_{a,b}' \frac{q_a q_b}{|\mathbf{r}_a - \mathbf{r}_b|}$$
$$+ \frac{1}{c}\sum_{a,k} q_a \mathbf{v}_a \cdot \mathbf{A}_k e^{i\mathbf{k}\cdot\mathbf{r}_a} + \frac{V}{8\pi c^2}\sum_k (\dot{\mathbf{A}}_k \cdot \dot{\mathbf{A}}_k^* - c^2 k^2 \mathbf{A}_k \cdot \mathbf{A}_k^*). \tag{79.9}$$

The expression for L deserves several comments. First, it is in fact gauge invariant, since it involves only the transverse part of $\mathbf{A}$, which is gauge invariant. Second, the independent degrees of freedom are the positions of the charges, $\mathbf{r}_a$, and coefficients, $\mathbf{A}_k$. The fields are *defined* in terms of these degrees of freedom by the formulas

$$\mathbf{E} = -\nabla\sum_a \frac{q_a}{|\mathbf{r} - \mathbf{r}_a|} - \frac{1}{c}\frac{\partial \mathbf{A}}{\partial t}, \quad \mathbf{B} = \nabla \times \mathbf{A}. \tag{79.10}$$

Thus, the Maxwell equations for $\nabla \cdot \mathbf{B}$, and Gauss's and Faraday's laws are automatic. The Euler-Lagrange equations for A_k amount to a single second-order differential equation for $\mathbf{A}(\mathbf{r}, t)$, equivalent to the Ampere-Maxwell law. The Euler-Lagrange equation for $\mathbf{r}_a$ is the Lorentz force equation (see below). Third, what happens if we work in, say, the Lorenz gauge? The formulation then has extra degrees of freedom from the scalar potential and the longitudinal part of $\mathbf{A}$. The Lorenz condition (or other gauge-fixing condition) appears as constraints on these degrees of freedom, which hold at all times if they hold at an initial time.[3] Thus, the true dynamical content of the theory is the same in all gauges.

---

[3] See, e.g., Heitler (1984), sec. I.6.

The Lorenz gauge is manifestly Lorentz invariant, but the Coulomb gauge is conceptually simpler.

Let us now obtain the Euler-Lagrange (EL) equations, beginning with the charges. We have

$$\frac{\partial L}{\partial \mathbf{v}_a} = m_a \mathbf{v}_a + \frac{q_a}{c} \sum_{\mathbf{k}} \mathbf{A_k} e^{i\mathbf{k} \cdot \mathbf{r}_a}, \tag{79.11}$$

$$\frac{\partial L}{\partial \mathbf{r}_a} = q_a \sum_{b \neq a} q_b \frac{\mathbf{r}_a - \mathbf{r}_b}{|\mathbf{r}_a - \mathbf{r}_b|^3} + \frac{q_a}{c} \sum_{\mathbf{k}} i\mathbf{k}(\mathbf{v}_a \cdot \mathbf{A_k}) e^{i\mathbf{k} \cdot \mathbf{r}_a}. \tag{79.12}$$

We recognize the sum in the first term in eq. (79.12) as the longitudinal electric field $\mathbf{E}_L(\mathbf{r}_a, t)$. Next, let us note that

$$\frac{d}{dt}\frac{\partial L}{\partial \mathbf{v}_a} = m_a \ddot{\mathbf{r}}_a + \frac{q_a}{c} \sum_{\mathbf{k}} \dot{\mathbf{A}}_{\mathbf{k}} e^{i\mathbf{k} \cdot \mathbf{r}_a} + \frac{q_a}{c} \sum_{\mathbf{k}} i\mathbf{A_k}(\mathbf{k} \cdot \mathbf{v}_a) e^{i\mathbf{k} \cdot \mathbf{r}_a}. \tag{79.13}$$

The first sum on the right is $-\mathbf{E}_T(\mathbf{r}_a, t)$. The EL equation is, therefore,

$$m_a \ddot{\mathbf{r}}_a = q_a(\mathbf{E}_T + \mathbf{E}_L) + i\frac{q_a}{c} \sum_{\mathbf{k}} [\mathbf{k}(\mathbf{v}_a \cdot \mathbf{A_k}) - \mathbf{A_k}(\mathbf{k} \cdot \mathbf{v}_a)] e^{i\mathbf{k} \cdot \mathbf{r}_a}. \tag{79.14}$$

The last term is

$$\frac{q_a}{c} \sum_{\mathbf{k}} \mathbf{v}_a \times (i\mathbf{k} \times \mathbf{A_k}) e^{i\mathbf{k} \cdot \mathbf{r}_a} = \frac{q_a}{c} \mathbf{v}_a \times \mathbf{B}(\mathbf{r}_a, t). \tag{79.15}$$

And so, the EL equation is

$$m_a \ddot{\mathbf{r}}_a = q_a \mathbf{E} + \frac{q_a}{c} \mathbf{v}_a \times \mathbf{B}, \tag{79.16}$$

with the fields evaluated at $\mathbf{r}_a$. This is, of course, the Lorentz force law.

Next, let us obtain the Euler-Lagrange equation for $\mathbf{A_k}$. Just as for eq. (78.18), we have

$$\frac{\partial L}{\partial \dot{\mathbf{A}}_{\mathbf{k}}} = \frac{V}{4\pi c^2} \dot{\mathbf{A}}_{\mathbf{k}}^*. \tag{79.17}$$

In calculating $\partial L/\partial \mathbf{A_k}$, however, we must remember that $\mathbf{A_k} \perp \mathbf{k}$. Thus, in the dot product $\mathbf{v}_a \cdot \mathbf{A_k}$, the component of $\mathbf{v}_a$ parallel to $\mathbf{k}$ simply does not appear, and

$$\frac{\partial}{\partial \mathbf{A_k}}(\mathbf{v}_a \cdot \mathbf{A_k}) = \mathbf{v}_a - (\mathbf{v}_a \cdot \hat{\mathbf{k}})\hat{\mathbf{k}} \equiv \mathbf{v}_{a,\perp}. \tag{79.18}$$

Hence,

$$\frac{\partial L}{\partial \mathbf{A_k}} = -\frac{V}{4\pi c^2} k^2 c^2 \mathbf{A}_{\mathbf{k}}^* + \frac{1}{c} \sum_a q_a \mathbf{v}_{a,\perp} e^{i\mathbf{k} \cdot \mathbf{r}_a}. \tag{79.19}$$

Taking the complex conjugate and rearranging, the Euler-Lagrange equation is

$$-\frac{1}{c^2}(\ddot{\mathbf{A}}_{\mathbf{k}} + c^2 k^2 \mathbf{A}) = -\frac{4\pi}{c}\frac{1}{V} \sum_a q_a \mathbf{v}_{a,\perp} e^{-i\mathbf{k} \cdot \mathbf{r}_a}. \tag{79.20}$$

The left-hand side is the FT of the D'Alembertian of $\mathbf{A}$. If we can show that the right-hand side is the FT of $-(4\pi/c)\mathbf{j}^T$, we will obtain the inhomogeneous wave equation (33.15) for $\mathbf{A}$ in the Coulomb gauge, equivalent to the Ampere-Maxwell law. We have, using eq. (33.15) and exercise 33.1(b),

$$\mathbf{j}^T(\mathbf{r}, t) = \sum_a q_a \mathbf{v}_a \delta(\mathbf{r} - \mathbf{r}_a) - \frac{1}{4\pi} \frac{\partial}{\partial t} \sum_a \nabla \frac{q_a}{|\mathbf{r} - \mathbf{r}_a|}. \tag{79.21}$$

Hence,

$$\begin{aligned}
\mathbf{j}_\mathbf{k}^T &= \frac{1}{V} \int \mathbf{j}^T(\mathbf{r}) e^{-i\mathbf{k}\cdot\mathbf{r}} d^3x \\
&= \frac{1}{V} \sum_a q_a \left[\mathbf{v}_a - \frac{1}{4\pi} \frac{\partial}{\partial t} \frac{4\pi i \mathbf{k}}{k^2} \right] e^{-i\mathbf{k}\cdot\mathbf{r}_a(t)} \\
&= \frac{1}{V} \sum_a q_a \left[\mathbf{v}_a - \frac{\mathbf{k}(\mathbf{k} \cdot \mathbf{v}_a)}{k^2} \right] e^{-i\mathbf{k}\cdot\mathbf{r}_a(t)} \\
&= \frac{1}{V} \sum_a q_a \mathbf{v}_{a,\perp} e^{-i\mathbf{k}\cdot\mathbf{r}_a}. \tag{79.22}
\end{aligned}$$

Thus, we have shown, as we wanted, that

$$\nabla^2 \mathbf{A} - \frac{1}{c^2} \frac{\partial^2 \mathbf{A}}{\partial t^2} = -\frac{4\pi}{c} \mathbf{j}^T. \tag{79.23}$$

Exercise 79.1 Show that the cross term on the right of eq. (79.6) vanishes. (*Hint*: Write $\mathbf{E}_L = -\nabla\phi$ and integrate by parts, or Fourier transform and use Parseval's theorem.)

Exercise 79.2 Fill in the steps leading to the derivation of L.

The Hamiltonian is obtained from the Lagrangian by the standard method:

$$\mathcal{H} = \sum_a \mathbf{p}_a \cdot \mathbf{v}_a + \sum_k \mathbf{P}_\mathbf{k} \cdot \mathbf{A}_\mathbf{k} - L, \tag{79.24}$$

where $\mathbf{p}_a$ and $\mathbf{P}_\mathbf{k}$ are the momenta conjugate to $\mathbf{r}_a$ and $\mathbf{A}_\mathbf{k}$. Omitting the details, we present the answer:

$$\mathcal{H} = \sum_a \frac{1}{2m_a} \left(\mathbf{p}_a - \frac{q_a}{c} \mathbf{A}(\mathbf{r}_a, t) \right)^2 + \frac{1}{2} {\sum_{a,b}}' \frac{q_a q_b}{|\mathbf{r}_a - \mathbf{r}_b|} + \frac{1}{8\pi} \int \left(E_T^2 + B^2 \right) d^3x. \tag{79.25}$$

The last two terms can be combined into the total energy of the field to yield

$$\mathcal{H} = \sum_a \frac{1}{2m_a} \left(\mathbf{p}_a - \frac{q_a}{c} \mathbf{A}(\mathbf{r}_a, t) \right)^2 + \int \frac{\mathbf{E}^2 + \mathbf{B}^2}{8\pi} d^3x. \tag{79.26}$$

$$\mathcal{H} = \sum_a \frac{1}{2m_a} \left(\mathbf{p}_a - q_a \mathbf{A}(\mathbf{r}_a, t) \right)^2 + \int \frac{1}{2} \left(\epsilon_0 \mathbf{E}^2 + \frac{\mathbf{B}^2}{\mu_0} \right) d^3x. \quad \text{(SI)}$$

We do not bother to examine Hamilton's equations, since it is obvious that they must lead to Maxwell's equations, as before.

Exercise 79.3 Fill in the steps leading to the derivation of $\mathcal{H}$.

Exercise 79.4 As we have seen, there is no explicit $q\phi$ term in the Hamiltonian (79.26), because the electrostatic energy is included in the field energy. Convince yourself that $d\mathcal{H}/dt = 0$ as another way to see that this is so.

Exercise 79.5 (inclusion of spin magnetic moment) If the charges in question are electrons, then at low speeds their spin can be incorporated via the Pauli Hamiltonian, which can be written as[4]

$$\mathcal{H} = \mathcal{H}_{\text{no spin}} - \sum_a \mu_a \cdot \mathbf{B}(\mathbf{r}_a, t), \tag{79.27}$$

where $\mu_a = -2(e\hbar/mc)\mathbf{S}_a$, and $\mathbf{S}_a$ is the spin operator. We can "classicalize" this Hamiltonian by turning the operators into classical dynamical variables. In particular, μ_a becomes a vector of fixed length. This Hamiltonian can then be used to modify Maxwell's equations to account for point magnetic dipoles. It is simpler, however, to use the corresponding Lagrangian, which is given by

$$L = L_{\text{no spin}} - \sum_a \left[S_a(1 - \cos\theta_a)\dot{\varphi}_a + \mu_a \cdot \mathbf{B}(\mathbf{r}_a, t) \right], \tag{79.28}$$

where θ_a and φ_a are the spherical polar coordinates of μ_a, and we have used the result of exercise 75.2 to add the kinetic term in L.

Show (a) that the Lagrangian (79.28) leads to the Hamiltonian (79.27), and (b) that Maxwell's equations are the same as before, provided we interpret $\mathbf{j}$ to include an extra contribution

$$\mathbf{j}_{\text{dip}} = c\nabla \times \sum_a \mu_a \delta(\mathbf{r} - \mathbf{r}_a). \tag{79.29}$$

This is in accord with our findings in section 26. Note, however, that it again rests crucially on $\mathbf{B}$ being the *curl* of $\mathbf{A}$.

80 Gauge invariance and charge conservation

The action formulation provides an effective way to see the connection between gauge invariance and charge conservation mentioned in section 2. Since a gauge change does not affect physical observables, it must produce only physically inconsequential changes in the action. To find this change, we consider a gauge transformation,

$$\phi \to \phi' = \phi - \frac{1}{c}\frac{\partial\psi}{\partial t}, \quad \mathbf{A} \to \mathbf{A}' = \mathbf{A} + \nabla\psi, \tag{80.1}$$

---

[4] See Shankar (1981), sec. 20.2.

where ψ is arbitrary. Using the form (79.1), we find the change in the Lagrangian to be

$$\delta L = \sum_a \frac{q_a}{c} \left[\frac{\partial}{\partial t} \psi(\mathbf{r}_a, t) + \mathbf{v}_a \cdot \nabla \psi(\mathbf{r}_a, t) \right]$$

$$= \frac{d}{dt} \sum_a \frac{q_a}{c} \psi(\mathbf{r}_a, t), \tag{80.2}$$

which is a total time derivative. Such an addition to a Lagrangian is completely immaterial, however, as the corresponding change in the action is

$$\delta S = \int_{t_1}^{t_2} \delta L \, dt = \sum_a \frac{q_a}{c} \left[\psi(\mathbf{r}_a, t_2) - \psi(\mathbf{r}_a, t_1) \right]. \tag{80.3}$$

To find the equations of motion, we must vary the coordinates $\mathbf{r}_a$, $\phi(\mathbf{r})$, and $\mathbf{A}(\mathbf{r}, t)$ *while keeping their boundary values at the initial and final times fixed.* Since the extra term in the action, δS, depends only on these boundary values, it follows that it cannot affect the equations of motion.

We now determine the change in the action using the continuum charge and current density form (79.2). We find

$$\delta S = \int \left[\mathbf{j} \cdot \nabla \psi + \frac{1}{c} \rho \frac{\partial \psi}{\partial t} \right] d^3x \, dt, \tag{80.4}$$

which turns, after integrations by parts, into

$$\delta S = - \int \left[\nabla \cdot \mathbf{j} + \frac{1}{c} \frac{\partial \rho}{\partial t} \right] \psi \, d^3x \, dt. \tag{80.5}$$

If the action is not to change for arbitrary $\psi(\mathbf{r}, t)$, then we must have

$$\nabla \cdot \mathbf{j} + \frac{1}{c} \frac{\partial \rho}{\partial t} = 0. \tag{80.6}$$

But this is the law of charge conservation. Of course, this law follows from Maxwell's equations in terms of the $\mathbf{E}$ and $\mathbf{B}$ fields. The relation between these fields and the potentials directly implies gauge invariance. If we regard the potentials as the basic fields, on the other hand, the Maxwell equations take the form of second-order partial differential equations for these fields. By themselves, these equations do not reveal that there are gauge degrees of freedom in the theory. Charge conservation follows by demanding that these degrees of freedom be physically redundant. In quantum electrodynamics, the potentials are basic, so gauge invariance becomes essential.

We can now answer why, in section 78, we did not fix the gauge via a constraint while varying the action. For a general variation of the potentials, the part that amounts to a change in gauge does not lead to any change in the action (modulo the endpoint terms). Hence, there is no point in restricting the variation to only nongauge changes.

13 | Electromagnetic fields in material media

With this chapter we begin our study of electromagnetic fields in material media. Materials come in a surprising variety: metals, insulators, electrolytic solutions, liquid crystals, magnets, superconductors, plasmas, transparent crystals, polar and nonpolar liquids, gases, polymer melts, etc., etc. Many of these distinctions are based on the differences in the way the material responds to electromagnetic fields. It might be felt that our study is premature and that one should first develop a theory of the relevant material. It is possible to make substantial progress, however, by using a heuristic approach that makes use only of certain macroscopic properties shared by all materials. The key property is electrical neutrality on the atomic or short-distance scale, of the order of angstroms. ($1\,\text{Å} = 10^{-8}\,\text{cm}$). This property is itself a consequence of the inverse-square nature and enormous strength of the Coulomb force. By and large, matter maintains its integrity extremely well under the electromagnetic fields that we can apply to it, and this property enables us to develop simple models for its response. Further, it means that we do not really need to consider the rapid atomic-scale variations in the fields. Rather, we can consider a spatially averaged or smoothed-out field. When we place the material in an external field, the changes in the smoothed-out field will be enough to describe laboratory measurements on the macroscopic scale. In its modern form, this idea is due to H. A. Lorentz.

Thus, the program of studying electromagnetism in media divides into two parts. The first part is to obtain the equations obeyed by the smoothed-out fields and is relatively easy. The second part is to describe how the medium responds. This part is much more difficult. It is here that the huge variety of materials must be confronted. There is no simple way to write down one or two neat equations that will cover every case. Further, we can construct only simplified models for this response, and it must be clearly understood that these models are approximations with limitations on their regimes of validity. This is the nature of equations such as $\mathbf{D} = \epsilon \mathbf{E}$, and they have a fundamentally different status from the basic Maxwell equations in vacuum.

81 Macroscopic fields

As already mentioned, the electric field inside any condensed form of matter varies on the scale of the Bohr radius, $a \sim 10^{-8}$ cm. The scale on which the magnitude of the field varies is also extremely large. Inside the core of an atom, we can estimate $E \sim e/a^2 \sim 10^{11}$ V/m, where e is the magnitude of the electron charge. We seek to average the microscopic equations over dimensions that are much larger than the interatomic spacing, a, but still small on the macroscopic scale, L. What we choose to call the "macroscopic scale" can still be very small compared to 1 cm. In most situations, choosing $L \simeq 100a \simeq 10^{-6}$ cm would serve to wash out all microscopic variations.

Let us denote the microscopic electric and magnetic fields by lowercase letters $\mathbf{e}(\mathbf{r}, t)$ and $\mathbf{b}(\mathbf{r}, t)$, and the microscopic charge and current densities by $\rho_{\text{micro}}(\mathbf{r}, t)$ and $\mathbf{j}_{\text{micro}}(\mathbf{r}, t)$. These fields obey the exact Maxwell equations

$$\nabla \cdot \mathbf{e} = 4\pi \rho_{\text{micro}}, \tag{81.1}$$

$$\nabla \cdot \mathbf{b} = 0, \tag{81.2}$$

$$\nabla \times \mathbf{e} + \frac{1}{c} \frac{\partial \mathbf{b}}{\partial t} = 0, \tag{81.3}$$

$$\nabla \times \mathbf{b} - \frac{1}{c} \frac{\partial \mathbf{e}}{\partial t} = \frac{4\pi}{c} \mathbf{j}_{\text{micro}}. \tag{81.4}$$

All the above fields vary on the length scale a. We denote their spatial averages (indicated by overbars) by uppercase letters $\mathbf{E}(\mathbf{r}, t)$ and $\mathbf{B}(\mathbf{r}, t)$ for the fields, and ρ and $\mathbf{j}$ without suffixes for the charge and current densities:

$$\mathbf{E}(\mathbf{r}, t) = \bar{\mathbf{e}}(\mathbf{r}, t), \quad \rho(\mathbf{r}, t) = \bar{\rho}_{\text{micro}}(\mathbf{r}, t), \tag{81.5}$$

etc. More precisely, we define these averages through the equations

$$\mathbf{E}(\mathbf{r}, t) = \int s(\mathbf{r} - \mathbf{r}') \mathbf{e}(\mathbf{r}', t) \, d^3x', \tag{81.6}$$

$$\rho(\mathbf{r}, t) = \int s(\mathbf{r} - \mathbf{r}') \rho_{\text{micro}}(\mathbf{r}', t) \, d^3x', \tag{81.7}$$

and so on. Here, $s(\mathbf{r})$ is any function that "smudges" the fields and densities over the length scale L (fig. 13.1). The exact choice of this function is unimportant, except that it should be smooth and approximately constant for $r < L$ and vanish for $r \gg L$. Further, since we wish it to represent an average, we should have

$$\int s(\mathbf{r}) \, d^3x = 1. \tag{81.8}$$

It is now extremely simple to average the microscopic equations. When applied to eq. (81.1), e.g., the averaging procedure yields

$$\nabla \cdot \mathbf{E} = 4\pi \rho, \tag{81.9}$$

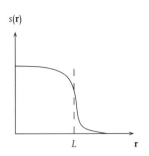
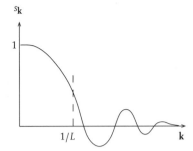

FIGURE 13.1. A smudge function and its Fourier transform.

which has the same formal appearance as the microscopic equation. One way to derive eq. (81.9) is by taking Fourier transforms. From eqs. (81.6) and (81.7) we obtain

$$\mathbf{E_k} = s_k \mathbf{e_k}, \quad \rho_k = s_k \rho_{\text{micro,k}}. \tag{81.10}$$

Thus,

$$\begin{aligned}
\text{F.T.}(\nabla \cdot \mathbf{E}) &= i\mathbf{k} \cdot \mathbf{E_k} \\
&= s_k(i\mathbf{k} \cdot \mathbf{e_k}) \\
&= s_k[\text{F.T.}(\nabla \cdot \mathbf{e})] \\
&= 4\pi s_k \rho_{\text{micro,k}} \\
&= 4\pi \rho_k.
\end{aligned} \tag{81.11}$$

Taking inverse Fourier transforms, we obtain eq. (81.9).

Note that the third of the chain of equalities (81.11) states that

$$\overline{\nabla \cdot \mathbf{e}} = \nabla \cdot \bar{\mathbf{e}}. \tag{81.12}$$

In the same way, we see that

$$\overline{\frac{\partial \mathbf{e}}{\partial t}} = \frac{\partial \bar{\mathbf{e}}}{\partial t}. \tag{81.13}$$

All other field derivatives can be averaged similarly, and we immediately obtain the averaged or macroscopic Maxwell equations

$$\nabla \cdot \mathbf{E} = 4\pi \rho, \tag{81.14}$$

$$\nabla \cdot \mathbf{B} = 0, \tag{81.15}$$

$$\nabla \times \mathbf{E} + \frac{1}{c}\frac{\partial \mathbf{B}}{\partial t} = 0, \tag{81.16}$$

$$\nabla \times \mathbf{B} - \frac{1}{c}\frac{\partial \mathbf{E}}{\partial t} = \frac{4\pi}{c}\mathbf{j}. \tag{81.17}$$

These have the same form as the equations in vacuum. However, the *meanings* of the symbols are very different, especially of ρ and $\mathbf{j}$. Indeed, these equations are not very useful in the form written and have no predictive power whatsoever, until we can say something about ρ and $\mathbf{j}$. We will turn to that in the succeeding sections. Here, we conclude with some general comments.

First, the k-space language not only allows us to average field derivatives efficiently, it also provides a valuable way of thinking about the averaging process. Because of the general properties of $s(\mathbf{r})$, $s_\mathbf{k}$ is exactly equal to 1 for $\mathbf{k} = 0$, is nearly unity for $k \leq 1/L$, and is very small for $k \gg 1/L$. From eq. (81.10), we see that the averaging corresponds to essentially throwing away the very high wave vector components of the fields and densities.

Secondly, what are the implications of quantum mechanics for macroscopic electromagnetism? and how should the above discussion be viewed in light of quantum theory? Since all physical observables are represented by operators in quantum mechanics, the microscopic fields $\mathbf{e}(\mathbf{r}, t)$ and $\rho_{\mathrm{micro}}(\mathbf{r}, t)$ must be replaced by operators. The same is true of the macroscopic fields $\mathbf{E}$ and ρ, since they are defined as linear combinations of $\mathbf{e}$ and ρ_{micro} via eqs. (81.6) and (81.7).[1] Now, the key feature of quantum mechanics is that physical quantities do not in general have definite values. The classical description ignores this indefiniteness and deals only with the mean values of the operators. In other words, classical equations such as eqs. (81.9) and (81.16) must be regarded as arising from quantum mechanically averaging some underlying operator equation, such as a Heisenberg equation of motion.

The real question, therefore, is the extent of the quantum mechanical indefiniteness or fluctuation. At the atomic and subatomic length scales, this is a very subtle and complex issue far outside the scope of our discussion, although quantum mechanics must clearly be taken into account in describing phenomena such as X-ray scattering, emission, and absorption, and the Pauli exclusion principle is critical to a proper understanding of electrical conduction in metals. For the spatially averaged fields appearing in the macroscopic theory, however, quantum fluctuations are very small and utterly negligible for most purposes. There is an extremely large range of phenomena that may be viewed as exactly described by the classical theory for all practical purposes. Secondly, even when quantum mechanics is needed, it is often only the matter fields—the charges and the currents—that must be treated quantum mechanically, and the electromagnetic fields can continue to be regarded as classical quantities. Quantum fluctuations of the electromagnetic field *can* be observed in the electromagnetic energy density in a transparent medium, and in phenomena such as the Casimir effect, but we shall not discuss them.

Exercise 81.1 (For readers averse to performing the spatial averages by Fourier transforms.) Derive the macroscopic equations by direct real-space integration.

---

[1] To be more explicit, recall that we must regard $\mathbf{e}(\mathbf{r}, t)$ as a separate operator at *each* space point $\mathbf{r}$. The right-hand side of eq. (81.6) thus adds together many different operators with different coefficients. Likewise for eq. (81.7).

Solution (partial): Consider the equation for $\nabla \cdot \mathbf{E}$. Since eq. (81.6) is a convolution, we may write

$$\mathbf{E}(\mathbf{r}, t) = \int s(\mathbf{r}')\mathbf{e}(\mathbf{r} - \mathbf{r}', t)\, d^3x. \tag{81.18}$$

The divergence can now be taken inside the integral. This yields

$$\nabla \cdot \mathbf{E} = \int s(\mathbf{r}')\nabla \cdot \mathbf{e}(\mathbf{r} - \mathbf{r}', t)\, d^3x \tag{81.19}$$

$$= \int s(\mathbf{r}')\rho_{\text{micro}}(\mathbf{r} - \mathbf{r}', t)\, d^3x \tag{81.20}$$

$$= \rho(\mathbf{r}, t), \tag{81.21}$$

where the last result again exploits the properties of convolutive integrals.

82 The macroscopic charge density and the polarization

Let us now take up the macroscopic charge density. There is no uniform way of discussing this for every type of material, but two broad contributions may be identified. In metals, some of the electrons are not bound to the atoms, i.e., they are delocalized. This is what gives metals their ability to conduct. By the same token, an isolated piece of metal may have electrons added to or removed from it, leaving it with a net charge. We define the macroscopic charge density due to such electrons to be ρ_{free}, where the word "free" means that the charges are free to move over macroscopic distances. This concept also applies to insulators, although in this case any free charge must be externally supplied.

The second contribution to ρ is from charges that are bound to the atoms or molecules. One can expect that the details have to be framed differently for different types of materials. To start with, therefore, we will present a simplified discussion for a system made out of a large number of molecules that are well separated on the microscopic scale. The key concept will turn out to be that of the *polarization* field. Later we will see that this concept is actually very general.

Each molecule is neutral. The next term in the moment expansion that can give rise to an electric field is the electric dipole moment. Let us suppose, therefore, that each molecule has a dipole moment $\mathbf{d}$. If there are n molecules per unit (macroscopic) volume, the material will have a *dipole moment per unit volume* given by

$$\mathbf{P} = n\mathbf{d}. \tag{82.1}$$

The vector $\mathbf{P}$ is known as the *polarization*. In general, $\mathbf{P}$ varies with $\mathbf{r}$, which in our simple model could come about because either n or $\mathbf{d}$ is spatially varying. Note that $\mathbf{P}$ is a macroscopic field, i.e., it varies on a macroscopic-length scale, because it is defined by averaging only over a macroscopic volume. By definition, $\mathbf{P}$ vanishes outside a material medium.

The existence of a polarization leads to a charge distribution in general. We can see this as follows. Let V be a (macroscopically) infinitesimal volume element, with surface S (fig. 13.2). The electric potential at a point $\mathbf{r}$ due to the polarization inside this volume

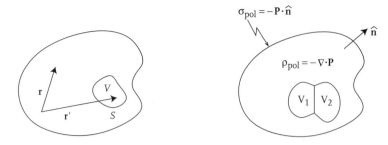

FIGURE 13.2. Volume and surface charges in a polarized medium.

element is given by

$$\phi(\mathbf{r}) = \int_V \frac{\mathbf{P}(\mathbf{r}') \cdot (\mathbf{r} - \mathbf{r}')}{|\mathbf{r} - \mathbf{r}'|^3} d^3 x'$$

$$= \int_V \mathbf{P}(\mathbf{r}') \cdot \nabla' \frac{1}{|\mathbf{r} - \mathbf{r}'|} d^3 x'$$

$$= \int_{S = \partial V} \frac{\mathbf{P}(\mathbf{r}') \cdot \hat{\mathbf{n}}'}{|\mathbf{r} - \mathbf{r}'|} d^2 s' - \int_V \frac{\nabla' \cdot \mathbf{P}(\mathbf{r}')}{|\mathbf{r} - \mathbf{r}'|} d^3 x', \tag{82.2}$$

where we use Gauss's theorem in the last step. The second term in this formula is the potential due to a volume charge density,

$$\rho_{\text{pol}} = -\nabla \cdot \mathbf{P}, \quad \text{(Gaussian and SI)} \tag{82.3}$$

while the first is the potential due to a surface charge density,

$$\sigma_{\text{pol}} = \mathbf{P} \cdot \hat{\mathbf{n}}. \quad \text{(Gaussian and SI)} \tag{82.4}$$

Of course, if the volume element is internal to the material, the surface charge density will be canceled by that of adjoining elements. For instance, if in fig. 13.2 we consider the surface charge densities on volume elements V_1 and V_2, the contributions from the common part of the surface will be equal and opposite and add to zero. At the true surface of the body, on the other hand, there is no such cancellation, and eq. (82.4) gives the net surface charge density.

The suffix "pol" in eqs. (82.3) and (82.4) signifies that the charges in question originate in the polarization and are bound to the molecules. The physical reason for ρ_{pol} is simple. Even though the material is neutral, there may be local accumulations of charge if the polarization vector varies in space. With this contribution, we can write the source term in Gauss's law as

$$\rho = \rho_{\text{pol}} + \rho_{\text{free}}. \tag{82.5}$$

Defining

$$\mathbf{D} = \mathbf{E} + 4\pi\mathbf{P}, \tag{82.6}$$

$$\mathbf{D} = \epsilon_0 \mathbf{E} + \mathbf{P}, \quad \text{(SI)}$$

and making use of eq. (82.3), we obtain

$$\nabla \cdot \mathbf{D} = 4\pi \rho_{\text{free}}. \tag{82.7}$$

$$\nabla \cdot \mathbf{D} = \rho_{\text{free}}. \quad \text{(SI)}$$

The field $\mathbf{D}$ is known as the *electric displacement*. Unlike $\mathbf{E}$, it does not have an operational meaning such as the force on a unit test charge.

Let us now discuss why the notion of a polarization field is a general one. It is clear that we can define the dipole moment for any microscopically neutral collection of charges. In a periodic solid, e.g., we could use the dipole moment for a unit cell of the crystal. By smoothing this we arrive at $\mathbf{P}$. One may ask why one should not also consider the quadrupole and higher moments of the microscopic charge collections. One reason is that these moments are rarely large enough to be important. A somewhat formal reason is that just as the divergence of the polarization gave us a local charge density, the divergence of a smoothed quadrupole moment density $D_{ij}(\mathbf{r})$ will give us a dipole moment density, $\mathbf{P}'(\mathbf{r})$:

$$P_i' = -\frac{\partial D_{ij}}{\partial x_j}. \tag{82.8}$$

Similar contributions will be given by still higher moments. There is no way to distinguish these contributions to the polarization from that previously considered by macroscopic measurements. Thus, as far as the form of eq. (82.7) goes, it is completely general. The problem is really one of how the polarization should be calculated and related to the microscopic charge distribution. This is a very difficult exercise, and the author is unaware of a situation in which the quadrupolar contributions were reliably included. More generally, one models this relationship in terms of a few parameters, which are then taken by fitting to experiment.

Equation (82.7) is the conventional way of writing Gauss's law for any material medium. At the interface between two media, it needs to be supplemented by a boundary condition, which can be derived by considering a Gaussian pillbox whose end caps lie on opposite sides of the interface. The condition is

$$(\mathbf{D}_2 - \mathbf{D}_1) \cdot \hat{\mathbf{n}}_{21} = 4\pi \sigma_{\text{free}}. \tag{82.9}$$

$$(\mathbf{D}_2 - \mathbf{D}_1) \cdot \hat{\mathbf{n}}_{21} = \sigma_{\text{free}}. \quad \text{(SI)}$$

The vector $\hat{\mathbf{n}}_{21}$ points from medium 1 to 2 (fig. 13.3). If medium 1 is a vacuum, then $\mathbf{D}_1 = \mathbf{E}_1$.

We also need a boundary condition for the tangential component of an $\mathbf{E}$-like field (either $\mathbf{E}$ or $\mathbf{D}$). This is obtained by integrating the macroscopic Faraday law around a rectangular circulation contour whose long sides are opposite sides of the interface and lie parallel to it. In this way we obtain

$$\mathbf{E}_{1t} = \mathbf{E}_{2t}. \tag{82.10}$$

The suffix t stands for the tangential component.

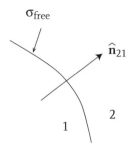

FIGURE 13.3. Convention for direction for $\hat{\mathbf{n}}_{21}$ in the boundary condition for the jump in **D** at an interface between two media.

It may help the reader to better understand the meaning of **E**, **P**, and **D** if we consider the problem of a piece of matter placed in an external electric field and ask for the field inside the matter. It follows from eq. (82.2) and the subsequent discussion that

$$\mathbf{E}_{\text{in}} = \mathbf{E}_0 + \mathbf{E}_d. \tag{82.11}$$

Here, $\mathbf{E}_0$ is the applied field, that which would exist if the body were not there. If the body is placed between the plates of a capacitor, the sources of this field are the charges on the capacitor. The field $\mathbf{E}_d$ is produced by the $\nabla \cdot \mathbf{P}$ and the $\mathbf{P} \cdot \hat{\mathbf{n}}$ sources. The subscript "d" in $\mathbf{E}_d$ stands for "depolarization," as this field generally tends to oppose the applied field, and we shall study it in more detail in section 96. Of course, to find $\mathbf{E}_d$ we need to know what **P** is, but that is an altogether separate problem, which we shall begin to address in section 84.

The field $\mathbf{D} = \mathbf{E} + 4\pi\mathbf{P}$ is an auxiliary field, which is introduced mainly to give the macroscopic Maxwell equations an appealing form. The equation $\nabla \cdot \mathbf{D} = 4\pi\rho_{\text{free}}$ suggests that its sources are the free charges (those on the capacitor in the example above) only, but that is not so, since $\nabla \times \mathbf{D}$ entails **P**.

Let us now discuss the conversion of the **P** and **D** fields from the SI to Gaussian systems and vice versa. The conversions are

$$\mathbf{E}_{\text{SI}} = \frac{1}{\sqrt{4\pi\epsilon_0}}\mathbf{E}_{\text{Gau}}, \quad \mathbf{P}_{\text{SI}} = \sqrt{4\pi\epsilon_0}\,\mathbf{P}_{\text{Gau}}, \quad \mathbf{D}_{\text{SI}} = \sqrt{\frac{\epsilon_0}{4\pi}}\mathbf{D}_{\text{Gau}}. \tag{82.12}$$

The conversion for **E** was given in chapter 1. The conversion for **P** is exactly the same as for charge, since the dipole moment is a charge times a separation. To find the conversion for **D**, we transform the SI system equation $\mathbf{D} = \epsilon_0\mathbf{E} + \mathbf{P}$ to Gaussian. This yields

$$\mathbf{D}_{\text{SI}} = \frac{\epsilon_0}{\sqrt{4\pi\epsilon_0}}\mathbf{E}_{\text{Gau}} + \sqrt{4\pi\epsilon_0}\,\mathbf{P}_{\text{Gau}} = \sqrt{\frac{\epsilon_0}{4\pi}}(\mathbf{E}_{\text{Gau}} + 4\pi\mathbf{P}_{\text{Gau}}). \tag{82.13}$$

Comparison with the corresponding Gaussian system equation yields the answer quoted in eq. (82.12). The result may also be obtained by starting from the equations for $\nabla \cdot \mathbf{D}$ and recalling the conversion for charge density.

Next, let us discuss the units. In the Gaussian system, **E**, **P**, and **D** all have the same dimensions and units of statcoulombs/cm² or statvolt/cm, although **P** is often quoted in esu/cm^3. Note that 1 statcoulomb/cm^2 = 1 statvolt/cm = 1 esu/cm^3. In the SI system, **P** and **D** have the same dimensions as each other and units of coulomb/m^2, but the dimensions of **E** are different, as, therefore, are the units (newton/coulomb, or volt/m). The conversion factors between the two systems are

$$E : 1 \text{ V/m} = (10^{-4}/3) \text{ statvolt/cm} ,$$

$$D : 1 \text{ C/m}^2 = 3 \times 4\pi \times 10^5 \text{ statvolt/cm} , \qquad (82.14)$$

$$P : 1 \text{ C/m}^2 = 3 \times 10^5 \text{ esu/cm}^3.$$

83 The macroscopic current density and the magnetization

Our next task is to determine the macroscopic current density, which is the source term in the macroscopic Ampere-Maxwell law. The development proceeds in parallel with what we did for the charge. We will see that the current divides into four parts:

$$\mathbf{j} = \mathbf{j}_{\text{free}} + \mathbf{j}_{\text{pol}} + \mathbf{j}_{\text{mag}} + \mathbf{j}_{\text{conv}}, \qquad (83.1)$$

which we call the *free, polarization, magnetization,* and *convection currents,* respectively.

The free current is due to the motion of electrons and ions that are not bound to each other, i.e., the charges that go into making up ρ_{free}. Accordingly, this part of the current separately obeys the continuity equation

$$\frac{\partial \rho_{\text{free}}}{\partial t} + \nabla \cdot \mathbf{j}_{\text{free}} = 0. \qquad (83.2)$$

This is the current that describes the transport of charge over macroscopic distances. In solids, this current is almost entirely due to the motion of electrons and is relevant only in conductors and semiconductors. In other cases, it includes the free motion of charges through space.

The polarization current $\mathbf{j}_{\text{pol}}$ arises from a time-varying polarization. Since a spatially varying polarization **P** produces a volume charge density $\rho_{\text{pol}} = -\nabla \cdot \mathbf{P}$, and since $\partial \rho / \partial t = -\nabla \cdot \mathbf{j}$ by charge conservation, it follows that a time-dependent polarization must result in a current

$$\mathbf{j}_{\text{pol}} = \frac{\partial \mathbf{P}}{\partial t}. \qquad (83.3)$$

In magnetostatics, where we study only steady-state situations, $\partial \mathbf{P}/\partial t = 0$, and so $\mathbf{j}_{\text{pol}}$ is also zero. In general situations, however, a polarization current will be present. Since the polarization can only vary between finite limits, this current is necessarily time dependent.

The convection current $\mathbf{j}_{\text{conv}}$ is present only in gases and liquids, such as molten metals and plasmas, and describes the charges that are bodily carried along by the mean macroscopic motion of the fluid. If the velocity of the fluid is **u**, then

$$\mathbf{j}_{\text{conv}} = (\rho_{\text{free}} + \rho_{\text{pol}})\mathbf{u} = (\rho_{\text{free}} - \nabla \cdot \mathbf{P})\mathbf{u}. \qquad (83.4)$$

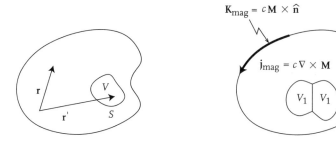

FIGURE 13.4. Volume and surface currents in a magnetized medium. The latter is indicated by the heavy arrow on the boundary.

The remaining part, $\mathbf{j}_{\mathrm{mag}}$, is the current due to electrons circulating in tiny atomic current loops, also known as *Amperean currents*.[2] From the equivalence between a current loop and a magnetic dipole sheet, we see that these currents will have the same effect as a volume distribution of magnetic moments, and this is why $\mathbf{j}_{\mathrm{mag}}$ is known as the *magnetization current*. To see this equivalence mathematically, let us suppose that a body has a *magnetic dipole moment per unit volume*, $\mathbf{M}(\mathbf{r})$. This quantity is also known as the *magnetization*. Referring to fig. 13.4, we see that the magnetization inside a volume element V gives rise to a vector potential

$$\mathbf{A}(\mathbf{r}) = \int_V \frac{\mathbf{M}(\mathbf{r}') \times (\mathbf{r} - \mathbf{r}')}{|\mathbf{r} - \mathbf{r}'|^3} d^3 x' \tag{83.5}$$

$$= \int_V \mathbf{M}(\mathbf{r}') \times \nabla' \frac{1}{|\mathbf{r} - \mathbf{r}'|} d^3 x'$$

$$= \int_{S = \partial V} \frac{\mathbf{M}(\mathbf{r}')}{|\mathbf{r} - \mathbf{r}'|} \times d\mathbf{s} + \int_V \frac{\nabla' \times \mathbf{M}(\mathbf{r}')}{|\mathbf{r} - \mathbf{r}'|} d^3 x', \tag{83.6}$$

using Gauss's theorem in the last step. The second term in this formula is the vector potential due to a volume current distribution,

$$\mathbf{j}_{\mathrm{mag}} = c \nabla \times \mathbf{M}, \tag{83.7}$$

$$\mathbf{j}_{\mathrm{mag}} = \nabla \times \mathbf{M}, \quad (\mathrm{SI})$$

while the first is the potential of a surface current density,

$$\mathbf{K}_{\mathrm{mag}} = c \mathbf{M}(\mathbf{r}) \times \hat{\mathbf{n}}. \tag{83.8}$$

$$\mathbf{K}_{\mathrm{mag}} = \mathbf{M}(\mathbf{r}) \times \hat{\mathbf{n}}. \quad (\mathrm{SI})$$

If the volume element is internal to the body, the surface current density will be canceled by that of the adjoining elements. At the true surface of the body, on the other hand, there is no such cancellation, and eq. (83.8) gives the true surface current.

---

[2] We are in a little bit of trouble here. In metals, the conduction electrons contribute not only to $\mathbf{j}_{\mathrm{free}}$ but also to $\mathbf{j}_{\mathrm{mag}}$. This is especially true in ferromagnetic metals, such as Fe and Ni. This contribution originates in the spin magnetic moment of the electron and is discussed in section 104.

The magnetization $\mathbf{M}(\mathbf{r})$ is necessarily zero outside a material body the way we have defined it. In the macroscopic theory there is no way to separate the spin and the orbital contributions to $\mathbf{M}$. Therefore, we must modify our initial remarks and regard $\mathbf{j}_{mag}$ as an effective current that includes the effect of spin. This is completely consistent with what we showed in section 26, namely, that the spin magnetic moment is also equivalent to a current. Indeed, for a point dipole, we should write $\mathbf{M}(\mathbf{r}) = \mathbf{m}\delta(\mathbf{r})$, and then eq. (83.7) reduces to eq. (26.14). We could also ask whether $\mathbf{j}_{mag}$ should not acquire contributions from higher magnetic multipole densities. The resolution of this question is entirely parallel to that of including electric quadrupole moments in ρ_{pol}.

It is convenient to define a new field $\mathbf{H}$,

$$\mathbf{H} = \mathbf{B} - 4\pi \mathbf{M}. \tag{83.9}$$

$$\mathbf{H} = \frac{1}{\mu_0}\mathbf{B} - \mathbf{M}. \quad \text{(SI)}$$

From eq. (83.7), it then follows that the Ampere-Maxwell law can be rewritten as

$$\nabla \times \mathbf{B} - \frac{1}{c}\frac{\partial \mathbf{E}}{\partial t} = \frac{4\pi}{c}\left(\mathbf{j}_{free} + \mathbf{j}_{conv} + c\nabla \times \mathbf{M} + \frac{\partial \mathbf{P}}{\partial t}\right),$$

$$\nabla \times \mathbf{H} - \frac{1}{c}\frac{\partial \mathbf{D}}{\partial t} = \frac{4\pi}{c}(\mathbf{j}_{free} + \mathbf{j}_{conv}). \tag{83.10}$$

For the most part we will not consider liquids, so $\mathbf{j}_{conv}$ is zero, and the equation becomes

$$\nabla \times \mathbf{H} - \frac{1}{c}\frac{\partial \mathbf{D}}{\partial t} = \frac{4\pi}{c}\mathbf{j}_{free}. \tag{83.11}$$

$$\nabla \times \mathbf{H} - \frac{\partial \mathbf{D}}{\partial t} = \mathbf{j}_{free}. \quad \text{(SI)}$$

As explained in chapter 4, we refer to $\mathbf{H}$ as the *magnetizing* field, reserving the term *magnetic field* for $\mathbf{B}$. The reason is given in section 107. $\mathbf{H}$ is like $\mathbf{D}$ in that its circulation is determined by the mobile or free currents, whereas that of $\mathbf{B}$ is determined by the total current including the bound currents.[3] Purcell makes the nice point that in electrical systems, the variable easiest to control is the potential, which amounts to controlling $\mathbf{E}$, not $\mathbf{D}$. In magnetic systems, on the other hand, we can control the free currents most easily, and thus $\mathbf{H}$ really is the directly controlled field.

At the interface between two media, eqs. (81.15) and (83.11) must be supplemented by boundary conditions, which can be derived by integrating these equations over a Gaussian pillbox or an Amperean current loop straddling the interface. The conditions are

$$B_{1n} = B_{2n}, \qquad \text{(Gaussian and SI)} \tag{83.12}$$

$$(\mathbf{H}_2 - \mathbf{H}_1) \times \hat{\mathbf{n}}_{21} = \begin{cases} (4\pi/c)\mathbf{K}_{free}, & \text{(Gaussian)} \\ \mathbf{K}_{free}, & \text{(SI)} \end{cases} \tag{83.13}$$

where $\mathbf{K}_{free}$ is the *free* surface current, and the vector $\hat{\mathbf{n}}_{21}$ points from medium 1 to medium 2 (fig. 13.5).

---

[3] But, since $\nabla \cdot \mathbf{H} \neq 0$, it would be incorrect to say that the mobile currents are the only sources of $\mathbf{H}$.

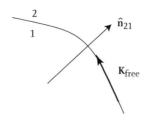

FIGURE 13.5. Convention for direction for $\hat{n}_{21}$ in the boundary condition for the jump in H at an interface between two media.

Let us now discuss the **B**, **H**, and **M** fields in analogy with the discussion surrounding eq. (82.11). A piece of matter is placed inside a large solenoid. What is **B** inside the solenoid? Answer:

$$\mathbf{B}_{\text{in}} = \mathbf{B}_0 + \mathbf{B}_d, \tag{83.14}$$

where $\mathbf{B}_0$ is the applied field whose sources are the currents in the solenoid, and $\mathbf{B}_d$ is the field due to the $\nabla \times \mathbf{M}$ and $\mathbf{M} \times \hat{\mathbf{n}}$ volume and surface Amperean currents.

So far, the discussion has paralleled the electric case completely. It pays, however, to add a slight twist in the magnetic case. We do not refer to $\mathbf{B}_d$ as the demagnetizing field. That term is reserved for $\mathbf{H}_d$, defined as $\mathbf{B}_d - 4\pi\mathbf{M}$. Because of the difference between the scalar and vector magnetic potentials, the sources of $\mathbf{H}_d$ are the volume and magnetic surface magnetic charges (not currents) $\nabla \cdot \mathbf{M}$ and $\mathbf{M} \cdot \hat{\mathbf{n}}$. With these definitions, we have

$$\mathbf{B}_{\text{in}} = \mathbf{B}_0 + 4\pi\mathbf{M} + \mathbf{H}_d. \tag{83.15}$$

The point of defining the demagnetizing field this way is that the mathematical problems of finding $\mathbf{H}_d$ and $\mathbf{P}_d$ are identical. Secondly, $\mathbf{H}_d$ generally opposes **M**, just as $\mathbf{E}_d$ opposes **P**, so the *de*magnetizing appellation is correct. We shall discuss demagnetization at greater length in section 108. That the sources of $\mathbf{H}_d$ are the charges $\nabla \cdot \mathbf{M}$ and $\mathbf{M} \cdot \hat{\mathbf{n}}$ is shown in eqs. (105.12) and (105.13).

Again, note that we need **M** to find $\mathbf{B}_{\text{in}}$ and we haven't yet said how to solve that problem. We begin on it in the next section. It should be noted though that both the $4\pi\mathbf{M}$ and the $\mathbf{H}_d$ terms in eq. (83.15) are due to the dipole fields of the magnetic moments (induced or permanently present) in the material.

Finally, let us find the conversion between SI and Gaussian fields **B**, **M**, and **H**. For **B**, this was done in chapter 1. The conversion for **M** is the same as that for the magnetic moment, and that for **H** follows from the macroscopic Ampere-Maxwell law. Summarizing the answers, we have

$$\mathbf{B}_{\text{SI}} = \sqrt{\frac{\mu_0}{4\pi}}\mathbf{B}_{\text{Gau}}, \quad \mathbf{M}_{\text{SI}} = \sqrt{\frac{4\pi}{\mu_0}}\mathbf{M}_{\text{Gau}}, \quad \mathbf{H}_{\text{SI}} = \frac{1}{\sqrt{4\pi\mu_0}}\mathbf{H}_{\text{Gau}}. \tag{83.16}$$

The SI and Gaussian units of B are the tesla and the gauss, as already mentioned. The SI units for H and M are the same, the ampere per meter (A/m). There are two easy ways to remember this. We found the B field due to an infinite straight wire to be

$\mu_0 I/2\pi r$, and since $H = B/\mu_0$ in vacuum, $H = I/2\pi r$, which has dimensions of current over length. Or, we recall that the magnetic moment of a current loop is current times area, so the dimensions of M (magnetic moment per unit volume) are current over length, which are the same as those of H. In the Gaussian system, H is given in oersteds (Oe), even though 1 Oe is exactly the same as 1 G. The different name, however, helps in distinguishing which quantity one is talking about. The Gaussian unit of M does not honor any great physicist and is the simple and direct emu/cm^3 (electromagnetic unit per cubic centimeter). Again, 1 emu/cm^3 = 1 G = 1 Oe. The conversions between the SI and Gaussian units are

$$B : 1\,\text{T} = 10^4\,\text{G} \,,$$

$$H : 1\,\text{A/m} = 4\pi \times 10^{-3}\,\text{Oe} \,, \tag{83.17}$$

$$M : 1\,\text{A/m} = 10^{-3}\,\text{emu/cm}^3.$$

Exercise 83.1 Take the curl of eq. (83.5) to find the magnetic field of a magnetized body, and show that

$$\mathbf{B}(\mathbf{r}) = 4\pi\mathbf{M}(\mathbf{r}) - \nabla \int \frac{\mathbf{M}(\mathbf{r}') \cdot (\mathbf{r} - \mathbf{r}')}{|\mathbf{r} - \mathbf{r}'|^3} d^3x'. \tag{83.18}$$

The integral may be taken as the scalar potential of the magnetization distribution. As we saw in exercise 26.1, its gradient does not yield $\mathbf{B}$, and we see now that it yields $\mathbf{H}$ instead.

84 Constitutive relations

We have now obtained macroscopic versions of the basic electromagnetic laws. We rewrite them all together for ready reference:

Law	Equation (Gaussian)	Equation (SI)	
Gauss's law	$\nabla \cdot \mathbf{D} = 4\pi\rho_{\text{free}}$	$\nabla \cdot \mathbf{D} = \rho_{\text{free}}$	
Faraday's law	$\nabla \times \mathbf{E} + \dfrac{1}{c}\dfrac{\partial \mathbf{B}}{\partial t} = 0$	$\nabla \times \mathbf{E} + \dfrac{\partial \mathbf{B}}{\partial t} = 0$	(84.1)
Ampere-Maxwell law	$\nabla \times \mathbf{H} - \dfrac{1}{c}\dfrac{\partial \mathbf{D}}{\partial t} = \dfrac{4\pi}{c}\mathbf{j}_{\text{free}}$	$\nabla \times \mathbf{H} - \dfrac{\partial \mathbf{D}}{\partial t} = \mathbf{j}_{\text{free}}$	
No magnetic monopoles	$\nabla \cdot \mathbf{B} = 0$	$\nabla \cdot \mathbf{B} = 0$	

These equations are useless, however, without a knowledge of $\mathbf{D}$ and $\mathbf{H}$, or, what is really the issue, a knowledge of the densities $\mathbf{P}$ and $\mathbf{M}$. In addition, we must also have some knowledge of ρ_{free} and $\mathbf{j}_{\text{free}}$. This information is provided by *constitutive relations*. As the name suggests, these relations reflect the constitution or makeup of the material under study. Hence, different constitutive relations apply to different materials, and they are not universal. At the risk of belaboring the point, we restate that they are also not fundamental in the sense of Coulomb's law. Ohm's law, e.g., may well be the oath that all licensed electrical engineers are required to take, but it is not true for every material, and even for one material it may be more or less accurate under different operating conditions. To understand a constitutive relation invariably requires some understanding, very crude perhaps, of the material. The microscopic derivation of these relationships makes up a large part of theoretical solid-state physics. From the point of view of a course in electromagnetism, this is a side issue, and one generally works with simplified, broadly applicable relations that are regarded as taken from experiment.

In the rest of this section, we survey some of the more common constitutive relations. The reader has probably encountered many of them earlier. We shall study them in more detail, along with their major implications, in succeeding chapters.

The first is Ohm's law, already mentioned. This law states that the *steady* free current in a conductor is given by

$$\mathbf{j} = \sigma \mathbf{E}. \tag{84.2}$$

The constant σ is known as the *conductivity*. The law fails at high enough fields, which may actually be quite low for semiconductors. This relationship is often generalized to time-dependent fields, in the form

$$\mathbf{j}_\omega = \sigma(\omega)\mathbf{E}_\omega. \tag{84.3}$$

The quantity $\sigma(\omega)$ is known as the *frequency-dependent* or *ac conductivity*. Its zero-frequency value $\sigma(0)$ is the static or dc conductivity σ in eq. (84.2).

In insulators or dielectrics, the displacement and electric fields are related by

$$\mathbf{D} = \epsilon \mathbf{E}. \quad \text{(Gaussian and SI)} \tag{84.4}$$

The quantity ϵ (in Gaussian units), or ϵ/ϵ_0 (in SI), is known as the *dielectric constant* of the material.[4] It is a dimensionless number with the same value in both systems. Equation (84.4) holds only if the field $\mathbf{E}$ is small enough, and the range of validity can be quite small for some materials. Indeed, most materials undergo *dielectric breakdown* at a characteristic field, arising from ionization of the material, and consequent formation of a conducting path. Air, for example, with $\epsilon = 1.00054$ (at 1 atmosphere), breaks down at approximately 1 statvolt/cm (3×10^4 V/m); $SrTiO_3$, with $\epsilon = 310$, breaks down at 2.7 statvolt/cm (8×10^4 V/m). Note that these fields are well below the internal electric fields in matter. Secondly, the dielectric constant of a substance is influenced by variables such as temperature and pressure. The dielectric constant of water, e.g., changes from 80.4 at 20°C to 78.5 at 25°C.

In addition to eq. (84.4), one also has a frequency-dependent generalization

$$\mathbf{D}_\omega = \epsilon(\omega)\mathbf{E}_\omega, \tag{84.5}$$

which is closely related to the frequency-dependent Ohm's law.

---

[4] The notation κ for the ratio ϵ/ϵ_0 is also common in the SI system.

The magnetic analog of eq. (84.4) is[5]

$$\mathbf{B} = \mu \mathbf{H}. \quad \text{(Gaussian and SI)} \tag{84.6}$$

The dimensionless number μ (in the Gaussian system), or μ/μ_0 (in SI), is known as the *permeability* of the medium. It has the same value in the two systems.

Equation (84.6) does not apply to two important classes of materials, ferromagnets and superconductors, though it is unfortunately rather common to see it written down for the former. At best one can talk only in terms of an effective permeability for a ferromagnet, and this is not a constant over the range of fields that can easily be investigated. As we shall see later, $\mathbf{M}$ and $\mathbf{H}$ are related intrinsically nonlinearly at a length scale of 50–100 Å. Nevertheless, it is useful to note that in some nickel–iron alloys ("supermalloy" and "MuMetal"), the ratio of B to H can be as large as 10^6. For obvious reasons, these are known as *high-permeability* materials.

In fact, for ferromagnets the constitutive relationship for ferromagnets cannot be written down in a single compact equation, as it can for dia- and paramagnets.[6] Instead, it is best embodied in a thermodynamic minimum-free-energy principle, along with a phenomenological expression for the free energy that takes into account the principal physical phenomena. The same remark applies to ferroelectrics.

Superconductors form another class of materials in which the constitutive $\mathbf{B}$ versus $\mathbf{H}$ relation is highly nonlinear. The fundamental property of a superconductor is that it expels magnetic flux. This is known as the *Meissner effect*. When the flux expulsion is complete, in the body of the superconductor we must have

$$\mathbf{B} = 0. \tag{84.7}$$

This is the basic constitutive relation. It holds only in the bulk, for small enough H and temperatures, in what are known as the type I superconductors, and in the so-called Meissner phase of type II superconductors.

We have written the relationships above for isotropic materials. In anisotropic substances, many of them have a tensorial generalization. For example, Ohm's law becomes $j_i = \sigma_{ij} E_j$, and eq. (84.4) becomes $D_i = \epsilon_{ij} E_j$. In superconductors, the field below which eq. (84.7) holds may be anisotropic.

A second general remark is that there are constraints on the constitutive laws. For example, thermodynamic stability forces σ and ϵ to both be positive. Similarly, the demands of causality impose significant restrictions on the frequency-dependent quantities $\sigma(\omega)$ and $\epsilon(\omega)$, which are embodied in the *Kramers-Kronig relations*, connecting the real and imaginary parts of these functions.

---

[5] For historical reasons, constitutive relations and many other formulas in magnetism are usually written as if $\mathbf{H}$ were the fundamental field and $\mathbf{B}$ the derived field, which is the opposite of what we now recognize.

[6] Or at least it cannot without considerable prior discussion of the meaning of the various terms.

85 Energy conservation

Let us now ask if we can make any general statements about energy conservation for macroscopic electromagnetic fields in material media. It is obvious from the outset that great care will have to be taken in doing this. First, any expression for the energy will have to be interpreted as a coarse-grained average, both in space and in time. Second, in thinking of the electromagnetic energy, the part that constitutes the cohesive energy of the material itself must be excluded. Mathematically, $\mathbf{E}^2$ and $\mathbf{B}^2$ are not the same as $\overline{\mathbf{e}^2}$ and $\overline{\mathbf{b}^2}$. An electromagnetic energy would have to be interpreted as the difference in the internal energy of the system, with and without the field. Third, we must allow for the possibility of dissipation or the evolution of heat. This means that the expression for the internal energy is sensible only when interpreted as an average over a sufficiently short time.

It is simplest to begin by calculating the work received by the free or mobile charges in a medium. The work received by the bound charges is hard to calculate and is, any case, expected to be converted either to the internal energy of the system or to heat. The work on the free charges is done at the rate $\mathbf{j}_{\text{free}} \cdot \mathbf{E}$. The macroscopic Ampere-Maxwell law enables us to write

$$\mathbf{j}_{\text{free}} \cdot \mathbf{E} = \frac{c}{4\pi} \left(\nabla \times \mathbf{H} - \frac{1}{c} \frac{\partial \mathbf{D}}{\partial t} \right) \cdot \mathbf{E}. \tag{85.1}$$

The $(\nabla \times \mathbf{H}) \cdot \mathbf{E}$ term is part of a total divergence:

$$\nabla \cdot (\mathbf{E} \times \mathbf{H}) = \mathbf{H} \cdot (\nabla \times \mathbf{E}) - \mathbf{E} \cdot (\nabla \times \mathbf{H}). \tag{85.2}$$

Hence,

$$\mathbf{j}_{\text{free}} \cdot \mathbf{E} = -\frac{c}{4\pi} \nabla \cdot (\mathbf{E} \times \mathbf{H}) - \frac{1}{4\pi} \mathbf{E} \cdot \frac{\partial \mathbf{D}}{\partial t} + \frac{c}{4\pi} \mathbf{H} \cdot (\nabla \times \mathbf{E}). \tag{85.3}$$

Using Faraday's law in the last term and rearranging, we get

$$-\nabla \cdot \frac{c}{4\pi} (\mathbf{E} \times \mathbf{H}) = \mathbf{j}_{\text{free}} \cdot \mathbf{E} + \frac{1}{4\pi} \left(\mathbf{E} \cdot \frac{\partial \mathbf{D}}{\partial t} + \mathbf{H} \cdot \frac{\partial \mathbf{B}}{\partial t} \right). \tag{85.4}$$

This is as far as one can go without invoking the properties of the medium or the constitutive relations. The term

$$\mathbf{S} = \frac{c}{4\pi} (\mathbf{E} \times \mathbf{H}) \tag{85.5}$$

is the generalization of the Poynting vector. It goes over into $\mathbf{E} \times \mathbf{B}$ in vacuum. The vectors $\mathbf{D} \times \mathbf{H}$ and $\mathbf{D} \times \mathbf{H}$ do that too, but the normal component of these quantities is discontinuous across the interface of a medium and a vacuum. If $\mathbf{S}$ is to be the energy density flux, $\mathbf{S} \cdot \hat{\mathbf{n}}$ must be continuous across such an interface.

The last term in eq. (85.4),

$$\frac{1}{4\pi} \left(\mathbf{E} \cdot \frac{\partial \mathbf{D}}{\partial t} + \mathbf{H} \cdot \frac{\partial \mathbf{B}}{\partial t} \right), \tag{85.6}$$

is necessarily equal to the rate of increase of energy *plus* the power evolved as heat. The key point is that, in general, one cannot obtain separate expressions for these two quantities. Thus, it cannot be written as the time derivative of any function of the fields alone, even for a linear dispersive medium, i.e., one for which $\epsilon(\omega)$ and $\mu(\omega)$ have significant dependence on frequency. If the medium is linear and dispersionless, i.e., if the dielectric constant and permittivity are constants, we can write it as

$$\frac{\partial}{\partial t}\left[\frac{1}{8\pi}(\epsilon \mathbf{E}^2 + \mu \mathbf{H}^2)\right],$$
(85.7)

and the quantity in square brackets can be interpreted as an energy density. However, it must always be remembered that this is an approximation. The formulas $\mathbf{D} = \epsilon \mathbf{E}$ and $\mathbf{B} = \mu \mathbf{H}$ presuppose that the fields have a time dependence that essentially excludes high frequencies. Approximate expressions for an energy can also be derived for nearly monochromatic fields in transparent media. This will be done in chapter 18.

14 Electrostatics around conductors

In this chapter we consider static electric fields around conducting bodies. This is the simplest type of problem in electromagnetism involving material media, as the macroscopic field does not penetrate into the media, and the question of constitutive relations essentially does not arise. The macroscopic response of a conductor to an applied field is confined to the surface and can be absorbed into the mathematical problem of solving Laplace's or Poisson's equation with appropriate boundary conditions on the conducting surfaces. Therefore, treatments of this subject traditionally tend to stress various methods of solving these equations in closed form, and the physical context becomes a surrogate for learning about the mathematics of linear partial differential equations, linear operators, Sturm-Liouville theory, and the "special functions" bearing the names of various giants of physical sciences dating back to the seventeenth century. Although we too shall follow this tradition, there is something illogical about it. The number of problems that can be solved with analytical methods is the proverbial set of measure zero, and, in many cases, the solutions appear in the form of series whose usefulness is not assured in advance. From this point of view, it would make more sense to simply give everybody a catalogue of these exactly solvable problems once and for all,[1] and spend much more time on numerical methods, which are the only methods of general applicability. However, one can learn much from the special cases that can be solved analytically, and the mathematical results are applicable in other areas of physics. We shall discuss the two simplest methods, that of images and of separating variables, in sections 89 and 90. Other methods, such as inversion and conformal mapping, are much too specialized, and are not discussed.[2] In sec. 91 we discuss the variational method, which is even less systematic, but which can provide quick approximate answers. Finally, in section 92, we discuss the numerical relaxation method.

---

[1] Such a catalogue would be rather thin, but the writer does not know of one. However, the books by Smythe (1968) and Jeans (1966) come close.

[2] For these see, e.g., Jeans (1966), chap. VIII.

The reader should not infer from the above comments and the title of this chapter that the electrostatics *inside* conductors is unworthy of study. Nonzero static electric fields can arise inside a conductor in the vicinity of impurity ions with a different valence than the host metal (Al in Cu, e.g.), and near metal surfaces. While it is true that in these two examples, the fields are limited to spatial regions of atomic dimensions, they can still have macroscopic manifestations. If two unlike metals are in contact, e.g., then there is a potential difference between them, known as the *contact potential*, of the order of 1 V. This potential gives rise to a macroscopic electric field outside them and can be measured with macroscopic instruments. We discuss such effects in section 93.

86 Electric fields inside conductors, and at conductor surfaces

The macroscopic Maxwell equations that determine the electric field are eqs. (81.9) and (81.16). In electrostatics we are concerned only with the fields produced by static charges. In particular, there are no steady currents, which would give rise to magnetic fields. For most metals, $\mathbf{B} = 0$ inside the conductor, and so $\mathbf{B} = 0$ everywhere. For ferromagnetic metals, such as iron and nickel, $\mathbf{B} \neq 0$ in the interior, and then because of the remaining Maxwell equations, $\mathbf{B} \neq 0$ in the exterior either.[3] For statics, however, $\mathbf{B}$ is independent of time, and so, even in these cases, the equations to be solved reduce to

$$\nabla \cdot \mathbf{E} = 4\pi\rho, \quad \nabla \times \mathbf{E} = 0. \tag{86.1}$$

We write Gauss's law in terms of $\mathbf{E}$ rather than $\mathbf{D}$ for reasons that will soon be clear. These equations hold both inside and outside a conductor. Outside the body, of course, the field $\mathbf{E}$ is essentially identical to the microscopic field $\mathbf{e}$, except in an atomically thick layer right next to it.

In a static situation, one cannot have any currents inside the conductor, since this would give rise to ohmic dissipation inside the conductor, as seen in the previous chapter. Since $\mathbf{j} = \sigma\mathbf{E}$, the electric field must vanish everywhere inside a conductor. This implies that $\nabla \cdot \mathbf{E}$ also vanishes in the conductor, except possibly at its surface. Hence,

$$\mathbf{E} = 0 \quad \text{and} \quad \rho = 0 \quad \text{inside a conductor.} \tag{86.2}$$

Therefore, any free charges in a conductor must lie entirely on its surface. If we consider a conductor in the presence of other charged bodies, it is these surface charges that must arrange themselves in such a way as to cancel the electric field produced by the charged bodies outside the conductor. This surface charge is directly related to the electric field $\mathbf{E}$ immediately outside the conductor by the following argument. Let us denote the surface charge per unit area by σ. Let us consider the flux of $\mathbf{E}$ through a small "pillbox"-like cylinder (see fig. 14.1), whose end caps, both of area $\mathcal{A}$, lie on either side of the conducting surface parallel to it, and whose height is of order $a \sim 10a_0$, where a_0 is the

---

[3] The problem of finding the steady magnetic field in these systems will be taken up in chapter 16.

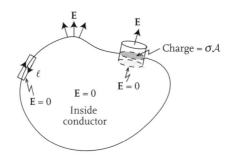

FIGURE 14.1. Diagram to show that $\mathbf{E} = 0$ inside a conductor and is normal to the surface right outside it, with $E_n = 4\pi\sigma$, where σ is the surface charge density.

Bohr radius. Only the end cap lying outside the conductor makes any contribution to the flux; denoting the component of $\mathbf{E}$ in the direction of the *outward* normal to the surface by E_n, this contribution is $E_n A$. The charge inside the pillbox is σA, and so by Gauss's law, we have

$$E_n = 4\pi\sigma. \tag{86.3}$$

The tangential component, $\mathbf{E}_t$, is similarly found by considering the circulation around a rectangular contour whose long sides (of length ℓ) lie on either side of the conductor parallel to it, and whose short sides are of order a (fig. 14.1). Only the long leg outside the conductor contributes to the circulation, the amount being $E_t \ell$. Since $\nabla \times \mathbf{E} = 0$, the circulation must be zero, and we conclude therefore that

$$\mathbf{E}_t = 0. \tag{86.4}$$

In other words, the electric field is normal to the surface of a conductor right next to it.[4]

A very interesting application of these ideas is to a *cavity* inside a solid conductor. If there is no net charge inside the cavity, then we can conclude that there can be no induced charges on the cavity surface. Suppose there are. We first argue by considering the flux of $\mathbf{E}$ through a Gaussian surface surrounding the cavity and lying inside the conducting material that there is no *net* charge on the cavity surface. If there are induced charges, there must be positively and negatively charged portions of the surface and, therefore, lines of $\mathbf{E}$ going from the former to the latter and terminating on the cavity. We now consider a loop that follows one of these lines and close the loop inside the conductor. The circulation of $\mathbf{E}$ on this loop is nonzero, contradicting the fact that $\nabla \times \mathbf{E} = 0$ everywhere. It follows that there is no electric field inside the cavity, no matter what fields and charges are present outside the conductor, and a test charge inside the cavity will feel no forces. This is the principle behind electrostatic screening, and tests of the validity of Coulomb's law based on looking for fields inside cavities.

---

[4] The location of the surface charge at a sharp surface of zero thickness and the abrupt jump in E_n at this surface are, of course, mathematical idealizations that are sensible only in the context of solving for the macroscopic fields. At the microscopic level, we expect the charges to be distributed in a thin layer of thickness $\sim a_0$, and the electric field to change rapidly from nearly zero inside to the appropriate value outside in a layer of similar thickness. This expectation is confirmed by model calculations.

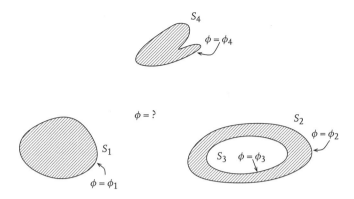

FIGURE 14.2. The type 1 electrostatic problem. The potential is specified on a number of conducting surfaces. What is the potential at a general point in space?

It is extremely useful to restate the boundary conditions on $\mathbf{E}$ in terms of the potential. We note first that since $\nabla \times \mathbf{E} = 0$ everywhere, we may indeed write $\mathbf{E} = -\nabla\phi$, as before. That $\mathbf{E} = 0$ inside a conductor then means that the potential is a constant in that region. In other words, it means that the entire body of a conductor is an *equipotential region*. Since $\mathbf{E}$ is finite at the surface, it follows that the potential is continuous and has the same value at the surface of the conductor as in the interior. The surface of a conductor is therefore an *equipotential surface*. This fact is consistent with the condition $\mathbf{E}_t = 0$. Next, the condition (86.3) on E_n becomes

$$\frac{\partial\phi}{\partial n} = -4\pi\sigma, \tag{86.5}$$

where $\partial\phi/\partial n$ denotes the gradient of ϕ in the direction of the outward normal. Lastly, in the space outside the conductors, the potential obeys Poisson's equation,

$$\nabla^2\phi = -4\pi\rho, \tag{86.6}$$

which is equivalent to eq. (86.1).

The typical electrostatic problem involving conductors is to determine the electric field everywhere for a given arrangement of conductors given either the potential (type 1 problem) or the total charge (type 2 problem) on each conductor (see fig. 14.2). We can formulate these mathematically as follows. In the region outside the conductors, with a specified source density ρ, the potential obeys Poisson's equation (86.6). In addition, it must obey boundary conditions at the conducting surfaces, which depend on the type of problem being solved. Suppose that we have a number of conducting surfaces S_a ($a = 1, 2, 3, \ldots$), disconnected from each other, and suppose that we have the first type of problem, in which the potential is specified on each surface:

$$\phi(\mathbf{r}) = \phi_a = \text{constant}, \quad \mathbf{r} \in S_a. \tag{86.7}$$

Note that if the conductor-free region is unbounded, the value of the potential (zero, for example) as $\mathbf{r} \to \infty$ can be specified by imagining one of the surfaces to be at infinity.

Physically, this condition corresponds to holding each surface at a certain fixed potential, which we may imagine as being achieved by connecting the surfaces by very thin wires to large reservoirs of charge that are very far away from the region of interest and that are maintained at fixed potentials relative to each other. Once ϕ is found everywhere, $\mathbf{E}$ may be found by differentiation. As a by-product of the calculation, one obtains from eq. (86.5) the charge density σ on each surface.

In the second type of problem, where we place a fixed total charge Q_a on each conductor, the surfaces S_a are still equipotentials, but the values ϕ_a are a priori unknown. However, we must have

$$\oint_{S_a} \frac{\partial \phi}{\partial n} d^2s = -4\pi Q_a. \tag{86.8}$$

At first sight, this corresponds to a rather different type of boundary condition on the potential than eq. (86.7). As we shall show in section 88, however, the potentials ϕ_a can be found from the charges Q_a, once we know certain quantities, known as *mutual capacitances*, that depend only on the spatial arrangement of the conductors. In this way, this problem can be transformed into a problem of the first type. We will develop methods for solving this problem in sections 89–92, after we have understood some general properties of the electrostatic field.

87 Theorems for electrostatic fields

Uniqueness theorem: This theorem is very similar to that proved in section 16, namely, that a vector field is uniquely determined by giving its curl and divergence everywhere. It states that given the source density ρ and the values of the potentials ϕ_a on the boundary surfaces, the solution to Poisson's equation is unique. To prove it, let us suppose the contrary and assume that there are two solutions $\phi^{(1)}$ and $\phi^{(2)}$ that meet the requirements, i.e.,

$$\nabla^2 \phi^{(i)} = -4\pi\rho, \quad \phi^{(i)} = \phi_a \text{ on } S_a, \quad i = 1, 2. \tag{87.1}$$

Now consider the function $\chi = \phi^{(1)} - \phi^{(2)}$. Clearly,

$$\nabla^2 \chi = 0, \tag{87.2}$$

$$\chi(\mathbf{r}) = 0, \quad \mathbf{r} \in S_a. \tag{87.3}$$

Let us denote by V the region free of the conductors, and denote by S its boundary. This is the formal sum of the conducting surfaces (including one at infinity if needed), $\sum_a S_a$. We now note that

$$\nabla \cdot (\chi \nabla \chi) = \nabla \chi \cdot \nabla \chi + \chi \nabla^2 \chi$$
$$= |\nabla \chi|^2,$$

using eq. (87.2). Integrating both sides of this equation over all V, using Gauss's theorem, and noting that the outward normal to V is in the opposite direction from the normal used to define the quantities E_n and $\partial\phi/\partial n$, we get

$$\int_V |\nabla\chi|^2 \, d^2x = -\oint_S \chi \frac{\partial\chi}{\partial n} d^2s. \tag{87.4}$$

But since from eq. (87.3), χ vanishes everywhere on S, the right-hand side vanishes. Since the integrand of the volume integral, $|\nabla\chi|^2$, can never be negative, it follows that $\nabla\chi = 0$ everywhere inside V, i.e., χ is a constant whose value must be zero because of the boundary condition eq. (87.3). Thus, $\phi^{(1)} = \phi^{(2)}$.

More general uniqueness theorem; Dirichlet and Neumann problems: The reader may have noticed that we have actually established a stronger result than we set out to prove, namely, the uniqueness of the solution to Poisson's equation under more general boundary conditions. For the χ in eq. (87.4) to vanish everywhere on S, it is not necessary that the potential be constant on each of the surfaces S_a; we could specify it to be *any* function. The proof is easy: the difference χ between two supposedly different solutions will again obey eqs. (87.2) and (87.3), and the rest of the argument will follow. Thus, the solution to the problem

$$\nabla^2\phi = -4\pi\rho, \quad \phi(\mathbf{r}) \text{ given on } S \quad \text{(Dirichlet)} \tag{87.5}$$

is also unique. This type of boundary condition is known as a *Dirichlet boundary condition*. Equation (86.7) is a special case of this.

In fact, the class of allowed boundary conditions can be extended still further. All that is needed to show uniqueness is that the right-hand side of eq. (87.4) should vanish. Clearly, this happens if $\chi = 0$ on S, but it also happens if $\partial\chi/\partial n = 0$ on S. This corresponds to specifying the normal derivative of the potential on the boundaries. In other words, the solution to the problem

$$\nabla^2\phi = -4\pi\rho, \quad \frac{\partial\phi}{\partial n} \text{ given on } S \quad \text{(Neumann)} \tag{87.6}$$

is also unique (up to an inconsequential additive constant). This boundary condition is known as a *Neumann boundary condition*.

Two more results follow relatively quickly. First, we could have mixed boundary conditions, i.e., Dirichlet on part of S and Neumann on the rest. Second, in general, we may *not* specify both ϕ and $\partial\phi/\partial n$ on any part of the boundary (even though this makes the right-hand side of eq. (87.4) vanish), for giving either ϕ or $\partial\phi/\partial n$ determines the solution uniquely; giving both *overdetermines* it. It is too much to hope that one would be so lucky as to pick values for both quantities that were consistent with each other.

The reader may also have noticed that in contrast to section 13, where we found an explicit solution to Poisson's equation in an unbounded region (see eq. (13.5)), here we have not done so. We have only shown uniqueness if a solution exists, but we have not shown that a solution exists in the first place. A proof of existence and a formal expression for the solution to the Dirichlet and Neumann problems can be constructed using Green functions,[5] but we shall skip over this, since in the specialized setups where a Green function can be found in closed or semiclosed form, the problem can usually be solved directly. Instead, we shall use physical arguments to show that a solution exists. This is done in the next section. In the same vein, we shall also ignore the question of what types of physical problems correspond to the more general boundary conditions. For now, let us simply take the mathematical results as bonuses at little extra cost.

The mean value theorem; equilibrium in electrostatic fields: Let us recall the mean value theorem (also known as *Earnshaw's theorem*) proved in section 19, namely, that in charge-free regions, the mean value of the potential over any spherical surface is equal to its value at the center of that sphere. The theorem continues to hold in the presence of conductors, and it shows that in charge-free regions, the potential cannot have a minimum or a maximum. If it did, then its mean value over a tiny sphere surrounding the supposed extremum would exceed or be exceeded by its value at the center. The theorem has several physical consequences. First, no configuration of charges can be in equilibrium under purely electrostatic forces. Second, a test charge cannot be in stable equilibrium in an electrostatic field in an otherwise charge-free region. Points of equilibrium *can* be found in charge-filled regions, for example, the center of a uniformly charged sphere. In the presence of boundary surfaces, equilibrium can be found at points on the surface.

Exercise 87.1 An interesting variation on the question of equilibrium arises in the presence of conductors. If we displace a charge in the space surrounding the conductors, is it possible that the induced charges on the conducting surfaces respond in such a way that the entire charge distribution tends to return to its original configuration? In other words, can an arrangement of charges and fixed conductors be in electrostatic equilibrium? Show that the answer to this question is also no.

88 Electrostatic energy with conductors; capacitance

Let us consider a system of conductors and suppose there are no charges in the space around them. Each conductor has a charge Q_a and is at a potential ϕ_a. For a system of point charges, the potential energy is given by $\frac{1}{2}\sum_a q_a \phi_a$. In the present case, all the charges at the same conducting surface are at the same potential, so we can add them together and write the potential energy as

$$U = \frac{1}{2}\sum_a Q_a \phi_a. \tag{88.1}$$

Note that the index a now labels the different conductors.

It is clear that the charges and the potentials cannot be independent, for if we add a positive charge to one of the conductors, the potential of all the other conductors goes up.

---

[5] See Jackson (1999), sec. 1.10.

To see what the relationship is, let us imagine changing the surface potentials by an amount $\delta\phi_a$. The original potentials ϕ_a give rise to the space potential $\phi(\mathbf{r})$. Since $\phi(\mathbf{r})$ already obeys Laplace's equation with the boundary values ϕ_a, the extra potential $\delta\phi(r)$ can be found by solving Laplace's equation with the boundary values $\delta\phi_a$. Thus, $\phi(\mathbf{r})$ is a linear functional of the boundary values ϕ_a. In other words,

$$\phi(\mathbf{r}) = \sum_a \phi_a F_a(\mathbf{r}), \tag{88.2}$$

where the functions F_a do not depend on the ϕ_a's but only on the geometry of the conductors. Since the charge Q_a is proportional to the integral of the normal derivative of $\phi(\mathbf{r})$ over the surface S_a, it follows that the charges are also linearly related to the potentials. We write this as

$$Q_a = \sum_b C_{ab}\phi_b, \quad \text{(Gaussian and SI)} \tag{88.3}$$

where the quantities C_{aa} are called *capacities* or *capacitances*, and the quantities C_{ab} $(a \neq b)$ are called *coefficients of electrostatic induction*.[6] Note that every coefficient depends on the geometry of the entire collection of conductors. If we consider C_{ab} as the elements of a matrix and denote the elements of the inverse matrix by C_{ab}^{-1} (note in particular that $C_{ab}^{-1} \neq 1/C_{ab}$), then we can invert eq. (88.3) and write

$$\phi_a = \sum_b C_{ab}^{-1} Q_b. \tag{88.4}$$

The quantities C_{ab} have dimensions of length and units of centimeters in the Gaussian system and, as we shall see, are equal in magnitude to the linear size of the bodies involved. In the SI system, the unit of capacitance is the *farad*, which is numerically equal to one coulomb per volt. A farad is an extraordinarily large unit, and one more commonly deals with picofarads (pf) or microfarads (μf). The conversion between the two systems is

$$1 \text{ farad (SI)} \equiv 9 \times 10^{11} \text{ cm (Gaussian).} \tag{88.5}$$

To convert a formula for capacitance in the Gaussian system to the SI system, one multiplies by $4\pi\epsilon_0$, i.e.,

$$C \text{ (SI)} = 4\pi\epsilon_0 C \text{ (Gaussian).} \tag{88.6}$$

This follows from eq. (88.3) and the conversions for charge and potential.

The words *capacity* and *capacitance* have specialized meanings in two contexts. If only one conductor is present at potential ϕ (relative to points infinitely far from the conductor), and with a total charge Q, we have $Q = C\phi$, and the single coefficient C is called the *capacity* of the body. If two conductors are present, with equal and opposite charges $\pm Q$, and the potential difference between them is V, one writes $Q = CV$, and the coefficient C is called the *capacitance*. Such a pair of conductors is called a *capacitor*.

The next few exercises establish some basic results about capacitances.

---

[6] We encountered similar relations when dealing with systems of current loops in magnetostatics in section 31 and defined the coefficients of self and mutual inductance without any additional "magnetostatic" qualifier.

Exercise 88.1 Find the capacity of a conducting sphere of radius a. [**Answer**: a (Gaussian), $4\pi\epsilon_0 a$ (SI).]

Exercise 88.2 Find the capacitance of a capacitor in terms of the coefficients C_{ab}. Assume that $C_{12} = C_{21}$. (We will prove this result below.)

$$C = \left[(C^{-1})_{11} + (C^{-1})_{22} - 2(C^{-1})_{12}\right]^{-1}$$
$$= (C_{11}C_{22} - C_{12}^2)/(C_{11} + C_{22} + 2C_{12}). \tag{88.7}$$

In many cases, such as those explored in the next three exercises, it is simpler to find the capacitance directly.

Exercise 88.3 Find the capacitance of a capacitor formed by two large parallel plates of area A separated by a distance d that is much smaller than the linear dimensions of the plates.

Solution: It is clear that far from the edges of the plates, the electric field $\mathbf{E}$ points normal to the plates and is uniform. Gauss's law gives $E \simeq 4\pi\sigma$, where $\sigma = Q/A$ is the surface charge density. The potential difference V is the work done in transporting a unit test charge between the plates, which equals Ed. Hence, $C \simeq A/4\pi d$ ($\epsilon_0 A/d$ in SI).

Exercise 88.4 (coaxial cable) Find the capacitance per unit length of two infinitely long coaxial cylinders of radii r_1 and $r_2 > r_1$. [**Answer** (in Gaussian): $1/(2\ln(r_2/r_1))$. Note that if $r_2 - r_1 \equiv d \ll r_1$, this becomes approximately $r_1/2d$, consistent with the previous exercise.]

Exercise 88.5 Find the capacitance of two concentric spheres of radii r_1 and $r_2 > r_1$. [**Answer**: $r_1 r_2/(r_2 - r_1)$ (Gaussian), $4\pi\epsilon_0 r_1 r_2/(r_2 - r_1)$ (SI).]

We now show that the capacitance matrix is symmetric, i.e., $C_{ab} = C_{ba}$. To do this, we first consider a system of point charges, q_1, q_2, etc., located at positions $\mathbf{r}_1$, $\mathbf{r}_2$, etc. Let the potential at the position of the ath point charge in this configuration be ϕ_a. Now, consider a second system q_1', q_2', etc. located at the *same* positions, and let the potentials in this case be ϕ_a'. Then,

$$\sum_a q_a \phi_a' = \sum_a q_a' \phi_a \tag{88.8}$$

since both sums are equal to

$$\sum_{a,b\neq a} \frac{q_a q_b'}{|\mathbf{r}_a - \mathbf{r}_b|}. \tag{88.9}$$

If we apply this result to our system of conductors, then we may add up all the charges at the same potential first. Thus, with the index a now referring to the various conducting surfaces, we deduce that if charges Q_a on the conductors produce potentials ϕ_a, and charges Q_a' produce potentials ϕ_a', then

$$\sum_a Q_a \phi_a' = \sum_a Q_a' \phi_a. \tag{88.10}$$

This is a form of *Green's reciprocation theorem*. If we now use eq. (88.3), we obtain

$$\sum_{a,b} C_{ab} \phi_b \phi_a' = \sum_{a,b} C_{ab} \phi_b' \phi_a. \tag{88.11}$$

Equating the coefficients of $\phi_b \phi_a'$ on both sides, we conclude that

$$C_{ab} = C_{ba}. \tag{88.12}$$

The total energy U can be written entirely in terms of the charges or the potentials using the above results:

$$U = \frac{1}{2} \sum_{a,b} C_{ab} \phi_a \phi_b = \frac{1}{2} \sum_{a,b} C_{ab}^{-1} Q_a Q_b. \tag{88.13}$$

Since U can also be written as the volume integral of $\mathbf{E}^2$, it is positive, and so the matrices formed by C_{ab} and C_{ab}^{-1} must both be positive definite. In particular, $C_{aa} > 0$ and $C_{aa}^{-1} > 0$. We have already argued that for $a \neq b$, $C_{ab}^{-1} > 0$, and one can similarly show that $C_{ab} < 0$.

The calculation of the capacitance matrix amounts to solving a series of Dirichlet problems. To find C_{12}, e.g., we solve for the potential $\phi(\mathbf{r})$, putting $\phi_2 = 1$, and $\phi_a = 0$ for all $a \neq 2$. The charge on conductor 1, which can be found by integrating $\partial \phi / \partial n$ over its surface, gives C_{12} directly.

We can now address the question raised in section 86: finding the potentials when the charges on each conductor are specified. Let us suppose first that there are no volume charges outside the conductors. We can immediately reduce the problem to the type we have solved before, i.e., one in which the potentials are specified on the conducting surfaces, by using eq. (88.4). Since finding the capacitance matrix also requires solving Dirichlet problems, it is plausible that both steps can be combined into one.

If there is a volume charge $\rho(\mathbf{r})$ outside the conductors, we approach the problem in two stages. In stage one, we solve the Dirichlet problem with this source density, but with the boundary condition that all the surfaces are at the same potential, zero. In other words, if we denote the solution to this problem by $\phi^{\mathrm{I}}(\mathbf{r})$, then

$$\nabla^2 \phi^{\mathrm{I}}(\mathbf{r}) = -4\pi\rho, \quad \phi^{\mathrm{I}}(\mathbf{r}) = 0, \mathbf{r} \in S_a. \tag{88.14}$$

Knowing ϕ^{I}, we can find the corresponding charges induced on each conductor. Let us denote these by Q_a^{I}. The problem now is to add the deficits $Q_a^{\mathrm{II}} = Q_a - Q_a^{\mathrm{I}}$ to each conductor. Since the volume charge is already balanced by the charges Q^{I}, it is necessary only for the deficit charges to balance each other. Thus, the problem now is of the type solved in the previous paragraph: finding ϕ given the surface charges and no volume charge. To be explicit, we calculate the surface potentials ϕ_a^{II} according to

$$\phi_a^{\mathrm{II}} = \sum_b C_{ab}^{-1} Q_b^{\mathrm{II}} \tag{88.15}$$

and then solve Laplace's equation with these boundary values:

$$\nabla^2 \phi^{II}(\mathbf{r}) = 0, \quad \phi^{II}(\mathbf{r}) = \phi_a^{II}, \mathbf{r} \in S_a. \tag{88.16}$$

The complete solution to the original problem is given by superposition: $\phi(r) = \phi^I(\mathbf{r}) + \phi^{II}(\mathbf{r})$.

We conclude this section with some exercises that bear on one class of tests of the inverse-square character of Coulomb's law. The tests are based on looking for potential differences between conductors enclosed inside other conductors or for a charge on the inner conductor when it is connected to the outer one by thin wires. The first such test was performed by Cavendish in 1772 and has been continually improved by many others since: Maxwell circa 1870, Plimpton and Lawton in 1936,[7] and Williams, Faller, and Hill in 1971.[8] The results are usually quoted in terms of limits on the parameter η in an assumed force law of the form $r^{-2-\eta}$, or the parameter μ in an assumed potential law of the form $e^{-\mu r}/r$. This latter form is especially attractive theoretically, as it fits into a Lagrangian scheme.[9] In this context, the quantity $\hbar\mu/c$ is known as the *photon mass*. The best present-day limits on η and μ are based on still other arguments and experiments, but a review from 1971 is still definitive.[10]

Exercise 88.6 Consider a conductor enclosed entirely inside another one. Number the surface of the innermost conductor 1, the interior and exterior surfaces of the enclosing conductor 2 and 3, respectively, and other conducting surfaces that may be present 4, 5, 6, etc. Show that

$$C_{13} = C_{14} = \cdots = 0. \tag{88.17}$$

That is, if the inner conductor is initially uncharged, no charge is induced on it if the other conductors are subsequently raised to any potentials whatsoever.

Exercise 88.7 If a conductor is enclosed inside another conductor, and the two are connected to each other by wires, an inverse-square force implies that there is no charge on the inner conductor, no matter what the shapes of the conductors. How much charge would we find if the force were not inverse square? The requisite calculation is difficult for any but spherical shapes. Consider, therefore, two conducting spherical shells of radii a and b, with $b > a$, connected together by a very thin wire. Let a charge Q be placed on the outer shell. Find the charge q on the inner shell assuming that the potential due to a point charge varies as (a) $r^{-1-\eta}$ with $\eta \ll 1$, (b) $e^{-\mu r}/r$ with $\mu > 0$. Give your answers to leading nonzero order as $\eta \to 0$ or $\mu \to 0$. (It pays to first find the potential due to a spherical shell for an arbitrary force law and to write the corresponding point charge potential in the form $r^{-1}df/dr$, where $f(r)$ is arbitrary.)

---

[7] S. J. Plimpton and W. E. Lawson, *Phys. Rev.* **50**, 1066 (1936).
[8] E. R. Williams, J. E. Faller, and H. A. Hill, *Phys. Rev. Lett.* **26**, 721 (1971). It should be noted that this is not a strict dc experiment.
[9] See exercise 162.3.
[10] A. S. Goldhaber and M. M. Nieto, *Rev. Mod. Phys.* **43**, 277 (1971).

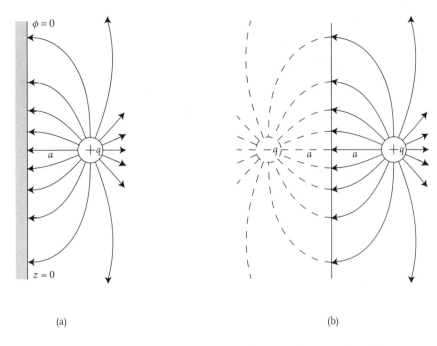

(a)   (b)

FIGURE 14.3. The electric field for a charge near a planar conducting surface (a) is obtained by restricting to the region outside the conductor, the field due to the original charge $+q$ plus an image charge $-q$ at the mirror image location inside the conductor (b).

Answer:

(a) : $\qquad \dfrac{q}{Q} = -\dfrac{\eta}{2(b-a)}\left[2a\ln(2b) - (b+a)\ln(b+a) + (b-a)\ln(b-a)\right].$ (88.18)

(b) : $\qquad \dfrac{q}{Q} = \dfrac{\mu^2}{6}a(b+a).$ (88.19)

It is a general rule that all observable effects are proportional to μ^2 as $\mu \to 0$.

Exercise 88.8 Reconsider the shells of the previous exercise, but do not take them to be connected. Instead, find the potential difference between them if a charge Q is placed on the outer shell.

89 The method of images

We now turn to solving problems in electrostatics. As stated earlier, the only methods of general applicability are numerical, and analytical methods work only in specialized situations. The simplest of these is the method of images.

Point charge near an infinite conducting plane: Consider a point charge in the presence of an infinite plane conducting surface (fig. 14.3). We take the conductor as occupying the half-space $z < 0$, and the point charge $+q$ at $(0, 0, a)$. The plane $z = 0$ is an equipotential, and the goal is to solve Poisson's equation for $z > 0$ subject to this boundary condition. We now observe that if we consider *two* charges, the original one plus another charge $-q$ at $(0, 0, -a)$, and remove the conductor, then the surface $z = 0$ is again an equipotential. By the uniqueness theorem proved in section 16, the solution for ϕ to the second problem must coincide with that for the first problem in the region $z > 0$ (apart from an additive constant). The solution in the region $z < 0$ is obviously not relevant to the original problem and can be discarded. The fictitious charge at $(0, 0, -a)$ is known as an *image* charge.

The potential in the region $z > 0$ is thus found to be

$$\phi(\mathbf{r}) = \frac{q}{|\mathbf{r} - a\hat{\mathbf{z}}|} - \frac{q}{|\mathbf{r} + a\hat{\mathbf{z}}|}. \tag{89.1}$$

We can use this to find several interesting results. First, the charge density on the conducting surface is

$$\sigma(x, y) = -\frac{1}{4\pi} \frac{\partial \phi}{\partial z}\bigg|_{z=0} = -\frac{q}{2\pi} \frac{a}{(x^2 + y^2 + a^2)^{3/2}}. \tag{89.2}$$

The total induced charge on the conductor is therefore

$$Q_{\text{ind}} = \iint \sigma(x, y) \, dx \, dy = -q, \tag{89.3}$$

as is easily shown. Using plane polar coordinates r and θ in the plane, the z component of the force on the point charge due to this charge distribution is seen to be

$$F_z = q \int_0^\infty \frac{\sigma(r)}{(r^2 + a^2)} \frac{a}{(r^2 + a^2)^{1/2}} 2\pi r \, dr = -\frac{q^2}{4a^2}, \tag{89.4}$$

exactly as we would (and should) find if we considered the force due to the image charge alone.

Point charge near a conducting sphere: The above example illustrates the general idea behind this method completely. We try and replace the conducting surface with fictitious charge(s) so that along with the true charges, the resulting equipotential surfaces are such that one of them coincides with the original conducting surface. For a general arrangement of conductors, this is, of course, a losing strategy, but one other successful application is to a point charge outside a spherical conductor. We first consider the conductor to be grounded, i.e., at zero potential, which is also the potential at infinity. We take the sphere to be of radius a, and the point charge to be at a distance $s > a$ from its center (see fig. 14.4). It is clear by symmetry that any image charge must lie on the same ray as this charge. Further, it must lie inside the sphere, since otherwise $\nabla^2 \phi$ would be singular at another point outside the sphere besides the true charge location. Denoting the image charge by q' and its distance from the center by s', and defining vectors as shown in the figure, the potential at a point $\mathbf{r}$ is given by

$$\phi(\mathbf{r}) = \frac{q'}{|\mathbf{r} - \mathbf{s}'|} + \frac{q}{|\mathbf{r} - \mathbf{s}|}. \tag{89.5}$$

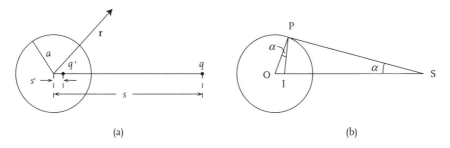

FIGURE 14.4. Method of images for a charge near a conducting sphere: (a) charges and locations, (b) geometrical construction.

Let us denote a general point on the surface of the sphere by $\mathbf{a}$. If this surface is to be an equipotential with $\phi = 0$, the ratio of the distances $|\mathbf{a} - \mathbf{s}'|$ and $|\mathbf{a} - \mathbf{s}|$ must be a constant $(= -q'/q)$ independent of the orientation $\hat{\mathbf{a}}$. To see that this is possible, refer to fig. 14.4b. Choose any point P on the surface of the sphere. Then choose point I so that the angles OPI and OSP are equal. Then the triangles OPI and OSP are similar, and IP/SP = OP/OS. But the latter ratio is a/s, which is a constant irrespective of where on the sphere the point P is chosen. Hence, the point I is uniquely fixed and exactly where the image charge should be placed. Choosing P on the line between I and S, we get $(a - s')/(s - a) = a/s$, which implies, firstly, that

$$ss' = a^2, \tag{89.6}$$

and, secondly, that, for a general point $\mathbf{a}$ on the sphere,

$$|\mathbf{a} - \mathbf{s}'| = \frac{a}{s}|\mathbf{a} - \mathbf{s}|. \tag{89.7}$$

It now follows that for $\phi(\mathbf{a})$ to vanish, q'/q must equal $-a/s$, i.e.,

$$q' = -q\frac{a}{s}. \tag{89.8}$$

The field in the region $r > a$ can now be written down immediately:

$$\mathbf{E}(\mathbf{r}) = \frac{q'(\mathbf{r} - \mathbf{s}')}{|\mathbf{r} - \mathbf{s}'|^3} + \frac{q(\mathbf{r} - \mathbf{s})}{|\mathbf{r} - \mathbf{s}|^3}. \tag{89.9}$$

In particular, we can find the electric field on the surface $r = a$, and the induced charge density $\sigma(\mathbf{a}) = \hat{\mathbf{a}} \cdot \mathbf{E}(\mathbf{a})/4\pi$. Using eq. (89.8), we find

$$\sigma(\mathbf{a}) = \frac{q}{4\pi a^2}\frac{a(a^2 - s^2)}{|\mathbf{a} - \mathbf{s}|^3}. \tag{89.10}$$

Note that this has the opposite sign to q everywhere on the surface. Writing θ for the angle between $\mathbf{a}$ and $\mathbf{s}$ and integrating over the entire surface, we find that the total induced charge is

$$Q_{\text{ind}} = \frac{q}{4\pi a^2}\int_0^\pi \frac{a(a^2 - s^2)}{(a^2 + s^2 - 2as\cos\theta)^{3/2}}2\pi\sin\theta\, d\theta$$

$$= -q\frac{a}{s} = q', \tag{89.11}$$

equal to the image charge. This is entirely expected, since the flux of **E** through a surface just surrounding the conductor can be calculated for the equivalent problem of the original plus image charge and is then $4\pi q'$ by Gauss's law.

The force on the charge q due to the induced surface charge is equal to that exerted by the image charge,

$$\mathbf{F} = -q^2 \frac{as}{(s^2 - a^2)^2} \hat{\mathbf{s}}. \tag{89.12}$$

The potential energy can be found by integrating this equation with respect to s,

$$U = -\frac{q^2 a}{2(s^2 - a^2)}. \tag{89.13}$$

The solution to the above problem is immediately generalized to two other problems. We note that if we add a second image charge at the center of the sphere, the surface of the sphere remains an equipotential, and the additional electric field satisfies Laplace's equation everywhere outside it. By choosing this charge to be $\phi_0 a$, we solve the problem where the sphere is held at a specified potential ϕ_0 different from zero, and by choosing this charge to be $Q - q'$, we solve the problem where the sphere is specified to have a total charge Q.

Exercise 89.1 Show that by interchanging the source and the image charges, we solve the problem where a point charge is placed *inside* a spherical conducting shell.

Exercise 89.2 Develop the method of images for a uniform line charge parallel to the axis of a conducting cylinder.

Exercise 89.3 (capacitance of two parallel cylinders)   Consider the field created by two parallel line charges of equal and opposite charge per unit length and show that the equipotential surfaces are (noncoaxial) cylinders. Use this fact to find the capacitance of two infinitely long conducting parallel cylinders (wires perhaps) of radii a and b, whose axes are separated by a distance d.

Solution: This problem crops up in a number of other situations, so we give the solution, even though it is not difficult to obtain. Consider two line charges along the lines $y = 0$, $x = \pm s$, with charges $\lambda_\pm = \pm 1/2$ per unit length, as shown in fig. 14.5. The potential at a point (x, y) due to a single line is $\pm \ln r_\pm$, so the total potential is

$$\phi = -\frac{1}{2} \ln \left[\frac{(x-s)^2 + y^2}{(x+s)^2 + y^2} \right]. \tag{89.14}$$

If we exponentiate both sides, we obtain a second-degree equation in x and y:

$$e^\phi \left[(x-s)^2 + y^2\right] = e^{-\phi} \left[(x+s)^2 + y^2\right]. \tag{89.15}$$

Expressing e^ϕ and $e^{-\phi}$ in terms of hyperbolic functions and rearranging terms, we get

$$(x - s \coth \phi)^2 + y^2 = s^2 \mathrm{cosech}^2 \phi. \tag{89.16}$$

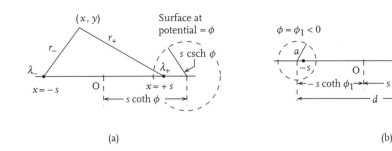

FIGURE 14.5. Method of images for finding the capacitance of two parallel cylinders: (a) The equipotential surfaces of two wires carrying charges $\lambda_\pm = \pm 1/2$ per unit length are cylinders; (b) diagram to accompany eq. (89.18), showing the distances s, a, b, and d, for which two equipotential surfaces coincide with the surfaces of the cylinders forming the capacitor.

This is the equation of a circle with center at $(s \coth \phi, 0)$ and radius $s|\operatorname{cosech}\phi|$. Note that $\phi = 0$ at $(0, 0)$, and that if $\phi > 0$, the center is to the right and at a distance greater than the radius.

Let us now choose the surfaces of the wires to be equipotentials, with values ϕ_0 and ϕ_1, as shown in fig. 14.5b. Since the radii of the equipotential circles are b and a, and the center-to-center separation is d, we have

$$s(\coth \phi_0 - \coth \phi_1) = d, \tag{89.17}$$

$$s \operatorname{cosech}\phi_0 = b, \tag{89.18}$$

$$s \operatorname{cosech}\phi_1 = a. \tag{89.19}$$

These three equations must be solved for s, ϕ_0, and ϕ_1 in terms of a, b, and d. We first eliminate s by multiplying the last two equations by $(\coth \phi_0 - \coth \phi_1)$ and using the first. This yields

$$d \sinh \phi_1 = b \sinh(\phi_1 - \phi_0), \tag{89.20}$$

$$d \sinh \phi_0 = -a \sinh(\phi_1 - \phi_0). \tag{89.21}$$

We now rewrite these equations in terms of the variables $V = (\phi_0 - \phi_1)/2$, $A = (\phi_0 + \phi_1)/2$, and solve for $\sinh A$ and $\cosh A$ to obtain

$$d \sinh A = (a - b) \sinh V, \quad d \cosh A = (a + b) \cosh V. \tag{89.22}$$

By squaring and subtracting, we eliminate A, and obtain

$$d^2 = (a^2 + b^2) + 2ab \cosh(2V). \tag{89.23}$$

This is the equation we want. Since $2V$ is the potential difference between the two wires, and since the charge per unit length is $1/2$, the capacitance per unit length is $1/2(2V)$, i.e.,

$$\frac{1}{C} = 2 \cosh^{-1} \left[\frac{d^2 - (a^2 + b^2)}{2ab} \right]. \tag{89.24}$$

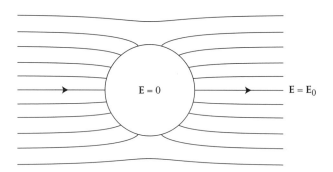

FIGURE 14.6. Electric field lines for a conducting sphere in a uniform electric field.

When $a = b$, we can also write $1/C = 4\cosh^{-1}(d/2a)$. If $d \gg a$, b, we have $C \approx 1/2\ln(d^2/ab)$.[11]

Exercise 89.4 Consider a uniform line charge of length $2d$. Show that the equipotential surfaces are prolate spheroids (a figure made by rotating an ellipse about its major axis). Use this result to find the capacity of a prolate spheroid with major and minor axes $2a$ and $2b$. The algebra is simplified by working in terms of variables $u = (r_1 + r_2)/2$ and $v = (r_1 - r_2)/2$, where r_1 and r_2 are distances from the ends of the line charge to a field point.

Conducting sphere in uniform external electric field; polarizability: A third interesting problem that involves image charges in the form of a dipole is that of a conducting sphere of radius a in a uniform field $\mathbf{E}_0$ (see fig. 14.6). Let us take the field to be along the z axis, and the origin of our coordinate system at the center of the sphere. The potential due to this field and a point dipole $\mathbf{d}\|\hat{\mathbf{z}}$ at the origin is given by

$$\phi(\mathbf{r}) = \frac{dz}{r^3} - E_0 z. \tag{89.25}$$

The surface $r = a$ will be an equipotential (with $\phi = 0$), if $d = E_0 a^3$. It follows immediately that the potential for our original problem is given by

$$\phi(\mathbf{r}) = -\left(1 - \frac{a^3}{r^3}\right)\mathbf{E}_0 \cdot \mathbf{r} = -\left(1 - \frac{a^3}{r^3}\right)E_0 r \cos\theta, \tag{89.26}$$

where θ is the angle between $\mathbf{E}_0$ and $\mathbf{r}$, or the usual polar angle in spherical polar coordinates. We thus see that the induced charges on the sphere produce an additional field that is exactly that of a point dipole of moment

$$\mathcal{P} = a^3 \mathbf{E}_0. \tag{89.27}$$

The net electric field at the surface is easily found to be $3(\mathbf{E}_0 \cdot \hat{\mathbf{r}})\hat{\mathbf{r}}$, from which it follows that the surface charge distribution is

$$\sigma = \frac{3E_0}{4\pi}\cos\theta. \tag{89.28}$$

---

[11] C is not precisely the same as c^2/L for two parallel wires (exercise 31.5) because the charge distribution is not uniform over the surface of the wires.

This example illustrates a more general phenomenon. If we place any uncharged insulated conductor in a uniform external electric field $\mathbf{E}_0$, the field due to the induced charges may be expanded in terms of multipoles. Since the net charge on the conductor must remain zero, this expansion will begin at the dipole term. The induced dipole moment, $\mathcal{P}$, will clearly be linear in the electric field because of the linearity of Laplace's equation. For the sphere, we found $\mathcal{P} = a^3 \mathbf{E}_0$, but, in general, $\mathcal{P}$ and $\mathbf{E}_0$ will not be parallel, and we must write

$$\mathcal{P}_i = \alpha_{ij} E_{0,j}, \tag{89.29}$$

$$\mathcal{P}_i = \alpha_{ij} \epsilon_0 E_{0,j}, \quad \text{(SI)}$$

where the quantities α_{ij} form a tensor known as the *polarizability tensor*. This tensor depends only on the geometry (shape and size) of the conductor. It has dimensions of volume in both SI and Gaussian systems, although its value in SI is 4π times its value in Gaussian:

$$\alpha \text{ (SI)} = 4\pi\alpha \text{ (Gaussian)}. \tag{89.30}$$

The energy of the conductor in the external field may be written in the approximation of keeping only the induced dipole moment as

$$U = -\frac{1}{2}\mathcal{P} \cdot \mathbf{E}_0 = -\frac{1}{2}\alpha_{ij} E_{0,i} E_{0,j}. \tag{89.31}$$

This may be derived by calculating the work done in increasing the electric field from zero to its final value. In this way we see that the energy is not $-\mathcal{P} \cdot \mathbf{E}_0$, but *half* that value.

The concept of polarizability is not limited to conductors and is of central importance in understanding the behavior of insulators or dielectrics. Any physical system that retains its basic integrity but in which electric charges can undergo limited displacement in response to an externally applied electric field is polarizable. We thus speak of *atomic* and *molecular polarizabilities*. The polarizability of the hydrogen atom, e.g., is $(9/2)a_0^3$ (Gaussian), where a_0 is the Bohr radius.[12] We will study this point further in section 96.

Exercise 89.5 Verify that eq. (89.31) is true for the conducting sphere in a uniform field in the following way. Consider a point charge q at a distance $r > a$ from the center of the sphere, and use the image solution to determine the energy of interaction between this charge and the induced surface charges on the sphere. As $r \to \infty$, the field at the sphere is essentially uniform, so the interaction energy should tend to $-\alpha E_0^2/2$, with $E_0 = q/r^2$.

Exercise 89.6 It is a general rule that the polarizability of a conductor is proportional to its volume. As an example that shows this, consider a cylinder of base radius a and height h in an electric field parallel to its long axis. Next, consider another cylinder with dimensions λa and λh, where λ is an arbitrary constant. Use scaling arguments and asymptotic behavior to relate the potentials for these two problems and thus deduce the proportionality of polarizability to volume.

---

[12] See, e.g., Shankar (1980), sec. 17.2. This result is sensible provided the electric field is not so large as to significantly alter the ground state, i.e., provided $E_0 \ll e/a_0^2$.

Exercise 89.7 Repeat the previous exercise for a conductor of general shape. Some readers may find it easier to do just this exercise.

90 Separation of variables and expansions in basis sets

A second method for solving potential problems is that of separation of variables in a suitable system of orthogonal coordinates. This technique is useful when the boundary conditions can be specified in simple form on one or more of the coordinate surfaces and is therefore largely limited to problems with high symmetry. It is closely related to the method of expanding the potential in a complete set of basis functions suited to the symmetry. We consider the most commonly used coordinate systems and illustrate the ideas with several examples.

Spherical polar coordinates: In spherical polar coordinates, Laplace's equation takes the form

$$\frac{1}{r}\frac{\partial^2}{\partial r^2}(r\phi) - \frac{1}{r^2}\mathbf{L}_{op}^2\phi = 0. \tag{90.1}$$

We have written the angular part of the Laplacian in terms of the angular momentum operator introduced in chapter 2. It is this writer's experience that it is preferable to do this if the range of angles or orientations is unrestricted and that the explicit differential form for $\mathbf{L}_{op}^2$ in terms of spherical polar coordinates[13] θ and φ is necessary only if this range is restricted. The latter situation is far less common, so we shall not consider it any further.[14] If we now seek a product form for the solution,

$$\phi(\mathbf{r}) = F(r)G(\hat{\mathbf{r}}), \tag{90.2}$$

we must take $G(\hat{\mathbf{r}})$ to be an eigenfunction of $\mathbf{L}_{op}^2$, i.e., a spherical harmonic. This is the only choice that is single valued for all $\hat{\mathbf{r}}$. In other words, we must choose

$$\phi(\mathbf{r}) = F(r)Y_{\ell m}(\hat{\mathbf{r}}). \tag{90.3}$$

Then, since $\mathbf{L}_{op}^2 Y_{\ell m}(\hat{\mathbf{r}}) = \ell(\ell+1)Y_{\ell m}(\hat{\mathbf{r}})$, the radial equation becomes

$$\frac{1}{r}\frac{d^2}{dr^2}(rF) - \frac{\ell(\ell+1)}{r^2}F = 0. \tag{90.4}$$

This equation has solutions $rF = r^{\ell+1}$ and $rF = r^{-\ell}$. Thus, the solutions to Laplace's equation are of the form

$$r^{\ell}Y_{\ell m}(\hat{\mathbf{r}}), \quad \frac{1}{r^{\ell+1}}Y_{\ell m}(\hat{\mathbf{r}}), \tag{90.5}$$

---

[13] We are in the unfortunate position of risking symbol confusion, a certain amount of which seems almost unavoidable. We denote the azimuthal angle by φ to distinguish it from the potential ϕ.

[14] An example of a restricted angular range arises if we consider a region bounded by a conical conducting surface. In this problem, only the range of θ is restricted and that of φ is not. Instead of $Y_{\ell m}$, the angular part of the solutions is a linear combination of $e^{im\varphi}P_\nu^m(\theta)$ and $e^{im\varphi}Q_\nu^m(\theta)$, where P_ν^m and Q_ν^m are associated Legendre *functions* (not Legendre polynomials) of the first and second kind.

and the general solution is a linear combination of these, i.e.,

$$\phi(\mathbf{r}) = \sum_{\ell,m} \left(A_{\ell m} r^\ell + B_{\ell m} \frac{1}{r^{\ell+1}} \right) Y_{\ell m}(\hat{\mathbf{r}}). \tag{90.6}$$

In fact, the above result is contained in our discussion of spherical harmonics and the multipole expansion. Specifically, it follows from eqs. (A.2) and (A.3) if we expand $P_\ell(\hat{\mathbf{r}} \cdot \hat{\mathbf{r}}')$ via the addition theorem and look at the coefficient of $Y_{\ell m}^*(\hat{\mathbf{r}}')$.

When the problem has azimuthal symmetry, the potential must be independent of φ. Only the terms with $m = 0$ in eq. (90.6) meet this requirement, and since $Y_{\ell 0}(\hat{\mathbf{r}}) \propto P_\ell(\cos\theta)$, we may write the general solution more simply as

$$\phi(\mathbf{r}) = \sum_{\ell=0}^{\infty} \left(A_\ell r^\ell + B_\ell \frac{1}{r^{\ell+1}} \right) P_\ell(\cos\theta). \tag{90.7}$$

As a simple application of this method, let us reconsider the problem of an uncharged conducting sphere of radius a in a uniform electric field $E_0\hat{\mathbf{z}}$. Far from the sphere, as $r \to \infty$, the potential is $-E_0 z$, i.e., $-E_0 r\, P_1(\cos\theta)$. Thus, in the expansion (90.7), we must have

$$A_1 = -E_0, \quad A_2 = A_3 = \cdots = 0. \tag{90.8}$$

On the surface $r = a$, on the other hand, ϕ is independent of θ. Hence, the coefficients of all $P_\ell(\cos\theta)$ terms with $\ell \neq 0$ must vanish. This means that $B_\ell = 0$ for $\ell \geq 2$, while for $\ell = 1$ we have $A_1 a + B_1/a^2 = 0$, i.e., $B_1 = -a^3 A_1 = a^3 E_0$. To fix B_0 we note that the total electric flux through the surface $r = a + \epsilon$ must vanish, as the net charge within this surface is zero. A term B_0/r in ϕ corresponds to an enclosed charge of B_0, so $B_0 = 0$. This leaves only A_0 undetermined, but this is an overall additive constant in ϕ, equal to the potential at infinity, which we may take as zero. Hence, the complete solution is

$$\phi(\mathbf{r}) = -\left(1 - \frac{a^3}{r^3} \right) E_0 r \cos\theta, \tag{90.9}$$

exactly as found above, in eq. (89.26).

Cylindrical coordinates: In cylindrical coordinates, which we take as r, φ, and z,[15] Laplace's equation takes the form

$$\frac{1}{r} \frac{\partial}{\partial r} \left(r \frac{\partial \phi}{\partial r} \right) + \frac{\partial^2 \phi}{\partial z^2} + \frac{1}{r^2} \frac{\partial^2 \phi}{\partial \varphi^2} = 0. \tag{90.10}$$

If we look for a product solution of the form

$$\phi(\mathbf{r}) = R(r)\, Z(z)\, F(\varphi), \tag{90.11}$$

---

[15] One frequently encounters the notation ρ for the distance from the axis of symmetry, but this risks confusion with the charge density. Of course, our choice risks some confusion with spherical polar variables, but usually one can tell fairly quickly from the form of the functions appearing which system one is dealing with.

the equations will separate, provided we choose parameters κ and m such that

$$\frac{d^2 F}{d\varphi^2} + m^2 F = 0, \tag{90.12}$$

$$\frac{d^2 Z}{dz^2} - \kappa^2 Z = 0, \tag{90.13}$$

$$\frac{d^2 R}{dr^2} + \frac{1}{r}\frac{dR}{dr} + \left(\kappa^2 - \frac{m^2}{r^2}\right) R = 0. \tag{90.14}$$

If the range of φ is unrestricted, then m must be an integer, and the solutions for F are of the form $e^{im\varphi}$ and $e^{-im\varphi}$. The solutions for Z are also exponentials, but the nature depends on the sign of κ^2, which in turn depends on the boundary conditions imposed on the problem. Let us suppose first that $\kappa^2 > 0$, so that κ is real. Then Z is of the form $e^{\pm\kappa z}$. The equation for R is then *Bessel's equation* of order m, and the solutions are $J_m(\kappa r)$ and $Y_m(\kappa r)$.[16] On the other hand, if the constant $\kappa^2 < 0$, we write $\kappa^2 = -k^2$, so that k is real. The function Z is then of the form $e^{\pm ikz}$, and the solutions for R are given by the *modified Bessel functions* $K_m(kr)$ and $I_m(kr)$. In summary,

$$\begin{aligned} &\text{if } Z = e^{\pm\kappa z}, \text{ then } R = AJ_m(\kappa r) + BY_m(\kappa r),\\ &\text{if } Z = e^{\pm ikz}, \text{ then } R = AK_m(kr) + BI_m(kr), \end{aligned} \tag{90.15}$$

where A and B are constants.

It is extremely useful to understand the qualitative behavior of the Bessel functions at the origin and at infinity. We list this for all four functions.[17]

Function	Behavior at 0	Behavior at ∞	
$J_m(x)$	regular	oscillatory	
$Y_m(x)$	power law/log divergent	oscillatory	(90.16)
$K_m(x)$	power law/log divergent	exponentially decaying	
$I_m(x)$	regular	exponentially growing	

For brevity, we may think of J_m and Y_m as oscillatory functions, and I_m and K_m as exponentially growing and decaying. The behavior in eq. (90.15) is then easily remembered as the rule that if the solutions are oscillatory in one of the variables, z or r, then they must be exponential (growing or decaying) in the other variable. That the solution cannot be oscillatory in all three factors, $F(\varphi)$, $Z(z)$, and $R(r)$, is directly connected with the impossibility of minima and maxima in a solution of Laplace's equation.

---

[16] The function Y_m is variously known as the *Bessel function of the second kind*, *Weber's function*, or *Neumann's function* and is sometimes denoted by N_m. We follow the conventions adopted by Abramowitz and Stegun (1965), and by Copson (1970), Whittaker and Watson (1927), and Watson (1951). A brief discussion of Bessel functions is contained in appendix B.

[17] The following qualifications should be added. The logarithmic divergence arises only for Y_0 and K_0, and the behavior at infinity contains a multiplicative factor of $x^{-1/2}$ in addition to the oscillatory or exponential factors.

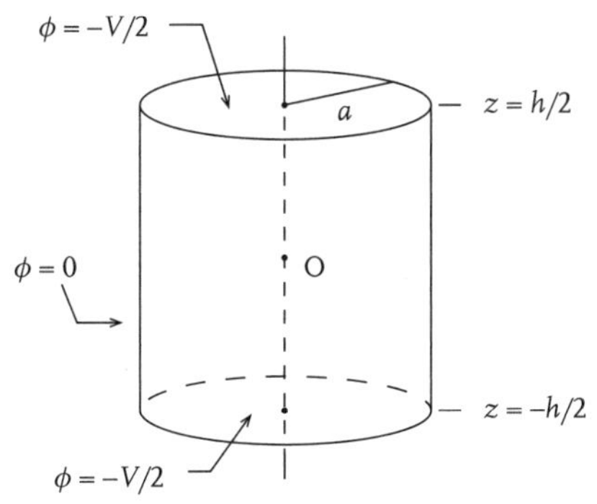

FIGURE 14.7. A can whose top, sides, and bottom are maintained at different potentials.

The general solution to Laplace's equation can therefore be written in either of the two forms,

$$\phi(\mathbf{r}) = \sum_{\kappa,m}\left(A_{\kappa m}J_m(\kappa r) + B_{\kappa m}Y_m(\kappa r)\right)e^{\kappa z}e^{im\varphi}, \tag{90.17}$$

$$\phi(\mathbf{r}) = \sum_{k,m}\left(A_{km}K_m(kr) + B_{km}I_m(kr)\right)e^{ikz}e^{im\varphi}. \tag{90.18}$$

Two notational points should be kept in mind here. First, we have displayed the superposition over the parameters κ and k as sums. In some cases, these parameters may take on a continuous range of values, and the sum then represents an integral. Second, both positive and negative values of κ (or k) and of m should be understood as included in this sum.

In fact, we have already solved a problem in cylindrical coordinates when we found the potential due to a linear array of point charges in section 15. There the solution was of the second type listed in eq. (90.15)—oscillatory in z, exponential in r. As another illustration of this method, let us find the potential inside a cylindrical can whose top and bottom are maintained at potentials $\pm V/2$, and whose sides are maintained at potential 0, being insulated from the top and bottom. (We may imagine this being accomplished by very thin O-ring spacers.) We take the radius of the can as a, and the top and bottom surfaces to be at $z = \pm h/2$ (fig. 14.7). We begin by noting that since the problem has azimuthal symmetry, F must be independent of φ, i.e., only $m = 0$ terms appear in the solution. Second, the solution must be odd in z because of the form of the boundary conditions. Hence, $Z(z)$ can be chosen to be of the form $\sinh(\kappa z)$. This forces $R(r)$ to be either $J_0(\kappa r)$ or $Y_0(\kappa r)$. The latter, however, is not regular on the z axis ($r = 0$) and so must be discarded. It follows that

$$\phi(\mathbf{r}) = \sum_{\kappa} A_{\kappa} \sinh(\kappa z) J_0(\kappa r), \tag{90.19}$$

where the A_κ are constants. To find these we must impose the boundary conditions. Since $\phi = 0$ at $r = a$, $J_0(\kappa a)$ must be zero. If we denote the successively larger zeros of $J_0(x)$ by

x_1, x_2, etc., it follows that κ must take on values

$$\kappa_i = x_i/a, \qquad (J_0(x_i) = 0, \quad i = 1, 2, 3, \ldots). \tag{90.20}$$

For reference, we mention that the first three zeros of $J_0(x)$ are 2.405, 5.520, and 8.654. Changing the summation index from κ to one that labels the zeros x_i, we rewrite eq. (90.19) as

$$\phi(\mathbf{r}) = \sum_i A_i \sinh(\kappa_i z) J_0(\kappa_i r). \tag{90.21}$$

We now impose the boundary condition ar $z = h/2$. (The condition at $z = -h/2$ is automatically satisfied because we have already chosen ϕ to be odd in z.) This gives

$$\frac{V}{2} = \sum_i A_i \sinh(\kappa_i h/2) J_0(\kappa_i r). \tag{90.22}$$

The problem is now reduced to finding coefficients A_i such that eq. (90.22) holds. Note that the right-hand side of this equation is explicitly dependent on r, and it is the demand that the equation hold for all $r \leq a$ that enables all the A_i to be found. This exercise entails some classical analysis belonging to the realm of Sturm-Liouville theory, in particular, the Sturm-Liouville eigenvalue problem.

The key step is to prove that the functions $J_0(\kappa_i r)$ may be used to expand any function of r vanishing at $r = a$. We shall give the standard physicist's argument for this, i.e., prove orthogonality but not completeness. To do this, we first observe that the $J_0(\kappa_i r)$ are in fact solutions $R_i(r)$ to the eigenvalue problem

$$\frac{1}{r}(r R_i')' = -\kappa_i^2 R_i, \tag{90.23}$$

where primes denote differentiation with respect to r, and the boundary conditions are $R_i(a) = 0$, and regularity at the origin. The quantity κ_i^2 is an eigenvalue, since unless this is specially chosen, there will be no solution. We now multiply this equation by r times another eigenfunction R_j, subtract the corresponding equation with the indices i and j interchanged, and integrate over r. This yields

$$\int_0^a [R_j(r R_i')' - R_i(r R_j')'] \, dr = (\kappa_j^2 - \kappa_i^2) \int_0^a r R_j R_i \, dr. \tag{90.24}$$

But the integrand on the left-hand side can be written as a total derivative,

$$[R_j(r R_i') - R_i(r R_j')]', \tag{90.25}$$

so

$$(\kappa_j^2 - \kappa_i^2) \int_0^a r R_j R_i \, dr = \left[R_j(r R_i') - R_i(r R_j') \right]_0^a. \tag{90.26}$$

The right-hand side vanishes at $r = a$ because R_i and R_j vanish there, and also vanishes at $r = 0$ because the factor r vanishes, while R and R' remain finite. Hence, if $\kappa_i \neq \kappa_j$, i.e., if $i \neq j$, the integral on the left-hand side vanishes:

$$\int_0^a r R_i R_j \, dr = 0, \quad i \neq j, \tag{90.27}$$

which is what we wanted. Explicitly written in terms of Bessel functions, this states that

$$\int_0^a r J_0(\kappa_i r) J_0(\kappa_j r) dr = 0, \quad i \neq j. \tag{90.28}$$

Hence, if we multiply eq. (90.22) by $r J_0(\kappa_j r)$ and integrate over r, we get

$$A_j = \frac{V}{2 \sinh(\kappa_j h/2)} \frac{\int_0^a r J_0(\kappa_j r) \, dr}{\int_0^a r \left(J_0(\kappa_j r) \right)^2 dr}. \tag{90.29}$$

At this point the problem is solved in a fundamental sense. We have found the expansion coefficients in terms of one-dimensional integrals, which is more or less the same thing as having found them as numbers. However, the integrals in eq. (90.29) can be expressed compactly in terms of the Bessel function J_1 of argument $\kappa_j r$, and it is useful to see how this is done.[18] To do this we return to eq. (90.26) and write $J_0(\kappa_i r)$ explicitly instead of R_i etc. We note that the derivation of this formula would continue to hold even if κ_j and κ_i were allowed to be arbitrary, except that the right-hand side would not vanish at the upper limit. With this in mind, we continue to let $\kappa_j a$ be a zero of J_0 but replace κ_i by an arbitrary κ. The equation then reads

$$\int_0^a r J_0(\kappa r) J_0(\kappa_j r) dr = \kappa_j a \frac{J_0(\kappa a) J_0'(\kappa_j a)}{\kappa^2 - \kappa_j^2}. \tag{90.30}$$

(Note that J_0' denotes the derivative of J_0 with respect to its argument. This accounts for the extra factor of κ_j on the right-hand side.) We now make use of two properties of Bessel functions,

$$J_0(0) = 1, \quad J_0'(x) = -J_1(x). \tag{90.31}$$

If we let $\kappa \to 0$ in eq. (90.30) and use these properties, we get

$$\int_0^a r J_0(\kappa_j r) dr = \frac{a}{\kappa_j} J_1(\kappa_j a), \tag{90.32}$$

while if we let $\kappa \to \kappa_j$ and use L'Hospital's rule, we get

$$\int_0^a r \left(J_0(\kappa_j r) \right)^2 dr = \frac{1}{2} a^2 [J_1(\kappa_j a)]^2. \tag{90.33}$$

Hence,

$$A_j = \frac{V}{\kappa_j a} \frac{1}{\sinh(\kappa_j h/2) J_1(\kappa_j a)} \tag{90.34}$$

---

[18] The results are special cases of what are known as *Lommel's integrals* for the products of two Bessel functions. See Watson (1944), sec. 5.11.

and

$$\phi(\mathbf{r}) = \sum_j \frac{V}{\kappa_j a} \frac{\sinh(\kappa_j z) J_0(\kappa_j r)}{\sinh(\kappa_j h/2) J_1(\kappa_j a)}. \tag{90.35}$$

This solution has a somewhat formal character. Just how useful is it? and would one be better off numerically solving Laplace's equation in the first place? Since there exist efficient methods of calculating Bessel functions, the key issue is how rapidly the sum converges. Since $\sinh x \approx e^x/2$ for large $x > 0$, the ratio of hyperbolic sines behaves as $\exp(-\kappa_j(\frac{1}{2}h - z))$ if $\kappa_j h/2 \gg 1$. Thus, provided the aspect ratio h/a of the can is large, successive terms fall approximately exponentially, and the convergence is rapid. If $h/a \ll 1$, the convergence is slow, and one should perhaps seek a different calculational scheme.

It is also useful to have an expansion of the potential due to a point charge, $|\mathbf{r} - \mathbf{r}'|^{-1}$, in cylindrical basis functions. There are two such expansions, analogous to eq. (19.17):[19]

$$\frac{1}{|\mathbf{r} - \mathbf{r}'|} = \frac{1}{\pi} \sum_{m=-\infty}^{\infty} \int_{-\infty}^{\infty} dk\, e^{im(\varphi-\varphi')} e^{ik(z-z')} I_m(kr_<) K_m(kr_>), \tag{90.36}$$

$$\frac{1}{|\mathbf{r} - \mathbf{r}'|} = \sum_{m=-\infty}^{\infty} \int_0^{\infty} dk\, e^{im(\varphi-\varphi')} e^{-k|z-z'|} J_m(kr) J_m(kr'). \tag{90.37}$$

In the first form, the functions are oscillatory in z and exponential in r, while in the second they are oscillatory in r and exponentials in z.

Plane polar coordinates; the wedge problem: In problems where all the charges and surfaces are extended infinitely in one direction, which we may take as z, the fields are independent of this variable, and plane polar coordinates may be used. Using the same notation as for cylindrical coordinates, we seek a separated solution of the form

$$\phi(r, \varphi) = R(r) F(\varphi). \tag{90.38}$$

The equations for F and R are

$$\frac{d^2 F}{d\varphi^2} + \nu^2 F = 0, \qquad r^2 \frac{d^2 R}{dr^2} + r \frac{dR}{dr} - \nu^2 R = 0. \tag{90.39}$$

The solutions can be written in complete generality:

$$\begin{aligned} F &= A\cos\nu\varphi + B\sin\nu\varphi, & R &= ar^\nu + br^{-\nu}, & (\nu > 0); \\ F &= A + B\varphi, & R &= a + b\ln r, & (\nu = 0). \end{aligned} \tag{90.40}$$

Naturally, if the full range of φ is allowed, single-valuedness of $F(\varphi)$ forces ν to be an integer.

Let us consider a specific problem as an illustration. Consider a wedge-shaped region of opening angle β, bounded by conducting half-planes (fig. 14.8), both held at potential V. To make the problem mathematically well posed, we must also give boundary conditions far from the edge, e.g., on some surface $r = R$, where R is large. We will see that

---

[19] The second of these is derived in appendix B, eq. (B.29).

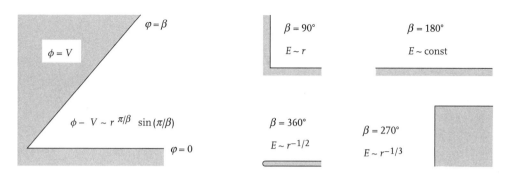

FIGURE 14.8. The potential near the inside edge of a wedge-shaped region of opening angle β inside a conductor. The smaller figures show how the electric field strength varies for chosen values of β.

irrespective of the details of the large-r conditions, the field varies in a characteristic manner near the edge of the wedge. It is easy to see that the conditions that ϕ be finite at $r = 0$, and $\tilde{\phi} \equiv \phi - V = 0$ for all r when $\varphi = 0$ or β, can be met only if we drop the $r^{-\nu}$ and $\ln r$ solutions in R, drop the $\cos \nu\varphi$ and φ solutions in F, and restrict ν so that $\sin \nu\beta = 0$, i.e., choose $\nu = m\pi/\beta$, where m is an integer. Thus, the general form for ϕ is

$$\phi(r, \varphi) = V + \sum_{m=1}^{\infty} A_m r^{m\pi/\beta} \sin(m\pi\varphi/\beta). \tag{90.41}$$

The coefficients A_m are determined by the conditions on ϕ at large r. The interesting point is that for general β, the fields are not analytic near $r = 0$. Keeping only the leading term, the electric field is easily shown to be

$$\mathbf{E} = -\frac{A_1 \pi}{\beta} r^{-1+(\pi/\beta)} \left(\cos(\pi\varphi/\beta)\mathbf{e}_\varphi + \sin(\pi\varphi/\beta)\mathbf{e}_r \right). \tag{90.42}$$

In particular, $|\mathbf{E}| \sim r^{-1+(\pi/\beta)}$, which vanishes as $r \to 0$ if $\beta < \pi$ but which diverges if $\beta > \pi$. The surface charge density behaves in the same way. For the case $\beta = 2\pi$, corresponding to the edge of a thin conducting plate, we find that $E \sim r^{-1/2}$. This rapid rise in E near sharp edges and corners[20] is so general and important a phenomenon that it is worth understanding in simple terms. Consider two surface charges a small distance apart in a region where the surface is highly curved, and compare this pair with another pair with the same separation in a region of small curvature. The force between the first pair will have a relatively larger component normal to the surface, and a smaller component parallel to the surface. Since the charges can move only in the surface, the force tending to separate the first pair will be smaller than that on the second pair. Consequently, we will get a higher surface charge density in the region of high curvature, and a higher electric field.

The above phenomenon is exploited in lightning rods. The large electric field increases the likelihood of an electrical breakdown of air near the tip of the rod, thus providing a

---

[20] For the very similar problem of a sharp conical point on the surface of a conductor, with the half-angle of the cone being β, the field varies as $r^{-1+\epsilon}$, where $\epsilon = 1/(2\ln\beta)$. See Landau and Lifshitz (1960), prob. 4 in sec. 3.

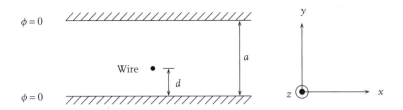

FIGURE 14.9. Figure for exercise 90.1. The wire has charge per unit length λ.

high-conductivity path for the lightning currents between the ground and thunderclouds. If the rod is properly grounded, this lowers the odds of nearby structures being hit by lightning.

Cartesian coordinates: The separation of Laplace's equation in this system is perhaps the simplest of all. We mention it for completeness. If we write the potential as $\phi = X(x)Y(y)Z(z)$, the equations for X, Y, and Z are all of the same form and have as solutions either oscillatory ($e^{\pm ik_x x}$) or nonoscillatory ($e^{\pm \kappa_x x}$) exponentials. All three factors cannot be of the same type, however. We illustrate the ideas with an exercise.

Exercise 90.1 Two infinite parallel conducting plates, at $y = 0$ and $y = a$, are both grounded (potential equal to zero). An infinite wire carrying a charge per unit length of λ runs parallel to the z axis a distance $d < a$ above the lower plate. (See fig. 14.9.) Find the potential in the region $0 < y < a$.

Solution: This is, in fact, a two-dimensional problem and the only new point worthy of note in it is the method of treating the singularity in ϕ near the wire. We take the wire to run through the point $x = 0$, $y = d$. Let us take $Y(y)$ in the form $\sin ky$ and $\cos ky$; then $X(x)$ must be of the form $e^{\pm kx}$. It is easy to show that the boundary conditions at $y = 0$, $y = a$, and $x = \pm\infty$ are met by taking

$$\phi(x, y) = \sum_{n=1}^{\infty} A_n^{\pm} e^{-k_n|x|} \sin k_n y, \tag{90.43}$$

where $k_n = n\pi/a$, and the coefficients are taken as A_n^+ for $x > 0$, and A_n^- for $x < 0$. Let us now consider ϕ for $y \neq d$. The continuity of ϕ in x across the plane $x = 0$ implies that $A_n^+ = A_n^-$. Accordingly, we write both as just A_n. To determine A_n, we consider Poisson's equation,

$$\nabla^2 \phi = -4\pi\lambda\delta(x)\delta(y - d), \tag{90.44}$$

and integrate it from $x = -\epsilon$ to $+\epsilon$, where ϵ is infinitesimally positive. This yields

$$\frac{\partial \phi}{\partial x}\bigg|_{x=-\epsilon}^{x=\epsilon} = -4\pi\lambda\delta(y - d),$$

$$\sum_n A_n k_n \sin k_n y = 2\pi\lambda\delta(y - d).$$

Using the orthogonality of the set $\sin k_n y$ over the interval $0 < y < a$, we find $A_n = (4\lambda/n) \sin k_n d$. Hence,

$$\phi(x, y) = 4\lambda \sum_{n=1}^{\infty} \frac{1}{n} \sin k_n d \sin k_n y \, e^{-k_n|x|}. \tag{90.45}$$

The sum is simple, and the final answer is

$$\phi(x, y) = -\lambda \ln \left(\frac{\cosh \frac{\pi}{a}x - \cos \frac{\pi}{a}(y - d)}{\cosh \frac{\pi}{a}x - \cos \frac{\pi}{a}(y + d)} \right). \tag{90.46}$$

The similarity to eq. (17.11) is more than coincidental. We could have approached the problem by the method of images, which would have given us two periodic arrays of oppositely charged wires, each of period $2a$ and displaced from each other by a distance $2d$.

Exercise 90.2 Find the potential and surface charge densities near a corner inside a cubical box. ·

91 The variational method*

A third method for solving potential problems is based on the variational theorem. This method is worth discussing largely because the theorem is very general, and variational methods are also used in many other branches of physics. It is difficult to assess the method's usefulness as a practical technique for potential problems per se.

We will consider only the Dirichlet problem (87.5). The argument for Neumann problems is very similar. Let the boundary values of $\phi(\mathbf{r})$ on the surface S be denoted by $\phi_0(\mathbf{r})$. The variational problem consists of minimizing the quantity

$$W[\psi] = \frac{1}{8\pi} \int_V (\nabla \psi(\mathbf{r}))^2 d^3x - \int_V \rho(\mathbf{r}) \psi(\mathbf{r}) d^3x, \tag{91.1}$$

where $\psi(\mathbf{r})$ is any smooth function[21] satisfying the condition $\psi(\mathbf{r}) = \phi_0(\mathbf{r})$ on S.

Suppose we have found the function $\psi(\mathbf{r})$ that minimizes W. If we consider a slightly different function $\psi(\mathbf{r}) + \eta(\mathbf{r})$, where $\eta(\mathbf{r})$ is very small everywhere, then to first order the value of W should not change, and to second order it should increase. Calling the first- and second-order changes in W, δW and $\delta^2 W$, respectively, we have

$$\delta W = \frac{1}{4\pi} \int_V \nabla \psi \cdot \nabla \eta \, d^3x - \int_V \rho \eta \, d^3x,$$

$$\delta^2 W = \frac{1}{8\pi} \int_V (\nabla \eta)^2 d^3x.$$

---

[21] More precisely, continuous and piecewise smooth.

The second variation $\delta^2 W$ is clearly positive, and so the goal is to see if the first variation δW vanishes. We integrate the first term by parts to obtain

$$\delta W = -\frac{1}{4\pi} \int_V (\nabla^2 \psi + 4\pi\rho)\eta(\mathbf{r})d^3x + \frac{1}{4\pi}\int_S \eta(\mathbf{r})\nabla\psi \cdot \hat{\mathbf{n}}\, d^2s. \tag{91.2}$$

We now note that to be in the admissible class of functions among which the minimum of W is to be sought, the varied function $\psi + \eta$ must also equal $\phi_0(r)$ on the boundary. Since ψ already equals ϕ_0 for $\mathbf{r}$ on S, η must vanish everywhere on S. Thus, the surface integral in the above equation vanishes identically. For arbitrary variations $\eta(\mathbf{r})$, the volume term vanishes only if

$$\nabla^2\psi(\mathbf{r}) + 4\pi\rho(\mathbf{r}) = 0, \tag{91.3}$$

i.e., if $\psi(\mathbf{r})$ is the solution to the original Dirichlet problem.

We thus have the theorem that the functional $W[\psi]$ is minimized by the true potential function $\phi(\mathbf{r})$. The simplest type of practical method that this gives rise to consists of guessing a function $\psi(\mathbf{r})$ that obeys the boundary conditions on ϕ and that has a few adjustable parameters, $\alpha_1, \alpha_2, \cdots, \alpha_n$. We then evaluate the quantity W and minimize with respect to the α_i by ordinary calculus. The optimal function ψ is the resulting approximation to $\phi(\mathbf{r})$. In some cases, ψ may be completely determined, i.e., there may be no adjustable parameters α_i. However, in other problems, one is able to restrict ψ in such a mild way that a complete functional form remains to be determined. One must then solve a modified variational problem. There is no cookbook for such an approach, and the literature of physics is full of clever and beautiful variational solutions to all sorts of problems.

Note that if there is no charge density ρ in the region V, the quantity W is the energy in the electric field. In particular, if there are two conducting surfaces S_1 and S_2 bounding V, we immediately get an upper bound for the capacitance of the setup. If we set the potentials to 0 and 1 on the two conductors, the energy is $C/2$. Therefore,

$$C = \text{Min}\frac{1}{4\pi}\int_V (\nabla\psi)^2 d^3x, \quad \phi = 0 \text{ on } S_1, \quad \phi = 1 \text{ on } S_2. \tag{91.4}$$

In practice, there are several hurdles to successful application of this method. First, one must be able to find a suitable approximation that builds in the correct physical features of the solution. Second, one must be able to evaluate the integrals for W. Last, one must be able to do the minimization of W. If one is looking for accurate answers, it may be better and more systematic to use a numerical method from the very start. For approximate answers, however, especially for the total energy, the method can be useful.

The next two exercises show how the method can be used to find approximate capacitances in simple geometries. The first exercise is a standard one where the exact answer is known. The next is related to a problem that we shall revisit in connection with the numerical relaxation method.

Exercise 91.1 A cylindrical capacitor is formed out of two coaxial pipes. The inner pipe has outer radius a, and the outer pipe has inner radius b. Find estimates for the capacitance

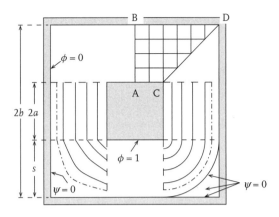

FIGURE 14.10. Cross section of square waveguide capacitor. The shaded areas are the conductors. Equipotential contours for the second variational solution, and that discussed beginning with eq. (91.15), are shown on the lower right and lower left, respectively. The contours are at intervals of 0.2 except for the dot-dashed contours, which are at $\phi = 0.1$. In the upper right corner, we show the grid to be used in exercise 92.3.

per unit length by using the following trial functions: (i) $\psi(r) = \alpha(r - a)$, (ii) $\psi(r) = \alpha(r - a) + \beta(r - a)^2$. [The functions depend only on the radial distance r, and α and β are to be adjusted so that $\phi(b) = 1$]. Compare with the exact result $1/2 \ln(b/a)$ for various values of the ratio b/a.

Answer: (i) $(b + a)/4(b - a)$, (ii) $(b^2 + 4ab + a^2)/6(b^2 - a^2)$ (Gaussian).

Exercise 91.2 Consider a capacitor similar to that in the previous exercise, except that the conductors are square in cross section. The squares are again concentric, with parallel sides $2a$ and $2b$ ($b > a$). It is convenient to choose $\phi = 1$ on the inner conductor and $\phi = 0$ on the outer. Consider general trial functions with the property that the equipotential lines are (i) themselves squares with the same centers as the pipes, (ii) "squares" with rounded corners consisting of circular arcs in the regions $a < |x| < b$, $a < |y| < b$ (fig. 14.10). For each type of potential, estimate the capacitance per unit length.[22]

Solution: We first present the answers: In Gaussian units, the capacitance per unit length is

$$(i) \quad \frac{2}{\pi \ln(b/a)}, \qquad (ii) \quad \left[2 \ln \left(1 + \frac{\pi}{4} \left(\frac{b}{a} - 1 \right) \right) \right]^{-1}. \tag{91.5}$$

For $b/a = 2$, these give (i) 0.9184, (ii) 0.8626, while the exact, numerically obtained answer is 0.8141.

---

[22] More details of the solutions may be found in A. Garg, *Am. J. Phys.* **75**, 509 (2007).

Since both types of potentials are treated very similarly, we give the solution only for type (ii). By symmetry,

$$C = \frac{1}{4\pi} \int_V (\nabla \psi)^2 d^3 x = \frac{1}{\pi} (w_s + w_c), \tag{91.6}$$

where w_s is the contribution to the integral from one rectangular side region, say $a < x < b$, $-a < y < a$, and w_c is the contribution from one square region in the corner. In the side region, the fact that the equipotentials are parallel to the y axis means that ψ depends only on x. If we define $\psi(\mathbf{r}) = f(x - a)$ in this region, then

$$w_s = 2a \int_0^{b-a} \left(\frac{df}{dx}\right)^2 dx. \tag{91.7}$$

To handle the corner contribution, we employ polar coordinates centered at $x = a$, $y = a$. Since the equipotentials are circular arcs, it follows that $\psi = f(r)$, with the same function f as above. Hence,

$$w_c = \frac{\pi}{2} \int_0^{b-a} \left(\frac{df}{dr}\right)^2 r \, dr. \tag{91.8}$$

Our estimate for the capacitance is, therefore,

$$C = \frac{1}{\pi} \int_0^{b-a} \left(2a + \frac{\pi}{2}x\right) \left(\frac{df}{dx}\right)^2 dx. \tag{91.9}$$

The function f must now be chosen so as to minimize C, subject to the boundary conditions

$$f(0) = 1, \quad f(b - a) = 0. \tag{91.10}$$

The method of minimization is standard. If we vary f to $f + \eta$, the first-order variation in C is

$$\delta C = \frac{2}{\pi} \int_0^{b-a} \left(\frac{d\eta}{dx}\right) \left(\frac{df}{dx}\right) \left(2a + \frac{\pi}{2}x\right) dx. \tag{91.11}$$

Since $f + \eta$ must also obey the boundary conditions if it is to be an acceptable potential, η must vanish at $x = 0$ and $x = b - a$. This fact means that if we integrate the expression for δC by parts, the integrated term will vanish, leaving us with

$$\delta C = -\frac{2}{\pi} \int_0^{b-a} \eta \frac{d}{dx} \left[\frac{df}{dx}\left(2a + \frac{\pi}{2}x\right)\right] dx. \tag{91.12}$$

Since $\eta(x)$ is arbitrary, its coefficient in the integrand must vanish. This gives us a very easy differential equation for f, whose solution is

$$f(x) = c \ln\left(2a + \frac{\pi}{2}x\right) + d. \tag{91.13}$$

Here c and d are constants that are fixed by the boundary conditions. If we feed the result for f back into the integral for C, we obtain the answer (91.5) quoted above.

Let us briefly discuss the characteristics of the two trial potentials. With the choice (i), the electric field has sharp discontinuities on the diagonals $y = \pm x$ and has a constant nonzero value everywhere on the outer conductor. The correct electric field must, by contrast, vary linearly with the distance from the corners close to the corners. Nevertheless, the error in the capacitance is only 11% for $b = 2a$. The type (ii) potential does not have discontinuities along the diagonals, but it leaves a dead space near the outer corners where the trial electric field is zero. Along the outer walls, the field is zero next to the corner and jumps to a constant value near the center. The error is reduced to 6%.

A trial potential that does even better is as follows. Let u and v be the Cartesian distances measured from one of the outer corners, the lower left one, say, and let

$$s = b - a. \tag{91.14}$$

In the square corner region, we choose

$$\phi(\mathbf{r}) = h(u)h(v), \quad (0 \le u \le s, \quad 0 \le v \le s), \tag{91.15}$$

where h is to be determined. Since we must have $\phi = 0$ at $u = v = 0$, and $\phi = 1$ at $u = v = s$, the function h must obey

$$h(0) = 0, \quad h(s) = 1. \tag{91.16}$$

In the rectangular region on the side, we choose

$$\phi(\mathbf{r}) = h(v), \quad (b - a \le u \le b + a, \quad 0 \le v \le b - a). \tag{91.17}$$

This ensures that ϕ is continuous.

Let us define w_s and w_c as for the previous trial solution. We now have

$$w_s = 2a \int_0^s \left(\frac{dh}{dv}\right)^2 dv, \tag{91.18}$$

$$w_c = \int_0^s \int_0^s \left[\left(\frac{dh(u)}{du}\right)^2 h^2(v) + u \leftrightarrow v\right] du\, dv. \tag{91.19}$$

The two terms in w_c are equal. Hence, we have

$$C = \frac{2}{\pi}\left[a \int_0^s (h')^2 + \int_0^s (h')^2 \int_0^s h^2\right]. \tag{91.20}$$

Here h' denotes the derivative of h, and we use an obvious and abbreviated notation for the integrals. If we vary h to $h + \eta$ and integrate the terms involving η' by parts, the boundary terms vanish once again, and we obtain

$$\delta C = -\frac{4}{\pi}\int_0^s \eta[K_1 h''(u) - K_2 h(u)]\, du, \tag{91.21}$$

where

$$K_1 = a + \int_0^s h^2(u)du, \quad K_2 = \int_0^s [h'(u)]^2 du. \tag{91.22}$$

For δC to vanish, the coefficient of η must vanish. This yields the following differential equation for h:

$$h'' - \kappa^2 h = 0, \quad \kappa^2 \equiv \frac{K_2}{K_1}. \tag{91.23}$$

The solution that obeys the required boundary conditions is

$$h(u) = A\sinh(\kappa u), \quad A = 1/\sinh(\kappa s). \tag{91.24}$$

It remains to evaluate κ. To do this, we insert the solution back into the definitions of K_1 and K_2. This yields

$$K_1 = a - \frac{A^2 s}{2}\left[1 - \frac{\sinh(2\kappa s)}{2\kappa s}\right], \tag{91.25}$$

$$K_2 = \frac{\kappa^2 A^2 s}{2}\left[1 + \frac{\sinh(2\kappa s)}{2\kappa s}\right]. \tag{91.26}$$

Writing $K_2 = \kappa^2 K_1$ and simplifying, we obtain $A^2 s = a$, i.e.,

$$\sinh^2 \kappa s = s/a. \tag{91.27}$$

This determines the optimal trial function completely. We have already evaluated the integrals needed to find C in the form of K_1 and K_2. Some straightforward analysis gives, finally,

$$C = \frac{a^2}{2\pi s^2}\left[\frac{\sqrt{s(s+a)}}{a} + \sinh^{-1}(\sqrt{s/a})\right]^2. \tag{91.28}$$

For $b = 2a$, this yields $\kappa s = 0.8814$ and $C = 0.8387$, which is off by 3%. This solution has the correct analytic behavior near the outer corner, but not the inner one.

92 The relaxation method

As mentioned earlier, the only methods for solving problems in electrostatics that are of general applicability are numerical. We now discuss what is perhaps the simplest of these, the *relaxation*, or *iterative finite difference method*. We will present the discussion in two spatial dimensions in order to keep it simple. The extension to three dimensions is obvious and easy. As before, we wish to solve Poisson's equation,

$$\nabla^2 \phi(x, y) = -4\pi\rho(x, y), \tag{92.1}$$

in some region, given the charge density ρ and the value of ϕ on the boundary (fig. 14.11). We cover the region by a square grid and suppose, to begin with, that the boundary is made up of segments that coincide with the grid lines. Let the grid points be

$$x_i = x_0 + ih, \quad y_j = y_0 + jh, \tag{92.2}$$

where h is the grid or mesh size, and the ranges of i and j are determined by the shape of the boundary. We denote $\phi(x_i, y_j)$ by $\phi_{i,j}$, and likewise for $\rho(x, y)$. The objective, clearly, is to approximate the Laplacian by some finite difference approximant, ∇^2_{FD}, i.e., a combination of differences between neighboring $\phi_{i,j}$, and thereby replace the differential

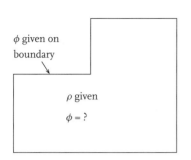

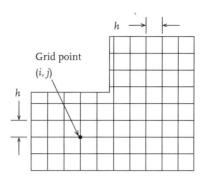

FIGURE 14.11. The relaxation method. The region where the potential is sought is divided into a grid, and the potential is found on the grid points.

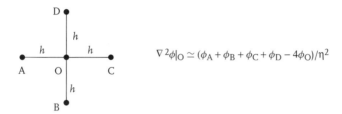

$$\nabla^2\phi|_O \simeq (\phi_A + \phi_B + \phi_C + \phi_D - 4\phi_O)/\eta^2$$

FIGURE 14.12. The simplest finite difference approximation to the Laplacian.

equation (92.1) by the linear algebraic equations

$$\nabla^2_{FD}\phi_{i,j} = -4\pi\rho_{i,j}. \tag{92.3}$$

The simplest and most useful choice for ∇^2_{FD} is obtained by noting that

$$\phi(x+h, y) + \phi(x-h, y) + \phi(x, y+h) + \phi(x, y-h) = 4\phi(x, y) + h^2\left(\frac{\partial^2\phi}{\partial x^2} + \frac{\partial^2\phi}{\partial y^2}\right)$$

$$+ \frac{h^4}{12}\left(\frac{\partial^4\phi}{\partial x^4} + \frac{\partial^4\phi}{\partial y^4}\right) + \cdots, \tag{92.4}$$

whence we have (see fig. 14.12).

$$\nabla^2_{FD}\phi_{i,j} = \frac{1}{h^2}(\phi_{i+1,j} + \phi_{i-1,j} + \phi_{i,j+1} + \phi_{i,j-1} - 4\phi_{i,j}). \tag{92.5}$$

With this choice, $\nabla^2_{FD}\phi = \nabla^2\phi + O(h^2)$. Note how $\nabla^2_{FD}\phi$ is proportional to the average of the function at the star of the four nearest neighbors minus its value at the point where the Laplacian is desired. (This is reminiscent of Earnshaw's theorem: if the Laplacian vanishes, the average over the star of nearest neighbors equals the value at the center.) There is a handy pictorial way of representing this equation:

$$\nabla^2_{FD}\phi_{i,j} = \frac{1}{h^2}\left\{\begin{matrix} & 1 & \\ 1 & -4 & 1 \\ & 1 & \end{matrix}\right\}\phi_{i,j}. \tag{92.6}$$

Exercise 92.1 Show that another approximant to the Laplacian is

$$\frac{1}{6h^2} \begin{Bmatrix} 1 & 4 & 1 \\ 4 & -20 & 4 \\ 1 & 4 & 1 \end{Bmatrix} \phi_{i,j}, \tag{92.7}$$

and that the leading correction to this is $(h^2/12)\nabla^2\nabla^2\phi$.

Equations (92.3) form a system of linear equations with as many unknowns as interior points in our grid. At the risk of belaboring the obvious, let us see how this is so when we use eq. (92.5) for ∇_{FD}^2. For any point (i, j) in the grid, on the left-hand side of eq. (92.3) we will have a linear combination of $\phi_{i,j}$ and its four nearest neighbors. If any of these neighbors happens to be on the boundary, its value is known, and we may move it over to the right-hand side. The resulting problem can be written in standard matrix form if we label the points inside the grid by a single index I, e.g., by starting from the point at the extreme lower-left corner and going along the row until we hit an edge, then going to the next row, and so on. In this way we make a column vector Φ with elements ϕ_I out of the $\phi_{i,j}$. We do the same for $\rho_{i,j}$. For each I, let us denote the set of nearest neighbors that lie inside the grid by $NN(I)$ and the set of those that lie on the boundary by $B(I)$. The equations then become

$$\phi_I - \frac{1}{4} \sum_{J \in NN(I)} \phi_J = \pi h^2 \rho_I + \frac{1}{4} \sum_{J \in B(I)} \phi_J. \tag{92.8}$$

This is of the standard form $\mathbf{A}\Phi = \mathbf{b}$. The matrix $\mathbf{A}$ has entries 1, 0, and $-1/4$ only. It is very sparse: in each row, it has an entry 1 on the diagonal, up to four other entries equal to $-1/4$, and all the rest are zero. If the boundary is a rectangle, the nonzero entries fall on neat diagonals, and the matrix is periodic except for boundary truncations. In this case, the linear equations can be solved directly by fast Fourier transform methods, reflecting the fact that vectors of $\mathbf{A}$ correspond to Bloch states of an electron hopping on a square crystal lattice with rectangular boundaries. In general, however, the boundary is not so simple, and even though the matrix is sparse, direct solution of the equations is hopeless if the grid is even moderately large. One must then resort to some kind of iterative method. The relaxation method is one such.

To understand this method, let us go back to the notation $\phi_{i,j}$ and, with eq. (92.5), rewrite eq. (92.3) as

$$\phi_{i,j} = \frac{1}{4} \begin{Bmatrix} & 1 & \\ 1 & & 1 \\ & 1 & \end{Bmatrix} \phi_{i,j} + \pi h^2 \rho_{i,j}. \tag{92.9}$$

We now implement this equation iteratively. That is, suppose we have some set of approximate values $\phi_{i,j}$ for the solution. We substitute these on the right-hand side and thus obtain new values. In other words,

$$\phi_{i,j}^{\text{new}} = \frac{1}{4} \left\{ \begin{array}{ccc} & 1 & \\ 1 & & 1 \\ & 1 & \end{array} \right\} \phi_{i,j}^{\text{old}} + \pi h^2 \rho_{i,j}. \tag{92.10}$$

Obviously, points (i, j) lying on the boundary are not updated, and the known boundary values are used if any such point appears on the right-hand side. The process is repeated until convergence is obtained.

The precise way in which the iteration is carried out depends on what one means by "old" and "new." In the *Jacobi scheme*, a complete set of old ϕ values is kept for all (i, j) and is replaced by the new values only when the latter are known for all (i, j). In the *Gauss-Seidel scheme*, which converges about twice as fast as the Jacobi and requires lesser storage, the new values are used as soon as they are available. In other words, in a given sweep through the grid, if any of the $\phi_{i,j}$ on the right-hand side of eq. (92.10) have been found in that very sweep, they are used. However, one makes a single complete sweep through the grid in every iteration, updating $\phi_{i,j}$ for every point (i, j) once and only once.

Programming this method is extremely simple. One starts by making a guess for the $\phi_{i,j}$ on the grid. Almost any guess will do, including a constant. One then applies eq. (92.10), going systematically over every lattice point, until some convergence criterion is attained. For example, one could demand that the solution not change at any lattice point by more than some amount. The bookkeeping is not onerous either. If one labels the points with a single index I as described above, all one needs is to make a $(i, j) \leftrightarrow I$ correspondence table at the start, along with lookup tables for the sets $NN(I)$ and $B(I)$.

Why does the relaxation method work, and how fast is it? To answer this, let us go back to the Jacobi scheme and indicate the sweep number by a superscript thus: $\phi_{i,j}^{(n)}$. Equation (92.10) may be rewritten as

$$\phi_{i,j}^{(n+1)} - \phi_{i,j}^{(n)} = \frac{1}{4} \left\{ \begin{array}{ccc} & 1 & \\ 1 & -4 & 1 \\ & 1 & \end{array} \right\} \phi_{i,j}^{(n)} + \pi h^2 \rho_{i,j}. \tag{92.11}$$

(Note the extra -4 inside the braces on the right.) If we now think of the sweep index as representing time, we may think of this equation as a finite difference approximation to a diffusion equation

$$\frac{1}{D} \frac{\partial \phi}{\partial t} = \nabla^2 \phi + 4\pi\rho, \tag{92.12}$$

where, with τ_0 as an arbitrary measure of the time step, the diffusion constant D is given by

$$D = h^2/4\tau_0. \tag{92.13}$$

The physical point is that this equation has an equilibrium solution, ϕ^*, and departures from equilibrium relax toward zero. Of course, since $\partial\phi^*/\partial t = 0$, ϕ^* is the solution to our initial problem. Let us write the distribution ϕ at some general time as

$$\phi = \phi^* + \eta, \quad \eta = 0 \text{ on boundary.} \tag{92.14}$$

Then,

$$\frac{\partial \eta}{\partial t} = D\nabla^2 \eta. \tag{92.15}$$

We can understand the evolution of η if we think in terms of the eigenfunctions

$$D\nabla^2 \eta_j = -\epsilon_j \eta_j, \quad \eta_j = 0 \text{ on boundary.} \tag{92.16}$$

This equation can be thought of as a Schrödinger equation for a particle moving in a box, and then the ϵ_j are the energy eigenvalues. Alternatively, one could think in terms of a vibrating drumhead that is clamped at the boundary. The ϵ_j are then the squares of the natural frequencies of the drumhead. The key point is that all the ϵ_j are positive. Thus, if we expand our general disturbance η in terms of the eigenfunctions η_j, we have

$$\eta(t) = \sum_j a_j \eta_j e^{-\epsilon_j t}, \tag{92.17}$$

where the a_j are expansion coefficients. Let us order the eigenvalues so that $0 < \epsilon_1 \leq \epsilon_2 \leq \cdots$. As $t \to \infty$ only the lowest eigenvalue ϵ_1 need be kept, and it is this which determines the rate of convergence. For a region of linear dimension L, we may estimate ϵ_1 as $2D(\pi/L)^2$, so that if want p-figure accuracy (relative error of order 10^{-p} or less), the number of sweeps required varies as

$$N_{\text{sweep}} \simeq \frac{2 \ln 10}{\pi^2} p \left(\frac{L}{h} \right)^2 \simeq \frac{1}{2} p \left(\frac{L}{h} \right)^2. \tag{92.18}$$

This is rather slower than other methods.[23] Indeed, Press et al. (1986) dismiss this kind of relaxation as being of only "academic interest." Describing the faster methods would, however, take us too far afield.

Exercise 92.2 Show that $\epsilon_j > 0$ by showing that

$$\epsilon_j = \frac{\int (\nabla \eta_j)^2 d^2 x}{\int \eta_j^2 d^2 x}. \tag{92.19}$$

The integrals extend over the region in which $\phi(x, y)$ is sought.

At this point, we should also discuss the *accuracy* of the relaxation method. From eq. (92.5), we expect that the error is of order h^4. This is correct, but, of course, we know nothing about the coefficient, nor can any general remarks be made about the uniformity of the approximation. We refer the reader to specialized texts on numerical

---

[23] Note that going to three dimensions does not change the scaling of N_{sweep} with h. Of course, the number of operations required per sweep now scales as $(L/h)^3$ as opposed to $(L/h)^2$ in two.

analysis for more detailed discussion of these matters, as well as for other methods of solution.[24]

Exercise 92.3 (modified version of Purcell, problem 3.30, Jackson, problem 1.23.) Consider the square pipe capacitor of excercise 91.2. Take $a = 1$ cm, $b = 2$ cm, and a potential difference of 1 statvolt (or 1 V, for SI) between the conductors.

(a) Use the relaxation method to solve for the potential. Choose a grid size h around 0.01 cm. (Much smaller values of h may take too long to converge.) It suffices to solve for ϕ in the trapezoidal region ABCD (fig. 14.10), and use mirror symmetry about the lines AB and CD.

(b) Find the charge densities and total charges on the inner and outer conductors. Are the total charges on the two conductors equal in magnitude? If not, which one is likely to be more accurate, and why? Near the corners, do the charge densities behave as found analytically in the wedge problem (section 90)—inverse cube root divergence at the inner corner, linear approach to zero at the outer corner?

(c) Find the capacitance per unit length. [**Answer**: 0.8141 (dimensionless, in Gaussian system), $10.43\epsilon_0$ F/m (SI).]

Finally, let us comment on the effect of an irregular boundary. We could choose either to keep our grid fixed and shift the boundary points so that they lie on the closest grid point, or to keep the boundary where it is and alter the grid in its immediate vicinity. Then, for points adjacent to the boundary, ∇^2_{FD} has to be constructed out of a star of points with unequal arms. In both cases, one introduces errors of order h near the boundary.

93 Microscopic electrostatic field at metal surfaces; work function and contact potential*

Let us now consider the field at a metal surface in more detail. Imagine, first, a large metal block, without any net charge. In the bulk of the block, the positive ions are periodically arrayed in the main, and the mobile electrons have a density that is uniform on the macro scale. We wish to find the energy for removing an electron from this block. When the question is posed in this language, however, it risks being confused for the energy in a *physical* process in which the electron is moved out of the bulk, through the surface, and thence to infinity. That is, in fact, the main question of this section, but for the moment, we do not wish to consider the surface. Let us, therefore, suppose that the block has N electrons when it is neutral, where N is a very large number (of order 10^{24}), and let the total energy of this system be $\mathcal{E}_N$.[25] Similarly, let the energy of the same block with $N-1$ electrons be $\mathcal{E}_{N-1}$. One less electron does not affect charge neutrality in a macroscopic block, so the change in energy is an intensive quantity, determined by the electrostatic

---

[24] See, e.g., Press, Flannery, Teukolsky, and Vetterling (1986), chap. 17, or Stoer and Bulirsch (1980), sec. 8.9.

[25] More precisely, the total energy should be replaced by the total free energy.

potential of the positive ions and the mean electron density. The energy difference

$$\mu = \mathcal{E}_N - \mathcal{E}_{N-1} \tag{93.1}$$

is known as the *chemical potential*. Since the positive ions attract the electrons, $\mathcal{E}_N < \mathcal{E}_{N-1}$, and $\mu < 0$.[26]

Now imagine cutting through the metal, at the section AB in fig. 14.13, for example. This creates two surfaces, one on each side of the cut. The key point is that the electron density near the surfaces is not obtained by rigidly transporting that which previously prevailed in the bulk. In particular, the density does not vanish abruptly at the surface, for the kinetic energy can be lowered by delocalizing the electrons a little and allowing them to "spill out" of the surface, to a distance that is a small fraction of a_0. Thus, even in a model in which the densities of the positive and negative charges are smeared out into jellies in the bulk, there is a net negative charge in a layer on the outside, followed by a net positive charge layer an atomic separation away (see fig. 14.13). Inside this characteristic double layer, there is a strong electric field pointing along the outward surface normal. Moving an electron from the bulk ($x \to -\infty$) through this double layer, far into the vacuum region ($x \to \infty$), requires electrostatic work in the amount of $-e\Delta\phi$, where $\Delta\phi = \phi(\infty) - \phi(-\infty)$. As is easily verified,

$$\Delta\phi = -4\pi e \int_{-\infty}^{\infty} x \big(n_-(x) - n_+(x) \big) \, dx, \tag{93.2}$$

where $n_\pm(x)$ are the number densities of positive and negative charges.

The electrostatic contribution $-e\Delta\phi$ arising from the rearrangement of the charges at the surface must be added to the bulk energy $-\mu$, in order to yield the work required to remove an electron through the surface of the metal. This work, known as the *work function*, is therefore given by

$$W = -\mu - e\Delta\phi. \tag{93.3}$$

The computation of $\Delta\phi$ (and even of μ) is a complex many-body problem far beyond our scope. We wish to stress, however, that this quantity includes the work that must be done against the field of the macroscopic image charge as the electron is removed, as well as all many-body effects. The argument for this is simple.[27] First, there is no macroscopic difference in the system with N and $N-1$ charges. Second, $\phi(x)$ is the

---

[26] For readers familiar with elementary treatments of metals, we offer this footnote. If we ignore the electron–electron interaction, the energy of the N electron system is obtained by filling single–electron energy levels, and the chemical potential (often replaced by the less accurate quantity, the Fermi energy, ϵ_F), lies at the highest such unoccupied level. In the simplest model, the potential of the ions is replaced by a uniform constant V_0, whose value is generally taken as zero. The problem is then that of a "particle in a box," with the box being the block of metal. The single-particle energies are all positive with the choice $V_0 = 0$, and $\mathcal{E}_N$ appears to exceed $\mathcal{E}_{N-1}$. Indeed, the standard formula for the Fermi energy in this model, $\epsilon_F = \hbar^2 k_F^2 / 2m$, seems to contradict our assertion that $\mu < 0$. The problem lies in the choice $V_0 = 0$. This is perfectly fine for studying processes that involve the bulk only, but if we wish to consider the metal's exterior also, then we must recognize that the positive ions provide a negative potential for the electrons relative to the outside, and the reference value V_0 must be taken as large and negative. The Fermi energy is then $V_0 + \hbar^2 k_F^2 / 2m$, which is correctly negative.

[27] A more careful proof is given by N. D. Lang and W. Kohn, *Phys. Rev. B* **3**, 1215 (1971). This reference also contains model calculations of the charge-density profiles.

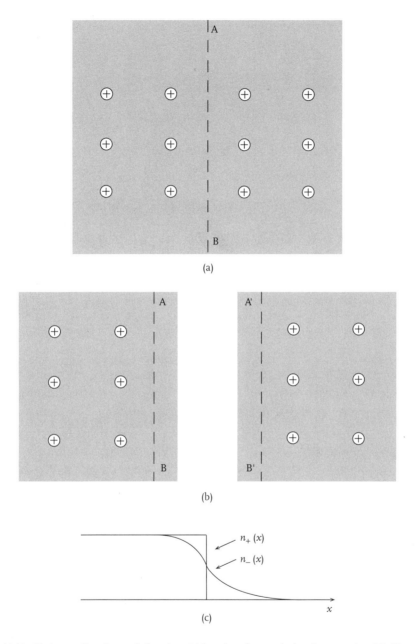

FIGURE 14.13. Understanding the work function. (a) Interior of a metal, showing a section AB. The negative charge density is spread into a jelly, shown by the shading. (b) If the metal is cut at the section AB, the negative charges spill out a small distance beyond the surfaces AB and A'B' so created. (c) Profiles of the charge densities $n_\pm(x)$ at a metal–vacuum interface.

electrostatic potential produced by the metal. Thus, if we start with the $N-1$ electron system, the additional energy of the Nth electron changes from $-e\phi(\infty)$ when it is at $x = \infty$ to $-e\phi(-\infty) + \mu$ when it is at $x = -\infty$. The precise way in which the remaining $N-1$ electrons adjust when the Nth electron is at intermediate positions (including those near the surface) is irrelevant.

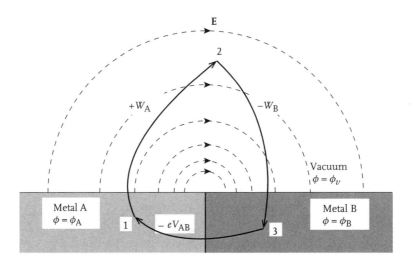

FIGURE 14.14. The contact potential. Two metals A and B with work functions W_A and W_B are in contact next to a vacuum. No net work must be done on an electron taken around the circuit 123. Hence, the work done on leg 31 must equal $W_B - W_A$. This work is $-eV_{AB}$, where $V_{AB} = \phi_A - \phi_B$ is the contact potential. The difference in the potentials of the two metals leads to a macroscopic electric field in the vacuum region as shown.

Work functions of typical metals lie in the 2- to 5-eV range. For a particular metal, the work function depends on whether the surface is atomically rough or smooth and, if the latter, on the particular crystallographic plane at the surface. These variations are of the order of 5.

The inequality of work functions for two dissimilar metals immediately implies that there will be a potential difference when these metals are in contact with each other. To see this, look at fig. 14.14. Let the work functions of metals A and B be W_A and W_B, and consider an electron transported around the closed circuit 123. Work W_A is done on leg 12, and $-W_B$ on leg 23, so work $W_B - W_A$ must be done on leg 32. This means that metal A must be at a potential V_{AB} relative to metal B, where

$$e V_{AB} = W_A - W_B. \tag{93.4}$$

The quantity V_{AB} is known as the *contact potential*. The need for its existence can also be seen by noting that the chemical potential for electrons must be the same in both metals when they are in contact. In the absence of the contact, we have $\mu_A = -W_A - e\Delta\phi_A$, and $\mu_B = -W_B - e\Delta\phi_B$. Setting $\mu_A = \mu_B$, we get

$$e(\Delta\phi_A - \Delta\phi_B) = W_A - W_B. \tag{93.5}$$

Writing $\Delta\phi_A = \phi_v - \phi_A$, and likewise for metal B, where ϕ_v, ϕ_A, and ϕ_B are the electrostatic potentials far in the vacuum, metal A, and metal B, asymptotically far from each other, we get

$$e(\phi_A - \phi_B) = W_A - W_B, \tag{93.6}$$

in full accord with eq. (93.4). Thus, the contact potential shifts the electronic energy levels of one metal relative to the other so as to align their chemical potentials (or Fermi energies).

Microscopically, the contact potential is brought about by a double layer of charge at the metal–metal interface, exactly like the double layer at a metal–vacuum interface. Macroscopically, it leads to an electric field in the vacuum region adjoining both metals, as shown in fig. 14.14. Although very small compared to the field in the double layer, it is nevertheless surprising that a field of order 1 V/cm extending over macroscopic distances can arise just by having two metals in contact.

15 | Electrostatics of dielectrics

94 The dielectric constant

Nonconducting materials, also known as insulators or dielectrics, differ from conductors in that (i) any free (mobile) charge must be externally supplied, and (ii) electric fields can penetrate inside them. For electrostatics, the macroscopic equations to be solved are Gauss's law

$$\nabla \cdot \mathbf{D} = 4\pi\rho_{\text{free}}, \tag{94.1}$$

$$\nabla \cdot \mathbf{D} = \rho_{\text{free}}, \quad \text{(SI)}$$

and the law that the electric field is irrotational,

$$\nabla \times \mathbf{E} = 0. \tag{94.2}$$

As discussed in chapter 13, to solve eqs. (94.1) and (94.2) one needs a constitutive relation connecting $\mathbf{D}$ and $\mathbf{E}$, or, equivalently, a knowledge of $\mathbf{P}$.[1] For most materials, it is found that $\mathbf{D}$ is linear in $\mathbf{E}$, at least for low enough electric fields. The reason is exactly the same as that behind the concept of polarizability (section 89): at low fields, the induced dipole moments are linear in $\mathbf{E}$, so $\mathbf{P}$ is linear in $\mathbf{E}$. For isotropic materials, we write[2]

$$\mathbf{D} = \epsilon\mathbf{E}. \quad \text{(Gaussian and SI)} \tag{94.3}$$

---

[1] Mathematically, we need the divergence and curl of the *same* vector field for the uniqueness theorem of section 15 to apply. Suppose we want to solve for $\mathbf{D}$. We know $\nabla \cdot \mathbf{D}$. If we knew $\mathbf{P}$, we would be in business, because $\nabla \times \mathbf{D} = 4\pi\nabla \times \mathbf{P}$, and so we would know $\nabla \times \mathbf{D}$ as well.

[2] We ignore here the class of crystals known as *pyroelectrics*, in which there is a nonzero polarization even in the absence of an applied field, so that the expansion of $\mathbf{D}$ in powers of $\mathbf{E}$ begins at the zeroth order: $D_i = D_{0,i} + \epsilon_{ij} E_{0,j}$. We discuss these very briefly at the end of section 99.

The quantity ϵ (in Gaussian units), or ϵ/ϵ_0 (in SI), is known as the *dielectric constant* of the material.[3] It is a dimensionless number with the same value in both systems. For anisotropic systems, such as solid or liquid crystals, a linear relationship must, in general, be tensorial:

$$D_i = \epsilon_{ij} E_j, \tag{94.4}$$

where ϵ_{ij} are the components of the dielectric tensor. In the special case of cubic crystals, the tensor is isotropic, $\epsilon_{ij} = \epsilon \delta_{ij}$, so one is back to eq. (94.3). Hence, such crystals behave like isotropic materials in their dielectric properties.

At the risk of being tiresome, we repeat that while eq. (94.3) is valid for a large variety of matter, it is not a fundamental law. Also, it holds only if the field **E** is small enough. We gave examples of the range of validity in chapter 13.

The linear relation (94.3) between **D** and **E** is equivalently parametrized in terms of a relation between **P** and **E**:

$$\mathbf{P} = \chi_e \mathbf{E}, \tag{94.5}$$

$$\mathbf{P} = \chi_e \epsilon_0 \mathbf{E}, \quad \text{(SI)}$$

where

$$\chi_e = \frac{\epsilon - 1}{4\pi}. \tag{94.6}$$

$$\chi_e = \frac{\epsilon - \epsilon_0}{\epsilon_0}. \quad \text{(SI)}$$

The quantity χ_e is known as the *electric susceptibility*. It is a dimensionless number in both Gaussian and SI systems, but not the same:

$$\chi_e(\text{SI}) = 4\pi \chi_e \text{ (Gaussian)}. \tag{94.7}$$

The following example from semiconductor physics provides a good illustration of the scope and usefulness of the concept of a dielectric constant.[4] Suppose we take a group IV semiconductor such as Si, and replace one of the Si atoms (4 valence electrons) by an atom of P (5 valence electrons). Four of the valence electrons of the impurity atom form covalent bonds with the neighboring Si atoms, but the fifth is de trop. It occupies one of the orbitals forming the conduction band and, as such, is free to hop from atom to atom in the crystal. It is not completely free, however, because the positive charge left behind on the P impurity pulls it back. The measured binding energy is rather weak: 0.045 eV for P in Si, and 0.0127 eV for As in Ge. We can construct a marvelously simple self-consistent model of this bound state by making use of only certain basic facts about semiconductors and using a dielectric constant to describe the electric field.

---

[3] The notation κ for the ratio ϵ/ϵ_0 is also common in the SI system.

[4] This example is discussed in all standard solid-state physics texts. See, e.g., Ashcroft and Mermin (1976), chap. 28; Kittel (1975), chap. 11; or Kittel (1987), chap. 14.

Table 15.1. Donor impurity states in semiconductors

System	m/m^*	ϵ	a_0^*	E_b^* (theory)	E_b^* (expt.)
As in Ge	10	15.8	84 Å	0.00531 eV	0.0127 eV
P in Si	5	11.7	31 Å	0.0199 eV	0.045 eV

If the extra electron is weakly bound to the impurity, we can expect the mean separation between it and the positive ion to be large. Let us ask what the potential energy $V(r)$ of interaction is for a fixed macroscopic separation, r. The $\mathbf{D}$ field of the P^+ ion is $e\hat{\mathbf{r}}/r^2$, so $\mathbf{E} = \mathbf{D}/\epsilon = e\hat{\mathbf{r}}/\epsilon r^2$, and integrating this we find that $V(r) = -e^2/\epsilon r$. The potential is reduced by a factor of ϵ from what we would have in free space because the atoms in the space between the ion and the electron get polarized and thus partially cancel the attraction. Of course, the electron is not static, but we may hope that as long its motion is slow (on some suitable time scale), the polarization of the atoms of the host medium will be able to follow along and that we may use the macroscopic static dielectric constant to describe it. The reduction of the attraction by the dielectric screening is one reason why the binding is weak.

The second fact is that the motion of conduction electrons is described not by the mass m of an electron moving in free space but by an effective mass, m^*, reflecting the energy versus wave-vector relation of electron Bloch states in the crystal. The ratio m^*/m is quite small, 0.1 for Ge, and 0.2 for Si. Since a lighter particle has less momentum, by the uncertainty principle, it must have a larger spatial spread, providing a second reason for the binding to be weak.

The picture that arises from these two points is exactly that of the hydrogen atom, except that the electron mass is replaced by m^*, and the coupling constant e^2 is replaced by e^2/ϵ. The hydrogen atom Bohr radius $a_0 = \hbar^2/me^2 = 0.529$ Å, and the ground-state binding energy $E_b = me^4/2\hbar^2 = 13.6$ eV. These get replaced by

$$a_0^* = \epsilon \frac{m}{m^*} a_0, \qquad E_b^* = \frac{1}{\epsilon^2} \frac{m^*}{m} E_b. \qquad (94.8)$$

The quantity a_0^* is an effective Bohr radius, giving the mean distance between the extra electron and the P^+ ion, and the quantity E_b^* is the energy with which the electron is bound. In table 15.1, we show the experimental values of these quantities for P in Si and for As in Ge, along with the answers given by our model.

The agreement between calculated and observed binding energies is fair at best. It can be improved by describing the motion of the electrons in the host via an anisotropic mass tensor. When this is done, one obtains binding energies of 0.009 eV for Ge and 0.030 eV for Si. From our perspective, the essential point is that the macroscopic concept of the dielectric constant is semiquantitatively accurate down to distances of the order of a_0^*—a few tens of angstroms. The true motion of the added electron and the response of the host atoms is in reality a fairly intricate many-body problem. It is pleasingly surprising that a simple model that lumps this response into a single macroscopic number works so well.

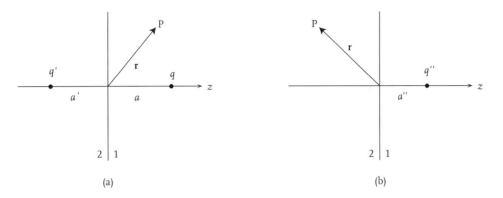

FIGURE 15.1. Method of images for a point charge q near a planar interface between dielectrics. Parts (a) and (b) show the image configurations for an observer (P) in regions 1 and 2, respectively. The distances a, a', and a'' all turn out to be equal, while the answers for the charges q' and q'' are given in eq. (95.7).

95 Boundary value problems for linear isotropic dielectrics

We now turn to the problem of solving for the fields produced by given charge configurations in the presence of dielectrics. As in the case of conductors, the only general methods are numerical, but the problem is even harder now because the boundary conditions are more complex. We will discuss only the simplest solvable situations.

Let us assume that space is filled by a number of dielectrics, each with its own dielectric constant ϵ. We will assume that each dielectric is homogeneous, so that except at the interface between two dielectrics, ϵ is a constant. Further, we will assume that there are no extraneous surface charges, so that $\sigma_{\text{free}} = 0$.

Since $\nabla \times \mathbf{E} = 0$, we may continue to define the potential ϕ via $\mathbf{E} = -\nabla\phi$, with the same meaning as before. The remaining equation, $\nabla \cdot \mathbf{D} = \rho_{\text{free}}$, therefore turns into Poisson's equation,

$$\nabla^2\phi = -4\pi\rho_{\text{free}}/\epsilon. \tag{95.1}$$

This equation holds piecewise with the appropriate value of ϵ in each dielectric, and the problem is to solve this set subject to the boundary conditions (82.10) and (82.9). In terms of ϕ, these conditions can be rewritten thus. First, since $\mathbf{E} = -\nabla\phi$, and the field is everywhere finite (except right on top of a point charge), it follows that the potential must be continuous. The continuity of $\mathbf{E}_t$ then follows automatically. This, along with eq. (82.9), then yields

$$\phi_1 = \phi_2, \tag{95.2}$$

$$\epsilon_1 \frac{\partial\phi_1}{\partial n} = \epsilon_2 \frac{\partial\phi_2}{\partial n}. \tag{95.3}$$

Point charge near plane interface; solution by images: As the first problem, we consider a point charge q near an infinite plane interface between two dielectrics. We take the interface as the plane $z = 0$, and the charge to lie at the point $(0, 0, a)$ (fig. 15.1). Let us

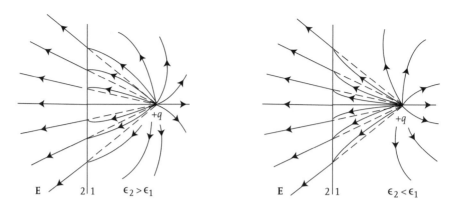

FIGURE 15.2. Electric field lines for a point charge q near a planar interface between dielectrics, for the cases $\epsilon_2 > \epsilon_1$ and $\epsilon_2 < \epsilon_1$, where the charge is placed in the region with dielectric constant ϵ_1.

denote the potential in regions 1 ($z > 0$) and 2 ($z < 0$) by ϕ_1 and ϕ_2, respectively. It is clear that if a method of images is to work, we must use different image charges to obtain ϕ_1 and ϕ_2. The reason is that ϕ_2 obeys Laplace's equation ($\nabla^2 \phi_2 = 0$) everywhere in region 2, so it must be described in terms of fictitious charges lying entirely in region 1. For ϕ_1, however, we must have one (and only one) source term in region 1, and so any image charges must lie in region 2. With this in mind, let us try the following *ansatz*:

$$\phi_1(\mathbf{r}) = \frac{1}{\epsilon_1} \left[\frac{q}{|\mathbf{r} - a\hat{\mathbf{z}}|} + \frac{q'}{|\mathbf{r} + a'\hat{\mathbf{z}}|} \right], \tag{95.4}$$

$$\phi_2(\mathbf{r}) = \frac{1}{\epsilon_2} \frac{q''}{|\mathbf{r} - a''\hat{\mathbf{z}}|}. \tag{95.5}$$

In other words, to find ϕ in region 1, we imagine all space to have a dielectric constant ϵ_1 and use an image charge q' at $(0, 0, -a')$, while to find it in region 2, we imagine all space to have a dielectric constant ϵ_2 and use an image charge q'' at $(0, 0, a'')$. Simple algebra shows that the conditions (95.2) and (95.3) can be met only if $a'' = a' = a$, and if

$$\frac{q + q'}{\epsilon_1} = \frac{q''}{\epsilon_2}, \qquad q - q' = q'', \tag{95.6}$$

i.e., if

$$q' = \frac{\epsilon_1 - \epsilon_2}{\epsilon_1 + \epsilon_2} q, \qquad q'' = \frac{2\epsilon_2}{\epsilon_1 + \epsilon_2} q. \tag{95.7}$$

The resulting electric field is sketched in fig. 15.2. For definiteness, we have taken $q > 0$. The interesting points to note are that (i) lines of $\mathbf{E}$ in region 2 are straight lines originating from the source in region 1, and (ii) $q' > 0$ if $\epsilon_2 < \epsilon_1$, while $q' < 0$ if $\epsilon_2 > \epsilon_1$. The case where medium 2 is a conductor can be formally obtained by taking the limit $\epsilon_2 \to \infty$. Naturally, the method of images works for any charge distribution near a plane dielectric interface. It also works for line charges parallel to dielectric cylinders. It does not work for a point charge near a sphere.

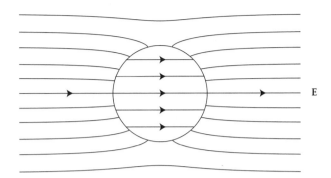

FIGURE 15.3. A dielectric sphere in a uniform external electric field. The **E** field inside the sphere is uniform but smaller than that outside. We show this by breaking off the lines of **E** and increasing the space between them inside the sphere.

Exercise 95.1 For a point charge near a flat dielectric interface, find the induced charge density at the interface, σ_{pol}, and show that it integrates to q'/ϵ_1. Physically interpret why q' has the same sign as $\epsilon_1 - \epsilon_2$. Also find the force on the charge $+q$ and the work required to bring it into final position from infinity. Compare with the answers when region 2 is a conductor and region 1 is vacuum.

Dielectric sphere in uniform external electric field: In many experimental situations a dielectric sample is placed in a uniform external electric field. It is of interest to know the field inside the sample if the field influences the sample property under study. Particularly important is whether the field inside the sample is homogeneous. This is so only for a few special shapes. The simplest of these is the sphere (fig. 15.3).[5] Choosing coordinates as for the corresponding problem with a conductor, we denote the potential inside and outside the sphere by ϕ_{in} and ϕ_{out}. Since there are no free charges, $\nabla^2 \phi_{\text{in}} = 0$, and $\nabla^2 \phi_{\text{out}} = 0$. Hence, in spherical polar coordinates, the solution is of the form

$$\phi_{\text{in}} = \sum_{\ell=0}^{\infty} A_\ell r^\ell P_\ell(\cos\theta), \tag{95.8}$$

$$\phi_{\text{out}} = -E_0 r \cos\theta + \sum_{\ell=0}^{\infty} \frac{B_\ell}{r^{\ell+1}} P_\ell(\cos\theta). \tag{95.9}$$

This incorporates the regularity of ϕ as $r \to 0$ and the behavior $\phi_{\text{out}} \approx -E_0 z$ for $r \to \infty$. From the continuity of ϕ at $r = a$, we get

$$A_1 = -E_0 + \frac{B_1}{a^3}, \tag{95.10}$$

$$A_\ell = \frac{B_\ell}{a^{2\ell+1}}, \qquad \ell \neq 1, \tag{95.11}$$

---

[5] The entire discussion that follows has a complete equivalent in magnetostatics, as we shall see in section 108. The case of ferromagnetic bodies, by themselves or in an external magnetic field, is perhaps of greatest practical importance.

while from the continuity of D_n, we get

$$\epsilon A_1 = -E_0 - 2\frac{B_1}{a^3},$$ (95.12)

$$\epsilon \ell A_\ell = -(\ell+1)\frac{B_\ell}{a^{2\ell+1}}, \qquad \ell \neq 1.$$ (95.13)

Solving these equations, we see that

$$A_1 = -\frac{3}{\epsilon+2}E_0, \qquad B_1 = \frac{\epsilon-1}{\epsilon+2}a^3 E_0.$$ (95.14)

All other A_ℓ and B_ℓ vanish. Hence,

$$\phi_{\text{in}} = -\frac{3}{\epsilon+2}E_0 z,$$ (95.15)

$$\phi_{\text{out}} = -E_0 z + \frac{\epsilon-1}{\epsilon+2}\frac{a^3}{r^2}E_0 \cos\theta.$$ (95.16)

The second term in ϕ_{out} is the potential due to an induced dipole of moment,

$$\mathbf{d}_{\text{ind}} = \frac{\epsilon-1}{\epsilon+2}a^3 \mathbf{E}_0.$$ (95.17)

$$\mathbf{d}_{\text{ind}} = 4\pi\epsilon_0\frac{\epsilon-\epsilon_0}{\epsilon+2\epsilon_0}a^3 \mathbf{E}_0. \quad \text{(SI)}$$

The electric field can now be easily obtained by taking gradients. The field outside the sphere is a sum of the uniform field $\mathbf{E}_0$ and that of the dipole $\mathbf{d}_{\text{ind}}$. More remarkable is the field inside the sphere:

$$\mathbf{E}_{\text{in}} = \frac{3}{\epsilon+2}\mathbf{E}_0,$$ (95.18)

$$\mathbf{E}_{\text{in}} = \frac{3\epsilon_0}{\epsilon+2\epsilon_0}\mathbf{E}_0, \quad \text{(SI)}$$

which is uniform. The polarization inside the sphere is also uniform and can be obtained either by dividing $\mathbf{d}_{\text{ind}}$ by the volume of the sphere or by using the relation (94.5) and (94.6) between $\mathbf{P}$ and $\mathbf{E}$. The result is

$$\mathbf{P} = \frac{3}{4\pi}\frac{\epsilon-1}{\epsilon+2}\mathbf{E}_0.$$ (95.19)

$$\mathbf{P} = 3\frac{\epsilon-\epsilon_0}{\epsilon+2\epsilon_0}\epsilon_0\mathbf{E}_0. \quad \text{(SI)}$$

96 Depolarization

The sphere in an external field can be profitably viewed in another way. The field inside the sphere is reduced from the field outside because of polarization charges induced on its surface. If we write

$$\mathbf{E}_{\text{in}} = \mathbf{E}_0 + \mathbf{E}_d,$$ (96.1)

where $\mathbf{E}_d$ is a *depolarization field*, then it is useful to relate $\mathbf{E}_d$ to $\mathbf{P}$, the polarization inside the dielectric. Using eqs. (95.18) and (95.19), we get

$$\mathbf{E}_d = -\frac{4\pi}{3}\mathbf{P}. \tag{96.2}$$

$$\mathbf{E}_d = -\frac{1}{3\epsilon_0}\mathbf{P}. \quad \text{(SI)}$$

The number $4\pi/3$ or $1/3$ is known as a *depolarization coefficient*.[6]

This point of view has the following advantage. Suppose the polarization of the sphere is somehow frozen in place. We can then regard eq. (96.1) as the superposition of two electric fields: $\mathbf{E}_0$, due to the external sources (capacitor plates, e.g.), and $\mathbf{E}_d$, due to the polarization $\mathbf{P}$. Equation (96.2) is then seen to be valid even if there is no external field, i.e., $\mathbf{E}_0 = 0$. Given its practical importance, it is useful to derive this result directly. The surface charge density $\sigma_{\text{pol}} = \mathbf{P} \cdot \hat{\mathbf{n}} = P\cos\theta$. The resulting "depolarization potential" ϕ_d inside the sphere is given by

$$\phi_d(\mathbf{r}) = \int \frac{P\cos\theta'}{|\mathbf{r} - a\hat{\mathbf{n}}'|} a^2 d\Omega', \tag{96.3}$$

where $\cos\theta' = \hat{\mathbf{z}} \cdot \hat{\mathbf{n}}'$. Expanding $|\mathbf{r} - a\hat{\mathbf{n}}'|^{-1}$ in spherical harmonics and writing $\cos\theta' = kY_{10}(\hat{\mathbf{n}}')$ (where k is a constant whose precise value is not needed), we obtain

$$\phi_d(\mathbf{r}) = a^2 P \int d\Omega' \sum_{\ell,m} \frac{4\pi}{2\ell+1} \frac{r^\ell}{a^{\ell+1}} Y_{\ell m}(\hat{\mathbf{r}}) Y_{\ell m}^*(\hat{\mathbf{n}}') k Y_{10}(\hat{\mathbf{n}}')$$

$$= a^2 P \frac{4\pi}{3} \frac{r}{a^2} k Y_{10}(\hat{\mathbf{r}})$$

$$= \frac{4\pi}{3} Pr\cos\theta. \tag{96.4}$$

(The integral over $\hat{\mathbf{n}}'$ follows by orthonormality, whence only the Y_{10} term survives, and in the last line we use $kY_{10}(\hat{\mathbf{r}}) = \cos\theta$ again.) Equation (96.2) follows from $\mathbf{E}_d = -\nabla\phi_d$.

The concept of depolarization extends to a dielectric body of any shape, not just a sphere. If we place the body between the plates of a capacitor, e.g., the matter of the dielectric will be polarized, so that a net negative charge will accumulate near the positive plate, and vice versa. These polarization charges can be regarded as forming a second capacitor that opposes the external one and leads to a reduced field inside the body. We show this schematically in fig. 15.4. In general, of course, we will also have volume polarization charges, and $\mathbf{P}$ and $\mathbf{E}$ will be nonuniform. Further, the larger the dielectric constant, the more polarizable the body, and the smaller the internal field.[7]

In fact, the depolarization field depends strongly on the shape of the body. We can see this by considering the exactly solvable cases of a long rod of any uniform cross-sectional shape in an external field parallel to the rod, and a uniformly thin slab of any shape in a field perpendicular to the slab. Ignoring end and edge effects, the answers are obtained

---

[6] Or, in the magnetic case, a *demagnetization coefficient*. See section 108.
[7] From this viewpoint, a conductor is infinitely polarizable.

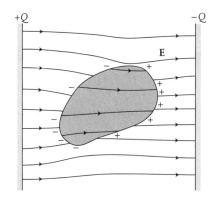

FIGURE 15.4. A dielectric body of general shape between the plates of a capacitor. Induced charges on the body's surface, as well as in its bulk (not shown), reduce the electric field inside the body, which will, in general, be nonuniform.

by using the continuity of E_t and D_n, respectively, and are

$$\mathbf{E}_{in} = \mathbf{E}_0, \qquad \text{(rod, Gaussian and SI)} \tag{96.5}$$

$$= \mathbf{E}_0/\epsilon. \qquad \text{(slab, Gaussian)} \tag{96.6}$$

$$= \epsilon_0 \mathbf{E}_0/\epsilon. \qquad \text{(slab, SI)}$$

Exercise 96.1 Find the electric field inside a cavity in a dielectric with a uniform applied electric field $\mathbf{E}_0$ far from the cavity, when the cavity is (a) spherical, (b) a thin cylinder (of any cross section) with axis parallel to the field, (c) a uniformly thin slab (of any shape) perpendicular to the field.

Exercise 96.2 A dielectric in the shape of an infinitely long right-circular cylinder is placed in a uniform external field $\mathbf{E}_0$ perpendicular to the axis of the cylinder. Find the electric field inside the cylinder. [**Answer**: $(2/(\epsilon + 1))\mathbf{E}_0$.]

We summarize our results in fig. 15.5, giving them in terms of the depolarization field. Formulas for material without a permanent polarization in a uniform external field are obtained by substituting $\mathbf{P} = (\epsilon - 1)\mathbf{E}_{in}/4\pi$ and $\mathbf{E}_d = \mathbf{E}_{in} - \mathbf{E}_0$. In general, the field inside the body will not be uniform even when placed in a uniform external field. The most general shape for which it *is* so is an ellipsoid. The calculations that establish this fact do not require any new physical principles, but they entail a laborious detour through ellipsoidal coordinates. Because of the unique nature of the result, and because researchers often try to approximate a general shape by an ellipsoid in order to estimate the internal field, we give the most important formulas for reference.[8]

---

[8] See Landau and Lifshitz (1960), vol. 8, sec. 8, for a succint derivation.

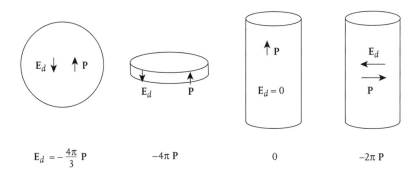

$$E_d = -\frac{4\pi}{3}\,P \qquad\qquad -4\pi\,P \qquad\qquad\qquad 0 \qquad\qquad\qquad -2\pi\,P$$

FIGURE 15.5. Depolarization fields for some special cases: a sphere, a thin slab, a long, narrow rod polarized along its axis, and a long, narrow, right-circular cylinder polarized perpendicular to its axis.

Let the ellipsoid surface be given by the equation

$$\frac{x^2}{a^2} + \frac{y^2}{b^2} + \frac{z^2}{c^2} = 1, \tag{96.7}$$

with a, b, and c all unequal, in general. The cross sections of this surface with the xy, yz, and zx planes are ellipses with semiaxes a, b, and c. The depolarization field is given by

$$E_{d,j} = -4\pi n_{jk}\,P_k, \tag{96.8}$$

$$E_{d,j} = -\frac{1}{\epsilon_0} n_{jk}\,P_k, \quad \text{(SI)}$$

where n_{jk} is the so-called *depolarization tensor*. This tensor is symmetric and has principal values $n^{(x)}$, $n^{(y)}$, and $n^{(z)}$ along the axes of the ellipsoid. The coefficient $n^{(z)}$ is given by

$$n^{(z)} = \frac{1}{2}abc \int_0^\infty \frac{ds}{(s+c^2)((s+a^2)(s+b^2)(s+c^2))^{1/2}}, \tag{96.9}$$

and the other two coefficients are obtained by permutation of a, b, and c. It is easy to see that these coefficients depend only on the ratios $a:b:c$. Thus, the tensor n_{jk} depends not on the volume but only on the shape of the ellipsoid. It also has the important property

$$\text{Tr}\,\mathbf{n} = n^{(x)} + n^{(y)} + n^{(z)} = 1, \tag{96.10}$$

which can be seen by adding eq. (96.9) and its permutates and using the substitution $u = (s+a^2)(s+b^2)(s+c^2)$. The results for the sphere, the infinite cylinder, and the infinitely thin circular disk can be obtained by using symmetry, the fact that there is no depolarization along a dimension that is effectively infinite—see eqs. (96.5) and (96.6), and this trace property.

97 Thermodynamic potentials for dielectrics

We now wish to investigate the energy of a dielectric in an electric field. Since the field penetrates into a dielectric, and since this will, in general, modify its other properties (entropy and energy, e.g.), the problem has to be posed as one of thermodynamics. There are several choices for the additional variable that must be used to specify the thermodynamic state of the dielectric ($\mathbf{E}$, $\mathbf{D}$, and $\mathbf{P}$, e.g.), and for each there is an appropriate thermodynamic potential (internal or free energy).[9] Different authors use different choices, and since there is no natural notational convention that would make it easy to remember the distinctions, it is hard to compare different treatments.[10] We will not discuss all the possibilities, just enough to enable the reader to see the interrelationships.[11]

As our starting point, let us consider the first law of thermodynamics and determine the electrostatic contribution to the energy accounting. We suppose that a dielectric is placed amidst a set of charged conductors, the charges on which are changed by infinitesimal amounts δQ_a while their potentials ϕ_a are held fixed (fig. 15.6). There are no free charges on the dielectric, and the entire system is thermally insulated. The adiabatic work received by the system (dielectric, conductors, and field) in this process is

$$\delta W = \sum_a \phi_a \delta Q_a. \tag{97.1}$$

Let us now denote by V the volume consisting of all space except that occupied by the conductors. We also denote the outward normal to this volume by $\hat{\mathbf{n}}$, so that $\hat{\mathbf{n}}$ points into the conductors. Then, since $\mathbf{D} \cdot \hat{\mathbf{n}} = -4\pi\sigma_a$ on the ath conductor, and since the surface S_a of this conductor is an equipotential, we can write

$$\phi_a \delta Q_a = -\frac{1}{4\pi} \oint_{S_a} \phi \, \delta\mathbf{D} \cdot \hat{\mathbf{n}} \, d^2s. \tag{97.2}$$

---

[9] We recall that a thermodynamic potential is a quantity that achieves its minimum value in thermodynamic equilibrium when its natural variables are held fixed and other variables are allowed to relax. Thus, the Helmholtz free energy of a fluid is a thermodynamic potential in terms of the volume, V, and the temperature, T. Further, all other thermodynamic quantities can be obtained if a potential is known as a function of its natural variables. As a counterexample, the internal energy, U, when given as a function of V and T, is *not* a thermodynamic potential.

[10] Nevertheless, excellent discussions of the subject exist. In addition to Jackson (1999), we have benefited from Landau and Lifshitz (1960), vol. 8, secs. 10 and 11; Balian (1991), vol. I, pp. 284–295; and Panofsky and Phillips (1955), chap. 6.

[11] In this section, we will write equations only in the Gaussian system. To write these in the SI system as well would add unnecessary clutter, since we would end up repeating a very large number of essentially identical expressions. To obtain SI expressions, replace a factor of $1/4\pi$ multiplying dot products $\mathbf{E} \cdot \mathbf{D}$, $\mathbf{E} \cdot \delta\mathbf{D}$, etc., by unity. Expressions involving dot products of $\mathbf{P}$ and $\mathbf{E}$ hold in both systems.

FIGURE 15.6. A dielectric body in the presence of charged conductors at fixed potentials.

Summing over all the conductors and using Gauss's law, we transform the total work into a volume integral,

$$\delta W = -\frac{1}{4\pi} \int_V \nabla \cdot (\phi \delta \mathbf{D}) d^3x. \tag{97.3}$$

Since there are no free volume charges, $\nabla \cdot \delta \mathbf{D} = 0$, so using $\nabla \phi = -\mathbf{E}$, this can be written as

$$\delta W = \frac{1}{4\pi} \int_V \mathbf{E} \cdot \delta \mathbf{D} \, d^3x. \tag{97.4}$$

This is the fundamental expression for the work in terms of the fields.

If we denote the total energy of the system by U, and the temperature and entropy of the dielectric by T and S, and add to eq. (97.4) the heat received in a nonadiabatic process, we obtain the first law,[12]

$$\delta U = T\delta S + \frac{1}{4\pi} \int_V \mathbf{E} \cdot \delta \mathbf{D} \, d^3x. \tag{97.5}$$

This energy is a thermodynamic potential with respect to entropy and $\mathbf{D}$. The potential with respect to temperature and $\mathbf{D}$ is the free energy,

$$F = U - TS, \tag{97.6}$$

$$\delta F = -S\delta T + \frac{1}{4\pi} \int_V \mathbf{E} \cdot \delta \mathbf{D} \, d^3x. \tag{97.7}$$

We can similarly define potentials $\tilde{U}$ and $\tilde{F}$, analogous to U and F, in which the natural field variable is $\mathbf{E}$ rather than $\mathbf{D}$. We tabulate the relevant formulas below along with those for dF and dU.

---

[12] We ignore here changes in the volume of the system. To include these, we add to δU, the mechanical work received by the system, a term $-p\delta V$, where p is the pressure.

Thermodynamic potential	Formula	Natural variables	Differential
U		$S, \mathbf{D}$	$T\delta S + \dfrac{1}{4\pi}\displaystyle\int_V \mathbf{E}\cdot\delta\mathbf{D}\, d^3x$
F	$U - TS$	$T, \mathbf{D}$	$-S\delta T + \dfrac{1}{4\pi}\displaystyle\int_V \mathbf{E}\cdot\delta\mathbf{D}\, d^3x$
$\tilde{U}$	$U - \dfrac{1}{4\pi}\displaystyle\int_V \mathbf{E}\cdot\mathbf{D}\, d^3x$	$S, \mathbf{E}$	$T\delta S - \dfrac{1}{4\pi}\displaystyle\int_V \mathbf{D}\cdot\delta\mathbf{E}\, d^3x$
$\tilde{F}$	$\tilde{U} - TS$	$T, \mathbf{E}$	$-S\delta T - \dfrac{1}{4\pi}\displaystyle\int_V \mathbf{D}\cdot\delta\mathbf{E}\, d^3x$

$$\tag{97.8}$$

To see the connection between $\tilde{F}$ and F (or $\tilde{U}$ and U), we rewrite the volume integral of $\mathbf{E}\cdot\mathbf{D}$ in terms of the potentials and charges on the conductors by essentially reversing the steps from eq. (97.1) to eq. (97.4):

$$
\begin{aligned}
\frac{1}{4\pi}\int_V \mathbf{E}\cdot\mathbf{D}\, d^3x &= -\frac{1}{4\pi}\int_V (\nabla\phi)\cdot\mathbf{D}\, d^3x \\
&= -\frac{1}{4\pi}\int_V \nabla\cdot(\phi\mathbf{D})\, d^3x \\
&= -\frac{1}{4\pi}\oint_{S=\partial V} \phi\mathbf{D}\cdot\hat{\mathbf{n}}\, d^2s \\
&= \sum_a \phi_a Q_a.
\end{aligned}
\tag{97.9}
$$

Denoting the change in a thermodynamic potential Φ at fixed X by $(\delta\Phi)_X$, eq. (97.1) is equivalent to asserting that

$$(\delta F)_T = (\delta U)_S = \sum_a \phi_a \delta Q_a. \tag{97.10}$$

Since $\tilde{U} - U = \tilde{F} - F = -\sum_a \phi_a Q_a$, it then follows that

$$(\delta\tilde{F})_T = (\delta\tilde{U})_S = -\sum_a Q_a \delta\phi_a. \tag{97.11}$$

The physical meaning of the free energies F and $\tilde{F}$ is contained in eqs. (97.10) and (97.11) and the principle that a thermodynamic potential is minimized in equilibrium if its natural variables are held fixed. Thus, if the temperature (T) and the charges on the conductors (the Q_a's) are held fixed, and other variables of the system are allowed to vary, the system will relax so as to minimize F in equilibrium. If, on the other hand, the temperature (T) and the *potentials* on the conductors (the ϕ_a's) are fixed and other variables of the system are allowed to vary, the system will relax so as to minimize $\tilde{F}$ in equilibrium. Similar remarks apply to U and $\tilde{U}$ in isentropic processes.

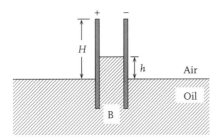

FIGURE 15.7. If a parallel plate capacitor is charged and dipped in an oil bath, the oil will be drawn up between the plates, irrespective of whether the plates are at a fixed charge or fixed potential. At which point does the force pulling up the oil act?

Up to now, we have only written relations for the differential changes in the energy when the charges or the potentials are changed. To integrate these expressions in general, we would need the constitutive relation $\mathbf{D(E)}$ or $\mathbf{E(D)}$. In the special case where this relationship can be assumed to be linear, this integration is simple, and we have

$$\mathbf{E} \cdot \delta\mathbf{D} = \delta\mathbf{E} \cdot \mathbf{D} = \tfrac{1}{2}\delta(\mathbf{E} \cdot \mathbf{D}). \tag{97.12}$$

Hence, if we denote the free energy in the absence of any externally applied electric field by F_0 etc., then

$$F - F_0 = \frac{1}{8\pi} \int_V \mathbf{E} \cdot \mathbf{D} d^3x = \frac{1}{2} \sum_a \phi_a Q_a, \tag{97.13}$$

$$\tilde{F} - \tilde{F}_0 = -\frac{1}{8\pi} \int_V \mathbf{E} \cdot \mathbf{D} d^3x = -\frac{1}{2} \sum_a \phi_a Q_a. \tag{97.14}$$

The first of these lines is easy to understand: the right-hand side is simply the work required to assemble the charges Q_a on the conductors. The second line is somewhat counterintuitive, but we must remember that now the potentials of the conductors are assumed to be held fixed as the state of the dielectric is altered. For this to happen, extra charges must flow from infinity onto the conductors, and the right-hand side also includes the work required for this charge transfer.

As an application of these ideas, let us consider the following problem. A parallel plate capacitor with fixed voltage drop V across its plates is dipped into a bath of oil with dielectric constant ϵ. We wish to find the height to which the oil rises inside the plates (fig. 15.7). Since the potentials on the capacitor plates are fixed, the appropriate potential that must be a minimum in equilibrium is $\tilde{U}$ (or $\tilde{F}$). Let the oil rise to a height h above the bath and the plates to a height H, and let the spacing between the plates be d and the transverse dimensions of the plates be L. There are two relevant contributions to $\tilde{U}$: the gravitational energy of the oil, and the volume integral of $-\mathbf{E} \cdot \mathbf{D}/8\pi$ inside the plates. Continuity of E_t implies that $E = V/d$ everywhere, but in the oil-filled region $D = \epsilon E$, while in the air-filled region, $D = E$. Thus,

$$\frac{\tilde{U}}{Ld} = \frac{1}{2}\rho_m g h^2 - \frac{E^2}{8\pi}(\epsilon h + (H - h)), \tag{97.15}$$

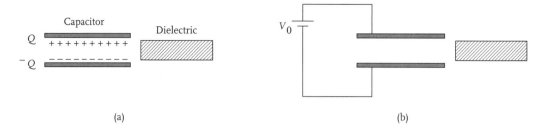

FIGURE 15.8. Is the dielectric slab pulled in or pushed out when the capacitor plates are (a) at a fixed charge, (b) at a fixed potential difference?

where ρ_m is the mass density of the oil, and g is the gravitational acceleration. Minimizing this with respect to h at fixed E (since that is equivalent to fixing V), we obtain the answer,

$$h = \frac{E^2}{8\pi\rho_m g}(\epsilon - 1). \tag{97.16}$$

Let us now consider the same problem except that this time let the charges $\pm Q$ on the plates be held fixed. The potential to minimize is now U. Introducing the electric field E between the capacitor as before, this is given by

$$\frac{U}{Ld} = \frac{1}{2}\rho_m g h^2 + \frac{E^2}{8\pi}(\epsilon h + (H - h)). \tag{97.17}$$

However, U must be minimized at fixed Q. Thus, we must reexpress E in terms of Q. To do this, let us write the charge density on the positively charged plate as σ_a and σ_o in the air and oil-filled regions, respectively. By the usual Gaussian pillbox argument, we find $\sigma_a = E/4\pi$ and $\sigma_o = \epsilon E/4\pi$. Since the total charge must add up to Q, we have

$$E = \frac{4\pi Q}{(\epsilon h + (H - h))L}. \tag{97.18}$$

Feeding this result into eq. (97.17) and minimizing the resulting expression for U, we find that if the final answer for h is reexpressed back in terms of E, it is once again given by eq. (97.16). In this case too, the oil is pulled in between the plates, not pushed down.

To properly appreciate these results, the reader should try and obtain them by considering the forces acting on the dielectric.

Exercise 97.1 The plates in a parallel plate capacitor have area A and are spaced by a distance d. The plates are charged to a potential difference V_0. A dielectric slab of exactly the shape required to fit completely between the plates is inserted between them (fig. 15.8), with the charging battery (a) disconnected and (b) left connected. In each case, find the work received by the dielectric slab, and state whether it is pulled in or must be pushed in.

Answer: In case (a) the energy stored between the plates is initially $U_i = AV_0^2/8\pi d$, and finally $U_f = AV_0^2/8\pi d\epsilon < U_i$. The difference $|\Delta U| = U_i - U_f$ is the (positive) work received by the slab. In case (b) $U_i = AV_0^2/8\pi d$, and $U_f = AV_0^2\epsilon/8\pi d$, which is *greater* than U_i. But, now, the charge on the capacitors increases from $Q_i = AV_0/4\pi d$ to

$Q_f = \epsilon A V_0/4\pi d$. This charge is supplied by the battery at a constant voltage of V_0. Consequently, the battery supplies an amount of energy $E_b = AV_0^2(\epsilon - 1)/4\pi d$. The dielectric slab receives an amount of work $\Delta W = E_b - (U_f - U_i) = AV_0^2(\epsilon - 1)/8\pi d$, which is again positive. In both cases, the slab is pulled in.

So far, we have discussed only *total* energies and free energies of a system of dielectrics and conductors. Our formulas for δU, δF, etc., also permit us, however, to discuss the energy or free energy densities, i.e., the quantities per unit volume. Denoting all densities by lowercase letters, u, f, s, etc., we have, for example,

$$\delta\tilde{u} = T\delta s - \frac{1}{4\pi}\mathbf{D}\cdot\delta\mathbf{E}. \tag{97.19}$$

Next, let us find the energy of a dielectric for given volume, temperature, and electric field, since these are common experimental conditions. We take a linear isotropic dielectric with $\mathbf{D} = \epsilon\mathbf{E}$, where ϵ depends on variables such as temperature and specific volume, but not on $\mathbf{E}$. Further, it is useful to imagine that the total volume of the dielectric is V, but only a part V_0 is subject to a nonzero and uniform electric field $\mathbf{E}$, created, perhaps, by a parallel plate capacitor. In that case, eq. (97.14) yields

$$\tilde{F}(V, T, E) = \tilde{F}(V, T, 0) - V_0\frac{E^2}{8\pi}\epsilon(V, T). \tag{97.20}$$

Then, since $S = -\partial\tilde{F}/\partial T$, and $\tilde{U} = \tilde{F} + TS$,

$$S(V, T, E) = S(V, T, 0) + V_0\frac{E^2}{8\pi}\frac{\partial\epsilon}{\partial T}, \tag{97.21}$$

$$\tilde{U}(V, T, E) = \tilde{U}(V, T, 0) - V_0\frac{E^2}{8\pi}\left(\epsilon - T\frac{\partial\epsilon}{\partial T}\right). \tag{97.22}$$

This shows that the field contribution to the internal energy density of a dielectric is not just $\epsilon E^2/8\pi$; one must also add the heat received, given by the term in $\partial\epsilon/\partial T$. Naturally, the same result is obtained if we work in terms of U rather than $\tilde{U}$.

Exercise 97.2 (electrostriction) In some materials, the dielectric constant depends on the pressure (or, what is the same thing, the specific volume), although the dependence is always weak. Show that in such cases, the application of an electric field to a dielectric of volume V causes a change in volume ΔV given by

$$\frac{\Delta V}{V} = -\frac{E^2}{8\pi}\left(\frac{\partial\epsilon}{\partial p}\right)_T. \tag{97.23}$$

This is known as the *Helmholtz-Lippman formula*, and the phenomenon is known as *electrostriction*. It may be of either sign.

98 Force on small dielectric bodies

All the energy expressions derived in the previous section included a part that would be present even if the dielectric were not there. For, even in the absence of the dielectric, an electric field $\mathbf{E}_0$ exists in the volume V. If we subtract the energy stored in this field from F, we obtain what may be called the total free energy of the dielectric body,

$$F_{\text{bod}} = F - \frac{1}{8\pi} \int_V E_0^2 \, d^3x. \tag{98.1}$$

This free energy may be said to pertain to the dielectric body alone. It should be noted that $\mathbf{E}_0(\mathbf{r})$ is not the same as the electric field $\mathbf{E}(\mathbf{r})$ that exists when the body is present. In particular, with a general conductor geometry and a body of general shape, $\mathbf{E}_0$ is not even the field outside the body. For this reason, designating it as an "external" field or even the "applied" field is a source of confusion, and we shall refer to it as the "vacuum" field. At the same time, since $\mathbf{E}_0$ is independent of the properties of the body, F_{bod} has the same thermodynamic derivatives as F, e.g.,

$$\left(\frac{\partial F_{\text{bod}}}{\partial T} \right)_{\mathbf{D}} = -S. \tag{98.2}$$

It is interesting to examine the change in F_{bod} arising from a variation in the charges Q_a on the conductors while the temperature of the body is held fixed. Making use of eq. (97.7), we obtain

$$\delta F_{\text{bod}} = \frac{1}{4\pi} \int_V (\mathbf{E} \cdot \delta \mathbf{D} - \mathbf{E}_0 \cdot \delta \mathbf{E}_0) \, d^3x. \tag{98.3}$$

We now write

$$(\mathbf{E} \cdot \delta \mathbf{D} - \mathbf{E}_0 \cdot \delta \mathbf{E}_0) = (\mathbf{D} - \mathbf{E}_0) \cdot \delta \mathbf{E}_0 + [\delta(\mathbf{D} - \mathbf{E}_0)] \cdot \mathbf{E} - (\mathbf{D} - \mathbf{E}) \cdot \delta \mathbf{E}_0, \tag{98.4}$$

and integrate each of the three terms in turn. We will show below that the first two terms integrate to zero. This leaves only the third term. But since $\mathbf{D} - \mathbf{E} = 4\pi \mathbf{P}$, we obtain for δF_{bod} the very simple expression

$$\delta F_{\text{bod}} = - \int_V \mathbf{P} \cdot \delta \mathbf{E}_0 \, d^3x. \quad \text{(Gaussian and SI)} \tag{98.5}$$

The integrand in this equation vanishes outside the dielectric. Yet, this expression cannot be regarded as the differential of a free energy in the same way as eq. (97.7), because it would give $\mathbf{P}$ as the derivative of the free energy with respect to a field $\mathbf{E}_0$ that is not a thermodynamic characteristic of the body.

We will now show that the integral of the first term in eq. (98.4) is zero. Let ϕ_0 be the potential corresponding to $\mathbf{E}_0$. Then,

$$\int_V (\mathbf{D} - \mathbf{E}_0) \cdot \delta \mathbf{E}_0 \, d^3x = \int_V (\mathbf{D} - \mathbf{E}_0) \cdot \nabla \delta \phi_0 \, d^3x = \int_S (\mathbf{D} - \mathbf{E}_0) \cdot \hat{\mathbf{n}} \delta \phi_0 \, d^2s - \int_V [\nabla \cdot (\mathbf{D} - \mathbf{E}_0)] \delta \phi_0 \, d^3x. \tag{98.6}$$

Arguments familiar from the previous section now show that both the surface and volume integral in the final expression vanish. The surface integral is a sum of surface integrals over the various conducting surfaces, each one of which vanishes because (i) $\delta\phi_0$, being constant on the conductor, can be pulled out of the integral, and (ii) the remaining integral for both D_n and $E_{0,n}$ is proportional to the same total charge. The volume integral vanishes because there are no free volume charges, and so $\nabla \cdot \mathbf{D} = \nabla \cdot \mathbf{E}_0 = 0$.

The same series of steps shows that the integral of the second term in eq. (98.4) also vanishes.

Exercise 98.1 Construct $\tilde{F}_{\text{bod}}$ from $\tilde{F}$ in analogy with eq. (98.1). Find $\delta\tilde{F}_{\text{bod}}$ and explain why this potential still does not yield a free energy density.

Equation (98.5) is useful in understanding how uncharged dielectric bodies behave in an inhomogeneous electric field. Let us suppose that the body is small enough so that the vacuum field $\mathbf{E}_0$ can be regarded as uniform over its dimensions. Let us also take the medium to be linear, so that the differential change δF_{bod} can be integrated. Then, with $\mathbf{d}_{\text{ind}}$ the total induced dipole moment of the body as before, we have

$$F_{\text{bod}} = F_0(V, T) - \tfrac{1}{2}\mathbf{d}_{\text{ind}} \cdot \mathbf{E}_0(\mathbf{r}). \tag{98.7}$$

Since $\mathbf{d}_{\text{ind}} = \alpha\mathbf{E}_0$, where α is the polarizability of the body, it follows that the free energy contains a term varying as $-\mathbf{E}_0^2$. Consequently, the body experiences a force toward regions of higher electric field intensity $|\mathbf{E}_0|$. This is why a glass rod that has been charged by rubbing attracts small pieces of paper.[13] The effect also provides additional insight into the problem of the capacitor dipped in oil (or of the slab pulled into the capacitor). We see, first of all, that the force depends only on the electric field, which is a local property. This is why the oil is pushed up irrespective of whether the battery that charged the capacitor is connected or not. Finally, by considering small volume elements of the oil, we see that the force that pushes the oil up acts at point B in fig. 15.7. This point can be better understood by considering an elementary dipole in the fringing field of a capacitor (fig. 15.9).

99 Models of the dielectric constant

We conclude this chapter by examining how the macroscopic dielectric constant is related to the microscopic properties of the medium. Naturally, this relationship is different for different types of matter, and we will consider only the simplest cases.

Nonpolar gases: Suppose a gas composed of nonpolar molecules, such as He or H_2, is placed in an external electric field. A dipole moment is thereby induced in each molecule.

---

[13] One can also understand, very qualitatively, the operation of laser tweezers for bodies smaller than the wavelength of light. However, the electric field is not static in this case, so, at the very least, one should use a frequency-dependent dielectric constant. Secondly, one must average the force over several time periods of oscillation of the field.

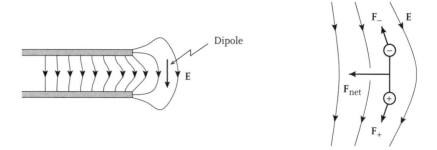

FIGURE 15.9. Force on a dipole in the fringing field of a capacitor. It pays to think of the dipole as made up of two opposite charges with a nonzero separation, as shown in the detail on the right. The dipole orients itself parallel to the local field and then feels a net force as shown because of unequal forces on its constituent charges.

The field experienced by each molecule consists of the applied field and the dipolar fields due to all the other induced dipoles. However, since the mean intermolecular distances in a gas are large, the dipolar contribution is negligible by comparison,[14] and we may take the applied and the actual fields to be the same, $\mathbf{E}$. If the number density of the gas is n and the polarizability of one molecule is α, the dipole moment per molecule is $\alpha\mathbf{E}$, the dipole moment per unit volume, or polarization, is $\mathbf{P} = \alpha n\mathbf{E}$, and the electric susceptibility is $\chi_e = \alpha n$. Hence, the dielectric constant is

$$\epsilon = 1 + 4\pi n\alpha. \tag{99.1}$$

$$\epsilon/\epsilon_0 = 1 + n\alpha. \quad \text{(SI)}$$

At STP, the number density n in a gas is 2.69×10^{19} cm^{-3}, and typical polarizabilities are ~ 1 Å^3 (Gaussian). For H, for instance, $\alpha = 9a_0^3/2 = 6.662 \times 10^{-25}$ cm^3, yielding $\epsilon = 1.00022$. Measured values for a variety of gases (table 15.2) all show that $\epsilon - 1 \simeq 10^{-4}$, in agreement with the general argument.

Polar gases: Certain molecules, such as H_2O and CO, have a permanent dipole moment due to the electronic charge distribution in the molecule and are said to be polar. (The value for H_2O is 1.86×10^{-18} esu·cm,[15] or 6.2×10^{-30} C·m (SI). The value for CO is 9×10^{-20} esu·cm, or 0.3×10^{-30} C·cm.) In a gas of such molecules the individual dipoles are completely randomly oriented in the absence of an applied electric field, and the net dipole moment per unit volume is zero. When a field is applied, there is a slight energetic preference for alignment of the dipoles parallel to the field, giving rise to an orientational imbalance and a nonzero net moment. In addition, there is an extra polarization of the electron distribution within the molecule, just as for nonpolar molecules. These two contributions to the polarization are labeled *orientational* and *intrinsic*, respectively.

However, as we shall see, under normal gaseous conditions, the intermolecular dipolar field is still negligible compared to the applied field, so this part of the argument is the

---

[14] The ratio is of order $n\alpha$ ($\sim 10^{-4}$) times the mean field.
[15] The unit 10^{-18} esu·cm is sometimes known as a *debye*.

Table 15.2. Dielectric constants of various nonpolar gases at STP

Gas	ϵ [ϵ/ϵ_0 (SI)]	Gas	ϵ [ϵ/ϵ_0 (SI)]
H_2	1.00026	Ar	1.000545
He	1.000068	CS_2	1.0029
O_2	1.000523	CCl_4	1.0030
Air	1.00054	NH_3	1.0072

same as that for a nonpolar gas. The only issue that remains, therefore, is to understand the orientational effect, to which we now turn.

Let the permanent moment of a molecule be denoted by $\mathbf{d}_0$, and the average per molecule by $\langle \mathbf{d} \rangle$. In an electric field $\mathbf{E}$, the potential energy of a molecule has a part

$$U(\theta) = -\mathbf{d}_0 \cdot \mathbf{E} = -d_0 E \cos\theta, \tag{99.2}$$

where θ is the angle between $\mathbf{d}_0$ and $\mathbf{E}$. At a temperature T, the probability that a molecule will be oriented at an angle θ is proportional to the Boltzmann factor $e^{-U(\theta)/k_B T}$, where k_B is Boltzmann's constant. Since the components of the dipoles transverse to $\mathbf{E}$ are still all equally probable, only the component parallel to $\mathbf{E}$ survives in the average, and is given by[16]

$$\langle \mathbf{d}_{\text{orient}} \rangle = \frac{\int d_0 \cos\theta\, e^{d_0 E \cos\theta/k_B T}\, d\Omega}{\int e^{d_0 E \cos\theta/k_B T}\, d\Omega} \hat{\mathbf{E}}, \tag{99.3}$$

the integration being over all orientations of $\mathbf{d}_0$. With $d_0 \simeq 10^{-18}$ esu·cm, $E \simeq 1$ statvolt/cm, and $T \simeq 400$ K, $\zeta \equiv d_0 E/k_B T \simeq 10^{-5}$, and we may evaluate the integrals in the small ζ approximation. This yields[17]

$$\langle \mathbf{d}_{\text{orient}} \rangle = \frac{d_0^2}{3 k_B T} \mathbf{E}. \tag{99.4}$$

Adding to this the intrinsic contribution $\alpha \mathbf{E}$, we obtain the net mean dipole moment per molecule,

$$\langle \mathbf{d} \rangle = \left(\alpha + \frac{d_0^2}{3 k_B T} \right) \mathbf{E}. \tag{99.5}$$

---

[16] The reason why such averaging, which is equivalent to adding together the dipole moments, is sensible is that if we consider a collection of arbitrarily oriented point dipoles confined to a finite region of space of size L, then the field at distances $\gg L$ is that of a dipole, given by vectorially adding the individual dipoles.

[17] The factor $d_0 E/3k_B T$ is the small ζ limit of $\coth\zeta - \zeta^{-1}$.

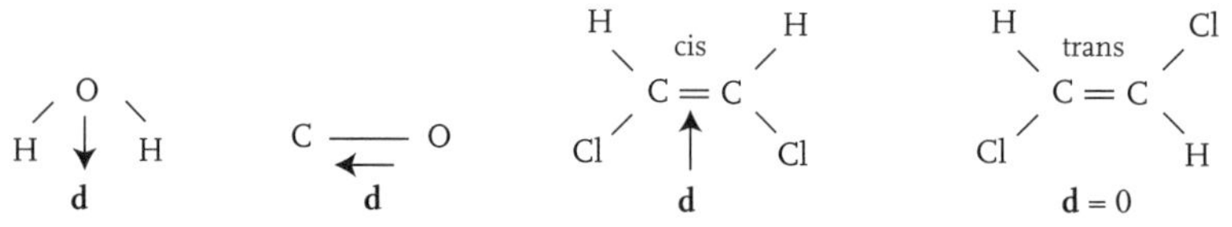

FIGURE 15.10. Dipole moments of a few simple molecules.

The rest of the argument runs exactly as for nonpolar molecules, and the final answer for the dielectric constant is

$$\epsilon = 1 + 4\pi n \left(\alpha + \frac{d_0^2}{3k_B T} \right). \tag{99.6}$$

$$\epsilon/\epsilon_0 = 1 + n \left(\alpha + \frac{d_0^2}{3\epsilon_0 k_B T} \right). \quad \text{(SI)}$$

Hence, a plot of the experimentally measured values of $\epsilon - 1$ versus $1/T$ (at fixed density) should be a straight line, and from the intercept and slope, both α and d_0 should be obtainable. This is indeed so, and this method gave structural information on many molecules before microwave spectroscopy became widely available. A historical anecdote may be of interest here. Of the cis and trans forms of dichloroethylene (CH_2Cl_2), only the former has a permanent dipole moment due to asymmetry (see fig. 15.10). Measurements of ϵ versus T were important in corroborating that the form believed to be trans from indirect evidence was indeed the one without a permanent moment.

To complete the argument, we still need to show that the intermolecular dipolar field is negligible. This follows from noting that the second term in the effective polarizability of a molecule, $d_0^2/3k_B T$, is also much smaller than $1/n$. For steam at 400 K, e.g., this term is $\simeq 2.1 \times 10^{-23}$ cm^3, whereas n is about 10^{19} cm^{-3}. Thus, the dielectric constant is still very close to unity and differs from it by a few parts per thousand.

Dense systems; the Clausius-Mossotti formula: It is clear from the discussion of gases that in dense systems such as liquids and solids, the dipolar fields due to other molecules will substantially modify the electric field felt by any one molecule. Since the dipolar fields due to neighboring molecules vary significantly in direction and magnitude on interatomic distance scales, it then follows that it is necessary to understand this variation. In other words, we must consider the microscopic field $\mathbf{e}(\mathbf{r})$ to understand the dielectric behavior of dense systems. The method for doing this is due to H. A. Lorentz.

Let us consider a dielectric body with a polarization $\mathbf{P}(\mathbf{r})$ and a macroscopic field $\mathbf{E}(\mathbf{r})$, both of which are slowly varying on the macroscopic length scale. By the definition of the dielectric constant,

$$\mathbf{P}(\mathbf{r}) = \frac{\epsilon - 1}{4\pi} \mathbf{E}(\mathbf{r}). \tag{99.7}$$

We now wish to find the *microscopic* field $\mathbf{e}$ felt by a molecule in this dielectric. (Whenever we write a field such as $\mathbf{e}$, $\mathbf{E}$, or $\mathbf{P}$ without an argument, we shall mean the field at this molecule.) It is convenient to regard this molecule as being at the origin of our coordinate

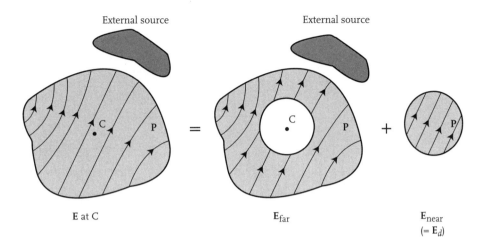

FIGURE 15.11. Argument for the macroscopic **E** field inside a dielectric.

system. We divide **e** into two parts, $\mathbf{e}_{near}$, the contribution from molecules within a sphere of radius, R, equal to 100 Å, say, centered at our origin, and $\mathbf{e}_{far}$, the contribution from molecules outside the sphere plus sources outside the dielectric:

$$\mathbf{e} = \mathbf{e}_{near} + \mathbf{e}_{far}. \tag{99.8}$$

We now argue that at points **r** within a few angstroms of the origin, the microscopic field $\mathbf{e}_{far}$ actually varies smoothly, since it is due to a collection of faraway molecules (plus the external sources). We can thus replace $\mathbf{e}_{far}$ by $\mathbf{E}_{far}$, the macroscopic field at the origin that is due to the external sources plus the part of the dielectric outside our sphere,

$$\mathbf{e}_{far} = \mathbf{E}_{far}. \tag{99.9}$$

In other words, $\mathbf{E}_{far}$ is the field that would exist at the center of a spherical cavity excised from the dielectric if the external sources and the polarization **P(r)** of the rest of the dielectric were frozen in place.

The third step in the argument is to note that, by superposition, the *macroscopic* field **E** at the origin is just the field $\mathbf{E}_{far}$ plus the field at the center of a uniformly polarized sphere of radius R (fig. 15.11). We call the latter field $\mathbf{E}_{near}$; it is just the depolarization field $-4\pi\mathbf{P}/3$ found in eq. (96.2). Hence,

$$\mathbf{E} = \mathbf{E}_{far} + \mathbf{E}_{near}, \tag{99.10}$$

and

$$\mathbf{E}_{far} = \mathbf{E} - \mathbf{E}_{near} = \mathbf{E} + \frac{4\pi}{3}\mathbf{P}. \tag{99.11}$$

The contribution $(4\pi/3)\mathbf{P}$ to the far-source field is also known as the *Lorentz field*.

From eqs. (99.8) and (99.9) we obtain

$$\mathbf{e} = \mathbf{E}_{far} + \mathbf{e}_{near} \tag{99.12}$$

$$= \frac{\epsilon + 2}{3} \mathbf{E} + \mathbf{e}_{near}, \tag{99.13}$$

using eq. (99.11) and $\mathbf{P} = (\epsilon - 1)\mathbf{E}/4\pi$ in the last step.

It remains to evaluate $\mathbf{e}_{near}$, the near-source microscopic field. As might be expected, this field depends on the actual geometry of the dipoles. If we consider an infinite cubic lattice of dipoles, all arranged parallel to one another, then it is easy to show that the field at any one dipole due to all the others vanishes. This can be done by grouping the source dipoles into sets with cubic symmetry, i.e., groups of six or eight at the vertices of octahedra or cubes centered about the origin, and showing that that the field due to any such group vanishes. We have already seen a special case of this result in exercise 19.7. More generally, the field vanishes at any site with cubic symmetry. Since the dipoles giving rise to $\mathbf{e}_{near}$ lie inside a sphere, it follows that they can be grouped into complete sets with cubic symmetry. Hence, $\mathbf{e}_{near}$ vanishes for a cubic crystal.

Exercise 99.1 Show that the electric field vanishes at the center of an octahedron with identical dipoles at the vertices, all pointing along the same but arbitrary direction. Repeat for a cube. Does the result hold for a tetrahedron? An icosahedron?

It should be carefully noted at this point that the device of using a spherical cavity is purely a convenience. Had we used a cavity with a different shape, both $\mathbf{E}_{far}$ and $\mathbf{e}_{near}$ would have been different (and difficult to calculate), but they still would have added up to the same microscopic field $\mathbf{e}$ (eq. (99.12)).

Exercise 99.2 Choose a Lorentzian cavity in the shape of a cube with one side parallel to the polarization $\mathbf{P}$. Since the cubic symmetry argument for $\mathbf{e}_{near} = 0$ still holds, it must be that eq. (99.11) also holds. Show that this is indeed so.

Solution: The problem amounts to showing that the depolarization field at the center of a cube uniformly polarized along one of the sides is $-4\pi\mathbf{P}/3$. Let the cube be of side $2a$, and let us choose Cartesian coordinates with the origin at the center of the cube, and $\hat{\mathbf{z}} \| \mathbf{P}$. We then have a surface charge density $\pm P$ on the faces $z = \pm a$, and the electric field at the origin is given by

$$\mathbf{E}_d = -2\hat{\mathbf{z}} \int_{-a}^{a} \int_{-a}^{a} \frac{aP}{(x^2 + y^2 + a^2)^{3/2}} dx \, dy. \tag{99.14}$$

The x integral is easily done via the substitution $x = (a^2 + y^2)^{1/2} \tan u$, and then the y integral via the successive substitutions $y = \sqrt{2}a \tan v$, $\sin v = \tan w$. The result, as expected, is $\mathbf{E}_d = -4\pi P\hat{\mathbf{z}}/3$.

Table 15.3. Test of the Clausius-Mossotti equation for nonpolar liquids. All numbers are in the Gaussian system

Compound	α (gas data) $(10^{-24}$ cm$^3)$	n (liquid) $(10^{22}$ cm$^{-3})$	$4\pi n\alpha/3$	ϵ (theory)	ϵ (experiment)
CS_2	8.61	1.02	0.368	2.75	2.64
O_2	1.547	2.238	0.145	1.509	1.507
CCl_4	12.47	0.622	0.325	2.44	2.24
Ar	1.617	2.171	0.147	1.517	1.54

Exercise 99.3 Let the cube of the previous exercise be uniformly polarized in an arbitrary direction. Find $\mathbf{E}_d$ at the center of the cube.

Exercise 99.4 Consider a simple cubic lattice with identical dipoles, all pointing along the $\hat{z}$ direction. Take the lattice sites as (n_x, n_y, n_z), where the n's are integers. Numerically find the electric field at the edge centers $(\frac{1}{2}, 0, 0)$ and $(0, 0, \frac{1}{2})$, the face centers $(\frac{1}{2}, \frac{1}{2}, 0)$ and $(0, \frac{1}{2}, \frac{1}{2})$, the body center $(\frac{1}{2}, \frac{1}{2}, \frac{1}{2})$, and the point $(\frac{1}{4}, \frac{1}{4}, \frac{1}{4})$.

For noncubic crystals, $\mathbf{e}_{\text{near}}$ does not in general vanish at any lattice site. For a liquid in which the dipoles have random locations and random orientations, it is not difficult to show that $\mathbf{e}_{\text{near}}$ vanishes on the average. From now on we will simply assume that this is so, so that

$$\mathbf{e} = \frac{\epsilon + 2}{3}\mathbf{E}. \tag{99.15}$$

Denoting the molecular polarizability by α, and the number density by n as before, we obtain

$$\mathbf{P} = n\alpha\mathbf{e} = n\alpha\frac{\epsilon + 2}{3}\mathbf{E}. \tag{99.16}$$

But $\mathbf{P} = (\epsilon - 1)\mathbf{E}/4\pi$, so

$$\frac{\epsilon - 1}{\epsilon + 2} = \frac{4\pi n\alpha}{3}. \tag{99.17}$$

$$\frac{\epsilon - \epsilon_0}{\epsilon + 2\epsilon_0} = \frac{n\alpha}{3}. \quad \text{(SI)}$$

This is known as the *Clausius-Mossotti* equation. It is also useful to write it as

$$\epsilon = 1 + \frac{4\pi n\alpha}{1 - (4\pi n\alpha/3)}. \tag{99.18}$$

In table 15.3, we show the extent to which this relation works for a few nonpolar liquids by using gas phase data on the same compound to obtain the molecular polarizability α and then using this same value of α for the liquid phase. As can be seen, the agreement is not bad.

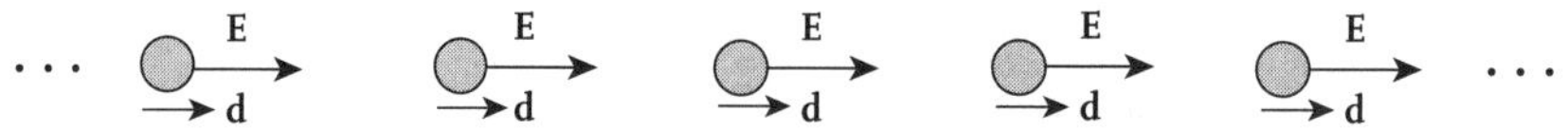

FIGURE 15.12. A one-dimensional model for a ferroelectric.

The Clausius-Mossotti equation also applies qualitatively to certain types of crystalline solids, such as the alkali halides, which have dielectric constants ranging from 4.68 (KCl) to 9.27 (LiF). However, the polarizability term $n\alpha$ must be appropriately modified. There are now at least two contributions to this term. First is the electronic polarizability of the various ions, i.e., the contribution due to the deformation of the electron distribution in an ion upon application of an electric field. These must be added for all the ions in the unit cell. Second is the ionic polarizability, which arises from the displacements of the ions relative to one another. However, the two contributions cannot be unambiguously separated, and the displacements also entail some distortion of the electron clouds. We refer readers to solid-state texts for more details.

The Clausius-Mossotti equation fails completely in polar liquids and solids. If we apply it to water at 300 K ($\epsilon = 81$), e.g., using an orientational polarizability of 2.8×10^{-23} cm^3 and a density $n = 3.34 \times 10^{22}$ cm^{-3}, we get $4\pi n\alpha/3 = 3.9$, implying $\epsilon < 0$! The reason, of course, is that there are strong orientational correlations among water molecules, and one can assume neither that the dipoles are all parallel nor that they are independently randomly oriented. It is clear that a full theory in this case would require a complete understanding of orientational and positional correlations among the molecules, a task far outside our scope and still the subject of active research.

In fact, we see that the Clausius-Mossotti equation leads to absurd results whenever $4\pi n\alpha/3 \geq 1$. This is known as the "polarization catastrophe" or the "$4\pi/3$ catastrophe." What happens in some cases is that a spontaneous fluctuation of polarization in one region of the medium polarizes nearby molecules, which polarize still more molecules, leading in this way to long-range polarization. The runaway predicted by the Clausius-Mossotti equation is eventually arrested, since one cannot distort electron distributions at the atomic or molecular level without bringing into play strong nonlinear intraatomic forces. A large class of solids, known as *pyroelectrics*, show spontaneous polarization. A simple model shows how this can happen. Suppose we have a chain of atoms, with interatomic spacing a and polarizability α. Suppose each atom spontaneously acquires a dipole moment d pointing along the chain (fig. 15.12). The electric field at any atom due to the dipoles on all the other atoms is given by

$$E = 2\frac{2d}{a^3} \sum_{n=1}^{\infty} \frac{1}{n^3}. \qquad (99.19)$$

The dipole moment that would be induced by this field is $d' = \alpha E - \alpha_3 E^3$, where α_3 is a positive constant. The second term is the next in an expansion of d in powers of the electric effect, and α_3 is a third-order polarizability. (There is no E^2 term, since $\mathbf{d}$ and $\mathbf{E}$ are both odd under parity.) We include it because we now wish to consider large fields.

Equating d' to d, we obtain (using $\sum_n n^{-3} = \zeta(3) = 1.202$)

$$d = 4d\alpha\zeta(3)/a^3 - Kd^3, \tag{99.20}$$

where $K > 0$ is a constant whose precise value does not concern us. This equation has a nontrivial solution if $\alpha > a^3/4\zeta(3)$. For $a = 2$ Å, this means $\alpha > 1.66 \times 10^{-24}$ cm^3, which is not an atypical value.

In most pyroelectrics, the spontaneous moment is too large to be influenced by laboratory electric fields, and the moment is masked by the accumulation of charges from the air that cancel the polarization charges on the surface of the crystal. In some, however, the arrangement of ions is such that the critical condition for a spontaneous polarization can be met by a slight change in some variable, temperature, e.g., and the moment can be reversed by applying a moderate electric field. It is not too difficult to imagine how such cases may arise from our simple model if we consider several ionic species with different polarizabilities and also allow for displacements of the ions. Such pyroelectric crystals are known as *ferroelectrics*. Examples are potassium dihydrogen phosphate (KH$_2$PO$_4$) and barium titanate (BaTiO$_3$), with polarizations of 16,000 and 78,000 (Gaussian units) at room temperature. The dielectric constant of these substances is large and strongly temperature dependent; for BaTiO$_3$ it is of order 5000 in the temperature range 200–500 K. Finally, in some crystals, called *antiferroelectrics*, each unit cell has a spontaneous moment, but the moments of different cells line up in some nonparallel ordered structure. For example, neighboring lines of moments can be antiparallel. Examples of antiferroelectrics are tungsten trioxide (WO$_3$) and lead zirconate (PbZrO$_3$).[18]

---

[18] For more about this subject, see Jona and Shirane (1993).

16 | Magnetostatics in matter

Our purpose in this chapter is to obtain a theory for the spatially averaged or macroscopic magnetic field in matter and to develop models for the response of certain types of materials to applied magnetic fields. While the mathematical development parallels that of electrostatics in dielectrics, the physics is quite different. The biggest difference is that, unlike the electric case, the magnetic susceptibility may be either positive or negative, i.e., the magnetic response can either oppose or reinforce the applied field. The two cases are known as *diamagnetic* and *paramagnetic*, respectively. This difference can be traced to the lack of any work done by the magnetic field on a moving charge. Second, except for ferromagnets and superconductors, the response is much weaker than in the electric case. Thus, there is often little penalty for confusing the magnetic field **B** with the magnetizing field **H**, and many authors even use the symbols interchangeably. This leads to much confusion, which we shall avoid with only partial success.

100 Magnetic permeability and susceptibility

In magnetostatics, where we study only steady-state situations, the macroscopic equations to be solved are Ampere's law,

$$\nabla \times \mathbf{H} = \frac{4\pi}{c} \mathbf{j}_{\text{free}}, \tag{100.1}$$

$$\nabla \times \mathbf{H} = \mathbf{j}_{\text{free}}, \quad \text{(SI)}$$

and the law that **B** is solenoidal,

$$\nabla \cdot \mathbf{B} = 0. \tag{100.2}$$

As in the electrostatic case, one needs a constitutive relation between **B** and **H** to solve eqs. (100.1) and (100.2). For many materials a linear relationship holds to very good approximation over a large range. If the material is isotropic, we write[1]

$$\mathbf{B} = \mu\mathbf{H}, \quad \text{(Gaussian and SI).} \tag{100.3}$$

The dimensionless number μ (in the Gaussian system), or μ/μ_0 (in SI), is known as the *permeability* of the medium. It has the same value in the two systems. In solid or liquid crystals, it must be replaced by a tensor.

We also define the *magnetic susceptibility* in parallel with the electric susceptibility:

$$\mathbf{M} = \chi_m\mathbf{H}. \quad \text{(Gaussian and SI)} \tag{100.4}$$

This is related to the permeability via

$$\chi_m = \begin{cases} (\mu - 1)/4\pi. & \text{(Gaussian)} \\ (\mu - \mu_0)/\mu_0. & \text{(SI)} \end{cases} \tag{100.5}$$

It is also a dimensionless number, but

$$\chi_m \text{ (SI)} = 4\pi\chi_m \text{ (Gaussian).} \tag{100.6}$$

For the majority of materials for which eq. (100.3) is a good approximation, χ_m is very small, of order 10^{-5}. As mentioned above, unlike χ_e, χ_m can be either positive (*paramagnets*) or negative (*diamagnets*). Because of the smallness of χ_m, if either a para- or diamagnetic body is placed in an external $\mathbf{B}$ field, the modification of the field is small and can be calculated perturbatively. For this reason, we do not consider solving the boundary value problem at this time.

We repeat, as stated in chapter 13, that eq. (100.3) does not apply to ferromagnets and superconductors.

101 Thermodynamic relations for magnetic materials

The thermodynamics of magnetic materials can be understood by an extension of the arguments leading to the magnetic field energy in vacuum (section 29), which, as we have seen, is rather more subtle than the electric field energy. The mathematical formalism, however, is very similar to the case of dielectrics in section 97. We shall therefore skip the detailed derivation of many of the formulas that follow.[2]

---

[1] For historical reasons, constitutive relations and many other formulas in magnetism are written as if $\mathbf{H}$ were the fundamental field and $\mathbf{B}$ the derived field, which is the opposite of what we now recognize. If logic were stronger than tradition, we would parallel the electric case by introducing two new symbols, say ξ and $\bar{\chi}_m$, to which we would give special names, and write the constitutive relations as $\mathbf{H} = \xi\mathbf{B}$ and $\mathbf{M} = \bar{\chi}_m\mathbf{B}$.

[2] Once again, we shall give formulas only in the Gaussian system in this section. To obtain SI formulas, replace any factor of $1/4\pi$ multiplying dot products such as $\mathbf{B}\cdot\mathbf{H}$, $\mathbf{B}\cdot\delta\mathbf{H}$, etc., by unity, factors of $1/c$ multiplying dot products of $\mathbf{j}$ and $\mathbf{A}$ also by unity, and leave dot products of $\mathbf{M}$ and $\mathbf{H}$ untouched.

As in the free-space case, let us imagine placing a piece of matter between a set of coils, and change the current in the coils by a small amount $\delta \mathbf{j}_{\text{free}}$ in a duration δt while holding the driving emf's fixed. The change in current causes a change in $\mathbf{B}$, which produces an induced $\mathbf{E}$ field, which, by Lenz's law opposes the change in currents. The work required to overcome this opposition is

$$\delta W = -\delta t \int_{\text{coils}} \mathbf{j}_{\text{free}} \cdot \mathbf{E}_{\text{ind}} \, d^3x. \tag{101.1}$$

Note that it is only the current that we can control, $\mathbf{j}_{\text{free}}$, whose change appears in this formula, since it is only useful to find the work done in an experimentally realizable situation. Second, only the induced electric field, $\mathbf{E}_{\text{ind}}$, appears in the formula, and not the field whose circulation gives us the emf.

The integral in eq. (101.1) extends only over the region of space filled by the coils. Since $\mathbf{j}_{\text{free}}$ vanishes outside this region, we can transform the integral to one over all space. In doing so, it pays to imagine that $\mathbf{j}_{\text{free}}$ is smoothed out at the surface of the coils so that it does not go to zero abruptly, so that various integrals may be evaluated by parts without fear of picking up any surface terms. We now transform the integral over all space as follows. First, we trade in $\mathbf{j}_{\text{free}}$ for $\nabla \times \mathbf{H}$. Second, we integrate by parts, turning the integrand into $\mathbf{H} \cdot (\nabla \times \mathbf{E}_{\text{ind}})$. Third, we use Faraday's law to replace $\delta t(\nabla \times \mathbf{E}_{\text{ind}})$ by $\delta \mathbf{B}$. In this way, we get

$$\delta W = \frac{1}{4\pi} \int \mathbf{H} \cdot \delta \mathbf{B} \, d^3x, \tag{101.2}$$

which is the fundamental expression for the work received by the material in terms of the fields. It should be compared with the expression in free space, where the integrand is $\mathbf{B} \cdot \delta \mathbf{B}$.

The expression for the first law is obtained if we add to eq. (101.2) the heat received by the material, which yields

$$\delta U = T\delta S + \frac{1}{4\pi} \int \mathbf{H} \cdot \delta \mathbf{B} \, d^3x. \tag{101.3}$$

Thus, U is a thermodynamic potential with respect to S and $\mathbf{B}$. Potentials with respect to T and/or $\mathbf{H}$ are now obtained via standard Legendre transforms. We tabulate the key formulas below as in the dielectric case.

Thermodynamic potential	Formula	Natural variables	Differential
U		$S, \mathbf{B}$	$T\delta S + \dfrac{1}{4\pi}\displaystyle\int \mathbf{H}\cdot\delta\mathbf{B}\, d^3x$
F	$U - TS$	$T, \mathbf{B}$	$-S\delta T + \dfrac{1}{4\pi}\displaystyle\int \mathbf{H}\cdot\delta\mathbf{B}\, d^3x$
$\tilde{U}$	$U - \dfrac{1}{4\pi}\displaystyle\int_V \mathbf{H}\cdot\mathbf{B}\, d^3x$	$S, \mathbf{H}$	$T\delta S - \dfrac{1}{4\pi}\displaystyle\int \mathbf{B}\cdot\delta\mathbf{H}\, d^3x$
$\tilde{F}$	$\tilde{U} - TS$	$T, \mathbf{H}$	$-S\delta T - \dfrac{1}{4\pi}\displaystyle\int \mathbf{B}\cdot\delta\mathbf{H}\, d^3x$

$$\text{(101.4)}$$

We also have

$$\tilde{F} = F - \frac{1}{4\pi}\int_V \mathbf{B}\cdot\mathbf{H}\, d^3x. \tag{101.5}$$

To further see the meaning of F, let us rewrite δF in terms of the current and vector potential. We hold the temperature fixed, since the contribution $-S\delta T$ is unaffected and can be added on at the end if wished. Using $\delta\mathbf{B} = \nabla \times \delta\mathbf{A}$, integrating by parts, and replacing $\nabla \times \mathbf{H}$ by $\mathbf{j}_{\text{free}}$, we get

$$(\delta F)_T = \frac{1}{c}\int_V \mathbf{j}_{\text{free}}\cdot\delta\mathbf{A}\, d^3x. \tag{101.6}$$

Similarly,

$$(\delta\tilde{F})_T = -\frac{1}{c}\int_V \mathbf{A}\cdot\delta\mathbf{j}_{\text{free}}\, d^3x. \tag{101.7}$$

Thus, the more convenient potential in the present case is $\tilde{F}$, because it is minimized in equilibrium with respect to changes at fixed currents. The Helmholtz free energy F, on the other hand, is minimized when the vector potential is held fixed. Experimentally it is much easier to control the current than the vector potential. Since the line integral of $\mathbf{A}$ around any loop is the magnetic flux threading that loop, the potential F can also be described as being minimum at fixed flux.

Integration of the differentials of F, U, etc., requires a constitutive relation. We shall not pause to address this issue.

Lastly, in analogy with the discussion in section 98, let us obtain the total magnetic free energy of a body by subtracting the field energy that would be present even if the body were not there. We imagine that the body is in a field created by current loops I_a. The energy to be subtracted is found keeping these currents fixed (fig. 16.1), so the potential to be used is $\tilde{F}$. Denoting by $\mathbf{B}_0$ the field that would be present in the absence of the body

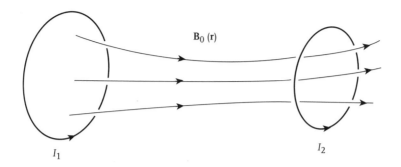

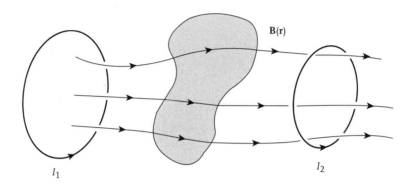

FIGURE 16.1. Alteration of the magnetic field by a body.

(the vacuum field), we have

$$\tilde{F}_{\text{bod}} = \tilde{F} + \frac{1}{8\pi} \int_V B_0^2 \, d^3x. \tag{101.8}$$

Hence,

$$\delta\tilde{F}_{\text{bod}} = -\frac{1}{4\pi} \int_V (\mathbf{B} \cdot \delta\mathbf{H} - \mathbf{B}_0 \cdot \delta\mathbf{B}_0) \, d^3x. \tag{101.9}$$

We write

$$(\mathbf{B} \cdot \delta\mathbf{H} - \mathbf{B}_0 \cdot \delta\mathbf{B}_0) = (\mathbf{H} - \mathbf{B}_0) \cdot \delta\mathbf{B}_0 + [\delta(\mathbf{H} - \mathbf{B}_0)] \cdot \mathbf{B} - (\mathbf{H} - \mathbf{B}) \cot \delta\mathbf{B}_0, \tag{101.10}$$

and integrate each term separately. Consider the first term first. Letting $\mathbf{A}_0$ be the vector potential that yields $\mathbf{B}_0$, and using Gauss's theorem, we have

$$\int_V (\mathbf{H} - \mathbf{B}_0) \cdot \delta\mathbf{B}_0 \, d^3x = \int_V (\mathbf{H} - \mathbf{B}_0) \cdot (\nabla \times \delta\mathbf{A}_0) \, d^3x = \int_V [\nabla \times (\mathbf{H} - \mathbf{B}_0)] \cdot \delta\mathbf{A}_0 \, d^3x. \tag{101.11}$$

However, the curls of $\mathbf{H}$ and $\mathbf{B}_0$ are both equal to the same current distribution $\mathbf{j}_{\text{free}}$, since these currents are outside the body (the loops I_a), which by hypothesis are not changed when the body is removed. Thus, the integral vanishes.

The integral of the second term in eq. (101.10) vanishes for the same reason. Thus, only the third term is left, and since $\mathbf{H} - \mathbf{B} = -4\pi\mathbf{M}$,

$$\delta\tilde{F}_{\text{bod}} = -\int_V \mathbf{M} \cdot \delta\mathbf{B}_0 \, d^3x. \quad \text{(Gaussian and SI)} \tag{101.12}$$

The integral in this formula can be taken over the volume of the body, since $\mathbf{M}$ vanishes outside it. This result should be compared to eq. (98.5).

Equation (101.12) allows us to understand how nonmagnetic bodies behave in an inhomogeneous magnetic field. The argument is identical to that for dielectrics. However, since χ_m need not always be positive, the behavior of para- and diamagnets is different. A paramagnet feels a force toward regions of higher B, while a diamagnet is attracted toward regions of lower B.

Exercise 101.1 A long but finite-length solenoid of diameter b and a single layer of windings is to be used to measure the susceptibility of a para- or diamagnetic material sample in the shape of a small cylinder of diameter much smaller than b. Where on the axis of the solenoid should the sample be placed in order to experience the greatest possible force?

Answer: A distance $b/\sqrt{15}$ inside one end of the solenoid.

102 Diamagnetism

When studying dielectric polarization, we divided the polarizability of atoms into an induced part and an orientational part. The induced part is due to the changes in the atomic charge distribution due to the applied electric field. Similarly, an applied magnetic field changes the atomic current distribution. This response gives rise to diamagnetism or, more generally, to a diamagnetic contribution to the net magnetic susceptibility. To see how this comes about, we consider the following simple model of an atom. We consider an electron in a circular orbit of radius r and angular frequency ω around a nucleus of charge Ze (fig. 16.2). A field $\mathbf{B}$ perpendicular to the orbit is also applied. In such an orbit, the angular momentum of the electron is $L = m_e\omega^2 r$, and by the result of section 75, the magnetic moment is

$$\mathcal{M} = \frac{-e}{2m_e c}L = -\frac{e}{2c}\omega r^2. \tag{102.1}$$

(The last formula also follows if we regard the orbit as a current loop of area πr^2 and current $-e\omega/2\pi$, since a charge $-e$ crosses any point on the orbit $\omega/2\pi$ times in unit time.) Equation (102.1) holds whatever the value of ω and r.

Next, let us obtain the relation between ω and r. Equating the centripetal force to the Coulomb plus Lorentz force, we obtain

$$m_e\omega^2 r = \frac{Ze^2}{r^2} + \frac{e\omega r B}{c}. \tag{102.2}$$

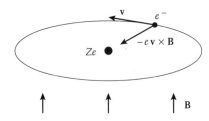

FIGURE 16.2. Classical model of an electron orbiting an atomic nucleus.

We rewrite this as

$$\omega^2 = \omega_0^2 + 2\omega_0 |\omega_L|, \tag{102.3}$$

where

$$\omega_0 = \left(\frac{Ze^2}{m_e r^3} \right)^{1/2} \tag{102.4}$$

is the frequency in the absence of the **B** field, and $\omega_L = -e\,B/2m_e c$ is the Larmor frequency. Before solving eq. (102.3), let us first calculate the typical numbers involved. Taking $r = a_0$ and $Z = 1$ for hydrogen, we obtain $\omega_0 \simeq 4 \times 10^{16} \sec^{-1}$, while even with $B = 10^6$ G, ω_L is only $8 \times 10^{12} \sec^{-1}$. Thus, for all laboratory scale fields, we may assume $\omega_L \ll \omega_0$, and the modified frequency of the atom's orbit is

$$\omega \approx \omega_0 + |\omega_L|. \tag{102.5}$$

This is in accord with Larmor's theorem, as it should be.

From eq. (102.1), it follows that in response to the applied B field, the atom acquires an additional magnetic moment

$$\Delta \mathcal{M} = -\frac{e}{2c} |\omega_L| r^2 = -\frac{e^2}{4 m_e c^2} r^2 B. \tag{102.6}$$

Let us now imagine that we have a collection of many such atoms, n per unit volume. When $\mathbf{B} = 0$, their moments point randomly in all directions. When $\mathbf{B} \neq 0$, each atom acquires an induced moment as shown above. The magnetization, i.e., the total moment per unit volume, is given by

$$M = -\frac{ne^2}{4 m_e c^2} \langle x^2 + y^2 \rangle B, \tag{102.7}$$

where we have replaced r^2 by the mean-square distance perpendicular to **B** to account for the random orientation of the orbits. Thus, the susceptibility is given by

$$\chi_m = -\frac{ne^2}{4 m_e c^2} \langle x^2 + y^2 \rangle, \tag{102.8}$$

which is negative, i.e., diamagnetic.

The typical magnitude of the diamagnetic susceptibility just computed can be found by putting $n = 10^{24}/cm^3$ and $\langle r^2 \rangle = a_0^2$. One finds, as asserted earlier, that $\chi_m \sim 10^{-6}$–10^{-5}, the higher number being achieved in many-electron atoms. If we recall that $a_0 = \hbar^2/me^2$, then we see that χ_m is smaller than its electrical counterpart, χ_e, by $\sim \alpha_{fs}^2$, where $\alpha_{fs} = e^2/c\hbar \simeq 1/137$ is the fine-structure constant. For example, for the H atom, the diamagnetic contribution to the magnetic polarizability is $-\alpha_{fs}^2 a_0^3/2$. The extra two powers of α_{fs} are what make the magnetic susceptibility so small.

The above calculation is quite standard and can be found in many texts. It is often accompanied by the caveat that it is, in fact, incorrect. A deficiency often noted is that using the Bohr radius for r is wrong, since classical mechanics would allow an orbit to exist at any radius in the absence of radiation. Indeed, it is a commonly cited result that classical theory, in which the only moments are due to moving charges, does not permit the existence of either paramagnetism or diamagnetism. (Spin is automatically excluded in this point of view as a quantum mechanical phenomenon.) These criticisms appear to this writer not to be fruitful. After all, in calculating dielectric susceptibilities, one considers displacement of the charges about a preexisting stable atomic or molecular structure, in which the typical interatomic distances are taken to be those given by quantum mechanics. This is just as indefensible as using a_0 for r in the calculation of χ_m, since classical mechanics does not allow for *any* stable atoms! The present calculation can be understood as sensible if we accept that quantum mechanics leads to relatively rigid bound electron orbits in atoms. A weak magnetic field cannot modify the electron state radically, and our quasiclassical approach correctly identifies the small changes in the orbits that do take place.

The calculation will clearly fail for metals and semiconductors, where the notion of an atomic center does not make sense for the delocalized conduction electrons. In this case, one does need a fully quantum mechanical calculation, first done by Landau. This shows that the orbital motion of the electrons always yields a diamagnetic response. If the periodic crystal structure of the metal and the electron–electron interactions are ignored, Landau's result is

$$\chi_m = -\frac{e^2}{4m_e c^2} \left(\frac{n}{9\pi^4} \right)^{1/3}, \tag{102.9}$$

where n is the conduction electron density.[3] We can understand this from the result (102.8) for atoms if we estimate $\langle x^2 + y^2 \rangle$ as $n^{-2/3}$. Essentially, because of the Pauli principle, each electron is confined to rattle in a cage of size $n^{-1/3}$, roughly equal to the average electron–electron separation. Since this separation is of the same order as the Bohr radius, the typical magnitude of χ_m is again 10^{-6}.

---

[3] Readers may wonder how this result can be quantum mechanical if it doesn't have an $\hbar$ in it. The answer is that $\hbar$ is hidden in n, the electron density. Like the sizes of atoms and molecules, interatomic distances in metals are determined by the interplay of Coulomb energy, kinetic energy, and the Pauli exclusion principle. The last is clearly quantal in character, and so is the kinetic energy, since by the uncertainty principle, the more delocalized the electrons are, the smaller it is. The same comment applies to eq. (103.3) for the Pauli paramagnetic susceptibility.

103 Paramagnetism

Paramagnetism is the magnetic equivalent of the orientational polarization of preexisting electric dipole moments along an applied electric field. The simplest case is that of insulating solids containing atomic or ionic magnetic moments whose interaction with each other can be neglected. Examples of such materials are $CuSO_4 \cdot K_2SO_4 \cdot 6H_2O$, in which the magnetic ion is Cu^{2+}, $Gd(C_2H_5SO_4)_3 \cdot 9H_2O$ (magnetic ion Gd^{3+}), and many other salts containing transition-metal or rare-earth ions. Solid ^3He at temperatures above 0.04 K forms another example. In this case the magnetic moments are those of the nuclei.

If we assume that each magnetic moment is $\mathcal{M}$, and the density of these moments is n, then a calculation identical to that in section 99 yields (compare with eq. (99.4))

$$\mathbf{M} = n\frac{\mathcal{M}^2}{3k_B T}\mathbf{B}. \tag{103.1}$$

The susceptibility χ_m now follows immediately. There are two points worth noting. First, since we will find $\chi_m \ll 1$, we may replace $\mathbf{B}$ by $\mathbf{H}$. Second, it is customary to write the answer in terms of the dimensionless atomic or ionic angular momentum J. By the usual rules of quantum mechanics, $\mathbf{J}^2 = J(J+1)$, so by eqs. (75.13) and (75.15), we get

$$\chi_m = n\frac{g^2\mu_B^2}{3k_B T}J(J+1). \tag{103.2}$$

The $1/T$ dependence on temperature is known as the *Curie law*.[4] For illustrative purposes, let us take $n = 10^{23}$ cm^{-3}, $g = 2$, $T = 300$ K, and $J = 2$. This yields $\chi_m = 1.7 \times 10^{-3}$, which is a typical order of magnitude. This is about 500 times larger than the diamagnetic contribution and completely dominates it in ions with partially filled electron shells. Systems such as the noble-gas and alkali halide solids have only filled shells and *are* diamagnetic.

The above calculation does not apply to metals. Here, reorientation of the conduction electron spin moment is strongly restricted by the Pauli exclusion principle, and the susceptibility is therefore much smaller. Ignoring the periodic crystal structure and electron–electron interactions, the result (known as *Pauli paramagnetism*) is

$$\chi_m = \frac{3e^2}{4m_e c^2}\left(\frac{n}{9\pi^4}\right)^{1/3}. \tag{103.3}$$

Note that this is exactly -3 times the Landau diamagnetic susceptibility, eq. (102.9).

104 The exchange interaction; ferromagnetism

In the next few sections, we discuss ordered magnetic materials, principally ferromagnets. This is a vast area and still a subject of active research from many points of view, ranging

---

[4] After Pierre Curie.

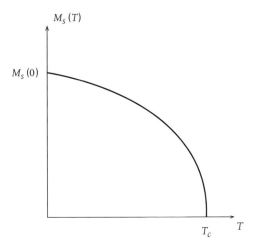

FIGURE 16.3. Temperature dependence of the spontaneous magnetization.

from basic questions about the origin of ferromagnetism, to the nature of magnetic phase transitions, to how to make better computer hard drives. From the point of view of macroscopic electrodynamics, perhaps the most important questions relate to domain theory, magnetization reversal, eddy currents, magnetostriction, hysteresis, etc. However, one cannot study these topics without bringing in concepts from thermodynamics and phenomenological solid-state physics. The next few sections therefore have a rather different flavor to them than what we have done up to now.

The defining property of a ferromagnet is that it has a nonzero or *spontaneous* magnetization, M_s, even in the absence of an applied magnetic field. This quantity is a function of the thermodynamic state of the material, especially its temperature, as shown in fig. 16.3. The temperature T_c is known as the *critical* or *Curie temperature*. Values of M_s and T_c for selected materials are given in table 16.1.

In connection with this table, it should be mentioned that the measurement and definition of M_s is a subtle affair. Let us note just two of the possible complications. First, the net moment of a macroscopic body will generally be close to zero because different regions, or *domains*, within the body, of size 100–1000 Å, will be magnetized in different directions, leading to an overall cancellation. Second, even if domains are eliminated, unless the body is properly shaped, the magnetization will not be uniform, nor will **H** be zero. The materials in table 16.1 are all chosen to be homogeneous and crystalline to avoid complications from defects and inhomogeneities.

Two important conclusions can be drawn from this data. First, if we take 10 Å^3 and 10^3 emu/cm^3 as rough mean values for the volume per magnetic atom and the magnetization, we obtain a magnetic moment per atom $\sim 10^{-20}$ emu, i.e., about 1 Bohr magneton. Since this is comparable to the magnitude of the magnetic moment of the free atom (or ion), we conclude that at low temperatures (say $T < T_c/2$), the atomic moments are substantially aligned parallel to each other. Let us ask if this alignment can be due to the direct interaction of the moments, as in the case of ferroelectrics. If neighboring ions

Table 16.1. Spontaneous magnetizations and critical
temperatures of selected materials. The magnetization
quoted is at zero temperature, except for that of Fe_3O_4,
which is at room temperature

Material	$M_s(0)$ (emu/cm^3)	T_c (K)
Fe	1740	1043
Co	1446	1388
Ni	510	627
Gd	2010	292
Fe_3O_4 (magnetite)	480*	858
γ-Fe_2O_3	370	973
$Y_3Fe_5O_{12}$ (YIG)	200	560
$Gd_3Fe_5O_{12}$	605	564
$SmCo_5$	780	993

are taken to be a distance a apart, the interaction energy between them is of order

$$\frac{\mu_B^2}{a^3} \simeq \frac{10^{-40}}{10 \times 10^{-24}} \text{ erg} \simeq 0.1 \text{ K.} \tag{104.1}$$

This is vastly smaller than the typical value of T_c. We conclude that the direct dipole–dipole interaction cannot be responsible for the alignment, for the thermal fluctuations would rapidly destroy it.[5] Another way to state this conclusion is in terms of an effective field H_{ex} that would stabilize a moment against the thermal disordering. Putting $\mu_B H_{ex} = k_B T_c$, we see that we need $H_{ex} \simeq 5 \times 10^6$ Oe. Whatever the agency behind the ordering is, it can be regarded as creating an effective field of $\sim 5 \times 10^6$ Oe on each moment. For reasons that will be clear shortly, this field is known as the *exchange field*. It is useful to remember that the energy density associated with this ordering is $M_s H_{ex} \simeq 10^{10}$ erg/cm^3.

The true reason for the alignment of the moments lies in the Pauli exclusion principle. According to this principle, two electrons with the same spin cannot simultaneously be at the same point in space and are automatically forced to be spatially far apart from each other. The Coulomb repulsion between the electrons is thereby minimized, and the total energy is lowered. This is the basic mechanism for magnetic ordering. Since the Coulomb energy is so much greater than the direct interaction between magnetic moments, we see that even reductions in the former at the level of 1% can be sufficient to explain why the spins in a ferromagnet order at a few hundred kelvins.

---

[5] Our discussion excludes *dipolar ferromagnets* (e.g., LiHoF$_4$) in which the ordering *is* due to the direct interaction and takes place at temperatures ~ 0.1–1 K.

This relative-spin-orientation-dependent part of the Coulomb interaction between electrons is known as the *exchange interaction*. The term reflects the fact that the many-electron wave function must be antisymmetric under exchange of the space and spin coordinates of any two electrons, and the relevant energy is, in the simplest cases, expressible as a two-electron integral in which the electron coordinates are interchanged.

A very simple illustration of exchange is provided by what might be called *atomic ferromagnetism*. Consider the ion Mn^{2+}, which has five $3d$ electrons in its outer shell. By filling the different $3d$ orbitals with the five electrons in all possible ways (always obeying the Pauli principle, of course), one obtains several possible values of total spin and orbital angular momenta. Since all single-electron $3d$ levels have the same energy, to a first approximation it does not matter how the orbitals are filled. This approximation is rather poor, however, as it ignores energy differences that are quite important at room temperature. Elementary chemistry courses emphasize that the filling must be done in accordance with *Hund's rules*. The first of these empirical rules puts the electrons into orbitals with parallel spins as far as possible. We can now see that this minimizes the Coulomb energy. To do this, we would like the electrons to be as far apart from each other as possible, and that as few orbitals be doubly occupied as possible. In the Mn^{2+} case, this implies that all five electrons have parallel spin. Thus, the ground state of this ion should have spin and orbital angular momenta of 5/2 and 0, respectively. In the atomic case, the energy difference between states due to this mechanism is of order 0.1 eV $\sim 10^3$ K.

Yet, it should be noted that parallel spin alignment is not always the least energetic. In the H_2 molecule, the Coulomb interaction with the protons favors a spatial electronic orbital that has greatest probability density at the midpoint between the protons. Both electrons prefer to be in this same spatial state, which is possible only if they have antiparallel spins. A certain amount of the attractive electron–proton energy is then offset by the electron–electron repulsion, but the former effect wins out. Indeed, the state wherein the electrons have parallel spin leads to dissociation of the molecule (hence the term *antibonding* state).

Because of the subtle interplay of positive and negative Coulomb energies, the calculation of exchange effects is still among the most challenging problems in the theory of solids. Even the sign can often be uncertain.

Both signs of exchange energy are found in solids. In MnF_2, e.g., the Mn^{2+} ions can be arranged into two sublattices, denoted A and B, such that the nearest neighbors of every A site are B sites, and vice versa. The spins on the A sites all point one way ("up"), but the spins on the B sites all point the opposite way ("down"). There is no net magnetic moment in this type of order, which is known as *antiferromagnetism*. An intermediate arrangement, known as *ferrimagnetism*, is also found. In this case, the "up" moments, say, have a greater magnitude or are more numerous than the "down" moments, leading to a net moment. Magnetite (Fe_3O_4), the oldest known magnet, is a ferrimagnet, as are many ferrites. The macroscopic properties of ferrimagnets are identical to those of ferromagnets, since the net moment is nonzero. We refrain from the joys of discussing the many other types of order encountered (canted ferro and antiferromagnets, helimagnets, speromagnets, etc.).

105 Free energy of ferromagnets

A common type of problem associated with ferromagnets is the following. A C-shaped piece of iron is to be used as an electromagnet. It is wrapped around with a current-carrying coil, and we are to find the magnetic field produced in the air gap. Clearly, one needs a constitutive or **B** versus **H** relationship to solve this and other boundary value problems. One approach is to use an experimentally measured curve for the **H** dependence of the total magnetization of the macroscopic body in question. As we shall see, however, such magnetization curves or hysteresis loops do not always describe a homogeneous relationship. In fact, the standard hysteresis loop of a large body (with dimensions greater than 1 mm, say) involves an average over a spatial structure on a mesoscale of 10^{-6}–10^{-3} cm. In this and the next section, we will try to describe some of the principles that go into determining this structure.

The constitutive relationship for ferromagnets cannot be written down in a single compact equation as it can for dia- and paramagnets.[6] Instead, it is best embodied in a thermodynamic minimum free energy principle, along with a phenomenological expression for the free energy that takes into account the principal physical phenomena. This was first done by Landau and Lifshitz. We shall follow their treatment but limit ourselves to low temperatures ($T < T_c/2$, say), far from the Curie point.

The thermodynamic potential most convenient to use turns out to be $\tilde{F}$, as defined in section 101. The natural variable for $\tilde{F}$ is **H**. We take $\tilde{F}$ to depend on **M**(**r**) as an additional variable. This may then be regarded as a *conditional* free energy, to be minimized with respect to **M**(**r**) in order to obtain the condition for equilibrium. Further, since **M** and **H** both vary spatially in many interesting situations, we will consider the free energy density, or free energy per unit volume, and denote this by the lowercase symbol $\tilde{f}$. The total free energy $\tilde{F}$ has many parts to it, which we call exchange, anisotropy, gradient, and magnetostatic:

$$\tilde{F} = \int_V (\tilde{f}_{\text{ex}}(\mathbf{M}, \mathbf{H}) + f_{\text{aniso}}(\mathbf{M}) + f_{\text{grad}}(\mathbf{M})) \, d^3x + E_{\text{ms}}. \tag{105.1}$$

We now discuss these terms one by one.

Exchange term: Let us first consider our magnet when $\mathbf{H} = 0$. The free energy then depends on **M** and other variables, such as temperature and pressure, which we do not show explicitly. The leading term in the free energy, $f_0(\mathbf{M})$, must then be such that it is minimum at $M = M_s$. Its value at this minimum must be of order $-M_s H_{\text{ex}}$. Second, it must depend only on the magnitude M and not the orientation of **M**, since exchange interactions are isotropic. Third, the energy must increase rapidly near $M_s(T = 0) = M_{\text{sat}}$, since the moments in the material are maximally aligned, saturated so to speak, at this value. Hence, $f_0(M)$ must have the general shape shown in fig. 16.4. Next, let us consider

---

[6] Or at least it cannot without considerable prior discussion of the meaning of the various terms.

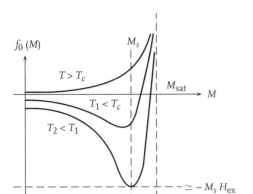
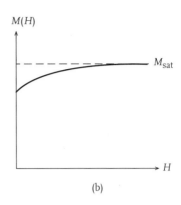

FIGURE 16.4. Schematics of the (a) conditional free energy $f_0(M)$ of an ideal ferromagnet and (b) the resulting equilibrium magnetization as a function of the magnetizing field **H**.

$\mathbf{H} \neq 0$.[7] To find the potential $\tilde{f}_{ex}(\mathbf{M}, \mathbf{H})$ (the subscript denoting exchange) corresponding to f_0, we note that

$$\frac{\partial \tilde{f}_{ex}}{\partial \mathbf{H}} = -\frac{\mathbf{B}}{4\pi} = -\frac{1}{4\pi}(\mathbf{H} + 4\pi\mathbf{M}),\qquad(105.2)$$

and that $\tilde{f}_{ex}(\mathbf{M}, 0) = f_0(M)$. Hence,

$$\tilde{f}_{ex}(\mathbf{M}, \mathbf{H}) = f_0(M) - \mathbf{M} \cdot \mathbf{H} - \frac{H^2}{8\pi}.\qquad(105.3)$$

Minimizing this with respect to **M** (with **H** fixed), we obtain

$$\frac{\partial f_0}{\partial \mathbf{M}} = \mathbf{H}.\qquad(105.4)$$

This can be regarded as an implicit constitutive equation. It determines **M(H)** in equilibrium in a completely homogeneous situation (ignoring of course, the other, smaller terms in the free energy). Note that it implies $\mathbf{M} \| \mathbf{H}$. Since $M(H)$ is not very different from M_s at low temperatures, we conclude that the curvature around the minimum of $f_0(M)$ must be quite large, of order H_{ex}/M_s.

Anisotropy energy: Next, we must incorporate the fact that because of the crystal structure, all directions of **M** are not equivalent. Thus, there is a slight anisotropy in the energy. For example, the solid may be more easily magnetizable along a high-symmetry direction in the crystal lattice. Such an anisotropy can originate in either the spin–spin or the spin–orbit interaction. Both these terms are of order $(v/c)^2$ relative to the exchange

---

[7] It is very important to distinguish the total **H** field, including the part due to an external current source, and the **H** field produced by the magnet itself. We will denote the latter by $\mathbf{H}_d$. The subscript "d" stands for "demagnetization," since, as we shall see in more detail in section 108, the field $\mathbf{H}_d$ tends to demagnetize the magnet.

energy, where v is the typical electron velocity. For the outer electrons responsible for magnetism, generally $v \ll c$, so the anisotropy energy is in this sense small. A proper justification of this statement is beyond the scope of our discussion, but one point is easily seen. Since $\mu_B = e\hbar/2mc$, the direct dipole–dipole interaction (which is the same as spin–spin) will contain a factor of $1/c^2$, which can only be compensated by a factor of v^2—ignoring constants of order unity, $\mu_B^2/a_0^3 \sim (e^2/a_0)(v/c)^2$, where v is the mean electron speed in hydrogen.

To write down a phenomenological form for the anisotropy energy, we note that it cannot be too violent a function of the orientation. Thus, we should be able to expand it in powers of the direction cosines of $\mathbf{M}$, i.e., in $\hat{\mathbf{M}}_x$, $\hat{\mathbf{M}}_y$, and $\hat{\mathbf{M}}_z$, where $\hat{\mathbf{M}}_i = M_i/M$, and keeping only a few terms in this expansion should be adequate. Second, we note that it must contain only even powers of $\hat{\mathbf{M}}_i$. The reason is that under time reversal $\mathbf{M} \to -\mathbf{M}$. This is easy to understand: if we reverse the flow of all the electrons in the atomic current loops, the moments of these loops, as well as the $\mathbf{B}$-field they produce, will all switch sign. However, since time reversal is a fundamental symmetry of nature, the energy must not change. Hence, it cannot contain odd powers, and we may write the general expansion as

$$f_{\text{aniso}} = \beta_{ij}\hat{\mathbf{M}}_i\hat{\mathbf{M}}_j + \zeta_{ijkl}\hat{\mathbf{M}}_i\hat{\mathbf{M}}_j\hat{\mathbf{M}}_k\hat{\mathbf{M}}_l + \cdots, \tag{105.5}$$

where β_{ij} and ζ_{ijkl} are tensors depending on the crystal symmetry. We will discuss only simple specific cases. For uniaxial crystals (tetragonal, rhombohedral, and hexagonal), we write

$$f_{\text{aniso}} = -K_1\hat{\mathbf{M}}_z^2 + K_2(\hat{\mathbf{M}}_x^2 + \hat{\mathbf{M}}_y^2)^2 + \cdots, \tag{105.6}$$

where the z axis is taken to be the principal symmetry axis of the crystal.[8] If $K_1 > 0$, the energy is minimized when $\mathbf{M}$ is along this axis, which is then said to be the *easy axis*. If $K_1 < 0$ (and K_2 is not too large), the energy is maximum when $\mathbf{M}$ is along $\hat{z}$ (which is now called the *hard* axis) and minimum in the xy plane (the easy plane), with the exact direction being determined by terms not shown in eq. (105.6).

For cubic crystals, the anisotropy energy may be written as

$$f_{\text{aniso}} = K_1(\hat{\mathbf{M}}_x^2\hat{\mathbf{M}}_y^2 + \hat{\mathbf{M}}_y^2\hat{\mathbf{M}}_z^2 + \hat{\mathbf{M}}_z^2\hat{\mathbf{M}}_x^2) + K_2\hat{\mathbf{M}}_x^2\hat{\mathbf{M}}_y^2\hat{\mathbf{M}}_z^2 + \cdots, \tag{105.7}$$

where x, y, and z are the axes along the edges of the cubic unit cell. Note that the expansion begins at the fourth order, since the only second-order term with cubic symmetry is $\hat{\mathbf{M}}_x^2 + \hat{\mathbf{M}}_y^2 + \hat{\mathbf{M}}_z^2$, which equals unity. If $K_1 > 0$ (as in Fe), the easy axes are

---

[8] Note that we do not include in this expression a term such as $K_1'(\hat{\mathbf{M}}_x^2 + \hat{\mathbf{M}}_y^2)$, since this could be rewritten as $K_1'(1 - \hat{\mathbf{M}}_z^2)$ and then absorbed into the K_1 term and an overall additive constant in the energy. Similar reductions limit the number of independent terms of higher orders.

cube axes x, y, and z, while if $K_1 < 0$ (as in Ni), the easy axes are along the cube body diagonals, e.g., along $(\hat{x} + \hat{y} + \hat{z})$. Note that in every case, if $\hat{n}$ is an easy direction, so is $-\hat{n}$.

The coefficients K_1, K_2, etc., are known as *anisotropy coefficients*. They have dimensions of energy density. By our earlier discussion we expect that $K_i / H_{\text{ex}} M_s \sim (v/c)^2 \sim 10^{-3} - 10^{-4}$. Since $H_{\text{ex}} M_s \sim 10^{10}$ erg/cm^3, we expect $K_i \sim 10^6$ erg/cm^3. This is actually the case.[9]

Gradient energy: If the magnetization is not uniform, the exchange energy is increased. If the variations in $\mathbf{M(r)}$ are slow on the atomic length scale, the increase can be represented in terms of gradients of $\mathbf{M}$. For cubic crystals, the simplest term that meets the requirements of time-reversal invariance, isotropy of exchange, and positivity is

$$f_{\text{grad}} = \frac{1}{2} \frac{C}{M_s^2} \left[\left(\frac{\partial \mathbf{M}}{\partial x} \right)^2 + \left(\frac{\partial \mathbf{M}}{\partial y} \right)^2 + \left(\frac{\partial \mathbf{M}}{\partial z} \right)^2 \right], \tag{105.8}$$

with $C > 0$. (The coefficient is written as C/M_s^2 instead of just C for later convenience.) For noncubic crystals a tensorial structure must be put in.

To estimate the size of C, we first note that it has dimensions of energy/length. Second, it must be proportional to the exchange energy density, $M_s H_{\text{ex}}$, and since exchange is a short-ranged phenomenon, we expect C to be of order $M_s H_{\text{ex}} a^2$, where a is an interatomic distance or lattice constant. Taking $M_s H_{\text{ex}} \sim 10^{10}$ erg/cm^3 and $a \sim 2$ Å, we get $C \simeq 10^{-6}$ erg/cm. This is indeed correct. For both Fe and Ni, $C \simeq 2 \times 10^{-6}$ erg/cm.

Magnetostatic energy: The last term in the energy is due to the direct dipole–dipole interaction. In terms of the macroscopic magnetization field $\mathbf{M(r)}$, this can be written as

$$E_{\text{ms}} = \frac{1}{2} \int_V \int_V \frac{(\mathbf{M}_1 \cdot \mathbf{M}_2) r_{21}^2 - 3(\mathbf{M}_1 \cdot \mathbf{r}_{21})(\mathbf{M}_2 \cdot \mathbf{r}_{21})}{r_{21}^5} d^3 x_1 \, d^3 x_2, \tag{105.9}$$

where $\mathbf{M}_i = \mathbf{M(r}_i)$, $\mathbf{r}_{21} = \mathbf{r}_2 - \mathbf{r}_1$, and the integrals extend over the volume of the body.[10]

This energy differs from the other contributions already considered (in particular, it is also much smaller in magnitude than the exchange energy) in that it is long-ranged. Indeed, because of it, statements involving idealized *infinite* ferromagnets are often incorrect and must always be examined with great care.

We shall see that E_{ms} is the driving force behind domain formation in ferromagnets. Its long-ranged character makes it generally the most difficult term to evaluate. Because of this, it is useful to rewrite it in a number of other ways that provide additional insight into its structure and enable one to estimate it qualitatively.

---

[9] Some characteristic values are, in units of 10^6 erg/cm^3, Co (hexagonal): $K_1 = 4.1$, $K_2 = 1.0$; SmCo$_5$ (hexagonal): $K_1 = 1100$; Fe (cubic): $K_1 = 0.48$, $K_2 = 0.05$; Ni (cubic): $K_1 = -0.045$, $K_2 = 0.023$.

[10] It should be noted that this energy is distinct from the *anisotropic* part of the spin–spin interaction. In particular, it makes no reference to the orientation of $\mathbf{M}$ relative to the crystal axes.

A second form for E_{ms} is obtained by noting that it is equivalent to the electrostatic energy of a charge distribution,

$$\rho_m = -\nabla \cdot \mathbf{M}. \tag{105.10}$$

That is,

$$E_{ms} = \frac{1}{2} \int_{V_+} \int_{V_+} \frac{\rho_m(\mathbf{r}_1)\rho_m(\mathbf{r}_2)}{|\mathbf{r}_1 - \mathbf{r}_2|} d^3x_1 \, d^3x_2. \tag{105.11}$$

The notation V_+ means that the integrals in this expression are to be extended to a volume slightly beyond the physical surface of the body. In this way ρ_m can be understood to include (in the form of a delta-function contribution) the surface charge $\mathbf{M} \cdot \hat{\mathbf{n}}$, and one does not have to write separate expressions for it.

Further forms follow by considering the magnetized body alone and supposing that no external magnetizing fields, i.e., current sources, are present. Let us denote the magnetizing field produced by the magnetization $\mathbf{M}(\mathbf{r})$ by $\mathbf{H}_d(\mathbf{r})$ to distinguish it from the thermodynamic variable $\mathbf{H}(\mathbf{r})$ in $\tilde{F}$, eq. (105.1).[11] Similarly, $\mathbf{B}_d \equiv \mathbf{H}_d + 4\pi\mathbf{M}$. The equations for $\nabla \cdot \mathbf{B}_d$ and $\nabla \times \mathbf{H}_d$ can be rewritten as

$$\nabla \cdot \mathbf{H}_d = -4\pi\nabla \cdot \mathbf{M} = 4\pi\rho_m, \tag{105.12}$$

$$\nabla \times \mathbf{H}_d = 0. \tag{105.13}$$

The field $\mathbf{H}_d$ obeys exactly the same equations as the electric field, with the volume and surface densities ρ_m and $\sigma_m(= \mathbf{M} \cdot \hat{\mathbf{n}})$ as the charge sources! (It is this analogy between electrostatics and static ferromagnetism that historically led to $\mathbf{H}$ being called the magnetic field. This nomenclature makes perfect sense when we realize that there were no current sources, and magnetic effects could be studied only with permanent magnets.) We can therefore write down two more formulas for E_{ms}:

$$E_{ms} = -\frac{1}{2} \int_V \mathbf{M} \cdot \mathbf{H}_d \, d^3x, \tag{105.14}$$

$$= \frac{1}{2} \int_{V_+} \rho_m \phi_d \, d^3x, \tag{105.15}$$

where ϕ_d is the magnetic scalar potential (vanishing at infinity) yielding $\mathbf{H}_d$, i.e., $\mathbf{H}_d = -\nabla\phi_d$. In these formulas, the integral is only over the volume of the body. The first form is particularly useful.

The concepts of magnetic charge density and magnetic scalar potential are extremely useful in solving magnetostatic boundary value problems involving ferromagnets. Some old textbooks used to shun their use to avoid implying the existence of magnetic monopoles. In the twenty-first century, there is no reason to give into this kind of timidness. We know that ρ_m is not a *true* magnetic charge but only a mathematically convenient construct.

---

[11] $\mathbf{H}_d$ is the demagnetizing field. See section 108.

One last form for E_{ms} follows by considering the integral of $\mathbf{B}_d \cdot \mathbf{H}_d$ over all space. We write

$$\mathbf{B}_d \cdot \mathbf{H}_d = -\mathbf{B}_d \cdot \nabla \phi_d$$

$$= -\nabla \cdot (\mathbf{B}_d \phi_d) + (\nabla \cdot \mathbf{B}_d) \phi_d$$

$$= -\nabla \cdot (\mathbf{B}_d \phi_d),$$

since $\nabla \cdot \mathbf{B}_d = 0$. Integrating both sides over all space, the right-hand side is transformed into a vanishing surface integral at infinity.[12] We thus obtain

$$\int \mathbf{B}_d \cdot \mathbf{H}_d \, d^3x = 0. \tag{105.16}$$

This is an interesting and important result in its own right. But if we note that $\mathbf{B}_d = \mathbf{H}_d + 4\pi\mathbf{M}$ and use eq. (105.14), we obtain an even more useful result,

$$E_{\mathrm{ms}} = \int \frac{H_d^2}{8\pi} \, d^3x. \tag{105.17}$$

Note that the integral extends over all space.

The last form leads to the *pole avoidance principle*. Since the integrand is always positive, the goal is clearly to keep $\mathbf{H}_d$ as small as possible everywhere. The only finite body in which $\mathbf{H}_d$ can be made zero everywhere is the torus (though see fig. 16.6b), but in other cases, one should try to minimize the creation of magnetic charges or poles (surface magnetic charges). Configurations in which $\nabla \cdot \mathbf{M} \neq 0$ inside the body are particularly bad. If charges are to be made, one should try to put opposite charges next to one another so that $\mathbf{H}_d$ is limited to small regions of space.

As an application of these ideas, let us consider a bar magnet with a rectangular cross section, uniformly magnetized along its long side. What do the $\mathbf{H}_d$ and $\mathbf{B}_d$ fields look like? The lines of $\mathbf{H}_d$ can be seen by considering the magnetic charges. Note that $\mathbf{H}_d$ is discontinuous at the surface. This is easy to remember if we remember that $\mathbf{H}_d$ is like an $\mathbf{E}$ field. Lines of $\mathbf{B}_d$, on the other hand, are continuous and are better visualized as arising from an equivalent solenoid. (See fig. 16.5.) Next, let half the bar be magnetized one way and the other half the other way. The fields now look as shown in fig. 16.6a. The lines of $\mathbf{H}$, in particular, are concentrated near the end faces of the bar. Compared with the uniformly magnetized bar, the magnetostatic energy is now lower. To see this, note that in the uniform $\mathbf{M}$ case, if the bar is not too long, $\mathbf{H}_d$ is approximately antiparallel to $\mathbf{M}$ throughout the body, so the product $\mathbf{M} \cdot \mathbf{H}$ is large and negative everywhere. Now $\mathbf{H}_d$ is large in magnitude only near the ends, but even there, it is frequently at an obtuse angle to $-\mathbf{M}$, so the product $-\mathbf{H}_d \cdot \mathbf{M}$ is reduced everywhere.

---

[12] There is no delta-function contribution at the surface of the body, since both ϕ_d and the normal component of $\mathbf{B}_d$ are continuous at the surface.

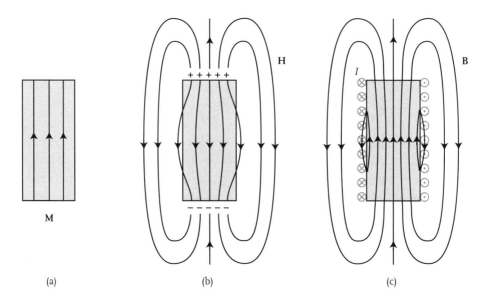

FIGURE 16.5. Lines of **M**, **H**, and **B** for a uniform bar magnet (a). **H** is best viewed as arising from surface magnetic charges (b), while **B** is best viewed as arising from an equivalent solenoid (c).

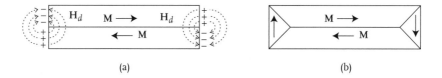

FIGURE 16.6. A bar magnet divided into two oppositely oriented domains along its length (a) leads to surface magnetic charges on its ends, which produce a sharply localized magnetic field H_d near these ends. This reduces the magnetostatic energy. This energy is further reduced by creating domains of closure (b). Now there are almost no surface charges and thus no H_d field.

In this way, we see that the magnetostatic energy favors division of a magnetic body into *domains*, such that **M** is uniform (or nearly so) within each domain, but changes abruptly from one domain to another. Structures in which magnetic charges are minimized are best. Especially good is an arrangement as in fig. 16.6b, showing so-called domains of closure, in which almost no H_d field is created.

Exact calculations of magnetostatic energies are generally difficult. The following textbook problem may seem artificial but is nevertheless instructive.

Exercise 105.1 Calculate the magnetostatic energy for a sphere of radius a magnetized (a) uniformly with $|\mathbf{M}| = M_s$ everywhere, (b) with $\mathbf{M} = \pm M_s \hat{x}$ in the hemispheres $z > 0$ and $z < 0$ (fig. 16.7).

Solution: For part (a), we can integrate $\mathbf{M} \cdot \mathbf{H}$ over the volume of the sphere to obtain $E_{\mathrm{ms}} = 8\pi^2 M_s^2 a^3/9$. For part (b), we use the method of a magnetic scalar potential ϕ, from

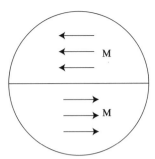

FIGURE 16.7. A magnetic sphere divided into two hemispherical domains separated by a 180° domain wall.

which we obtain the field via $\mathbf{H}_d = -\nabla\phi$. Let us denote this function for $r < a$ and $r > a$ by ϕ_{in} and ϕ_{out}, respectively. The potential must obey Laplace's equation, and, since

$$\mathbf{M} = M_s \, \text{sgn}(z)\hat{\mathbf{x}}, \tag{105.18}$$

it must obey the following boundary conditions on the surface $r = a$:

$$\phi_{\text{in}} - \phi_{\text{out}} = 0, \tag{105.19}$$

$$\frac{\partial \phi_{\text{in}}}{\partial r} - \frac{\partial \phi_{\text{out}}}{\partial r} = 4\pi M_s \hat{\mathbf{x}} \cdot \hat{\mathbf{n}} \, \text{sgn}(z). \tag{105.20}$$

The general solutions of Laplace's equation can be taken as

$$\phi_{\text{in}} = \sum_{\ell,m} A_{\ell m} \frac{r^\ell}{a^\ell} Y_{\ell m}(\hat{\mathbf{n}}), \tag{105.21}$$

$$\phi_{\text{out}} = \sum_{\ell,m} A_{\ell m} \frac{a^{\ell+1}}{r^{\ell+1}} Y_{\ell m}(\hat{\mathbf{n}}), \tag{105.22}$$

where we have written the formulas so that the continuity of ϕ at $r = a$ is obeyed from the outset. The condition on the jump of the normal gradient now gives

$$\frac{1}{a} \sum_{\ell,m} (2\ell + 1) A_{\ell m} Y_{\ell m}(\hat{\mathbf{n}}) = 4\pi M_s \hat{\mathbf{x}} \cdot \hat{\mathbf{n}} \, \text{sgn}(z), \tag{105.23}$$

i.e.,

$$A_{\ell m} = \frac{4\pi a}{2\ell + 1} M_s B_{\ell m}^*, \tag{105.24}$$

where

$$B_{\ell m} = \int d^2\hat{\mathbf{n}} \, [\text{sgn}(\hat{\mathbf{z}} \cdot \hat{\mathbf{n}})\hat{\mathbf{x}} \cdot \hat{\mathbf{n}}] Y_{\ell m}(\hat{\mathbf{n}}). \tag{105.25}$$

Once the coefficients $B_{\ell m}$ are known, we proceed as follows. The surface magnetic charge is given by

$$\sigma_m = \mathbf{M} \cdot \hat{\mathbf{n}} = M_s \hat{\mathbf{x}} \cdot \hat{\mathbf{n}} \, \mathrm{sgn}(\hat{\mathbf{z}} \cdot \hat{\mathbf{n}}). \tag{105.26}$$

We now use the form (105.15) for the magnetostatic energy:

$$
\begin{aligned}
E_{\mathrm{ms}} &= \frac{1}{2} \int \sigma_m(\hat{\mathbf{n}}) \phi(\hat{\mathbf{n}}) \, a^2 d^2 \hat{\mathbf{n}} \\
&= \frac{1}{2} a^2 M_s \sum_{\ell,m} A_{\ell m} \int d^2 \hat{\mathbf{n}} \, \mathrm{sgn}(\hat{\mathbf{z}} \cdot \hat{\mathbf{n}}) \, \hat{\mathbf{x}} \cdot \hat{\mathbf{n}} \, Y_{\ell m}(\hat{\mathbf{n}}) \\
&= 2\pi M_s^2 a^3 \sum_{\ell m} \frac{|B_{\ell m}|^2}{2\ell + 1}.
\end{aligned}
\tag{105.27}
$$

The problem is, therefore, to find $B_{\ell m}$. We present the answer for this first and give the analysis later. We have (i) $B_{\ell m} = 0$ unless ℓ is even and $m = \pm 1$, (ii) $B_{\ell,-1} = -B_{\ell 1}$, and (iii)

$$B_{\ell 1} = (-1)^{\ell/2} \frac{\sqrt{\ell(\ell+1)(2\ell+1)\pi}}{2^{\ell-1}(\ell+2)(\ell-1)} \binom{\ell-1}{\ell/2}. \tag{105.28}$$

Hence,

$$E_{\mathrm{ms}} = 4\pi^2 M_s^2 a^3 \sum_{\ell=2,4,\ldots}^{\infty} \frac{\ell(\ell+1)}{2^{2(\ell-1)}(\ell+2)^2(\ell+1)^2} \binom{\ell-1}{\ell/2}^2. \tag{105.29}$$

Numerical evaluation gives the sum (the summand drops as $1/\ell^3$ for large ℓ) to be 0.106, so

$$E_{\mathrm{ms}} = 4.18 \, M_s^2 a^3, \tag{105.30}$$

which is about half the value for the uniformly magnetized sphere.

To evaluate $B_{\ell m}$, we first expand $\mathrm{sgn}(\hat{\mathbf{z}} \cdot \hat{\mathbf{n}})$ in Legendre polynomials $P_\ell(\hat{\mathbf{z}} \cdot \hat{\mathbf{n}})$. If we write

$$\mathrm{sgn}(u) = \sum_\ell \frac{2\ell+1}{2} C_\ell P_\ell(u), \tag{105.31}$$

then

$$C_\ell = \int_{-1}^{1} \mathrm{sgn}(u) P_\ell(u) \, du. \tag{105.32}$$

To find C_ℓ, we multiply both sides of the generating function (A.72) (writing u for z) by $\mathrm{sgn}(u)$ and integrate over $(-1, 1)$. The right-hand side yields $\sum_\ell C_\ell t^\ell$, while the left-hand side equals

$$\int_{-1}^{1} \frac{\text{sgn}(u)}{(1+t^2 - 2tu)^{1/2}} du = \frac{2}{t}\left[\sqrt{1+t^2} - 1\right]$$

$$= t + \sum_{k=2}^{\infty} \frac{(-1)^{k-1}}{2^{2k-3}} \frac{(2k-3)!}{k!(k-2)!} t^{2k-1}. \tag{105.33}$$

Equating the two power series in t term by term, we see that C_ℓ vanishes for even ℓ, while for odd ℓ, we have

$$C_\ell = \frac{(-1)^{(\ell-1)/2}}{2^{\ell-2}} \frac{(\ell-2)!}{\left(\frac{\ell+1}{2}\right)! \left(\frac{\ell-3}{2}\right)!}. \tag{105.34}$$

To complete the first step, we use eq. (A.71) to rewrite eq. (105.31) as

$$\text{sgn}(\hat{\mathbf{z}} \cdot \hat{\mathbf{n}}) = \sum_{\ell} \frac{2\ell+1}{2} C_\ell \sqrt{\frac{4\pi}{2\ell+1}} Y_{\ell 0}(\hat{\mathbf{n}}). \tag{105.35}$$

The second step is to expand $(\hat{\mathbf{x}} \cdot \hat{\mathbf{n}}) Y_{\ell m}(\hat{\mathbf{n}})$ in terms of other $Y_{\ell m}$'s. We have already done this in the form of the recursion formulas (A.81). Since we eventually must integrate the product of this expansion with eq. (105.35), we need only consider the values $m = \pm 1$ to begin with. (This is why $B_{\ell m}$ vanishes for $m \neq \pm 1$.) Equation (A.81) gives

$$(\hat{\mathbf{x}} \cdot \hat{\mathbf{n}}) Y_{\ell,1} = \frac{1}{2}\left(\frac{\ell(\ell+1)}{2\ell+1}\right)^{1/2}\left(\frac{Y_{\ell+1,0}}{\sqrt{2\ell+3}} - \frac{Y_{\ell-1,0}}{\sqrt{2\ell-1}}\right) + Y_{\ell\pm1,2} \text{ terms.} \tag{105.36}$$

Multiplying this with eq. (105.35) and integrating over all orientations, we get

$$B_{\ell 1} = \frac{1}{2}\sqrt{\frac{\ell(\ell+1)}{2\ell+1}}\left[\frac{1}{\sqrt{2\ell+3}}\sqrt{\frac{4\pi}{2\ell+3}}\frac{2\ell+3}{2}C_{\ell+1} - \frac{1}{\sqrt{2\ell-1}}\sqrt{\frac{4\pi}{2\ell-1}}\frac{2\ell-1}{2}C_{\ell-1}\right]$$

$$= \frac{\sqrt{\pi}}{2}\sqrt{\frac{\ell(\ell+1)}{2\ell+1}}(C_{\ell+1} - C_{\ell-1}). \tag{105.37}$$

Substituting eq. (105.34) in this result, we obtain the answer (105.28) quoted before. Further, since

$$(\hat{\mathbf{x}} \cdot \hat{\mathbf{n}}) Y_{\ell,-1} = -(\hat{\mathbf{x}} \cdot \hat{\mathbf{n}}) Y_{\ell,1}, \tag{105.38}$$

we immediately obtain $B_{\ell,-1} = -B_{\ell 1}$, as also quoted.

106 Ferromagnetic domain walls*

In discussing domain structure in the previous section, we have ignored the gradient energy that will have to be paid at the surfaces separating the domains, which are known as *domain walls*. An infinitely sharp wall is clearly a mathematical idealization, and the

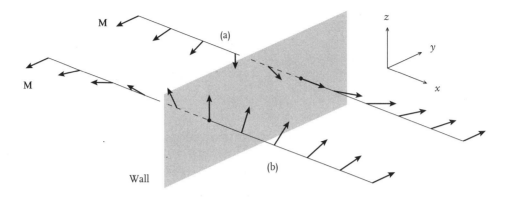

FIGURE 16.8. Domain wall structure. The structure in (a) leads to a large magnetostatic energy, as $\nabla \cdot \mathbf{M} \neq 0$. The structure in (b) is better. It is the Landau-Lifshitz 180° domain wall.

true transition must be smoothed out over some microscopic length. We now examine the details of this transition.

Let us consider a uniaxial material with easy axis z and take the wall to be the yz plane, with $\mathbf{M} = \pm M_s \hat{\mathbf{z}}$ for $x > 0$ and $x < 0$ far from the wall. The first priority is to minimize the exchange term $f_0(M)$, as this is the dominant part of the free energy. This is done by keeping $|\mathbf{M}| = M_s$ everywhere. Effectively, $\mathbf{M}$ can only rotate from one direction to another. Two possible profiles are shown in fig. 16.8. In (a) $\mathbf{M}$ rotates in the xy plane, being perpendicular to the wall at $x = 0$, while in (b) $\mathbf{M}$ rotates in the yz plane, with M_n, the component normal to wall being zero. The profile (a) leads to a large magnetostatic energy, since $\nabla \cdot \mathbf{M} = -\nabla \cdot \mathbf{H}_d/4\pi$ is nonzero and leads to a $\mathbf{H}_d$ field along x. Profile (b) is clearly the best in this regard, since $\nabla \cdot \mathbf{M} = 0$ everywhere.[13]

The problem is therefore reduced to minimizing the sum of the gradient and anisotropy energies. Writing the magnetization as

$$\mathbf{M} = M_s (\cos \theta(x) \hat{\mathbf{y}} - \sin \theta(x) \hat{\mathbf{z}}), \tag{106.1}$$

these energies become

$$f_{\text{grad}} = \frac{1}{2} C \left(\frac{d\theta}{dx} \right)^2,$$

$$f_{\text{aniso}} = K_1 \sin^2 \theta,$$

where we have added a constant K_1 to f_{aniso} so that it vanishes far from the wall.

We denote the integral of the sum of f_{grad} and f_{aniso} per unit area of the wall by σ_{dw}, which may be thought of as a domain wall surface tension. The mathematical problem is

---

[13] Of course, we must eventually consider how $\mathbf{M}$ and $\mathbf{H}$ behave far away from the region considered, either at another domain wall or at the surface of the body. Here we are concerned only with the immediate vicinity of one domain wall, and how to find the overall magnetostatic energy of the body as a whole is not at issue.

thus to minimize

$$\sigma_{dw} = \int_{-\infty}^{\infty} \left[\frac{1}{2} C \left(\frac{d\theta}{dx} \right)^2 + K_1 \sin^2 \theta \right] dx, \tag{106.2}$$

subject to the boundary conditions $\theta \to \pi$ as $x \to -\infty$, and $\theta \to 0$ as $x \to \infty$.

The solution to this problem can be found by a simple mechanical analogy. We think of θ as the position coordinate of a particle moving in one dimension, x as the time, and C as the mass. The integrand in eq. (106.2) is then like the Lagrangian $T - V$, with kinetic energy $T = \frac{1}{2} C (d\theta/dx)^2$ and potential energy $V = -K_1 \sin^2 \theta$. The integral is like the action, and minimizing it leads to the usual mechanical equations of motion. It follows immediately that for the optimal solution for $\theta(x)$, the total energy $T + V$ is a constant, whose value is found to be zero by examining the boundary conditions at $x \to \pm\infty$. Thus, $T = -V$, i.e.,

$$\frac{d\theta}{dx} = - \left(\frac{2K_1}{C} \right)^{1/2} \sin \theta. \tag{106.3}$$

(We choose the square root leading to a negative $d\theta/dx$. The other is physically equivalent.) Integration of this equation gives

$$\cos \theta = \tanh(x/\xi), \quad \sin \theta = \text{sech}(x/\xi), \tag{106.4}$$

where

$$\xi = (C/2K_1)^{1/2} \tag{106.5}$$

may be regarded as the domain wall width, since it gives the distance scale over which **M** varies significantly. The energy σ_{dw} may be found by putting $T = -V$ in the integral (106.2), i.e.,

$$\sigma_{dw} = 2K_1 \int_{-\infty}^{\infty} \text{sech}^2 \left(\frac{x}{\xi} \right) dx = 2\sqrt{2K_1 C}. \tag{106.6}$$

The solution we have found above is known as the *Landau-Lifshitz 180° wall*, since **M** turns by 180°. A very similar 90° wall is obtained in cubic ferromagnets like Fe, which have $K_1 > 0$, in which the wall is, say, normal to the $\hat{\mathbf{x}} + \hat{\mathbf{y}}$ direction, and **M** rotates from $\hat{\mathbf{x}}$ on one side to $\hat{\mathbf{y}}$ on the other. Many other types of domain wall geometries and solutions are known,[14] and this is an active area of research at present. For our purpose, the key point is that the estimates (106.5) and (106.6) are generally valid up to factors of order unity, since these are the only dimensionally correct combinations of anisotropy and gradient energy coefficients. Since, in CGS units, $C \sim 10^{-6}$ and $K_1 \sim 10^6$, $\xi \sim 10^{-6}$ cm, or 100 Å, while $\sigma_{dw} \sim 1$ erg/cm^2. For the most common walls in Fe, Co, and Ni, one finds experimentally that the width is 400, 150, and 100 Å, respectively, while the surface tension is 3, 8, and 1 erg/cm^2.

---

[14] See, e.g., Aharoni (2000).

Exercise 106.1 Solve for the magnetization profile, width, and surface tension of a 90° wall in a cubic material with $K_1 > 0$.

Exercise 106.2 Solve the problem of a 180° wall in a uniaxial material, keeping the K_2 term in the anisotropy energy (105.6).

107 Hysteresis in ferromagnets

In many devices such as transformer cores, electric motors, and relays, a ferromagnetic material is subjected to a time-dependent magnetic field. In all cases, it is observed that the magnetization of the material is *hysteretic*, i.e., that **M** is dependent not only on **H** but also on the past history of the material. This behavior is graphically depicted in the form of a curve of **M** versus **H** (or sometimes of **B** versus **H**) known as a *hysteresis loop*. See fig. 16.9. We shall discuss how this curve is to be read shortly, but let us first ask why **M** should be hysteretic in the first place. To do this, let us first consider uniform magnetization and neglect the gradient and magnetostatic energies. To determine the value of **M** that is energetically most favorable, we must effectively minimize the energy

$$f_{\text{aniso}}(\hat{\mathbf{M}}) - \mathbf{M} \cdot \mathbf{H} \tag{107.1}$$

for $M = M_s$. For general nonzero **H**, this energy landscape will have more than one minimum on account of f_{aniso}. As **H** is varied, these minima will move up and down, and a global minimum for some values of **H** may turn into a local minimum for other values. The key point is that if **M** starts out in a minimum that turns from the global minimum into a local one, *it need not immediately switch to the new global minimum*. In other words, **M** can get stuck in a local minimum. The actual magnetization of a bulk ferromagnetic body is determined not only by the energetics but also by whether it is dynamically possible for the minimum energy to be achieved. The problem is analogous to that of a particle moving in a time-varying, asymmetric, two-dimensional double well in which the wells move up and down relative to each other. It is clear that the dynamical position of the particle need not always be at the deeper minimum, especially if there is friction.

Exercise 107.1 Find the range of magnetic fields for which the energy (107.1) has multiple minima for a uniaxial ferromagnet.

When the gradient and magnetostatic energies are included, the problem of determining the evolving magnetization profile in a macroscopic body of general shape becomes very complicated. In practical situations, the waters are muddied even further by the fact that a real magnet is likely to be polycrystalline. Not only do the different crystallites have different easy and hard axes, but the grain boundaries separating them tend to act as pinning centers for domain walls. Other defects, local strains, inhomogeneities, and impurities play a similar role, and empirical evidence suggests that they all affect hysteresis in significant ways.

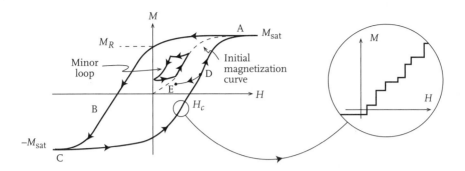

FIGURE 16.9. Schematic of a hysteresis loop for a ferromagnet, showing the remnant magnetization, M_R, and the coercive field, H_c. Also shown are the initial magnetization curve and a minor loop. The enlargement shows the stepped nature of $M(H)$ as revealed by Barkhausen noise.

Let us now return to the hysteresis loop of fig. 16.9. The simplest way of determining $M(H)$ is to wind a solenoid around a body in the shape of a torus. This geometry minimizes the tendency to form domains, and by applying the macroscopic Ampere law, we obtain

$$H = 2NI/cR \tag{107.2}$$

$$H = \mu_0 NI/2\pi R \quad \text{(SI)}$$

for the H field inside the body. Here, N is the total number of windings around the torus, I is the current through them, and R is the radius of the large circle of the torus. We have neglected the variation of the circumference of the Amperean loop; in effect, the torus is assumed to be thin. Equation (107.2) gives us H in terms of the experimentally controllable I, and M can be obtained by purely inductive measurements. If a sufficiently large positive field is applied, the magnetization is saturated in the same direction, and one reaches the point A on the curve. If the field is now decreased, one follows the curve ABC. In particular, M does not vanish when H reaches 0, but only when H equals a negative value $-H_c$. As H is decreased further, M saturates in the negative direction. If H is now increased, passed through zero, and eventually made positive again, one traces out the curve CDA, which is obtained by inverting ABC through the origin.

The curve in fig. 16.9 is in fact the *limiting* hysteresis loop. If we imagine ramping up H, but we turn back at the point D before saturation is attained, we do not retrace the prior evolution of M but, rather, follow the path DE. In fact, by appropriate cycling of H, any point inside the outermost loop can be reached. If the sample is completely demagnetized, i.e., brought to the point $M = H = 0$, and the field is then increased, one obtains the so-called *magnetization curve* or the *virgin magnetization curve*. If, as in many applications, one starts from a demagnetized state and cycles H back and forth over a smaller range of values than that needed to saturate the sample, one obtains a *minor loop*, shown by the dashed curve in the figure.

Let us pause to note that eq. (107.2), and others like it, are the basis for calling **H** the magnetizing field. The current in the windings is clearly the physical agency that magnetizes the initially unmagnetized piece of iron, and since I and H are so directly related, we may just as well regard this agency as the field **H**.

It is evident from our description that the hysteresis curve represents a large length scale phenomenon and involves averaging over many stochastic aspects, especially the orientation of crystallites in a polycrystalline sample. Much more importantly, hysteresis is clearly an irreversible phenomenon. This irreversibility can arise only if the energy $-\mathbf{M} \cdot \mathbf{H}$ is somehow dissipated into other degrees of freedom. A complete theory for hysteresis does not at present exist, but it is evident that it must entail not only the dynamical equations of motion for the atomic moments but also their coupling to the dissipative degrees of freedom. The most plausible candidates for the latter are lattice vibrations and, additionally, eddy currents in the case of metallic magnets. It is generally believed that the rotation of moments inside a domain is reversible, while the motion of domain walls is irreversible.

One of the clearest demonstrations of the irreversible aspects of hysteresis can be done as follows. If we employ not a toroidal but a cylindrical magnet, and bring a second coil near the magnet, then changes in M in the magnet will produce a small current in the second coil because of the inductive coupling (mutual inductance) between the magnet and the coil. If this current is amplified and fed to a loudspeaker, one hears distinct clicks at irregular intervals. This is known as *Barkhausen noise*. It arises from domain walls moving in jerky fashion from one pinning center to another. A blowup of the hysteresis loop will have the irregular stepped appearance shown in fig. 16.9.

Engineers have developed a large number of parameters for characterizing the hysteresis loop of a material for practical applications. The two most important are the *coercivity*, H_c, which is the reverse field required to drive the magnetization to zero starting from a saturated state, and the *remanent magnetization* M_R, which is the magnetization at zero field, again starting from a saturated state. In addition, one finds the *initial permeability*, μ_i, which equals the ratio B/H as one proceeds up the virgin magnetization curve. The initial permeability and the coercivity are generally related inversely. If the minor loops are narrow and long, as they are for materials like iron and permalloy, one can represent the **B** versus **H** relation approximately as $\mathbf{B} = \mu_{\text{eff}} \mathbf{H}$, where μ_{eff} is known as an effective permeability. This relation can then be used to perform calculations provided H lies within the appropriate range. It should be noted, however, that this formula is really a rule of thumb and does not have the status as the constitutive relation $B = \mu H$ for a paramagnet, where it does imply a truly linear relation for small fields. Finally, one has the *hysteresis loss*, W_h, which is the energy lost as heat in one complete hysteresis cycle. It is evident that W_h is the magnetic work done in going around the hysteresis loop once, i.e.,

$$W_h = \frac{1}{4\pi} \oint B \, dH. \tag{107.3}$$

The hysteresis properties of ferromagnets greatly influence how they are used. Ferromagnetic materials are divided into two broad classes: *soft* and *hard*. Soft materials typically have coercivities less than 15 Oe and high initial permeabilities. These materials have both ac applications (power generation and transmission elements, such as motors, generators, and transformers, waveguides for microwaves, and radio receivers) and dc applications (electromagnets, magnetic shielding). In the ac uses, low hysteresis loss is important, while in the dc uses, high permeability is the main consideration. Examples of soft magnets are iron, iron–silicon alloys, nickel–iron alloys, all of which are metallic, and ferrites, such as magnetite, cobalt–iron ferrite, and YIG, which are insulating. Hard magnetic materials are generally used where a permanent magnet is needed. Here, one desires both the coercivity and the remanent magnetization to be high. In $Nd_2Fe_{14}B$, e.g., $H_c = 14$ kOe, and $M_R = 1050$ emu/cm^3. These materials are also used in motors and generators, galvanometers, loudspeakers, and magnetic storage (tapes, disks, credit card stripes, etc.). The high coercivity is often obtained by using compounds containing a rare earth, which has intrinsically higher anisotropy energy, and a transition-metal element to provide a large moment. In addition to $Nd_2Fe_{14}B$ already mentioned, the compounds $SmCo_5$ and Sm_2Co_{17} use the same idea. Other hard magnets are made by deliberately adding impurities to or introducing defects into iron, nickel, and cobalt so as to provide extra sites for domain wall pinning.

108 Demagnetization

In the last few sections, we have encountered several times the idea that volume and surface magnetic charges ($\nabla \cdot \mathbf{m}$ and $\mathbf{M} \cdot \hat{\mathbf{n}}$, respectively) increase the magnetostatic energy of the body and produce a (de)magnetizing $\mathbf{H}_{demag}$ inside it that is opposite to $\mathbf{M}$ on the average. This concept is not limited to ferromagnets and applies whether the magnetization is induced or spontaneous. This effect, known as *demagnetization*, is completely analogous to the depolarization of dielectrics studied in section 96. Indeed, the calculation of the demagnetization field $\mathbf{H}_{demag}$ (or just $\mathbf{H}_d$ as before) is mathematically identical to that of $\mathbf{E}_d$. The conceptual aspects of the problem have already been discussed in connection with the magnetostatic energy of ferromagnets (section 105). However, enormous confusion arises from abuse of notation (overworking the poor μ) and SI versus Gaussian units. In this section, we therefore give some key formulas explicitly.

Let us first consider a para- or diamagnetic body in an external field $\mathbf{B}_0$. The equations to be solved are

$$\nabla \times \mathbf{H} = 0, \quad \nabla \cdot \mathbf{B} = 0, \quad \mathbf{B} = \mathbf{H} + 4\pi\mathbf{M}, \quad \mathbf{B} = \mu\mathbf{H}, \tag{108.1}$$

along with continuity of H_t and B_n. The corresponding equations for a dielectric body in an external field $\mathbf{E}_0$ are

$$\nabla \times \mathbf{E} = 0, \quad \nabla \cdot \mathbf{D} = 0, \quad \mathbf{D} = \mathbf{E} + 4\pi\mathbf{P}, \quad \mathbf{D} = \epsilon\mathbf{E}, \tag{108.2}$$

along with continuity of E_t and D_n. Thus, the problems are identical with the equivalences

$$\mathbf{E} \leftrightarrow \mathbf{H}, \quad \mathbf{D} \leftrightarrow \mathbf{B}, \quad \mathbf{P} \leftrightarrow \mathbf{M}, \quad \epsilon \leftrightarrow \mu. \tag{108.3}$$

Hence, for a spherical body,

$$\mathbf{B}_{\text{in}} = \frac{3\mu}{\mu+2}\mathbf{B}_0, \qquad \mathbf{H}_{\text{in}} = \frac{3}{\mu+2}\mathbf{B}_0, \qquad \mathbf{M} = \frac{3}{4\pi}\frac{\mu-1}{\mu+2}\mathbf{B}_0. \quad \text{(Gaussian)}$$

$$\mathbf{B}_{\text{in}} = \frac{3\mu}{\mu+2\mu_0}\mathbf{B}_0, \quad \mathbf{H}_{\text{in}} = \frac{3}{\mu+2\mu_0}\mathbf{B}_0, \quad \mathbf{M} = 3\frac{\mu-\mu_0}{\mu+2\mu_0}\frac{\mathbf{B}_0}{\mu_0}. \quad \text{(SI)}$$

(108.4)

Note that in SI, if we want to write the external field as $\mathbf{H}_0$ instead of $\mathbf{B}_0$, we must remember that $\mathbf{H}_0 = \mathbf{B}_0/\mu_0$.

Standalone formulas for the extra or demagnetization fields $\mathbf{H}_d = \mathbf{H}_{\text{in}} - \mathbf{H}_0$ and $\mathbf{B}_d = \mathbf{B}_{\text{in}} - \mathbf{B}_0$ in terms of the magnetization are

$$\mathbf{H}_d = -\frac{4\pi}{3}\mathbf{M}, \quad \mathbf{B}_d = \frac{8\pi}{3}\mathbf{M}. \quad \text{(Gaussian)}$$

(108.5)

$$\mathbf{H}_d = -\frac{1}{3}\mathbf{M}, \quad \mathbf{B}_d = \frac{2}{3}\mu_0\mathbf{M}. \quad \text{(SI)}$$

Thus, $\mathbf{H}_d$ is negative (opposite to $\mathbf{M}$), while the extra $\mathbf{B}$ field, $\mathbf{B}_d$, is positive (parallel to $\mathbf{M}$).

For a ferromagnet, only the formulas (108.5) can be used without reservations. If the magnetization is regarded as given (for a hard ferromagnet, we might, for example, take it as equal to its saturation value in magnitude), these formulas give the magnet's contribution to $\mathbf{H}$ and $\mathbf{B}$, and adding them to $\mathbf{H}_0$ and $\mathbf{B}_0$ gives the total fields inside the body.

The formulas (108.4) can be used for soft, high-permeability materials such as permalloy and MuMetal, where hysteresis may be neglected, and provided the applied field is not too high, it is meaningful to write

$$\mathbf{B} = \mu_{\text{eff}}\mathbf{H}. \quad \text{(Gaussian and SI)} \tag{108.6}$$

Equation (108.4) then applies with $\mu \to \mu_{\text{eff}}$. Taking $\mu \gg 1$ (or $\mu \gg \mu_0$ for SI), we get

$$\mathbf{B}_{\text{in}} \simeq 3\mathbf{B}_0. \quad \text{(Gaussian and SI)} \tag{108.7}$$

Thus, the magnetic field inside the sphere is three times larger than the applied field. The ferromagnet essentially draws in the $\mathbf{B}$ field lines. Naturally, this property holds only if the applied field is weak enough. To see this, note that $M \simeq 3B_0/4\pi$, and since M cannot exceed the saturation magnetization M_{sat}, eq. (108.7) applies only if $B_0 \ll 4\pi M_{\text{sat}}/3$ (or $B_0 \ll \mu_0 M_{\text{sat}}/3$ in SI). Otherwise, one can only write $\mathbf{B}_{\text{in}} = \mathbf{B}_0 + 8\pi\mathbf{M}/3$ (or $\mathbf{B}_{\text{in}} = \mathbf{B}_0 + 2\mu_0\mathbf{M}/3$ in SI).

Exercise 108.1 The ability of a soft and highly permeable ferromagnet to draw in the field lines is exploited in magnetic shielding, when it is desirable to have a region of space essentially free of magnetic fields. Consider a spherical shell of such a material of radius a and thickness $\Delta \ll a$, and place it in a uniform applied field $\mathbf{B}_0$. Find the field in the hollow space inside the shell, assuming the effective permeability is much greater than 1. (This is a standard if lengthy calculation which may be found in many texts.)

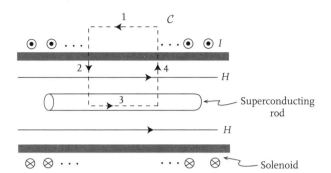

FIGURE 16.10. A superconducting rod inside a solenoid. Since $\mathbf{B} = 0$ inside the rod, application of Ampere's law to the loop C shows that there must be screening currents flowing at the surface of the rod in a sense opposite to the currents in the solenoid windings.

Answer:

$$\mathbf{B}_{\text{in}} \simeq \frac{3a}{2\Delta} \frac{\mathbf{B}_0}{\mu_{\text{eff}}}. \tag{108.8}$$

(For SI, multiply the right-hand side by μ_0.) Since μ_{eff} may be as large as 10^5, the interior field is greatly reduced even for quite thin shells.

109 Superconductors*

Superconductors form another class of materials in which the magnetic response is highly nonlinear. At a basic level, in fact, one cannot define a magnetization in the sense of a magnetic moment density distributed throughout the medium, so it is improper to speak of a B vs. H relation. The fundamental property of a superconductor is that it expels magnetic flux. This is known as the *Meissner effect*. When the flux expulsion is complete, in the body of the superconductor we must have

$$\mathbf{B} = 0. \tag{109.1}$$

This is the basic constitutive relation. It holds in what are known as the type I superconductors and in the so-called Meissner phase of type II superconductors. Discussion of the distinctions between these two types and the full range of magnetic behavior of the type IIs is beyond our scope, and we shall limit ourselves to studying the consequences of eq. (109.1).

To this end, let us consider a long, thin superconducting rod in a solenoid (fig. 16.10). This shape has the advantage that we may ignore end effects or, what is the same thing, shape demagnetization effects. By symmetry, the field can point only along the axis of the solenoid. On the contour C shown in the figure, $\mathbf{B} = 0$ on both legs 1 and 3; it is zero on leg 1, because that leg is outside the solenoid, and it is zero on leg 3 because of eq. (109.1). Thus, the circulation of $\mathbf{B}$ on C is zero. By Ampere's law, the total current passing through the surface spanning C must then vanish. Since this vanishing must happen no matter how deep in the superconductor we take the loop C to lie, we conclude that a current $\mathbf{K}_s$ must flow at the surface of the superconductor in a sense opposite to that in the windings. If we let the current in the windings be I and let there be n turns per unit length, we obtain $K_s = nI = cH_s/4\pi$, where $H_s = 4\pi nI/c$ is the field at the surface, which in this case is the field everywhere in the solenoid except for the region occupied by the superconductor. More generally,

$$\mathbf{K}_s = \frac{c}{4\pi}\hat{\mathbf{n}} \times \mathbf{H}_s, \tag{109.2}$$

where $\hat{\mathbf{n}}$ is the outward normal to the superconductor.[15]

Although we have used the same symbol $\mathbf{K}$ for the surface current as in section 83, the current here flows not in a mathematical layer of zero thickness but in a real sheath of thickness λ_L, known as the *London penetration depth*. This depth is a material property of the superconductor and ranges from 10^2 to 10^3 Å. It is clear that the magnetic field must also penetrate into the superconductor over the same length scale, and so in this surface layer, eq. (109.1) does not in fact hold. These surface currents are also known as *screening currents*, since they screen the interior of the superconductor from the magnetic field.

Since the state of expelled flux is thermodynamically stable, it follows that the superconductor must be able to sustain the surface currents in equilibrium, i.e., without the dissipation of heat that would occur in a normal conductor. This is, of course, the property that gives superconductors their name, but we now see that it is a consequence of the flux expulsion. For this reason, it is often said that superconductors would be better described as superdiamagnets.

The surface currents induced by the field $\mathbf{H}$ cause the superconductor to have a net nonzero moment, $\mathcal{M}$. Let us return to the setup of fig. 16.10, and let the superconducting rod have a length L and cross-sectional radius R. The total moment is

$$\mathcal{M} = \frac{1}{c}(K_s L)(\pi R^2) = \frac{1}{4\pi}HV, \tag{109.3}$$

where V is the volume of the rod, and we have used $K_s = cH/4\pi$. Note that this moment points against the field. Its presence means that the total free energy of a superconductor in a field H, $F_{sc}(T, H)$, is given by $F_{sc}(T, 0)$, plus the work required to build up the magnetic moment $\mathcal{M}$, which in the present case equals $-\mathcal{M}H/2$, i.e., $H^2V/8\pi$. Thus,

$$F_{sc}(T, H) = F_{sc}(T, 0) + \frac{H^2 V}{8\pi}. \tag{109.4}$$

Expelling the field raises the free energy of a superconductor.

Equation (109.4) shows that the superconducting state will eventually become unstable under application of a sufficiently strong field. The *critical field*, H_c, can be found by

---

[15] Since $\mathbf{B} = 0$ vanishes deep in the superconductor, so does the current $\mathbf{j}$ and the magnetization $\mathbf{M}$. It is therefore pointless to talk of the H field inside the superconductor. Outside the superconductor, of course, $\mathbf{B} = \mathbf{H}$. Therefore, all formulas in this section could be written using B everywhere instead of H. It is traditional in superconductivity to use H, however. For example, one sees the notation B_c for the critical field only rarely.

equating $F_{sc}(T, H_c)$ to the total free energy $F_n(T, H_c)$ of a normal metal. Typical values of H_c are of the order of 1000 Oe. At such fields, the state of most (nonferromagnetic) metals is hardly affected at all, so we may replace $F_n(T, H_c)$ by $F_n(T, 0)$. This gives the relation[16]

$$\frac{H_c^2}{8\pi} = \frac{F_n(T, 0) - F_{sc}(T, 0)}{V}. \tag{109.5}$$

Since the total free energies for $H = 0$ are extensive quantities, proportional to V, the right-hand side can be written as the difference of the free energy *densities*, and H_c is seen to be an intensive quantity. Essentially, H_c arises from a maximum sustainable screening surface current density K_s. Any attempt to drive a larger current at any point through the superconductor surface drives the adjoining region normal.

We conclude this section and chapter by considering a spherical superconductor in an external field, for which explicit calculations can easily be done. The main ideas hold for any other shape, however.

Let the sphere have a radius a, and let the field $\mathbf{H} = H_0 \hat{\mathbf{z}}$ far from it. Let us write the general field $\mathbf{H}(\mathbf{r})$ as $\mathbf{H}_0 + \mathbf{H}'(\mathbf{r})$, where $\mathbf{H}'$ is the additional field due to the screening currents on the surface of the sphere. From the analogous problem for a ferromagnetic sphere, we know that the extra field $\mathbf{H}'$ for $r > a$ will be that of a dipole with magnetic moment $\mathcal{M}$:

$$\mathbf{H}' = \frac{3(\hat{\mathbf{n}} \cdot \mathcal{M})\hat{\mathbf{n}} - \mathcal{M}}{r^3}. \tag{109.6}$$

At the surface, $\mathbf{H} \cdot \hat{\mathbf{n}}$ must vanish, from which it follows that

$$\mathcal{M} = -\frac{a^3}{2}\mathbf{H}_0. \tag{109.7}$$

But now, the tangential component H_t of $\mathbf{H}$ at the surface of the sphere depends on the polar angle θ. Specifically,

$$H_t = H_0 \sin\theta + \tfrac{1}{2} H_0 \sin\theta = \tfrac{3}{2} H_0 \sin\theta. \tag{109.8}$$

This is largest at the equator, $\theta = \pi/2$, and on that line it equals H_c when $H_0 = (2/3) H_c$. Therefore, when the applied field H_0 just exceeds $(2/3) H_c$, the portion of the surface near the equator will tend to go normal, as it will then have a lower free energy. At the same time, the entire sphere cannot enter the normal state, since it would then be permeated with a field less than H_c throughout its volume and thus be in a state of energy higher than that of the superconducting state for every small volume element. It follows that the sphere must break up into superconducting and normal domains (see fig. 16.11). This problem shares many elements in common with that for a ferromagnet divided into domains, in that we must minimize the total free energy, consisting of the sum of the

---

[16] We are neglecting effects such as a change in volume or lattice structure of the material due to the transition to superconductivity or due to the application of the field (magnetostriction).

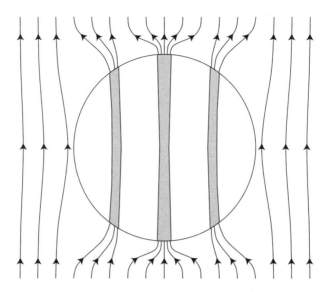

FIGURE 16.11. A superconducting sphere placed in an external magnetic field develops normal regions (shaded) through which the field passes in order to satisfy the constitutive relation (109.1).

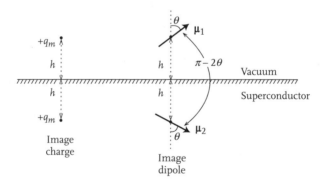

FIGURE 16.12. Method of images for a superconductor.

free energies of the superconducting and normal regions, the magnetostatic energy, and the free energy associated with the surfaces or "domain walls" separating the two types of regions. The actual execution of this program is complicated, and we shall not investigate it further.

Exercise 109.1 (superconducting levitation) A point magnetic dipole is placed above a superconductor with a horizontal surface, as shown fig. 16.12. The dipole has mass m and is pulled down by the earth's gravity. Using a method of images, find the potential energy of this system as a function of θ and h, and minimize this energy to find the stable configurations if any. Find and sketch the current distribution at the surface of the superconductor.

Solution: At the surface, $B_n = 0$. For a monopole, therefore, the image charge must be of the same sign. By thinking of the dipole as two closely spaced monopoles, we see that the image dipole is obtained by reflecting in the surface. The potential energy is then readily calculated. Minimization gives the stable point as

$$\theta = \pi/2, \quad h = (3\mu^2/16mg)^{1/4}, \tag{109.9}$$

where μ is the dipole magnetic moment. Taking a small iron magnet with a mass of 10 g and $\mu \simeq 1700$ emu (guessing the volume as 1 cm^3), we get $h \simeq 2.7$ cm.

The surface current is given by $\mathbf{K} = (c/4\pi)\hat{\mathbf{z}} \times \mathbf{B}$. Taking the surface to be the xy plane, the moment to lie in the xz plane, and adding the **B** fields due to the source and image dipoles to obtain the total **B**, we get

$$K_x = -\frac{3c\mu}{2\pi R^5}xy, \quad K_y = -\frac{c\mu}{2\pi R^5}(R^2 - 3x^2), \tag{109.10}$$

where $R^2 = x^2 + y^2 + h^2$.

17 | Ohm's law, emf, and electrical circuits

We turn in this chapter from basic laws of electromagnetism to the practical subject of electrical circuits. This subject is invariably discussed in elementary courses, and we expect that readers are aware of Kirchhoff's laws, RC circuits, resonant circuits, etc. Our goal, therefore, is not to revisit these topics but to show how their central concepts relate to the fundamental laws, and some of the approximations involved.

The chapter has three themes. The first is Ohm's empirical law relating current and electric potential in a conductor. From this law, one is naturally tempted to solve for the current distribution in extended conductors, but the difficulty of this problem soon becomes apparent. Fortunately, the problem is not very important, so there is no great loss in abandoning it.

It follows from Ohm's law that a current through any material body (other than a superconductor) is accompanied by the dissipation of energy as heat. Hence, to sustain a current through a material body, one needs a source of electrical energy to overcome the dissipation, or, as is more generally said, a source of emf (electromotive force). The practical problem of making such a source was not solved till 1800, when Volta made the first battery by stacking alternating disks of silver and zinc separated by pieces of cardboard soaked in brine. Since then, a huge variety of sources have been devised, but they all share one essential feature. Inside any such source, one must move charge *against* a potential gradient. The energy required to achieve this motion can take any form—chemical (as in the battery), mechanical, thermal, radiative, etc.—but it cannot ultimately be electromagnetic.

Our second theme, therefore, will be to describe a few selected sources of emf. Our description will have very much of a "how things work" flavor, since the physics of the devices is highly varied and outside the theory of electromagnetism per se. An understanding of these sources is nevertheless very useful, as the notion of an emf tends to remain somewhat mysterious without it.

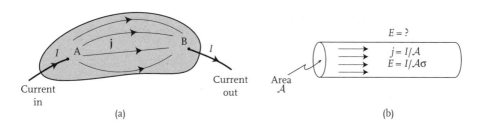

FIGURE 17.1. Current through (a) a general conductor, (b) a uniform wire.

The discussion of emf and Ohm's law naturally leads to our last theme, linear circuits. We will discuss these very briefly, mainly to highlight the approximations and the limits of their validity.

110 Ohm's law

The fundamental feature of metals is the existence of free or delocalized electrons which can conduct electricity. To solve Maxwell's equations in the presence of current-carrying bodies, one needs a constitutive equation connecting the current to the fields in the body. This is provided by Ohm's law, which Ohm discovered through extensive experiments. For a body carrying a steady current, the law takes the local form

$$\mathbf{j} = \sigma \mathbf{E}. \quad \text{(Gaussian and SI)} \tag{110.1}$$

Here, $\mathbf{j}$ and $\mathbf{E}$ are the current density and electric field at any point in the body, and σ is known as the *conductivity* of the material in question. It is necessarily a positive quantity, since the electric field inside the body accelerates the electrons and increases their kinetic energy. In steady state, this increased energy must be dissipated as heat. The rate of dissipation per unit volume is easily shown to be $\dot{Q} = \mathbf{j} \cdot \mathbf{E} = \sigma E^2$, and its positivity implies $\sigma > 0$.

Conductivity has dimensions of inverse time in the Gaussian system and may therefore be quoted in $\sec^{-1}$. It is equally common to give the inverse of the conductivity, known as the resistivity:

$$\rho = \frac{1}{\sigma}. \tag{110.2}$$

This has units of sec in the Gaussian system and ohm m in the SI system. One ohm m is a rather high resistivity, however, so one often sees the derived unit micro-ohm-centimeter, written μohm cm. The conversion factor for resistivity is

$$1 \sec \equiv 9 \times 10^{17} \, \mu\text{ohm cm}. \tag{110.3}$$

Ohm's law also applies to materials other than metals, e.g., semiconductors and electrolytic solutions. When these are included, resistivities can vary over a wide range.

At 273 K, the resistivities of pure copper, tin, and bismuth are 1.6, 11, and 110, all in μohm cm, that of germanium is 200 ohm cm, and of seawater is $\sim$25 ohm cm (note the missing micro). By varying the temperature and doping, semiconductor resistivities can be made to vary over 13 orders of magnitude, from 10^{-2} to 10^{11} ohm cm. We shall develop a physical model of conduction in chapter 18 that will enable us to understand some of these values. Our present goal is to understand the macroscopic consequences of Ohm's law if we take it as given.

To that end, suppose a current I enters and leaves a body through thin wires at two points A and B as shown in fig. 17.1a. This current distributes itself in some way as it flows through the body. The local electric field at a point inside the body is given by $\mathbf{E}(\mathbf{r}) = \mathbf{j}(r)/\sigma$. The line integral of $\mathbf{E}$ from A to B gives the potential or voltage difference between these points:

$$V \equiv \phi_A - \phi_B = \int_A^B \mathbf{E}(\mathbf{r}) \cdot d\boldsymbol{\ell}. \tag{110.4}$$

Since Maxwell's equations are linear, it follows that if we double I, we will double $\mathbf{j}(\mathbf{r})$ and $\mathbf{E}(\mathbf{r})$ everywhere, and thus double V. The ratio

$$R = V/I \tag{110.5}$$

is known as the *resistance* of the conductor. It depends on the shape of the conductor and the location of the points where the current enters and leaves.

The issue thus arises of determining the current and field distribution in the above setup. The obvious first problem to solve is that of a long, straight wire of uniform circular cross section of area A, carrying a current I—fig. 17.1b. The requirements that $\mathbf{j}$ have zero divergence and no normal component at the surface of the wire can be met by having a uniform current density $j = I/A$. This gives rise to an electric field $E = I/A\sigma$ everywhere in the wire parallel to the current. If we measure the potential difference between two points a distance ℓ apart along the wire, we will get $V = I\ell/A\sigma$. The resistance of this length of wire is therefore

$$R = \rho\ell/A. \tag{110.6}$$

The reader has undoubtedly seen the result (110.6) before. It is a rather good approximation for wires of macroscopic length. However, the discussion is seriously lacking from a logical point of view. The vanishing of $\nabla \times \mathbf{E}$ requires $\mathbf{E}_t$ to be continuous at the surface of the wire. If idealize to an infinitely long wire, we reach the absurd result that the space surrounding the wire also be filled with the same electric field as that within it. Another way of seeing the quandary is to ask for the charge sources that produce the electric field (since $\nabla \cdot \mathbf{E} = 4\pi\rho$). In an infinitely long wire, we would expect the charge density to be the same along the length of the wire. But this produces a field perpendicular to the wire, not along it! To get a sensible result, we must consider that current can flow only in a closed circuit, which requires the wire to curve somewhere.

Thus, the problem of finding the true current density and field in a conductor is a rather difficult one for almost any geometry approximating an actual circuit, and we know of

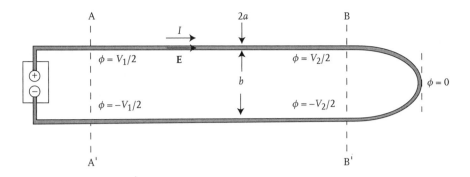

FIGURE 17.2. Discharging a capacitor through a very long wire loop.

almost none that have been solved. In the next section, we solve a toy problem that will help us understand the essential ideas.

111 Electric fields around current-carrying conductors—a solvable example*

A geometry for which we can approximately find the surface charges and electric fields in the space outside the conductors is shown in fig. 17.2. Let a wire of uniform circular cross section be connected across the plates of a charged capacitor. We take an extremely long wire, so it has enough resistance that the capacitor discharges very slowly. The induced magnetic field can then be ignored, and we can solve for the electric field at every instance as if the problem were one of statics. In particular, we can find the field in the region far from the cell and far from the U-bend, where we need only focus on the two long, parallel wires.

Let the potential difference across the capacitor be V, and the total length of the wire be ℓ_0. To a first approximation, eq. (110.6) implies that the potential drops uniformly along the length of the wire, so the electric field has a magnitude $E_0 = V/\ell_0$ everywhere inside the wire, and the current density is $j_0 = \sigma E_0$.[1] It will be convenient to take the zero of potential at the end of the U-bend, so that the potential is $\pm V_1/2$ at the points A and A', and $\pm V_2/2$ at B and B'. If the length of the segment AA'–BB' is ℓ, then $(V_1 - V_2)/2 = E_0\ell$.

Outside the wires, we have $\nabla \cdot \mathbf{E} = \nabla \times \mathbf{E} = 0$, so we can write $\mathbf{E} = -\nabla\phi$, where ϕ satisfies Laplace's equation. Let us take the z axis in the direction of the wires, with the $z = 0$ plane chosen so that $\phi = -E_0 z$ on the surface of the wire AB, and $\phi = E_0 z$ on the surface of A'B'. With these boundary conditions, we have a problem of the Dirichlet type. To solve for ϕ, let us try the ansatz

$$\phi(\mathbf{r}) = z F(x, y). \tag{111.1}$$

---

[1] We ignore the magnetic field produced by the current, and the redistribution of the current due to the resulting Lorentz force. The errors incurred are of order $(v/c)^2$, where v is the drift velocity of the electrons, to be discussed in chapter 18. For typical current densities, $v \sim 0.01$–1.0 cm/s, so the error is utterly negligible.

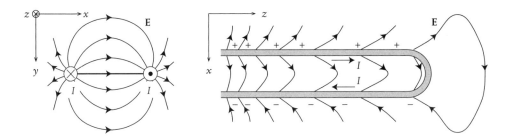

FIGURE 17.3. Sketch of the electric field for the setup of fig. 17.2.

Then,

$$\nabla^2\phi = z\left(\frac{\partial^2}{\partial x^2} + \frac{\partial^2}{\partial y^2}\right) F(x, y). \tag{111.2}$$

Further, ϕ obeys the boundary conditions if $F = -E_0$ on AB, and $F = E_0$ on A'B'. The problem is thus reduced to solving Laplace's equation in two dimensions with boundary conditions given on two nonconcentric circles. We solved this already in exercise 89.3, when we found the capacitance of parallel wires.[2] If the radius of each wire is a and the separation between their centers is c, then

$$F(x, y) = -\lambda \ln\left[\frac{(x-s)^2 + y^2}{(x+s)^2 + y^2}\right], \tag{111.3}$$

where

$$s = \frac{1}{2}(c^2 - 4a^2)^{1/2}, \tag{111.4}$$

$$\lambda = \frac{E_0}{2\cosh^{-1}(c/a)}. \tag{111.5}$$

The electric field is given by

$$E_z = -F(x, y), \tag{111.6}$$

$$\mathbf{E}_\perp = 2\lambda z\left[\left(\frac{x-s}{(x-s)^2 + y^2} - \frac{x+s}{(x+s)^2 + y^2}\right)\hat{\mathbf{x}} + \left(\frac{y}{(x-s)^2 + y^2} - \frac{y}{(x+s)^2 + y^2}\right)\hat{\mathbf{y}}\right]. \tag{111.7}$$

which we sketch approximately in fig. 17.3. Note that the z component of $\mathbf{E}$ is uniform along the wire, but the transverse components grow linearly with distance as we approach the capacitor.

Lastly, we can find the surface charge on the wires from the formula $\Sigma_s = E_n/4\pi$. Denoting the angle around the cross section of the wire by θ, for the wire on the positive side, we get

$$\Sigma_s = \frac{\lambda z}{\pi} \frac{s}{a(c - 2a\cos\theta)}. \tag{111.8}$$

---

[2] $F(x, y)$ here is the same as $\phi(x, y)$ of that problem with $a = b$. We also take the charge density on the image wires to be $\pm\lambda$ by replacing ϕ by $\phi/2\lambda$ everywhere.

This also increases linearly with z, but it is never greater than $\sim V/8\pi a \ln(c/a)$ for $c \gg a$. As mentioned before, one can regard these surface charges as the source of the electric field that drives the current.

This example illustrates several important points. First, the electric field does not vanish outside the conductor. Indeed, its magnitude can exceed that of the interior field. Second, the field and surface charge distribution is determined by the geometry of the entire circuit, not just "local" properties such as the curvature of the wire. In the example, the surface charge is least right at the U-bend. The entire wire acts as a distributed capacitor. In real circuits however, the electromagnetic fields are much larger near specific circuit elements, as we shall discuss in section 116, and the fields due to the wires are generally negligible. In that case, the entire problem of finding the external field can be sidestepped in favor of the much simpler one of finding the potential at different points of the circuit.

Exercise 111.1 In the above example, consider the magnetic field produced by the current, and thus find the Poynting vector. Show that right at the surface of the wire, the energy flux precisely balances the heat produced. Sketch, approximately, the global pattern of the Poynting vector, and thus convince yourself of the consistency of the overall energy flow.

112 van der Pauw's method*

In this section, we discuss a remarkable method for measuring the resistivity of thin laminar samples or films, discovered by van der Pauw in 1958.[3] Although it is somewhat specialized and relies on the theory of conformal mapping, it provides an interesting case where we can find the current distribution, and it is actively used by researchers. Our treatment follows van der Pauw's closely.

Before discussing the method, it is useful to introduce the concept of *sheet resistance*. Consider a thin conducting sheet of resistivity ρ of uniform thickness d, in the shape of a rectangle of length ℓ and width w. If current enters through the face with dimensions $w \times d$, flows with uniform distribution through the sheet, and exits through the opposite face, the formula (110.6) gives a resistance

$$R = \frac{\rho\ell}{wd}. \tag{112.1}$$

The combination

$$R_s \equiv \rho/d \tag{112.2}$$

is known as the sheet resistance. It determines the resistance of any geometry as long as the current flow is parallel to the sheet and uniform across its thickness. In some cases (such as a two-dimensional electron gas at the interface of two semiconductors) d may also be difficult to measure separately, and R_s is then a more basic sample parameter.

---

[3] L. J. van der Pauw, *Philips Res. Rep.* **13**, 1–9 (1958).

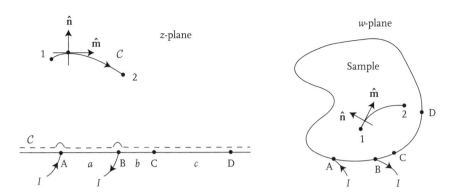

FIGURE 17.4. van der Pauw's method.

The quantity R_s has dimensions of resistance and is therefore given in ohms, or in ohms/square. The latter usage signals that one is talking about a sheet resistance and comes from the fact that for the rectangular sheet with uniform current distribution, $R = R_s(\ell/w)$. If the sheet is square, $\ell = w$, and $R = R_s$ no matter how big the length and width are separately.

Returning to van der Pauw's result, it is as follows. Consider a laminar conductor of any shape. The conductor must be uniform in thickness and be simply connected (annuluses are excluded). Let four *point* contacts, A, B, C, and D, be made at the edge of the conductor. Let a current I enter at A and leave at B. The resulting potential difference at D and C defines the resistance

$$R_{AB,CD} = (\phi_D - \phi_C)/I. \tag{112.3}$$

By letting the current leads be B and C and the voltage leads be A and D, we get $R_{BC,DA}$, and so on. van der Pauw's formula states that

$$\exp(-\pi R_{AB,CD}/R_s) + \exp(-\pi R_{BC,DA}/R_s) = 1. \tag{112.4}$$

If $R_{AB,CD}$ and $R_{BC,DA}$ are measured, this formula enables us to solve for R_s.

The mathematical problem is to solve the time-independent equations $\nabla \cdot \mathbf{E} = 0$, and $\nabla \times \mathbf{E} = 0$, with $\mathbf{E} = \rho \mathbf{j}$, such that $\mathbf{j}$ is in the plane of the sheet. Thus, we are ignoring the magnetic field produced by the current. We also ignore the electric field outside the sheet. The only boundary condition is $\mathbf{j} \cdot \hat{\mathbf{n}} = 0$ everywhere at the edge except at A and B, where we have a source and a sink. The problem is thus purely two dimensional. Writing $\mathbf{E} = -\nabla\phi$, we get $\nabla^2\phi = -\rho\nabla \cdot \mathbf{j}$. Taking account of the source and sink, we get

$$\nabla^2\phi = -I R_s \left(\delta(\mathbf{r} - \mathbf{r}_A) - \delta(\mathbf{r} - \mathbf{r}_B)\right), \tag{112.5}$$

where $\mathbf{r} = (x, y)$, and the delta functions are two dimensional.

We first solve this problem for a half-plane (see fig. 17.4). If the current enters at A and leaves at B, the potential is that of two line charges parallel to the z axis through these points. Taking the origin at A and the coordinates of B as $(a, 0)$, we have

$$\phi(x, y) = -\frac{I R_s}{\pi} \left[\ln(x^2 + y^2)^{1/2} - \ln((x - a)^2 + y^2)^{1/2} \right]. \tag{112.6}$$

The boundary condition $\mathbf{j} \cdot \hat{\mathbf{n}} = 0$ is also satisfied, as $\nabla\phi \| \hat{\mathbf{x}}$ when $y = 0$. Hence, with the spacings a, b, and c as shown in fig. 17.4, we have

$$R_{AB,CD} = \frac{R_s}{\pi} \ln \frac{(a + b)(b + c)}{b(a + b + c)}. \tag{112.7}$$

In the same way,

$$R_{BC,DA} = \frac{R_s}{\pi} \ln \frac{(a + b)(b + c)}{ca}. \tag{112.8}$$

Since

$$b(a + b + c) + ac = (a + b)(b + c), \tag{112.9}$$

eq. (112.4) follows straightforwardly.

To extend the result to our odd-shaped laboratory sample, we proceed in two steps. In step 1, we use the theorem from complex analysis that if $f(z)$ is an analytic function of $z = x + iy$, with real and imaginary parts $u(x, y)$ and $v(x, y)$, then u and v obey the Cauchy-Riemann conditions,

$$\frac{\partial u}{\partial x} = \frac{\partial v}{\partial y}, \quad \frac{\partial u}{\partial y} = -\frac{\partial v}{\partial x}, \tag{112.10}$$

and also Laplace's equation except at points where the second partial derivatives do not exist. We may therefore choose f such that u coincides with the potential ϕ in the half-space. To see the physical meaning of v, let $\hat{\mathbf{m}}$ and $\hat{\mathbf{n}}$ be any two orthogonal unit vectors obtained by rigidly rotating the dyad $(\hat{\mathbf{x}}, \hat{\mathbf{y}})$. Then, using the Cauchy-Riemann conditions and $m_x = n_y$, $m_y = -n_x$, we get

$$\hat{\mathbf{m}} \cdot \nabla v = m_x \frac{\partial v}{\partial x} + m_y \frac{\partial v}{\partial y} = n_y \left(-\frac{\partial u}{\partial y} \right) - n_x \frac{\partial u}{\partial x} = -\hat{\mathbf{n}} \cdot \nabla u. \tag{112.11}$$

But, since $\mathbf{j} = -\nabla\phi / R_s$ and $\phi = u$, this means that

$$\hat{\mathbf{m}} \cdot \nabla v = R_s \mathbf{j} \cdot \hat{\mathbf{n}}. \tag{112.12}$$

It is useful to restate this law in integral form. Let $\mathcal{C}$ be any open contour with endpoints 1 and 2 and tangent $\hat{\mathbf{m}}$. Then, with ds being the arc length element,

$$v_2 - v_1 = \int_1^2 \hat{\mathbf{m}} \cdot \nabla v \, ds = R_s \int_1^2 \mathbf{j} \cdot \hat{\mathbf{n}} \, ds. \tag{112.13}$$

Let $\mathcal{C}$ now run along the real axis, infinitesimally above it, except for semicircular indentations around A and B. Choosing 1 at $-\infty$, and 2 as a variable point along $\mathcal{C}$, we see that v vanishes all the way to the left of A, jumps up by $I R_s$ as we cross A, stays at this value till B, and then jumps back to zero as we cross B, staying zero till $+\infty$.

In step 2, we use another theorem from complex analysis that states that the upper half-plane can be mapped into the interior of any simply connected domain by a suitable analytic function $w(z)$.[4] We take this domain to coincide with our sample and let the points A, B, C, and D on the sample boundary be the images of the corresponding points on the half-plane. Further, with $z(w)$ being the inverse mapping, let $F(w) = f(z(w))$ have real and imaginary parts g and h. That is, if the mapping takes a point z in the half-plane into a point ζ inside the sample,

$$g(\zeta) = u(z), \quad h(\zeta) = v(z). \tag{112.14}$$

The functions g and h obey the Cauchy-Riemann conditions and Laplace's equation. Further, h vanishes on the boundary except between A and B where it equals $I R_s$. Thus, h has the same physical meaning in the sample as v did in the half-plane: $\hat{\mathbf{m}} \cdot \nabla h = R_s \hat{\mathbf{n}} \cdot \mathbf{j}$. It follows that g is the potential field in the sample, in particular that $g_D - g_C = u_D - u_C$ is the voltage difference between the points C and D on the sample. Hence, all the ratios $R_{AB,CD}/R_s$, $R_{BC,AD}/R_s$, etc., are invariant under the mapping, and eq. (112.4) holds for the sample too.

Let us repeat the conditions for van der Pauw's method to work. The sample must be uniformly thin and simply connected. The contacts must be on the boundary and essentially point-like. Other practical considerations may be found in Wikipedia.

113 The Van de Graaff generator

A discharging capacitor is too transitory a source of emf for most experiments or applications. One needs, instead, a source of *steady* emf. The Van de Graaff machine is one of the simplest such, as its energy input is mechanical. The operation is shown in fig. 17.5. An insulating belt is driven around two pulleys. Near the lower pulley, positive charge is generated on the belt by means of a brush that rubs lightly against it. This charge is transported upward, where another brush removes the charge and distributes it over a metallic dome. In this way positive charge accumulates on the dome, which serves as one electrode, and negative charge on another electrode connected to the lower brush. The two electrodes can now serve as the terminals of an emf source and be used to drive a current through a load connected to them. When the Van de Graaff is running, there exists an electric field whose lines run from the dome to the lower electrode. This field exists both outside and inside the insulating housing containing the belt. Note that inside the housing, this electric field is in a direction that *opposes* the motion of the charges on the belt.

---

[4] See, e.g., Copson (1970), chap. VIII.

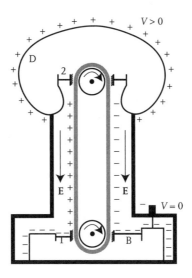

FIGURE 17.5. Schematic of a Van de Graaff generator.

The emf of a Van de Graaff machine is the potential difference between its electrodes. The precise definition of emf is the energy released per unit charge transported around a circuit connected to the terminals, and it is therefore given in volts.

114 The thermopile

Consider a bar of metal, whose ends are maintained at temperatures T_h and T_l, with $T_h > T_l$. Heat is transported from the hot to the cold end of the bar by means of electrons. The electrons from the hot end have more kinetic energy and thus higher speeds than the electrons at the cold end. By itself, this implies that the electrons have a nonzero mean velocity opposite to ∇T. Immediately after the temperature gradient is established, this nonzero velocity will lead to a pileup of electrons at the cold end, until an opposing electric field $\mathbf{E} \parallel -\nabla T$ is set up (see fig. 17.6a), such that in steady state, there is no net charge flow. The electric field is related to the temperature gradient by

$$\mathbf{E} = Q\nabla T, \tag{114.1}$$

where Q is known as the *thermopower* of the metal. It is a negative coefficient as defined. It follows that in steady state, the hot end of the bar is at a higher electric potential than the cold end. This potential difference can be used to drive an electric current in an external circuit, so the device can serve as a source of emf. Note that inside the bar, a current will flow from the cold end to the hot end, uphill, against the electric field.

If the ends of the bar of fig. 17.6a are directly connected to an external circuit, however, a temperature gradient will arise in the external leads that will generate an opposing

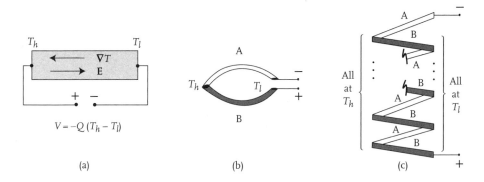

FIGURE 17.6. (a) The thermoelectric effect. (b) A thermocouple. (c) A thermopile.

thermoelectric voltage. There is no way to avoid this offsetting effect, so to obtain a device with a definite emf, one connects a second, different, metal to the first one at the hotter end, as in fig. 17.6b. The cold ends of the two metals then serve as the terminals of the device. If we label the two metals A and B and denote their thermopowers by Q_A and Q_B, then the emf of the device is given by

$$\mathcal{E} = -(Q_A - Q_B)(T_h - T_l). \tag{114.2}$$

We have written this equation assuming that A serves as the negative terminal, which will be the case if $|Q_A| > |Q_B|$.

Thermopowers are generally $\sim 10\ \mu\text{V/K}$, so a single bimetallic pair (called a thermo-couple) is hardly a great source of emf. By attaching many such pairs in series, however, and maintaining one set of junctions at T_h and the other at T_l, as in fig. 17.6c, one obtains a thermopile, and quite respectable emf's can be had. Thermopiles were commonly used around 1900 and were Ohm's preferred source over batteries because of their stability. One use today is in certain spacecraft.

In passing, let us note that although it is not a viable emf source, a thermocouple makes a useful thermometer. If a sensitive voltmeter is connected across the open ends, which are at, say, room temperature, the temperature at the hot junction is directly given by eq. (114.2).

115 The battery

Since their discovery in 1800, batteries have grown ever more varied and sophisticated and are vital pieces of electrical technology today, with uses ranging from cardiac pacemakers with an energy storage capacity of 0.1 Wh (1 watt-hour = 3600 J) to power utility load levelers with capacities of 10^7 Wh and more. In this section, we describe

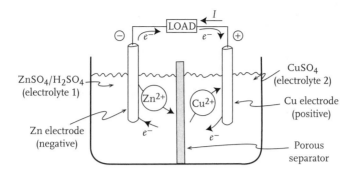

FIGURE 17.7. Schematic of a Daniell cell.

the basic elements of a battery.[5] In fig. 17.7, we show the essential components of an electrochemical cell, using the so-called Daniell cell for specificity. The essence of its operation lies in two chemical reactions, one at each electrode. At the negative electrode zinc dissolves into the $ZnSO_4$ electrolyte: [6]

$$Zn \longrightarrow Zn^{2+} + 2e^-, \quad \mathcal{E}_1 = 0.763\,V, \tag{115.1}$$

while at the positive electrode, cupric ions from the $CuSO_4$ electrolyte leave the solution and are deposited onto the copper electrode:

$$Cu^{2+} + 2e^- \longrightarrow Cu, \quad \mathcal{E}_2 = 0.345\,V. \tag{115.2}$$

The net result is that electrons accumulate on the negative electrode and are depleted from the positive. If the circuit is open, the reactions will quickly stop due to an opposing potential from this charge buildup. If, however, the electrodes are connected by a resistive load and an ammeter, current will be seen to flow from the positive to the negative electrode, i.e., electrons will flow from the negative to the positive through the external circuit. For continuous operation, an equal positive current must flow from the negative to the positive *within* the cell, and this is achieved by diffusional exchange of ions in the electrolytes. However, direct mixing of the electrolytes must not occur; if $CuSO_4$ came in contact with the Zn, Cu^{2+} ions would be driven out of solution and plate onto the Zn electrode, and the battery would cease functioning very quickly. This mixing is prevented by a separator, the cardboard in Volta's original pile, a porous pot, or gel, or pasteboard, as in modern dry cells.[7]

The driving force behind a battery is the chemical energy[8] that is released (or absorbed) in the reaction at each electrode. Recall that the emf of the cell is the energy released

---

[5] For a more detailed and excellent pedagogical account, see W. Saslow, *Am. J. Phys.* **67**, 574 (1999). More comprehensive discussion of the various types of batteries and their actual specifications may be found in Vincent, Bonino, Lazzari, and Scrosati (1984), and in Dell and Rand (2001). A very clear discussion of the underlying principles of physical chemistry is given by Mahan (1965), chaps. 7 and 8.

[6] The quantities $\mathcal{E}_1$ and $\mathcal{E}_2$ will be explained shortly.

[7] Daniell, in 1836, first employed the gullet of an ox!

[8] More precisely, the appropriate thermodynamic potential—at fixed temperature and pressure, the Gibbs free energy.

per unit charge transported across a circuit. This quantity depends on the operating conditions, the temperature and concentration of the electrolytes, and the particular reactions involved in each "half-cell." Since a battery could, in principle, be made by pairing any two half-cells, it is convenient to give for each half-cell reaction the voltage or the *standard cell potential*, as it is called, by pairing it with a standard gas electrode, a half-cell consisting of a piece of platinum in water of pH 7, over which H_2 gas is bubbled at a pressure of 1 atm. The reaction at this electrode is

$$H_2(g) = 2H^+(aq) + 2e^-. \tag{115.3}$$

(The aq stands for aqueous.) Further, the standard operating conditions are 25°C temperature, 1 atm pressure, and 1 molal concentration (1 mole of solute in 1 kg of solvent) for all solutions. We have given the standard cell potentials of the Daniell cell half-reactions in eqs. (115.1) and (115.2). The positive signs of both potentials mean that both reactions tend to spontaneously proceed in the directions shown.[9] The net emf $\mathcal{E}$ of the Daniell cell is therefore 1.11 V (0.763 + 0.345), and under ideal conditions this is the value that would be measured by a voltmeter connected across its terminals. Since chemical bonds entail energies of a few eV at most, it follows that the emf of any electrochemical cell will be at most a few volts. However, by connecting cells in series, sources of larger emf may be produced.

Let us now consider the energetics of a battery during operation. If we follow an electron around the circuit, it gains an amount of energy $e\mathcal{E}_1'$ in the reaction at the negative electrode and another amount $e\mathcal{E}_2'$ at the positive electrode, such that $\mathcal{E}_1' + \mathcal{E}_2' = \mathcal{E}_1 + \mathcal{E}_2$. (One cannot separately equate $\mathcal{E}_1'$ and $\mathcal{E}_2'$ to $\mathcal{E}_1$ and $\mathcal{E}_2$, since the latter are the net energy changes when each electrode is used along with a standard gas electrode.) This energy is then available for use as work or heat in the external circuit. In reality, a certain amount of energy is also lost in the ion diffusion and in the process of charge transfer at the electrodes. This makes up the so-called internal resistance of the cell, typically $\sim$0.1 ohm. We again note that inside the cell, charge moves against the electric potential gradient.

The total amount of energy that a battery can store is limited by the least plentiful reactant, i.e., the one that runs out first. If the total charge that can be transported across an external circuit is Q_T, the total energy available is at most $E_T = \mathcal{E} Q_T$. It is clear that since E_T depends on the total amount of a material, it will scale with the volume of a battery. Thus, the commonly used AA and D alkaline cells both have an emf of 1.5 V but have volumes of 8.3 and 56.5 cm^3 and rated charge capacities of 2.7 and 18 Ah (1 amp-hour = 3600 C), in direct proportion. It should also be noted that in a battery such as a Daniell cell, the emf depends on the concentration of the ionic species, and runs down gradually as these reactants are exhausted. By contrast, a zinc–silver oxide battery

---

[9] Tables of standard potentials usually show all reactions as reductions, i.e., as gaining electrons. Thus, reaction (115.1) would be written in the reverse direction, and its potential shown as −0.763 V.

is based on the reaction

$$Zn(s) + Ag_2O(s) \longrightarrow ZnO(s) + 2Ag(s), \quad \mathcal{E} = 1.59\,V, \tag{115.4}$$

involving only solid reactants, as indicated by the parenthetical s's. (A KOH or NaOH electrolyte is also present, but not used up.) Such batteries have very flat discharge curves, i.e., a nearly constant emf, followed by an abrupt battery death. They are commonly found in the button-shaped cells used in watches and calculators.

In connection with battery life, mention must be made of rechargeable batteries, in which the reactions can be reversed by applying a large external negative emf, i.e., by driving current through the battery in the direction opposite to that of discharge. The most well-known example of such a battery is the lead–acid cell used in cars. The positive and negative electrodes are made of PbO_2 and Pb, respectively, and the electrolyte is H_2SO_4. The half-cell reactions are

$$PbO_2 + HSO_4^- + 3H^+ + 2e^- \longrightarrow PbSO_4 + 2H_2O, \quad \mathcal{E}_1 = 1.690\,V, \tag{115.5}$$

$$Pb + HSO_4^- \longrightarrow PbSO_4 + H^+ + 2e^-, \quad \mathcal{E}_2 = 0.358\,V, \tag{115.6}$$

at the positive and negative electrodes, respectively. The directions shown are those of discharge, during which sulfuric acid is consumed and water is produced. The cell voltage is thus $\sim2\,V$, and the standard 12-V pack is made of six cells in series. When the car is driven, the engine drives an alternator. The resulting ac output is rectified, and this voltage is then used to recharge the battery as needed.

The actual calculation of electric fields, charge densities, and currents inside a battery is very complex. Qualitatively, we can say that most of the potential change occurs near the electrodes, and that these are therefore the "seats" of the emf. In an open circuit, there must therefore be a layer of opposite charge close to each electrode, forming a dipole layer, similar to the interface between two dissimilar metals. The thickness of this layer is a few angstroms. Its detailed structure is complex, and it affects the kinetics of charge transfer near the electrode significantly. Further discussion of these matters is too digressive and beyond the author's expertise.

116 Lumped circuits

The lumped circuit approximation enables one to get around the problem of finding the actual fields in the vicinity of a circuit. The key idea is to regard the circuit as made of elements—resistors, capacitors, inductors, sources of emf, etc.[10] Each of these is a two-terminal device, connected to other devices by perfectly conducting wires. In fig. 17.8, we depict a generic such element by a black box, with terminals A and B. The black box can also stand for a small spatial region surrounding the element, such that for any path C

---

[10] An emf source is an active element, and the others are passive. These four elements can be used to make approximate representations of still other circuit elements, such as transistors. Discussing these opens the door to the wonderful world of electronics.

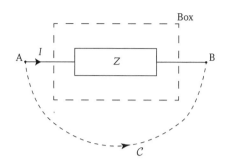

FIGURE 17.8. A generic lumped circuit element, with impedance Z. The line integral of $\mathbf{E}$ along any curve C lying outside the box is assumed to be independent of the path. The current is taken to be positive when flowing from A to B.

lying *outside* the box, the line integral

$$\int_C \mathbf{E}(\mathbf{r}) \cdot d\boldsymbol{\ell} \tag{116.1}$$

is independent of the path. Furthermore, this path independence is assumed to hold even for circuits with time-dependent currents. Naturally, this is an approximation, but if it is accepted, one can associate potentials $V_A(t)$ and $V_B(t)$ with the terminals of the element and call

$$V(t) = V_A(t) - V_B(t) = \int_C \mathbf{E}(\mathbf{r}, t) \cdot d\boldsymbol{\ell} \tag{116.2}$$

the *potential drop* across the element. We stress that V is now uniquely defined.

With these approximations, one discovers that there is a definite relation between the drop across the element and the current flowing through it. For the simple elements we have mentioned this relation is linear. That is, if we take

$$V(t) = V_\omega e^{-i\omega t}, \quad I(t) = I_\omega e^{-i\omega t}, \tag{116.3}$$

then[11]

$$V_\omega = Z(\omega) I_\omega. \tag{116.4}$$

The frequency-dependent quantity $Z(\omega)$ is known as the *impedance* of the element. Connecting wires are taken to be ideal, i.e., to have zero impedance, so that the potential drop across them is zero.

We now examine, following Feynman, how these ideas are implemented for the three simplest elements—the resistor, the inductor, and the capacitor (fig. 17.9).

The resistor: We take wires made of a very good conductor and connect them to a piece of a poorer conductor. Current I enters at the terminal A and leaves at B. We now ignore the magnetic field created by the current. This automatically means that we ignore

---

[11] We alert readers that the standard convention in electrical engineering is to take the time factor as $e^{i\omega t}$. This leads to impedances that are the complex conjugates of ours.

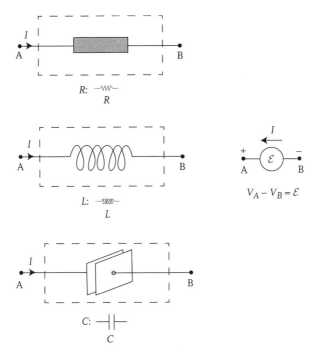

FigURE 17.9. The three basic lumped circuit elements. Under each element we show the symbol for it. We also show the convention for the potential drop across a source of emf.

any induced electric fields. The circulation of **E** is then zero, and the line integral for V can be taken to run *inside* the resistor. Secondly, the entire potential drop is taken to occur within the resistor and none within the connecting wires. This is tantamount to ignoring all charges on the wires and confining them within the resistor. Exactly as in our discussion of Ohm's law, it then follows that

$$V(t) = RI(t). \tag{116.5}$$

The impedance of a resistor is therefore just the resistance:

$$Z_R = R. \tag{116.6}$$

In writing this, we are limiting ourselves to low frequencies, for which the conductivity continues to be given by its steady-state value. This in itself is not a limitation of the lumped circuit element description, however, and we could take Z_R to be frequency dependent. We discuss frequency-dependent conductivity in section 119.

The inductor: We take a coil or solenoid made of a very good conducting wire, so when a current flows through it, a large magnetic field **B** is produced. We justify the path indepence of $\int_C \mathbf{E} \cdot d\boldsymbol{\ell}$ by assuming **B** to lie almost entirely within the coil, i.e., setting $\mathbf{B} = 0$ outside the black box. This implies, as for the resistor, that there is no induced **E** outside the box. Adding and subtracting the line integral of **E** from B to A *through* the wire

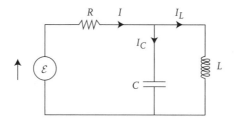

FIGURE 17.10. An example of a lumped circuit.

loop, we get

$$V = \oint \mathbf{E} \cdot d\boldsymbol{\ell} + \int_{\substack{\text{along} \\ \text{wire}}} \mathbf{E} \cdot d\boldsymbol{\ell}. \tag{116.7}$$

Now the integral along the wire is taken as zero on the grounds that its conductivity is very high, so $\mathbf{E} = \mathbf{j}/\sigma \approx 0$. By Faraday's law, the closed-loop integral equals $d\Phi/dt$, where Φ is the magnetic flux through the coil. If the coil has self-inductance L, we thus get

$$V(t) = L\, d\, I(t)/dt. \tag{116.8}$$

In terms of complex amplitudes, this reads, $V_\omega = -i\omega L\, I_\omega$, i.e.,

$$Z_L = -i\omega L. \tag{116.9}$$

We leave it to the reader to show that our signs are correct.

The capacitor: We take the conducting wires to be connected to a closely spaced pair of parallel plates, so that the fringing field is small. By the same argument as for the resistor, we can deform the integral along C to run between the plates instead:

$$V(t) = \int_{\text{plate A}}^{\text{plate B}} \mathbf{E} \cdot d\boldsymbol{\ell}. \tag{116.10}$$

The last integral is given by $Q(t)/C$, where $\pm Q(t)$ is the charge on the two plates, and C is their capacitance. Since $I = d\,Q/dt$, this implies that $-i\omega V_\omega = I_\omega/C$, or

$$Z_C = i/\omega C. \tag{116.11}$$

To study a general circuit, we apply Kirchhoff's laws. That is, we require the total potential drop around a closed loop (including sources of emf if necessary) to be zero and the current to be conserved at every node (charge conservation). In this way we can solve for the potential and the current at every point in the circuit. The fields are never explicitly mentioned. For example, for the circuit of fig. 17.10, we have the three equations

$$\mathcal{E} = IR - i\omega L\, I_L, \quad \mathcal{E} = IR + iI/\omega C, \quad I = I_L + I_C, \tag{116.12}$$

which can be solved for I, I_L, and I_C. We skip this topic entirely, except to mention that (i) impedances in series and parallel obey the same rules as for resistances, and (ii) any combination of passive elements can be replaced by a single impedance Z.

If one has more than one circuit in proximity, one writes a similar set of equations for each circuit, except that there may be coupling terms between the currents in one circuits and the potentials in another, arising from mutual inductances and capacitances. Such couplings may be desirable, as in a transformer, or undesirable, as in integrated circuit chips, where they lead to cross talk.

Let us now examine the approximations made in this section more closely. First, each of the elements has parasitic contributions of the other two types. For the resistor, e.g., we have ignored stray charges which lead to some capacitance, and we have neglected the magnetic field of the current, which means there is some inductance also. For the inductor, we have clearly neglected the resistance of the coil. In addition, surface charges on the wire lead to a capacitance between successive turns of the wire, which we have neglected. Analogous arguments apply to the capacitor. We can improve the lumped circuit description by including these parasitic elements. The inductor could be modeled, e.g., by adding a resistor in series and a small capacitor in parallel with the L-R combination. These improvements work only up to a point, however. The most significant approximation that we have made is in neglecting induced electric fields. This is known as the *quasistatic* approximation, which we discuss further in chapter 19. It is equivalent to neglecting displacement currents. This, in turn, means that there is no radiation of energy, and electric fields adjust instantaneously everywhere in the circuit to changes in the driving emf. A necessary condition for this to be valid is that the dimensions of the circuit, d, be much less than c/ω.

In fact, at low frequencies there is a much more restrictive condition on d. Information about changes in the emf is propagated in a time d/c. This time must be much less than the characteristic times RC, L/R, and $(LC)^{1/2}$ associated with the response of the circuit as per the lumped element description. Thus, we require that

$$\frac{d}{c} \ll RC, \ \frac{L}{R}, \ (LC)^{1/2}. \tag{116.13}$$

Exercise 116.1 In exercise 35.2 we considered a capacitor with circular plates of radius R and spacing d. Reexamine the electric field, keeping the first correction due to induced fields. Use this approximation to define a frequency-dependent capacitance via $Q = C(\omega)V$, with Q being the charge on the plates, and V being the line integral of $\mathbf{E}$ between the centers of the circular plates. Show that this approximation is equivalent to adding an inductor in parallel to the original capacitor, and find the value of the inductance. For what frequencies is the correction small?

Answer: $C(\omega) = C_0(1 - \omega^2 R^2/8c^2)$, where C_0 is the ordinary capacitance.

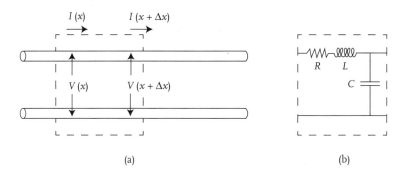

FIGURE 17.11. (a) A transmission line. A small segment of length Δx is equivalent to the lumped circuit combination shown in (b), with $R = R' \Delta x$, $L = L' \Delta x$, and $C = C' \Delta x$, where the primed quantities are the parameters per unit length of the line.

117 The telegrapher's equation*

An interesting extension of the lumped circuit idea is to the telegraph or transmission line, which consists of two long, parallel conductors with a fixed cross section. In the usual power lines, the conductors are wires, but they could be of any shape. In the coaxial cable, e.g., one conductor surrounds the other.

Any such line will have a resistance, capacitance, and inductance per unit length. Let these be denoted by R', C', and L'.[12] Let us denote the current through one of the conductors at a distance x along the line by $I(x)$, and its potential with respect to the other by $V(x)$, as in fig. 17.11. A small segment of the line of length Δx then has resistance $R = R' \Delta x$ (sum of the corresponding resistances of the two conductors), and likewise for the inductance and capacitance, and may be replaced by the lumped circuit combination shown in part (b). It then follows that

$$V(x + \Delta x) - V(x) = -R' \Delta x I(x) - L' \Delta x \dot{I}(x), \tag{117.1}$$

$$I(x + \Delta x) - I(x) = -C' \Delta x \dot{V}(x). \tag{117.2}$$

We have written these equations in the time domain, and the overdots denote time derivatives. Taking $\Delta x \to 0$, we get

$$\frac{\partial V}{\partial x} = -R' I(x) - L' \frac{\partial I}{\partial t}, \quad \frac{\partial I}{\partial x} = -C' \frac{\partial V}{\partial t}. \tag{117.3}$$

---

[12] It is also possible for current to leak from one conductor to the other through the intervening medium. In this case, we should include a conductance per unit length, G'. In most practical cases, $G' \ll R'/L'C'$, but it is easy to extend the analysis to allow for $G' \neq 0$.

Differentiating the second equation with respect to x once more and eliminating V, we obtain

$$\frac{\partial^2 I}{\partial x^2} - R'C'\frac{\partial I}{\partial t} - L'C'\frac{\partial^2 I}{\partial t^2} = 0. \tag{117.4}$$

This is the telegrapher's equation. It is a lossy wave equation. Assuming that the losses are small, signals travel along the transmission line with a speed $\tilde{c} = (L'C')^{-1/2}$. The attenuation coefficient for the power is $\alpha = \tilde{c}R'C'$, the inverse of which gives the attenuation length.

We found L' and C' for a two-wire line in exercises 31.5 and 89.3. At the end of the latter, we remarked that C' was only approximately equal to c^2/L. The reason is that the charges on the cross section of the wires are not distributed uniformly, which affects the capacitance, and the current need not flow only on the surface of the wire, which affects the inductance. We shall show in section 126 that for frequencies of order 10^2 Hz or greater, the current in a conductor is confined to a thin layer or skin near the surface. Further, if the separation of the wires is much greater than their radii, the charges are also distributed uniformly. Thus, $L'C' \approx 1/c^2$, and $\tilde{c} \approx c$, so that the signal speed is almost the speed of light. If the dielectric constant and permittivity of the insulating medium separating the conductors is very different from 1, though, then the speed is reduced. The same comments apply to the coaxial cable (exercises 31.4 and 88.4).

Exercise 117.1 Let the voltage and current in a transmission line be written as $V = V_0 e^{i(kx-\omega t)}$, $I = I_0 e^{i(kx-\omega t)}$. (a) Find the characteristic impedance of the line, $Z_0 = V_0/I_0$. (b) Find the time-averaged energy loss per unit length, assuming R' is small, and explain why a larger inductance L' reduces this loss. (c) Suppose the line is terminated by an impedance Z_T. In that case, the signals on the line can be written as a sum of an incident wave $V_i e^{i(kx-\omega t)}$ and a reflected wave $V_r e^{i(kx-\omega t)}$. Find the reflection coefficient V_r/V_i, and find the condition for this coefficient to vanish. This is known as the impedance-matching condition.

Exercise 117.2 Find the characteristic impedance in SI units for an air-filled coaxial cable whose inner and outer conductors have radii a and b.

Our analysis of the transmission line distributes the inductance, etc., of the line over its entire length, so the fields do not propagate down the line instantaneously. The quasistatic approximation is still made with respect to the transverse dimensions, however, so these dimensions must be much less than c/ω. For standard 60-Hz power lines, this condition is very well met. The other approximation is that the **E** and **B** fields are purely transverse to the line, since the expressions for the capacitance and inductance are derived by finding the respective field energies per unit length and treating the problem as two dimensional. The inclusion of resistance means that the fields are not purely transverse. One way of stating the necessary condition on the resistance is that the attenuation length $L'\tilde{c}/R'$ should be much greater than the transverse dimensions of the line. For

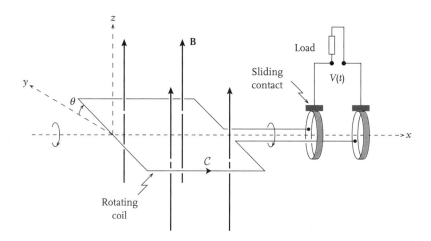

FIGURE 17.12. The basic elements of an ac generator. The positive directions around the loop and of a current in the external circuit are as shown. Readers should convince themselves that these are consistent.

low frequencies this is more stringent than the quasistationarity condition. At sufficiently high frequencies, a different complication sets in. We shall see in section 126 that an ac current in a conductor is confined to a *skin depth* near the surface of the conductor, given by $(c^2/2\pi\sigma\omega)^{1/2}$, where σ is the conductivity of the conductor. If this depth is much less than the radius of the wires, the resistance of the wire varies as $\omega^{1/2}$. We can then use the telegraph equation with a frequency-dependent resistance only for signals made by synthesizing a narrow range of frequencies.

Finally, we have ignored the losses due to radiation. For 60-Hz power lines, these are negligible, but for higher frequencies, the losses can be significant. One can reduce them by using twisted pairs ($\omega \sim 1\,\text{kHz}$) or coaxial cables (MHz) and waveguides (GHz). In fact, both the coaxial cable and the two-wire line can be regarded as waveguides operating in what is known as the TEM mode (see section 143).

118 The ac generator

We complete our discussion of ac circuits by describing the most pervasive and important source of emf today, the ac generator (see fig. 17.12). If the area of the coil is A and it rotates at a frequency ω, the magnetic flux through the coil is given by $AB\cos(\omega t)$. By Faraday's law, there is then an emf in the coil, given by

$$\mathcal{E} = \omega A B \sin(\omega t). \tag{118.1}$$

Let us now discuss energy transformation in this generator. We do this at sufficiently low frequencies, where radiative effects may be neglected. With the choice of positive direction shown for the loop (coil), the emf (118.1) may be regarded as the electrostatic potential difference between the terminals a and b of the generator. Let a load with impedance $Z = R + iX$ be attached across these terminals. This results in a current (in complex notation)

$$I_\omega = \frac{1}{Z}\mathcal{E}_\omega. \tag{118.2}$$

The quantities R and X are known as the *resistance* and the *reactance* of the load. The direction of this current in the external circuit is as shown.

Since a unit charge loses an energy $\mathcal{E}$ when it travels from a to b in the external circuit, the instantaneous power delivered is given by $I(t)\mathcal{E}(t)$, and its time average therefore equals

$$P_{\text{elec}} = \frac{1}{2}\text{Re } I_\omega\mathcal{E}_\omega^* = \text{Re } \frac{|\mathcal{E}_\omega|^2}{2Z} = (\omega\mathcal{A}B)^2\frac{R}{2|Z|^2}. \tag{118.3}$$

The suffix shows that this power is delivered in electrical form.

To complete the circuit, current must flow in the generator coil in the *negative* direction, i.e., *against* the motional emf induced by Faraday's law. With the simple setup we have drawn, this current results in forces on the legs of the coil parallel to the rotational axis, such that the net instantaneous torque on the coil is (with $\theta = \omega t$)

$$N = I(t)\mathcal{A}B\sin\omega t. \tag{118.4}$$

We leave it to the reader to confirm that this torque tends to oppose the sense of rotation shown. To keep the coil turning at a steady frequency, we must supply an equal and opposite torque by some external mechanical agency (a turbine, e.g.). Hence, the instantaneous power supplied must be $P(t) = N\omega$. Since this is exactly algebraically identical to the power delivered to the external circuit, its time average is also equal:

$$P_{\text{mech}} = (\omega\mathcal{A}B)^2\frac{R}{2|Z|^2}. \tag{118.5}$$

The suffix indicates the mechanical nature of the supply. The generator converts mechanical power to electrical power. In reality, of course, the conversion entails inefficiencies, the minimization of which is an enormously important engineering concern, which we skip over.

Since the power delivered to the external circuit must always be positive, we conclude that the resistance R (which may depend on the frequency ω) must also be positive, while the reactance X may have any sign. This is a simple form of the Kramers-Kronig relationships that we shall encounter later.

Because it is so similar to the generator, we also present a very brief discussion of the ac motor and the key concept of the *back emf*.[13] The motor is simply a generator run in reverse. That is, if we send a current through the coil in fig. 17.12 in the positive direction, we will generate a torque (by the Lorentz force law) that turns the coil in the direction shown. This rotation will now produce a motional emf (of amplitude $\omega\mathcal{A}B$) that opposes the emf being employed to turn the coil. This is the back emf, and it is an

---

[13] There are, in fact, several types of ac motors. We are discussing the simplest, a synchronous ac motor.

important consideration in the workings of all motors. Let the driving emf be $\mathcal{E}_0$, the back emf be $\mathcal{E}_b$, and let us take the impedance of the coil and associated circuitry to be purely resistive with a value R. Then the current amplitude when the motor is running will be $(\mathcal{E}_0 - \mathcal{E}_b)/R$. Since the back emf is proportional to ω, however, it is very small when the motor is not turning or turning very slowly, as happens when the motor is first started or if it is used to overcome too large a mechanical load—a large moment of inertia or a large rotational damping. The current in the coil then rises to $\mathcal{E}_0/R$. At start-up, motors usually achieve their operational rotational speed quickly, so the large current does not persist for long, but if the motor jams due to a heavy mechanical load, the large current can cause the coil to overheat and burn out.

18 Frequency-dependent response of materials

119 The frequency-dependent conductivity

The simplest constitutive equation for the current carried by mobile charges in a conductor in time-dependent situations is a generalization of Ohm's great law for steady currents:

$$\mathbf{j}_\omega = \sigma(\omega)\mathbf{E}_\omega, \quad \text{(Gaussian and SI)} \tag{119.1}$$

where $\mathbf{j}_\omega$ and $\mathbf{E}_\omega$ are the current and electric field in the frequency domain. That is,

$$\mathbf{E}(t) = \int \frac{d\omega}{2\pi}\mathbf{E}_\omega e^{-i\omega t}, \quad \mathbf{j}(t) = \int \frac{d\omega}{2\pi}\mathbf{j}_\omega e^{-i\omega t}. \tag{119.2}$$

The quantity $\sigma(\omega)$ is the known as the frequency-dependent or ac conductivity. Its units and dimensions are the same as those of the static conductivity.

As in the static case, eq. (119.1) implies that current flow in a metal is accompanied by the evolution and dissipation of heat. This dissipation arises because the electric field inside the conductor accelerates the electrons, which then collide with impurities, ions, or other electrons. The same two processes—acceleration due to the field, and degradation of the current due to collisions—determine the response of a conductor to a time-dependent electric field. To describe this response, let us try to write Newton's second law for a free electron. The force due to the electric field is just $-e\mathbf{E}$. The collisions are much harder to describe. The simplest way is to adopt a probabilistic point of view and posit that each electron has a probability $1/\tau$ of suffering a collision per unit time. Now suppose that at time t the electrons have an *average* velocity $\bar{\mathbf{v}}(t)$. At time $t + \Delta t$, where Δt is very small, a fraction $\Delta t/\tau$ of the electrons will have collided. We assume that these electron velocities will be completely randomized right after the collision and therefore contribute nothing to the average velocity. Hence,

$$\bar{\mathbf{v}}(t + \Delta t) - \bar{\mathbf{v}}(t) \simeq -\frac{\Delta t}{\tau}\bar{\mathbf{v}}(t) - \frac{e\mathbf{E}(t)}{m}\Delta t, \tag{119.3}$$

where m is the mass of an electron. The first term gives the change in the average velocity due to the collisions, and the second the change due to the acceleration due to the electric field. We now recall that $\mathbf{j} = -n_f e \bar{\mathbf{v}}$, where n_f is the free electron number density. Multiplying the above equation by $n_f e$, dividing by Δt, and replacing increments by differentials, we obtain

$$\frac{d\mathbf{j}}{dt} = -\frac{\mathbf{j}}{\tau} + \frac{n_f e^2}{m} \mathbf{E}(t). \tag{119.4}$$

This is the response or constitutive equation we are after. In the absence of an applied electric field, we obtain $\mathbf{j}(t) = \mathbf{j}(0) \exp(-t/\tau)$, i.e., any current dies off in a time of order τ.

If we rewrite eq. (119.4) in the frequency domain, we get

$$-i\omega \mathbf{j}_\omega = -\frac{\mathbf{j}_\omega}{\tau} + \frac{n_f e^2}{m} \mathbf{E}_\omega. \tag{119.5}$$

Solving for $\mathbf{j}_\omega$, we get $\mathbf{j}_\omega = \sigma(\omega)\mathbf{E}_\omega$, with

$$\sigma(\omega) = \frac{n_f e^2 \tau}{m} \frac{1}{1 - i\omega\tau}. \quad \text{(Gaussian and SI)} \tag{119.6}$$

The model leading to this particular form is known as the *Drude model* of the conductivity. The time τ is known as the collision or relaxation time and is of order 10^{-14}–10^{-15} sec for most metals at room temperature. This model cannot describe many aspects of the electromagnetic behavior of metals, especially at optical frequencies, but it gets many gross features right. Thus, at low frequencies,

$$\sigma(\omega) \approx \sigma_0 \equiv \frac{n_f e^2 \tau}{m}, \quad \omega \ll \tau^{-1}, \tag{119.7}$$

so the response is mainly dissipative. The quantity σ_0 is the static conductivity. With $\tau \sim 10^{-14}$–10^{-15} sec and $n_f \sim 10^{22}$–10^{23} cm^{-3}, we get $\sigma_0 \sim 10^{18}$ sec^{-1}, or a resistivity of approximately 1 μohm cm.[1] At high frequencies, on the other hand,

$$\sigma(\omega) \approx i \frac{n_f e^2}{m\omega}, \quad \omega \gg \tau^{-1}. \tag{119.8}$$

The factor of i signifies that the response is mainly reactive, or inertial. Further, it drops as $1/\omega$, indicating that as the frequency becomes very large, the electrons cannot follow the rapidly changing electric field and get paralyzed.

In fact, the result (119.8) is completely general and model independent, as signaled by the absence of τ. The reason is that at high frequencies, the details of the relaxation mechanism do not enter the picture because the motion of the electrons changes before they can collide. Secondly, the answer is proportional to n_f because *all* mobile electrons must respond. We shall use eq. (119.8) in section 121 to establish an important sum rule on $\sigma(\omega)$.

---

[1] To take a specific example, for Cu at 273 K, one has, $n_f = 8.47 \times 10^{22}$ cm^{-3} and $\sigma = 5.78 \times 10^{17}$ sec^{-1} $(6.42 \times 10^5 \text{ (ohm cm)}^{-1}$ in SI), which imply $\tau = 2.7 \times 10^{-14}$ sec. Seawater has a conductivity of 3.6×10^{10} sec^{-1} and two types of charge carriers, Na$^+$ and Cl$^-$ ions, each with a density of 3×10^{20} cm^{-3}. If we recall that the ions are much heavier than an electron, we estimate a collision time of about 10^{-14} sec again.

Exercise 119.1 Show that if one extends the Drude model to include a magnetic field, Ohm's law for static fields is modified to read

$$\mathbf{j} = \sigma_0 \mathbf{E} - \tilde{\mu}\mathbf{j} \times \mathbf{B}, \tag{119.9}$$

where $\tilde{\mu} = e\tau/mc$. We caution that this result is grossly inadequate for real metals, but it does capture some of the observed phenomena. **(a)** Consider first the geometry of the Hall effect, in which $\mathbf{B} \perp \mathbf{j}$. Show that in this case we must have $\mathbf{B} \perp \mathbf{E}$. Taking $\mathbf{B}\|\hat{\mathbf{z}}$ and $\mathbf{j}\|\hat{\mathbf{x}}$, find the *Hall coefficient*, $R_H = E_y/j_x B$, and *transverse magnetoresistance*, $\rho(B) = E_x/j_x$. You should find that R_H depends on the sign of the charge carriers and that $\rho(B)$ is independent of B. What is the average velocity, $\bar{\mathbf{v}}$? **(b)** By solving for $\mathbf{j}$ in the form $j_i = \sigma_{ij}(\mathbf{B})E_j$, find the conductivity tensor $\sigma_{ij}(\mathbf{B})$ and show that $\sigma_{ij}(-\mathbf{B}) = \sigma_{ji}(\mathbf{B})$. The last result is an example of an Onsager relation, a very general symmetry principle for transport coefficients.

120 The dielectric function and electric propensity

In addition to the conduction electrons, the bound electrons in a metal also respond to an electric field. Their response can be described by a frequency-dependent dielectric function just as for insulators,

$$\mathbf{D}_\omega = \epsilon_b(\omega)\mathbf{E}_\omega. \tag{120.1}$$

The suffix b indicates that only the bound electrons are involved. The two functions $\sigma(\omega)$ and $\epsilon_b(\omega)$ together describe the electric response of a conductor completely. There is, however, another frequently used way to describe this response, which is completely equivalent. To understand this point, let us consider the Maxwell equations with source terms:

$$\nabla \times \mathbf{H} - \frac{1}{c}\frac{\partial \mathbf{D}}{\partial t} = \frac{4\pi}{c}\mathbf{j}^{\text{free}} \tag{120.2}$$

$$\nabla \cdot \mathbf{D} = 4\pi\rho^{\text{free}}. \tag{120.3}$$

We have explicitly reattached the "free" suffix, which we now put as a superscript. If we make use of eq. (120.1) and switch to the frequency domain, we obtain

$$\nabla \times \mathbf{H}_\omega + \frac{i\omega}{c}\epsilon_b(\omega)\mathbf{E}_\omega = \frac{4\pi}{c}\mathbf{j}_\omega^{\text{free}} \tag{120.4}$$

$$\epsilon_b(\omega)\nabla \cdot \mathbf{E}_\omega = 4\pi\rho_\omega^{\text{free}}. \tag{120.5}$$

In a conductor, we may further divide the free current into parts that are spatially internal and external to the conductor,

$$\mathbf{j}^{\text{free}} = \mathbf{j}^{\text{free,int}} + \mathbf{j}^{\text{free,ext}}, \tag{120.6}$$

and do the same for the charge. Now for the internal part of the current, we have

$$\mathbf{j}_\omega^{\text{free,int}} = \sigma(\omega)\mathbf{E}_\omega. \tag{120.7}$$

Combining this with the continuity equation gives

$$\rho_\omega^{\text{free,int}} = -\frac{i}{\omega}\nabla\cdot\mathbf{j}_\omega^{\text{free,int}} = -i\frac{\sigma(\omega)}{\omega}\nabla\cdot\mathbf{E}_\omega. \tag{120.8}$$

Using the last two results to eliminate the internal part of the current and charge densities from eqs. (120.4) and (120.5), we obtain

$$\nabla\times\mathbf{H}_\omega + \frac{i\omega}{c}\zeta(\omega)\mathbf{E}_\omega = \frac{4\pi}{c}\mathbf{j}_\omega^{\text{free,ext}} \tag{120.9}$$

$$\zeta(\omega)\nabla\cdot\mathbf{E}_\omega = 4\pi\rho_\omega^{\text{free,ext}}, \tag{120.10}$$

where

$$\zeta(\omega) = \epsilon_b(\omega) + \frac{4\pi i\sigma(\omega)}{\omega} \tag{120.11}$$

is another response function, which may be called the *electric propensity* of the metal. Further, in parallel with eq. (120.1), it is useful to define a field $\mathbf{Z}(\mathbf{r}, t)$ such that

$$\mathbf{Z}_\omega = \zeta(\omega)\mathbf{E}_\omega. \tag{120.12}$$

Finally, for completeness, let us rewrite eqs. (120.9) and (120.10) in the time domain, in terms of the fields $\mathbf{H}$ and $\mathbf{Z}$:

$$\nabla\times\mathbf{H} - \frac{1}{c}\frac{\partial\mathbf{Z}}{\partial t} = \frac{4\pi}{c}\mathbf{j}^{\text{free,ext}} \tag{120.13}$$

$$\nabla\cdot\mathbf{Z} = 4\pi\rho^{\text{free,ext}}. \tag{120.14}$$

We now see that the equations (120.9) and (120.10) have exactly the same form as in a dielectric, except that $\epsilon_b(\omega)$ is replaced by $\zeta(\omega)$. For this reason, $\zeta(\omega)$ is also referred to as the *full dielectric function* of the metal. Physically, there is no fundamental difference between conductors and dielectrics in an AC field, since *all* charges carry out an oscillatory and bounded motion. Thus, one cannot really separate the current into a polarization part and a free part, and the only combination which can ever matter is the sum

$$\mathbf{j} = \mathbf{j}^{\text{free,int}} + \frac{\partial\mathbf{P}}{\partial t}. \tag{120.15}$$

Indeed, if we write eq. (120.12) in full in the time domain, we get

$$\frac{\partial\mathbf{Z}}{\partial t} = \frac{\partial\mathbf{E}}{\partial t} + 4\pi\left(\mathbf{j}^{\text{free,int}} + \frac{\partial\mathbf{P}}{\partial t}\right), \tag{120.16}$$

which should be compared with

$$\frac{\partial\mathbf{D}}{\partial t} = \frac{\partial\mathbf{E}}{\partial t} + 4\pi\frac{\partial\mathbf{P}}{\partial t} \tag{120.17}$$

for a dielectric. (Recall $\mathbf{D} = \mathbf{E} + 4\pi\mathbf{P}$.)

Unfortunately, with few exceptions, the quantities that we denote by $\zeta(\omega)$ and $\epsilon_b(\omega)$ are both denoted by the *same* symbol, $\epsilon(\omega)$, and are both referred to as "the dielectric function". The term "propensity" is completely nonstandard. Similarly, the field we call $\mathbf{Z}$ is generally denoted $\mathbf{D}$ as for a dielectric. Needless to say, this double duty served by the symbols can lead to a great deal of confusion. The use of a dielectric function that incorporates the conduction electron response is especially common in discussions of the optical properties of metals, i.e., at frequencies of the order 10^{15} Hz, since the distinction between free and bound electrons then ceases to be useful even at a practical level. One way to decipher which dielectric function is involved is to note that while the bound response $\epsilon_b(\omega)$ is finite at $\omega = 0$, the full response $\zeta(\omega)$ is singular at zero frequency.

The inclusion of $\epsilon_b(\omega)$ is also important for poor conductors such as seawater and semiconductors even at low ω, since it can easily be as large as 10 or more.

We also need to know the magnetic response to solve the macroscopic Maxwell equations. For most materials at the frequencies under consideration, we may take $\mathbf{B}_\omega = \mu(\omega)\mathbf{H}_\omega$ with constant (frequency-independent) $\mu(\omega)$ to good approximation. In fact, μ is very close to unity in most cases. We discuss the dispersion of the magnetic response further in section 124.

121 General properties of the ac conductivity*

The ac conductivity obeys a number of important, model-independent properties. We tabulate these below and then discuss them. In the table, we have separated $\sigma(\omega)$ into its real and imaginary parts,[2]

$$\sigma(\omega) = \sigma'(\omega) + i\sigma''(\omega). \tag{121.1}$$

	Property	Reason
1.	$\sigma^*(\omega) = \sigma(-\omega)$	reality of $\mathbf{E}(t)$ and $\mathbf{j}(t)$.
2.	$\sigma'(\omega) > 0$	2nd law of thermodynamics
3.	$\sigma(\omega)$ analytic in upper half ω plane	causality
4.	Kramers-Kronig relations	causality
5.	$\sigma(\omega)$ nonzero in upper half ω plane	all the above
6.	f-sum rule	high-frequency behavior of $\sigma(\omega)$

$$\tag{121.2}$$

These properties are of great generality, and they also apply (with minor changes) to the frequency-dependent dielectric function, frequency-dependent susceptibility, or any other linear response function.

---

[2] The notations σ_1 and σ_2 are also common.

Property 1 is merely a statement of reality. For real $\mathbf{E}(t)$ and $\mathbf{j}(t)$, we must have $\mathbf{E}_\omega^* = \mathbf{E}_{-\omega}$ and $\mathbf{j}_\omega^* = \mathbf{j}_{-\omega}$. It follows at once that

$$\sigma^*(\omega) = \sigma(-\omega). \tag{121.3}$$

Or, separating this equation into its real and imaginary parts, we get

$$\sigma'(-\omega) = \sigma'(\omega), \quad \sigma''(-\omega) = -\sigma''(\omega). \tag{121.4}$$

The real part of the conductivity is even in ω, and the imaginary part is odd. (More generally, for complex ω we have $\sigma(-\omega^*) = \sigma^*(\omega)$.)

The second property follows from the second law of thermodynamics. The energy dissipated per unit time and volume is $\dot{Q} = \mathbf{j}(t) \cdot \mathbf{E}(t)$. This must be positive. If we consider a purely harmonic electric field $\mathbf{E}(t) = \mathrm{Re}\,\mathbf{E}_\omega e^{-i\omega t}$, then, by the usual rule for bilinear quantities, the time-averaged dissipation is

$$\dot{Q} = \tfrac{1}{4}(\mathbf{j}_\omega \cdot \mathbf{E}_\omega^* + \mathbf{j}_\omega^* \cdot \mathbf{E}_\omega) = \tfrac{1}{2}\sigma'(\omega)|\mathbf{E}_\omega|^2. \tag{121.5}$$

Positivity of $\dot{Q}$ requires that

$$\sigma'(\omega) > 0. \tag{121.6}$$

Property 3 is a consequence of causality. In the time domain, a linear relation such as eq. (119.1) can be written as

$$\mathbf{j}(t) = \int_{-\infty}^{t} R(t - t')\mathbf{E}(t')\, dt', \tag{121.7}$$

where R is some kernel. The important point here is that the integral does not extend to times greater than t. In other words, the current at time t can depend on the electric field only at prior times. If we write this equation in the frequency domain and compare with eq. (119.1), we obtain

$$\sigma(\omega) = \int_{0}^{\infty} R(t)e^{i\omega t}\, dt, \tag{121.8}$$

the important point being that the lower limit of integration is zero. This equation can be used to define $\sigma(\omega)$ for *complex* ω. In the upper half ω plane, we obtain immediately

$$|\sigma(\omega)| \leq \int_{0}^{\infty} |R(t)|e^{-(\mathrm{Im}\,\omega)t}\, dt, \tag{121.9}$$

which is strongly convergent, and hence finite. Further, we may safely assume that $\sigma(\omega)$ is smooth and differentiable. In short, $\sigma(\omega)$ is analytic in the entire upper half-plane. One should note here that it is the sign convention for time Fourier transforms that determines the half-plane—upper or lower—in which $\sigma(\omega)$ is analytic.

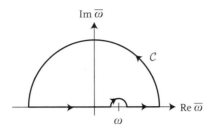

FIGURE 18.1. Contour for Kramers-Kronig relations.

We now exploit the analyticity of $\sigma(\omega)$ to prove property 4, the Kramers-Kronig formulas. Consider the integral

$$\oint_{\mathcal{C}} \frac{\sigma(\overline{\omega})}{\overline{\omega} - \omega} d\overline{\omega}, \tag{121.10}$$

where the contour $\mathcal{C}$ is as shown in fig. 18.1. Since the integrand is analytic everywhere inside $\mathcal{C}$, the integral is zero by Cauchy's theorem. Now the contribution from the large semicircle also vanishes as this semicircle tends to infinity. This is because $|\sigma(\overline{\omega})| \to 0$ as $\overline{\omega} \to \infty$ in the upper half-plane, so the integrand vanishes faster than $1/\overline{\omega}$. We are therefore left with the contribution from the indented straight line along the real axis. The contribution from the semicircular indentation is evaluated in the standard way by letting its radius go to zero, and we obtain

$$\mathrm{P} \int_{-\infty}^{\infty} \frac{\sigma(\overline{\omega})}{\overline{\omega} - \omega} d\overline{\omega} - i\pi\sigma(\omega) = 0, \tag{121.11}$$

where the P denotes a principal value. Separating the real and imaginary parts of this equation, we obtain

$$\sigma''(\omega) = -\frac{1}{\pi} \mathrm{P} \int_{-\infty}^{\infty} \frac{\sigma'(\overline{\omega})}{\overline{\omega} - \omega} d\overline{\omega}, \tag{121.12}$$

$$\sigma'(\omega) = \frac{1}{\pi} \mathrm{P} \int_{-\infty}^{\infty} \frac{\sigma''(\overline{\omega})}{\overline{\omega} - \omega} d\overline{\omega}. \tag{121.13}$$

These formulas are an example of *Kramers-Kronig relations*.[3] They can be used to find both the real and imaginary parts of $\epsilon_b(\omega)$ (or $\zeta(\omega)$) for a real material if only one part is known or if the reflectivity of the material is known over a sufficiently large frequency range. We discuss these points further in sections 123 and 137.

We shall forgo a proof of property 5, as it is rather formal. It follows from a larger theorem that states that $\sigma(\omega)$ is real in the upper half-plane only on the imaginary axis, and on this axis, it decreases monotonically from $\sigma(0)$ at $\omega = 0$ to its value at $\omega = i\infty$ (which is zero in this case, but need not be for other response functions).[4]

---

[3] From the viewpoint of the theory of functions of a complex variable, the functions σ' and σ'' are known as a *Hilbert transform pair* after Hilbert, who analyzed such transforms in 1904. Kramers and Kronig's work dates to 1927.

[4] See Landau and Lifshitz (1969), sec. 125.

To derive property 6, we consider the first Kramers-Kronig relation (18.39), and divide the integral into the negative- and positive-frequency regions:

$$\sigma''(\omega) = -\frac{1}{\pi} P \int_0^\infty \frac{\sigma'(\overline{\omega})}{\overline{\omega} - \omega} d\overline{\omega} - \frac{1}{\pi} P \int_{-\infty}^0 \frac{\sigma'(\overline{\omega})}{\overline{\omega} - \omega} d\overline{\omega}. \tag{121.14}$$

In the second integral, we change the variable of integration to $-\overline{\omega}$, and use the fact that $\sigma'(\overline{\omega})$ is even in $\overline{\omega}$. The two integrals may then be combined into one to yield

$$\sigma''(\omega) = -\frac{2\omega}{\pi} P \int_0^\infty \frac{\sigma'(\overline{\omega})}{\overline{\omega}^2 - \omega^2} d\overline{\omega}. \tag{121.15}$$

Let us now let $\omega \to \infty$. We know that the response is reactive at high frequencies, so $\sigma'(\overline{\omega})$ vanishes for very large $\overline{\omega}$. Thus, the contribution to the integral arises from $\overline{\omega} \ll \omega$, and the denominator of the integrand may be replaced by $-\omega^2$. Thus,

$$\sigma''(\omega) \approx \frac{2}{\pi\omega} \int_0^\infty \sigma'(\overline{\omega}) d\overline{\omega}, \quad (\omega \to \infty). \tag{121.16}$$

Comparing with eq. (119.8), we find the general result

$$\int_0^\infty \sigma'(\omega) d\omega = \frac{\pi n_f e^2}{2m}. \tag{121.17}$$

This is known as the *conductivity sum rule*, or the *f-sum rule*. The latter terminology is historical: in a completely equivalent sum rule for the dielectric function, the analog of $\sigma'(\omega)$ was known as the oscillator strength and denoted by the symbol f. (See sections 123 and 147.)

The value of the *f-sum rule* lies in that for any particular material, $\sigma(\omega)$ depends on the details of the relaxation mechanisms in that material. The sum rule is an important constraint on any theory for that material and may provide valuable clues to the relaxation mechanisms and the nature of excitations and charge carriers in the interpretation of experimental data.

Exercise 121.1 Verify properties 1–5 for the Drude model, eq. (119.6).

Comment: Except, perhaps, for the Kramers-Kronig relations, these properties hardly need showing. For example, property 2, $\sigma'(\omega) > 0$, is built into the model in the tacit assumption that $\tau > 0$. Otherwise, a spontaneous fluctuation in the current would grow with time, which is another way of saying that the second law would be violated.

Exercise 121.2 Find the kernel $R(t)$ for the Drude model.

Exercise 121.3 Consider the model where $\sigma'(\omega) = \sigma_0$ for $|\omega| < \gamma$, and vanishes otherwise. Use the Kramers-Kronig relations to find $\sigma''(\omega)$, and compare your answers for σ' and σ'' with those from the Drude model. (The requisite integrals can be done using the fundamental definition of a principal value integral.)

Exercise 121.4 Discuss the evenness and oddness of $\epsilon_b'(\omega)$ and $\epsilon_b''(\omega)$.

122 Electromagnetic energy in material media*

We mentioned briefly in section 85 that one cannot obtain separate expressions for the internal energy and the heat evolved in a material medium. There is a physical reason for this. As long as the permittivity $\epsilon(\omega)$ and permeability $\mu(\omega)$ have some dispersion, the Kramers-Kronig formulas imply that the imaginary parts of these quantities, ϵ'' and μ'', cannot vanish for all frequencies. These imaginary parts are due to dissipative processes in the medium, as a result of which the electromagnetic energy in the wave is gradually converted to heat. Thus, strictly speaking, there can be no thermodynamic quantity that we can interpret as the macroscopic energy density of the electromagnetic field. However, approximate expressions for this quantity and the heat evolution can be found if the fields are quasimonochromatic, and if the dispersion is weak.

In order not to interrupt the analysis that follows, we now discuss a particular Fourier representation of the fields that proves convenient. We divide the standard decomposition,

$$\mathbf{E}(t) = \int_{-\infty}^{\infty} \frac{d\omega}{2\pi} \mathbf{E}_\omega e^{-i\omega t}, \tag{122.1}$$

into its positive- and negative-frequency parts,

$$\mathbf{E}(t) = \int_0^{\infty} \mathbf{E}_\omega e^{-i\omega t} \frac{d\omega}{2\pi} + \int_0^{\infty} \mathbf{E}_\omega^* e^{i\omega t} \frac{d\omega}{2\pi}, \tag{122.2}$$

where the second term is obtained by the substitution $\omega \to -\omega$ in the original integral over the range $-\infty < \omega < 0$, and the result $\mathbf{E}_{-\omega} = \mathbf{E}_\omega^*$, which is a consequence of the reality of $\mathbf{E}(t)$. Let us now suppose the field is quasimonochromatic with a mean frequency ω_0. This means that the amplitude $\mathbf{E}_\omega$ is large only in a small range of frequencies around ω_0. Writing

$$\mathbf{E}_{\omega_0+\alpha} = \mathbf{a}_\alpha, \tag{122.3}$$

we get

$$\mathbf{E}(t) = e^{-i\omega_0 t} \int_{-\omega_0}^{\infty} \mathbf{a}_\alpha e^{-i\alpha t} \frac{d\alpha}{2\pi} + \text{c.c.} \tag{122.4}$$

Since the range of frequencies α that matter is much smaller than ω_0, the lower limit of the α integral may be extended to $-\infty$, after which the integral can be equated to its FT, $\mathbf{a}(t)$. In other words, we may write

$$\mathbf{E}(t) = \mathbf{a}(t)e^{-i\omega_0 t} + \mathbf{a}^*(t)e^{i\omega_0 t}. \tag{122.5}$$

The envelope functions $\mathbf{a}(t)$ and $\mathbf{a}^*(t)$ vary little over one time period $2\pi/\omega_0$. Hence, if we average $\mathbf{E}^2$ over several such periods, the rapidly oscillating terms at frequencies $\pm 2\omega_0$ get washed out. Denoting this average by an overbar, we have

$$\overline{E^2(t)} = 2\mathbf{a}(t) \cdot \mathbf{a}^*(t). \tag{122.6}$$

The magnetic field is treated the same way.

We now turn to the main argument. We showed in section 85 that (eq. (85.4))

$$-\nabla \cdot \frac{c}{4\pi}(\mathbf{E} \times \mathbf{H}) = \mathbf{j}_{\text{free}} \cdot \mathbf{E} + \frac{1}{4\pi}\left(\mathbf{E} \cdot \frac{\partial \mathbf{D}}{\partial t} + \mathbf{H} \cdot \frac{\partial \mathbf{B}}{\partial t}\right). \tag{122.7}$$

This is an exact result. If we integrate this equation over a volume V and multiply by a time interval Δt, the first term on the right is the mechanical work received by the free charges. The left-hand side is the energy inflow. It leads us to identify the vector

$$\mathbf{S} = c(\mathbf{E} \times \mathbf{H})/4\pi \tag{122.8}$$

as the energy flux density, or the generalization of the Poynting vector to fields in media. The last term on the right must therefore represent the rate of increase of the internal energy *plus* the heat evolved. To analyze it, we write

$$\mathbf{E} \cdot \dot{\mathbf{D}} = \int_{-\infty}^{\infty} \frac{d\omega}{2\pi} \mathbf{E}_\omega^* e^{i\omega t} \cdot \int_{-\infty}^{\infty} \frac{d\omega'}{2\pi}(-i\omega')\epsilon(\omega')\mathbf{E}_{\omega'} e^{-i\omega' t}. \tag{122.9}$$

In this formula, we have written the complex conjugate of the standard FT expression for $\mathbf{E}(t)$, since $\mathbf{E}(t)$ is real. We could, however, conjugate the integral for $\dot{\mathbf{D}}$ instead, and write

$$\mathbf{E} \cdot \dot{\mathbf{D}} = \int_{-\infty}^{\infty} \frac{d\omega}{2\pi}(i\omega)\epsilon^*(\omega)\mathbf{E}_\omega^* e^{i\omega t} \cdot \int_{-\infty}^{\infty} \frac{d\omega'}{2\pi}\mathbf{E}_{\omega'} e^{-i\omega' t}. \tag{122.10}$$

Adding these two expressions, we get

$$2\mathbf{E} \cdot \dot{\mathbf{D}} = \iint \frac{d\omega d\omega'}{4\pi^2} i[\omega\epsilon^*(\omega) - \omega'\epsilon(\omega')]\mathbf{E}_\omega^* \cdot \mathbf{E}_{\omega'} e^{i(\omega-\omega')t}. \tag{122.11}$$

Let us now average this formula over several periods of the rapid oscillation at ω_0. Only the two regions $\omega \approx \omega' \approx \omega_0$ and $\omega \approx \omega' \approx -\omega_0$ survive this averaging. Further, the contributions of the two regions are equal, so we need only double the first contribution. (The equality can be seen via the substitution $\omega \to -\omega'$, $\omega' \to -\omega$, and the identity $\epsilon(-\omega) = \epsilon^*(\omega)$.) Writing $\omega = \omega_0 + \alpha$, etc., and the amplitudes in terms of $\mathbf{a}_\alpha$, we get

$$2\overline{\mathbf{E} \cdot \dot{\mathbf{D}}} \approx 2\int_{-\infty}^{\infty}\int_{-\infty}^{\infty} \frac{d\alpha d\alpha'}{4\pi^2} i[\omega\epsilon^*(\omega) - \omega'\epsilon(\omega')]_0 \mathbf{a}_\alpha^* \cdot \mathbf{a}_{\alpha'} e^{i(\alpha-\alpha')t}. \tag{122.12}$$

Here, the suffix 0 on the term in square brackets means that it is to be evaluated for frequencies near ω_0, and we have pushed the lower limits of the integrals from $-\omega_0$ to $-\infty$.

Let us now examine the integrand of eq. (122.12) more carefully. We have

$$\epsilon(\omega) = \epsilon'(\omega) + i\epsilon''(\omega). \tag{122.13}$$

So, with $\omega = \omega_0 + \alpha$, $\omega' = \omega_0 + \alpha'$, we obtain

$$\text{Re}\left[\omega\epsilon^*(\omega) - \omega'\epsilon(\omega')\right]_0 \approx (\alpha - \alpha')\frac{d}{d\omega}(\omega\epsilon'(\omega))\bigg|_{\omega_0}, \qquad \text{(Term 1)} \tag{122.14}$$

$$\text{Im}\left[\omega\epsilon^*(\omega) - \omega'\epsilon(\omega')\right]_0 \approx -2\omega_0\epsilon''(\omega_0). \qquad \text{(Term 2)} \tag{122.15}$$

We can now undo the frequency integrals. It is easy to do this for term 2, and we can do term 1 by noting that

$$i(\alpha - \alpha')e^{i(\alpha-\alpha')t} = \frac{d}{dt}e^{i(\alpha-\alpha')t}. \tag{122.16}$$

Proceeding in this way, and noting that $2|\mathbf{a}(t)|^2 = \overline{E^2(t)}$, we get

$$\overline{\mathbf{E}\cdot\mathbf{D}} \text{ (Term 1)} \approx \frac{1}{2}\frac{d}{d\omega}\left(\omega\epsilon'(\omega)\right)\frac{d}{dt}\overline{E^2(t)}, \tag{122.17}$$

$$\overline{\mathbf{E}\cdot\mathbf{D}} \text{ (Term 2)} \approx \omega\epsilon''(\omega)\overline{E^2(t)}. \tag{122.18}$$

We have removed the suffix 0 in ω_0, as there is now no need for it.

The magnetic terms can be handled in exactly the same way. Putting together all the pieces, the equation for energy conservation becomes

$$-\nabla\cdot\overline{\mathbf{S}} = \mathbf{j}_{\text{free}}\cdot\mathbf{E} + \frac{d}{dt}\overline{U} + \dot{Q}, \tag{122.19}$$

where

$$\overline{U} = \frac{1}{8\pi}\left[\frac{d}{d\omega}\left(\omega\epsilon'(\omega)\right)\overline{E^2} + \frac{d}{d\omega}\left(\omega\mu'(\omega)\right)\overline{H^2}\right] \tag{122.20}$$

is the internal energy, and

$$\dot{Q} = \frac{1}{4\pi}\left[\omega\epsilon''(\omega)\overline{E^2} + \omega\mu''(\omega)\overline{H^2}\right] \tag{122.21}$$

is the rate of heat evolution per unit volume. Since we must have $\dot{Q} > 0$ by thermodynamics, we obtain the important results

$$\epsilon''(\omega) > 0, \quad \mu''(\omega) > 0 \quad (\omega > 0). \tag{122.22}$$

The above results are largely due to Brillouin, especially the form (122.20) for the internal energy. He argues that because the term $\epsilon' E^2/8\pi$ represents the mechanical work done in polarizing the medium when the electric field is varied slowly and adiabatically, the remaining part, $\omega(d\epsilon'/d\omega) E^2/8\pi$, must be the average kinetic energies of the bound charges. Similar arguments apply to the magnetic terms.

We reemphasize that the above analysis is meaningful only when absorption and dispersion are both weak, i.e., $\epsilon''(\omega) \ll 1$, and $\epsilon'(\omega)$ varies only weakly with frequency. We can see that our expression for $\overline{U}$ is but the first term of a Taylor series in frequency derivatives of ϵ', and if absorption is strong, this expansion is not useful, and the very concept of an internal energy is shaky.

123 Drude-Lorentz model of the dielectric response

We discussed classical models of the atom and of mobile electrons in conductors in sections 67 and 119. We now discuss a very similar model for the dielectric constant because of its pedagogic value. This model is generally named for Drude and Lorentz, but the names of Kramers and Lorenz are also associated with it.

We consider an atom or molecule with electrons bound to a center like harmonic oscillators and also include a linear damping term to represent absorption and radiation damping. The equation of motion for an electron is then

$$m\ddot{\mathbf{r}} + m\gamma\dot{\mathbf{r}} + m\omega_0^2\mathbf{r} = -e\mathbf{E}(t), \tag{123.1}$$

where $\mathbf{E}(t)$ is the electric field in the medium. Taking FTs, we get

$$(-m\omega^2 - im\gamma\omega + m\omega_0^2)\mathbf{r}_\omega = -e\mathbf{E}_\omega. \tag{123.2}$$

If there are n_b such bound electrons per unit volume, then the polarization of the medium is given by

$$\mathbf{P}_\omega = -n_b e\mathbf{r}_\omega = \frac{n_b e^2}{m}\frac{1}{-\omega^2 + \omega_0^2 - i\omega\gamma}\mathbf{E}_\omega. \tag{123.3}$$

The electric susceptibility $\chi_e(\omega)$ is defined via $\mathbf{P}_\omega = \chi_e(\omega)\mathbf{E}_\omega$, so our model gives

$$\chi_e(\omega) = \frac{n_b e^2}{m}\frac{1}{-\omega^2 + \omega_0^2 - i\omega\gamma}. \tag{123.4}$$

Finally, since $\epsilon_b = 1 + 4\pi\chi_e$, we get

$$\epsilon_b(\omega) = 1 + \frac{4\pi n_b e^2}{m}\frac{1}{(\omega_0^2 - \omega^2) - i\omega\gamma}. \tag{123.5}$$

We have taken the local field at the electron in our model to be the same as the macroscopic field. If we improve upon this approximation in the same spirit as in the Clausius-Mossotti model, we obtain[5]

$$\frac{\epsilon_b(\omega) - 1}{\epsilon_b(\omega) + 2} = \frac{4\pi}{3}\chi_e(\omega). \tag{123.6}$$

Feeding in the form (123.4) for $\chi_e(\omega)$ and simplifying, we get

$$\epsilon_b(\omega) = 1 + \frac{4\pi n_b e^2}{m}\frac{1}{(\omega_1^2 - \omega^2) - i\omega\gamma}, \tag{123.7}$$

where

$$\omega_1^2 = \omega_0^2 - \frac{4\pi n_b e^2}{3m}. \tag{123.8}$$

Because of the extreme simplicity of this model, however, one should not attach too much weight to the exact meanings of the frequencies ω_0 and ω_1, and the main value of the

---

[5] If we rewrite eq. (123.6) in terms of the refractive index $n(\omega) = \sqrt{\epsilon_b(\omega)}$, which we shall encounter in chapter 20, we obtain the so-called Lorenz-Lorentz formula.

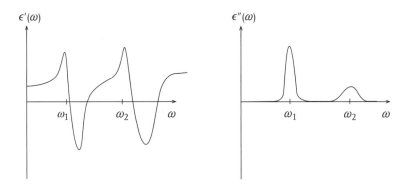

FIGURE 18.2. Schematic real and imaginary parts of the dielectric constant of a material, showing two absorption peaks. The region between ω_1 and ω_2 is a "transparency window."

model is in the functional form for $\epsilon(\omega)$, which captures a large amount of the qualitative behavior of many dielectrics. Henceforth, we shall stick with the simpler eq. (123.5).[6]

A common improvement of the model (123.5) is founded in the fact that electrons in real atoms can resonate at many frequencies. An atom that has energy levels $\mathcal{E}_i$ can resonate at the transition frequencies $\omega_i = (\mathcal{E}_i - \mathcal{E}_0)/\hbar$ if it is initially in the ground level $\mathcal{E}_0$. However, not all possible transition frequencies occur, or even do so with equal effectiveness. To each transition we assign an *oscillator strength*, f_i, a dimensionless number of order unity, which is a measure of the amplitude of the dipole moment of the atom when it is oscillating between levels $\mathcal{E}_i$ and $\mathcal{E}_0$. The *f-sum rule* dictates that

$$\sum_i f_i = Z, \tag{123.9}$$

where Z is the number of electrons per atom. See exercise 123.5 and section 147 for more on it.

With these definitions, equation (123.5) is generalized to

$$\epsilon_b(\omega) = 1 + \frac{4\pi n_b e^2}{mZ} \sum_i \frac{f_i}{(\omega_i^2 - \omega^2) - i\omega\gamma_i}, \tag{123.10}$$

where γ_i is the damping of the resonance at ω_i. Once again, if we do not attach too literal a meaning to the material constants such as n_b, Z, etc., this functional form provides a rough model for a medium having many absorption bands. In fig. 18.2 we show a schematic for a case with two such bands. In the regions where absorption is strong and $d\epsilon'/d\omega < 0$, the dispersion is said to be "anomalous." The reason for this terminology will become clear in section 134.

For poor conductors, in which there may be both bound and mobile electrons, we can combine the above model for the bound response (with a single absorption peak for

---

[6] Indeed, essentially the same functional form describes ionic insulators such as ZnS and NaCl, even though the oscillations in question are very different, namely, the long-wavelength phonons of the crystal. The poles in ϵ then occur at $\sim 10^{12}$ Hz, as opposed to 10^{14}–10^{15} Hz for atomic resonances. See *Ashcroft and Mermin* (1976), chap. 27.

brevity) with the Drude model for the conductivity to give a propensity (see section 119)

$$\zeta(\omega) = \epsilon_b(\omega) + \frac{4\pi i \sigma(\omega)}{\omega}$$

$$= 1 + \frac{4\pi n_b e^2}{m} \frac{1}{(\omega_0^2 - \omega^2) - i\omega\gamma} - \frac{4\pi n_f e^2}{m\omega(\omega + i\tau^{-1})}. \tag{123.11}$$

This viewpoint has the advantage that the free-electron contribution can be viewed as just another term in eq. (123.10), obtained by setting ω_0 to 0, and γ to the inverse collision time τ^{-1}.

Although eqs. (123.10) and (123.11) are both simple-minded models over the full range of frequencies, they correctly predict that at very high frequencies (X-ray or extreme ultraviolet and above), well above electronic transition frequencies in a medium,

$$\zeta(\omega) \approx 1 - \frac{4\pi n e^2}{m\omega^2}, \tag{123.12}$$

where n is the total number of electrons per unit volume. The physical content of this behavior is the same as that underlying eq. (119.8): at large ω, the electrons cannot follow the rapidly changing electric field. They are, so to speak, paralyzed, and the polarization is very small. Hence, $\zeta \approx 1$, and eq. (123.12) gives the leading departure from unity. We can also understand eq. (123.12) by arguing that because the distances the electrons can move in one period of the imposed electric field are much smaller than the wavelength for that field, the electrons can be treated as essentially free in calculating this response, and, indeed, eq. (123.12) is precisely the dielectric or propensity function for plasmas.

Exercise 123.1 The oddness or evenness of $\epsilon'(\omega)$ was investigated in exercise 121.4. Verify these properties for the Drude-Lorentz model for $\epsilon(\omega)$.

Exercise 123.2 It was stated in section 121 that a response function such as $\epsilon_b(\omega)$ has no singularities or zeros in the upper half ω plane. Verify that this is so for the Drude-Lorentz model.

Exercise 123.3 Show that the Kramers-Kronig relations for $\epsilon_b(\omega)$ can be written in the form

$$\epsilon_b''(\omega_0) = -\frac{2\omega_0}{\pi} P \int_0^\infty \frac{(\epsilon_b'(\omega) - 1)}{\omega^2 - \omega_0^2} d\omega, \quad \epsilon_b'(\omega_0) = 1 + \frac{2}{\pi} P \int_0^\infty \frac{\omega\epsilon_b''(\omega)}{\omega^2 - \omega_0^2} d\omega, \tag{123.13}$$

paying particular attention to why the first integrand involves $\epsilon_b'(\omega) - 1$ and not just $\epsilon_b'(\omega)$.

Exercise 123.4 Extend the previous exercise to a poor conductor that has some free electrons. Show that now

$$\zeta''(\omega_0) = -\frac{2\omega_0}{\pi} P \int_0^\infty \frac{(\zeta'(\omega) - 1)}{\omega^2 - \omega_0^2} d\omega + 4\pi \frac{\sigma'(0)}{\omega_0}, \tag{123.14}$$

$$\zeta'(\omega_0) = 1 + \frac{2}{\pi} P \int_0^\infty \frac{\omega\zeta''(\omega)}{\omega^2 - \omega_0^2} d\omega. \tag{123.15}$$

Note that in the first equation, the last term involves the zero-frequency conductivity, $\sigma'(0)$, not $\sigma'(\omega)$, and that both integrals are convergent near the $\omega = 0$ limit.

Exercise 123.5 (f-sum rule) Use the Kramers-Kronig relations to examine the form of $\epsilon'(\omega)$ [or $\zeta'(\omega)$] at very high frequencies, and compare with eq. (123.12) to deduce that

$$\int_0^\infty \omega \epsilon''(\omega)\, d\omega = \frac{2\pi^2 e^2 n}{m}. \tag{123.16}$$

This is the f or dipole sum rule. Apply it to the model (123.10) to obtain eq. (123.9).

124 Frequency dependence of the magnetic response*

We conclude this chapter with a qualitative look into the dispersion or frequency dependence of the magnetic susceptibility in nonmagnetic materials. Although the smallness of this susceptibility diminishes the importance of the issue, it is nevertheless interesting.

We introduced the magnetization density, **M**, as a way of thinking about the atomic currents in the macroscopic Ampere-Maxwell law. Ultimately, this is what we care about—understanding the source terms in this law. In the time-dependent case, the bound current can be written as

$$\mathbf{j}_{\text{bound}} = c\nabla \times \mathbf{M} + \frac{\partial \mathbf{P}}{\partial t}, \tag{124.1}$$

where the two terms are the Amperean or magnetization current and the polarization current. The magnetization is a useful concept as long as the polarization current is much smaller than the Amperean current. We therefore try to estimate the relative sizes of the two terms. Estimating M as $\chi_m H \sim \chi_m B$, where χ_m is the magnetic susceptibility, and demanding that **M** vary over lengths much bigger than an atomic size a, we have

$$c\nabla \times \mathbf{M} \ll c\chi_m B/a. \tag{124.2}$$

For the polarization current, we first appeal to Faraday's law to conclude that $E \gg a\omega B/c$. Taking the dielectric susceptibility to be of order 1, we then find that

$$\frac{\partial \mathbf{P}}{\partial t} \gg a\omega^2 B/c. \tag{124.3}$$

We have overestimated the magnetization current and underestimated the polarization current. Hence, for the former to dominate, it should certainly be the case that

$$\chi_m \gg a^2 \omega^2/c^2. \tag{124.4}$$

We estimated after eq. (102.8) that $\chi_m \sim (e^2/c\hbar)^2$. Hence, our condition for the magnetization to be meaningful becomes

$$\omega \ll (e^2/\hbar a). \tag{124.5}$$

The right-hand side of this inequality is of the order of an optical frequency. Thus, we see that in contrast to the dielectric case, the magnetic response becomes negligible at relatively low frequencies. At higher frequencies, the difference bewteen **B** and **H** is unimportant, as the polarization currents induced by a time-dependent magnetic field overshadow the Amperean currents. One says that the magnetic susceptibility shows dispersion at low frequencies, sometimes in the microwave. Alas, there is no analog of the Drude-Lorentz model that is both simple and general enough that captures the low-frequency dispersion.

The condition (124.5) can also be written as $\omega \ll v/c$, where v is a typical electronic velocity in an atom. This result is in line with the picture of diamagnetism developed in section 102. The orbital frequency of atomic electrons is $\sim v/c$, and if the applied field varies much more rapidly than this, the orbital magnetic moment will not be able to respond. It will get paralyzed, just as the electrons do in response to electric fields at X-ray and higher frequencies.

The same conclusion applies to the paramagnetic susceptibility with even greater force, since the time scales for paramagnetic relaxation are much longer than atomic time periods. In this connection, we mention in passing that since the ambient magnetic fields in matter are essentially zero, in general, one often applies a static magnetic field and a transverse time-dependent magnetic field, as in nuclear or electronic spin resonance. One then defines a dynamic susceptibility to the transverse field, wherein the static magnetic field is regarded as part of the unperturbed system. The questions that arise here take us far beyond electromagnetism into the realm of condensed matter physics.[7]

Exactly the opposite situation holds in ordered systems (ferromagnets and antiferro-magnets). Here the ambient exchange field is much stronger than applicable laboratory fields, and the dominant excitations are *spin waves*, normal modes of the system quite analogous to phonons in crystals, in which the moments execute small precessions around their mean directions, coupled together by the exchange interaction. To discuss these further would again take us into condensed matter physics territory.[8]

The response of superconductors to time-dependent fields is a fully quantum mechanical problem, well beyond our scope.

---

[7] For a good discussion, see White (1983), chap. 5.
[8] For more, see White (1983), chap. 6.

19 | Quasistatic phenomena in conductors

125 Quasistatic fields

Consider a metallic body of linear size a, and suppose the fields vary at frequencies $\omega \ll c/a$. Even for fairly large bodies, this can still allow a considerable range of frequencies. For example, for $a \sim 1$ m, the upper limit is of the order of 10 MHz. We shall also assume that the frequencies are low compared to the inverse of the collision time. We may then ignore the frequency dispersion of the conductivity and take simply $\sigma(\omega) = \sigma_0$ for the most part. For a good metal, this limits us to frequencies less than $\sim 10^{13}$ Hz (in the infrared). The condition $\omega \ll c/a$ automatically implies $\omega \ll \sigma_0$ (or $\omega \ll \sigma_0/4\pi\epsilon_0$ in SI), unless the body is very small or the conductor is poor.

Let us see how the Maxwell equations simplify under these conditions. The source-free equations stay unchanged:

$$\nabla \cdot \mathbf{B} = 0, \quad \nabla \times \mathbf{E} + \frac{1}{c}\dot{\mathbf{B}} = 0. \tag{125.1}$$

Next, we take up the Ampere-Maxwell law inside the body:

$$\nabla \times \mathbf{H} - \frac{1}{c}\dot{\mathbf{D}} = \frac{4\pi}{c}\mathbf{j}. \tag{125.2}$$

We estimate $\dot{\mathbf{D}} \sim \omega \mathbf{D} = \omega\epsilon_b\mathbf{E}$, and write $\mathbf{j} = \sigma\mathbf{E}$. Since $\epsilon_b = O(1)$ and $\sigma \gg \omega$, we may neglect the $\dot{\mathbf{D}}$ term, obtaining simply[1]

$$\nabla \times \mathbf{H} = \frac{4\pi\sigma}{c}\mathbf{E}. \tag{125.3}$$

$$\nabla \times \mathbf{H} = \sigma\mathbf{E}. \quad \text{(SI)}$$

---

[1] For SI users, although we will give the main equations in SI too in the next few sections, it will help to remember the Gaussian-to-SI conversion: replace μ by μ/μ_0, σ by $\sigma(\mu_0 c^2/4\pi)$, E by $E(4\pi/\mu_0 c^2)^{1/2}$, and H by $H(4\pi\mu_0)^{1/2}$.

The form of Gauss's law for quasistatics follows from the divergence of this equation. We get

$$\nabla \cdot \mathbf{E} = 0. \tag{125.4}$$

But since $\mathbf{j} = \sigma \mathbf{E}$ and $\nabla \cdot \mathbf{E} = 4\pi\rho$, we also get

$$\nabla \cdot \mathbf{j} = 0, \quad \rho = 0. \tag{125.5}$$

We thus see that in quasistatics a conductor continues to expel bulk charges from its interior, as in statics. Finally, the equation of continuity is automatically satisfied.

We can eliminate $\mathbf{E}$ from this system of equations by taking the curl of eq. (125.3), making use of $\nabla \cdot \mathbf{H} = \mu^{-1}\nabla \cdot \mathbf{B} = 0$, and substituting for $\nabla \times \mathbf{E}$ from Faraday's law, to get

$$\nabla^2 \mathbf{H} = \frac{4\pi\sigma\mu}{c^2} \frac{\partial \mathbf{H}}{\partial t}. \tag{125.6}$$

$$\nabla^2 \mathbf{H} = \sigma\mu \frac{\partial \mathbf{H}}{\partial t}. \quad \text{(SI)}$$

By solving this equation along with $\nabla \cdot \mathbf{H} = 0$, we obtain $\mathbf{H}$ inside the metal. The electric field is then found by using eq. (125.3), and the current follows at once from $\mathbf{j} = \sigma \mathbf{E}$.

Equation (125.6) has the form of a diffusion equation. By taking its curl and using eq. (125.3), we see that $\mathbf{E}$ obeys the same equation,

$$\nabla^2 \mathbf{E} = \frac{4\pi\sigma\mu}{c^2} \frac{\partial \mathbf{E}}{\partial t}. \tag{125.7}$$

$$\nabla^2 \mathbf{E} = \sigma\mu \frac{\partial \mathbf{E}}{\partial t}. \quad \text{(SI)}$$

Thus, alternatively, we can solve this equation along with $\nabla \cdot \mathbf{E} = 0$ and then use Faraday's law or eq. (125.3) to find $\mathbf{H}$.

To solve the problem completely, we also need boundary conditions. If we suppose that the only interface is between the conductor and vacuum (or what is practically the same thing, air), then at the interface, by arguments now familiar to the reader, we must have continuity of the tangential components of $\mathbf{E}$ and $\mathbf{H}$ and of the normal component of $\mathbf{B}$. It remains to examine the normal component of $\mathbf{E}$. For simplicity, let us only consider conductors not carrying any externally supplied current. Then, the condition (120.14) becomes

$$\nabla \cdot \mathbf{Z} = 0. \tag{125.8}$$

This is valid everywhere with $\mathbf{Z} = \mathbf{E}$ outside the conductor. Integrating it over a pillbox-shaped Gaussian surface, we find that $\mathbf{Z} \cdot \hat{\mathbf{n}}$ must be continuous, or

$$E_{\text{out},n} = \zeta E_{\text{in},n} \approx (4\pi i \sigma/\omega) E_{\text{in},n}, \tag{125.9}$$

where the subscripts in and out refer to the inside and the outside of the conductor, and we have again invoked $\sigma \gg \omega$ to approximate ζ. In the time domain, eq. (125.9) reads

$$\frac{\partial}{\partial t} E_{\text{out},n} = 4\pi\sigma E_{\text{in},n}. \tag{125.10}$$

Equation (125.9) can also be obtained by integrating Gauss's law and the equation of continuity over the same pillbox. Denoting the surface charge density by Σ,[2] we get

$$E_{\text{out},n} = \left(4\pi - i(\omega/\sigma)\right)\Sigma \approx 4\pi\Sigma, \tag{125.11}$$

$$E_{\text{in},n} = -i(\omega/\sigma)\Sigma. \tag{125.12}$$

Eliminating Σ, we get eq. (125.9). The physical point to note is that the normal component of the electric field is very strongly attenuated inside the conductor.

We can draw two important qualitative conclusions from the general nature of the solution to the diffusion problem. First, let us suppose that a conducting body of size a is placed in an external magnetic field, which is then suddenly removed. The field in the body decays on a time scale

$$\tau_d \simeq \frac{4\pi\sigma\mu a^2}{c^2}. \tag{125.13}$$

$$\tau_d \simeq \sigma\mu a^2. \quad \text{(SI)}$$

For $a = 1$ m, $\sigma \sim O(10^{17})$ sec, and $\mu \simeq 1$, $\tau_d \sim 10$ sec. The physical reason the field within the conductor cannot vanish right away is simple. When the magnetic field is removed, the induced electric field sets up currents, which give rise to a magnetic field in the same direction as the initial field. (Recall Lenz's rule.) The higher the conductivity, the longer it takes for the currents to die away.

Second, suppose the fields vary at a frequency ω (which might be externally imposed). By dimensional analysis of the diffusion equation, all the fields, $\mathbf{H}$, $\mathbf{E}$, and $\mathbf{j} = \mathbf{E}/\sigma$, will then exist in a layer of thickness $\delta \sim (c^2/\sigma\mu\omega)^{1/2}$ next to the surface of the metal (unless $a < \delta$). Thus, one obtains dynamic shielding of the electromagnetic fields in the bulk of the conductor. We will investigate this phenomenon more carefully in the following sections.

The induced currents set up in a metal due to the changing magnetic field are known as *eddy currents*. They are an undesirable source of heating and energy loss in the iron cores used to channel the magnetic field in transformers, motors, etc. They are also put to good use, however, in induction heaters, magnetic brakes, and eddy current maglev (magnetic levitation).

126 Variable magnetic field: eddy currents and the skin effect in a planar geometry

Suppose that a conductor is placed in a periodically varying external magnetic field. We wish to find the electromagnetic fields, especially inside the conductor. We will suppose that the conditions for a quasistatic field are satisfied. This automatically means that the body has finite size; however, attempting to solve for the fields for any finite body requires

---

[2] In this chapter we use Σ for the surface charge density, as σ is being used for the conductivity.

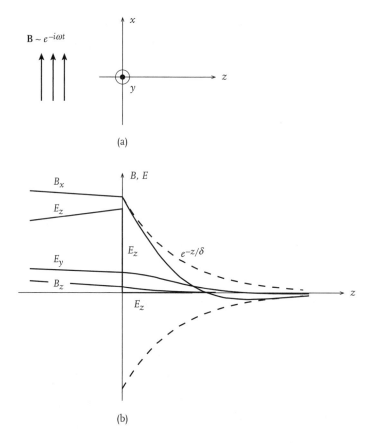

(a)

(b)

FIGURE 19.1. The skin effect. (a) Geometry used in the text. (b) Fields near the surface. As discussed in the text, outside the conductor, **B** must be essentially parallel to the surface, and **E** must be perpendicular to it. We also show the much smaller components: $\mathbf{E} \times \hat{\mathbf{n}}$ and $\mathbf{B} \cdot \hat{\mathbf{n}}$, where $\hat{\mathbf{n}}$ is the surface normal. Note that $\mathbf{E} \cdot \hat{\mathbf{n}}$ is extremely small inside the conductor, as discussed in subsequent sections.

solving a rather involved boundary value problem in even the simplest case. We shall therefore approach the task in stages and solve three problems of increasing difficulty.

Throughout this and the next section, we shall assume that $\mu = 1$ (or $\mu = \mu_0$ in SI). Except for the ferromagnetic metals at low frequencies, this is a very good approximation, and nothing qualitatively important is lost by it. This means that we can take $\mathbf{H} = \mathbf{B}$ everywhere, and **B** to be continuous at the interface.

In this section, we solve the simplest problem, that of a conductor occupying the half-space $z > 0$ and subjected to a time-dependent magnetic field

$$\mathbf{B} = B_0 e^{-i\omega t}\hat{\mathbf{x}}, \quad (z = 0-), \tag{126.1}$$

at the surface. (See fig. 19.1.) The frequency is such that the field may regarded as quasistatic. We wish to find the fields inside the conductor. The first step is to note that there is no surface current, by which we mean that there is no current in a sheet of idealized zero thickness, because the tangential component of **B** must be continuous.

This continuity also implies that (omitting the $e^{-i\omega t}$ factor)

$$\mathbf{B} = B_0 \hat{\mathbf{x}} \quad (z = 0+) \tag{126.2}$$

just inside the conductor. The diffusion equation (125.6) for **B** now reads

$$\frac{\partial^2 B_x}{\partial z^2} = -\frac{4\pi\sigma\omega}{c^2} i B_x. \tag{126.3}$$

The solutions to this equation are of the form $e^{\pm i\kappa z}$, where

$$\kappa^2 = \frac{4\pi\sigma\omega}{c^2} i. \tag{126.4}$$

Writing $i = e^{i\pi/2}$ and taking the square root, we obtain

$$\kappa = \sqrt{\frac{2\pi\sigma\omega}{c^2}}(1 + i). \tag{126.5}$$

$$\kappa = \sqrt{\mu_0\sigma\omega/2}(1 + i). \quad \text{(SI)}$$

The solution $e^{-i\kappa z}$ diverges as $z \to \infty$, and must be discarded. Imposing the boundary condition (126.2) at $z = 0$, we obtain

$$\mathbf{B} = B_0 e^{-(1-i)z/\delta}\hat{\mathbf{x}}, \quad \text{(Gaussian and SI)} \tag{126.6}$$

where

$$\delta = \sqrt{\frac{c^2}{2\pi\sigma\omega}} \tag{126.7}$$

$$\delta = \sqrt{\frac{2}{\mu_0\sigma\omega}} \quad \text{(SI)}$$

is the *skin depth*, so called because the field penetrates into the conductor only to distances of order δ from the surface. It is generally a small length that decreases as the inverse square root of the frequency of the applied field. For Cu at room temperature, e.g., it is 8.5 mm at 60 Hz, 7 μm at 100 MHz, and 0.7 μm at 10 GHz. For seawater, it is 30 m at 60 Hz, and 2.5 cm at 100 MHz.

Next, let us calculate the electric field inside the conductor. We have

$$\mathbf{E} = \frac{1}{\sigma}\mathbf{j} = \frac{c}{4\pi\sigma}\nabla \times \mathbf{B}, \tag{126.8}$$

which gives

$$\mathbf{E} = E_0 e^{-(1-i)z/\delta}\hat{\mathbf{y}}, \tag{126.9}$$

where

$$E_0 = -\frac{c}{4\pi\sigma}\frac{(1-i)}{\delta} B_0. \tag{126.10}$$

$$E_0 = -\frac{(1-i)}{\sigma\delta} B_0. \quad \text{(SI)}$$

We see that the electric field is also tangential to the surface of the conductor. It is useful to restore the time dependence and write the fields in real form explicitly:

$$\mathbf{B} = B_0 e^{-z/\delta} \cos\left(\frac{z}{\delta} - \omega t\right) \hat{\mathbf{x}}, \tag{126.11}$$

$$\mathbf{E} = -|E_0| e^{-z/\delta} \cos\left(\frac{z}{\delta} - \omega t - \frac{\pi}{4}\right) \hat{\mathbf{y}}. \quad \text{(Gaussian and SI)} \tag{126.12}$$

We see that the electric field leads the magnetic field by $45°$ in phase and is much smaller in magnitude:

$$\left|\frac{E_y}{B_x}\right| \approx \sqrt{\frac{\omega}{2\pi\sigma}} \ll 1. \tag{126.13}$$

(Recall that $\sigma \gg \omega$ in the frequency range being considered.)

Third, we note that since the tangential component of the electric field must be continuous, at the surface we must have (reverting to complex form, and omitting the $e^{-i\omega t}$ factor again)

$$\mathbf{E} \equiv \mathbf{E}_0 = -E_0 \hat{\mathbf{y}}. \tag{126.14}$$

This leads to a nonzero power flowing into the metal, as can be seen by calculating the time-averaged Poynting vector:

$$\overline{\mathbf{S}} = \tfrac{1}{4} \sigma \delta |\mathbf{E}_0|^2 \hat{\mathbf{z}}. \quad \text{(Gaussian and SI)} \tag{126.15}$$

This power is dissipated as ohmic heat in a thickness of the order of the skin depth. This is because $\mathbf{j} = \sigma \mathbf{E}$, and so, by eq. (126.14), there is a current flowing near the surface in the direction $\mathbf{B} \times \hat{\mathbf{n}}$, where $\hat{\mathbf{n}}$ is the outward normal from the conductor (in this case $-\hat{\mathbf{z}}$). It is often convenient to regard this current in terms of an equivalent or effective surface current

$$\mathbf{K}_s = \int_0^\infty \mathbf{j}(z) dz = -\frac{c}{4\pi} B_0 \hat{\mathbf{y}}. \tag{126.16}$$

Note that the conductivity has dropped out of the result, and that this current is of precisely the amount required to satisfy a gross-length-scale version of the jump condition on the tangential magnetic field. If we denote the field in the vacuum just outside the conductor by $\mathbf{B}_{\text{out}}$ and the field deep inside it by $\mathbf{B}_{\text{in deep}}$, we have

$$(\mathbf{B}_{\text{out}} - \mathbf{B}_{\text{in deep}}) \times \hat{\mathbf{n}} = -\frac{4\pi}{c} \mathbf{K}_s. \tag{126.17}$$

Since $\mathbf{B}_{\text{in deep}} = 0$, we get

$$\mathbf{K}_s = -\frac{c}{4\pi} \mathbf{B}_{\text{out}} \times \hat{\mathbf{n}} = -\frac{c}{4\pi} B_0 \hat{\mathbf{x}} \times (-\hat{\mathbf{z}}) = -\frac{c}{4\pi} B_0 \hat{\mathbf{y}}, \tag{126.18}$$

as before.

The effective surface current can be used to define a *surface impedance* Z_s in complete parallel with the bulk resistivity,

$$\mathbf{K}_s = \frac{1}{Z_s}\mathbf{E}_0. \quad \text{(Gaussian and SI)} \tag{126.19}$$

For the skin effect,

$$Z_s = (1-i)/\sigma\delta. \quad \text{(Gaussian and SI)} \tag{126.20}$$

The power dissipated per unit area of the surface is

$$\dot{Q} = \frac{1}{2}\int_0^\infty \mathbf{j}\cdot\mathbf{E}^* dz = \frac{1}{4}\sigma\delta|\mathbf{E}_0|^2. \tag{126.21}$$

This is in precise agreement with the power flowing in, eq. (126.15), as it should be. We can also write

$$\dot{Q} = \frac{1}{2}\text{Re}(\mathbf{K}_s\cdot\mathbf{E}_0^*) = \frac{1}{2}\text{Re}\frac{1}{Z_s}|\mathbf{E}_0|^2 = \frac{1}{2}\left(\frac{c}{4\pi}\right)^2 \text{Re}\, Z_s|\mathbf{B}_0|^2. \tag{126.22}$$

The last result is valid whenever a surface impedance can be defined. This concept proves of use when talking of dissipative losses in transmission lines, waveguides, etc.

Exercise 126.1 Consider a half-wave center-fed linear antenna as discussed in section 64, made of cylindrical Al rods of total length 1 m and diameter 0.5 cm. Find the resistance due to ohmic losses in the rods, and compare with the radiation resistance 73 ohms. Take $\rho = 0.3\,\mu$ohm cm for Al.

Let us finish the discussion of the above problem by noting two other points. First, what is the electric field in the region $z < 0$? This is given by solving the equation $\nabla\times\mathbf{E} = i(\omega/c)\mathbf{B}$ with the boundary condition $\mathbf{E}(z=0) = -E_0\hat{\mathbf{y}}$. The solution is

$$E_y = -E_0 + ik\,B_0 z, \tag{126.23}$$

but this is valid only as long the magnetic field it in turn induces is smaller than B_0, i.e., for $|z| < c/\omega$. This is because we have effectively ignored the $\dot{\mathbf{E}}$ term in the Ampere-Maxwell law. If the variable magnetic field is being produced by a solenoid, for example, then provided the dimensions of the solenoid are much smaller than the wavelength c/ω, the magnetic field is hardly affected within the solenoid. Far from it, of course, the fields must be such as to describe electromagnetic radiation.

The second point is that as $\omega\to 0$, the skin depth becomes very large. This might seem to contradict the fact that a static electric field cannot penetrate into a conductor, except that $|\mathbf{E}|\sim(\omega/\sigma)^{1/2}|\mathbf{B}|$, so that, in the limit, the magnitude of the electric field vanishes altogether.

Next, let us ask what happens if we consider a flat interface as above, but require the magnetic field to be normal to it. Suppose the conductor occupies the half-space $z\geq 0$, and suppose $\mathbf{B} = B_0\hat{\mathbf{z}}$ for $z < 0$. This is also the value of the field on the surface. In the conductor, $\mathbf{B}$ obeys the diffusion equation, and the only solution to this equation that equals $B_0\hat{z}$ at $z = 0$, and is physically well behaved as $z \to \infty$, is

$$\mathbf{B} = B_0\hat{z}e^{i\kappa z}. \tag{126.24}$$

But now, $\nabla \cdot \mathbf{B} = i\kappa B_0 e^{i\kappa z} \neq 0$. The only way out would appear to be to put $B_0 = 0$, and it would seem that it is not allowed to have a magnetic field perpendicular to the conductor. At first sight this conclusion seems absurd, since we can surely place a conductor inside a solenoid with a surface normal to the solenoid axis. We will analyze this situation in more detail shortly, but let us continue with our example and also allow a field in the xy plane in the conductor. If the magnitude of this field is $\sim B_\perp$ and it varies on a length scale a in the xy plane, the divergence condition implies that $B_\perp \sim (a/\delta) B_0$. If the length scale a is set by the dimensions or radius of curvature of the conductor, and $a \gg \delta$, then $B_\perp \gg B_0$, i.e., the field is largely parallel to the interface in the skin-depth layer. By continuity, it must be so at the surface and above it as well. This field configuration *can* be sustained by creating surface currents in the metal. The correct conclusion is that no matter what the shape of the conductor, surface currents will be induced in such a way that the magnetic field will in essence be locally parallel to the surface everywhere, and any normal component will be canceled.

127 Variable magnetic field: eddy currents and the skin effect in finite bodies*

To firm up the conclusions of the previous section, we now consider finite bodies and solve the problem for the cylinder and the sphere.

Cylinder in periodically varying external magnetic field: Let us consider a metal cylinder of length L and radius $a \ll L$ placed inside a larger solenoid with its axis parallel to that of the solenoid. (See fig. 19.2.) Let the solenoid be much longer than L, so that in the absence of the conductor, the field can be taken to be uniform over the entire region of interest. As usual, we take this as $\mathbf{B} = B_0\hat{z}e^{-i\omega t}$. We will consider the problem only at frequencies high enough that $a \gg \delta$, i.e., when the skin effect is strong. This is the most interesting situation. In this case, for the current and field along the sides of the cylinder, we can ignore the curvature of the cylinder and use our solution for the flat interface. The induced current is given by

$$\mathbf{j}(\mathbf{r}) \approx -\frac{c}{4\pi}\frac{(1-i)}{\delta}B_0 e^{-(1-i)(a-r)/\delta}\mathbf{e}_\varphi. \tag{127.1}$$

The magnetic field produced by this current sheet can be computed using the integral form of Ampere's law for points inside the bulk of the cylinder (i.e., far from the surface on the scale of δ), far from the ends. At these points it clearly equals $-B_0\hat{z}$ to very good approximation, and it cancels the external field.

However, the cancellation would appear to be partial near the ends. If we think of the induced current sheet as a smaller solenoid inside the larger one, then by the results of

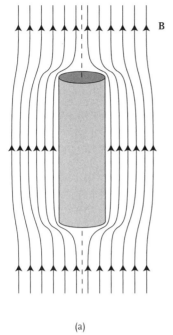

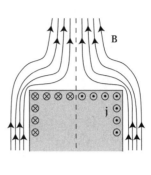

(a)   (b)

FIGURE 19.2. (a) A conducting cylinder in a periodically varying magnetic field. (b) Detail of fields and currents under the end cap.

exercise 23.5, we find that on the center of the upper end surface, say, the field produced by it is only $-\frac{1}{2}B_0\hat{\mathbf{z}}$. This leaves a net magnetic field of order $\frac{1}{2}B_0\hat{\mathbf{z}}$ near the upper end cap, which must necessarily penetrate a distance of order a into the cylinder along its length.

This state of affairs is not acceptable because it implies $\nabla^2\mathbf{B} \sim \mathbf{B}/a^2$ near the ends, which contradicts the diffusion equation. Hence, there must also be a current sheet flowing just under the upper surface that screens the field in this region too. Let us try to investigate this issue more closely. We take the upper surface to be the plane $z = 0$ and assume that the field near and above it can be written as

$$\mathbf{B}(\mathbf{r}) = b_0\hat{\mathbf{z}} - \frac{B_1}{a}(x\hat{\mathbf{x}} + y\hat{\mathbf{y}} - 2z\hat{\mathbf{z}}) \quad (z > 0, r < a). \tag{127.2}$$

We have taken the uniform, or dipolar, part of the field to be b_0, different from B_0. The B_1 term is a locally quadrupolar field compatible with the symmetry, and B_1 has the dimensions of magnetic field.

On the $z = 0$ surface, $\mathbf{B} = b_0\hat{\mathbf{z}} - (B_1/a)(x\hat{\mathbf{x}} + y\hat{\mathbf{y}})$. We can write down the solution to the diffusion equation immediately:

$$\mathbf{B} = \left[b_0\hat{\mathbf{z}} - \frac{B_1}{a}(x\hat{\mathbf{x}} + y\hat{\mathbf{y}})\right]e^{-i\kappa z}, \quad (z < 0). \tag{127.3}$$

Then,

$$\nabla \cdot \mathbf{B} = -\left[i\kappa b_0 + 2\frac{B_1}{a}\right]e^{-i\kappa z}, \tag{127.4}$$

which vanishes if

$$b_0 = \frac{2i\,B_1}{\kappa a} = O\left(\frac{\delta}{a}B_1\right). \tag{127.5}$$

Therefore, $b_0 \ll B_1$ and, to good approximation, the field outside the conductor is tangential to it, as deduced earlier. If we neglect the b_0 term in eq. (127.2), then, since the field must be of order B_0 at a distance of order a from the end cap, we conclude that $B_1 = O(B_0)$.

We can now find the eddy current under the end cap. Switching to cylindrical coordinates and letting r be the distance from the axis, we have

$$\mathbf{j} = \frac{c}{4\pi}\nabla \times \mathbf{B} = \frac{i\kappa c}{4\pi a}B_1 r e^{-i\kappa z}\mathbf{e}_\varphi. \tag{127.6}$$

What is the field produced by this current distribution in the $z > 0$ region? For points on the z axis, this is relatively easy to find. Using the basic formula for the field due to an arbitrary distribution of current, we get

$$\mathbf{B}_{\mathrm{ind}} = 2\pi\frac{i\kappa}{4\pi a}B_1\hat{\mathbf{z}}\int r'dr'\int_{-\infty}^{0} dz'\frac{r'^2}{[r'^2+(z-z')^2]^{3/2}}e^{-i\kappa z'}. \tag{127.7}$$

Since $e^{-i\kappa z'} \ll 1$ for $|z'| \gg \delta$, we can put $z' = 0$ in the term in the denominator. The z' integral is then easy. For the r' integral, let us assume that the simplified form for the current is valid all the way to a. Then

$$\begin{aligned}
\mathbf{B}_{\mathrm{ind}} &= -\hat{\mathbf{z}}\frac{B_1}{2a}\int_0^a\frac{r'^3}{(r'^2+z^2)^{3/2}}dr' \\
&= -\hat{\mathbf{z}}\frac{B_1}{2a}\left[\frac{(a^2+2z^2)}{(a^2+z^2)^{1/2}}-2z\right].
\end{aligned} \tag{127.8}$$

For $z \gg a$, this equals $-(B_1/8)(a/z)^3\hat{\mathbf{z}}$. For $z \approx 0$, we get $-(B_1/2)\hat{\mathbf{z}}$. With $B_1 = O(B_0)$, this is of exactly the order needed to cancel the field that was left over after we took the currents on the side of the cylinder into account.

Sphere in periodically varying external magnetic field: As our final example, we consider a sphere of radius a placed in an external field $\mathbf{B}_0 e^{-i\omega t}$. This problem can be solved in closed form and allows us to verify in detail the conclusions we have reached qualitatively.[3]

As usual, we take $\mathbf{B}_0 \| \hat{\mathbf{z}}$. We wish to find the eddy currents, and the fields inside and outside the sphere. It is simplest to first find the current, $\mathbf{j}$, which equals $(c/4\pi)\nabla \times \mathbf{B}$. If we take the curl of eq. (125.6), we see that $\mathbf{j}$ also obeys the same diffusion equation, i.e.,

$$\nabla^2\mathbf{j}+\kappa^2\mathbf{j}=0, \tag{127.9}$$

where κ is defined by eq. (126.5). This equation, along with $\nabla \cdot \mathbf{j} = 0$, determines $\mathbf{j}$ fully. Note that κ is of the order of an inverse skin depth, since we may write $\kappa = (1+i)/\delta$.

---

[3] A solution is also given in Landau and Lifshitz (1960), vol. 8, sec. 45. It was first solved by H. Lamb, *Phil. Trans. Roy. Soc. (London)* **174**, 519 (1883).

It follows by symmetry that $\mathbf{j}$ can flow only along the $\mathbf{e}_\varphi$ direction. This demand, along with $\nabla \cdot \mathbf{j} = 0$, can be met by writing

$$\mathbf{j} = (\nabla f) \times \mathbf{B}_0 = (\nabla \times f \mathbf{B}_0), \tag{127.10}$$

where f depends on z and r. It now follows that

$$(\nabla^2 + \kappa^2)\mathbf{j} = \nabla \times [(\nabla^2 + \kappa^2) f]\mathbf{B}_0, \tag{127.11}$$

which will vanish provided $(\nabla^2 + \kappa^2) f = 0.$[4] We now argue that f can depend only on r, since there is no natural parameter in this problem that sets the scale of variation with z relative to that with r.[5] The equation is then easy to solve, and the solution, which is regular at the origin, is

$$f(r) = b \frac{\sin \kappa r}{\kappa r}, \tag{127.12}$$

where b is a constant. From this result we get $\mathbf{j} = f'(\hat{\mathbf{r}} \times \mathbf{B}_0)$. In particular, $\mathbf{j} \approx -b\kappa^2(\mathbf{r} \times \mathbf{B}_0)/3$ near the origin, which is well behaved, as it should be.

Next, let us find the total magnetic moment $\mathcal{M}$ of the sphere due to the induced currents. We have

$$\mathcal{M} = \frac{1}{2c} \int (\mathbf{r} \times \mathbf{j}) d^3 x. \tag{127.13}$$

The integration is straightforward to perform, and we get $\mathcal{M} = V\alpha \mathbf{B}_0$, where $V = 4\pi a^3/3$ is the volume of the sphere, and

$$\alpha = -\frac{b}{c}\left[\frac{\sin \kappa a}{\kappa a} + 3\frac{\cos \kappa a}{(\kappa a)^2} - 3\frac{\sin \kappa a}{(\kappa a)^3}\right]. \tag{127.14}$$

The quantity α has the meaning of the magnetic polarizability per unit volume.

It is now easy to find the magnetic field $\mathbf{B}_{\text{out}}$ outside the sphere. In addition to the external field $\mathbf{B}_0$, we have the field due to the induced currents. The latter can only be the field of a dipole with moment $\mathcal{M}$. Hence,

$$\mathbf{B}_{\text{out}} = \mathbf{B}_0 + \frac{4\pi a^3}{3}\alpha\left(\frac{3(\mathbf{B}_0 \cdot \hat{\mathbf{r}})\hat{\mathbf{r}} - \mathbf{B}_0}{r^3}\right). \tag{127.15}$$

Lastly, let us find the magnetic field inside the sphere, $\mathbf{B}_{\text{in}}$. By combining Faraday's law with Ohm's law for $\mathbf{E}$, we obtain

$$\mathbf{B}_{\text{in}} = -\frac{ic}{\omega\sigma}\nabla \times \mathbf{j}. \tag{127.16}$$

[4] In fact, we need require only $(\nabla^2 + \kappa^2) f = A$, where A is a constant. The solution to this inhomogeneous equation differs from the solution to the homogeneous one by the constant A/κ^2. This last constant makes no contribution to ∇f, and thus to the current, so we may as well set $A = 0$.

[5] To put it another way, a dependence on z in addition to that on r would imply that the current distribution had multipoles higher than the dipole. While the orientation of the dipole is clearly fixed by $\mathbf{B}_0$, there is no equivalent tensor that is linear in $\mathbf{B}_0$ to fix the structure of any higher multipole.

The differentiation is straightforward but lengthy. A page or two of algebra gives

$$\mathbf{B}_{\text{in}} = \frac{4\pi}{c\kappa^2}\left(\frac{f'}{r} + \kappa^2 f\right)\mathbf{B}_0 - \frac{4\pi}{c\kappa^2}\left(\frac{3f'}{r} + \kappa^2 r\right)(\mathbf{B}_0 \cdot \hat{\mathbf{r}})\hat{\mathbf{r}} \tag{127.17}$$

$$= \frac{4\pi b}{c}\left[\frac{\sin\kappa r}{\kappa r} + \frac{\cos\kappa r}{(\kappa r)^2} - \frac{\sin\kappa r}{(\kappa r)^3}\right]\mathbf{B}_0 - \frac{4\pi b}{c}\left[\frac{\sin\kappa r}{\kappa r} + 3\frac{\cos\kappa r}{(\kappa r)^2} - 3\frac{\sin\kappa r}{(\kappa r)^3}\right](\mathbf{B}_0 \cdot \hat{\mathbf{r}})\hat{\mathbf{r}}. \tag{127.18}$$

To find b we enforce continuity at the interface. Equating the coefficients of the $(\mathbf{B}_0 \cdot \hat{\mathbf{r}})\hat{\mathbf{r}}$ terms in eqs. (127.15) and (127.18), we get

$$\alpha = -\frac{b}{c}\left[\frac{\sin\kappa a}{\kappa a} + 3\frac{\cos\kappa a}{(\kappa a)^2} - 3\frac{\sin\kappa a}{(\kappa a)^3}\right], \tag{127.19}$$

which is precisely the same as before. Similarly, equating the coefficients of $\mathbf{B}_0$, we get

$$1 - \frac{4\pi}{3}\alpha = \frac{4\pi b}{c}\left[\frac{\sin\kappa a}{\kappa a} + \frac{\cos\kappa a}{(\kappa a)^2} - \frac{\sin\kappa a}{(\kappa a)^3}\right]. \tag{127.20}$$

Solving eqs. (127.19) and (127.20), we obtain

$$b = \frac{3c}{8\pi}\frac{\kappa a}{\sin\kappa a}, \tag{127.21}$$

$$\alpha = -\frac{3}{8\pi}\left[1 + 3\frac{\cot\kappa a}{\kappa a} - \frac{3}{(\kappa a)^2}\right]. \tag{127.22}$$

This completes the solution.

Let us analyze our results in the case where the skin effect is strong, i.e., $\delta \ll a$. The current and fields are significant only in a thin layer near the surface. These are easily found by noting that in this region,

$$\sin\kappa r \approx \frac{i}{2}e^{(1-i)r/\delta}, \quad \cos\kappa r \approx \frac{1}{2}e^{(1-i)r/\delta} \quad (a - r \ll a). \tag{127.23}$$

The current is given by

$$\mathbf{j}(\mathbf{r}) = \frac{3c}{8\pi\delta}(1-i)e^{-(1-i)(a-r)/\delta}(\hat{\mathbf{r}} \times \mathbf{B}_0). \tag{127.24}$$

(The electric field is given by the same form, since $\mathbf{E} = \mathbf{j}/\sigma$.) This is confined to a layer of width $O(\delta)$, but it becomes smaller in magnitude as we approach the poles because of the $\hat{\mathbf{r}} \times \mathbf{B}_0$ factor. The field right at the poles is nevertheless small, of order $(\delta/a)\mathbf{B}_0$, because the large currents near the equator and the rest of the sphere still manage to cancel most of the applied field. A short calculation shows that in the surface layer, the normal and tangential components of the field are given by

$$\mathbf{B}_n = -\frac{3\delta}{4a}(1+i)(\mathbf{B}_0 \cdot \hat{\mathbf{r}})\hat{\mathbf{r}}e^{-(1-i)(a-r)/\delta}, \tag{127.25}$$

$$\mathbf{B}_t = \frac{3}{2}\hat{\mathbf{r}} \times (\mathbf{B}_0 \times \hat{\mathbf{r}})e^{-(1-i)(a-r)/\delta}. \tag{127.26}$$

Thus, the normal component of the field is everywhere of order $(\delta/a)\mathbf{B}_0$, as asserted earlier.[6] On the equator, the field is $(3/2)\mathbf{B}_0$. This is larger than $\mathbf{B}_0$ because of the additional field due to the induced dipole $\mathcal{M}$. This dipole moment is found by first calculating the polarizabilty from eq. (127.22). We get $\alpha \approx -3/8\pi$, so

$$\mathcal{M} = -\frac{1}{2}a^3\mathbf{B}_0. \tag{127.27}$$

Since this is negative, the response is strongly diamagnetic.

128 Variable electric field, electrostatic regime

To discuss the response of a conductor to a low-frequency *electric* field one must distinguish two frequency regimes, based on whether the skin depth, δ, is larger or smaller than the size of the conductor, a. If $\delta \gg a$, the response is essentially electrostatic and rather different from that to a variable magnetic field. Surface charges are induced that shield the interior of the conductor from the electric field, but because these charges are now time dependent, the shielding is imperfect, and we get currents in the bulk of the conductor. These currents then act as sources of quasistatic magnetic fields. This completes the solution to the problem. The key difference with the magnetic case is that while the magnetic polarizability is diamagnetic or negative—weakly if $\delta \gg a$, and strongly if $\delta \ll a$—the electric polarizability is always strongly positive.

To illustrate these ideas, we again take a conducting sphere of radius a in an external field:

$$\mathbf{E} = E_0 e^{-i\omega t}\hat{\mathbf{z}} \quad (r \gg a). \tag{128.1}$$

We also take $\mu = 1$. Then, with $\hat{\mathbf{r}}$ denoting the direction on the sphere, let the surface charge density be $\Sigma(\hat{\mathbf{r}}, t) = \Sigma(\hat{\mathbf{r}})e^{-i\omega t}$. The key point now is that since we are in the limit $\delta \gg a$, the diffusion equation for $\mathbf{E}$ reduces to Laplace's equation. Thus, we may seek $\mathbf{E}$ as $-\nabla\phi$, where ϕ is the potential of the external field, plus the instantaneous potential due to the surface charges:

$$\phi(\mathbf{r}) = -E_0 z + 4\pi a^2 \int \frac{\Sigma(\hat{\mathbf{r}}')}{|\mathbf{r} - a\hat{\mathbf{r}}'|}d\Omega'. \tag{128.2}$$

(We have omitted the $e^{-i\omega t}$ factors.) We now argue that the induced field can only be that of a dipole, as there is no tensor linear in $\mathbf{E}_0$ available to fix the orientation of any multipole. Since this symmetry argument also holds in the static problem, $\Sigma(\hat{\mathbf{r}})$ must take the same functional form,

$$\Sigma(\hat{\mathbf{r}}) = \Sigma_0 \hat{\mathbf{r}} \cdot \hat{\mathbf{z}}, \tag{128.3}$$

---

[6] The electric field is even smaller, of order $(\delta/\lambda)\mathbf{B}_0$.

where Σ_0 is a constant. The integral over directions is known from the static case, and we get

$$\phi(\mathbf{r}) = -E_0 z + \frac{4\pi}{3}\Sigma_0 \times \begin{cases} z, & r < a, \\ (a^3/r^3)z, & r > a. \end{cases} \tag{128.4}$$

The electric field is now easily found. Imposing the boundary conditions (125.11) and (125.12) at $r = a$, we get

$$\Sigma_0 = \frac{3E_0}{4\pi - 3i\omega/\sigma} \approx \frac{3E_0}{4\pi}\left(1 + \frac{3i\omega}{4\pi\sigma}\right), \tag{128.5}$$

up to linear order in (ω/σ). To the same order, the field inside the sphere is given by

$$\mathbf{E} = -\frac{3i\omega}{4\pi}E_0\hat{\mathbf{z}}, \tag{128.6}$$

and outside by

$$\mathbf{E} = E_0\hat{\mathbf{z}} + E_0\left(1 + \frac{3i\omega}{4\pi\sigma}\right)\frac{a^3}{r^3}\left(3(\hat{\mathbf{r}}\cdot\hat{\mathbf{z}})\hat{\mathbf{r}} - \hat{\mathbf{z}}\right). \tag{128.7}$$

In fact, it is not legitimate to keep the higher-order terms in the expansion for Σ_0 or these fields. Essentially, since $\delta \sim \omega^{-1/2}$ and $\delta \gg a$, we are developing the fields, currents, and charge densities as series in ω.

It remains to find the magnetic field. This can be done using Ampere's law with $\mathbf{j} = \sigma\mathbf{E}$ for $r < a$, and Laplace's equation for $r > a$, with the requirement of continuity at $r = a$. The calculation is straightforward, and we just give the answer:

$$\mathbf{B} = -\frac{i\omega E_0}{c} \times \begin{cases} \frac{3}{2}\hat{\mathbf{z}} \times \mathbf{r} & r < a, \\ \left(\frac{a^3}{r^3} + \frac{1}{2}\right)\hat{\mathbf{z}} \times \mathbf{r}, & r > a. \end{cases} \tag{128.8}$$

Exercise 128.1 Fill in the missing steps in the above calculation.

Exercise 128.2 Consider a charge q moving parallel to the surface of a large conducting slab, at a velocity $v \ll c$ and height h above the surface. (Take the conductor as occupying the half-space $z < 0$.) Find the fields, currents, and charges, as corrections about the instantaneous values, to first order in v.[7]

Exercise 128.3 Repeat the previous exercise for a charge moving normal to the conductor.

---

[7] For the solution, see T. H. Boyer, *Phys. Rev. A* **9**, 68 (1974).

129 Variable electric field, skin-effect regime

We now consider a conductor in an electric field at higher frequencies, such that $\delta \ll a$ (but with $\omega \ll c/a$ and $\omega \ll \sigma$ still). As in the magnetic case, let us first examine the special cases of an external field parallel and perpendicular to the surface of a semi-infinite conductor occupying the half-space $z \leq 0$. Let us first take the external field parallel to the surface:

$$\mathbf{E} = E_0 e^{-i\omega t}\hat{\mathbf{x}}. \tag{129.1}$$

The continuity of the tangential component of $\mathbf{E}$ implies $\mathbf{E} = E_0\hat{\mathbf{x}}$ at $z = 0-$, and the diffusion equation yields

$$\mathbf{E} = E_0 e^{\kappa z}\hat{\mathbf{x}}. \tag{129.2}$$

The equation for $\nabla \times \mathbf{H}$ reduces to

$$-\frac{\partial H_y}{\partial z} = \frac{4\pi}{c}\sigma E_0 e^{\kappa z}, \tag{129.3}$$

which has the solution

$$H_y = \frac{4\pi\sigma}{c\kappa} E_0 e^{\kappa z}. \tag{129.4}$$

Right at the surface, the ratio H_y/E_0 is of order $(\sigma/\omega)^{1/2}$, as in the case of a variable magnetic field. This is a tenable solution.

Next, let us take the external field to be normal to the surface:

$$\mathbf{E} = E_0 e^{-i\omega t}\hat{\mathbf{z}}. \tag{129.5}$$

To first order in ω/σ, the surface conditions (125.11) and (125.12) now imply

$$E_{z-} \approx -\frac{i\omega}{4\pi\sigma} E_0, \quad \Sigma \approx \frac{E_0}{4\pi}. \tag{129.6}$$

But there is now no way to satisfy both the diffusion equation for $\mathbf{E}$ and the $\nabla \cdot \mathbf{E} = 0$ requirement. The resolution is that to the degree of approximation being carried out, we should put $\mathbf{E}_{in} = 0$, since a very small transverse component in $\mathbf{E}_{out}$ could give rise to a transverse part to $\mathbf{E}_{in}$ that would be much greater than the normal part. The correct conclusion to draw from this toy calculation is that for any finite conductor, the induced charges and currents will arrange themselves in such a way as to make $\mathbf{E}_{in}$ and $\mathbf{j}$ essentially parallel to the surface over most of the conductor.

To see these points in detail, we return to the problem of the sphere. The solution exactly parallels the magnetic case[8] and can be done for any δ/a, but because it is lengthy, we just give the answer. For the exterior field we have

$$\mathbf{E}_{out} \approx E_0\hat{\mathbf{z}} + E_0\frac{a^3}{r^3}\left(3(\hat{\mathbf{z}}\cdot\hat{\mathbf{r}})\hat{\mathbf{r}} - \hat{\mathbf{z}}\right), \tag{129.7}$$

---

[8] For details, see A. Garg, *Am. J. Phys.* **76**, 615 (2008).

omitting a tangential term of relative order δ/a, while for the internal field we have

$$\mathbf{E}_{\text{in}} \approx \frac{1}{\zeta_s}\left[\frac{\sin \kappa r}{\kappa r} + \frac{\cos \kappa r}{(\kappa r)^2} - \frac{\sin \kappa r}{(\kappa r)^3}\right]\mathbf{E}_0 - \frac{1}{\zeta_s}\left[\frac{\sin \kappa r}{\kappa r} + 3\frac{\cos \kappa r}{(\kappa r)^2} - 3\frac{\sin \kappa r}{(\kappa r)^3}\right](\mathbf{E}_0 \cdot \hat{\mathbf{r}})\hat{\mathbf{r}},$$

(129.8)

with

$$\zeta_s = \frac{2}{3}\zeta\left(\frac{\sin \kappa a}{(\kappa a)^3} - \frac{\cos \kappa a}{(\kappa a)^2}\right).$$

(129.9)

For $\delta \ll a$, in the skin layer near $r = a$, eq. (129.8) yields

$$\mathbf{E}_{\text{in},\,n} = -\frac{3i\omega}{4\pi\sigma}(\mathbf{E}_0 \cdot \hat{\mathbf{r}})\hat{\mathbf{r}}\, e^{-(1-i)(a-r)/\delta},$$

(129.10)

$$\mathbf{E}_{\text{out},\,t} = -\frac{3\omega a}{8\pi\sigma\delta}(1+i)\big(\hat{\mathbf{r}} \times (\mathbf{E}_0 \times \hat{\mathbf{r}})\big)e^{-(1-i)(a-r)/\delta}.$$

(129.11)

Thus, $\mathbf{E}_{\text{out}}$ is essentially the same as in the low-frequency case and is almost perfectly normal to the surface everywhere ($E_t/E_n \sim \kappa a/\zeta$). The surface charge density is still given by eq. (128.5). $\mathbf{E}_{\text{in}}$, on the other hand, is very different. It is mainly tangential to the surface, except in a very narrow region of area $\sim\delta^2$ near the poles. The same holds for the current, since $\mathbf{j} = \sigma\mathbf{E}$. This verifies our earlier remarks about the field inside the conductor.

The magnetic field can now be found from Faraday's law. In the skin layer, we get

$$\mathbf{B}_{\text{in}} = -\frac{3i}{2}\frac{\omega a}{c}(\mathbf{E}_0 \times \hat{\mathbf{r}})e^{-(1-i)(a-r)/\delta}.$$

(129.12)

Outside the conductor, we get

$$\mathbf{B}_{\text{out}} = -\frac{i\omega a}{c}\left(\frac{a^3}{r^3} + \frac{1}{2}\right)(\mathbf{E}_0 \times \mathbf{r}).$$

(129.13)

We can understand the essential aspects of the solution as follows. The surface charge is fixed by eq. (125.11) to be $\sim E_0/4\pi$, independent of ω. Thus, a charge of order $E_0 a^2$ must flow from the lower hemisphere to the upper every half-period π/ω, giving a net current $\sim E_0 a^2\omega$. This current is confined to a layer of thickness δ near the surface, so the current density is of magnitude $E_0 a^2\omega/a\delta \sim E_0\omega\kappa a$. Since the magnetic field must vanish deep in the sphere, it follows that $|\mathbf{B}_{\text{out}}| \sim (4\pi/c)j\delta \sim E_0(\omega a/c)$. This argument applies to conductors of any shape and gives the order of magnitude of all the fields in the problem.

It is useful to summarize the magnitudes of the various fields near the surface of the conductor, both inside and outside. We tabulate them below in order of decreasing strength, expressing the ratios entirely in terms of the lengths λ, a, and δ:

$E_{\text{out},n}$	B	$E_{\text{out},t}$	$E_{\text{in},t}$	$E_{\text{in},n}$	
E_0	$\dfrac{a}{\lambda}E_0$	$\dfrac{a\delta}{\lambda^2}E_0$	$\dfrac{a\delta}{\lambda^2}E_0$	$\dfrac{\delta^2}{\lambda^2}E_0$	(129.14)

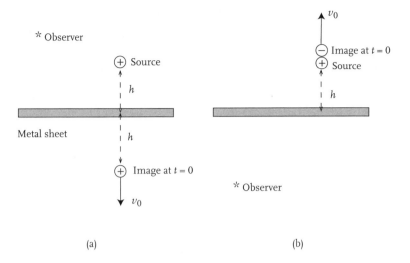

FIGURE 19.3. Maxwell's receding image method. A monopole is created above a thin metal sheet at $t = 0$. Depending on whether the field is desired (a) above or (b) below the sheet, it is obtained by superposing the field of the source monopole with that of a receding image monopole as shown.

130 Eddy currents in thin sheets, Maxwell's receding image construction, and maglev*

As the last problem in our study of eddy currents, let us consider a conducting sheet that is thin compared to the skin depth, and let the externally imposed magnetic field be created by a time-dependent current or magnetic moment distribution. The simplest problem is that of a flat metallic sheet, with permeability $\mu = 1$, further idealized to be of infinite extent, and effectively of zero thickness. All such problems can be solved if we can solve the problem of a magnetic monopole of strength q_m that materializes abruptly at a height h above the sheet at time $t = 0$. A general distribution can then be described by creating and destroying monopoles at appropriate points and times.

The monopole problem was considered by Maxwell in 1872, who also gave an ingenious solution in terms of moving image monopoles. Specifically, he stated that for a sheet of thickness d, the magnetic field below the sheet is that of the original monopole plus another of strength $-q_m$ that is created at the same location as the original monopole at $t = 0$ and thereafter moves vertically away from the sheet at the velocity

$$v_0 = \frac{c^2}{2\pi\sigma d}. \tag{130.1}$$

In the region above the sheet, the field is that of the original monopole plus an image monopole of strength q_m that is created at the mirror image point below the sheet at $t = 0$ and that then recedes vertically away from the sheet at the velocity v_0 (see fig. 19.3). To prove the correctness of Maxwell's solution, we will find the magnetic scalar potential for the field, which obeys Laplace's equation. The main tool is the uniqueness theorem for the solution of Laplace's equation with given boundary conditions.[9] We take the sheet to lie in the xy plane and let the monopole be at the point $(0, 0, h)$. We will use the variables **R** and **r** to distinguish functions defined over all space and over just the sheet.

The first step is to find the boundary conditions on **B** at the metal sheet. Let the surface current, which is the eddy current density $\mathbf{j}(\mathbf{r})$ integrated across the thickness of the sheet, be $\mathbf{K}(\mathbf{r}, t)$. Ampere's law gives the discontinuity in the magnetic field to be

$$\Delta \mathbf{B} = -\frac{4\pi}{c}\hat{\mathbf{z}} \times \mathbf{K}, \tag{130.2}$$

where $\Delta \mathbf{B} = \mathbf{B}(z = 0+) - \mathbf{B}(z = 0-)$. We now note that if we divide **B** into a direct part $\mathbf{B}^{\text{dir}}$ due to the source monopole and a part $\mathbf{B}^{\text{ind}}$ due to the induced currents, the discontinuity is entirely due to the induced part. Further, the tangential components of the induced field must be equal and opposite just above and below the sheet. Hence, we may write

$$\mathbf{B}_t^{\text{ind}}(\mathbf{r}) = \mp\frac{2\pi}{c}\hat{\mathbf{z}} \times \mathbf{K}(\mathbf{r}), \quad (z = 0\pm). \tag{130.3}$$

Next, let us consider the normal component of $\mathbf{B}^{\text{ind}}$. This must be continuous, but its normal derivative must be equal and opposite on the two sides of the sheet. (To see this, note that $\nabla \cdot \mathbf{B} = 0$ implies $\partial_z B_z = -\nabla_t \cdot \mathbf{B}_t$. Since the induced part of $\mathbf{B}_t$ is equal and opposite for $z = 0\pm$, so are its derivatives with respect to x and y. The oddness of $\partial_z B_z^{\text{ind}}$ then follows.) Taking the z component of the diffusion equation obeyed by the field inside the sheet, we get

$$\nabla^2 B_z = \frac{4\pi\sigma}{c^2}\frac{\partial B_z}{\partial t}. \tag{130.4}$$

Now the transverse derivatives in $\nabla^2 B_z$ are of order $h^{-2}B_z$, while the normal derivatives are much bigger, of order $d^{-2}B_z$. Hence, $\nabla^2 B_z \approx \partial^2 B_z/\partial z^2$. Integrating the equation across the sheet and recalling that the direct part $\mathbf{B}^{\text{dir}}$ is continuous, on the left-hand side we get

$$\Delta\left[\partial_z B_z^{\text{ind}}\right] = \pm 2\left[\partial_z B_z^{\text{ind}}\right]_{z=0\pm}, \tag{130.5}$$

while on the right we get

$$\frac{4\pi\sigma d}{c^2}\frac{\partial B_z}{\partial t}\bigg|_{z=0\pm}. \tag{130.6}$$

Rearranging terms, the boundary condition on the normal component is, therefore,

$$\left(\frac{\partial}{\partial t} \mp v_0\frac{\partial}{\partial z}\right)B_z^{\text{ind}} = -\frac{\partial}{\partial t}B_z^{\text{dir}}, \quad (z = 0\pm). \tag{130.7}$$

---

[9] An alternative proof that is less formal is given in appendix H. The proof here is essentially that of James Jeans (1966), sec. 538. See also W. M. Saslow, *Am. J. Phys.* **60**, 693 (1992), who adds many interesting physical considerations and also discusses several related problems.

Let us now consider the scalar potential ϕ, in terms of which $\mathbf{B} = -\nabla\phi$, and let us separate it into direct and induced parts ϕ^{dir} and ϕ^{ind}. The direct part is known:

$$\phi^{\text{dir}}(\mathbf{R}, t) = \frac{q_m}{|\mathbf{R} - h\hat{\mathbf{z}}|}\theta(t); \tag{130.8}$$

the objective clearly is to find ϕ^{ind}. To do this, we first rewrite eq. (130.7) in terms of ϕ:

$$\frac{\partial^2}{\partial t \partial z}\left(\phi^{\text{ind}} + \phi^{\text{dir}}\right) = \pm v_0 \frac{\partial^2}{\partial z^2}\phi^{\text{ind}} \quad (z = 0\pm). \tag{130.9}$$

Next, we note that the right-hand side is clearly finite for all $\mathbf{r}$ and t. Hence, the *rate of change* of the z component of the total field is finite at all t. In particular, since $B_z = 0$ on the surfaces of the sheet at $t = 0-$, we must also have $B_z = 0$ at $t = 0+$. In other words,

$$\frac{\partial\phi^{\text{ind}}}{\partial z} = -\frac{\partial\phi^{\text{dir}}}{\partial z}, \quad (z = 0\pm, t = 0+). \tag{130.10}$$

This equation determines ϕ^{ind} at $t = 0+$, not only on the surfaces $z = 0\pm$ but for all $\mathbf{R}$. For, ϕ^{ind} obeys Laplace's equation everywhere and vanishes as $z \to \pm\infty$. Equation (130.10) is a boundary condition on its normal derivative. The choice (upper and lower signs to be applied in the regions with $z > 0$ and $z < 0$, respectively)

$$\phi^{\text{ind}}(\mathbf{R}, t = 0+) = \pm\frac{q_m}{|\mathbf{R} \pm h\hat{\mathbf{z}}|} \tag{130.11}$$

meets all these criteria and, so, by the uniqueness theorem, must be the solution. Note that the solution for $z > 0$ is that of an image monopole in the region $z < 0$ (and vice versa) and is, therefore, nonsingular everywhere.

To find the induced potential for all t, we note that the ϕ^{dir} term in eq. (130.9) vanishes for all $t > 0$. We therefore consider the functions

$$F_{\pm}(\mathbf{R}) = \left(\frac{\partial^2}{\partial t \partial z} \mp v_0 \frac{\partial^2}{\partial z^2}\right)\phi^{\text{ind}}, \tag{130.12}$$

defined in the half-spaces $z > 0$ and $z < 0$, respectively. These functions vanish on the sheets $z = 0\pm$. They also vanish as $|z| \to \infty$. Since they obey Laplace's equation, by the uniqueness theorem, they must vanish everywhere in their respective half-spaces.[10] Hence, if we take F_- first, and integrate the equation $F_- = 0$ with respect to z from $-\infty$ to z, we obtain

$$\frac{\partial\phi^{\text{ind}}}{\partial t} = -v_0 \frac{\partial\phi^{\text{ind}}}{\partial z} \quad (z < 0). \tag{130.13}$$

This is a first-order differential equation with the general solution

$$\phi^{\text{ind}}(\mathbf{R}, t) = f(x, y, z - v_0 t) \quad (z < 0), \tag{130.14}$$

where f is an arbitrary function. In the same way, by considering F_+, we see that for $z > 0$, ϕ^{ind} is a general function of $z + v_0 t$. But, since we know ϕ^{ind} at $t = 0+$, these functions

---

[10] Or note that if they did not, they would have a maximum or minimum somewhere, which is impossible.

are completely determined, and the solution is

$$\phi^{\text{ind}}(\mathbf{R}, t) = \pm \frac{q_m}{|\mathbf{R} \pm (h + v_0 t)\hat{\mathbf{z}}|}. \tag{130.15}$$

This is precisely the potential due to the image monopoles.

Although the problem is at this point solved, it is also useful to find the induced current distribution. By symmetry $\mathbf{K}(\mathbf{r}) = K_\varphi(r)\mathbf{e}_\varphi$. Recall from section 23 that the scalar potential is discontinuous across a surface spanning a current loop. If the current is I, the discontinuity is $4\pi I/c$. In our problem, the current through the annular loop lying between radii r and $r + dr$ is $K_\varphi dr$. The total discontinuity in $\phi^{\text{ind}}(\mathbf{r})$ is the sum of the discontinuities due to all the infinitesimally wide loops extending from r to infinity. Equating this sum to the discontinuity found from the explicit solution (130.15), we obtain

$$\Delta\phi^{\text{ind}} = \frac{4\pi}{c} \int_r^\infty K_\varphi(r', t)\,dr' = 2\frac{q_m}{\sqrt{r^2 + (h + v_0 t)^2}}. \tag{130.16}$$

Differentiation with respect to r yields

$$K_\varphi(r, t) = \frac{q_m c}{2\pi} \frac{r}{\left(r^2 + (h + v_0 t)^2\right)^{3/2}}. \tag{130.17}$$

The above results find application in the idea of magnetic levitation, or maglev for short. In its simplest form, we imagine a train moving over a track consisting of a thin metal sheet. The train carries current-carrying coils in its floor, as shown in fig. 19.4. The eddy currents induced in the track will then tend to oppose the magnetic field of the coils and thereby lift the train off the track. Since such a train would not suffer the frictional forces between the wheels and the track of a conventional one, it is hoped that it could form the basis of high-speed (400 km/h or more) rail transportation. The calculation of the forces on an extended moving current loop is an involved problem (although see exercise 130.2). A simpler problem is to consider a monopole q_m moving at some velocity v parallel to the sheet. Since we want only the induced field above the sheet, only the images below the sheet are of interest. To find these, it is simplest to think of the monopole as moving in jerks, abruptly jumping a distance $v\Delta t$ in a short time Δt. Each jerk may be conceived of as the creation of an antimonopole at the previous location and a monopole at the new one. At any given instant, the source monopole will see the freshly created image and the previously created train of monopoles and antimonopoles, all receding at the velocity v_0. By taking the limit as $\Delta t \to 0$, we find that the image source will consist of a line of *dipoles*, with dipole moment per unit length given by

$$\boldsymbol{\mu}(\mathbf{r}) = q_m \frac{v}{(v^2 + v_0^2)^{1/2}}\hat{\mathbf{x}}. \tag{130.18}$$

If the source monopole is taken to be at $(0, 0, h)$ at a given instant, the dipole line starts at the point $(0, 0, -h)$ and trails off to infinity in the negative x and z directions, at an angle $\tan^{-1}(v_0/v)$ to the x axis. As the source monopole moves, the dipole string drags along, rather like a runner holding a ribbon in a headwind.

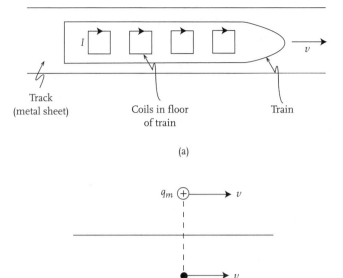

(a)

(b)

FIGURE 19.4. (a) Schematic of a maglev train in an overhead view. (b) Simpler toy problem of a monopole moving above a metal sheet. The image now consists of a moving dipole ribbon at the angle shown.

The magnetic potential of the dipole string is given by

$$\phi(\mathbf{R}) = \int_C \frac{\mu(\mathbf{R}') \cdot (\mathbf{R} - \mathbf{R}')}{|\mathbf{R} - \mathbf{R}'|^3} dl', \tag{130.19}$$

where the line integral is taken over the string. We parametrize the string as

$$\mathbf{R}' = (-v\tau, 0, -(h + v_0\tau)). \tag{130.20}$$

Then $\mu dl' = q_m(v/v_0)d\tau$, and we get

$$\phi(x, y, z) = q_m v \int_0^\infty d\tau \frac{x + v\tau}{[(x + v\tau)^2 + y^2 + (z + h + v_0\tau)^2]^{3/2}}$$

$$= \frac{q_m v}{V\tilde{R} + \mathbf{V} \cdot \tilde{\mathbf{R}}} \left(\frac{x}{\tilde{R}} + \frac{v}{V} \right), \tag{130.21}$$

where we have introduced the vectors

$$\mathbf{V} = (v, 0, v_0), \quad \tilde{\mathbf{R}} = (x, y, z + h). \tag{130.22}$$

Note that these are merely shorthand to achieve a compact writing of the answer. In particular, $\mathbf{V}$ is not the velocity of anything.

The force on the source is $-q_m \nabla \phi$, evaluated at $(0, 0, h)$. The components along the z and $-x$ directions are known as the *lift* and *drag* for obvious reasons. When the calculation is done, we find

$$F_L = \frac{q_m^2}{4h^2} \left(1 - \frac{v_0}{\sqrt{v_0^2 + v^2}} \right), \tag{130.23}$$

$$F_D = \frac{q_m^2}{4h^2} \frac{v_0}{v} \left(1 - \frac{v_0}{\sqrt{v_0^2 + v^2}} \right). \tag{130.24}$$

There are several interesting points about these results. The drag is proportional to v at low velocities, passes through a maximum at $v \sim v_0$, and falls as $1/v$ for $v \gg v_0$. The lift is proportional to v^2 for low v, rises monotonically with v, and becomes independent of v for $v \gg v_0$.

The low-speed behavior can be understood as follows. The magnetic field at a fixed point in the region above which the source is passing varies at a frequency $\omega \sim v/h$ and may be taken in the first approximation to be the field of the source itself. The electric field induced in the sheet is then given by solving Faraday's equation, $\nabla \times \mathbf{E} \sim \dot{B}$, and hence $\mathbf{E} \sim \omega$. The ohmic losses in the sheet are given by σE^2 and hence vary as ω^2 or v^2. To keep the source moving at speed v, an equal amount of power must be supplied, and since this power is $F_D v$, $F_D \sim v$. The lift force is due to the induced magnetic field, and since we can obtain this iteratively by solving the Ampere-Maxwell equation, $\nabla \times \mathbf{B} \sim \dot{\mathbf{E}}$, this contains an additional power of ω or v.[11] Hence, $F_L \sim v^2$. This is a very general argument that applies to any moving source.

The high-speed behavior, on the other hand, is best understood in terms of the Maxwell construction itself. Because the speed is so high, at any given instant, we can regard the monopole as having materialized at its location at that instant out of thin air. This is just the problem solved at the beginning of this section. The instantaneous response of the sheet is to produce an identical monopole at the image point, which repels the source monopole with a force $q_m^2/4h^2$. This is exactly what eq. (130.23) gives. This idea also applies to any moving source and shows that the lift tends to a value independent of the velocity for $v \gg v_0$.

The most remarkable feature of eqs. (130.23) and (130.24), however, is the drag-to-lift ratio,

$$\frac{F_D}{F_L} = \frac{v_0}{v}, \tag{130.25}$$

which holds for *all* velocities. This is, in fact, a general result for any distribution of sources that moves rigidly with a velocity v. To discuss this would take us too far afield, and we refer the reader to Saslow. Once it is accepted, however, we see that the drag at high speeds must vary as $1/v$ in all cases.

---

[11] We can also see these ω dependences in the exactly solved problem of the sphere in section 127. The induced magnetic field is proportional to the polarizability α. In the limit $\delta \gg a$, we find that the out-of-phase part of α, which is responsible for the dissipation, is proportional to $(a/\delta)^2$, i.e., to ω, while the in-phase part is proportional to $(a/\delta)^4$, i.e., to ω^2.

Exercise 130.1 Find the torque on an arbitrarily oriented magnetic dipole moving at a speed $v \gg v_0$. Find out if this torque ever vanishes, and discuss the stability of the equilibrium with respect to reorientation.

Exercise 130.2 Find the lift on a square current loop of side a, carrying a current I, at a height $h \ll a$, if it is moving at a speed $v \gg v_0$ above a conducting sheet.

Solution: The lift is approximately that due to an oppositely circulating identical current loop at a distance h below the sheet. Since $h \ll a$, we may ignore the corners and use the formula for the force per unit length between two parallel wires. This gives $F_L = 4I^2a/c^2h$.

Let us conclude this section by calculating a few representative numbers. For a 1-cm-thick track of Al, which has a conductivity of 3.2×10^{17} sec^{-1} at 20°C, the Maxwell recession velocity, v_0, is 16 km/h. At the speeds at which maglev would start to get useful, about 400 km/h, one would appear to be well in the high-velocity regime of our analysis, with a small drag-to-lift ratio. On the other hand, if we assume that the current coils are a distance of 10 cm above the track, then with $v = 400$ km/h, $\omega \sim v/h = 1100$ sec^{-1}. At this frequency, the skin depth $\delta = 0.63$ cm, which is less than the assumed 1-cm thickness of the track, so the assumption $d \ll \delta$ breaks down. In the opposite limit, $\delta \ll d$, the lift is unaffected at high speeds, but the drag increases, falling only as $v^{-1/2}$.

131 Motion of extended conductors in magnetic fields*

Let us now consider a conductor moving in a magnetic field. Let a volume element d^3x of the conductor have a velocity $\mathbf{v}$, and let the macroscopically averaged fields in this element be $\mathbf{E}$ and $\mathbf{B}$. The force on any mobile charges will then be given by $q(\mathbf{E} + \mathbf{v} \times \mathbf{B}/c)$, and since it is the net force that accelerates the charges, Ohm's law should be modified to read

$$\mathbf{j} = \sigma \left(\mathbf{E} + \frac{\mathbf{v}}{c} \times \mathbf{B} \right). \tag{131.1}$$

The same conclusion can be reached by transforming to an inertial frame instantaneously comoving with the volume element. The field in this frame is given by eq. (38.3), $\mathbf{E}' = \mathbf{E} + \mathbf{v} \times \mathbf{B}/c$, to first order in v/c. In its own rest frame, $\mathbf{j}' = \sigma\mathbf{E}'$. Since $\mathbf{j} = \mathbf{j}'$ under a Galilean transformation, we again obtain eq. (131.1).

If the conductor is in the form of a wire loop, and the external field is purely magnetic, the above conclusions reproduce those of section 28. This is easily seen by integrating eq. (131.1) around the loop and noting that the current flowing in the loop is given by $\mathcal{E}/R$, where R is the total loop resistance.

Exercise 131.1 Derive the flux rule when both the magnetic field and the shape of the wire loop are changing with time.

More interesting effects arise in steady-state situations, for example, if the body moves at a uniform velocity in a uniform magnetic field, or if it rotates uniformly in a static magnetic field. In these cases, no currents can flow in steady state, and we must conclude that an electric field exists inside the body. Such a field can arise only if there is a redistribution of the mobile charges so that local charge neutrality is disturbed.

As an example, let us consider a rotating metal sphere of radius a in a uniform magnetic field parallel to the rotation axis. Let this axis be denoted by z, and let the angular velocity of the sphere be ω. Since $\mathbf{j} = 0$ in steady state, the interior electric field is given by

$$\mathbf{E} = -\frac{\mathbf{v}}{c} \times B\hat{\mathbf{z}}. \tag{131.2}$$

Since $\mathbf{v} = \omega \times \mathbf{r}$, we get

$$\mathbf{E} = -\frac{\omega B}{c}(\hat{\mathbf{z}} \times \mathbf{r}) \times \hat{\mathbf{z}} = -\frac{\omega B}{c}(x\hat{\mathbf{x}} + y\hat{\mathbf{y}}) \quad (r < a). \tag{131.3}$$

We can now use Gauss's law to find the interior charge density:

$$\rho = \frac{1}{4\pi}\nabla \cdot \mathbf{E} = -\frac{\omega B}{2\pi c}. \tag{131.4}$$

Since the sphere is neutral as a whole, there must be a positive surface charge on it. This is as it should be, since the Lorentz force pushes the electrons toward the axis of rotation.

To find the surface charge density and the exterior electric field, we proceed as follows. Let the contributions of the volume and surface charges to the interior field (131.3) be denoted by $\mathbf{E}_v$ and $\mathbf{E}_s$. Since the volume charge density is uniform, we know that

$$\mathbf{E}_v = -\frac{2\omega B}{3c}r\hat{\mathbf{r}} = -\frac{2\omega B}{3c}(x\hat{\mathbf{x}} + y\hat{\mathbf{y}} + z\hat{\mathbf{z}}). \tag{131.5}$$

Hence,

$$\mathbf{E}_s = \mathbf{E} - \mathbf{E}_v = -\frac{\omega B}{3c}(x\hat{\mathbf{x}} + y\hat{\mathbf{y}} - 2z\hat{\mathbf{z}}). \tag{131.6}$$

We now observe that

$$\mathbf{E}_s = -\frac{\omega B}{6c}\nabla(x^2 + y^2 - 2z^2), \tag{131.7}$$

which shows the potential from which this field derives. This potential is a spherical harmonic of order $\ell = 2$, and the field is a quadrupole type. More importantly, the potential outside must have the same spherical harmonic structure and must die as $r^{-(\ell+1)}$ or r^{-3}. Further, it must be continuous at $r = a$. Thus,

$$\mathbf{E}_s = -\frac{\omega B}{6c}\nabla\frac{a^5}{r^5}(x^2 + y^2 - 2z^2) \quad (r > a). \tag{131.8}$$

The corresponding surface charge density can be found from the discontinuity in the normal component of the field:

$$\Sigma_s = \frac{1}{4\pi}\,\hat{\mathbf{r}} \cdot \mathbf{E}_s\Big|_{r=a-}^{r=a+} = \frac{a\omega B}{12\pi c}(1 - 3\cos^2\theta). \tag{131.9}$$

We have written the final answer for Σ_s in terms of the usual polar angle.

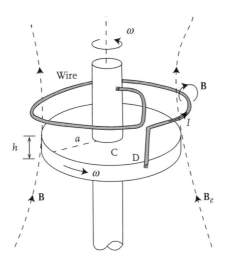

FIGURE 19.5. A dynamo. The rotating disk and its axle must be metallic and make contact with the wire by means of sliding contacts or brushes.

However, Σ_s cannot be the total surface charge density, because it integrates to zero and so does not cancel the negative volume charge. The problem is easily fixed by noting that a uniform surface charge does not produce a field inside the sphere and so may be added on without affecting our previous interior decomposition of $\mathbf{E}$ into $\mathbf{E}_v$ and $\mathbf{E}_s$. Since the net negative charge is $-2\omega Ba^3/3c$, the required surface charge is

$$\bar{\Sigma}_s = \frac{a\omega B}{6\pi c},\tag{131.10}$$

and the total surface charge is

$$\Sigma = \bar{\Sigma}_s + \Sigma_s = \frac{a\omega B}{4\pi c}\sin^2\theta.\tag{131.11}$$

Finally, by Gauss's law and symmetry, the contribution of the volume charge density and the surface density $\bar{\Sigma}_s$ to the electric field outside the sphere vanishes. Thus, the exterior field is entirely given by $\mathbf{E}_s$.

132 The dynamo*

Consider a conducting disk that is rotating at an angular velocity ω as shown in fig. 19.5, in a magnetic field $\mathbf{B}_e$ parallel to the axis of the disk. As found in the previous section, positive and negative charges will accumulate on the rim and axle of the disk, respectively. If a wire is connected between the rim and the axle by means of sliding contacts or brushes, then a current will flow through the wire as long as the rotation is maintained. The current in the wire will produce a magnetic field $\mathbf{B}$. If this wire is looped around the axle (possibly more than once) as shown, the field $\mathbf{B}$ will reinforce the external field $\mathbf{B}_e$. This raises the question of whether the external field $\mathbf{B}_e$ can be dispensed with entirely.

In other words, can the magnetic field producing the inductively generated emf be produced by a current driven by that very emf? To answer this question, let us assume that the conductivity of the wire and the axle is much greater than that of the disk and that the magnetic field **B** is cylindrically symmetric. Then, since $\mathbf{E} = -(\mathbf{v} \times \mathbf{B})/c$ and $\mathbf{v} = \omega \times \mathbf{r}$, the emf is given by

$$\mathcal{E} = -\int_C^D \mathbf{E} \cdot d\ell = \frac{\omega}{c} \int_C^D Br \, dr = \frac{\omega}{2\pi c} \Phi, \tag{132.1}$$

where Φ is the flux through a loop at the rim of the disk. If we denote the mutual inductance of this loop with the circuit comprising the wire and the radius CD by L_{wd}, and the resistance and self-inductance of the circuit by R and L, then Kirchhoff's voltage law for the circuit reads

$$L\frac{dI}{dt} + RI = \frac{\omega L_{wd}}{2\pi} I. \tag{132.2}$$

This equation has an exponentially growing solution (representing a self-sustaining magnetic field) if

$$\omega > 2\pi R / L_{wd}. \tag{132.3}$$

It is useful to express this condition in more geometrical terms. If we take the radius of the disk as a and estimate $R \sim 1/\sigma a$ and $L_{wd} \sim 2\pi a/c^2$, the condition becomes

$$\omega \gtrsim c^2 / \sigma a^2. \tag{132.4}$$

Such a device is known as a *dynamo*, or a homopolar or self-excited dc generator. Note that there is an asymmetry in the device. Once the wire is connected, it is necessary for the disk to spin in the sense shown. To see this, let us reintroduce the external field $\mathbf{B}_e$. If we now reverse the rotation, **B** reverses, and it cancels $\mathbf{B}_e$ instead of reinforcing it. The device is an anti-dynamo! The direction of $\mathbf{B}_e$ does not matter, however. If we reverse $\mathbf{B}_e$ (keeping ω as drawn in the figure), the emf reverses, so do I and **B**, and **B** ends up reinforcing $\mathbf{B}_e$ once again. This shows that if the dynamo condition (132.3) is met, the magnetic field is generated via an instability. This instability is seeded by fluctuations or stray fields, so **B** can have either direction. Eventually, of course, the growth of **B** will be arrested by nonlinearities not included in our simple model.

The dynamo instability is believed to underlie the earth's magnetic field. Seismic evidence indicates that the outer core of the earth has a radius of 3500 km and is made of molten iron with a conductivity of $\sim 3 \times 10^3$ mho/cm. The decay time (125.13) is then of order 2×10^5 years, far shorter than the $\sim 10^9$ years for which the geomagnetic field is inferred to have existed on the basis of frozen-in magnetization in various geological rocks. On the other hand, the estimate (132.4) leads to easily attainable velocities, so a dynamo could explain how the geomagnetic field is maintained. The situation is much more complicated, however, since the roles of the conducting disk and the connecting wire must both be played by the molten iron. If we feed Ohm's modified law (131.1) in

Ampere's law and take the curl, instead of eq. (125.6), we obtain

$$\nabla^2 \mathbf{H} = \frac{4\pi\sigma\mu}{c^2}\left(\frac{\partial \mathbf{H}}{\partial t} - \mathbf{v} \times (\nabla \times \mathbf{H})\right),$$

(132.5)

where $\mathbf{v}(\mathbf{r})$ is the velocity field of the conducting fluid. Because of the Lorentz force, this velocity field is in turn influenced by $\mathbf{H}(\mathbf{r})$, and so is determined by a modified Navier-Stokes equation in which the magnetic pressure and stresses are included. The coupled problem of determining both $\mathbf{v}(\mathbf{r})$ and $\mathbf{H}(\mathbf{r})$ is very complicated and still the subject of active research. Even the far simpler problem of solving eq. (132.5) for a *prescribed* $\mathbf{v}(\mathbf{r})$ is beset with several pitfalls in the form of various no-go theorems. The first and most important of these is due to Cowling and states that cylindrically symmetric $\mathbf{v}(\mathbf{r})$ and $\mathbf{H}(\mathbf{r})$ fields cannot exist. (Note that our disk dynamo breaks cylindrical symmetry in the way the wire is attached to the disk and axle.) Thus, simple, highly symmetric solutions are ruled out from the start. Nevertheless, fully three-dimensional solutions do exist, as demonstrated by Ponomarenko in an explicit solution for a model in which the flow is restricted to a circular cylinder inside an infinite sea of conducting fluid. There is even some evidence that the full set of equations is capable of explaining the magnetic field reversals known to have happened in the earth's history at intervals of $\sim 10^6$ years.[12]

---

[12] For more on these fascinating topics, we refer the reader to the very readable text by Fitzpatrick, *Introduction to Plasma Physics: A Graduate Level Course* (2006). The book is also available on http://farside.ph. utexas.edu/teaching/plasma/380.pdf. Sections 5.12 and 5.13 discuss Cowling's theorem and the Ponomarenko dynamo and are well within the reach of readers of the present text.

20 | Electromagnetic waves in insulators

133 General properties of EM waves in media

We now consider the propagation of electromagnetic waves in matter, limiting ourselves to linear isotropic media, in which it is permissible to write the constitutive equations as

$$\mathbf{D}_\omega = \epsilon(\omega)\mathbf{E}_\omega, \quad \mathbf{B}_\omega = \mu(\omega)\mathbf{H}_\omega. \tag{133.1}$$

We consider waves of a single frequency, ω, in a homogeneous medium, and assume that there are no charge or current sources in the region of interest. Then the macroscopic Faraday and Ampere-Maxwell laws can be written as

$$\nabla \times \mathbf{E}_\omega = \frac{i\omega}{c}\mu(\omega)\mathbf{H}_\omega, \tag{133.2}$$

$$\nabla \times \mathbf{H}_\omega = -\frac{i\omega}{c}\epsilon(\omega)\mathbf{E}_\omega. \tag{133.3}$$

For plane wave solutions of the type

$$\mathbf{E}_\omega, \mathbf{H}_\omega \sim e^{i\mathbf{k}\cdot\mathbf{r}}, \tag{133.4}$$

these equations reduce to

$$\mathbf{k} \times \mathbf{E}_\omega = \frac{\omega}{c}\mu(\omega)\mathbf{H}_\omega, \tag{133.5}$$

$$\mathbf{k} \times \mathbf{H}_\omega = -\frac{\omega}{c}\epsilon(\omega)\mathbf{E}_\omega. \tag{133.6}$$

These two equations automatically imply that

$$\mathbf{k}\cdot\mathbf{E}_\omega = \mathbf{k}\cdot\mathbf{H}_\omega = 0. \tag{133.7}$$

Hence, if we take the cross product of $\mathbf{k}$ with eq. (133.5) and use eq. (133.6) for $\mathbf{k} \times \mathbf{H}_\omega$, we get

$$k^2\mathbf{E}_\omega = \frac{\omega^2}{c^2}\mu(\omega)\epsilon(\omega)\mathbf{E}_\omega. \tag{133.8}$$

To get a nonzero solution for $\mathbf{E}_\omega$, we must have

$$k^2 = \frac{\omega^2}{c^2} \mu(\omega)\epsilon(\omega). \tag{133.9}$$

This is the general *dispersion relation* for EM waves in a linear medium.

Let us now consider the spatial dependence of the fields. In general, for a real frequency ω, the wave vector $\mathbf{k}$ will not be real. If we write its real and imaginary parts as $\mathbf{k}'$ and $\mathbf{k}''$,

$$\mathbf{k} = \mathbf{k}' + i\mathbf{k}'', \tag{133.10}$$

then eq. (133.9) becomes

$$k'^2 - k''^2 + 2i\mathbf{k}' \cdot \mathbf{k}'' = \frac{\omega^2}{c^2} \mu(\omega)\epsilon(\omega). \tag{133.11}$$

This equation does not determine k', k'', and the angle between $\mathbf{k}'$ and $\mathbf{k}''$ uniquely, since we get only two conditions by equating the real and imaginary parts. Secondly, the wave is not truly plane, since the surfaces of constant phase are perpendicular to $\mathbf{k}'$, while the surfaces of constant amplitude are perpendicular to $\mathbf{k}''$. The formulas $\mathbf{k} \cdot \mathbf{E}_\omega = \mathbf{k} \cdot \mathbf{H}_\omega = 0$ do not have a simple geometrical interpretation either, since $\mathbf{k}$, $\mathbf{E}_\omega$, and $\mathbf{H}_\omega$ are all complex. Such solutions are known as *inhomogeneous plane waves*.

Far simpler are homogeneous plane waves, in which $\mathbf{k}'$ and $\mathbf{k}''$ are parallel. If we take this mutual direction to be $\hat{\mathbf{x}}$, say, then we can write $\mathbf{k} = k\hat{\mathbf{x}}$, where k is the complex wave vector

$$k = \frac{\omega}{c} \left(\mu(\omega)\epsilon(\omega) \right)^{1/2}. \tag{133.12}$$

It is customary to separate k into its real and imaginary parts and write

$$k = \frac{\omega}{c}(n + i\kappa). \tag{133.13}$$

The quantities n and κ are known as the *refractive index* and *attenuation coefficient*, respectively, and the combination

$$\tilde{n} = n + i\kappa \tag{133.14}$$

is known as the *complex refractive index*.[1]

Exercise 133.1 Consider a nonmagnetic dielectric, for which $\mu \simeq 1$. Show that in terms of the real and imaginary parts of the dielectric function, $\epsilon = \epsilon' + i\epsilon''$,

$$n = \frac{1}{\sqrt{2}} \left[\epsilon' + (\epsilon'^2 + \epsilon''^2)^{1/2} \right]^{1/2}, \tag{133.15}$$

$$\kappa = \frac{1}{\sqrt{2}} \left[-\epsilon' + (\epsilon'^2 + \epsilon''^2)^{1/2} \right]^{1/2}. \tag{133.16}$$

---

[1] It is also common to see n stand for both the complex refractive index and its real part.

134 Wave propagation velocities

Let us consider a quasimonochromatic plane wave traveling through the medium:

$$\mathbf{E}(\mathbf{r}, t) = \int_{\omega \approx \omega_0} \frac{d\omega}{2\pi} \mathbf{E}_\omega e^{i(\mathbf{k}\cdot\mathbf{r} - \omega t)}. \tag{134.1}$$

An analogous expression can be written for $\mathbf{H}(\mathbf{r}, t)$. We assume that absorption can be neglected in the frequency window being considered, so $\epsilon(\omega) \approx \epsilon'(\omega)$, and likewise for $\mu(\omega)$. In that case, the dispersion relation between wave vector and frequency is

$$k = \frac{\omega}{c}(\epsilon\mu)^{1/2}. \tag{134.2}$$

We wish to investigate the velocity of the wave.

As the reader probably knows, the presence of dispersion (by which we mean that the ratio k/ω varies with frequency) implies that one cannot assign a frequency-independent velocity to the wave. One must distinguish between the *phase velocity*, v_p, at which the phase fronts advance, and which therefore enters into all calculations of interference, and the group velocity, v_g, at which the envelope of a wave packet composed of a narrow band of frequencies advances. These velocities are given by

$$v_p = \frac{\omega(k)}{k}, \tag{134.3}$$

$$v_g = \frac{d\omega(k)}{dk}. \tag{134.4}$$

It should be stressed that the notion of a group velocity makes sense only when the width of the band of frequencies making up the wave packet is much smaller than the characteristic frequency scale on which the dispersion occurs. Otherwise, the wave packet undergoes significant distortion, and one cannot say that it is moving with a definite velocity. Secondly, the damping must also be small. These are precisely the conditions we are considering.

It is also possible to ask, however, for the velocity, v_E, at which energy is transported by the wave. If $\overline{S}$ is the time-averaged Poynting vector or energy flux density, and $\overline{U}$ is the time-averaged energy density in the wave, then, clearly, we should have

$$\overline{S} = v_E \overline{U}, \tag{134.5}$$

Since $\overline{S}$ and $\overline{U}$ can be calculated using the formulas in section 122, v_E can be found directly.

Let us find $\overline{S}$ first. Appealing to the rule for time averages of bilinear quantities, we have for a general wave,

$$\overline{\mathbf{S}} = \frac{c}{16\pi}(\mathbf{E}_\omega \times \mathbf{H}_\omega^* + \mathbf{E}_\omega^* \times \mathbf{H}_\omega)e^{-2\mathbf{k}''\cdot\mathbf{r}}. \tag{134.6}$$

We now use the relations (133.5) and (133.6) and set k'', ϵ'', and μ'' all to zero (since absorption is being neglected). This yields $\overline{\mathbf{S}} = \overline{S}\hat{\mathbf{k}}$, with

$$\overline{S} \simeq \frac{c}{16\pi}\left(\sqrt{\frac{\epsilon}{\mu}}\mathbf{E}_\omega \cdot \mathbf{E}_\omega^* + \sqrt{\frac{\mu}{\epsilon}}\mathbf{H}_\omega \cdot \mathbf{H}_\omega^*\right). \tag{134.7}$$

Again remembering that $\epsilon'' = \mu'' \approx 0$, squaring either of eqs. (133.5) and (133.6), and using the dispersion relation, we obtain

$$\epsilon \mathbf{E}_\omega \cdot \mathbf{E}_\omega^* = \mu \mathbf{H}_\omega \cdot \mathbf{H}_\omega^*. \tag{134.8}$$

The two contributions to $\overline{S}$ are then equal.

To find $\overline{U}$, we use eq. (122.20):

$$\overline{U} = \frac{1}{16\pi} \left(\frac{d(\omega\epsilon)}{d\omega} \mathbf{E}_\omega \cdot \mathbf{E}_\omega^* + \frac{d(\omega\mu)}{d\omega} \mathbf{H}_\omega \cdot \mathbf{H}_\omega^* \right). \tag{134.9}$$

Equations (134.5)–(134.9) now imply

$$\frac{1}{v_E} = \frac{1}{2c} \sqrt{\frac{\mu}{\epsilon}} \left(\frac{d(\omega\epsilon)}{d\omega} + \frac{\epsilon}{\mu} \frac{d(\omega\mu)}{d\omega} \right)$$

$$= \frac{1}{c} \left(\sqrt{\mu\epsilon} + \frac{1}{2}\sqrt{\frac{\mu}{\epsilon}} \omega \frac{d\epsilon}{d\omega} + \frac{1}{2}\sqrt{\frac{\epsilon}{\mu}} \omega \frac{d\mu}{d\omega} \right). \tag{134.10}$$

The three terms can be combined into a single derivative, yielding

$$\frac{1}{v_E} = \frac{1}{c} \frac{d}{d\omega} (\omega\sqrt{\epsilon\mu}). \tag{134.11}$$

But this is exactly the formula for $1/v_g$ also, since $k = \omega\sqrt{\epsilon\mu}/c$. Hence,

$$v_E = v_g, \tag{134.12}$$

i.e., the energy also travels with the group velocity.

This happy state of affairs prevails only when the dispersion and absorption are both weak. For a counterexample that makes this point clear, consider the Lorentz model (123.5) for the dielectric constant, and assume that $\mu(\omega) = 1$. For quick understanding, let us first neglect absorption (set $\gamma = 0$). The refractive index is then given by $n(\omega) = \sqrt{\epsilon(\omega)}$, where

$$\epsilon(\omega) = \frac{\omega^2 - (\omega_0^2 + \omega_b^2)}{\omega^2 - \omega_0^2} \tag{134.13}$$

with

$$\omega_b^2 = 4\pi n_b e^2/m. \tag{134.14}$$

Since $\epsilon(\omega) < 0$ in the range $\omega_0 < \omega < \omega_b$, $n(\omega)$ is not even defined in this range, and neither are the group or the phase velocities. The situation is not really improved if we do not neglect absorption and write $c/v_p = n$, $c/v_g = d(\omega n)/d\omega$, where n is the real part of the refractive index. Now v_p is defined for all ω, but v_g is larger than c in a narrow band of frequencies around ω_0, and even becomes negative for some ω. Clearly, the concept of group velocity is inadequate to describe wave propagation in regions of anomalous dispersion.

We can, however, calculate v_E explicitly for this model, following Brillouin. The idea is to find $\overline{S}$ and $\overline{U}$, and then $v_E = \overline{S}/\overline{U}$. For $\overline{S}$, we have

$$\overline{S} = \frac{c}{8\pi} |\mathbf{E}_\omega|^2 n(\omega), \tag{134.15}$$

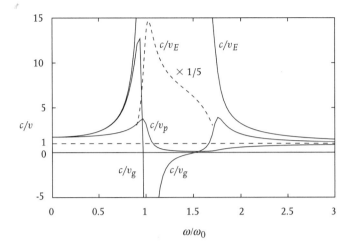

FIGURE 20.1. The inverse of the phase, group, and energy transport velocities as a function of frequency for the Lorentz model, with $\omega_b^2 = 2\omega_0^2$, $\gamma = 0.1\omega_0$. Note the unphysical values of v_g (greater than c or negative), and the very small values of v_E in the region of anomalous dispersion. The dashed curve shows the central peak in c/v_E reduced by a factor of 5.

where $n(\omega)$ is given by the formulas

$$n(\omega) = \frac{1}{\sqrt{2}} \left[\epsilon' + (\epsilon'^2 + \epsilon''^2)^{1/2} \right]^{1/2}, \tag{134.16}$$

$$\epsilon'(\omega) = 1 + \frac{\omega_b^2(\omega_0^2 - \omega^2)}{(\omega_0^2 - \omega^2)^2 + \omega^2\gamma^2}, \tag{134.17}$$

$$\epsilon''(\omega) = \frac{\omega\gamma\omega_b^2}{(\omega_0^2 - \omega^2)^2 + \omega^2\gamma^2}. \tag{134.18}$$

We can find the energy density $\overline{U}$ by adding the time-averaged energy of the oscillators to the vacuum energy:

$$\overline{U} = \frac{1}{16\pi}\left(|\mathbf{E}_\omega|^2 + |\mathbf{H}_\omega|^2\right) + \frac{1}{2}n_b m\left(\overline{\mathbf{v}^2} + \omega_0^2\overline{\mathbf{r}^2}\right). \tag{134.19}$$

Here, $\mathbf{v} = \dot{\mathbf{r}}$. The averages for the oscillator are easily computed, and combining the results with that for the magnetic energy, $|\mathbf{H}_\omega|^2 = |\epsilon(\omega)|\,|\mathbf{E}_\omega|^2$, we get

$$\overline{U} = \frac{|\mathbf{E}_\omega|^2}{16\pi}\left[1 + |\epsilon(\omega)| + \frac{(\omega_0^2 + \omega^2)\omega_b^2}{(\omega_0^2 - \omega^2)^2 + \omega^2\gamma^2}\right]. \tag{134.20}$$

Thus,

$$\frac{c}{v_E} = \frac{1}{n(\omega)}\left[1 + \frac{1}{2}\frac{\omega_b^2}{[(\omega_0^2 - \omega^2)^2 + \omega^2\gamma^2]^{1/2}} + \frac{1}{2}\frac{(\omega_0^2 + \omega^2)\omega_b^2}{(\omega_0^2 - \omega^2)^2 + \omega^2\gamma^2}\right]. \tag{134.21}$$

In fig. 20.1 we plot the inverse of the various velocities, essentially the transit times for a fixed distance. As can be seen, $v_E \leq c$ for all frequencies. It is essentially equal to

v_g far away from the resonance, but it drops sharply in its vicinity. Thus, this velocity appears to be sensibly defined for all frequencies and may be a better characterization of the speed of wave propagation. It must be kept in mind, however, that in regions of strong absorption and large $|d\epsilon/d\omega|$, the entire notion of wave propagation is much more qualified. The wave loses amplitude as it propagates, and the wave form is rapidly distorted, so a characterization by a single number or velocity is necessarily of limited value. These issues were investigated in great detail by Sommerfeld and Brillouin in back-to-back papers in *Annalen der Physik* in 1914.[2] They examine the propagation of a wave form with an initially sharp front and identify a "signal velocity," v_s, which characterizes the arrival of the dominant contribution to the wave form in a steepest descent analysis. They find that

$$v_E \leq v_s \leq c \tag{134.22}$$

for all ω. In the region of anomalous dispersion, v_E and v_s are significantly different, but far from this region, both velocities approach v_g. The definition of v_s is mathematical in character, however, and not based on a physical principle, so its significance is again limited. Further, our general remarks above on wave propagation when the dispersion is anomalous should be kept in mind.

One general point, however, can be stated unequivocally. If the initial wave form has a sharp front, i.e., the amplitude ahead of a certain point ($x > 0$, say) is strictly zero prior to a certain time ($t < 0$, say), then the wave form continues to have a sharp front as it evolves. This front can never travel at a speed greater than c. This physically obvious fact is confirmed by the mathematical analysis of Sommerfeld and Brillouin, and is almost tautological, since it is built into the properties of $\epsilon(\omega)$ that follow from causality, namely, the analyticity in the upper half of the frequency plane. Claims of superluminal transmission appear from time to time despite the fact that they violate this theorem, but so far, no such claim has withstood careful scrutiny.

Exercise 134.1 Consider a steady monochromatic plane wave propagating through a medium. In this case, the internal energy cannot increase with time, and the average over many cycles of the energy inflow, $-\nabla \cdot \overline{\mathbf{S}}$, must yield the dissipation, $\dot{Q}$. Use this fact to rederive eq. (122.21) for the heat evolved.

135 Reflection and refraction at a flat interface (general case)

We now consider what happens when a homogeneous plane EM wave at a single frequency is incident from medium 1 at a flat interface with medium 2 (see fig. 20.2). Both media are assumed to be homogeneous. As might be expected, the wave will be partially reflected and partially transmitted (or refracted). We will denote the material quantities μ and ϵ pertaining to the two media by suffixes 1 and 2, and quantities pertaining to

---

[2] The original papers are translated into English and reprinted in Brillouin (1960).

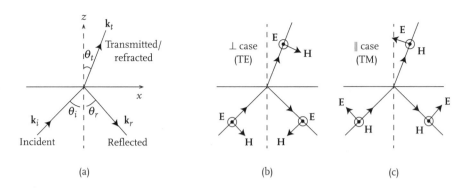

FIGURE 20.2. Reflection and refraction at an interface. (a) Plane of incidence showing the directions of the incident, reflected, and transmitted wave vectors, and the associated angles. (b) Instantaneous directions of **E** and **H** at the interface for the ⊥ or TE case, in which the light is polarized perpendicular to the plane of incidence. (c) Same as (b) for the ∥ or TM case.

the incident, reflected, and transmitted waves by suffixes i, r, and t, respectively. We will begin by allowing for attenuation in both media, but we will then take medium 1 to be transparent (negligible attenuation). The case where both media are transparent is special enough to deserve its own section (the next one). Let us choose coordinates as shown in the figure, i.e., take the interface to be the xy plane, and the z axis to point along the normal from medium 1 to medium 2. Further, we take the wave vector $\mathbf{k}_i$ of the incident wave to be in the xz plane. This can be done even when k_i is complex, as the unit vector $\hat{\mathbf{k}}_i = \mathbf{k}_i / k_i$ is real. With these conventions, the fields can be written down without further ado:

$$
\mathbf{E} = \begin{cases} \mathbf{E}_i e^{i\mathbf{k}_i \cdot \mathbf{r}} + \mathbf{E}_r e^{i\mathbf{k}_r \cdot \mathbf{r}}, & z \le 0, \\ \mathbf{E}_t e^{i\mathbf{k}_t \cdot \mathbf{r}}, & z \ge 0 \end{cases},
\tag{135.1}
$$

$$
\mathbf{H} = \begin{cases} \mathbf{H}_i e^{i\mathbf{k}_i \cdot \mathbf{r}} + \mathbf{H}_r e^{i\mathbf{k}_r \cdot \mathbf{r}}, & z \le 0, \\ \mathbf{H}_t e^{i\mathbf{k}_t \cdot \mathbf{r}}, & z \ge 0 \end{cases}.
\tag{135.2}
$$

As usual, we have omitted the $e^{-i\omega t}$ factor and the suffix ω on the complex field amplitudes. The amplitude relations (133.5) and (133.6) imply that these various amplitudes are connected by

$$
\mathbf{k}_{i,r} \times \mathbf{E}_{i,r} = \frac{\omega}{c}\mu_1 \mathbf{H}_{i,r}, \quad \mathbf{k}_t \times \mathbf{E}_t = \frac{\omega}{c}\mu_2 \mathbf{H}_t,
\tag{135.3}
$$

$$
\mathbf{k}_{i,r} \times \mathbf{H}_{i,r} = -\frac{\omega}{c}\epsilon_1 \mathbf{E}_{i,r}, \quad \mathbf{k}_t \times \mathbf{H}_t = -\frac{\omega}{c}\epsilon_2 \mathbf{E}_t.
\tag{135.4}
$$

Finally, the wave vectors in each medium obey the appropriate dispersion relation:

$$
k_{i,r} = k_1 \equiv \frac{\omega}{c}(\mu_1\epsilon_1)^{1/2},
\tag{135.5}
$$

$$
k_t = k_2 \equiv \frac{\omega}{c}(\mu_2\epsilon_2)^{1/2}.
\tag{135.6}
$$

We must now impose the conditions on continuity of the various field components. It is clear that because these conditions are the same for every point in the xy plane, all wave vectors must have the same projections onto this plane. That is, we must have

$$k_{tx} = k_{rx} = k_{ix}, \quad k_{ty} = k_{ry} = 0. \tag{135.7}$$

Thus, all three vectors lie in the one plane. This plane, which may be defined by the vectors $\mathbf{k}_i$ and the normal to the interface, is known as the *plane of incidence*.

The equations (135.7) along with the dispersion relations, determine the vectors $\mathbf{k}_r$ and $\mathbf{k}_t$ completely. For the reflected wave in particular, we obtain

$$k_{rz} = -k_{iz}. \tag{135.8}$$

Geometrically, this relation, along with eq. (135.7) implies that the angles of reflection and incidence, as defined in fig. 20.2, are equal:

$$\theta_r = \theta_i. \tag{135.9}$$

This is, of course, the law of reflection familiar from elementary treatments. Note that we have shown this quite generally, without neglecting attenuation.

For the transmitted wave, on the other hand,

$$k_{tz} = (k_2^2 - k_1^2 \sin^2 \theta_i)^{1/2} = \frac{\omega}{c}(\epsilon_2 \mu_2 - \epsilon_1 \mu_1 \sin^2 \theta_i)^{1/2}, \tag{135.10}$$

and there is, in general, no simple relation between k_{tx} and k_{tz}. Thus, the transmitted wave will, in general, be inhomogeneous.

The continuity conditions on the fields now reduce to linear equations relating the amplitudes of the three waves. The number of equations to be solved simultaneously can be halved by recalling that $\mathbf{E}$ is odd under parity, while $\mathbf{H}$ is even. Under a reflection in the xz plane, the components E_y, H_x, and H_z change sign, while E_x, E_z, and H_y do not. Thus, these two sets of components must decouple from each other, and we can think of the solutions in which one or the other set vanishes as normal modes of the problem. If E_y is the only nonzero component of $\mathbf{E}$, the waves are polarized perpendicular or transverse to the plane of incidence, so this mode is known as the TE (transverse electric) mode. In the case where E_x and E_z are nonzero, the wave is polarized in the plane of incidence. Since $\mathbf{H}$ is now transverse to the plane of incidence, this mode is known as the TM (transverse magnetic) mode.[3]

Polarization perpendicular to plane of incidence (the TE case): The only nonzero field components are E_y, H_x, and H_z. The relevant equations are obtained by imposing continuity of the tangential components of $\mathbf{E}$ and $\mathbf{H}$, i.e., E_y and H_x. Continuity of the normal components of $\mathbf{B}$ and $\mathbf{D}$ does not have to be imposed separately. Recall that these conditions are consequences of the laws $\nabla \cdot \mathbf{B} = \nabla \cdot \mathbf{D} = 0$. The latter conditions

---

[3] The TE and TM modes are often denoted by the symbols $\perp$ and $\parallel$ for obvious reasons. In ellipsometry, the TE mode is known as the s mode (s for surface), since $\mathbf{E}$ lies in the interface, and the TM mode is known as the p mode (p for perpendicular), since $\mathbf{E}$ is perpendicular to the interface.

are automatically assured by eqs. (135.3) and (135.4). Thus, we have

$$E_{iy} + E_{ry} = E_{ty}, \quad H_{ix} + H_{rx} = H_{tx}. \tag{135.11}$$

The x components of the H's can be related to the various E_y's by eq. (135.3), which leads to the simultaneous equations

$$E_{iy} + E_{ry} = E_{ty}, \tag{135.12}$$

$$\mu_1^{-1} k_{iz}(E_{iy} - E_{ry}) = \mu_2^{-1} k_{tz} E_{ty}. \tag{135.13}$$

These equations can be solved for the *amplitude* reflection and transmission coefficients to yield

$$\frac{E_r}{E_i} = -\frac{\mu_2^{-1} k_{tz} - \mu_1^{-1} k_{iz}}{\mu_2^{-1} k_{tz} + \mu_1^{-1} k_{iz}}, \tag{135.14}$$

$$\frac{E_t}{E_i} = \frac{2\mu_1^{-1} k_{iz}}{\mu_2^{-1} k_{tz} + \mu_1^{-1} k_{iz}}. \tag{135.15}$$

Polarization in plane of incidence (the TM case): The only nonzero field components this time are H_y, E_x, and E_z. The formulas for this case can be obtained from the previous one by the symbolic replacements $\mathbf{E} \to \mathbf{H}$, $\mathbf{H} \to -\mathbf{E}$, and $\mu \leftrightarrow \epsilon$, for, as is easily seen, these leave eqs. (135.3)–(135.6) collectively unchanged.[4] The analysis of this case is therefore more easily done in terms of the H's than the E's. The analogs of eqs. (135.14) and (135.15) are

$$\frac{H_r}{H_i} = -\frac{\epsilon_2^{-1} k_{tz} - \epsilon_1^{-1} k_{iz}}{\epsilon_2^{-1} k_{tz} + \epsilon_1^{-1} k_{iz}}, \tag{135.16}$$

$$\frac{H_t}{H_i} = \frac{2\epsilon_1^{-1} k_{iz}}{\epsilon_2^{-1} k_{tz} + \epsilon_1^{-1} k_{iz}}. \tag{135.17}$$

The ratios of the electric field components can be obtained by making use of eq. (135.4).

These formulas simplify somewhat in the special but common case where the light is incident from a transparent medium (vacuum or air, e.g.). In this case, using eqs. (134.7) and (134.8) for the intensity, we can obtain the *reflectivity*, R, defined as the ratio of the intensities in the reflected and incident light, by squaring the appropriate amplitude ratios. Thus,

$$R_{TE} = \left| \frac{\mu_2^{-1} k_{tz} - \mu_1^{-1} k_{iz}}{\mu_2^{-1} k_{tz} + \mu_1^{-1} k_{iz}} \right|^2, \tag{135.18}$$

$$R_{TM} = \left| \frac{\epsilon_2^{-1} k_{tz} - \epsilon_1^{-1} k_{iz}}{\epsilon_2^{-1} k_{tz} + \epsilon_1^{-1} k_{iz}} \right|^2. \tag{135.19}$$

---

[4] For this reason, it is sometimes advantageous to leave the magnetic permeabilities in the general formulas, even when one is dealing with materials where the permeabilities are close to unity.

136 More reflection and refraction (both media transparent and nonmagnetic)

We now specialize the discussion of the previous section to the important case where both media are transparent (i.e., the attenuation is negligible) and nonmagnetic, i.e., $\mu_1 \approx \mu_2 \approx 1$. A large number of additional physical phenomena can be analyzed in this case. It pays to keep in mind that now ϵ, $\tilde{n} = \sqrt{\epsilon}$, and k are all real, so we may write $\tilde{n} = n$, $k = n(\omega/c)$. Secondly, the intensity in the wave is proportional to $\epsilon^{1/2}|\mathbf{E}|^2 = \epsilon^{-1/2}|\mathbf{H}|^2$.

From eqs. (135.7) and (135.10), and $k_{ix} = k_1 \sin \theta_i$, it follows that

$$\mathbf{k}_t = (k_1 \sin \theta_i, 0, \sqrt{k_2^2 - k_1^2 \sin^2 \theta_i}). \tag{136.1}$$

We shall assume first that $k_1 \sin \theta_i < k_2$, in which case the wave vector $\mathbf{k}_t$ of the transmitted wave is completely real, so that this wave is also homogeneous.[5] We can then write the x and z components of $\mathbf{k}_t$ as $k_2 \sin \theta_t$ and $k_2 \cos \theta_t$, where θ_t is the angle between the direction of the transmitted wave and the normal to the interface. (See fig. 20.2 on page 476.) Equating the x components, we get

$$n_1 \sin \theta_i = n_2 \sin \theta_t. \tag{136.2}$$

This is the basic law of refraction, also known as the *law of sines*.[6] It, along with the law of reflection, holds irrespective of the polarization state of the incident light.

Next, let us consider the ratios of the fields and the reflectivities for the two polarizations.

The TE mode: From eq. (135.14) we get

$$\frac{E_r}{E_i} = -\frac{n_2 \cos \theta_t - n_1 \cos \theta_i}{n_2 \cos \theta_t + n_1 \cos \theta_i}. \tag{136.3}$$

Using the law of refraction, we can express this entirely in terms of the angles of incidence and refraction:

$$\frac{E_r}{E_i} = -\frac{\sin(\theta_i - \theta_t)}{\sin(\theta_i + \theta_t)}. \tag{136.4}$$

The reflectivity can be written as

$$R_{TE} = \left(\frac{n_1 \cos \theta_i - \sqrt{n_2^2 - n_1^2 \sin^2 \theta_i}}{n_1 \cos \theta_i + \sqrt{n_2^2 - n_1^2 \sin^2 \theta_i}} \right)^2 \tag{136.5}$$

$$= \frac{\sin^2(\theta_i - \theta_t)}{\sin^2(\theta_i + \theta_t)} \quad (\theta_i \neq 0). \tag{136.6}$$

---

[5] If $k_1 \sin \theta_i > k_2$, we get the phenomenon of total internal reflection, which we will discuss separately.

[6] This law is generally named after Snell or Descartes, who used it in the 17th century. Recently, however, R. Rashed, *Isis* **81**, 464 (1990), has documented clear knowledge and publication of this law by Ibn Sahl in 964.

The TM mode: From eq. (135.16) we get

$$\frac{H_r}{H_i} = -\frac{n_2^{-1}\cos\theta_t - n_1^{-1}\cos\theta_i}{n_2^{-1}\cos\theta_t + n_1^{-1}\cos\theta_i}. \tag{136.7}$$

This can also be expressed entirely in terms of the angles:

$$\frac{H_r}{H_i} = \frac{\sin 2\theta_i - \sin 2\theta_t}{\sin 2\theta_i + \sin 2\theta_t}. \tag{136.8}$$

The reflectivity is therefore given by

$$R_{TM} = \left(\frac{n_2^2 n_1 \cos\theta_i - n_1^2\sqrt{n_2^2 - n_1^2\sin^2\theta_i}}{n_2^2 n_1 \cos\theta_i + n_1^2\sqrt{n_2^2 - n_1^2\sin^2\theta_i}}\right)^2, \tag{136.9}$$

$$= \left(\frac{\sin 2\theta_i - \sin 2\theta_t}{\sin 2\theta_i + \sin 2\theta_t}\right)^2. \tag{136.10}$$

Exercise 136.1 Show that for the TM mode, the reflected magnetic field amplitude ratio and the reflectivity can also be written as

$$\frac{H_r}{H_i} = \frac{\tan(\theta_i - \theta_t)}{\tan(\theta_i + \theta_t)}, \tag{136.11}$$

$$R_{TM} = \frac{\tan^2(\theta_i - \theta_t)}{\tan^2(\theta_i + \theta_t)} \quad (\theta_i \neq 0). \tag{136.12}$$

Normal incidence: For normal incidence ($\theta_i = 0$) the distinction between TE and TM modes disappears, and all the formulas for the reflectivity collapse into

$$R = \left(\frac{n_2 - n_1}{n_2 + n_1}\right)^2 \quad (\theta_i = 0). \tag{136.13}$$

For visible light incident from air onto water ($n = 1.33$) or glass ($n = 1.5$), the reflectivity is 2% and 4%, respectively.

Polarization by reflection: The above analysis shows that the reflectivity depends strongly upon the angle of incidence, especially for the TM mode. This dependence is shown in fig. 20.3. We can see from these figures that $R_{TM} \leq R_{TE}$ for all angles of incidence. This means that if the light incident upon an interface is unpolarized, the reflected light will be *partially polarized perpendicular to the plane of incidence*. In particular, for sunlight reflecting off a pool of water, or any other surface on the ground, the reflected light is predominantly horizontally polarized, since the plane of incidence is mainly vertical. Polaroid sunglasses exploit this effect by selectively transmitting vertically polarized light.[7] Further, as can be seen from eq. (136.10) or (136.12), R_{TM} vanishes

---

[7] In this connection, one should note that sunlight scattered from the atmosphere is also predominantly horizontally polarized, but one cannot think of this as due to reflection from a flat interface. Rather, the correct description is in terms of Rayleigh scattering from air molecules, section 149.

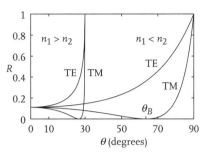

FIGURE 20.3. Reflectivity of light scattered from an interface between two media, as a function of the angle of incidence, for both TE and TM modes. The light is incident from the medium with refractive index n_1 onto the medium with index n_2. Both cases $n_1 > n_2$ and $n_1 < n_2$ are shown. The figure is drawn assuming a $2 : 1$ ratio of refractive indices. Note that whether n_1 is bigger than or smaller than n_2, the reflectivity for the TM mode vanishes at a certain angle of incidence, known as the Brewster angle, given by eq. (136.14).

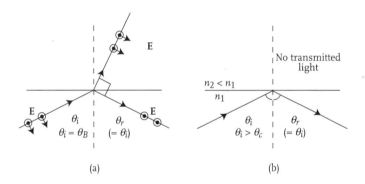

FIGURE 20.4. Two special angles of incidence. (a) At Brewster's angle, the reflected light is polarized perpendicular to the plane of incidence. Note that only the electric field vector is shown, and the incident and refracted waves contain both polarization components. (b) Light incident from a dense medium onto a rarer one at an angle of incidence exceeding the critical angle is totally reflected; there is only an evanescent wave in the rare medium, and no energy is transmitted.

completely if $\theta_i + \theta_t = \pi/2$, i.e., if the reflected and transmitted light rays are at right angles. From the law of refraction, we can see that this happens at the special angle of incidence

$$\theta_B = \tan^{-1}(n_2/n_1), \tag{136.14}$$

known as *Brewster's angle*. Light reflected at Brewster angle incidence is perfectly polarized perpendicular to the plane of incidence (see fig. 20.4). Conversely, if the incident light is polarized in the plane of incidence, there will be no reflected light. These properties are exploited to produce polarized light in some devices and in Brewster angle windows in gas laser cavities to kill unwanted reflections. Let us now suppose that the incident beam is linearly polarized in an arbitrary direction. This means that the components of $\mathbf{E}_i$ along the $\hat{\mathbf{y}}$ and $\hat{\mathbf{k}}_i \times \hat{\mathbf{y}}$ directions may both be taken to be real. Since the linear equations that determine the reflected and transmitted electric fields have real coefficients, the polarization components of these waves will also be real. Hence, these waves will also be linearly polarized.

Exercise 136.2 Let the incident light be linearly polarized at angle α_i to the plane of incidence, with $0 \le \alpha_i \le \pi/2$. Find the corresponding angles α_r and α_t for the reflected and transmitted beams.

Phase changes upon reflection: We now examine the phase of the reflected wave relative to that of the incident wave, limiting ourselves to the case of normal incidence. Then,

$$\frac{E_r}{E_i} = -\frac{n_2 - n_1}{n_2 + n_1}. \tag{136.15}$$

Hence, if the reflection is off an optically denser medium ($n_2 > n_1$), the reflected wave is $180°$ out of phase with the incident one. Conversely, there is no phase change if the reflection is off an optically rarer medium ($n_2 < n_1$). The transmitted wave never suffers a phase shift. These facts must be kept in mind when considering interference in reflection by thin films, nonreflecting coatings, etc.

Total internal reflection: Let us now take up the situation where $n_2 < n_1$, i.e., the light is incident from an optically denser medium onto a rarer one, and $\sin \theta_i > n_2/n_1$. In this case, there is no solution to the law of refraction, eq. (136.2), with a real value of θ_t. The angle

$$\theta_c = \sin^{-1}(n_2/n_1) \tag{136.16}$$

is known as the critical angle for total internal reflection. As we shall see, whenever the angle of incidence exceeds this critical angle, no energy is transmitted across the interface, and the incident wave is said to be totally internally reflected (fig. 20.4). This phenomenon is exploited in numerous optical devices.

If we introduce the real quantity

$$\kappa = (k_1^2 \sin^2 \theta_i - k_2^2)^{1/2} = k_1(\sin^2 \theta_i - \sin^2 \theta_c)^{1/2}, \tag{136.17}$$

then the wave vector in medium 2 can be written as

$$\mathbf{k}_t = (k_1 \sin \theta_i, 0, i\kappa). \tag{136.18}$$

Since the real and imaginary parts of $\mathbf{k}_t$ are not parallel, we have an inhomogeneous wave in this medium, and since the intensity dies exponentially along z, the wave is said to be *evanescent*. The electric and magnetic fields take the form

$$\begin{pmatrix} \mathbf{E} \\ \mathbf{H} \end{pmatrix} = \begin{pmatrix} \mathbf{E}_t \\ \mathbf{H}_t \end{pmatrix} e^{ik_1 \sin \theta_i x - \kappa z} e^{-i\omega t}, \tag{136.19}$$

with $\mathbf{H}_t = (c/\omega)\mathbf{k}_t \times \mathbf{E}_t$. The time-averaged Poynting vector in medium 2 is therefore given by

$$\overline{\mathbf{S}} = \frac{c^2}{8\pi\omega} \text{Re} \left(\mathbf{E}_t \times (\mathbf{k}_t^* \times \mathbf{E}_t^*) \right) e^{-2\kappa z}$$

$$= \frac{c^2}{8\pi\omega} \text{Re} \left[(\mathbf{E}_t \cdot \mathbf{E}_t^*)\mathbf{k}_t^* - (\mathbf{E}_t \cdot \mathbf{k}_t^*)\mathbf{E}_t^* \right] e^{-2\kappa z}. \tag{136.20}$$

In the first term, we write $\text{Re}\, \mathbf{k}_t^* = k_{ix}\hat{\mathbf{x}}$. In the second, because $\mathbf{E}_t \cdot \mathbf{k}_t = 0$, we may write

$$\mathbf{E}_t \cdot \mathbf{k}_t^* = \mathbf{E}_t \cdot (\mathbf{k}_t^* - \mathbf{k}_t) = -2i\kappa\, E_{tz}. \tag{136.21}$$

Secondly, we note that $\text{Re}\, i\, E_{tz} E_{tz}^* = 0$. Hence,

$$\overline{\mathbf{S}} = \frac{c^2}{8\pi\omega} \left[|\mathbf{E}_t|^2 k_{ix} - 2\kappa \,\text{Im}\, (E_{tz} E_{tx}^*) \right] e^{-2\kappa z}\hat{\mathbf{x}}. \tag{136.22}$$

Since this has no z component, there is no energy flowing into medium 2, as asserted. Energy does flow *parallel* to the interface in a thin layer of thickness $1/\kappa$, but this does not present any logical problem, since we are considering an incident wave that is plane and of infinite lateral extent.

Since $\overline{S}_z = 0$, it must follow from energy conservation that the intensity of the totally internally reflected light is equal to that of the incident one. We can see this from our formulas for the field amplitude ratios eqs. (135.14) and (135.16), which now become

$$\frac{E_r}{E_i} = \frac{k_{iz} - i\kappa}{k_{iz} + i\kappa} \quad \text{(TE case)}, \tag{136.23}$$

$$\frac{H_r}{H_i} = \frac{\epsilon_1^{-1}k_{iz} - i\epsilon_2^{-1}\kappa}{\epsilon_1^{-1}k_{iz} + i\epsilon_2^{-1}\kappa} \quad \text{(TM case)}. \tag{136.24}$$

Both of these are quantities of the form $(a - ib)/(a + ib)$, with a and b real. Such a quantity always has unit modulus, and so

$$R_{TE} = |E_r/E_i|^2 = 1, \quad R_{TM} = |H_r/H_i|^2 = 1. \tag{136.25}$$

The reflectivity is unity in both cases, showing once again that all the energy is reflected. Note, however, that the reflected wave suffers a phase shift, which is, in general, unequal for the two polarizations.

137 Reflection from a nonmagnetic opaque medium[*]

Let us now specialize the results of section 135 in a different direction. As before, we take medium 1 as the vacuum ($\epsilon_1 = \mu_1 = 1$), and medium 2 to be nonmagnetic ($\mu_2 = 1$), but we do not assume it to be transparent. That is, writing just ϵ for ϵ_2, we do not assume ϵ or $\tilde{n} = \sqrt{\epsilon}$ to be real. This is also a very common situation, and its importance lies in that measurements on the reflected light permit us to find the complex dielectric function of the reflecting medium. The techniques by which this is done are rather different when the light is normally or obliquely incident, so we discuss them separately.

Normal incidence: The extraction of ϵ' and ϵ'' (or n and κ) relies on the use of Kramers-Kronig relations. The mathematical analysis itself is straightforward, so we shall present most of it as a guided exercise for the reader and add physical commentary as necessary.

Exercise 137.1 Show that under the conditions described above, the reflectivity is given by (there is clearly no meaning now to a TM or TE assignment)

$$R(\omega) = \left| \frac{n(\omega) + i\kappa(\omega) - 1}{n(\omega) + i\kappa(\omega) + 1} \right|^2 = \frac{(n(\omega) - 1)^2 + \kappa^2(\omega)}{(n(\omega) + 1)^2 + \kappa^2(\omega)}, \tag{137.1}$$

where $\kappa(\omega)$ is the imaginary part of the refractive index.

Next, consider the function

$$f(\omega) = \frac{\sqrt{\epsilon(\omega)} - 1}{\sqrt{\epsilon(\omega)} + 1}, \tag{137.2}$$

in terms of which the reflectivity is given by $|f(\omega)|^2$. Show that if we write

$$f^2(\omega) = R(\omega)e^{i\theta(\omega)}, \tag{137.3}$$

then

$$\tan\theta = \frac{4n\kappa}{n^2 - \kappa^2 - 1}. \tag{137.4}$$

Since $\epsilon(\omega)$ is analytic and without zeros in the upper half-plane (see exercise 121.1), $\sqrt{\epsilon(\omega)}$ is analytic in this half-plane, and so are $f(\omega)$ and $\ln f(\omega)$. (The function $\sqrt{\epsilon(\omega)}$ possesses branch points at the zeros and poles of $\epsilon(\omega)$, but these all lie in the lower half-plane.) Thus, we may apply the Kramers-Kronig method to the real and imaginary parts of $\ln f(\omega)$. Show that this yields

$$\theta(\omega) = \frac{1}{\pi} P \int_0^\infty \frac{d\ln R}{d\omega} \ln \left| \frac{\omega + \omega_0}{\omega - \omega_0} \right| d\omega. \tag{137.5}$$

If one has measurements of $R(\omega)$, one can numerically evaluate this integral to find the phase $\theta(\omega)$. One may then invert eqs. (137.1) and (137.4) to find $n(\omega)$ and $\kappa(\omega)$, from which one may find $\epsilon'(\omega)$ and $\epsilon''(\omega)$. In practice this exercise requires quite a lot of care. The Kramers-Kronig integrals extend over all frequencies, but experimental data are necessarily available over some finite range. One must then use some sensible extrapolation of the data to the remaining frequency range.[8] For example, $R(\omega)$ is often available only for $\omega < \omega_p$. At very high frequencies, one has the general rule that $R(\omega) \sim \omega^{-4}$ (why?), but one cannot assume this form right away for $\omega > \omega_p$. In the intermediate range, however, one often has data for $\kappa(\omega)$ from transmission measurements on thin samples. One technique is to combine these data with models of κ at lower ω to perform a preliminary Kramers-Kronig evaluation of $n(\omega)$. These rough values of $n(\omega)$ and $\kappa(\omega)$ are then used to find an approximate $R(\omega)$ in the missing range, after which eq. (137.5) is employed.

---

[8] For an early illustration of the dangers of casual extrapolation, see H. R. Philipp and E. A. Taft, *Phys. Rev.* **136**, A1445 (1964).

The process is iterated until it converges. However, one must always ensure that certain physical constraints are met. For example, θ must vanish at $\omega = 0$ and also be nearly zero in any transparency ranges. Essential sum rules, such as the f-sum rule, must also be obeyed. For more on these matters, we refer the reader to the literature.[9]

Oblique incidence; ellipsometry: We have already seen in the phenomenon of Brewster's angle that the reflectivity can depend dramatically on the polarization state of the incident light. More generally, the polarization of the reflected and incident light beams may be quite different. In general, if the incident light is linearly polarized in a combination of TE and TM modes, the reflected light will be elliptically polarized. Measurements of the polarization ellipse can yield both parts of $\epsilon(\omega)$ without the need to measure the absolute reflectivity or use Kramers-Kronig relations. This makes this technique, known as ellipsometry, very attractive.

Let us introduce the ratios[10]

$$r_{TE} = \frac{E_r}{E_i} = -\frac{\sqrt{\epsilon - \sin^2\theta} - \cos\theta}{\sqrt{\epsilon - \sin^2\theta} - \cos\theta}, \tag{137.6}$$

$$r_{TM} = \frac{H_r}{H_i} = -\frac{\sqrt{\epsilon - \sin^2\theta} - \epsilon\cos\theta}{\sqrt{\epsilon - \sin^2\theta} - \epsilon\cos\theta}, \tag{137.7}$$

which follow easily from the formulas in section 135. We have written θ instead of $\theta_i (= \theta_r)$, since we focus on the reflected light only. Further, let us define

$$\hat{\mathbf{u}}_i = \hat{\mathbf{y}} \times \hat{\mathbf{k}}_i, \quad \hat{\mathbf{u}}_r = \hat{\mathbf{y}} \times \hat{\mathbf{k}}_r, \tag{137.8}$$

so that $(\hat{\mathbf{u}}_i, \hat{\mathbf{y}}, \hat{\mathbf{k}}_i)$ and $(\hat{\mathbf{u}}_r, \hat{\mathbf{y}}, \hat{\mathbf{k}}_r)$ form orthonormal right triads. We then have the following short table of field orientations in the TE and TM modes:

Mode	$\mathbf{E}_i$	$\mathbf{E}_r$	$\mathbf{H}_i$	$\mathbf{H}_r$	
TE	$\hat{\mathbf{y}}$	$\hat{\mathbf{y}}$	$-\hat{\mathbf{u}}_i$	$-\hat{\mathbf{u}}_r$	(137.9)
TM	$\hat{\mathbf{u}}_i$	$\hat{\mathbf{u}}_r$	$\hat{\mathbf{y}}$	$\hat{\mathbf{y}}$	

Let us now consider incident light linearly polarized at an angle α to the TM mode direction, i.e.,

$$\mathbf{E}_i \sim E_0(\cos\alpha\,\hat{\mathbf{u}}_i + \sin\alpha\,\hat{\mathbf{y}}). \tag{137.10}$$

The reflected light vector will then be given by

$$\mathbf{E}_r \sim E_0(r_{TE}\cos\alpha\,\hat{\mathbf{u}}_r + r_{TM}\sin\alpha\,\hat{\mathbf{y}}). \tag{137.11}$$

---

[9] See, e.g., E. Shiles, T. Sasaki, M. Inokuti, and D. Y. Smith, *Phys. Rev.* **B22**, 1612 (1980).
[10] Ellipsometrists often write r_s for r_{TE} and r_p for r_{TM}.

If we define

$$\frac{r_{TM}\cos\alpha}{r_{TE}\sin\alpha} = \tan\Psi e^{i\Delta}, \tag{137.12}$$

we can write the reflected field as

$$\mathbf{E}_r \sim (\sin\Psi e^{i\Delta}\hat{\mathbf{u}}_r + \cos\Psi\,\hat{\mathbf{y}}). \tag{137.13}$$

The point of this last representation is the following. From measurements on the polarization ellipse (orientation and ratio of semiaxes), one can obtain the parameters Ψ and Δ using eqs. (42.17) and (42.18). Thus, Ψ and Δ can be regarded as experimentally known quantities. Since α is also at the experimenter's control, the complex parameter

$$w = \tan\Psi\tan\alpha e^{i\Delta} \tag{137.14}$$

may be regarded as known. If we recast eq. (137.12) as

$$\frac{r_{TM}}{r_{TE}} = w, \tag{137.15}$$

the right-hand side is known. The left-hand side can be written as a function of ϵ using eqs. (137.6) and (137.7), and the task is to invert this equation and get ϵ in terms of w. One can transform it into a quartic equation for ϵ, but that is not much simpler, so the inversion is best done numerically. Since the angles θ and α can be varied (although α is often fixed at the single value, $\pi/4$), one can achieve some redundancy in one's data, which can be used to reduce errors and perform cross-checks.

Ellipsometry is often employed to study reflection from a thin film on a substrate, and in this case one can measure the thickness of the film in addition to its dielectric properties.[11]

---

[11] See, e.g., *Principles of Semiconductor Devices* by B. Van Zeghbroeck, http://ece-www.colorado.edu/bart/ellipsom·htm.

21 | Electromagnetic waves in and near conductors

138 Plasma oscillations

Suppose that a local charge imbalance is set up inside a metal, as shown in fig. 21.1. Electrons are removed from one region and placed in another, leading to local charge densities as shown. This gives rise to an electric field inside the metal, which tends to pull the electrons in such a way as to undo the charge imbalance. To understand how this comes about, let us take the Drude model equation (119.3) for $\mathbf{j}(t)$,

$$\left(\frac{\partial}{\partial t} + \frac{1}{\tau}\right)\mathbf{j} = \frac{n_f e^2}{m}\mathbf{E}, \tag{138.1}$$

and take its divergence. Using $\nabla \cdot \mathbf{E} = 4\pi\rho$, and $\nabla \cdot \mathbf{j} = -\dot{\rho}$ from the equation of continuity, we obtain

$$\left(\frac{\partial^2}{\partial t^2} + \frac{1}{\tau}\frac{\partial}{\partial t} + \omega_p^2\right)\rho = 0, \tag{138.2}$$

where

$$\omega_p^2 = \frac{4\pi n_f e^2}{m}. \tag{138.3}$$

$$\omega_p^2 = \frac{n_f e^2}{\epsilon_0 m}. \quad \text{(SI)}$$

Equation (138.2) describes a damped harmonic oscillator. If, as in most metals, $\omega_p \gg \tau^{-1}$, the oscillations are underdamped. We can write $\rho \sim e^{-i\omega t}$, with

$$\omega \approx \pm\omega_p - \frac{i}{2\tau}. \tag{138.4}$$

The quantity ω_p is known as the *plasma frequency*, and these charge density oscillations are known as *plasma oscillations*. They arise whenever there are mobile charges that can move inertially, i.e., without suffering too many collisions. They also occur in the earth's

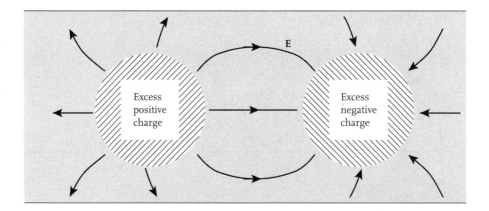

FIGURE 21.1. Electric fields set up by charge imbalance in a conductor.

ionosphere, which starts at a height of about 100 km above the earth's surface. Now, $n_f \sim 10^4$–10^6 cm^{-3}, so $\omega_p/2\pi \sim 1$–10 MHz, and $\lambda_p \sim 1$ km.

The ratio $\omega_p\tau$ provides a way to classify conductors. If $\omega_p\tau \gg 1$, the conductor can support plasma oscillations, and the conductor is said to be good. If $\omega_p\tau \ll 1$, plasma oscillations are not possible (even though we still refer to ω_p as the plasma frequency), and the conductor is said to be poor. In Cu, $\omega_p = 1.6 \times 10^{16}$ sec^{-1}, i.e., the frequency is 2.55×10^{15} Hz. The corresponding wavelength λ_p is 1200 Å, in the ultraviolet. Since $\tau = 2.5 \times 10^{-14}$ sec, $\omega_p\tau$ is indeed much greater than 1—there are about 60 oscillation periods in the decay time τ—and copper is a good conductor. These numbers are typical of good metals. For seawater, $\omega_p \sim 5 \times 10^{12}$ sec^{-1} and $\omega_p\tau \sim 1/20$, so this is a poor conductor.

Exercise 138.1 Show that plasma oscillations entail longitudinal oscillations in the electric field.

Exercise 138.2 Express the condition for plasma oscillations generally in terms of the response functions of a conductor.

Solution: The charge density inside the conductor obeys

$$\zeta(\omega)\rho_\omega = 4\pi\rho_\omega^{\text{free,ext}}. \tag{138.5}$$

In a plasma oscillation, we get a nonzero ρ_ω in the conductor even when no external charge is imposed. This is possible only if $\zeta(\omega) = 0$. The desired condition is, therefore, $\zeta(\omega_p) = 0$, or $\omega_p = -4\pi i\sigma(\omega_p)/\epsilon_b(\omega_p)$.

139 Dispersion of plasma waves*

Our discussion in section 138 neglected dispersion, i.e., the k dependence of ω. This is because in writing eq. (138.1), we ignored the interelectron forces. This is correct as long as *all* the electrons in the medium oscillate in phase (tantamount to putting $k = 0$), for then the electron density remains uniform and is perfectly canceled by the background density of positive charge. If the density variation is nonuniform, though, there will also be nonuniform pressure variations, providing a second restoring force in addition to the electric force, which tends to undo the density variations, as in a sound wave. Finding the electron pressure lies in the realm of quantum theory, as it is essential to incorporate Fermi statistics. It is still possible, however, to give a semiclassical hydrodynamic treatment.[1]

Let us consider an electron density of the form

$$n(\mathbf{r}, t) = n_f + \delta n(\mathbf{r}, t). \tag{139.1}$$

This has an FT $n_{\mathbf{k}\omega} = \delta n_{\mathbf{k}\omega}$, since we need not consider $k = 0$. We now imagine the time dependence to be on the scale ω_p^{-1} and average the energy of the electron gas over a time $T_0 \gg \omega_p^{-1}$. Denoting this average by angular brackets, we find the electrostatic or Coulombic contribution to be

$$\mathcal{E}_{\text{Coul}} = \frac{1}{2} \int\int e^2 \frac{\langle \delta n(\mathbf{r}, t) \delta n(\mathbf{r}', t) \rangle}{|\mathbf{r} - \mathbf{r}'|} d^3x \, d^3x' = \frac{1}{2T_0} \int \frac{4\pi}{k^2} e^2 |n_{\mathbf{k}\omega}|^2 \frac{d^3k \, d\omega}{(2\pi)^4}. \tag{139.2}$$

The hydrodynamic contribution can be written as

$$\mathcal{E}_{\text{hyd}} = \int \frac{1}{2} \frac{\partial^2 u}{\partial n_f^2} \langle [\delta n(\mathbf{r}, t)]^2 \rangle d^3x, \tag{139.3}$$

where $u(n_f)$ is the internal energy density of the gas. Fourier transformation yields

$$\mathcal{E}_{\text{hyd}} = \frac{\partial^2 u}{\partial n_f^2} \frac{1}{2T_0} \int |n_{\mathbf{k}\omega}|^2 \frac{d^3k \, d\omega}{(2\pi)^4}. \tag{139.4}$$

Lastly, let us find the time average of the kinetic energy. If $\mathbf{v}(\mathbf{r}, t)$ is the velocity field, then

$$\mathcal{E}_{\text{kin}} = \frac{1}{2} m n_f \int \langle v^2(\mathbf{r}, t) \rangle d^3x = \frac{m n_f}{2T_0} \int |\mathbf{v}_{\mathbf{k}\omega}|^2 \frac{d^3k \, d\omega}{(2\pi)^4}. \tag{139.5}$$

Now, since $\mathbf{j} = -ne\mathbf{v}$, and $\nabla \cdot \mathbf{j} = e\partial n/\partial t$ by the equation of continuity, and since the oscillation is longitudinal,

$$v_{\mathbf{k}\omega} = \frac{1}{n_f} \frac{\omega}{k} n_{\mathbf{k}\omega}. \tag{139.6}$$

Thus,

$$\mathcal{E}_{\text{kin}} = \frac{m}{n_f} \frac{1}{2T_0} \int \frac{\omega^2}{k^2} |n_{\mathbf{k}\omega}|^2 \frac{d^3k \, d\omega}{(2\pi)^4}. \tag{139.7}$$

In a wave, the average kinetic and potential energies must be equal. This yields the dispersion relation as

$$\omega^2 = \omega_p^2 + \frac{n_f}{m} \frac{\partial^2 u}{\partial n_f^2} k^2. \tag{139.8}$$

---

[1] Our treatment follows Feynman (1972), section 9.5. The deeper question of why this semiclassical argument should work at all is far outside our scope. The answer is due to David Bohm and David Pines, who showed in the 1950s (Phys. Rev. **82**, 625 (1951); **85**, 338 (1952); **92**, 609 (1953)) that a long wavelength density fluctuation in an electron gas behaves as a collective excitation that does not decay into a mess of particle–hole pairs, because the dominant force on it is the Coulomb force. The long-range nature of the Coulomb force is essential to this argument.

For a system of volume V, we have $\partial(uV)/\partial V = -p$, where p is the pressure. Hence,

$$\frac{\partial u}{\partial n_f} = \frac{u}{n_f} + \frac{p}{n_f}, \tag{139.9}$$

$$\frac{\partial^2 u}{\partial n_f^2} = \frac{1}{n_f}\frac{\partial p}{\partial n_f} = \frac{1}{n_f^2 \kappa}, \tag{139.10}$$

where the very last result is written in terms of the compressibility, κ, which must be found quantum mechanically. In the simplest approximation, we get $\kappa^{-1} = n_f m v_F^2/3$, where v_F is the Fermi velocity, given by $v_F = (\hbar/m)(3\pi^2 n_f)^{1/3}$. Hence,

$$\omega^2 = \omega_p^2 + a v_F^2 k^2, \tag{139.11}$$

where $a = 1/3$. The dispersion is small, since in typical metals, $k v_F \ll \omega_p$ as long $k \ll a_0^{-1}$, i.e., as long as the wavelength is much greater than the interatomic spacing.[2]

140 Transverse EM waves in conductors

As we have seen above, plasma oscillations are longitudinal EM waves. We now consider transverse EM waves in a metal. In this case, $\nabla \cdot \mathbf{E} = 0$, so there is no net charge density in the metal. Since the macroscopic Faraday and Ampere-Maxwell laws [eq. (120.9)] now take the form

$$\nabla \times \mathbf{E}_\omega = \frac{i\omega}{c}\mu(\omega)\mathbf{H}_\omega, \tag{140.1}$$

$$\nabla \times \mathbf{H}_\omega = -\frac{i\omega}{c}\zeta(\omega)\mathbf{E}_\omega, \tag{140.2}$$

all the formulas of section 133 can be used by writing $\zeta(\omega)$ in lieu of $\epsilon(\omega)$. In particular, the dispersion relation is

$$k^2 = \frac{\omega^2}{c^2}\mu(\omega)\zeta(\omega). \tag{140.3}$$

To get a physical feeling for this result, let us set $\mu(\omega)$ and $\epsilon_b(\omega)$ both equal to 1 and use the Drude model conductivity. This yields

$$c^2 k^2 = \omega^2 + \omega_p^2 \frac{i\omega\tau}{1 - i\omega\tau}. \tag{140.4}$$

The regime of greatest interest is $\omega\tau \gg 1$. In that case, this relation becomes

$$\omega^2 \approx \omega_p^2 + c^2 k^2. \tag{140.5}$$

---

[2] The more accurate treatment of Bohm and Pines (see footnote 1) gives $a = 3/5$ for a dense electron gas instead of $1/3$. The hydrodynamic argument is off the mark because it presupposes that the frequency at which the density varies is low, whereas the exact opposite is true. We are interested in variations at the frequency ω_p. The characteristic response frequency of an electron gas at a wave vector k is $v_F k$. The hydrodynamic treatment requires $\omega_p \ll v_F k$, but, in fact, $\omega_p \gg v_F k$.

This implies that if $\omega < \omega_p$, k is imaginary, which, in turn, means that the fields must die in the interior of the conductor. We get a real solution for k only if $\omega > \omega_p$. In this case the wave *can* propagate through the medium.

Thus, an electromagnetic wave can propagate through a conducting medium only if its frequency is greater than the plasma frequency of the medium. We have already mentioned that the plasma wavelength of Cu is 1200 Å, which is a typical value for metals. This means that light in the ultraviolet and lower wavelength range can pass through metals. For this reason, this phenomenon is sometimes called *ultraviolet transparency*. The earth's ionosphere (altitude 100 km and above) is opaque to waves of frequency less than a few MHz, and this fact is exploited in shortwave radio, where long-distance transmission is achieved by bouncing the signal off the ionosphere.

Let us also use the Drude model along with $\epsilon_b = \mu = 1$ to find the refractive index and attenuation coefficient. If $\omega\tau \ll 1$,

$$n + i\kappa = \left(\frac{2\pi\sigma}{\omega}\right)^{1/2}(1+i), \quad k = \frac{1+i}{\delta}, \tag{140.6}$$

where δ is the skin depth. As expected, the metal is opaque at low frequencies. At very high frequencies, $\omega \gg \omega_p$ (which necessarily implies $\omega\tau \gg 1$), on the other hand,

$$k = \frac{\omega}{c}\left(1 - \frac{\omega_p^2}{2\omega^2}\right) + i\frac{\omega_p^2}{2\omega^2 c\tau}. \tag{140.7}$$

In particular, using eq. (138.3) for the plasma frequency, we have for the refractive index

$$n = 1 - \frac{4\pi n_f e^2}{2m\omega^2}, \tag{140.8}$$

which is slightly less than unity. In fact, if we replace n_f by a suitable electron density that includes the bound electrons, this result also applies to dielectrics at X-ray frequencies, since at such frequencies there is no difference between a conduction electron and a bound electron.

Exercise 140.1 Seawater is both a dielectric and a weak conductor, well described by the constitutive relations $\mathbf{B}_\omega = \mathbf{H}_\omega$, $\mathbf{D}_\omega = \epsilon\mathbf{E}_\omega$, and $\mathbf{j}_\omega = \sigma(\omega)\mathbf{E}_\omega$, with $\sigma(\omega) = \sigma_0/(1 - i\omega\tau)$, where ϵ, σ_0, and τ are constants. Further, $\tau^{-1} \gg \sigma_0$. Derive the dispersion relation for transverse EM waves assuming $\omega\tau \ll 1$.

Answer: $k^2 c^2 \approx \omega^2\epsilon + 4\pi i\omega\sigma_0$. If the frequency lies in the range $\tau^{-1} \gg \omega \gg \sigma_0$, $kc \approx \omega\sqrt{\epsilon} + 2\pi i\sigma_0/\sqrt{\epsilon}$. Thus, the waves have an attenuation length $c\sqrt{\epsilon}/2\pi\sigma_0$ [$2\epsilon_0 c\sqrt{\epsilon}/\sigma_0$ in SI], which is about 1 cm, independent of frequency. At very low frequencies, $\omega \ll \sigma_0$, the wave is overdamped, and the attenuation length varies as $(\omega\sigma_0)^{-1/2}$.

Exercise 140.2 (plasma in a magnetic field) Consider a plasma in the presence of a uniform magnetic field $\mathbf{B}_0$, and consider trasnverse EM waves traveling parallel to this field. Show that the eigenmodes are circularly polarized, and that within the Drude model, for

$\mathbf{E} \sim (\hat{\mathbf{x}} \pm i\hat{\mathbf{y}})$, the dispersion relation is

$$\omega^2 = c^2 k^2 + \frac{\omega_p^2 \omega}{(\omega \mp \omega_c) + i\tau^{-1}}, \tag{140.9}$$

where $\omega_c = e B_0 / mc$. Examine this dispersion relation assuming that $\omega_c \tau \gg 1$, and show that one branch (which one?) always has a propagating solution at arbitrarily low frequencies. Find the limiting form of the dispersion as $\omega \to 0$. Such waves are found both in the ionosphere and in pure enough metals, where they are known as whistlers and helicons, respectively. Find the typical wavelength for a 10-Hz frequency both for the ionosphere (take B_0 as 0.5 G) and for a good metal (take B_0 as 1 T).

Exercise 140.3 Extend the previous problem to allow for an arbitrary propagation direction. Ignore damping from the outset. (*Caution*: The analysis is straightforward, but lengthy.)

141 Reflection of light from a metal

Let us now consider what happens when light is incident from a vacuum onto a flat metallic surface. We expect that if the frequency of the light exceeds the plasma frequency of the metal, the light will be partially transmitted and partly reflected as for transparent media. The most interesting case is the opposite one, where we expect the light to be reflected. This is in accord with the everyday experience that metals are excellent reflectors of visible light. We shall therefore assume that $\omega\tau \ll 1$, where τ is the collision time in the metal. The problem can then be studied using the formulas of section 135.

As before, let the metal be medium number 2. We shall continue to assume that $\mu_2 \approx 1$. At the frequencies of interest, the wave vector in this medium then obeys

$$k^2 = \frac{\omega^2}{c^2}\left(\epsilon_b + \frac{4\pi i \sigma_0}{\omega}\right), \tag{141.1}$$

and we recall that the second term is much bigger than the first in magnitude. Now,

$$k_{tx} = k_1 \sin\theta_i, \tag{141.2}$$

and $k_1 = \omega/c$. Hence,

$$k_{tz}^2 = \frac{\omega^2}{c^2}\left((\epsilon_b - \sin^2\theta_i) + \frac{4\pi i \sigma_0}{\omega}\right). \tag{141.3}$$

Again, the second term dominates, so there is very little dependence of k_{tz} on θ_i, and we may write

$$k_{tz} \approx \frac{1+i}{\delta}\left(1 + O(k_1\delta)^2\right), \tag{141.4}$$

where δ is the skin depth, and $k_1\delta = \omega\delta/c \ll 1$.

It is now straightforward to calculate the reflectivity. Consider the TE mode first. Equation (135.14) simplifies to

$$\frac{E_r}{E_i} = -\frac{k_{tz} - k_{iz}}{k_{tz} + k_{iz}}. \tag{141.5}$$

But $k_{iz} = k_1 \cos \theta_i \ll k_{tz}$. Hence,

$$\frac{E_r}{E_i} \approx -\left(1 - 2\frac{k_{iz}}{k_{tz}}\right) \approx -\left(1 - \frac{\omega\delta}{c}\cos\theta_i\right), \tag{141.6}$$

where the imaginary part has been neglected in comparison with the real part. The departure from unity in the latter is, however, important. Thus,

$$R_{TE} \approx 1 - \frac{2\omega\delta}{c}\cos\theta_i = 1 - \sqrt{\frac{2\omega}{\pi\sigma_0}}\cos\theta_i. \tag{141.7}$$

For TM polarization, on the other hand, eq. (135.16) implies

$$\frac{H_r}{H_i} = -\frac{k_{tz} - \zeta(\omega)k_{iz}}{k_{tz} + \zeta(\omega)k_{iz}}. \tag{141.8}$$

Now, $\zeta \approx 2i(c/\omega\delta)^2 \gg k_{tz}$, so (except for extreme grazing angle incidence when $\cos\theta_i \simeq \omega\delta/c$),

$$\frac{H_r}{H_i} \approx 1 - 2\frac{k_t z}{\zeta k_{iz}} \approx -\left(1 - \frac{\omega\delta}{c\cos\theta_i}\right). \tag{141.9}$$

The reflectivity is therefore given by

$$R_{TM} \approx 1 - \sqrt{\frac{2\omega}{\pi\sigma_0}}\frac{1}{\cos\theta_i}, \quad (\cos\theta_i \gg \omega\delta/c). \tag{141.10}$$

At normal incidence ($\theta_i = 0$), this coincides with R_{TE}.

For both polarizations, we see that R is very close to 1. For example, for a wavelength of 5×10^4 Å, in the infrared, at normal incidence, $1 - R \simeq 0.02$ for Cu at room temperature ($\sigma = 5.8 \times 10^{17}$ sec^{-1}). Nevertheless, the $\omega^{1/2}$ dependence of $1 - R$ is confirmed experimentally.

142 Surface plasmons[*]

In the phenomenon of total internal reflection we have seen an evanescent wave that travels along the interface in the optically rarer medium. Can we have a wave that is evanescent in both media? Such a wave would be a self-sustaining mode of oscillation of charges or polarizations near the interface along with the associated fields. Naturally, we expect some damping, but we would like it to be small. This means that the motion of the charges should be dominated by inertia, and there should be some restoring force. This requirement is met in metals (plasmas) and possibly in dielectrics if the frequency is near one of the natural frequencies in a Drude-Lorentz type model. Secondly, we expect charges at the interface to give rise to electric fields normal to the interface. We now examine these expectations in a more careful analysis.

We shall study this problem only when one medium is the vacuum and the other medium is nonmagnetic. To perform the analysis, we use the notation and set up of section 135, with medium 2 as the vacuum, except that quantities in this medium shall be denoted by a suffix 0 instead of 2. There is now no incident wave, and one can think of the fields as as being made up of purely reflected and transmitted waves. Since both of these waves must be evanescent, we write the wave vectors in each medium as

$$\mathbf{k}_1 = k_{1x}\hat{\mathbf{x}} - i\kappa_1\hat{\mathbf{z}}, \tag{142.1}$$

$$\mathbf{k}_0 = k_{0x}\hat{\mathbf{x}} + i\kappa_0\hat{\mathbf{z}}, \tag{142.2}$$

where κ_1 and κ_0 are both real and positive. The components of the wave vectors in the interface, k_{1x} and k_{0x}, should be largely real. The dispersion relations then read

$$c^2(k_{1x}^2 - \kappa_1^2) = \omega^2\zeta(\omega)\mu(\omega), \tag{142.3}$$

$$c^2(k_{0x}^2 - \kappa_0^2) = \omega^2. \tag{142.4}$$

We write just $\zeta(\omega)$ and $\mu(\omega)$ instead of $\zeta_1(\omega)$ and $\mu_1(\omega)$, and delay setting $\mu(\omega)$ to 1.

The next step is to enforce the continuity conditions on the fields. This can be simplified by noting that the parity argument of section 135 still holds, so we can look for TE and TM mode solutions separately.

TE mode: Now only E_y is nonzero. Omitting the $e^{-i\omega t}$ factor, we write

$$\mathbf{E}_1 = E_1 e^{i\mathbf{k}_1 \cdot \mathbf{r}}\hat{\mathbf{y}}, \tag{142.5}$$

$$\mathbf{E}_0 = E_0 e^{i\mathbf{k}_0 \cdot \mathbf{r}}\hat{\mathbf{y}}. \tag{142.6}$$

The $\mathbf{H}$ fields can be found from the formula $\mathbf{H} = (c/\omega\mu)\mathbf{k} \times \mathbf{E}$, which yields

$$\mathbf{H}_1 = \frac{c}{\omega\mu(\omega)} E_1 e^{i\mathbf{k}_1 \cdot \mathbf{r}}(k_{1x}\hat{\mathbf{z}} + i\kappa_1\hat{\mathbf{x}}), \tag{142.7}$$

$$\mathbf{H}_0 = \frac{c}{\omega} E_0 e^{i\mathbf{k}_0 \cdot \mathbf{r}}(k_{0x}\hat{\mathbf{z}} - i\kappa_0\hat{\mathbf{x}}). \tag{142.8}$$

It suffices to enforce continuity of $\mathbf{E}_t$ and $\mathbf{H}_t$, i.e., E_y and H_x. The former leads to $E_1 = E_0$, and

$$k_{1x} = k_{0x}. \tag{142.9}$$

That is, the wave vectors in the plane are equal, as expected from momentum conservation. Continuity of H_x leads to

$$\kappa_1 = -\mu(\omega)\kappa_0. \tag{142.10}$$

If $\mu(\omega) = 1$, $\kappa_0 = -\kappa_1$, and since $k_{0x} = k_{1x}$ also, the dispersion relations (142.3) and (142.4) cannot both be satisfied if $\zeta \neq 1$. Hence, there is no TE mode confined entirely to the interface. This is in keeping with our earlier argument for the electric field to have a normal component.

TM mode: The results for this mode can be found from the TE mode results by the substitutions $\mathbf{E} \to \mathbf{H}$, $\mathbf{H} \to -\mathbf{E}$, $\zeta(\omega) \leftrightarrow \mu(\omega)$. The analog of eq. (142.10) is thus

$$\kappa_1 = -\zeta(\omega)\kappa_0. \tag{142.11}$$

This equation immediately shows that ζ must be real and negative for κ_1 and κ_0 to both be positive. This requirement is only met for plasmas, and for dielectrics in a frequency range very close to an absorption peak, but in that case, the damping is not weak. We therefore consider only plasmas from now on.

Let us now solve for the wave vectors fully. Setting $\mu(\omega) = 1$ and using $k_{0x} = k_{1x}$ along with eq. (142.11) in the dispersion relations (142.3) and (142.4), some simple algebra yields

$$ck_{0x} = \omega \left(\frac{\zeta(\omega)}{1 + \zeta(\omega)} \right)^{1/2}, \tag{142.12}$$

$$c\kappa_1 = \omega \left(\frac{-\zeta^2(\omega)}{1 + \zeta(\omega)} \right)^{1/2}, \tag{142.13}$$

$$c\kappa_0 = \omega \left(\frac{-1}{1 + \zeta(\omega)} \right)^{1/2}. \tag{142.14}$$

These equations show that, in fact, we need the stronger condition $\zeta(\omega) < -1$ for a solution. Using the Drude model in the limit $\omega\tau \gg 1$, i.e., using

$$\zeta(\omega) = 1 - \omega_p^2/\omega^2, \tag{142.15}$$

we get

$$c^2 k_{0x}^2 = \frac{\omega^2(\omega^2 - \omega_p^2)}{(2\omega^2 - \omega_p^2)}. \tag{142.16}$$

Thus, surface plasmons (as these waves are called) exist for $0 < \omega < \omega_p/\sqrt{2}$. For low ω, $k_{0x} \approx \omega/c$.

Exercise 142.1 Complete the analysis of the TM mode and find the fields, charge densities (both bulk and surface), and currents.

Answer (partial): The charge density vanishes in the bulk. On the surface it is given (in the Drude model) by

$$\Sigma_s = \frac{H_0}{4\pi} \frac{\omega^2}{[(\omega_p^2 - \omega^2)(2\omega_p^2 - \omega^2)]^{1/2}} e^{ik_{0x}x}. \tag{142.17}$$

Exercise 142.2 Find the Poynting vector for the surface plasmon, and thus normalize the amplitude H_0 in terms of the energy flux along the interface.

Exercise 142.3 If a semiconductor is suitably doped, the plasma frequency of the mobile electrons can be made to lie in the infrared. At such low frequencies, the bound electrons

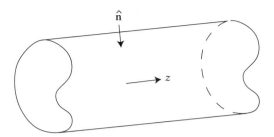

FIGURE 21.2. A section of waveguide, showing the inward-pointing normal. If the ends are capped with conductors, this becomes a cavity resonator.

respond as if the field were static and the propensity can be modeled as

$$\zeta(\omega) = \epsilon_\infty \left(1 - \frac{\omega_p^2}{\omega^2}\right),$$ (142.18)

where ϵ_∞ is the bound electron dielectric constant. Find the dispersion relation for surface plasmons for this model, and the limiting frequency.[3]

143 Waveguides

For frequencies in the microwave regime or higher, ordinary ac circuits cannot be used to send electrical signals for any great distance, as the dimensions of the circuit are then larger than the wavelength, and radiative losses become very large. In such cases, waveguides—hollow metallic pipes—provide an alternative. We show in this section how EM waves may be propagated down such pipes. We consider a pipe of arbitrary but uniform cross section as shown in fig. 21.2. In the hollow region, assuming a harmonic time dependence of the form $e^{-i\omega t}$, the fields obey the wave equation[4]

$$\left(\nabla^2 + \frac{\omega^2}{c^2}\right) \begin{pmatrix} \mathbf{E} \\ \mathbf{B} \end{pmatrix} = 0$$ (143.1)

and the Faraday and Ampere-Maxwell laws

$$\nabla \times \mathbf{E} - \frac{i\omega}{c}\mathbf{B} = 0,$$ (143.2)

$$\nabla \times \mathbf{B} + \frac{i\omega}{c}\mathbf{E} = 0.$$ (143.3)

---

[3] See, e.g., N. Marschall, B. Fischer, and H. J. Queisser, *Phys. Rev. Lett.* **27**, 95 (1971), and B. Fischer, N. Marschall, and H. J. Queisser, *Surf. Sci.* **34**, 50 (1973). These authors study InSb with the parameters $\omega_p/2\pi = 12.79$ THz and $\epsilon_\infty = 15.68$.

[4] Here, we are assuming that the waveguide is evacuated or, what is practically the same thing, filled with air. Very little is lost by this specialization. If the waveguide is filled with a linear dielectric of dielectric constant $\epsilon(\omega)$ and permeability $\mu(\omega)$, we simply replace ω^2 by $\mu\epsilon\omega^2$.

These equations determine the fields completely, when supplemented with boundary conditions. For the latter, we will assume that the waveguide is made with a good conductor, so the analysis of quasistatic fields from sections 126–129 is applicable. We found that in the strong skin-effect limit surface charges and currents in the conductor essentially kill any normal component of **H** or tangential component of **E** outside the conductor. Neglecting quantities of first and higher order in δ/λ, we obtain the idealized or perfect conductor boundary conditions,

$$\mathbf{B} \cdot \hat{\mathbf{n}} = 0, \quad \mathbf{E} \times \hat{\mathbf{n}} = 0, \tag{143.4}$$

where $\hat{\mathbf{n}}$ is the normal to the walls of the waveguide.

Let us choose the z axis along the axis of the waveguide and denote components of vectors transverse to this axis by the suffix t. We are interested in wavelike solutions of the form

$$\mathbf{E}(\mathbf{r}, t) = \mathbf{E}(x, y)e^{i(kz-\omega t)}, \quad \mathbf{B}(\mathbf{r}, t) = \mathbf{B}(x, y)e^{i(kz-\omega t)}. \tag{143.5}$$

With this form,

$$\hat{\mathbf{z}} \times (\nabla \times \mathbf{E}) = \nabla_t E_z - ik\mathbf{E}_t. \tag{143.6}$$

Hence, taking the cross product of the Faraday law, eq. (143.2), with $\hat{\mathbf{z}}$, we obtain

$$ik\mathbf{E}_t + i\frac{\omega}{c}\hat{\mathbf{z}} \times \mathbf{B}_t = \nabla_t E_z. \tag{143.7}$$

Similarly,

$$(\nabla \times \mathbf{B})_t = \hat{\mathbf{z}}\nabla_z \times \mathbf{B}_t + \nabla_t \times B_z\hat{\mathbf{z}} = ik\hat{\mathbf{z}} \times \mathbf{B}_t - \hat{\mathbf{z}} \times \nabla_t B_z, \tag{143.8}$$

so the transverse component of the Ampere-Maxwell law, eq. (143.3), reads

$$i\frac{\omega}{c}\mathbf{E}_t + ik\hat{\mathbf{z}} \times \mathbf{B}_t = \hat{\mathbf{z}} \times \nabla_t B_z. \tag{143.9}$$

Equations (143.7) and (143.9) may be solved for $\mathbf{E}_t$ and $\hat{\mathbf{z}} \times \mathbf{B}_t$ to yield

$$\begin{pmatrix} \mathbf{E}_t \\ \hat{\mathbf{z}} \times \mathbf{B}_t \end{pmatrix} = -\frac{1}{k^2 - (\omega/c)^2} \begin{pmatrix} ik & -i\omega/c \\ -i\omega/c & ik \end{pmatrix} \begin{pmatrix} \nabla_t E_z \\ \hat{\mathbf{z}} \times \nabla_t B_z \end{pmatrix}. \tag{143.10}$$

Noting that $\mathbf{B}_t = -\hat{\mathbf{z}} \times (\hat{\mathbf{z}} \times \mathbf{B}_t)$, these two equations show that the transverse components of the fields are determined once the z components are known. For the latter, the wave equation implies

$$(\nabla_t^2 + \gamma^2) \begin{pmatrix} E_z \\ B_z \end{pmatrix} = 0, \tag{143.11}$$

with

$$\gamma^2 \equiv (\omega/c)^2 - k^2. \tag{143.12}$$

Lastly, the boundary conditions can also be cast entirely in terms of E_z and B_z. For E_z, this is already done. For B_z, we take the dot product of eq. (143.9) with $\hat{\mathbf{n}} \times \hat{\mathbf{z}}$. This yields

$$\hat{\mathbf{n}} \cdot \nabla_t B_z \equiv \frac{\partial B_z}{\partial n} = 0. \tag{143.13}$$

It follows from eq. (143.11) that there are two types of field configurations, or modes. In the first, $E_z = 0$, while in the second, $B_z = 0$. In some cases, there is also a mode in which both E_z and B_z vanish.

Transverse electric (TE) modes: In these modes, also known as B modes, the electric field is purely transverse. That is, $E_z = 0$, but $B_z \neq 0$. The problem is to solve the two-dimensional Helmholtz equation

$$(\nabla_t^2 + \gamma^2) B_z(x, y) = 0, \tag{143.14}$$

with the boundary condition $\partial B_z / \partial n = 0$. A solution exists only for certain discrete values of γ^2, so this is an eigenvalue problem. The transverse fields are given by

$$\mathbf{E}_t = -i \frac{\omega}{c\gamma^2} \hat{\mathbf{z}} \times \nabla_t B_z, \quad \mathbf{B}_t = i \frac{k}{\gamma^2} \nabla_t B_z. \tag{143.15}$$

Transverse magnetic (TM) modes: In these modes, also known as E modes, the magnetic field is purely transverse: $B_z = 0$, but $E_z \neq 0$. Now we must solve

$$(\nabla_t^2 + \gamma^2) E_z(x, y) = 0, \tag{143.16}$$

with the boundary condition $E_z = 0$. This is also an eigenvalue problem. The transverse fields are given by

$$\mathbf{E}_t = i \frac{k}{\gamma^2} \nabla_t E_z, \quad \mathbf{B}_t = i \frac{\omega}{c\gamma^2} \hat{\mathbf{z}} \times \nabla_t E_z. \tag{143.17}$$

Exercise 143.1 Show that the eigenvalue γ^2 is necessarily positive for both TE and TM modes.

Transverse electromagnetic (TEM) mode: If we seek a solution with $E_z = B_z = 0$, we see from eq. (143.10) that we must have $k^2 = (\omega/c)^2$, i.e., the dispersion relation must be that of free space (or the unbounded medium if the waveguide is filled with a dielectric). To find the transverse fields, we must solve the Maxwell equations directly. For $\mathbf{E}_t$, the z component of the Faraday law and Gauss's law yield

$$\nabla_t \times \mathbf{E}_t = 0, \quad \nabla_t \cdot \mathbf{E}_t = 0. \tag{143.18}$$

The magnetic field can be found from $\mathbf{E}_t$ via

$$\mathbf{B}_t = (\text{sgn } k)\hat{\mathbf{z}} \times \mathbf{E}_t. \tag{143.19}$$

Equation (143.18), along with the boundary condition $\mathbf{E}_t \| \hat{\mathbf{n}}$, constitutes a problem in two-dimensional electrostatics. If the boundary of the waveguide is simply connected, then since this boundary is an equipotential, one must have $\mathbf{E}_t = 0$, i.e., the mode cannot exist.

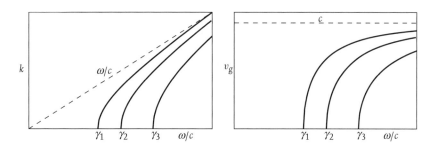

FIGURE 21.3. Dispersion relation and group velocity for waveguide modes. Three different modes are shown, with cutoffs γ_1, γ_2, and γ_3.

To have a TEM mode, one needs to have two or more disconnected surfaces bounding the nonconducting region. Two important cases where this is realized are the coaxial cable and the two-wire transmission line, which we discussed in connection with the telegrapher's equation (section 117). As this discussion showed, the speed of the wave is not quite c, nor is dispersion completely absent. The calculation of the fields itself is a simple problem: this is done (implicitly if not explicitly) for the two-wire line in Exercises 31.4 and 89.3, and for the coaxial cable in Exercises 31.4 and 88.4.

Dispersion relation for TE/TM modes: For both the TE and TM modes, the dispersion relation is

$$\omega^2 = (k^2 + \gamma^2)c^2. \tag{143.20}$$

As can be seen from fig. 21.3, a particular mode exists only above a threshold or cutoff frequency, γc. Secondly, the group velocity is given by

$$v_g = \frac{d\omega}{dk} = c\frac{k}{(k^2 + \gamma^2)^{1/2}} < c. \tag{143.21}$$

Rectangular waveguide: Let the cross section of the waveguide be a rectangle with sides a and b, and choose Cartesian coordinates with the origin at one corner. The modes are easily found to be given by

$$\text{TE}_{mn}: \quad B_z(x, y) = \cos\left(\frac{m\pi}{a}x\right)\cos\left(\frac{n\pi}{b}y\right), \quad m, n \text{ not both } 0, \tag{143.22}$$

$$\text{TM}_{mn}: \quad E_z(x, y) = \sin\left(\frac{m\pi}{a}x\right)\sin\left(\frac{n\pi}{b}y\right), \quad m \geq 1, \ n \geq 1. \tag{143.23}$$

For both sets of modes,

$$\gamma_{mn}^2 = \left(\frac{m\pi}{a}\right)^2 + \left(\frac{n\pi}{b}\right)^2, \tag{143.24}$$

but the lowest TE mode is TE_{10}, while the lowest TM mode is TM_{11}. If m and n are both nonzero, TE_{mn} and TM_{mn} are degenerate. The lowest mode overall is TE_{10}, which means that for radiation to propagate down such a waveguide, its free-space wavelength must be less than twice the long side of the rectangle. It is left for the reader to find the transverse

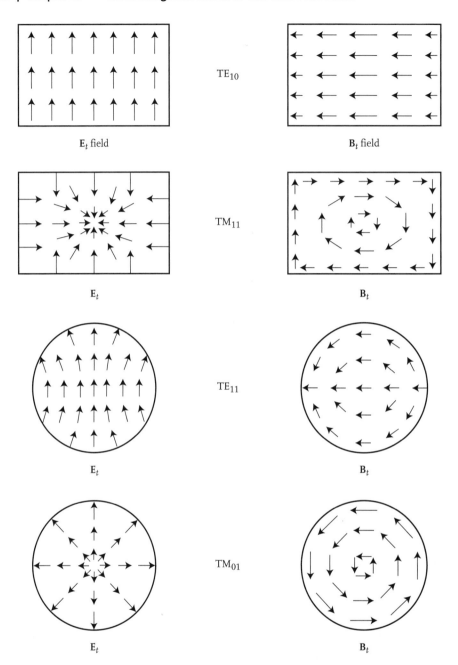

FIGURE 21.4. Some low-order modes for rectangular and circular waveguides.

fields and confirm that they obey the boundary conditions. These fields are shown in fig. 21.4 for the TE$_{10}$ and TM$_{11}$ modes.

Circular waveguide: Let the cross section of the waveguide be a circle of radius a. The wave equation for the longitudinal component of the fields is

$$(\nabla^2 + \gamma^2)\Psi = 0, \tag{143.25}$$

where Ψ stands for either E_z or B_z. Writing $\Psi = \psi(r)e^{im\varphi}$, we get

$$\left(\frac{d^2}{dr^2} + \frac{1}{r}\frac{d}{dr} - \frac{m^2}{r^2} + \gamma^2\right)\psi(r) = 0. \tag{143.26}$$

The solution that is regular at the origin is the Bessel function $J_m(\gamma r)$. The boundary conditions require the vanishing at $r = a$ of either the derivative J'_m (for TE modes) or J_m (for TM modes). Further, for each type, except when $m = 0$, the two modes with angular dependence $e^{\pm im\varphi}$ or, what is the same thing, $\cos(m\varphi)$ and $\sin(m\varphi)$, are degenerate. Hence, the modes can be given as

$$\text{TE}_{mn}: \quad B_z(r, \varphi) = J_m(\gamma r)\cos(m\varphi), \quad J'_m(\gamma^{\text{TE}}_{mn}a) = 0, \tag{143.27}$$

$$\text{TM}_{mn}: \quad E_z(r, \varphi) = J_m(\gamma r)\cos(m\varphi), \quad J_m(\gamma^{\text{TM}}_{mn}a) = 0. \tag{143.28}$$

In both cases, the labeling is such that $n \geq 1$. The first few modes are

$$\text{TE}: \quad \gamma_{11}a = 1.841, \quad \gamma_{21}a = 3.054, \quad \gamma_{01}a = 3.832, \tag{143.29}$$

$$\text{TM}: \quad \gamma_{01}a = 2.405, \quad \gamma_{11}a = 3.832, \quad \gamma_{02}a = 5.520. \tag{143.30}$$

Because $J'_0(x) = -J_1(x)$, $\gamma^{\text{TE}}_{0n} = \gamma^{\text{TM}}_{1n}$, but there are no other degeneracies. Once again the lowest mode is of type TE. The transverse fields for the TE_{11} and TM_{01} modes are shown in fig. 21.4.

Exercise 143.2 Construct a variational principle for TE and TM modes in a waveguide, and use it to find approximations for the $m = 0$ and $m = 1$ TE and TM modes in a circular waveguide.

Energy losses: Because of the nonzero conductivity of the walls, there will be some ohmic losses in every waveguide. The rate at which energy is lost per unit length of the waveguide is given by the last member of eq. (126.22) as

$$\dot{Q} = \frac{1}{2}\left(\frac{c}{4\pi}\right)^2 \text{Re } Z_s \oint_C |\mathbf{B}_t|^2 d\ell, \tag{143.31}$$

where Z_s is the surface impedance, and the integral is taken around the circumference of the cross section of the waveguide. On the other hand, the power flowing through any cross section is given by

$$P = \frac{c}{8\pi} \text{Re} \int_S \hat{\mathbf{z}} \cdot (\mathbf{E}_t \times \mathbf{B}_t^*)d^2s. \tag{143.32}$$

Thus, the power flow at a distance z along the waveguide can be written as

$$P(z) = P_0 e^{-\alpha z}, \tag{143.33}$$

where α, the power attenuation coefficient, is given by $\dot{Q}/P$. The problem is thus to evaluate these quantities. Now,

$$\hat{\mathbf{z}} \cdot (\mathbf{E}_t \times \mathbf{B}_t^*) = \frac{\omega k}{c\gamma^4}|\nabla_t\psi|^2, \tag{143.34}$$

where ψ is either B_z (TE) or E_z (TM). By integrating by parts and using the wave equation, we get

$$\int_S |\nabla_t \psi|^2 d^2 s = \gamma^2 \int_S |\psi|^2 d^2 s. \tag{143.35}$$

Hence,

$$P = \frac{\omega k}{8\pi \gamma^2} \int_S |\psi|^2 d^2 s. \tag{143.36}$$

To find $\dot{Q}$, we must integrate $|\mathbf{B} \times \hat{\mathbf{n}}|^2$ around the circumference. Now,

$$|\mathbf{B} \times \hat{\mathbf{n}}|^2 = \begin{cases} \dfrac{k^2}{\gamma^4} |\nabla_t B_z|^2 + |B_z|^2 & \text{(TE)}, \\[2ex] \dfrac{\omega^2}{c^2 \gamma^4} \left| \dfrac{\partial E_z}{\partial n} \right|^2 & \text{(TM)}. \end{cases} \tag{143.37}$$

From the Helmholtz equation, we can estimate $|\nabla_t \psi|^2$ as roughly $\gamma^2 |\psi|^2$, so

$$\oint_C |\nabla_t \psi|^2 d\ell \simeq \frac{\eta}{\ell} \gamma^2 \int_S |\psi|^2 d^2 s, \tag{143.38}$$

where ℓ is a characteristic transverse dimension of the waveguide, and η is a number of order unity. Thus, for the power attenuation coefficient, we have,

$$\alpha \simeq \frac{1}{\ell} \frac{c^2 \gamma^2}{4\pi \sigma \delta \omega k} \begin{cases} \dfrac{k^2}{\gamma^2} \eta_1 + \eta_2 & \text{(TE)}, \\[2ex] \dfrac{\omega^2}{c^2 \gamma^2} \eta_3 & \text{(TM)}. \end{cases} \tag{143.39}$$

The η_i are again numbers of order unity. For very high frequencies $\omega \gg c\gamma$, $\alpha \sim (\omega/\sigma)^{1/2} \ell^{-1}$.

144 Resonant cavities

Any metallic body that is hollow on the inside can serve as a resonant cavity that can support specific modes of oscillatory electromagnetic fields. These modes are analogous to the characteristic vibrations of a string or a drumhead. For a general shape, these can be found by solving for the potentials in a suitable gauge. We shall consider only the simplest type of cavity, obtained by putting metallic end caps at $z = 0$ and $z = L$ on the cylindrical waveguide of the previous section. Our previous discussion can be taken over almost entirely. The main change is that we must now have standing waves along the z direction, so in the z dependence of the modes,

$$e^{ikz} \to a \cos kz + b \sin kz, \tag{144.1}$$

where a and b are constants depending on the boundary conditions at the end caps. These are different for TE and TM modes, as we now show.

TE modes: Now, $B_z = 0$ at $z = 0$ and $z = L$, so

$$B_z(\mathbf{r}) = \psi(x, y) \sin\left(\frac{\pi z}{L} p\right), \quad p = 1, 2, \ldots, \tag{144.2}$$

where $\psi(x, y)$ is the same solution to the two-dimensional Helmholtz equation as discussed before.

TM modes: The relevant boundary condition now is $\mathbf{E}_t = 0$ at $z = 0$ and $z = L$. To relate $\mathbf{E}_t$ to E_z, we must rewrite the first of eq. (143.17) as

$$\mathbf{E}_t = \frac{1}{\gamma^2} \nabla_t \frac{\partial E_z}{\partial z}. \tag{144.3}$$

Hence, $\partial E_z / \partial z$ must vanish on the end caps, which implies that

$$E_z(\mathbf{r}) = \psi(x, y) \cos\left(\frac{\pi z}{L} p\right), \quad p = 0, 1, \ldots. \tag{144.4}$$

Note that now $p = 0$ is a possibility, i.e., that the fields need have no z dependence.

Since the z variation of the modes is also discretized, it follows that the spectrum of oscillation frequencies is completely discrete. For a cavity with circular cross section, we write these as

$$\omega_{mnp} = \left[\gamma_{mn}^2 + \left(\frac{\pi p}{L}\right)^2\right]^{1/2} c. \tag{144.5}$$

By far the hardest thing about this formula is remembering what the suffixes m, n, p mean: they are the azimuthal, radial, and longitudinal "quantum" numbers, respectively.

***Q*-factor:** The lossiness of a cavity is specified in terms of the Q-factor, defined as

$$Q = 2\pi \frac{\mathcal{E}}{\Delta\mathcal{E}}, \tag{144.6}$$

where $\mathcal{E}$ is the energy stored in the cavity, and $\Delta\mathcal{E}$ is the energy lost in one cycle. This loss can be found from the power loss in a waveguide as follows. We may think of the standing wave as a running wave that is reflected back and forth by the end caps. If we multiply the attenuation coefficient α for a waveguide by $2L$, it is clear that we will get the fractional energy loss per oscillation period due to the side walls of the cavity. The contribution from the end caps must be of the same order of magnitude. The key dependence in the formula (143.39) is that on σ and δ. Recalling that $\delta \sim (\omega\sigma)^{-1/2}$, we thus obtain the estimate

$$Q \simeq \frac{\ell}{\delta} \eta, \tag{144.7}$$

where ℓ is a linear dimension of the cavity, and η is a geometry-dependent factor of order unity. If we consider a cavity with $\ell \sim 5$ cm and take $\delta \sim 1 \, \mu$m as appropriate at microwave frequencies, we see that Q factors of 10^3–10^4 are attainable. Still higher Q's, of order 10^9, can be obtained with cavities with superconducting walls.

It is important to note that the losses also shift the resonance frequency. For large Q's this shift can be found in exactly the same way as for a very weakly damped linear harmonic oscillator, and is given by

$$\Delta\omega_0 = -\frac{\omega_0}{2Q}. \tag{144.8}$$

Exercise 144.1 Show that for a cavity of any shape, (a) the electric and magnetic energies are equal, and (b) the fields for two distinct modes labeled a and b obey

$$\int \mathbf{E}_a \cdot \mathbf{E}_b \, d^3x = \int \mathbf{B}_a \cdot \mathbf{B}_b \, d^3x = 0. \tag{144.9}$$

22 | Scattering of electromagnetic radiation

When electromagnetic radiation falls on a medium, the charges in the medium are accelerated and themselves radiate electromagnetic waves. In some cases, dielectrics, e.g., this reradiation combines with the original radiation in such a way as to lead to wave propagation, and the problem is analyzed using the language of the refractive index and the frequency-dependent dielectric function. If instead of a bulk medium, we have a small piece of matter, or even a single charge, the reradiation is said to arise from *scattering* of the incident wave. Scattering is an enormously important tool in the analysis of material media, as evidenced by the sheer variety of names: Thomson, Rayleigh, Compton, X-ray, Brillouin light scattering, Raman, etc. In this chapter, we study some of these.

Our task is complicated by the fact that the ultimate description of matter, and even of radiation, must be quantum mechanical. This aspect is especially important when the frequency of the radiation becomes comparable to the internal frequencies of atoms. Absorption and emission then take place at definite spectral frequencies, a fact that classical mechanics cannot explain. Even at low frequencies, the polarizabilites of atoms and molecules must eventually be found by quantum mechanics. At very high frequencies, the quantal nature of light becomes important, as in the Compton effect. Nevertheless, there is a wide range of phenomena where classical physics is applicable, and even where it fails in the details, it provides a valuable way of thinking about the problem.

145 Scattering terminology

The basic descriptor of scattering is the *cross section*, σ, defined as the power scattered divided by the incident energy flux:

$$\sigma = \frac{\text{Scattered power}}{\text{Incident energy flux}}. \tag{145.1}$$

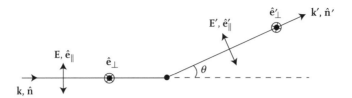

FIGURE 22.1. Scattering descriptors.

Since the energy flux is the energy passing unit area in unit time, the cross section has dimensions of area. By considering the scattered radiation in an infinitesimal solid angle around a given direction, we obtain the *differential scattering cross section*, denoted $d\sigma/d\Omega$, in analogy with the radiated power spectrum. The cross section may be further specialized with regard to the frequency of the incident or scattered radiation.

We shall denote the wave vector of the incoming light by **k**, its direction (equal to $\hat{\mathbf{k}}$) by $\hat{\mathbf{n}}$, electric field by **E**, and polarization by $\hat{\mathbf{e}}$, and attach primes to denote the same quantities for the scattered light. The angle between the vectors $\hat{\mathbf{n}}$ and $\hat{\mathbf{n}}'$ is the scattering angle, θ, and the plane containing them is known as the *scattering plane* (see fig. 22.1). It is often interesting to study the scattering of polarized light, or the polarization of the scattered light. The cross section is then still further specialized with regard to the polarization of the incident or scattered radiation. We denote polarization in and out of the scattering plane by the vectors $\hat{\mathbf{e}}_\parallel$, $\hat{\mathbf{e}}_\perp$ $\hat{\mathbf{e}}'_\parallel$, and $\hat{\mathbf{e}}'_\perp$. Unpolarized incident light may be regarded as an incoherent equal sum of polarizations $\hat{\mathbf{e}}_\parallel$ and $\hat{\mathbf{e}}_\perp$, and the scattering cross section is obtained by averaging the corresponding cross sections.

146 Scattering by free electrons

The simplest type of scattering is by a free electron. We assume that in the lab frame, we may ignore any displacement of the electron, which, as we shall see, is justified when the incident radiation is weak enough. This is equivalent to assuming that the motion of the electron is nonrelativistic, so the Lorentz force due to the magnetic field of the incident light may be neglected.

Let the mass and velocity of the electron be m and $\mathbf{v}$. If the incident electric field is $\mathbf{E}(t)$, the acceleration $\dot{\mathbf{v}}$ is $-e\mathbf{E}/m$. By the Larmor formula (58.12), the power radiated is

$$P_{\mathrm{rad}} = \frac{2}{3}\frac{e^4 E^2}{m^2 c^3}. \tag{146.1}$$

The incident energy flux is given by the Poynting vector and equals $c E^2/4\pi$. Hence, the total scattering cross section is

$$\sigma = \frac{8\pi}{3}\left(\frac{e^2}{mc^2}\right)^2 = \frac{8\pi}{3}r_e^2, \tag{146.2}$$

where the last form is written using the classical radius of the electron, $r_e = e^2/mc^2$. This is known as the *Thomson cross section*. It is independent of frequency and is of magnitude 6.65×10^{-25} cm^2.

If the incident light is polarized along $\hat{\mathbf{e}}$, the scattered radiation will be that of a dipole oscillating along $\hat{\mathbf{e}}$, leading to a differential scattering cross section,

$$\frac{d\sigma}{d\Omega} = \frac{e^4}{m^2c^4}(\hat{\mathbf{n}}' \times \hat{\mathbf{e}})^2. \tag{146.3}$$

Integration over all final directions $\hat{\mathbf{n}}'$ reproduces eq. (146.2).

We also have (see fig. 22.1)

$$(\hat{\mathbf{n}}' \times \hat{\mathbf{e}})^2 = \begin{cases} \cos^2\theta, & \hat{\mathbf{e}} = \hat{\mathbf{e}}_\parallel, \\ 1, & \hat{\mathbf{e}} = \hat{\mathbf{e}}_\perp. \end{cases} \tag{146.4}$$

Thus, for unpolarized light, the differential cross section is

$$\left(\frac{d\sigma}{d\Omega}\right)_{\text{unpol}} = \frac{e^4}{m^2c^4}\frac{1+\cos^2\theta}{2}. \tag{146.5}$$

The scattering is a minimum at right angles to the incident light and has forward–backward symmetry, i.e., it is equal for scattering angles θ and $\pi - \theta$. Also, the scattered light has the same frequency as the incident light.

The characteristic velocity of the electron in this scattering process is $eE/m\omega$, where ω is the incident light frequency. Our analysis holds only if this velocity is much less than c, i.e., if

$$E \ll m\omega c/e. \tag{146.6}$$

For lightof wavelength 5000 Å, this limits us to $E \ll 6 \times 10^{12}$ V/m. The limiting fields are quite large, but they can be realized with picosecond laser pulses with power 10^{13} W, focused down to a 10-μm-diameter spot.

A more serious failure of the classical analysis occurs due to quantum effects at high frequencies, when $\hbar\omega \sim mc^2$, i.e., at photon energies ~ 0.5 MeV. In fact, the failure is significant even at $\hbar\omega/mc^2 \simeq 0.2$—the differential scattering cross section agrees with the classical formula at $\theta = 0$ but is reduced to well less than half that value at $\theta = \pi$. The classical cross section has forward–backward symmetry, so the scattered light carries off no net momentum, and all the momentum lost by the incident light is given to the electron. Since the scattered power is σI, with I being the incident light intensity, the electron acquires momentum at a rate $\sigma I/c$, i.e., it feels a force of this magnitude. In reality, each *individual* photon scatters conserving energy and momentum, so that light scattered at an angle θ in the lab frame has a wavelength

$$\lambda' = \lambda + \frac{2\pi\hbar}{mc}(1 - \cos\theta), \tag{146.7}$$

where λ is the wavelength of the incident light. This shift in wavelength is known as the *Compton effect*, and the quantity $2\pi\hbar/mc$ ($= 0.024$ Å for the electron) is known as the *Compton wavelength*. If the spin of the electron is neglected, the cross section (146.5) is multiplied by a factor of $(\lambda/\lambda')^2$, which leads to a reduction of backscattering that is progressively more important at higher frequencies. In fact, scattering by the spin magnetic moment is quite important for $\hbar\omega \gtrsim mc^2$. The full formula for Compton scattering, named after Klein and Nishina, agrees very well with experiments and was one of the early successes of quantum electrodynamics.

Exercise 146.1 Find the differential Thomson scattering cross section for right- or left-circularly polarized incident light, and in this way obtain the formula for unpolarized light.

Exercise 146.2 Consider the scattering of unpolarized light by a free charge. Find the Stokes parameters and the degree of polarization at a scattering angle θ.

Answer: $a_1 = a_2 = 0$, $P = |a_3| = \sin^2\theta/(1 + \cos^2\theta)$. In particular, $90°$ scattered light is completely polarized perpendicular to the scattering plane.

147 Scattering by bound electrons

Let us now consider scattering by a bound charge, as in an atom or molecule. As a model, we consider an isotropic charged harmonic oscillator, with natural frequency ω_0, and linear damping with a small damping constant $\gamma_T \ll \omega_0$. The equation of motion for the charge is then

$$m\ddot{\mathbf{r}} + m\gamma_T\dot{\mathbf{r}} + m\omega_0^2\mathbf{r} = -e\mathbf{E}_0 e^{-i\omega t}. \tag{147.1}$$

The steady-state solution to this equation is

$$\mathbf{r}(t) = \frac{e\mathbf{E}_0}{m}\frac{1}{\omega^2 - \omega_0^2 + i\omega\gamma_T}e^{-i\omega t}. \tag{147.2}$$

Differentiating twice to obtain the acceleration, and feeding that result into the Larmor formula, we obtain the power radiated as

$$P_{\text{rad}} = \frac{2e^2}{3c^3}\frac{1}{2}\frac{e^2}{m^2}|\mathbf{E}_0|^2\frac{\omega^4}{(\omega^2 - \omega_0^2)^2 + \omega^2\gamma_T^2}. \tag{147.3}$$

Dividing by the incident energy flux, $\bar{\mathbf{S}} = c|\mathbf{E}_0^2|/8\pi$, we obtain the scattering cross section,

$$\sigma_{\text{sc}} = \frac{8\pi}{3}r_e^2\frac{\omega^4}{(\omega^2 - \omega_0^2)^2 + \omega^2\gamma_T^2}. \tag{147.4}$$

This tends to the free charge cross section for $\omega \gg \omega_0$. At low frequencies, $\omega \ll \omega_0$, it is reduced by a factor $(\omega/\omega_0)^4$. The proportionality to ω^4 is a universal feature, which we shall see again when we discuss Rayleigh scattering. At resonance, $\omega = \omega_0$, the cross section is greatly enhanced over the value for a free charge:

$$\sigma_{sc} = \sigma_{free} \left(\frac{\omega_0}{\gamma_T} \right)^2 \qquad (\omega = \omega_0). \tag{147.5}$$

Let us now inquire into the origin of the damping rate γ_T. The part of this rate due to radiation reaction is (eq. (67.14))

$$\gamma = \frac{2}{3} \omega^2 \frac{r_e}{c}. \tag{147.6}$$

In addition, there may be other dissipative processes, such as collisions, lattice vibrations (if the atom is in a solid), etc. Let the damping rate due to all these processes be γ'. Then,

$$\gamma_T = \gamma + \gamma'. \tag{147.7}$$

It follows that not all the energy absorbed by the oscillator is reemitted as radiation. The power absorbed is given by the time average of $-e\mathbf{E} \cdot \mathbf{v}$, i.e.,

$$\begin{aligned} P_{abs} &= \frac{1}{2} \mathrm{Re} \, \frac{(-e)(-ie\omega)\mathbf{E}_0 \cdot \mathbf{E}_0^*}{m(\omega^2 - \omega_0^2 + i\omega\gamma_T)} \\ &= \frac{1}{2} \frac{e^2 \omega^2 \gamma_T}{m[(\omega^2 - \omega_0^2)^2 + \omega^2\gamma_T^2]} |\mathbf{E}_0|^2. \end{aligned} \tag{147.8}$$

Dividing this by the incident intensity $c|\mathbf{E}_0|^2/8\pi$ gives the total absorption cross section,

$$\sigma_{abs} = \frac{4\pi e^2}{mc} \frac{\omega^2 \gamma_T}{(\omega^2 - \omega_0^2)^2 + \omega^2 \gamma_T^2}. \tag{147.9}$$

Thus,

$$\frac{\sigma_{abs}}{\sigma_{sc}} = \frac{3}{2} \frac{c}{\omega^2 r_e} \gamma_T = \frac{\gamma_T}{\gamma}, \tag{147.10}$$

which is clearly greater than unity. At resonance, it is also revealing to write σ_{abs} as follows:

$$\sigma_{abs}(\omega_0) = 4\pi \frac{e^2}{mc} \frac{1}{\gamma_T} = \frac{6\pi}{k_0^2} \frac{\gamma}{\gamma_T}, \tag{147.11}$$

where $k_0 = c/\omega_0$ is the wave vector. Since $\gamma \le \gamma_T$, we have the upper bound $\sigma_{abs}(\omega_0) \le 6\pi/k_0^2$.

Our model for an atom or molecule is clearly not realistic. Nevertheless, the formulas for the cross sections hold reasonably well with small modifications. An atom that has energy levels $\mathcal{E}_i$ can resonate at the transition frequencies $\omega_i = (\mathcal{E}_i - \mathcal{E}_0)/\hbar$ if it is initially in the ground level, $\mathcal{E}_0$. However, not all possible frequencies occur, or even do so with equal effectiveness. To each transition, we assign an *oscillator strength*, f_i, a dimensionless number of order unity, which is a measure of the amplitude of the dipole moment of the atom when it is oscillating between levels $\mathcal{E}_i$ and $\mathcal{E}_0$. Equation (147.4) is generalized to

$$\sigma_{sc} = \frac{8\pi}{3} r_e^2 \sum_i \frac{f_i \omega^4}{(\omega^2 - \omega_i^2)^2 + \omega^2 \gamma_{i,T}^2}, \tag{147.12}$$

where $\gamma_{i,T}$ is the total damping or *width* of the resonance at ω_i.

It is interesting to apply our discussion to the scattering of X-rays from atoms. X-ray energies are 10–100 keV, whereas the energy level differences $\hbar \omega_i$ are ~ 10 eV. Physically, $\omega_i \sim v/a$, where v is the electron velocity and a is the dimension of the atom. The X-ray wavelength is comparable to this dimension, i.e., $\omega \sim c/a$. Thus, $\omega_i/\omega \sim v/c \ll 1$. In one X-ray period, the electrons hardly move at all, so the force on an electron due to the nucleus and the other electrons may be neglected. The scattering is then essentially that from a free electron, and the Thomson formula may be used, provided the X-ray energy is not so high as to make the Compton effect important. If there are Z electrons in one atom, each one scatters independently, and the net cross section is

$$\sigma = \frac{8\pi}{3} Z r_e^2. \tag{147.13}$$

When $\omega \gg \omega_i$, the quantum mechanical answer (147.12) reduces to $\sigma_{sc} = (8\pi r_e^2/3) \Sigma_i f_i$. This must agree with the classical answer, so we conclude that

$$\sum_i f_i = Z. \tag{147.14}$$

This is the f-sum rule (also known as the conductivity or dipole sum rule) introduced in chapter 18. It requires the oscillator strengths to vanish sufficiently rapidly for high-frequency transitions, and it played an important role in the historical development of quantum mechanics.[1]

148 Scattering by small particles

Next, let us consider scattering from particles of dimension $a \ll \lambda$, the wavelength of the light. The fields acting on the particle may then be considered to be uniform, and the particle acquires a time-dependent electric dipole moment, $\mathbf{d}(t)$, that oscillates at the frequency of the incident wave. It also acquires a magnetic dipole as well as higher moments, but these are often smaller and may at first be neglected. The power radiated by this induced dipole is

$$P_{rad} = \frac{2}{3c^3} \ddot{\mathbf{d}}^2 = \frac{\omega^4}{3c^3} |\mathbf{d}_\omega|^2. \tag{148.1}$$

The scattering thus varies as ω^4, just as we saw in the previous section for low-frequency scattering from bound charges. In fact, that behavior is now seen to also be due to oscillations of the induced dipole moment.

---

[1] A clear discussion of the quantum theory of absorption, emission, and scattering of light by matter, and of the dipole sum rule, is given by Baym (1969), chap. 13. See especially pp. 286–287.

As an example, let us take the scatterer to be a dielectric sphere of radius a, with real dielectric constant $\epsilon(\omega)$ (absorption is assumed negligible). Then,

$$\mathbf{d}_\omega = \frac{\epsilon(\omega) - 1}{\epsilon(\omega) + 2} a^3 \mathbf{E}_0, \tag{148.2}$$

as found in chapter 15 (see eq. (95.17)). Thus, the scattering cross section is

$$\sigma = \frac{8\pi}{3} \left(\frac{\epsilon(\omega) - 1}{\epsilon(\omega) + 2} \right)^2 \frac{\omega^4 a^6}{c^4}. \tag{148.3}$$

When the scatterer is highly conducting, magnetic scattering cannot be neglected and, in fact, interferes with the electric scattering. The fields may be regarded as quasistatic, and both electric and magnetic dipoles are induced. For a sphere, the induced moments are (see eqs. (89.27) and (127.27))

$$\mathbf{d}_\omega = a^3 \mathbf{E}_0, \tag{148.4}$$

$$\mathbf{m}_\omega = -\frac{1}{2} a^3 \mathbf{B}_0. \tag{148.5}$$

The radiated electric field at a distance R is given by

$$\mathbf{E}' = \frac{1}{c^2 R} [(\ddot{\mathbf{d}} \times \hat{\mathbf{n}}') \times \hat{\mathbf{n}}' - (\ddot{\mathbf{m}} \times \hat{\mathbf{n}}')]. \tag{148.6}$$

Going to the frequency domain, feeding in the formulas for $\mathbf{d}_\omega$ and $\mathbf{m}_\omega$, and writing $\mathbf{B}_0 = \mathbf{E}_0 \times \hat{\mathbf{n}}$, we get

$$\mathbf{E}'_\omega = -\frac{\omega^2 a^3}{c^2 R} \left(\mathbf{E}_0 \times \hat{\mathbf{n}}' - \frac{1}{2} \mathbf{E}_0 \times \hat{\mathbf{n}} \right) \times \hat{\mathbf{n}}'. \tag{148.7}$$

Taking the absolute square and multiplying by $(c R^2 / 8\pi) d\Omega$, we obtain the power scattered into the solid angle $d\Omega$ around $\hat{\mathbf{n}}'$:

$$\frac{d P_{\text{rad}}}{d\Omega} = \frac{\omega^4 a^6}{8\pi c^3} \left[|\mathbf{E}_0 \times \hat{\mathbf{n}}'|^2 - \text{Re} \, (\mathbf{E}_0 \times \hat{\mathbf{n}}) \cdot (\mathbf{E}_0^* \times \hat{\mathbf{n}}') + \frac{1}{4} |(\mathbf{E}_0 \times \hat{\mathbf{n}}) \times \hat{\mathbf{n}}'|^2 \right]. \tag{148.8}$$

Dividing by $c|\mathbf{E}_0|^2 / 8\pi$ and introducing the incident light polarization $\hat{\mathbf{e}}$, we obtain the scattering cross section,

$$\frac{d\sigma}{d\Omega} = \frac{\omega^4 a^6}{c^4} \left[(\hat{\mathbf{e}} \times \hat{\mathbf{n}}')^2 - \hat{\mathbf{n}} \cdot \hat{\mathbf{n}}' + \frac{1}{4} ((\hat{\mathbf{e}} \times \hat{\mathbf{n}}) \times \hat{\mathbf{n}}')^2 \right]. \tag{148.9}$$

For the particular cases of $\hat{\mathbf{e}}$ in the scattering plane and perpendicular to it, this simplifies to

$$\frac{d\sigma}{d\Omega} = \frac{\omega^4 a^6}{c^4} \times \begin{cases} \cos^2 \theta - \cos \theta + \frac{1}{4}, & \hat{\mathbf{e}} = \hat{\mathbf{e}}_\parallel, \\ 1 - \cos \theta + \frac{1}{4} \cos^2 \theta, & \hat{\mathbf{e}} = \hat{\mathbf{e}}_\perp. \end{cases} \tag{148.10}$$

Averaging these two expressions and integrating over all solid angles, we obtain the total cross section for scattering of unpolarized light:

$$\sigma_{\text{unpol}} = \frac{10\pi}{3} \frac{\omega^4 a^6}{c^4}. \tag{148.11}$$

This is 25% larger than the $\epsilon \to \infty$ limit of the answer for a dielectric sphere. Note, however, that because of the interference, the scattering is not forward–backward symmetric but has a pronounced backward peak.

Exercise 148.1 Find the force due to radiation pressure on a small dielectric sphere by considering energy–momentum conservation. Find its value for a 1-μm radius, and refractive index 1.05 (a model of a biological cell).

Exercise 148.2 Find the degree of polarization as a function of scattering angle for scattering of unpolarized light by a dielectric sphere.

Exercise 148.3 Consider the scattering of unpolarized light by a conducting sphere, and find the differential cross sections for the *scattered* light to be linearly polarized in and out of the scattering plane. Also find the degree of polarization.

Exercise 148.4 Write the various cross sections for scattering from a conducting particle in terms of the electric and magnetic polarizabilities α_e and α_m.

149 Scattering by dilute gases, and why the sky is blue

In a dilute gas, the molecules may be regarded as scattering independently of each other, since there are no correlations between them on the scale of the wavelength of the light, λ.[2] We assume that the frequency of the light is not high enough to excite electronic transitions in the molecules, so there is no resonant absorption. At the same time, we assume that it is much larger than the frequencies of molecular vibrations or rotations. These conditions are well satisfied for light in the visible range of the spectrum. In this situation, the nuclei in the molecule hardly move at all (including center of mass motion) in one time period of the light, and may be regarded as frozen. The electrons, however, are so agile that they carry out many orbits in the same period. The treatment of the previous section is then immediately applicable. Let the electric polarizability of one molecule at the light frequency be α. (We neglect the magnetic polarizability, as it is usually very small.) The induced dipole moment is $\alpha \mathbf{E}$, and the scattering cross section per molecule is

$$\sigma = \frac{8\pi}{3} \frac{\omega^4}{c^4} \alpha^2. \tag{149.1}$$

The single-molecule cross section is an inconvenient descriptor of scattering from a gas. It is better to give the absorption coefficient (also called the *extinction coefficient*), β, defined as the ratio of the total amount of energy scattered per unit volume of gas per unit time, to the incident energy flux. This quantity is clearly obtained by multiplying σ by ρ_n,

---

[2] In the atmosphere at STP, the volume per molecule is 3.7×10^{-20} cm^3, i.e., there are 2.7×10^{19} molecules per cm^3, and the molecular mean free path, $\ell \sim 10^4$ Å.

the number of gas molecules per unit volume:[3]

$$\beta = \frac{8\pi}{3}\frac{\omega^4}{c^4}\rho_n\alpha^2. \tag{149.2}$$

To further understand the meaning of β, note that in a distance Δx traveled by the beam, a fraction $\rho_n\sigma\Delta x$ of the energy is lost, i.e., $\Delta I/\Delta x = -\rho_n\sigma I = -\beta I$. Integrating, we get $I(x) = I_0 e^{-\beta x}$. Hence, β^{-1} gives the length in which the intensity of the incident beam drops to $1/e$ of its initial value.

It is traditional to express β in terms of the refractive index of the gas. From section 99, we have the dielectric constant for a nonpolar gas,

$$\epsilon = 1 + 4\pi\rho_n\alpha. \tag{149.3}$$

Since the imaginary part of ϵ is very small in comparison to the real part, we have $n = \sqrt{\epsilon}$, and since ϵ is very close to unity, we have

$$n = 1 + 2\pi\rho_n\alpha. \tag{149.4}$$

Thus, finally,

$$\beta = \frac{2}{3\pi\rho_n}\frac{\omega^4}{c^4}(n-1)^2. \tag{149.5}$$

Equation (149.5) was obtained by Rayleigh in 1871 and used to explain the blue color of the sky and the red color of the sunset. The ω^4 dependence implies that higher frequencies are scattered much more than lower ones. With $n = 1.000293$ and $\rho_n = 2.69 \times 10^{19}$ cm^{-3}, we obtain β^{-1} equal to 39 km for $\lambda = 4500$ Å (blue) and 170 km for $\lambda = 6500$ Å (red). The blue light is scattered four times more strongly than the red. At sunset, the western sky appears red because the higher frequencies are removed from the forward direction of propagation. Finally, the polarization of scattered sunlight is also explained. We expect the light to be partially polarized perpendicular to the scattering plane. This plane has a major vertical component, so we expect the polarization to be roughly horizontal, as is observed.

Our treatment neglects multiple scattering events. For the atmosphere, this is justified by the large value of β^{-1}, which may be regarded as the mean free path of a photon. A bigger shortcoming is that the molecules O_2 and N_2, which make up the air almost entirely, are linear and not spherical, and their rotational frequencies are much less than those of visible light. Thus, their polarizabilities along the molecular axis, $\alpha_{\parallel}$, and perpendicular to it, $\alpha_{\perp}$, are unequal. The result is that the above formulas are correct provided α^2 is replaced by the rotational average of the tensor α_{ik}^2:

$$\alpha^2 \rightarrow \frac{1}{3}(\alpha_{\parallel}^2 + 2\alpha_{\perp}^2). \tag{149.6}$$

---

[3] The choice of ρ_n to denote the molecular number density is inelegant, but there is no other good way out of symbol confusion here. We wish to reserve n for the refractive index of the gas, and ρ for charge density. Likewise, we wish to reserve α for the polarizability, which many authors use for the extinction coefficient. In terms of the attenuation coefficient κ defined in section 133, $\beta = 2(\omega/c)\kappa$.

To see this, let us assume that the molecular axis is oriented along $\hat{\mathbf{u}}$, and the incident light is linearly polarized. Then, the induced dipole moment of the molecule is

$$\mathbf{d} = \alpha_{\parallel}(\mathbf{E} \cdot \hat{\mathbf{u}})\hat{\mathbf{u}} + \alpha_{\perp}[\mathbf{E} - (\mathbf{E} \cdot \hat{\mathbf{u}})\hat{\mathbf{u}}]. \tag{149.7}$$

The scattered power is proportional to $\ddot{\mathbf{d}}^2$, so

$$\beta \propto [\alpha_{\perp}^2 E_i E_i + (\alpha_{\parallel}^2 - \alpha_{\perp}^2) E_i E_j \hat{u}_i \hat{u}_j]. \tag{149.8}$$

We now average over all molecular orientations. This has the effect of replacing $\hat{u}_i \hat{u}_j$ by the tensor $\delta_{ij}/3$. Hence,

$$\beta \propto \left[\alpha_{\perp}^2 + \frac{1}{3}(\alpha_{\parallel}^2 - \alpha_{\perp}^2) \right] E^2. \tag{149.9}$$

The term in square brackets is precisely the average in eq. (149.6). The main effect of the molecular anisotropy is on the polarization of scattered light. In particular, even for 90° scattering, there is some polarization in the scattering plane, as the following exercise shows.

Exercise 149.1 Consider scattering of unpolarized light from a gas of linear molecules. Find the differential cross sections for scattered light polarized in and perpendicular to the scattering plane. What is the degree of polarization for scattering through 90°?

Solution: For the scattered light, $\mathbf{E}' \propto \mathbf{d} - (\mathbf{d} \cdot \hat{\mathbf{k}}')\hat{\mathbf{k}}'$. For a fixed molecular orientation $\hat{\mathbf{u}}$ and a fixed initial polarization $\mathbf{E}$, the scattered power in either of the polarization components is obtained by projecting $\mathbf{E}'$ onto $\hat{\mathbf{e}}'_{\parallel}$ or $\hat{\mathbf{e}}'_{\perp}$ and squaring. The desired cross sections are obtained by averaging the resulting expressions over $\hat{\mathbf{u}}$ and $\mathbf{E}$ directions.

The essential quantity that must be averaged is the tensor $d_i d_j$. The averaging over $\hat{\mathbf{u}}$ is easily shown to yield

$$d_i d_j \rightarrow \left(\bar{\alpha}^2 + \frac{1}{45}\Delta^2 \right) E_i E_j + \frac{1}{15}\Delta^2 E^2 \delta_{ij}, \tag{149.10}$$

where $\bar{\alpha} = (\alpha_{\parallel} + 2\alpha_{\perp})/3$, and $\Delta = (\alpha_{\parallel} - \alpha_{\perp})$. The subsequent averaging over the orientations of $\mathbf{E}$ is carried out by the formulas

$$\overline{(\mathbf{E} \cdot \hat{\mathbf{k}}')^2} = \overline{(\mathbf{E} \cdot (\hat{\mathbf{k}} \times \hat{\mathbf{k}}'))^2} = \frac{1}{2}\sin^2\theta\, E^2. \tag{149.11}$$

The resulting cross sections have the angular dependence

$$\left(\frac{d\sigma}{d\Omega} \right)_{\parallel,\perp} \propto \frac{1}{15}\Delta^2 + \frac{1}{2}\left(\bar{\alpha}^2 + \frac{1}{45}\Delta^2 \right) \times \begin{pmatrix} \cos^2\theta \\ 1 \end{pmatrix}. \tag{149.12}$$

The polarization is easily found at any angle from these expressions. In particular, at 90°, the component polarized in the scattering plane has a fraction of the intensity equal to approximately $2\Delta^2/15\bar{\alpha}^2$, when Δ is small.

150 Raman scattering

In the previous section, we treated the molecular polarizability as a fixed quantity and neglected the nuclear motions. When this approximation is relaxed, a new type of scattering appears, in which the frequency of the scattered light is shifted from that of the incident light. This effect was discovered by Raman in 1928, in scattering from water and other liquids.

For specificity, let us consider a diatomic molecule and let the deviation of the internuclear distance from its mean value be denoted by ξ. If we somehow hold ξ fixed at a nonzero value, the *electronic* polarizability of the molecule will differ from its $\xi = 0$ value. The difference is extremely small, however, so we may expand the polarizability in powers of ξ:

$$\alpha = \alpha_0 + \alpha_1 \xi. \tag{150.1}$$

On dimensional grounds, we expect $\alpha_1 \sim \alpha_0/a_0$, where a_0 is an atomic size, of the order of the Bohr radius. Since vibrational excursions in molecules are much smaller than this size, i.e., $\xi \ll a_0$, the expansion is justified.

Let us now suppose that the molecule is vibrating at its natural vibrational frequency ω_v, and that it is illuminated by light at a frequency $\omega_0 \gg \omega_v$. Since vibrational frequencies are typically in the infrared, while electronic transitions are in the visible or ultraviolet, it is safe to assume that the electronic distribution can respond more rapidly than the time scale $2\pi/\omega_v$. On this time scale, therefore, eq. (150.1) may be regarded as holding instantaneously, and we may write the induced dipole moment as

$$\begin{aligned}
\mathbf{d}(t) &= \left(\alpha_0 + \alpha_1 \xi(t)\right)\mathbf{E}(t) \\
&= \frac{1}{2}\operatorname{Re}\left(\alpha_0 e^{-i\omega_0 t} + \tfrac{1}{2}\alpha_1\left(\xi_0 e^{-i(\omega_0+\omega_v)t} + \xi_0^* e^{-i(\omega_0-\omega_v)t}\right)\right)\mathbf{E}_0.
\end{aligned} \tag{150.2}$$

It is immediately evident that the dipole moment oscillates at the frequencies $\omega_0 \pm \omega_v$ in addition to ω_0. The power scattered into each frequency component is given by a simple extension of the Rayleigh formula. In particular, the ratio of the intensity in one of the Raman lines to the unshifted, or Rayleigh, line is given by

$$\frac{I_{\omega_0 \pm \omega_v}}{I_{\omega_0}} = \frac{\alpha_1^2}{4\alpha_0^2}\langle \xi_0^2 \rangle. \tag{150.3}$$

We have replaced the ratios $(\omega_0 \pm \omega_v)^4/\omega_0^4$ by unity and put angular brackets around ξ_0^2 to indicate a thermal averaging. If M is the reduced mass of the vibrational coordinate, and the liquid or gas is at a temperature T, the equipartition theorem gives $\langle \xi_0^2 \rangle = k_B T/M\omega_v^2$. Replacing α_1/α_0 by a_0^{-1}, we obtain

$$\frac{I_{\omega_0 \pm \omega_v}}{I_{\omega_0}} \simeq \frac{k_B T}{2M\omega_v^2 a_0^2}. \tag{150.4}$$

Taking $M \sim 10^{-23}$ g, $\omega_v \sim 10^{14}$ sec^{-1}, and $a_0 \sim 10^{-8}$ cm^{-1}, we see that this ratio is 10^{-3}–10^{-2}.

It is apparent that our treatment can be generalized to cover vibrations or rotations of polyatomic molecules, by letting ξ stand for an appropriate normal mode or rotational coordinate. All that is necessary is that an excitation of this mode should lead to a modulation of the polarizability of the system. If we again consider our simple example, we see that if the molecule is homonuclear, the vibration of the molecule does not produce a dipole moment. Rather, it produces a time-dependent quadrupole moment. Hence, such a mode can be excited only by the gradient of the electric field, and not the electric field itself, and is said to be *dipole inactive* in consequence. The absorption cross section for such transitions is smaller than that of dipole active modes by the factor $\sim(ka_0)^2$, so they appear as very weak *forbidden lines* in ordinary absorption or emission spectra. The great importance of the Raman effect is that it allows one to observe such modes much more easily. Dipole inactive modes are often Raman active. In fact, the basic idea extends to electronic states also, and one then speaks of electronic Raman transitions. As we shall see in the next section, scattering with a shift in frequency also occurs in liquids and dense gases.

Our classical treatment leads to equal intensities for the lines at $\omega_0 \pm \omega_v$, proportional to the temperature. A proper quantum mechanical treatment shows that the intensity of the higher sideband at $\omega_0 + \omega_v$ (known as the *anti-Stokes line*) is obtained by replacing the factor $k_B T/\omega_v$ by $\hbar/(e^{\hbar\omega_v/k_B T} - 1)$ and that of the lower sideband (*Stokes line*) exceeds that of the higher sideband by the factor $e^{\hbar\omega_v/k_B T}$. This result for the ratio of the intensities is very general and follows from detailed balance.

151 Scattering by liquids and dense gases*

The propagation of light in a transparent medium can be regarded as a process in which an incident wave scatters from the molecules in the medium, and the incident and scattered waves combine coherently to produce a wave that has the dispersion characteristics of the medium.[4] Since the medium is dense, its response is captured by the equations of macroscopic electrodynamics, and, as we saw in chapter 20, the wave equation that describes the wave in the medium can be derived directly from these equations. For this derivation to work, however, the dielectric constant and permeability of the medium must be uniform. In liquids and gases, deviations from uniformity arise spontaneously due to thermodynamic fluctuations. For example, in a volume V, the mean-square deviation of the number density of the fluid from its mean value is given by

$$\langle(\delta\rho_n)^2\rangle_V = \frac{1}{V}\rho_n^2 k_B T \kappa_T, \tag{151.1}$$

---

[4] An elementary treatment following from this point of view may be found in Feynman et al. (1964), vol. I, chap. 31. In reality, for a dense substance, one would need to consider not just the once-scattered waves but the response of the medium to the *total* electric field, including that of the scattered waves. In other words, one would have to consider multiple-order scattering and sum the fields from all orders to get the net behavior.

where κ_T is the isothermal compressibility of the fluid. The fluctuations that scatter light of wavelength λ most effectively are those that occur in a volume of linear size $\sim \lambda$. For visible light, there are a large number of molecules in such a volume, so these fluctuations can be regarded as creating fluctuations in the dielectric constant and permeability of the medium. The latter fluctuations then give rise to small changes in the electromagnetic field that are not part of the main wave and are described in terms of scattering.

One especially interesting aspect of this scattering is that it may take place with a change in frequency. Of course, if the fluid is composed of polyatomic molecules, the frequency may change because of the Raman effect, i.e., the excitation of intramolecular vibrations.[5] Let us recall that the frequency shift arises because these vibrations modulate the time dependence of the molecular dipole moment. In the same way, density fluctuations also modulate the time dependence of the local dipole moment of a small volume element of the fluid. These modulations take place on much longer time scales than the intramolecular vibrations, and the frequency shifts they give rise to are correspondingly smaller. It is this effect that we wish to study in this section.

Let us divide the displacement field $\mathbf{D}$ into a sum of the incident and scattered fields $\mathbf{D}^{(0)}$ and $\mathbf{D}^{sc}$. Likewise for $\mathbf{E}$, $\mathbf{B}$, and $\mathbf{H}$. We expect that $|\mathbf{D}^{sc}| \ll |\mathbf{D}^{(0)}|$, and so on. The ability to determine the scattered fields now rests on the possibility of being able to separate motions with vastly different time scales. Among the slow degrees of freedom, let us include the centers of mass of the molecules, and let us denote their configuration at time t by the symbol $X(t)$. All other degrees of freedom, the electrons in particular, are regarded as fast. Since the configuration $X(t)$ changes slowly on the time scale on which the electrons and the other fast degrees of freedom move, we may describe the medium in terms of a dielectric constant $\epsilon(\mathbf{r}, t)$ that includes an averaging over the fast variables but regards the center of mass configuration as frozen in time. Since the density fluctuations are small, we may write this dielectric function as a position- and time-independent average plus a small deviation,

$$\epsilon(\mathbf{r}, t) = \bar{\epsilon} + \delta\epsilon(\mathbf{r}, t). \tag{151.2}$$

For simplicity, we have taken the average dielectric function as isotropic. This assumption holds unless the medium is macroscopically anisotropic, as e.g., in a nematic liquid crystal. We have also taken the deviation $\delta\epsilon$ to be isotropic. This assumption is just slightly more restrictive, and it needs to be questioned only if the liquid is made of large anisotropic molecules. In this case, one needs to think about fluctuations in the *anisotropy*, in addition to those in density and energy. In general, however, anisotropy fluctuations are rapid enough to be counted among the fast degrees of freedom and are not included in the set $X(t)$. Further, the fluctuation $\delta\epsilon(\mathbf{r}, t)$ should be regarded as a stochastic variable, since the time variation of the $X(t)$ cannot be specified definitely but only on the average.

---

[5] The fact that the medium is dense does not change the frequency shift very much if the molecules maintain their integrity. The main effect of the intermolecular coupling of the excitations is a broadening of the Raman lines.

We can make a similar decomposition for the magnetic susceptibility, but there is little gained by doing so, as no new physical ideas are introduced. We shall assume that the magnetic scattering is negligible, so we can ignore all fluctuations and take $\mu = 1$ everywhere, i.e., $\mathbf{B} = \mathbf{H}$.

Let us now seek the modified wave equation obeyed by the total field $\mathbf{D}$. We have

$$\mathbf{D}(\mathbf{r}, t) = \epsilon(\mathbf{r}, t)[\mathbf{E}^{(0)}(\mathbf{r}, t) + \mathbf{E}^{sc}(\mathbf{r}, t)]$$

$$= \bar{\epsilon}\, \mathbf{E}(\mathbf{r}, t) + \delta\epsilon(\mathbf{r}, t)\mathbf{E}^{(0)}(\mathbf{r}, t), \tag{151.3}$$

where in the second line we have discarded the second-order term $\delta\epsilon\,\mathbf{E}^{sc}$. We now write the last equation in the frequency domain:

$$\mathbf{D}_\omega(\mathbf{r}) = \bar{\epsilon}(\omega)\mathbf{E}_\omega(\mathbf{r}) + \int \frac{d\omega'}{2\pi}\delta\epsilon(\mathbf{r}, \omega - \omega')\mathbf{E}^{(0)}_{\omega'}(\mathbf{r}). \tag{151.4}$$

Note that we have shown the average dielectric function as frequency dependent. There is no contradiction in this, as at optical frequencies, the slow variables are essentially static. Therefore, we can think of a frequency-dependent dielectric constant that is varying very slowly with time. The situation is not unlike that of a flickering candle, where we can say that the flame is blue (a certain frequency), but the intensity is changing with time. Essentially, this means that in eq. (151.2), we should think of $\epsilon(\mathbf{r}, t)$ as a slowly varying kernel or operator rather than just a function of $\mathbf{r}$ and t.

We now take the curl of the curl of eq. (151.4). We assume that there are no free charges or currents in the medium, so $\nabla \cdot \mathbf{D}_\omega(\mathbf{r}) = 0$, and $\nabla \times \mathbf{H}_\omega(\mathbf{r}) = \nabla \times \mathbf{B}_\omega(\mathbf{r}) = -i\omega\mathbf{D}_\omega(\mathbf{r})/c$. Thus,

$$\nabla \times (\nabla \times \mathbf{E}_\omega(\mathbf{r})) = \frac{i\omega}{c}\nabla \times \mathbf{B}_\omega(\mathbf{r}) = \frac{\omega^2}{c^2}\mathbf{D}_\omega(\mathbf{r}), \tag{151.5}$$

and $\nabla \times (\nabla \times \mathbf{D}_\omega(\mathbf{r})) = -\nabla^2\mathbf{D}_\omega(\mathbf{r})$. Hence,

$$\nabla^2\mathbf{D}_\omega(\mathbf{r}) + \frac{\bar{\epsilon}(\omega)\omega^2}{c^2}\mathbf{D}_\omega(\mathbf{r}) = -\nabla \times \left[\nabla \times \int \frac{d\omega'}{2\pi}\delta\epsilon(\mathbf{r}, \omega - \omega')\mathbf{E}^{(0)}_{\omega'}(\mathbf{r})\right]. \tag{151.6}$$

Equation (151.6) possesses the homogeneous solution $\mathbf{D}^{(0)}_\omega(\mathbf{r})$ corresponding to the unscattered wave. The scattered wave is immediately written down using the Green function obtained in section 54. Writing

$$k^2_\omega = \bar{\epsilon}(\omega)\omega^2/c^2, \tag{151.7}$$

we have

$$\mathbf{D}^{sc}_\omega(\mathbf{r}) = \frac{1}{4\pi}\int d^3x' \frac{e^{ik_\omega|\mathbf{r}-\mathbf{r}'|}}{|\mathbf{r}-\mathbf{r}'|}\nabla' \times \left[\nabla' \times \int \frac{d\omega'}{2\pi}\delta\epsilon(\mathbf{r}', \omega - \omega')\mathbf{E}^{(0)}_{\omega'}(\mathbf{r}')\right]. \tag{151.8}$$

For points far away from the scattering region, we can write the exponential phase factor as $e^{ik_\omega r}$ times $e^{-i\mathbf{k}_\omega \cdot \mathbf{r}'}$, where $\mathbf{k}_\omega = k_\omega\hat{\mathbf{r}}$, and approximate $1/|\mathbf{r}-\mathbf{r}'|$ by $1/r$. Also, we take the incident wave to be monochromatic with frequency ω_0 and wave vector $\mathbf{k}_0$. Integrating twice by parts and keeping only the leading order in powers of $1/r$, we get

$$\mathbf{D}^{\text{sc}}_{\omega}(\mathbf{r}) = \frac{k_{\omega}^2 e^{ik_{\omega}r}}{4\pi r} \int d^3x' \, \delta\epsilon(\mathbf{r}', \omega - \omega_0) \mathbf{E}_{0\perp} e^{i(\mathbf{k}_0 - \mathbf{k}_{\omega})\cdot\mathbf{r}'}, \tag{151.9}$$

where the suffix $\perp$ denotes the component perpendicular to the direction of the scattered light, i.e., to $\mathbf{k}_{\omega}$. It is more convenient to write this result in terms of $\mathbf{E}^{\text{sc}}_{\omega}$ and at the same time undo the Fourier transform of $\delta\epsilon$ with respect to time. Defining

$$\mathbf{q}_{\omega} = \mathbf{k}_{\omega} - \mathbf{k}_0, \tag{151.10}$$

and recalling that $k_{\omega}^2 = \bar{\epsilon}(\omega)\omega^2/c^2$, and $\mathbf{E}^{\text{sc}}_{\omega}(\mathbf{r}) = \mathbf{D}^{\text{sc}}_{\omega}(\mathbf{r})/\bar{\epsilon}(\omega)$ to leading order, we obtain

$$\mathbf{E}^{\text{sc}}_{\omega}(\mathbf{r}) = \frac{\omega^2 e^{ik_{\omega}r}}{4\pi c^2 r} \int dt \, d^3x' \delta\epsilon(\mathbf{r}', t) \mathbf{E}_{0\perp} e^{-i[\mathbf{q}_{\omega}\cdot\mathbf{r}' - (\omega - \omega_0)t]}. \tag{151.11}$$

This is the scattered field for a given configuration $X(t)$ of the slow coordinates. To find the scattered power, we must square this field and then average over $X(t)$. It is simplest to just repeat the heuristic procedure by which we related the power radiated by a stochastic source to the current–current correlation function (see eq. (61.14)). The average of $|\mathbf{E}^{\text{sc}}|^2$ will involve the autocorrelation function of $\delta\epsilon$ at two different times, t_1 and t_2, followed by a double integral over t_1 and t_2. We argue that the fluctuations are stationary if the medium is in thermodynamic equilibrium, so this correlation function depends only on $t_1 - t_2$, enabling one of the time integrals to be evaluated. This yields an overall factor of time. Dividing by this factor converts the total energy to the power. We find

$$\frac{d^2 P_{\text{sc}}}{d\omega d\Omega} = \frac{\omega^4}{256\pi^4 c^3} \sqrt{\bar{\epsilon}(\omega)} \int d\tau \iint d^3x_1 d^3x_2 \, \langle\delta\epsilon(\mathbf{r}_1, \tau)\delta\epsilon(\mathbf{r}_2, 0)\rangle \, e^{-i[\mathbf{q}_{\omega}\cdot(\mathbf{r}_1 - \mathbf{r}_2) - (\omega - \omega_0)\tau]} |\mathbf{E}_{0\perp}|^2. \tag{151.12}$$

In a large volume, the correlation function depends only on the difference $\mathbf{r}_1 - \mathbf{r}_2$, and one of the spatial integrations can be performed to give a factor of V, the volume of the system. This is exactly as it should be, since the total scattered power should increase linearly with the volume of the scattering system. Dividing by V and by $c[\bar{\epsilon}(\omega_0)]^{1/2}|\mathbf{E}_0|^2/8\pi$, we obtain the scattered power per unit volume per unit incident energy flux, i.e., the differential extinction coefficient. At the same time, we may average over the polarizations of the incident beam, which we take to be unpolarized. We thus obtain

$$\frac{d^2\beta}{d\omega d\Omega} = \frac{\omega^4}{64\pi^3 c^4} \left(\frac{\bar{\epsilon}(\omega)}{\bar{\epsilon}(\omega_0)}\right)^{1/2} (1 + \cos^2\theta) \iint \langle\delta\epsilon(\mathbf{r}, \tau)\delta\epsilon(0, 0)\rangle \, e^{-i[\mathbf{q}_{\omega}\cdot\mathbf{r} - (\Delta\omega)\tau]} d^3x \, d\tau, \tag{151.13}$$

where

$$\Delta\omega = \omega - \omega_0. \tag{151.14}$$

The next step is to relate the fluctuations in the dielectric constant to those in the density and temperature. To that end, we write

$$\delta\epsilon = \left(\frac{\partial\epsilon}{\partial\rho_n}\right)_T \delta\rho_n + \left(\frac{\partial\epsilon}{\partial T}\right)_{\rho_n} \delta T. \tag{151.15}$$

For most systems, the dependence of the dielectric function on temperature is much weaker than that on density. Hence, we shall ignore the last term in eq. (151.15). Introducing the *structure factor*,

$$S_{\rho\rho}(\mathbf{q}, \omega) = \int\int \langle\delta\rho_n(\mathbf{r}, \tau)\delta\rho_n(0, 0)\rangle\, e^{-i(\mathbf{q}\cdot\mathbf{r}-\omega\tau)} d^3x\, d\tau, \tag{151.16}$$

we get

$$\frac{d^2\beta}{d\omega d\Omega} = \frac{\omega^4}{64\pi^3 c^4} \left(\frac{\bar\epsilon(\omega)}{\bar\epsilon(\omega_0)}\right)^{1/2} \left(\frac{\partial\epsilon}{\partial\rho_n}\right)_T^2 (1+\cos^2\theta)\, S_{\rho\rho}(\mathbf{q}_\omega, \omega - \omega_0). \tag{151.17}$$

The problem is now reduced to understanding the density fluctuations. The detailed calculation of the structure factor belongs to the subject of hydrodynamics, which is too far afield for us, so we merely give the answer and present a heuristic justification for it.[6] In any fluid, the density fluctuations have two principal sources, entropic and mechanical. A fluctuation in the local entropy density of the fluid relaxes by thermal diffusion. Such fluctuations produce a broad peak in the structure factor at $q = 0$, $\omega = 0$.[7] Fluctuations in the local pressure, however, generate sound waves. Suppose such a fluctuation occurs at the origin at time 0. The resulting sound pulse travels outward at speed c_s. Therefore, it will produce a strong fluctuation at a point a distance r away at time t if $r \simeq c_s t$. In terms of the Fourier transform this means that $S_{\rho\rho}(\mathbf{q}, \omega)$ has peaks at $\omega = \pm c_s q$. Because $c_s \ll c$, the frequency shift is very small, i.e., $|\Delta\omega| \ll \omega_0$. The scattering is essentially elastic, and $\omega \approx \omega_0$, $k_\omega \approx k_0$. Then (see fig. 22.2),

$$q_\omega = 2k_0 \sin(\theta/2). \tag{151.18}$$

We may also take $\bar\epsilon(\omega_0) \approx \bar\epsilon(\omega)$ and replace the ratio of the two dielectric constants in eq. (151.17) by unity.

It follows that if we examine the spectrum of the light at a scattering angle θ, we will see three peaks (see fig. 22.2). The central peak, known as the *Rayleigh peak*, is unshifted in frequency, and its width reflects the spectral density of entropy fluctuations. The two side peaks centered at $\omega_0 \pm c_s q$ are known as the *Brillouin peaks*. Their width reflects the damping of sound.

---

[6] An excellent discussion of this matter is given by L. P. Kadanoff and P. C. Martin, Hydrodynamic equations and correlation functions, *Ann. Phys.* **24**, 419–469 (1963). See especially sec. III.

[7] In a multicomponent fluid, we must also consider composition fluctuations. These also relax by diffusion and add to the Rayleigh peak at $q = 0$, $\omega = 0$. In what follows, we consider only single-component fluids.

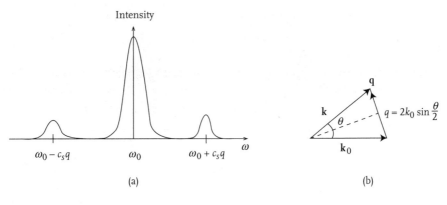

FIGURE 22.2. (a) The three-peak spectrum of Rayleigh-Brillouin scattering, and (b) the geometrical relation between the scattering vector and the wave vectors of the incident and scattered light.

The detailed form of $S_{\rho\rho}$ is a sum of three Lorentzians,

$$S_{\rho\rho}(\mathbf{q}, \omega)$$

$$= \rho_n^2 k_B T \kappa_T \left[\frac{2 D_T q^2 ((\gamma - 1)/\gamma)}{\omega^2 + (D_T q^2)^2} + \frac{(\Gamma q^2/2\gamma)}{(\omega - c_s q)^2 + \left(\frac{1}{2}\Gamma q^2\right)^2} + \frac{(\Gamma q^2/2\gamma)}{(\omega + c_s q)^2 + \left(\frac{1}{2}\Gamma q^2\right)^2} \right].$$

(151.19)

Here, $D_T = \kappa_T/c_v$ is the thermal conductivity, $\gamma = C_p/C_v$, where C_p and C_v are the specific heats at constant pressure and constant volume respectively, and

$$\Gamma = \frac{1}{\rho_n} \left(\eta_V + \frac{4}{3}\eta \right) + (\gamma - 1) D_T$$

(151.20)

is the sound attenuation coefficient. In this formula, η_V and η are the bulk and shear viscosities of the fluid. Recall further that $c_s^2 = \gamma/\rho_n \kappa_T$. The interesting point here is that all the quantities in the formula for $S_{\rho\rho}$ can be found by making macroscopic thermodynamic and transport measurements. That the same quantities should appear in an experiment that measures equilibrium fluctuations only is a reflection of *Onsager's regression hypothesis*, according to which the processes that govern such fluctuations are the same ones by which a system perturbed from equilibrium relaxes back to equilibrium. A measurement of the positions and widths of the Rayleigh and Brillouin peaks gives c_s, D_T, and Γ. A measurement of the absolute peak heights would also give us the index γ, but this is difficult. The same information can also be found in the following way. Let us denote the total intensity in the Rayleigh peak by I_R, and that in the Brillouin peaks by $I_{B\pm}$. Then, integrating each of the three Lorentzians in $S_{\rho\rho}$, we get

$$\mathcal{L} \equiv \frac{I_R}{I_{B+} + I_{B-}} = \frac{2(\gamma - 1)/\gamma}{2 \times (1/\gamma)} = \gamma - 1.$$

(151.21)

This ratio, known as the *Landau-Placzek ratio*, is easier to measure, as it requires knowing only the relative intensities of the peaks. In this connection, let us also note that the intensity of the two Brillouin peaks is equal in our classical treatment. The actual ratio, $\exp(\hbar\Delta\omega/k_B T)$, is very close to unity, as $\Delta\omega$ is of the order of a few gigahertz in most liquids.

Let us now find the total extinction coefficient. For this, we first integrate eq. (151.17) over all frequencies. Since the important range of shifts $\omega - \omega_0$ is very small, we may take $\mathbf{q}_\omega$ as a constant in this integration. We may also replace the ratio of the two dielectric constants by unity. This substitution yields

$$\frac{d\beta}{d\Omega} = \frac{\omega^4}{64\pi^3 c^4} \left(\frac{\partial\epsilon}{\partial\rho_n}\right)_T^2 (1 + \cos^2\theta) \int d\omega\, S_{\rho\rho}(\mathbf{q}, \omega - \omega_0). \tag{151.22}$$

We now observe that

$$\int_{-\infty}^{\infty} d\omega\, S_{\rho\rho}(\mathbf{q}, \omega) = 2\pi \int \langle \delta\rho_n(\mathbf{r}, 0)\delta\rho_n(0, 0)\rangle\, e^{-i\mathbf{q}_\omega \cdot \mathbf{r}} d^3x, \tag{151.23}$$

and that

$$\langle \delta\rho_n(\mathbf{r}, 0)\delta\rho_n(\mathbf{r}', 0)\rangle = \rho_n^2 k_B T \kappa_T \delta(\mathbf{r} - \mathbf{r}'). \tag{151.24}$$

The last result can be understood by noting that the equal-time correlator must vanish unless $\mathbf{r} = \mathbf{r}'$, since the fluctuations cannot propagate at all in zero time. The strength of the delta function is found by integrating both sides with respect to $\mathbf{r}$ and $\mathbf{r}'$ over an arbitrary macroscopic volume V and using eq. (151.1) for $\langle(\delta\rho_n)^2\rangle_V$.

Combining the last three formulas and integrating over all directions, we find

$$\begin{aligned}
\beta &= \frac{\omega^4}{32\pi^2 c^4} \left(\frac{\partial\epsilon}{\partial\rho_n}\right)_T^2 \rho_n^2 k_B T \kappa_T \int d\Omega\, (1 + \cos^2\theta) \\
&= \frac{\omega^4}{6\pi c^4} \left(\frac{\partial\epsilon}{\partial\rho_n}\right)_T^2 \rho_n^2 k_B T \kappa_T.
\end{aligned} \tag{151.25}$$

This result was found in somewhat more general form by Einstein. It is useful to verify that it reduces to Rayleigh's formula for a dilute gas. In this case, we have $\epsilon = 1 + 4\pi\rho_n\alpha$, so $\partial\epsilon/\partial\rho_n = 4\pi\alpha$, where α is the polarizability. Further, $p = \rho_n k_B T$ from the ideal-gas equation of state, so $\kappa_T = \rho_n^{-1}(\partial\rho_n/\partial p)_T = 1/\rho_n k_B T$. Thus,

$$\beta = \frac{8\pi\omega^4}{3c^4} \rho_n \alpha^2, \tag{151.26}$$

which is the same formula as eq. (149.2).

We conclude this section by commenting on Brillouin scattering in solids. The underlying principles are identical, but since a solid has three acoustic phonon branches (one longitudinal, two transverse), one observes three pairs of Brillouin doublets symmetrically placed about the Rayleigh line. This can be a valuable method of measuring phonon dispersion relations, which, in general, are not linear. One can also study optical phonons by light scattering, although one then tends to speak of the effect as Raman scattering. Secondly, although we have neglected magnetic scattering in our discussion, it is obvious that the same ideas will apply. Brillouin light scattering can therefore be used to study magnetic excitations or spin waves in solids with magnetic order, provided the solid is transparent.

Exercise 151.1 The Landau-Placzek ratio can be understood in the following way, without knowing the detailed form of the structure factor. The density fluctuation $\delta\rho_n$ can be divided into entropic and mechanical parts:

$$(\delta\rho_n)_{\text{ent}} = \left(\frac{\partial\rho_n}{\partial s}\right)_p \delta s, \tag{151.27}$$

$$(\delta\rho_n)_{\text{mech}} = \left(\frac{\partial\rho_n}{\partial p}\right)_s \delta p, \tag{151.28}$$

where s and p are the entropy density and pressure, respectively. Show that the L-P ratio is the ratio of the mean-square values of these two parts, i.e.,

$$\mathcal{L} = \left[\frac{(\delta\rho_n)_{\text{ent}}}{(\delta\rho_n)_{\text{mech}}}\right]^2. \tag{151.29}$$

23 | Formalism of special relativity

We have already mentioned that electromagnetism is incompatible with Galilean relativity, by citing the wave equation as an example. To take another example, which also incorporates the Lorentz force law, consider two parallel wires, each carrying a static electric charge density λ per unit length and separated by a distance d. The wires then repel each other with a force $\lambda^2/2d$ per unit length. Let us now consider an observer moving at a uniform speed v parallel to the wires. Suppose Galilean relativity holds. Then, in this observer's frame, the charge density on the wires is unchanged, so the electric field is the same. In addition, however, the wires now carry a current λv, so there is an additional repulsive force between the wires of order $(v/c)^2$ times λ^2/d. The conclusion clearly violates Galilean relativity, so the initial assumption must be incorrect. Many other similar examples can be constructed.

In this chapter we examine some of the formal aspects of relativity. In the next chapter we will see how these ideas relate to electromagnetism. We assume that readers have already seen elementary treatments of relativity for mechanics and are aware of the chief implications of relativity for electromagnetism at a qualitative level.[1]

152 Review of basic concepts

In discussing modern relativity, it proves extremely convenient to measure time in terms of the distance traveled by light. That is, we adopt units such that, c, the speed of light is unity:

$$c = 1. \tag{152.1}$$

---

[1] In writing this and the next chapter, I have made copious use of my lecture notes from a course by Saul Teukolsky at Cornell University. It is a pleasure to have this opportunity to thank Professor Teukolsky. He is, of course, not responsible for any errors or infelicities in my discussion.

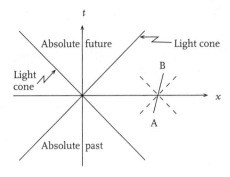

FIGURE 23.1. A space–time diagram. The line AB is the world line of a uniformly moving particle.

We shall consistently follow this convention, except in displaying certain results where the explicit appearance of c is useful. In checking the dimensional correctness of any equation, it must then be remembered that distance and time have the same dimensions, and that speeds are dimensionless numbers.

In Galilean relativity, space and time are absolutely separate concepts. Modern relativity eliminates this distinction to some extent. One is therefore led to consider a space consisting of ordinary three-dimensional space, plus a fourth dimension, time. We refer to this four-dimensional space as *space–time* and represent it diagrammatically by showing one of the spatial dimensions (since that is all one can do on paper) on the horizontal axis, and the time dimension on the vertical axis (see fig. 23.1). The lines $x = \pm t$ on this diagram are referred to as the *light cones*. The region inside these lines including the positive t axis, i.e., $t > |x|$, is known as the *absolute future*, and that including the negative t axis, $t < -|x|$, the *absolute past*. We can similarly define the future and past light cones around any other space–time point.

A uniformly moving particle with speed v appears as a straight line on this diagram, making an angle θ with the time axis, where $\tan \theta = \pm v$. This is known as the *world line* of the particle. For a nonuniformly moving particle, the world line is curved, but since its speed must not exceed 1 at any time, this line must not make an angle greater than 45° with the time axis at any point, i.e., the world line must lie inside the past and future light cones at any point on the line. The world line of a particle at rest is straight and parallel to the time axis.

A point in space–time is called an *event*. To every event, an observer assigns four coordinates: t, the time, and x, y, and z, the space coordinates. The assignment of time requires some discussion, since we must be able to do this for spatially separated points. The key point is that we should be able to synchronize an entire array of clocks distributed throughout space. Consider two points and suppose we have an observer at each point, each armed with a clock. They arrange for a light flash to go off at the exact halfway point between them, and each records his clock reading when the light reaches him. They then communicate the clock readings to each other and move one of the clocks forward or backward as necessary by the difference in the readings. In this way, we can synchronize as many clocks as needed.

Observers in motion relative to one another will assign different coordinates to the same event, just as in ordinary three-space, observers who choose to orient their Cartesian axes differently will assign different Cartesian components to the same space point. In this sense, the relationship between the space–time coordinates used by the two observers is that of a rotation. However, as we shall see, the angle of this rotation is imaginary.

We now recall for the reader a number of elementary results that follow from the postulate of the constancy of the speed of light in all inertial frames.

1. Consider two events with coordinate differences Δx, Δy, Δz, and Δt in the inertial frame K, and $\Delta x'$, etc., in another frame K'. Then, the quantity,

$$(\Delta s)^2 = (\Delta t)^2 - (\Delta x)^2 - (\Delta y)^2 - (\Delta z)^2, \tag{152.2}$$

known as the square of the *interval* between the events, is the same in the two frames. In other words,

$$(\Delta s)^2 = (\Delta s')^2 = (\Delta t')^2 - (\Delta x')^2 - (\Delta y')^2 - (\Delta z')^2. \tag{152.3}$$

A quantity such as $(\Delta s)^2$ which has the same value in all inertial reference frames is said to be a *Lorentz invariant* or a scalar.

2. Suppose K' moves with a velocity V along the x direction relative to K (which we shall also refer to as the lab frame), and that the Cartesian axes and clocks of the two observers coincide when $t = t' = 0$. Then, the coordinates of a general event as recorded by the two observers are related by[2]

$$t' = \gamma(t - Vx), \quad x' = \gamma(x - Vt), \quad y' = y, \quad z' = z, \tag{152.4}$$

where we have introduced the convenient and common abbreviation

$$\gamma = (1 - V^2)^{-1/2}. \tag{152.5}$$

The transformation law (152.4) is known as a *Lorentz transformation* (LT). More generally this term also includes spatial rotations. The special transformations involving observers moving relative to one another without any rotations are known as *boosts*.[3]

---

[2] The logical deduction of eq. (152.4) from the postulates of relativity is a simple and standard exercise. See, e.g., probs. 11.1 and 11.2 of Jackson (1999).

[3] The operations of parity and time reversal are sometimes included in the Lorentz transformations; these operations are then called *improper*, and pure rotations and boosts are called *proper*. A combination of an improper operation and a proper one is also improper, as is one involving both parity and time reversal. A second qualifier, *orthochronous* or *nonorthochronous*, may optionally be added to show whether time is reversed or not. Thus, the improper orthochronous transformations are those including a spatial inversion, but not time reversal. The set of all Lorentz transformations makes up the *full Lorentz group*. Adjoining translations in space and time gives the *Poincare group*.

Exercise 152.1 One check on the validity of eq. (152.4) is as follows. As seen by K', the frame K moves with velocity $-V\hat{x}$. Thus, it should be the case that

$$t = \gamma(t' + Vx'), \quad x = \gamma(x' + Vt'), \quad y = y', \quad z = z'. \tag{152.6}$$

Show that this is indeed the inverse of the transformation (152.4).

Exercise 152.2 Derive the handy formulas

$$\gamma^2 - 1 = \gamma^2 V^2, \quad \gamma - 1 = \frac{\gamma^2}{\gamma + 1} V^2. \tag{152.7}$$

Exercise 152.3 Show that the interval is invariant under an LT.

3. We can show that eq. (152.4) is a rotation by introducing the so-called *rapidity parameter*

$$\theta = \tanh^{-1} V. \tag{152.8}$$

Then,

$$\gamma = \cosh\theta, \quad V\gamma = \sinh\theta, \tag{152.9}$$

so we can write

$$\begin{pmatrix} t' \\ x' \end{pmatrix} = \begin{pmatrix} \cosh\theta & -\sinh\theta \\ -\sinh\theta & \cosh\theta \end{pmatrix} \begin{pmatrix} t \\ x \end{pmatrix}. \tag{152.10}$$

The matrix on the right-hand side is that of a rotation through an imaginary angle $i\theta$ if instead of t we use as a coordinate it. Such conventions are indeed found in some texts, but we shall avoid them.

The rapidity parameter allows us to combine two parallel boosts very easily. Suppose a third frame K'' moves at velocity $V'\hat{x}$ relative to K'. Then, with $V' \equiv \tanh\theta'$, we have

$$\begin{pmatrix} t'' \\ x'' \end{pmatrix} = \begin{pmatrix} \cosh\theta' & -\sinh\theta' \\ -\sinh\theta' & \cosh\theta' \end{pmatrix} \begin{pmatrix} \cosh\theta & -\sinh\theta \\ -\sinh\theta & \cosh\theta \end{pmatrix} \begin{pmatrix} t \\ x \end{pmatrix}$$

$$= \begin{pmatrix} \cosh\theta'' & -\sinh\theta'' \\ -\sinh\theta'' & \cosh\theta'' \end{pmatrix} \begin{pmatrix} t \\ x \end{pmatrix}, \tag{152.11}$$

where

$$\theta'' = \theta + \theta'. \tag{152.12}$$

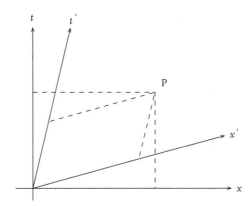

FIGURE 23.2. Space–time coordinates in inertial frames in relative motion, showing that a boost appears as a skew rotation. The coordinates of an event P as
measured in the primed and unprimed frames are the intercepts on the axes as
indicated.

Thus, the velocity of the frame K'' with respect to K is given by $V'' = \tanh \theta''$, i.e.,

$$V'' = \frac{\tanh \theta + \tanh \theta'}{1 + \tanh \theta \tanh \theta'}$$

$$= \frac{V + V'}{1 + VV'}. \tag{152.13}$$

If both V and V' are less than 1, it follows that $V'' < 1$. If $V = 1$ or $V' = 1$ (speed of light),
then $V'' = 1$. Thus, the velocity of frame K'' as measured in K can never exceed that of
light.

The rule that the rapidities of parallel boosts add linearly is particularly easy to
remember. The law for nonparallel boosts is derived in exercise 152.7.

4. The geometry of space–time is rather different from that of ordinary Euclidean three-
space, and is known as *Minkowskian*. We see this in the appearance of minus signs in the
definition of the interval. It also shows up in the way we must draw the space and time
axes for a moving observer. If we consider eq. (152.4), the x' axis is the set of points on
which $t' = 0$, i.e., it is the line $x = t/V$. Similarly, the t' axis is the set on which $x' = 0$, so
it is the line $x = Vt$ (fig. 23.2). Instead of lying in the second quadrant, as we would have
for an ordinary rotation, the t' axis is turned in the opposite sense from the x' axis. Note
in particular that the time axis of any inertial observer is simply that observer's world line.

The Minkowskian nature of space–time geometry gives rise to three physically distinct
types of intervals. Suppose that between two events, the squared interval is positive in
the K frame: $(\Delta s)^2 > 0$. We assert that in this case, there is another inertial frame K' in
which these events occur at the same spatial location. For, in such a frame, the squared
interval would be $(\Delta t')^2$, and since the interval is a Lorentz invariant, we would have

$$(\Delta t')^2 = (\Delta s)^2 > 0. \tag{152.14}$$

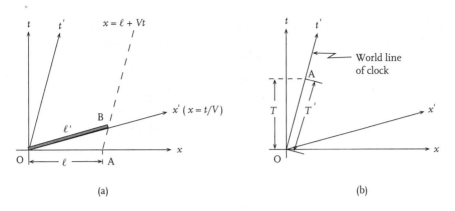

FIGURE 23.3. Space-time diagrams showing (a) Lorentz contraction and (b) time dilation.

Whatever the value of $(\Delta s)^2$, as long as it is positive, we can find a solution for $\Delta t'$. The velocity of the frame K' relative to K is simply that of a uniformly moving particle whose world line passes through the two events in question.

Events for which $(\Delta s)^2 > 0$ are said to have a *time-like separation*, and the interval is said to be time-like. One of the events lies in the future light cone of the other and can therefore be causally influenced by the latter.

By contrast, events for which $(\Delta s)^2 < 0$ are said to be *space-like* separated and cannot be causally connected. In this case, one can find a frame in which these events are simultaneous. The interval in the primed frame is $-(\Delta \mathbf{r}')^2$ (where $\Delta \mathbf{r}'$ is the ordinary three-dimensional spatial vector from one point to the other), and since the interval is an invariant, we must have

$$-(\Delta \mathbf{r}')^2 = (\Delta s)^2 < 0, \tag{152.15}$$

which always has a solution. It is worth noting that the temporal order of space-like separated events is *not* invariant.

Finally, if $\Delta s = 0$, the events are said to have a *light-like* or *null* separation, and lie *on* each other's light cones. In summary, we have

$$(\Delta s)^2 > 0: \quad \text{time-like separation}$$
$$(\Delta s)^2 < 0: \quad \text{space-like separation} \tag{152.16}$$
$$(\Delta s)^2 = 0: \quad \text{light-like separation}$$

5. We can now derive the expression for *length contraction* very easily. We consider a rod lying parallel to the x' axis, at rest in the K' frame. The world lines of the ends of the rod are shown in fig. 23.3. The essential point now is to understand the definition of length. To determine the length, an observer must measure the spatial coordinates of the ends of the rod *simultaneously*. Thus, the K' observer measures the rod's length as ℓ', the distance OB along the x' axis, since the events O and B both occur at $t' = 0$. But the K observer measures the length to be ℓ, the distance OA along the x axis, since to him, it is events O and A that are simultaneous. The problem is thus reduced to evaluating the interval OB in the two frames and equating the answers. Clearly,

$$(OB)^2 = \begin{cases} -\ell'^2 & \text{in } K' \\ t_B^2 - x_B^2 & \text{in K,} \end{cases} \tag{152.17}$$

where t_B and x_B are the coordinates of B. This point is determined by the intersection of the lines $x = \ell + Vt$, and $t' = 0$, i.e., $t = Vx$. Solving these equations, we get

$$x_B = \frac{\ell}{1 - V^2}, \qquad t_B = \frac{\ell V}{1 - V^2}. \tag{152.18}$$

Combining this with the previous two equations and simplifying, we obtain

$$\ell = \ell'\sqrt{1 - V^2} = \ell'/\gamma. \tag{152.19}$$

Thus, the rod appears to be shorter or contracted in a frame in which it is moving. It appears to be longest in the comoving frame, i.e., its own rest frame, and the length in this frame is known as the *proper length*.

Note that there is no contraction perpendicular to the direction of motion, since $y = y'$ and $z = z'$ in eq. (152.4).

6. The phenomenon of *time dilation* can be studied similarly. Suppose a clock is at rest in the K' frame, and it ticks at a regular time interval T'. The world line of the clock is just the t' axis. Suppose two successive ticks occur at the points O and A in fig. 23.3, so that T' is just the interval OA. In the K frame, the second tick occurs at a time T, as indicated. The point A has $x' = 0$ and $t' = T'$. Hence,

$$T = \gamma(t' + Vx') = \gamma T'. \tag{152.20}$$

In other words, the interval between ticks is longer in the lab frame, and the moving clock appears to run slow. The time elapsed in the rest frame is known as the *proper time*.

It is worth noting that in the above discussion, the moving clock must be compared with two different (and synchronized) clocks in the K frame, one at $x = 0$ and the other at $x = VT$. We see explicitly in this example the need for the ability to measure time at every spatial point in one inertial frame.

Exercise 152.4 Let K' carry two clocks a distance D apart along the direction of motion, synchronized in her frame. Show that K sees the clocks as out of sync; the leading clock, i.e., the one that is farther along the direction of motion, lags by an amount $\gamma V D$.

The concept of proper time is particularly important, as it can be generalized to an arbitrarily moving frame, not just an inertial one. Suppose a particle is moving along some trajectory with speed $v(t)$ as measured in the lab frame. At every instant of time, we can imagine some inertial frame that is comoving with the particle. The time elapsed in the particle's rest frame is equal to that elapsed in the comoving frame in an infinitesimal duration over which the particle may be considered to be moving uniformly.

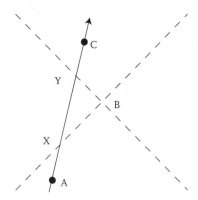

FIGURE 23.4. How to define the interval without using coordinates.

By introducing a succession of comoving inertial frames, we can thus determine the elapsed time in a large interval. In this way we see that in an infinitesimal interval dt of lab time, an interval $dt' = (1 - v^2)^{1/2} dt$ elapses in the rest frame of the particle. Hence, in an interval from time t_1 to time t_2 in the lab frame, the time elapsed in the particle frame is

$$\tau = \int_{t_1}^{t_2} \left(1 - v^2(t)\right)^{1/2} dt. \tag{152.21}$$

We shall use the symbol τ for the proper time consistently.

Exercise 152.5 Suppose AC is the world line of a uniformly moving particle, and B is an event not on this world line (see fig. 23.4). Let the forward and backward light cones from the event B intersect the world line at points X and Y. Show that the interval s_{AB} is given by

$$s_{AB}^2 = \tau_{AY} \tau_{AX}, \tag{152.22}$$

where τ_{AX} and τ_{AY} are the proper time intervals. This relationship can be used to provide a coordinate-free definition of the interval.

Exercise 152.6 A rod lies at rest at an angle θ' to the x axis in a frame K' moving along the x axis at a speed V relative to another observer K. At what angle to the x axis does the rod appear to K?

Answer: $\tan \theta = \gamma \tan \theta'$. (Hence, $\theta > \theta'$, opposite to the headlight effect, discussed below.)

Exercise 152.7 Derive the law for addition of nonparallel velocities.

Solution: Let K, K', and K'' be three inertial observers, where K' has a velocity $\mathbf{V}$ relative to K, and K'' has a velocity $\mathbf{V}'$ relative to K'. We want the velocity, $\mathbf{V}''$, of K'' relative to K. We take $\mathbf{V}$ to lie along $\hat{\mathbf{x}}$, and $\mathbf{V}'$ to lie in the xy plane. Then, according to K', in a short time

dt', the coordinates of K'' change by

$$dx' = V'_x dt', \quad dy' = V'_y dt'. \tag{152.23}$$

The corresponding increments measured by K are related to these by $dx = \gamma(dx' + Vdt')$, $dy = dy', dt = \gamma(dt' + Vdx')$. Hence,

$$V''_x = \frac{dx}{dt} = \frac{V'_x + V}{1 + V V'_x}, \quad V''_y = \frac{dy}{dt} = \frac{V'_y}{\gamma(1 + V V'_x)}. \tag{152.24}$$

In general vector notation, we have

$$\mathbf{V}'' = \frac{\gamma(\mathbf{V} \cdot \mathbf{V}' + V^2)\mathbf{V} + \mathbf{V} \times (\mathbf{V}' \times \mathbf{V})}{\gamma V^2 (1 + \mathbf{V} \cdot \mathbf{V}')}. \tag{152.25}$$

There are three points about eq. (152.25) worth noting. First, it reduces to eq. (152.13) when $\mathbf{V} \| \mathbf{V}'$. Second, it is not symmetric when $\mathbf{V} \not\| \mathbf{V}'$. That is, boosts do not commute. Third, since light always travels at speed c, it should be true that if $V' = 1$, then $V'' = 1$. From the component form, we have

$$(V'')^2 = \frac{(V'_x + V)^2 + (1 - V^2)(V'_y)^2}{(1 + V V'_x)^2}, \tag{152.26}$$

$$= \frac{(\mathbf{V} + \mathbf{V}')^2 - (\mathbf{V} \times \mathbf{V}')^2}{(1 + \mathbf{V} \cdot \mathbf{V}')^2}, \tag{152.27}$$

where the last result is in general vector form. Using $(V'_x)^2 + (V'_y)^2 = 1$, we obtain $(V'')^2 = 1$, as expected.

Exercise 152.8 A particle K' moving at velocity $\mathbf{V}$ decays. In the comoving frame of the particle, one of the decay particles is emitted at velocity $\mathbf{V}'$ at an angle θ' to the forward direction of motion. In what direction does this decay particle travel in the lab frame?

Solution: Use eq. (152.24). The decay particle will appear to travel at an angle θ to $\mathbf{V}$, where

$$\tan\theta = \frac{V''_y}{V''_x} = (1 - V^2)^{1/2} \frac{V' \sin\theta'}{V + V' \cos\theta'}. \tag{152.28}$$

This differs from the Galilean formula in the prefactor of $(1 - V^2)^{1/2}$. Since $\theta < \theta'$, the decay particle appears to travel in a direction closer to that of the parent particle's motion. This is also true in the Galilean case, but the effect is now more pronounced. When the decay particle is a photon, in which case $V' = 1$, this phenomenon is known as the *headlight effect*.

153 Four-vectors

In the previous section, we saw that an LT may be regarded as a rotation in space–time. Just as in three-dimensional Euclidean space, it is therefore highly advantageous to introduce the concept of a vector in Minkowski space, along with dot products, scalars, tensors, etc. The extensions required are straightforward, and our main task is to develop a notation that enables easy computation and makes the physics transparent.

Let us return to eq. (152.2) for the interval between two events. Let us take one of the events to be the origin of our space–time coordinate system, and the other to have coordinates t, x, y, z. The squared interval (which we denote by just s^2 now) equals

$$s^2 = t^2 - x^2 - y^2 - z^2. \tag{153.1}$$

This quantity is a Lorentz invariant, or a scalar. Except for the minus signs, it looks very much like the expression for the length of a vector in Euclidean space. We introduce a vector with four components, x^μ, where $\mu = 0, 1, 2, 3$ (note that the suffixes are to be read as superscripts and not exponents), and

$$x^0 = t, \quad x^1 = x, \quad x^2 = y, \quad x^3 = z. \tag{153.2}$$

We would like to write the squared "length" of this vector (s^2) as $x^\mu x^\mu$ using an implicit summation-over-repeated-indices convention, but this would not account for the minus signs. We get around this difficulty by writing

$$s^2 = g_{\mu\nu} x^\mu x^\nu, \tag{153.3}$$

where the 16 numbers $g_{\mu\nu}$ are given by

$$g_{00} = 1, \quad g_{11} = g_{22} = g_{33} = -1, \quad g_{\mu\nu} = 0 \text{ if } \mu \neq \nu, \tag{153.4}$$

and there is an implicit double sum on the indices μ and ν.

The quantity $g_{\mu\nu}$ is known as the *metric*, and eq. (153.3) is the formula for the squared length of a vector. We can write it in another form if we define a second "version" of the vector x^μ:

$$x_\mu = g_{\mu\nu} x^\nu. \tag{153.5}$$

Note that the suffix on this second vector is a subscript. Its components are related to those of x^μ by

$$x_0 = x^0, \quad x_1 = -x^1, \quad x_2 = -x^2, \quad x_3 = -x^3. \tag{153.6}$$

We can now write

$$s^2 = x^\mu x_\mu = x_\mu x^\mu. \tag{153.7}$$

The learned words for the vectors x^μ and x_μ are *contravariant* and *covariant*, respectively, but we shall often speak of them more informally as vectors with upper and lower indices. The location of an index is important, and in eq. (153.3) the indices on g were written as subscripts with this in mind. We call the process of summing over a repeated index *contraction*, but the index must appear in the lower position once and in the upper position once. Thus, in eq. (153.7) we are contracting over the index μ, and in eq. (153.3) we are contracting over both μ and ν. Contraction with the metric, as in eq. (153.5), lowers the index on a vector. The converse operation, raising an index, can also be carried out if we define the contravariant metric tensor (that it is a tensor will be shown later) $g^{\mu\nu}$ with components

$$g^{00} = 1, \quad g^{11} = g^{22} = g^{33} = -1, \quad g^{\mu\nu} = 0 \text{ if } \mu \neq \nu, \tag{153.8}$$

which are, in fact, identical to those of $g_{\mu\nu}$. With this object in hand, we have

$$x^{\mu} = g^{\mu\nu} x_{\nu}. \tag{153.9}$$

Let us now reexamine the Lorentz invariance of s^2. That is, if the event (t, x, y, z) has coordinates (t', x', y', z') in another inertial frame,

$$t^2 - x^2 - y^2 - z^2 = t'^2 - x'^2 - y'^2 - z'^2. \tag{153.10}$$

This is obviously true for a pure space rotation, for then $t' = t$, and $x^2 + y^2 + z^2$ and $x'^2 + y'^2 + z'^2$ are equal. We have also seen it to be true for a pure boost (in exercise 152.3, and in the algebraic resemblance of eq. (152.4) to a rotation). Let us write a general LT as

$$x'^{\mu} = \Lambda^{\mu}_{\nu} x^{\nu}, \tag{153.11}$$

where Λ^{μ}_{ν} is a 4×4 *matrix*, whose row index is μ and column index is ν. It should be carefully noted that the first index is raised, and the second index is lowered. The location of the indices is chosen to preserve the rule that summation over a repeated index must involve one upper and one lower index. For purposes of explicit algebraic evaluation, there is no difference between writing Λ^{μ}_{ν} and $\Lambda_{\mu\nu}$. In four-vector notation, eq. (153.10) reads

$$g_{\alpha\beta} x^{\alpha} x^{\beta} = g_{\mu\nu} x'^{\mu} x'^{\nu}. \tag{153.12}$$

Using eq. (153.11), this becomes

$$g_{\alpha\beta} x^{\alpha} x^{\beta} = g_{\mu\nu} \Lambda^{\mu}_{\alpha} x^{\alpha} \Lambda^{\nu}_{\beta} x^{\beta}. \tag{153.13}$$

The equality of the two sides for all vectors x^{μ} requires that

$$g_{\alpha\beta} = g_{\mu\nu} \Lambda^{\mu}_{\alpha} \Lambda^{\nu}_{\beta} = \Lambda^{\mu}_{\alpha} g_{\mu\nu} \Lambda^{\nu}_{\beta}. \tag{153.14}$$

We can write this in matrix notation as

$$g = \Lambda^{\mathrm{T}} g \Lambda, \tag{153.15}$$

where the superscript T denotes a transpose. (Readers who are not clear on the meaning of this expression should think about how they would evaluate the right-hand sides of eqs. (153.14) and (153.15) on a computer.) Equation (153.15) is the Minkowski-space analog of the formula $R^T R = 1$ for rotation matrices in three-space.

Let us verify that eq. (153.15) is correct for two special types of LTs. First consider a three-space rotation by an angle θ about the x axis. In this case, the matrix Λ is given by

(the order of the row and column labels is t, x, y, z)

$$\Lambda = \begin{pmatrix} 1 & 0 & 0 & 0 \\ 0 & 1 & 0 & 0 \\ 0 & 0 & \cos\theta & \sin\theta \\ 0 & 0 & -\sin\theta & \cos\theta \end{pmatrix},$$

(153.16)

and

$$\Lambda^T g \Lambda = \begin{pmatrix} 1 & 0 & 0 & 0 \\ 0 & 1 & 0 & 0 \\ 0 & 0 & \cos\theta & -\sin\theta \\ 0 & 0 & \sin\theta & \cos\theta \end{pmatrix} \begin{pmatrix} 1 & 0 & 0 & 0 \\ 0 & -1 & 0 & 0 \\ 0 & 0 & -1 & 0 \\ 0 & 0 & 0 & -1 \end{pmatrix} \begin{pmatrix} 1 & 0 & 0 & 0 \\ 0 & 1 & 0 & 0 \\ 0 & 0 & \cos\theta & \sin\theta \\ 0 & 0 & -\sin\theta & \cos\theta \end{pmatrix},$$

$$= \begin{pmatrix} 1 & 0 & 0 & 0 \\ 0 & -1 & 0 & 0 \\ 0 & 0 & -1 & 0 \\ 0 & 0 & 0 & -1 \end{pmatrix},$$

(153.17)

as is easily verified. But the right-hand side is g, so eq. (153.15) holds in this case.

Second, let us consider a boost along the x axis. Now,

$$\Lambda = \begin{pmatrix} \gamma & \gamma V & 0 & 0 \\ \gamma V & \gamma & 0 & 0 \\ 0 & 0 & 1 & 0 \\ 0 & 0 & 0 & 1 \end{pmatrix},$$

(153.18)

as implied by eq. (152.6). In this case,

$$\Lambda^T g \Lambda = \begin{pmatrix} \gamma & \gamma V & 0 & 0 \\ \gamma V & \gamma & 0 & 0 \\ 0 & 0 & 1 & 0 \\ 0 & 0 & 0 & 1 \end{pmatrix} \begin{pmatrix} 1 & 0 & 0 & 0 \\ 0 & -1 & 0 & 0 \\ 0 & 0 & -1 & 0 \\ 0 & 0 & 0 & -1 \end{pmatrix} \begin{pmatrix} \gamma & \gamma V & 0 & 0 \\ \gamma V & \gamma & 0 & 0 \\ 0 & 0 & 1 & 0 \\ 0 & 0 & 0 & 1 \end{pmatrix},$$

$$= \begin{pmatrix} 1 & 0 & 0 & 0 \\ 0 & -1 & 0 & 0 \\ 0 & 0 & -1 & 0 \\ 0 & 0 & 0 & -1 \end{pmatrix},$$

(153.19)

as is again easily verified. Since the right-hand side is g, eq. (153.15) holds in this case too.

Since any (proper) LT can be built up by a combination of three-space rotations and boosts, it follows that eq. (153.15) is true in general. We shall have very little occasion to write down explicit LTs as 4×4 matrices.

From eq. (153.15) we can also obtain the matrix inverse of the Lorentz transformation matrix. If we left multiply by g^{-1} (which is identical with g) and right multiply by Λ^{-1}, we obtain

$$\Lambda^{-1} = g\Lambda^{\mathrm{T}}g. \tag{153.20}$$

Exercise 153.1 Show that corresponding to eq. (153.11), the transformation law for covariant components of a vector is

$$x'_\mu = g_{\mu\nu}\Lambda^\nu{}_\alpha g^{\alpha\beta}x_\beta. \tag{153.21}$$

The matrix on the right is precisely the inverse of Λ^{T}, i.e., its $\mu\beta$ component is the $\beta\mu$ component of Λ^{-1}.

Exercise 153.2 Show that the most general LT requires six parameters for its specification.

Just as by combining time and space coordinates, we formed the event four-vector x^μ, we can form four-vector extensions of three-vectors such as velocity and momentum also. In general, a four-vector is any set of four quantities A^μ that transform according to the rule (153.11):

$$A'^\mu = \Lambda^\mu{}_\nu A^\nu. \tag{153.22}$$

Given a contravariant vector A^μ, we define a covariant vector via the relation[4]

$$A_\mu = g_{\mu\nu}A^\mu. \tag{153.23}$$

The *scalar product* of any two four-vectors A^μ and B^μ is given by

$$A^\mu B_\mu = A_\mu B^\mu = g_{\mu\nu}A^\mu B^\nu = g^{\mu\nu}A_\mu B_\nu. \tag{153.24}$$

All four forms are clearly equivalent. That the product is a Lorentz invariant follows immediately from the proof of the invariance of s^2, since we did not make use of the physical nature of the event four-vector, only its transformation law. Alternatively, the invariance can be shown algebraically by using eq. (153.15).

A vector A^μ is said to be time-, space-, or light-like (or null) depending on the sign of $A^\mu A_\mu$ in the same way as the interval.

It is useful at this point to introduce some other ways of writing four-vectors. The first is

$$A^\mu = (A^0, \mathbf{A}), \tag{153.25}$$

----

[4] In more modern language, the A^μ are said to be the components of a vector, and A_μ are the components of the one-form *dual* to that vector.

where **A** is the three-vector (A^1, A^2, A^3) formed from the spatial components. We refer to A^0 and **A** as the time and space parts of A^μ. In the same way,

$$A_\mu = (A^0, -\mathbf{A}). \tag{153.26}$$

(Recall that $A^0 = A_0$.) We also write

$$A^\mu = (A_0, A_i), \quad A_\mu = (A_0, -A_i), \tag{153.27}$$

where the A_i are the Cartesian components of the space part **A**. There is no need to distinguish contra- and covariant components for three-vectors. We shall use Greek indices for the components of a four-vector, and Latin indices for a three-vector. The latter run over the set $(1, 2, 3)$. With these usages, we can write the dot product as

$$A^\mu B_\mu = A_0 B_0 - \mathbf{A} \cdot \mathbf{B} = A_0 B_0 - A_i B_i, \tag{153.28}$$

where there is an implicit sum on i in the last form.

154 Velocity, momentum, and acceleration four-vectors

Let us now turn to constructing some physical four-vectors. Suppose a particle is moving with velocity **v** at some instant in the lab frame. Let the particle's coordinates be x^μ. The proper time $\Delta\tau$ elapsed in the particle's rest frame in an infinitesimal duration is clearly a Lorentz invariant.[5] Hence, the quantity

$$\frac{\Delta x^\mu}{\Delta\tau} \tag{154.1}$$

is a four-vector. Taking the limit as $\Delta\tau \to 0$, we obtain the four-velocity:

$$u^\mu = \frac{dx^\mu}{d\tau}. \tag{154.2}$$

In the lab frame, $\Delta\mathbf{r} = \mathbf{v}\Delta t$ and $\Delta t = \gamma\Delta\tau$, so

$$u^\mu = (\gamma, \gamma\mathbf{v}), \quad \gamma = (1 - v^2)^{-1/2} \quad \text{(lab frame)}. \tag{154.3}$$

In an inertial frame instantaneously comoving with the particle, on the other hand, $\Delta t = \Delta\tau$ and $\Delta\mathbf{r} = 0$, so

$$u^\mu = (1, 0) \quad \text{(comoving frame)}. \tag{154.4}$$

The four-velocity has the important property

$$u^\mu u_\mu = 1, \tag{154.5}$$

---

[5] In general, if a physical quantity can be measured as a *number* by an observer, that number is a Lorentz invariant, since the observer could record it in a notebook, say. There can be no dispute about this recorded entry among other observers. This point of view, which may be termed the *notebook principle*, is often very useful.

as can be verified by evaluating the scalar product in either the lab or the comoving frame. Hence, the four-velocity is a time-like vector.

The four-momentum p^μ is given in terms of the four-velocity by

$$p^\mu = mu^\mu. \tag{154.6}$$

Here m is the *rest mass* of the particle. We shall never use the concept of a velocity-dependent mass. Since $u^0 = \gamma = (1 - v^2)^{-1/2}$, $p^0 = m/\sqrt{1-v^2}$, which is just the energy of the particle, $\mathcal{E}$, and $\mathbf{p} = m\mathbf{v}/\sqrt{1-v^2}$, which is the relativistic definition of the three-momentum. Hence,

$$p^\mu = (\mathcal{E}, \mathbf{p}) \tag{154.7}$$

$$= \frac{(m, m\mathbf{v})}{\sqrt{1-v^2}} \quad \text{(massive particles)}. \tag{154.8}$$

As indicated, the last result works for massive particles only. For massless particles, we still have the notion of a world line. For such particles, the four-momentum is *defined* by eq. (154.7) in terms of the energy $\mathcal{E}$ of the particle (a basic physical property) and the direction of travel; $\mathbf{p}$ is parallel to this direction, and $|\mathbf{p}| = \mathcal{E}$.

The separate conservation laws of energy and momentum in Newtonian mechanics get replaced by the single law of conservation of four-momentum in Einsteinian relativity. Secondly, the relativistic relation between energy and three-momentum for one particle follows from the fact that $u^\mu u_\mu = 1$. Since $p^\mu = mu^\mu$,

$$p^\mu p_\mu = m^2, \tag{154.9}$$

i.e.,

$$\mathcal{E}^2 = m^2 + \mathbf{p}^2. \tag{154.10}$$

This relation holds for massless particles as well, in which case, $\mathcal{E}^2 = \mathbf{p}^2$.

Lastly, we define the four-acceleration in parallel with u^μ:

$$a^\mu = \frac{d^2 x^\mu}{d\tau^2}. \tag{154.11}$$

In the lab frame, since $d/d\tau = \gamma d/dt$, we get

$$a^\mu = \gamma^2(0, \dot{\mathbf{v}}) + \gamma\dot{\gamma}(1, \mathbf{v}). \tag{154.12}$$

We now wish to relate this four-vector to the acceleration measured in the momentarily comoving frame. The calculation is subtle. We first differentiate eq. (154.5) with respect to τ. This yields

$$0 = \frac{1}{2}\frac{d}{d\tau}(u^\mu u_\mu) = u^\mu a_\mu. \tag{154.13}$$

Evaluating the last dot product in the comoving frame, we conclude that in this frame a^μ must have the form

$$a^\mu = (0, \mathbf{a}_{co}). \tag{154.14}$$

The quantity $\mathbf{a}_{co}$ is just the Newtonian acceleration of the particle that the momentarily comoving inertial observer would measure. For, in his frame, the four-velocities at the exact instant of comobility (call that $t = 0$) and an infinitesimal time interval Δt later are given by

$$u(t = 0) = (1, \mathbf{0}), \quad u(\Delta t) = (1 - (\Delta v)^2)^{-1/2}(1, \Delta \mathbf{v}). \tag{154.15}$$

Now, noting that in this frame, $\Delta \tau / \Delta t \to 1$ as $t \to 0$, and $\Delta \mathbf{v} = O(\Delta t)$, we get

$$a^\mu = \lim_{\Delta t \to 0} (0, \Delta \mathbf{v}/\Delta t) = (0, \mathbf{a}), \tag{154.16}$$

which completes the proof.

To find $\mathbf{a}_{co}$ explicitly, we boost the lab-frame acceleration. Equation (154.12) gives $a^0 = \gamma \dot{\gamma}$, $a_\| = \gamma \dot{\gamma} v + \gamma^2 \dot{\mathbf{v}}_\|$, and $\mathbf{a}_\perp = \gamma^2 \dot{\mathbf{v}}_\perp$, where the suffixes denote the components parallel and perpendicular to $\mathbf{v}$. Hence,

$$\mathbf{a}_{co,\|} = \gamma(a_\| - v a^0) = \gamma^3 \mathbf{v}(\mathbf{v} \cdot \dot{\mathbf{v}})/v^2,$$

$$\mathbf{a}_{co,\perp} = \mathbf{a}_\perp = \gamma^2 (\dot{\mathbf{v}} - \mathbf{v}(\mathbf{v} \cdot \dot{\mathbf{v}})/v^2). \tag{154.17}$$

That is,

$$\mathbf{a}_{co} = \gamma^2 \dot{\mathbf{v}} - \frac{\gamma^2(1 - \gamma)}{v^2}(\mathbf{v} \cdot \dot{\mathbf{v}})\mathbf{v}. \tag{154.18}$$

The magnitude a_{co} can be found more simply by evaluating the invariant $a^\mu a_\mu$ in the lab and comoving frames and equating. In this way, we find that

$$a_{co}^2 = \frac{\dot{\mathbf{v}}^2 - (\mathbf{v} \times \dot{\mathbf{v}})^2}{(1 - v^2)^3}. \tag{154.19}$$

Exercise 154.1 Show that the sum of two time-like vectors is itself time-like. Thus, adding two momenta yields a momentum. Is the difference of two momenta a momentum?

Exercise 154.2 Let the four-velocities of the three observers in exercise 152.7 be u_α, u'_α, and u''_α. By evaluating the invariant $u^\alpha u''_\alpha$ in the frames K' and K'' and equating, obtain eq. (152.27).

Comment: If we replace $\mathbf{V}$ by $-\mathbf{V}$ and let $\mathbf{V}' = (\mathbf{V} + d\mathbf{V})$, then relative to K', K and K'' have infinitesimally different velocities $\mathbf{V}$ and $(\mathbf{V} + d\mathbf{V})$. The quantity $(V'')^2$ then gives the square of the relative velocity between K and K''—a length element in velocity space:

$$ds_v^2 = \frac{(d\mathbf{V})^2 - (\mathbf{V} \times d\mathbf{V})^2}{(1 - V^2)^2}. \tag{154.20}$$

This is manifestly non-Euclidean, showing that velocity space is curved. This fact is useful in understanding Thomas precession (section 168).

Exercise 154.3 Consider a particle subject to a constant acceleration g along a fixed direction (x, say) in the comoving frame. If the particle is at rest at $t = 0$, find the coordinates and velocity at all times.

Answer: Parametrically, in terms of the proper time, $x = g^{-1}\cosh(g\tau)$, $t = g^{-1}\sinh(g\tau)$, $u^x = \sinh(g\tau)$, $u^t = \cosh(g\tau)$. Such motion is known as *uniform hyperbolic motion*.

155 Four-tensors

Consider the 16 quantities $A^\mu B^\nu$, where A^μ and B^μ are four-vectors. Under an LT, these transform as follows:

$$A^\mu B^\nu \to A'^\mu B'^\nu = \Lambda^\mu_{\ \alpha} \Lambda^\nu_{\ \beta} A^\alpha B^\beta. \tag{155.1}$$

We say that an object $T^{\mu\nu}$ is a four-tensor of the second rank if it transforms according to

$$T'^{\mu\nu} = \Lambda^\mu_{\ \alpha} \Lambda^\nu_{\ \beta} T^{\alpha\beta}. \tag{155.2}$$

The product $A^\mu B^\nu$ is thus a second-rank tensor and is said to be formed by taking the *direct product* of the four-vectors A^μ and B^μ.

Tensors of higher rank can be defined similarly. Further, from a tensor such as $T^{\mu\nu}$, which is contravariant in both indices, we can obtain three related tensors by contracting with the metric:

$$T_{\alpha\beta} = g_{\alpha\mu} g_{\beta\nu} T^{\mu\nu}, \quad T^\alpha_{\ \beta} = g_{\beta\nu} T^{\alpha\nu}, \quad T_\alpha^{\ \beta} = g_{\alpha\mu} T^{\mu\beta}. \tag{155.3}$$

The first tensor is covariant in both indices, while the second and third are mixed. It should be noted that the latter two tensors will generally not be the same, so it is necessary to keep track of which index of a tensor is lowered or raised. It is also possible to write down the Lorentz transformation for these tensors, which we leave to the reader. In fact, we shall generally refrain from giving covariant and contravariant versions of the same formula.

A good example of a second-rank tensor is the angular momentum. In three-space, **L** has components like $xp_y - yp_x$. The natural extension to four-space is the tensor with components

$$J^{\mu\nu} = x^\mu p^\nu - x^\nu p^\mu. \tag{155.4}$$

Let us now discuss a few examples of purely geometric tensors. The first of these is the mixed second-rank "unit tensor," which we specify by giving its components in one frame:

$$\delta^\alpha_\beta = \begin{cases} 1, & \alpha = \beta, \\ 0, & \alpha \neq \beta. \end{cases} \tag{155.5}$$

In other words, this is merely the Kronecker symbol. In this case, we do not distinguish which index is the first or the second, because it is symmetric. This tensor has the important property that its components are the same in all inertial frames. To show this, we make use of eq. (153.21) and the remark following it. This yields

$$
\begin{aligned}
\delta'^{\alpha}_{\beta} &= \Lambda^{\alpha}{}_{\mu}(\Lambda^{-1})^{\nu}{}_{\beta}\delta^{\mu}_{\nu} \\
&= \Lambda^{\alpha}{}_{\mu}(\Lambda^{-1})^{\mu}{}_{\beta} \\
&= \delta^{\alpha}_{\beta}.
\end{aligned}
\tag{155.6}
$$

A second such tensor is the Levi-Civita, or fully antisymmetric tensor in four dimensions. This is now of fourth rank. As in three-space, we define it to be fully antisymmetric in all its indices:

$$
\epsilon^{\alpha\beta\gamma\delta} = -\epsilon^{\beta\alpha\gamma\delta} = -\epsilon^{\gamma\beta\alpha\delta}, \text{ etc.}
\tag{155.7}
$$

Its only nonzero components are therefore ϵ^{0123} and 23 others obtained by permuting the indices. We specify it fully by setting

$$
\epsilon^{0123} = 1.
\tag{155.8}
$$

To show its frame invariance, we perform an LT and get

$$
\epsilon'^{\alpha\beta\gamma\delta} = \Lambda^{\alpha}{}_{\bar{\alpha}}\Lambda^{\beta}{}_{\bar{\beta}}\Lambda^{\gamma}{}_{\bar{\gamma}}\Lambda^{\delta}{}_{\bar{\delta}}\epsilon^{\bar{\alpha}\bar{\beta}\bar{\gamma}\bar{\delta}}.
\tag{155.9}
$$

The right-hand side is clearly fully antisymmetric. Secondly,

$$
\epsilon'^{0123} = \Lambda^{0}{}_{\alpha}\Lambda^{1}{}_{\beta}\Lambda^{2}{}_{\gamma}\Lambda^{3}{}_{\delta}\epsilon^{\alpha\beta\gamma\delta}.
\tag{155.10}
$$

But the right-hand side is precisely the determinant of the matrix $\Lambda^{\alpha}{}_{\beta}$. From eq. (153.15), we have $(\det \Lambda)^2 = 1$. Since we are concerned only with proper LTs here, we must have $\det \Lambda = 1$. Hence, $\epsilon'^{0123} = 1$, and the frame invariance is proved.

From $\epsilon^{\alpha\beta\gamma\delta}$, we can construct $\epsilon_{\alpha\beta\gamma\delta}$. The latter is also fully antisymmetric and has

$$
\epsilon_{0123} = -1.
\tag{155.11}
$$

Next, let us show that the metric $g_{\mu\nu}$ is a tensor. The simplest way to do this is to introduce the notion of *basis vectors* $\vec{e}_{\mu}$ ($\mu = 0, 1, 2, 3$) in four-space.[6] Note that the index μ here labels the different vectors, each of which has four components. We then define an abstract four-vector $\vec{A}$ as

$$
\vec{A} = A^{\mu}\vec{e}_{\mu},
\tag{155.12}
$$

where the A^{μ} are now seen to be components with respect to this basis.[7] The dot product

---

[6] We use arrows to distinguish four-vectors from three-vectors, which we continue to denote by boldface type. The arrow notation is limited to this chapter and the next. In writing vectors by hand, one common practice is to use an arrow for four-vectors and a squiggle under the symbol for three-vectors.

[7] We mentioned above that in modern learned language the A_{μ} are the components of a one-form. These components are defined with respect to basis one-forms that are dual to the basis vectors. We shall say no more about forms, since for our purposes, doing so does not provide sufficient payoff to develop the required machinery.

of two vectors $\vec{A}$ and $\vec{B}$ is given by

$$\vec{A} \cdot \vec{B} = A^\mu B^\nu (\vec{e}_\mu \cdot \vec{e}_\nu). \tag{155.13}$$

For this to agree with our previous definition, we must have

$$\vec{e}_\mu \cdot \vec{e}_\nu = g_{\mu\nu}. \tag{155.14}$$

We now adopt the point of view that under an LT, the vectors $\vec{A}$ and $\vec{B}$ themselves remain fixed, and that it is the basis vectors and the components that change. Thus, in the new frame

$$\vec{A} = A'^\mu \vec{e}_\mu', \quad \vec{A} \cdot \vec{B} = A'^\mu B'^\nu (\vec{e}_\mu' \cdot \vec{e}_\nu'). \tag{155.15}$$

It is now obvious that the basis vectors must transform like the covariant components of a vector, i.e.,

$$\vec{e}_\mu = \Lambda^{\mu'}_{\ \mu} \vec{e}_{\mu'}. \tag{155.16}$$

Equation (155.14) then implies that $g_{\mu\nu}$ transforms like a covariant tensor of rank two. We have already shown that the components of this tensor are also frame invariant.

In abstract notation, we denote the direct product tensor by the symbol $\otimes$. Thus, for the example at the beginning of this section, we write

$$\overset{=}{T} = \vec{A} \otimes \vec{B} = A^\mu B^\nu (\vec{e}_\mu \otimes \vec{e}_\nu). \tag{155.17}$$

The products $\vec{e}_\mu \otimes \vec{e}_\nu$ form the basis tensors of rank two. For an antisymmetric product tensor such as the angular momentum, we write

$$\overset{=}{J} = \vec{x} \otimes \vec{p} - \vec{p} \otimes \vec{x} \equiv \vec{x} \wedge \vec{p}. \tag{155.18}$$

The wedge product symbol ($\wedge$) is reserved for antisymmetric products.

Exercise 155.1 Show that $\epsilon_{\delta\alpha\beta\gamma} = -\epsilon_{\alpha\beta\gamma\delta}$.

Exercise 155.2 (permutation tensors) Show that

$$\epsilon^{\alpha\beta\gamma\delta}\epsilon_{\alpha\beta\gamma\delta} = -4!,$$

$$\epsilon^{\mu\beta\gamma\delta}\epsilon_{\alpha\beta\gamma\delta} = -(3!)\delta^\mu_\alpha,$$

$$\epsilon^{\mu\nu\gamma\delta}\epsilon_{\alpha\beta\gamma\delta} = -(2!)\delta^{\mu\nu}_{\alpha\beta},$$

$$\epsilon^{\mu\nu\zeta\delta}\epsilon_{\alpha\beta\gamma\delta} = -\delta^{\mu\nu\zeta}_{\alpha\beta\gamma}. \tag{155.19}$$

Here, $\delta^{\mu\nu\zeta}_{\alpha\beta\gamma}$ is nonzero only if the three indices μ, ν, and ζ are all unequal, and likewise for α, β, and γ. Further, it is $+1$ if the triple $(\mu\nu\zeta)$ is an even permutation of the triple $(\alpha\beta\gamma)$, and -1 if it is an odd permutation. The tensor $\delta^{\mu\nu}_{\alpha\beta}$ is defined similarly. Clearly, these tensors are also frame invariant.

Exercise 155.3 (dual tensors) From any completely antisymmetric tensor T of rank $n = 1, 2,$ or 3, one can construct the *dual* tensor $*T$, which is also completely antisymmetric, and has rank $4 - n$, as follows:

$$*T_{\nu\eta\sigma} = \epsilon_{\mu\nu\eta\sigma} T^{\mu}, \quad *T_{\mu\nu} = \frac{1}{2!}\epsilon_{\mu\nu\eta\sigma} T^{\eta\sigma}, \quad *T_{\sigma} = \frac{1}{3!}\epsilon_{\mu\nu\eta\sigma} T^{\mu\nu\eta}. \tag{155.20}$$

These definitions are generalizations of the cross product in three dimensions. Show that (i) the dual contains exactly the same information as the original tensor, (ii) $**T = \pm T$, where the sign is positive if T is of rank 1 or 3, and negative if T is of rank 2.

Exercise 155.4 In section 153 we referred to $\Lambda^{\mu}_{\ \nu}$ as the LT *matrix*. Is $\Lambda^{\mu}_{\ \nu}$ also a tensor? If not, why not? If yes, how would you think of it as a physically measurable quantity?

156 Vector fields and their derivatives in space–time

From vectors and tensors, we move on to vector and tensor *fields*. The simplest such object is a scalar field, $\psi(x^{\mu})$, by which we mean that if an event P has coordinates x^{μ} and x'^{μ} in two different frames, the value of ψ at P is the same in the two frames:

$$\psi(x'^{\mu}) = \psi(x^{\mu}). \tag{156.1}$$

The components of a vector field, on the other hand, transform according to

$$A'^{\alpha}(x'^{\mu}) = \Lambda^{\alpha}_{\ \beta} A^{\beta}(x^{\mu}). \tag{156.2}$$

From A^{μ}, we can construct the associated covariant vector field. The transformation rules for this field and for higher-rank tensor fields are easily written down.

Just as in three-space, the basic derivative in space–time is the gradient. Given a scalar field $\psi(x^{\mu})$, we define

$$\psi_{,\mu} = \partial_{\mu}\psi = \frac{\partial \psi}{\partial x^{\mu}} = \left(\frac{\partial \psi}{\partial t}, \nabla\psi\right). \tag{156.3}$$

(The last entry is in 1+3 form.) To see how this transforms under an LT, consider ψ at two points x^{μ} and $x^{\mu} + w^{\mu}$, where w^{μ} is infinitesimal. Then,

$$\psi(x^{\mu} + w^{\mu}) - \psi(x^{\mu}) = w^{\mu}\frac{\partial \psi}{\partial x^{\mu}} + O(w^2). \tag{156.4}$$

Since the left-hand side is a scalar, so must the right-hand side be. This is possible only if it is a dot product. Hence, $\partial\psi/\partial x^{\mu}$ is a covariant vector. The notations in eq. (156.3) anticipate this fact. The comma notation is particularly handy.

If we raise the index in the gradient, we get

$$\partial^{\mu}\psi = g^{\mu\nu}\partial_{\nu}\psi. \tag{156.5}$$

This is clearly a contravariant vector field. We can also define it as a derivative with respect to the covariant components of x, but we shall avoid this route.

In the same way, derivatives of vector fields generate tensor fields. For example,

$$\partial_\mu A_\nu = A_{\nu,\mu} = \frac{\partial A_\nu}{\partial x^\mu} \tag{156.6}$$

is a second-rank tensor field, covariant in both indices. Contracting on the two indices gives us the four-divergence

$$A^\mu{}_{,\mu} = \partial_\mu A^\mu = \partial^\mu A_\mu = \frac{\partial A_\mu}{\partial x^\mu}. \tag{156.7}$$

This object is clearly a scalar. In the same way, we can take the divergence of higher-rank tensor fields, obtaining a tensor field of one lower rank. For example,

$$T^{\mu\nu}{}_{,\mu} \quad \text{and} \quad T^{\nu\mu}{}_{,\mu} \tag{156.8}$$

are both rank-1 fields. The two obviously need not be the same.

Of the higher derivatives, the *D'Alembertian* is the only one worth singling out:

$$\partial^\mu \partial_\mu \psi = \partial_\mu \partial^\mu \psi = \left(\frac{\partial^2}{\partial t^2} - \nabla^2 \right) \psi, \tag{156.9}$$

which we encountered before while studying electromagnetic waves.

As in three-space, it is also useful to introduce the concept of a directional derivative. The derivative of a scalar field ϕ along a vector $\vec{A}$ is written in several equivalent ways and defined by

$$\partial_{\vec{A}}\phi = \nabla_{\vec{A}}\phi = (\vec{A} \cdot \nabla)\phi \equiv A^\mu \frac{\partial \phi}{\partial x^\mu}. \tag{156.10}$$

It should be apparent that we can also take the directional derivative of vector and tensor fields. The directional derivative along the four-velocity is particularly important. By evaluating this in the comoving frame, we see that it is just the partial derivative with respect to the proper time:

$$\nabla_{\vec{u}}\phi = \frac{\partial \phi}{\partial \tau}. \tag{156.11}$$

157 Integration of vector fields*

Next, let us consider integrals of vector fields. In three dimensions, we can have line, surface, and volume integrals. In four-space, we can integrate over curves, two-dimensional surfaces, three-dimensional surfaces (also called *hypersurfaces*), and four-dimensional volumes. The last is perhaps the simplest. The volume element is

$$d^4\Omega = dt\, dx\, dy\, dz. \tag{157.1}$$

This is clearly a scalar. On changing frames, it will be multiplied by the determinant of the LT matrix, $\det \Lambda$, which is 1 for proper transformations. This result also shows that the four-dimensional delta function is a scalar:

$$\delta^{(4)}(x) = \delta^{(4)}(x'). \tag{157.2}$$

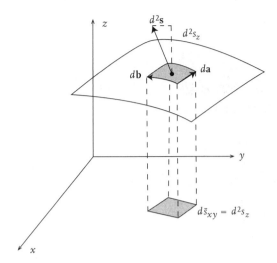

FIGURE 23.5. Surface integrals in three dimensions, and the meaning of the vectorial surface area element $d^2\mathbf{s}$.

For, integrating over all four-space, we get

$$\int \delta^{(4)}(x)d^4\Omega = 1 = \text{invariant}. \tag{157.3}$$

The line integral along a curve in four-space is also easy to understand. The integration element is just the differential dx^μ. If the curve is parametrized by a variable ξ, i.e., the coordinates x^μ are given as functions of ξ: $x^\mu(\xi)$,

$$dx^\mu = \frac{\partial x^\mu}{\partial \xi}d\xi. \tag{157.4}$$

Under reparametrization from ξ to a new parameter $\bar{\xi}$, this formula is form invariant, as is easily verified.

Surface integrals are more subtle. In three dimensions, given a surface, we can draw a tiny parallelogram on it spanned by the vectors $d\mathbf{a}$ and $d\mathbf{b}$. (See fig. 23.5.) The projection of this parallelogram onto the xy plane has an area $d\bar{s}_{xy} = da_x db_y - da_y db_x$. Likewise for the other coordinate planes. Instead of working with these projections, it proves advantageous to employ the vector defined by the cross product, $d^2\mathbf{s} = d\mathbf{a} \times d\mathbf{b}$. The component d^2s_z is the area of the projection onto the plane normal to $\hat{\mathbf{z}}$, i.e., the xy plane. More generally,

$$d^2s_i = \tfrac{1}{2}\epsilon_{ijk}d\bar{s}_{jk}. \tag{157.5}$$

In four-space, we define the projections of the parallelogram spanned by the four-vectors $d\vec{a}$ and $d\vec{b}$ in complete parallel:

$$d\bar{s}_{\alpha\beta} = da_\alpha db_\beta - da_\beta db_\alpha, \tag{157.6}$$

and construct the area element as

$$d^2 s_{\mu\nu} = \frac{1}{2!} \epsilon_{\mu\nu\alpha\beta} d\bar{s}^{\alpha\beta}. \tag{157.7}$$

Note that there are two outward normal directions to the surface, i.e., the normal is a tensor of rank two. Note also that the surface must not be nonorientable like a Möbius strip, for then one cannot integrate globally over it.

Exercise 157.1 Suppose a surface is parametrized by the variables ξ and η. Write the surface element in both three- and four-space in terms of these variables. Show that the expression is invariant under reparametrization.

Solution:

$$d^2 s_i = \frac{1}{2} \epsilon_{ijk} \frac{\partial(x^j, x^k)}{\partial(\xi, \eta)} d\xi d\eta, \tag{157.8}$$

$$d^2 s_{\mu\nu} = \frac{1}{2!} \epsilon_{\mu\nu\alpha\beta} \frac{\partial(x^\alpha, x^\beta)}{\partial(\xi, \eta)} d\xi d\eta. \tag{157.9}$$

Exercise 157.2 Find the surface element for a sphere of radius a in three dimensions in terms of the spherical polar coordinates θ and φ.

Answer: $d^2 \mathbf{s} = (\sin\theta\cos\varphi, \sin\theta\sin\varphi, \cos\theta) a^2 \sin\theta d\theta d\varphi$.

Exercise 157.3 Find the invariant magnitude of the surface element for a sphere of radius a in three dimensions in terms of Cartesian coordinates x and y.

Solution: Let us regard the sphere as embedded in a three-dimensional space, for which the three-volume element $dx\,dy\,dz$ is invariant under rotation. The condition that a point lie on the two-sphere can also be expressed in terms of the invariant delta function $\delta(x^2 + y^2 + z^2 - a^2)$. Hence, the invariant surface element is

$$d^2 s = \int 2a\delta(x^2 + y^2 + z^2 - a^2) dx\,dy\,dz. \tag{157.10}$$

(The proportionality factor of $2a$ will be seen to be necessary below.) Integrating over z, we obtain

$$d^2 s = \frac{a}{\sqrt{a^2 - (x^2 + y^2)}} dx\,dy. \tag{157.11}$$

That the overall factor is correct can be seen by looking at the north pole, where $x = y = 0$.

What this result shows is that the Cartesian element $dx\,dy$, which a north-pole-based observer unaware of the curvature of the sphere might use, is not invariant. If another observer uses a second Cartesian system x', y', and z', the two observers will agree on the surface element for a small patch only if they insert the square root prefactors.

The volume element for an integral over a three-dimensional hypersurface in four-space is now easy to construct. We give the answer only in terms of an assumed parametrization of the hypersurface, ξ, η, and ζ:

$$d^3 S_\mu = \frac{1}{3!} \epsilon_{\mu\alpha\beta\gamma} \frac{\partial(x^\alpha, x^\beta, x^\gamma)}{\partial(\xi, \eta, \zeta)} d\xi d\eta d\zeta. \tag{157.12}$$

As an example, consider the integral one over the subspace $x^0 = t = \text{const}$, i.e., a snapshot of all three-space. We choose the spatial coordinates x, y, and z as parameters. The only nonvanishing element of $d^3 S_\mu$ is $d^3 S_0$, with $d^3 S_0 = dx\, dy\, dz$. Hence, $d^3 S^0 = -dx\, dy\, dz = -d^3 x$, and

$$d^3 S^\mu = (-dV, \mathbf{0}). \tag{157.13}$$

The outward normal to the surface points toward the past.

A hypersurface is said to be time- or space-like if its normal is everywhere time- or space-like.

Exercise 157.4 Express the spatial volume element dV of an inertial observer moving with velocity $\mathbf{v}$ in four-vector form, i.e., as a hypersurface element.

Solution: Let u^α be the four-velocity of the observer, and let dV_p be the proper volume element. The former is a vector and the latter is a Lorentz scalar, so the product $-u^\alpha dV_p$ is a four-vector. In the rest frame of the particle, $u^\alpha = (1, \mathbf{0})$, so this product is $-(dx\, dy\, dz, 0)$, which is just the surface element in that frame. It follows that

$$d^3 S^\alpha = -u^\alpha dV_p = -(1, \mathbf{v}) dV \tag{157.14}$$

in any frame.

Exercise 157.5 A fluid of particles of mass m is described by the phase space density $f(\mathbf{r}, \mathbf{p})$, i.e., in a small volume element $d^3 r$ centered around the point $\mathbf{r}_A$ and a small momentum space volume element $d^3 p$ centered around $\mathbf{p}_A$, there are $f(\mathbf{r}_A, \mathbf{p}_A) d^3 r d^3 p$ particles. Show that $f(\mathbf{r}, \mathbf{p})$ is a Lorentz invariant.

Solution: Consider a group of particles with momentum $\mathbf{p}_A$, and let $\mathbf{v}_A$ be the velocity with which they move. The volume element in the lab frame is foreshortened from the proper volume element in the comoving frame of the particles by a factor of $\gamma_A = (1 - \mathbf{v}_A^2)^{-1/2}$. The proper volume is a Lorentz invariant, so

$$\gamma_A d^3 r = \text{invariant.} \tag{157.15}$$

To find how the momentum space volume element transforms, we proceed in complete analogy with exercise 157.3 on the surface element of the two-sphere. The four-volume element $dp^x dp^y dp^z dp^0$ is a Lorentz invariant. However, p_A^0 is fixed once $\mathbf{p}_A$ is given, via the mass-shell constraint $(p^0)^2 = m^2 + \mathbf{p}^2$. Hence, the following volume element on this mass-shell is also an invariant:

$$\int \delta \left(\sqrt{p^\alpha p_\alpha} - m \right) dp^x dp^y dp^z dp^0. \tag{157.16}$$

Performing the integration over p^0 and noting that $p_A = m\gamma_A$, we see that

$$\gamma_A^{-1} dp^x dp^y dp^z = \text{invariant}. \tag{157.17}$$

Multiplying the real and momentum space volume element invariants, we see that the phase space volume element is Lorentz invariant:

$$d^3 r\, d^3 p = \text{invariant}. \tag{157.18}$$

Since the number of particles in this element is clearly invariant, we conclude that the phase space *density* $f(\mathbf{r}, \mathbf{p})$ is an invariant.

Note that a frontal attack on this problem is very cumbersome. If we consider a primed observer moving at a velocity $\mathbf{V}$ relative to the unprimed one, there are two subtleties. First, the volume element $d^3 r$ in the lab frame is being transported with the particles. The best way to see this is to imagine painting the group of selected particles. Since it is ultimately the number of painted particles that is invariant, the volume in which they are contained is moving. Hence, $d^3 r' \neq (1 - V^2)^{1/2} d^3 r$. The correct relation is in fact found by going to the comoving frame. The second subtlety is that $\mathbf{p}'_A$ must be found from $\mathbf{p}_A$ by using the correct law for adding velocities, and differentials must then be taken to find $d^3 p'$. If we performed these maneuvers, we would discover that $d^3 p' = (\gamma'_A/\gamma_A) d^3 p$.

The theorems of Gauss and Stokes have natural analogs in four-space. Gauss's theorem takes the form

$$\int_{\partial V} A^\alpha d^3 S_\alpha = \int_V A^\alpha_{,\alpha} d^4 \Omega. \tag{157.19}$$

Here, V is some space–time region with boundary ∂V, and A_α is some vector field. But the result also applies to tensor fields, e.g.,

$$\int_{\partial V} T^{\alpha\beta} d^3 S_\alpha = \int_V T^{\alpha\beta}_{,\alpha} d^4 \Omega. \tag{157.20}$$

Likewise, Stokes's theorem reads

$$\int_{\partial \Sigma} A^{\alpha\beta} d^2 s_{\alpha\beta} = 2 \int_\Sigma A^{\alpha\beta}_{,\gamma} d^3 S^\gamma. \tag{157.21}$$

Now, Σ is a three-surface with boundary $\partial \Sigma$. Proof and verification are left to the reader.

158 Accelerated observers*

The example of uniform hyperbolic motion (exercise 154.3) illustrates one of the central characteristics of an accelerated observer in special relativity. From fig. 23.6a we see that no signals can reach the observer from regions B and C. Likewise, he can send no signals

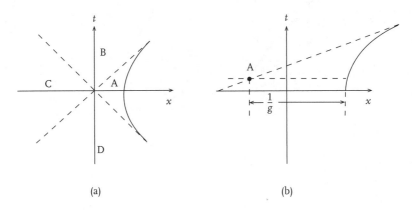

FIGURE 23.6. (a) An observer undergoing uniform hyperbolic motion. (b) An observer accelerated for a finite time only cannot avoid a coordinate singularity.

to regions C and D. There is no way for this observer to assign unique coordinates to events in region C, since he cannot communicate with it. To see that this problem afflicts all accelerated motion, consider an observer who undergoes some acceleration for a finite time and is moving uniformly before and after. It is clearly possible to assign coordinates in the states of uniform motion, but if we try to make a coordinate patch in between, we see from the figure that a singularity must develop somewhere near point A, at a distance of order $1/g$, where g is the acceleration.

The best an accelerated observer can do is make a local coordinate system. The first step in doing this is to construct a tetrad $\vec{e}_\mu(\tau)$ at each point on its world line. Since the tetrads for infinitesimally apart times must connect smoothly onto one another, let us examine how the tetrad changes with τ. The most general form for its time derivative is

$$\nabla_{\vec{u}}\vec{e}_\mu = \partial_\tau\vec{e}_\mu = \Omega_\mu{}^\nu\vec{e}_\nu, \tag{158.1}$$

since the left-hand side is a vector and hence expandable in the basis itself. We proceed to derive the properties of the 16 numbers $\Omega_\mu{}^\nu$.

First, it is clearly desirable to maintain orthonormality of the basis vectors with τ, i.e., we want

$$\vec{e}_\mu(\tau) \cdot \vec{e}_\nu(\tau) = g_{\mu\nu}. \tag{158.2}$$

Differentiating, we get

$$\begin{aligned}
0 &= \partial_\tau(\vec{e}_\mu(\tau) \cdot \vec{e}_\nu(\tau)) \\
&= \Omega_\mu{}^\sigma\vec{e}_\sigma \cdot \vec{e}_\nu + \vec{e}_\mu \cdot \Omega_\nu{}^\sigma\vec{e}_\sigma \\
&= \Omega_{\mu\nu} + \Omega_{\nu\mu}. \tag{158.3}
\end{aligned}$$

Hence, the object Ω is antisymmetric.

It pays to rewrite the time derivative of the tetrad as follows. If we observe that $(\vec{e}_\nu \otimes \vec{e}_\sigma) \cdot \vec{e}_\mu = \vec{e}_\nu g_{\sigma\mu}$, we have

$$\Omega_\mu{}^\nu \vec{e}_\nu = \Omega^{\sigma\nu} \vec{e}_\nu g_{\sigma\mu} = (\Omega^{\sigma\nu} \vec{e}_\nu \otimes \vec{e}_\sigma) \cdot \vec{e}_\mu. \tag{158.4}$$

Thus, with

$$\bar{\bar{\Omega}} = \Omega^{\nu\sigma} \vec{e}_\nu \otimes \vec{e}_\sigma, \tag{158.5}$$

we get (exploiting the antisymmetry of $\bar{\bar{\Omega}}$)

$$\nabla_{\vec{u}} \vec{e}_\mu = -\bar{\bar{\Omega}} \cdot \vec{e}_\mu. \tag{158.6}$$

This equation shows that $\bar{\bar{\Omega}}$ is a tensor.

A second condition on $\bar{\bar{\Omega}}$ arises from the requirement that the first member of the tetrad, $\vec{e}_0$, must be the same as the four-velocity $\vec{u}$ at all times. Since this singles out the time direction $\vec{u}$, let us decompose $\bar{\bar{\Omega}}$ into its time and space parts,

$$\Omega^{\mu\nu} = -u^\mu w^\nu + w^\mu u^\nu + \omega^{\mu\nu}. \tag{158.7}$$

In other words,

$$\bar{\bar{\Omega}} = -\vec{u} \otimes \vec{w} + \vec{w} \otimes \vec{u} + \bar{\bar{\omega}}. \tag{158.8}$$

Here, w^μ is some undetermined vector, and $\omega^{\mu\nu}$ is a purely spatial tensor, i.e., $u_\mu \omega^{\mu\nu} = \omega^{\mu\nu} u_\nu = 0$. In fact, the vector w^μ can also be taken to be purely spatial if we note that any temporal part must be parallel to u^μ, and the corresponding contributions to $\Omega^{\mu\nu}$ from the first two terms will cancel each other. We can then fix w^μ completely by noting that

$$\nabla_{\vec{u}} \vec{e}_0 = -\bar{\bar{\Omega}} \cdot \vec{e}_0. \tag{158.9}$$

But $\vec{e}_0 = \vec{u}$, and $\nabla_{\vec{u}} \vec{e}_0 = \partial_\tau \vec{u} = \vec{a}$. Hence, exploiting the facts $\bar{\bar{\omega}} \cdot \vec{u} = \vec{w} \cdot \vec{u} = 0$, we get

$$\vec{a} = -\bar{\bar{\Omega}} \cdot \vec{u}$$
$$= -(-\vec{u} \otimes \vec{w} + \vec{w} \otimes \vec{u} + \bar{\bar{\omega}}) \cdot \vec{u}$$
$$= -\vec{w}. \tag{158.10}$$

We thus determine $\bar{\bar{\Omega}}$ as

$$\bar{\bar{\Omega}} = \vec{u} \otimes \vec{a} - \vec{a} \otimes \vec{u} + \bar{\bar{\omega}}. \tag{158.11}$$

This is all one can say about $\bar{\bar{\Omega}}$ by requiring that the tetrad change smoothly. However, the tensor ω describes the rotation of the spatial components of the tetrad as one moves along the world line. The most natural choice is not to allow these components to spin needlessly. (For example, we would like $\bar{\bar{\Omega}} = 0$ if there is no acceleration, i.e., $\vec{a} = 0$.) Therefore, we set $\bar{\bar{\omega}}$ to zero, which yields

$$\bar{\bar{\Omega}} = \vec{u} \wedge \vec{a}. \tag{158.12}$$

Then,

$$\partial_{\vec{u}}\vec{e}_\mu = -(\vec{u} \wedge \vec{a}) \cdot \vec{e}_\mu. \tag{158.13}$$

This is the desired equation for how the basis vectors must change as we move along the world line of an accelerated observer. This is also known as *Fermi-Walker transporting the tetrad*, and it keeps the basis vectors as unchanged in direction as possible.

Exercise 158.1 Equation (158.13) may be easier to understand by analogy with parallel transport on a curved two-dimensional surface such as the earth. Let us embed the surface in three-dimensional space and move along a path r(t) on the surface. Let n̂(t) be the unit normal as we move along, and let $e_1(t)$ and $e_2(t)$ be two other unit vectors in the surface. We wish to find the law for parallel transporting e_i.

Show that we must have

$$\frac{d\mathbf{e}_i}{dt} = \mathbf{\Omega} \times \mathbf{e}_i, \tag{158.14}$$

where $\mathbf{\Omega}$ is some vector (the angular velocity). The important physical point now is that the e_i must not spin around n̂ (consider motion on a flat surface), so $\mathbf{\Omega}$ must obey $\mathbf{\Omega} \cdot \hat{\mathbf{n}} = 0$. Show that this implies

$$\frac{d\mathbf{e}_i}{dt} = \left(\hat{\mathbf{n}} \times \frac{d\hat{\mathbf{n}}}{dt} \right) \times \mathbf{e}_i, \tag{158.15}$$

which should be compared with eq. (158.13).

Let us resume the attempt to assign coordinates. The tetrad $\vec{e}^\mu(\tau)$ can be regarded as an infinitesimal coordinate system at some point along the observer's world line, denoted by $\vec{h}(\tau)$. We now wish to extend these coordinate systems for different τ and patch them up as smoothly as possible. Consider a point A on the world line, and the hypersurface of simultaneity passing through A for a comoving *inertial* observer. Let B be a point on this hypersurface, to which we try to assign coordinates as measured by the comoving observer at A. Let Euclidean coordinates in the simultaneity surface through A be denoted by ξ^i ($i = 1, 2, 3$). These are clearly acceptable spatial coordinates to be used by the comoving observer. The vector assigned to point B is therefore

$$\vec{x} = \vec{h}(\tau) + \xi^i \vec{e}_i(\tau). \tag{158.16}$$

The meaning of this equation is as follows. To find the coordinates for B, we first find the point $\vec{h}(\tau)$ on the world line that lies on the hypersurface of simultaneity through B (as measured by some inertial observer). Relative to the point $\vec{h}(\tau)$ as an origin, the point B has time coordinate zero and spatial coordinates ξ^i.

The question now is whether this assignment can be made globally. To answer this, let us find the metric in the vicinity of B. We have

$$(ds)^2 = g_{\mu\nu}dx^\mu dx^\nu = d\vec{x} \cdot d\vec{x}. \tag{158.17}$$

We next note that

$$d\vec{x} = \frac{d\vec{h}}{d\tau}d\tau + d\xi^i\vec{e}_i + \xi^i\frac{d\vec{e}_i}{d\tau}d\tau$$

$$= \vec{u}d\tau + d\xi^i\vec{e}_i - \xi^i(\vec{u}\otimes\vec{a} - \vec{a}\otimes\vec{u})\cdot\vec{e}_i d\tau$$

$$= (1 - \xi^i a_i)d\tau\vec{u} + d\xi^i\vec{e}_i. \tag{158.18}$$

Here, in the second line, we used the Fermi-Walker transport law. The last line follows from $\vec{u}\cdot\vec{e}_i = 0$, and $\vec{a}\cdot\vec{e}_i = a_i$. Identifying the coordinate ξ^0 with τ, we finally obtain

$$(ds)^2 = (1 - \xi^i a_i)^2(d\xi^0)^2 - \delta_{ij}d\xi^i d\xi^j. \tag{158.19}$$

The scale factor for the ξ^0 coordinate vanishes when $\xi^i a_i = 1$. At this point a coordinate singularity develops. As noted before, this breakdown happens at length scales of order $1/a$. Secondly, the scale factor $(1 - \xi^i a_i)$ shows that proper time elapses nonuniformly at different spatial locations in the accelerated frame. Suppose that we have synchronized clocks at the points A and B introduced above. If we now consider two points A′ and B′ at the same spatial separation ξ^i, but at a later time, clocks at these points will not be synchronized. Thus, the time coordinate cannot be thought of as the result of measurements made on rods and clocks. The same is true of the other coordinates too. This shows the intrinsic limitations of special relativity in describing accelerated observers.

24 | Special relativity and electromagnetism

In this chapter we will see how electromagnetism fits in with relativity. We first show that Maxwell's equations and the Lorentz force law obey modern relativity by showing that these equations can be written covariantly. We then examine some problems where relativistic considerations are prominent.

159 Four-current and charge conservation

One way to show that electromagnetism obeys special relativity is to show that its laws are form invariant under a Lorentz transformation. We will do this by writing all of them in terms of four-vectors, four-tensors, etc., and their four-gradients, four-divergences, etc. The invariance of the laws is then automatic, and written in this way they are said to be *manifestly covariant*.

The first four-object we study is the current. Consider a charge q moving along a trajectory $\mathbf{s}(t)$, with velocity $\mathbf{v} = d\mathbf{s}/dt$. We will show that the charge and current densities

$$\rho(\mathbf{r}, t) = q\delta(\mathbf{r} - \mathbf{s}(t)), \quad \mathbf{j}(\mathbf{r}, t) = q\mathbf{v}(t)\delta(\mathbf{r} - \mathbf{s}(t)), \tag{159.1}$$

are the time and space components of a four-vector j^μ. To do this, we think in terms of the world line of the particle rather than its trajectory. We parametrize this line by the proper time τ and, in addition to the space components $\mathbf{s}(\tau)$, give the time in the lab frame as $t(\tau) \equiv s^0(\tau)$. The world line is then given by $s^\alpha(\tau)$. Now consider the four-vector

$$j^\mu(x^\alpha) = q \int u^\mu \delta^{(4)} \big[x^\alpha - s^\alpha(\tau) \big] d\tau, \tag{159.2}$$

where $u^\mu = ds^\mu/d\tau$ is the four-velocity of the particle. Since $\delta^{(4)}$ and $d\tau$ are scalars, this object is a four-vector, as asserted. We can evaluate the τ integral explicitly as follows:

$$j^\mu = q \int u^\mu \delta^{(3)}\left[\mathbf{r} - \mathbf{s}(\tau)\right]\delta\left[t - s^0(\tau)\right]d\tau$$

$$= q u^\mu \frac{d\tau}{dt}\delta^{(3)}\left[\mathbf{r} - \mathbf{s}(\tau)\right]\Big|_{t=s^0(\tau)} . \tag{159.3}$$

But $dt/d\tau = ds^0/d\tau = u^0$, so

$$j^\mu = q \frac{u^\mu}{u^0}\delta^{(3)}\left[\mathbf{r} - \mathbf{s}(t)\right]. \tag{159.4}$$

Finally, we note that $u^\mu = (\gamma, \gamma\mathbf{v})$ (see eq. (154.3)) and $u^0 = \gamma$, so that

$$j^\mu = (\rho, \mathbf{j}). \tag{159.5}$$

By adding up the individual contributions to the current for a system of particles, we define the four-current in general. Equation (159.5) clearly continues to hold.

The equation of continuity,

$$\frac{\partial \rho}{\partial t} + \nabla \cdot \mathbf{j} = 0, \tag{159.6}$$

which expresses the conservation of charge, can be written as

$$\partial_\mu j^\mu = 0. \tag{159.7}$$

The left-hand side is clearly a scalar, so this equation is true in all inertial frames.

Exercise 159.1 One can verify that ρ and $\mathbf{j}$ transform as the time and space parts of a four-vector by an elementary argument. Consider a uniform wire lying parallel to the x axis, filled with a uniform density of positive charges. In a frame K, let the charge density be ρ, and let all the charges move at a velocity $v_0\hat{\mathbf{x}}$. Consider another frame K' moving at a velocity $V\hat{\mathbf{x}}$ relative to K. Find the charge and current densities in K'.

Solution: Let the velocity of the charges in K' be $v_0'\hat{\mathbf{x}}$. By the rule for addition of velocities,

$$v_0' = \frac{v_0 - V}{1 - v_0 V/c^2} = c\frac{\beta_0 - \beta}{1 - \beta_0\beta}, \tag{159.8}$$

where $\beta_0 = v_0/c$, $\beta_0' = v_0'/c$, and $\beta = V/c$. Define $\gamma_0 = (1 - \beta_0^2)^{-1/2}$, $\gamma_0' = (1 - \beta_0'^2)^{-1/2}$, and $\gamma = (1 - \beta^2)^{-1/2}$. Consider a segment of wire of proper length L_0. Its length in K and K' is L_0/γ_0 and L_0/γ_0'. Since the number of charges in this segment is invariant, $\rho' = \rho\gamma_0'/\gamma_0$. Expressing γ_0' in terms of v_0 and V, we find $\gamma_0' = \gamma\gamma_0(1 - \beta\beta_0)$, so

$$\rho' = \rho\gamma(1 - Vv_0/c^2). \tag{159.9}$$

To find j', we first note that in K, $j = \rho v$. Likewise, in K', $j' = \rho'v'$. Feeding in eqs. (159.8) and (159.9), we obtain

$$j' = \rho\gamma(v_0 - V). \tag{159.10}$$

It is easy to verify that $\rho' = \gamma(\rho - jV/c^2)$, $j' = \gamma(j - V\rho)$.

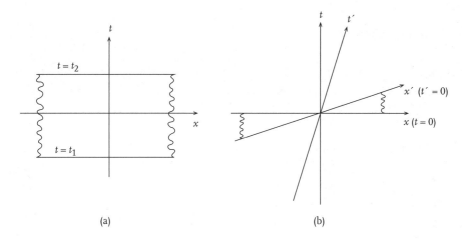

FIGURE 24.1. Four-volumes over which to integrate the charge density to prove that the total charge is (a) conserved and (b) a Lorentz scalar. The wavy lines represent space-like surfaces at infinity.

It is also useful to examine the four-space integral form of the conservation law. This states that the total charge in the universe,

$$Q = \int_{t=\text{const}} j^0 d^3 x, \tag{159.11}$$

is (a) independent of time and (b) a Lorentz scalar. The results follow by Gauss's theorem, according to which

$$\int_{\partial V} j^\mu d^3 S_\mu = \int_V j^\mu_{,\mu} d^4 \Omega = 0, \tag{159.12}$$

where V is any four-volume. To show the constancy of Q, we choose V to be bounded by the time-like surfaces $t = t_1$ and $t = t_2$ (see fig. 24.1). The contribution to the surface integral will vanish from the part that has a space-like normal. (We must assume that the current density dies off faster than r^{-2} as $r \to \infty$, which is physically reasonable.) The remaining surface integral gives

$$\int_{t=t_2} j^0 d^3 x - \int_{t=t_1} j^0 d^3 x = 0. \tag{159.13}$$

In other words, Q is constant in time.

To show that Q is a Lorentz scalar, we consider two inertial observers K and K' and choose V to be the volume bounded by the surfaces $t = 0$ and $t' = 0$. Once again, the contributions from the space-like surfaces at infinity vanish, and evaluating the integral over $t' = 0$ in x', t' coordinates, we get

$$\int_{t=0} j^0 d^3 x - \int_{t'=0} j'^0 d^3 x' = 0. \tag{159.14}$$

In other words, $Q = Q'$. Note that since we have shown the time independence of Q, we could have used any $t = \text{const}$, $t' = \text{const}$ surfaces to bound V.

This argument is easily generalized to show that any divergenceless tensor density of rank n generates a conserved quantity of rank $n - 1$.

160 The four-potential

We showed earlier that the D'Alembertian $\partial^\mu \partial_\mu = c^{-2}\partial^2/\partial t^2 - \nabla^2$. Hence, the wave equations for the scalar and vector potentials in the Lorenz gauge can be written as

$$\partial^\mu \partial_\mu \phi = 4\pi\rho, \quad \partial^\mu \partial_\mu \mathbf{A} = 4\pi\mathbf{j}. \tag{160.1}$$

We can combine these into the single equation

$$\partial^\mu \partial_\mu (\phi, \mathbf{A}) = 4\pi(\rho, \mathbf{j}). \tag{160.2}$$

Since the right-hand side is a four-vector and since $\partial^\mu \partial_\mu$ is a scalar operator, it follows that the *four-potential*, defined as

$$A^\mu = (\phi, \mathbf{A}), \tag{160.3}$$

must be a four-vector.

This conclusion relies on the specific use of the Lorenz gauge. However, note that under a general gauge transformation,

$$\phi \to \phi - \frac{\partial \chi}{\partial t}, \quad \nabla \mathbf{A} \to \mathbf{A} + \nabla\chi, \tag{160.4}$$

where χ is any scalar field. In other words,

$$A^\mu = (\phi, \mathbf{A}) \to (\phi, \mathbf{A}) - \left(\frac{\partial}{\partial t}, -\nabla\right)\chi = A^\mu - \partial^\mu \chi. \tag{160.5}$$

Since $\partial^\mu \chi$ is a four-vector, it follows that A^μ is a four-vector in *any* gauge.

Finally, we note that the Lorenz gauge condition $\nabla \cdot \mathbf{A} + \partial\phi/\partial t = 0$ can be written as

$$\partial_\mu A^\mu = 0, \tag{160.6}$$

which is itself Lorentz invariant. This proves the assertion made in section 33 that if the potentials are in the Lorenz gauge in one frame, they are in the same gauge in a boosted frame.

Exercise 160.1 Find the $\mathbf{E}$ and $\mathbf{B}$ fields of a charge in uniform motion by Lorentz transforming the four-potential of a charge at rest, and then differentiating these potentials. Compare with the answers in section 55.

Exercise 160.2 Repeat the previous exercise for point electric and magnetic dipoles.

161 The electromagnetic field tensor

Next, let us consider the electromagnetic fields themselves. What kind of four-objects are they? Neither $\mathbf{E}$ nor $\mathbf{B}$ can be the space part of a four-vector, since, as shown in section 38, they mix with one another under a boost. Secondly, they are obtained by taking derivatives of the potentials, which suggests a tensor of rank two. Let us therefore consider the tensor $\partial_\mu A_\nu$. This is not quite correct, as can be seen by looking at its space–space components. These are of the form $\partial_x A_y$, whereas to get the magnetic field we need the combinations $\partial_x A_y - \partial_y A_x$, etc. We are therefore led to examine the antisymmetric tensor

$$F_{\mu\nu} = \partial_\mu A_\nu - \partial_\nu A_\mu, \tag{161.1}$$

and its contra sibling

$$F^{\mu\nu} = g^{\mu\bar{\mu}} g^{\nu\bar{\nu}} F_{\bar{\mu}\bar{\nu}}. \tag{161.2}$$

It is straightforward to evaluate the components of F. We find that

$$F_{\mu\nu} = \begin{pmatrix} 0 & E_x & E_y & E_z \\ -E_x & 0 & -B_z & B_y \\ -E_y & B_z & 0 & -B_x \\ -E_z & -B_y & B_x & 0 \end{pmatrix}, \quad F^{\mu\nu} = \begin{pmatrix} 0 & -E_x & -E_y & -E_z \\ E_x & 0 & -B_z & B_y \\ E_y & B_z & 0 & -B_x \\ E_z & -B_y & B_x & 0 \end{pmatrix}. \tag{161.3}$$

Thus, F, known as the electromagnetic field tensor, is indeed the object we seek.

Exercise 161.1 Show that $^*F_{\mu\nu}$, the dual of the field tensor, is obtained from $F_{\mu\nu}$ by $\mathbf{E} \to \mathbf{B}$, $\mathbf{B} \to -\mathbf{E}$.

Exercise 161.2 From the transformation law for tensors, we can derive the transformation for $\mathbf{E}$ and $\mathbf{B}$. If K' is a frame moving at a velocity $\mathbf{V}$ relative to K, show that

$$\mathbf{E}'_\| = \mathbf{E}_\|, \quad \mathbf{E}'_\perp = \gamma(\mathbf{E}_\perp + \mathbf{V} \times \mathbf{B}_\perp/c), \tag{161.4}$$

$$\mathbf{B}'_\| = \mathbf{B}_\|, \quad \mathbf{B}'_\perp = \gamma(\mathbf{B}_\perp - \mathbf{V} \times \mathbf{E}_\perp/c), \tag{161.5}$$

where the subscripts indicate the components parallel and perpendicular to $\mathbf{V}$.

Exercise 161.3 One can verify the transformation laws for the fields by considering simple setups. To verify that the fields along the boost velocity are unchanged, consider (a) the electric field in an ideal parallel plate capacitor, and boost perpendicular to the plates, and (b) the magnetic field in an ideal solenoid, and boost along the axis of the solenoid. To study the transverse components, consider two infinite sheets of charge parallel to the xy plane, spaced a certain distance apart. In a particular frame K, let the charge densities on the two sheets be $\pm\sigma_0$, and let the charges all move with velocity $v_0\hat{\mathbf{x}}$. Find the sheet currents, and $\mathbf{E}$ and $\mathbf{B}$ as measured in K. Let K' be a frame moving at a velocity $V\hat{\mathbf{x}}$ with respect to K. Find the charge densities and sheet currents in K', and in this way find $\mathbf{E}'$ and $\mathbf{B}'$ and verify the transformations.

From the fact that $F_{\mu\nu}$ is a tensor, we can obtain the Lorentz invariants of the electromagnetic field, i.e., quantities that are the same in all Lorentz frames. Such invariants must be functions of $\mathbf{E}$ and $\mathbf{B}$, i.e., of $F_{\mu\nu}$. The simplest such scalar, $F_{\mu\mu}$, is useless, as it is identically zero. Next, we can try to multiply F with itself and contract all the indices in all possible ways, including the geometrically invariant tensors $g_{\mu\nu}$ and $\epsilon_{\mu\nu\sigma\tau}$. There are two such constructions possible, and it is not difficult to show that they are

$$F_{\mu\nu}F^{\mu\nu} = 2(B^2 - E^2), \tag{161.6}$$

$$ ^*F_{\mu\nu}F^{\mu\nu} = -4\mathbf{E}\cdot\mathbf{B}. \tag{161.7}$$

And this exhausts the list, for all Lorentz invariants must also be rotational invariants, all of which can be built out of three fundamental ones: E^2, B^2, and $\mathbf{E}\cdot\mathbf{B}$. A third Lorentz invariant besides $E^2 - B^2$ and $\mathbf{E}\cdot\mathbf{B}$, if it existed could be taken as $E^2 + B^2$. This quantity, however, is proportional to the energy density and is therefore not a Lorentz invariant—it is multiplied by γ^{-2} under a boost. Finally, note that $\mathbf{E}\cdot\mathbf{B}$ is a pseudoscalar, i.e., it is odd under either parity or time reversal.

Exercise 161.4 Using the transformation laws for the fields, (161.4) and (161.5), verify directly that $E^2 - B^2$ and $\mathbf{E}\cdot\mathbf{B}$ are Lorentz invariants.

Exercise 161.5 Suppose the field is purely electric (i.e., $\mathbf{B} = 0$) in some frame K. Show that boosts along $\mathbf{E}$ leave the field completely unchanged, i.e., $\mathbf{B}' = 0$, $\mathbf{E}' = \mathbf{E}$. Discuss the case of a purely magnetic field similarly.

Exercise 161.6 Suppose the fields are orthogonal ($\mathbf{E}\cdot\mathbf{B} = 0$) in some frame K. Find a frame (there are many) in which the field is purely electric or purely magnetic.

Solution: If $E = B \neq 0$, the invariant $E^2 - B^2$ is zero, so there can be no frame in which either E vanishes or B vanishes. If $E > B$, $E^2 - B^2 > 0$, so the field cannot be made purely magnetic. To go to a frame where it is purely electric, the required boost velocity is $\mathbf{V} = c(\mathbf{E}\times\mathbf{B})/E^2 + V'\mathbf{E}/E$, where V' is arbitrary, as long as $V < c$. Likewise, if $B > E$, a boost by $\mathbf{V} = c(\mathbf{E}\times\mathbf{B})/B^2 + V'\mathbf{B}/B$ renders the field purely magnetic.

Exercise 161.7 Suppose $\mathbf{E}$ and $\mathbf{B}$ are not orthogonal in a frame K. Show that one can always find a frame in which they are parallel.

Solution: Suppose we can boost to such a frame. A further boost along the direction of the fields leaves them unchanged, so the initial boost can be sought perpendicular to both $\mathbf{E}$ and $\mathbf{B}$. Equations (161.4) and (161.5) then simplify to

$$\mathbf{E}' = \gamma(\mathbf{E} + \mathbf{V}\times\mathbf{B}/c), \quad \mathbf{B}' = \gamma(\mathbf{B} - \mathbf{V}\times\mathbf{E}/c). \tag{161.8}$$

To find $\mathbf{V}$, we set $\mathbf{E}'\times\mathbf{B}'$ to 0, remembering that $\mathbf{V}\cdot\mathbf{E} = \mathbf{V}\cdot\mathbf{B} = 0$. This yields an easy quadratic equation for V. The solution with $V < c$ is

$$\frac{V}{c} = \frac{(E^2 + B^2) - \left[(E^2 - B^2)^2 - 4(\mathbf{E}\cdot\mathbf{B})^2\right]^{1/2}}{2|\mathbf{E}\times\mathbf{B}|}, \quad \mathbf{V} \parallel (\mathbf{E}\times\mathbf{B}). \tag{161.9}$$

The above exercises confirm the statements made in section 71 when we discussed the motion of charged particles in static **E** and **B** fields.

Exercise 161.8 Solve the equations of motion for a charge in a uniform **E** field. Compare with hyperbolic motion.

162 Covariant form of the laws of electromagnetism

With the field tensor in hand, we can write Maxwell's equations in covariant form. Consider first the equations with source terms:

$$\nabla \cdot \mathbf{E} = 4\pi\rho, \quad \nabla \times \mathbf{B} - \frac{\partial \mathbf{E}}{\partial t} = 4\pi\mathbf{j}. \tag{162.1}$$

The right-hand sides are the pieces of the four-current, so it is to be expected that the left-hand sides are the pieces of a four-vector obtained by taking the divergence of the field tensor. This obvious guess turns out to be correct, for

$$\partial_\nu F^{\mu\nu} = \partial_\nu(\partial^\mu A^\nu - \partial^\nu A^\mu)$$
$$= \partial^\mu(\partial_\nu A^\nu) - (\partial_\nu \partial^\nu) A^\mu. \tag{162.2}$$

In the Lorenz gauge $\partial_\nu A^\nu = 0$ and $\partial_\nu \partial^\nu A^\mu = 4\pi j^\mu$. Hence,

$$\partial_\nu F^{\mu\nu} = -\frac{4\pi}{c} j^\mu. \tag{162.3}$$

Since F and j are gauge invariant, this result is correct in any gauge. This equation is equivalent to eq. (162.1).

Now consider the equations without source terms:

$$\nabla \cdot \mathbf{B} = 0, \quad \nabla \times \mathbf{E} + \frac{\partial \mathbf{B}}{\partial t} = 0. \tag{162.4}$$

The left-hand sides of these equations are obtained from those of eq. (162.1) by replacing **E** by **B**, and **B** by −**E**. Since this is the operation that takes F to $^*F$ (see exercise 161.1), the desired four-vector form must be

$$\partial_\mu {}^*F^{\mu\nu} = 0. \tag{162.5}$$

To see this directly, we note that

$$\partial_\mu {}^*F^{\mu\nu} = \epsilon^{\mu\nu\sigma\tau}(\partial_\mu \partial_\sigma A_\tau - \partial_\mu \partial_\tau A_\sigma). \tag{162.6}$$

Consider the first term on the right. It vanishes because it is the contraction of a tensor (ϵ) antisymmetric in the indices μ and σ with $\partial_\mu \partial_\sigma$, which is symmetric. The second term vanishes for the same reason.

Exercise 162.1 Show that eq. (162.4) can also be written as

$$F_{\alpha\beta,\gamma} + F_{\gamma\alpha,\beta} + F_{\beta\gamma,\alpha} = 0. \tag{162.7}$$

Lastly, let us rewrite in covariant form the Lorentz force law,

$$\frac{d\mathbf{p}}{dt} = q(\mathbf{E} + \mathbf{v} \times \mathbf{B}). \tag{162.8}$$

It is clear that the generalization must involve the momentum and velocity four-vectors

$$p^{\mu} = (\mathcal{E}, \mathbf{p}), \quad u^{\mu} = (\gamma, \gamma\mathbf{v}), \tag{162.9}$$

and must be an equation for the four-vector $dp^{\mu}/d\tau$. The space part of this vector is

$$\frac{d\mathbf{p}}{d\tau} = \gamma\frac{d\mathbf{p}}{dt} = q(u^0\mathbf{E} + \mathbf{u} \times \mathbf{B}), \tag{162.10}$$

using the facts $u^0 = \gamma$, $\mathbf{u} = \gamma\mathbf{v}$. The time part is

$$\frac{d\mathcal{E}}{d\tau} = \gamma\frac{d\mathcal{E}}{dt} = \gamma\mathbf{v} \cdot \frac{d\mathbf{p}}{dt} = q\mathbf{u} \cdot \mathbf{E}. \tag{162.11}$$

It is not difficult to verify that eqs. (162.10) and (162.11) are together equivalent to

$$\frac{dp^{\mu}}{d\tau} = q\,F^{\mu\nu}u_{\nu}. \tag{162.12}$$

This is the form that we seek. It can also be written as

$$mc^2\frac{d^2x^{\mu}}{d\tau^2} = q\,F^{\mu\nu}\frac{dx_{\nu}}{d\tau}. \tag{162.13}$$

It is also instructive to reconsider the action formulation of the laws of EM in covariant terms. We derived the action in section 79. Let us consider the free-field action first. This is given by the space and time integral of the Lagrangian density $(E^2 - B^2)/8\pi$. The latter, however, is proportional to the invariant $F_{\mu\nu}F^{\mu\nu}$, so

$$S_{\text{field}} = -\frac{1}{16\pi}\int F_{\mu\nu}F^{\mu\nu}d^4\Omega. \tag{162.14}$$

This form makes it manifest that the action is a Lorentz invariant, as it must be.

Next, let us consider the action for a free charge. This is given by the time integral of the Lagrangian, $-mc^2(1 - v^2/c^2)^{1/2}$. However, $(1 - v^2/c^2)^{1/2}dt$ is the proper time interval $d\tau$, so

$$S_{\text{charge}} = -\int mc^2 d\tau. \tag{162.15}$$

Once again, the Lorentz invariance of this action is manifest.

Finally, the interaction term (for a single charge) is given by

$$S_{\text{int}} = \int \left(-q\phi + \frac{q}{c}\mathbf{v} \cdot \mathbf{A}\right)dt. \tag{162.16}$$

Since $\mathbf{v}dt = d\mathbf{r}$, this can be written in terms of the four-potential (160.3) as

$$S_{\text{int}} = -\frac{q}{c}\int A_{\mu}dx^{\mu}. \tag{162.17}$$

Alternatively, we can write this in terms of the four-current as

$$S_{\text{int}} = -\frac{1}{c}\int j^{\mu}A_{\mu}d^4\Omega. \tag{162.18}$$

Adding together the three parts, we get the complete action for a collection of charges

$$S = -\sum_a \int m_a c^2 d\tau_a - \frac{1}{c} \int j^\mu A_\mu d^4\Omega - \frac{1}{16\pi} \int F_{\mu\nu} F^{\mu\nu} d^4\Omega. \tag{162.19}$$

It is also useful to keep in mind the alternative form of the interaction term.

Exercise 162.2 Use the principle of least action to derive the covariant form of the Maxwell equations with source terms directly from eq. (162.19).

Exercise 162.3 (photon mass) Add to the Lagrangian a "photon mass term" $\mu^2 A^\mu A_\mu/8\pi$. (a) How is the wave equation, in particular, the dispersion relation for EM waves, modified? (b) Show that the field of a static point charge takes the form $e^{-\mu r}/r$. (c) Is gauge invariance still true?

163 The stress–energy tensor

Just as the local conservation of charge is expressed in the continuity equation, $\partial_\mu j^\mu = 0$, we expect that the conservation of four-momentum will have a local form in terms of a divergenceless four-tensor. This expectation is indeed correct, and the tensor in question is known as the stress–energy or energy–momentum tensor, denoted $T^{\mu\nu}$. This generalizes the stress tensor defined in section 36 for the electromagnetic field, and definable for other systems, such as fluids and elastic bodies in mechanics.

It turns out that the components of $T^{\mu\nu}$ have the following physical meaning. For any observer, T^{00} is the energy density measured by her, T^{i0} the momentum density, T^{0i} the energy flux, and T^{ij} the three-tensor of stress. If we start with these four physical quantities, however, we have no a priori guarantee that they make up a four-tensor. To show this is exceedingly cumbersome. We will, therefore, take an indirect route. We define $T^{\mu\nu}$ so that for any hypersurface element $d^3 S^\alpha$,

$$T^{\mu\nu} d^3 S_\nu = \text{four-momentum } P^\mu \text{ in } d^3 S. \tag{163.1}$$

The right-hand side of this equation is a four-vector, since it is just the sum of four-momenta. The element $d^3 S_\mu$ is also a four-vector. Hence, the left-hand side of the equation can be a four-vector only if $T^{\mu\nu}$ is a four-tensor.

The meanings of the elements of $T^{\mu\nu}$ now follow by considering particular choices for $d^3 S$. Let us first take the volume on a $t = \text{const}$ hypersurface. In this case, $d^3 S_\alpha = (dV, 0)$, so

$$T^{\mu\nu} d^3 S_\nu = T^{\mu 0} dV = P^\mu \text{ in } dV. \tag{163.2}$$

In other words,

$$T^{00} = \text{energy density}, \tag{163.3}$$

$$T^{i0} = \text{momentum density}. \tag{163.4}$$

Next, let us take an element of the hypersurface $x = $ const. Then,

$$d^3 S_\alpha = (0, dt\, dy\, dz, 0, 0),$$
(163.5)

and

$$T^{\mu\nu} d^3 S_\nu = T^{\mu x} dt\, dy\, dz = P^\mu \text{ in } dt\, dy\, dz.$$
(163.6)

For the $\mu = 0$ component of this equation, the right-hand side is an energy. So, the middle expression must also be an energy, but this can be so only if T^{0x} is the energy flux in the x direction, i.e., the energy flowing per unit time through a unit area normal to the x direction. Likewise, T^{jx} is the flux in the x direction of the j component of momentum. Since the rate at which momentum is transferred is just the force, the flux of momentum is just the force per unit area, i.e., the stress. Generalizing to the other spatial components, we get

$$T^{0i} = \text{energy flux in } i\text{th direction},$$
(163.7)

$$T^{ji} = j\text{-stress in } i\text{th direction}.$$
(163.8)

This confirms the meanings of the $T^{\mu\nu}$.

The stress–energy tensor possesses the important property that it is symmetric. Let us first see this for the space–space subtensor T^{ij}. To this end, let us consider the torque on all objects inside a cubical volume element of side L, and let us evaluate this torque about the center of the cube as the negative of the torque on the surroundings. For the z components, we have

$$N_z = - \left[\tfrac{1}{2} L L^2 T^{yx} + \left(-\tfrac{1}{2} L \right) L^2 (-T^{yx}) + 2 \left(-\tfrac{1}{2} L \right) L^2 T^{xy} \right].$$
(163.9)

The first term arises from the front face, i.e., that with normal $\hat{x}$, the second from the rear face, and the last term from the side faces (with normals $\pm\hat{y}$). The factor of $L/2$ is the lever arm, and that of L^2 is the area of a face. Thus,

$$N_z = (T^{yx} - T^{xy}) L^3.$$
(163.10)

As $L \to 0$, the moment of inertia of the objects inside the cube is of order $T^{00} L^3 L^2$, so the angular acceleration is of order $(T^{yx} - T^{xy})/L^2$. This will diverge unless $T^{xy} = T^{yx}$. The same symmetry requirement applies to all the space–space components.

In the same way, asymmetry of the time–space components such as T^{0x} would lead to infinite accelerations. It follows that the stress–energy tensor must be symmetric in general,[1]

$$T^{\mu\nu} = T^{\nu\mu}.$$
(163.11)

---

[1] We note here that the stress–energy tensor can also be obtained from the action by employing Noether's theorem. The tensor so obtained is known as the *canonical stress–energy tensor*, and it may not be symmetric. However, it can be always be symmetrized by adding the total divergence of an auxiliary field. This added term does not alter the total momentum of the system, since the integral of a divergence can be converted into a surface integral at infinity, which vanishes if the auxiliary field is sensibly chosen. The symmetrized stress–energy tensor is unique.

In particular, restoring the dimensional factors of c, we get

$$T^{i0} = \frac{1}{c^2} T^{0i}. \tag{163.12}$$

This is the argument for the general relationship between momentum density and energy flux asserted without proof in section 36 (there denoted $\mathbf{g}$ and $\mathbf{S}$).

Let us now consider a few examples of the stress–energy tensor, beginning with the EM field. We already have expressions for the various components of this tensor: ($E^2 + B^2$)/8π for the energy density, $\mathbf{S}$ (the Poynting vector) and $\mathbf{g} = \mathbf{S}/c^2$ for the energy flux and momentum density, and the Maxwell stress tensor for the space–space components. To write this tensor in four-dimensional form, we need only construct the small number of tensors of rank two that are quadratic in the fields. We leave it to the reader to show that

$$T^{\mu\nu} = \frac{1}{4\pi} \left(\frac{1}{4} g^{\mu\nu} F^{\alpha\beta} F_{\alpha\beta} - F^{\mu\alpha} F^{\nu}_{\alpha} \right). \tag{163.13}$$

This contribution to the stress–energy tensor has the special property of being traceless:

$$T^{\mu}_{\mu} = \frac{1}{4\pi} \left(\frac{1}{4} \delta^{\mu}_{\mu} F^{\alpha\beta} F_{\alpha\beta} - F^{\mu\alpha} F_{\mu\alpha} \right) = 0. \tag{163.14}$$

This property is related to the masslessness of the photon.

Next, let us consider the beam of particles of mass m discussed in section 36. Let n be the density of particles as *measured* in the comoving frame. (Note that this is a scalar.) In this frame, there is no transport of energy or momentum. Hence, the only nonzero component of $T^{\mu\nu}$ is T^{00}, and this is given by mn. We can now write the result in a general frame. We continue to use n, the proper density. The only vector from which a tensor could be constructed is u^{α}, the four-velocity of the particles. Hence,

$$T^{\mu\nu} = mn u^{\mu} u^{\nu}. \tag{163.15}$$

Another way to see that this is correct is to note that both sides are tensors and that the equation is true is in the rest frame. Hence, it must be true in all frames. We can rewrite this equation by noting that $m u^{\mu}$ is the momentum and that $n u^{\mu}$ is the number current j_n. Hence,

$$T^{\mu\nu} = p^{\mu} j^{\nu}_n. \tag{163.16}$$

Lastly, for a single particle, the stress–energy tensor can be obtained from eq. (163.15). If the trajectory of the particle is written as $\mathbf{s}(t)$, the density at a point $\mathbf{r}$ is given by $\gamma^{-1}\delta(\mathbf{r} - \mathbf{s}(t))$. Hence,

$$\begin{aligned}
T^{00} &= m\gamma \delta^{(3)}\big[\mathbf{r} - \mathbf{s}(t)\big], \\
T^{0i} &= m\gamma v^i \delta^{(3)}\big[\mathbf{r} - \mathbf{s}(t)\big], \\
T^{ij} &= m\gamma v^i v^j \delta^{(3)}\big[\mathbf{r} - \mathbf{s}(t)\big].
\end{aligned} \tag{163.17}$$

We can write this in four-dimensional form using the same method that we employed for the charge current. Parametrizing the world line as $s^\alpha(\tau)$, we have

$$T^{\mu\nu}(x^\alpha) = m \int u^\mu u^\nu \delta^{(4)}\big[x^\alpha - s^\alpha(\tau)\big]d\tau, \tag{163.18}$$

This is a four-tensor for reasons now familiar: $u^\mu u^\nu$ is a four-tensor, and $\delta^{(4)}$ and $d\tau$ are scalars. The integration over τ turns the four-dimensional delta function into $\delta^{(3)}(\mathbf{r} - \mathbf{s})$ and produces an additional factor of γ^{-1}. The space-plus-time forms (163.17) then follow. The tensor for a collection of particles follows by adding the contributions of the individual particles.

Exercise 163.1 Show that for a plane EM wave traveling in the direction $\hat{\mathbf{n}}$,

$$T^{\mu\nu} = \mathcal{E} n^\mu n^\nu, \tag{163.19}$$

where $\mathcal{E}$ is the energy density of the radiation, and $n^\mu = (1, \hat{\mathbf{n}})$.

164 Energy–momentum conservation in special relativity

In Newtonian mechanics, the notions of a system of particles interacting via action at a distance, or of rigid bodies, cause no difficulty. Neither is meaningful in relativity. Consider a system of particles. The interactions between them must be via fields (such as the EM field), which also carry energy, momentum, etc., which must then be included in any conservation law. The stress–energy tensor is the way to do this.

The law of energy–momentum conservation takes the form

$$T^{\mu\nu}_{\ ,\nu} = 0. \tag{164.1}$$

The formal proof is identical to that for the conservation of charge. The total four-momentum is the integral of $T^{\mu\nu}d^3 S_\nu$, which is conserved by Gauss's law. It is more interesting to verify it explicitly for a system of charges interacting electromagnetically. (We already know this result physically. The goal here is to see how it comes out of the stress–energy tensor.) We denote the electromagnetic and particle contributions to $T^{\mu\nu}_{\ ,\nu}$ by the superscripts EM and P preceding the letter T. Dealing with the former first, we have

$$4\pi\big({}^{EM}T_\mu^{\ \nu}{}_{,\nu}\big) = -F_{\mu\alpha}F^{\nu\alpha}{}_{,\nu} - F_{\mu\alpha,\nu}F^{\nu\alpha} + \tfrac{1}{2}\delta^\nu_\mu F^{\alpha\beta}F_{\alpha\beta,\nu}$$

$$= -4\pi F_{\mu\alpha}j^\alpha + \frac{1}{2}F^{\alpha\nu}(F_{\mu\alpha,\nu} - F_{\mu\nu,\alpha}) + \frac{1}{2}F^{\alpha\nu}F_{\alpha\nu,\mu} \tag{164.2}$$

$$= -4\pi F_{\mu\alpha}j^\alpha + \frac{1}{2}F^{\alpha\nu}(F_{\mu\alpha,\nu} + F_{\nu\mu,\alpha} + F_{\alpha\nu,\mu}). \tag{164.3}$$

In eq. (164.2), the first term is obtained by the Maxwell equation (162.1), the second term arises by antisymmetrizing the tensor $F_{\mu\alpha,\nu}$ in the indices ν and α (since it is multiplied by the antisymmetric tensor $F^{\nu\alpha}$), and the last term is obtained by first trading in the index ν for μ using the Kronecker delta and then writing ν as the dummy index instead of β. The last three terms are combined together in eq. (164.3), and since this combination vanishes by virtue of the Maxwell equation (162.7), we have

$$^{EM}T^{\mu\nu}{}_{,\nu} = -F^{\mu\alpha}j_\alpha. \tag{164.4}$$

We now consider the particle part. To avoid index clutter, we consider just one particle first:

$$\begin{aligned}
^{P}T^{\mu\nu}{}_{,\nu} &= m\int u^\mu u^\nu \frac{\partial}{\partial x^\nu}\delta^{(4)}\big[x^\alpha - s^\alpha(\tau)\big]d\tau, \\
&= -m\int u^\mu \frac{ds^\nu}{d\tau}\frac{\partial}{\partial s^\nu}\delta^{(4)}\big[x^\alpha - s^\alpha(\tau)\big]d\tau, \\
&= -m\int u^\mu \frac{d}{d\tau}\delta^{(4)}\big[x^\alpha - s^\alpha(\tau)\big]d\tau, \\
&= m\int \frac{du^\mu}{d\tau}\delta^{(4)}\big[x^\alpha - s^\alpha(\tau)\big]d\tau.
\end{aligned} \tag{164.5}$$

In the third line, we used the chain rule, and in the last line, we integrated by parts. We now note that $mdu^\mu/d\tau = dp^\mu/d\tau$ and make use of the four-vector form of the Lorentz force law (162.12). This yields

$$\begin{aligned}
^{P}T^{\mu\nu}{}_{,\nu} &= \int q\,F^{\mu\nu}u_\nu\delta^{(4)}\big[x^\alpha - s^\alpha(\tau)\big]d\tau \\
&= F^{\mu\nu}j_\nu,
\end{aligned} \tag{164.6}$$

where the last line follows from eq. (159.2), the four-vector expression for the current. Although derived for only one particle, the final result holds for an arbitrary collection, since the individual currents add to give the total current j_ν. Adding this to eq. (164.4), we see that the divergence of the total stress energy does vanish:

$$(^{EM}T^{\mu\nu} + {}^{P}T^{\mu\nu})_{,\nu} = 0. \tag{164.7}$$

165 Angular momentum and spin*

We now consider the conservation of angular momentum for a closed system of particles with interactions invariant under rotations and boosts. Since angular momentum changes under both boosts and a shift of the origin, we first define a special frame that generalizes the concept of center of mass.

Let the total momentum of the system be denoted P^μ. Since this is a time-like vector, and it is conserved, the quantity

$$m_{\text{tot}} = (P_\mu P^\mu)^{1/2} \tag{165.1}$$

is a conserved scalar, which we call the total "mass" of the system. It follows that

$$U^\mu \equiv P^\mu/m_{\text{tot}} \tag{165.2}$$

behaves like a four-velocity. In particular, since $U^\mu U_\mu = 1$, we can write it as

$$U^\mu = \gamma_{ci}(1, \mathbf{V}_{ci}), \tag{165.3}$$

where $\mathbf{V}_{ci}$ is known as the (three-)velocity of the *center of inertia* (COI), and $\gamma_{ci} = (1 - V_{ci}^2)^{-1/2}$. In the COI frame, i.e., the frame in which the center of inertia is stationary, we have

$$P^\mu = (m_{\text{tot}}, \mathbf{0}). \tag{165.4}$$

This is also known as the *center-of-momentum* frame, and it is the sought for generalization of the center-of-mass frame or the rest frame of a composite particle.

We now turn to the angular momentum. For noninteracting particles this is

$$J^{\mu\nu}(t) = \sum_a (x_a^\mu(t)\, p_a^\nu(t) - x_a^\nu(t)\, p_a^\mu(t)), \tag{165.5}$$

where t is the lab time. Since $dp_a^\nu/dt = 0$ (recall, no interactions), and $dx_a^\nu/dt = p_a^\nu/m_a\gamma_a$, $dJ^{\mu\nu}/dt = 0$, and conservation is proved. More generally, we have

$$J^{\mu\nu} = \int_\Sigma (x^\mu T^{\nu\alpha}(x^\beta) - x^\nu T^{\mu\alpha}(x^\beta)) d^3 S_\alpha, \tag{165.6}$$

where Σ is any time-like hypersurface. This is conserved, since the integrand is divergenceless.

Next, let us see how the angular momentum changes with change of origin. Consider another lab-frame observer $\overline{K}$ with origin at X^α, so that $\bar{x}^\alpha = x^\alpha - X^\alpha$, and $\overline{T}^{\mu\nu}(\bar{x}^\alpha) = T^{\mu\nu}(x^\alpha - X^\alpha)$. Adding bars to the integrand in eq. (165.6), and shifting the variable of integration, we see that

$$\overline{J}^{\mu\nu} = J^{\mu\nu} + \int (X^\mu T^{\nu\alpha}(x^\beta) - X^\nu T^{\mu\alpha}(x^\beta)) d^3 S_\alpha$$

$$= J^{\mu\nu} + (X^\mu P^\nu - X^\nu P^\mu). \tag{165.7}$$

However, in the COI frame, $P^i = 0$, and the space–space parts do not change, for the extra terms vanish. This leaves the time–space parts, for which we have

$$\overline{J}^{0i} = J^{0i} - X^i m_{\text{tot}}. \tag{165.8}$$

By a suitable choice of X^i, $\overline{J}^{0i}$ can be made to vanish. Thus, with this origin in the COI frame, the angular momentum has only space–space parts J^{ij}, and we may characterize it completely by the three-vector

$$s_i = \tfrac{1}{2}\epsilon_{ijk} J_{jk}, \tag{165.9}$$

which is known as the *spin* of the system or the body. This concept is especially useful when applied to a particle such as the electron and will hold even if the electron turns out to be a composite particle.

Instead of $\mathbf{s}$, it is often easier to work with the four-vector,

$$\mathcal{S}^\alpha = -\tfrac{1}{2}\epsilon^{\alpha\beta\gamma\delta} U_\beta J_{\gamma\delta}. \tag{165.10}$$

In the rest frame, $U_\beta = (1, \mathbf{0})$, so

$$\mathcal{S}^\alpha = \tfrac{1}{2}\epsilon^{0\alpha\gamma\delta} J_{\gamma\delta}. \tag{165.11}$$

The indices α, γ, and δ must all be spatial for a nonvanishing result, so we get

$$\mathcal{S}^\alpha = (0, \mathbf{s}). \tag{165.12}$$

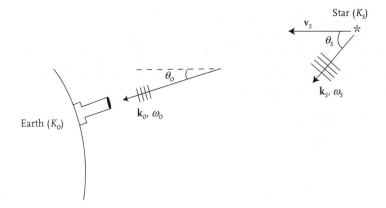

FIGURE 24.2. Light emitted by a moving source appears to a stationary observer to have different frequency (Doppler effect) and direction of travel (aberration/headlight effect).

Note that $\mathcal{S}^\alpha U_\alpha = 0$ by construction, and

$$\mathcal{S}^\alpha \mathcal{S}_\alpha = -\mathbf{s} \cdot \mathbf{s} = \text{invariant.} \tag{165.13}$$

Exercise 165.1 Consider a closed system of noninteracting particles and show that the vector $\mathbf{R}_{ci} = \sum \mathcal{E}_a \mathbf{r}_a / \sum \mathcal{E}_a$ moves at a uniform velocity $\mathbf{V}_{ci}$. Show further that $\mathbf{R}_{ci}$ is not the space part of any four-vector, in that although different observers will agree that it moves uniformly, they will not agree on its position within the body or system of particles. Generalize these results to a system of interacting particles.

166 Observer-dependent properties of light

Although light has the same speed in all inertial frames, its other characteristics do vary from frame to frame. We begin by discussing the changes in the frequency (Doppler shift) and the apparent direction of propagation (aberration). Let K_o be the earth, let K_s be one of the "fixed" stars, and let the latter be moving at a velocity $\mathbf{v}_s$ relative to K_o (fig. 24.2). (The suffixes s and o denote quantities measured in inertial frames comoving with the source and observer, but the specific example of a star and an earth-bound observer are useful in fixing one's ideas. Note, however, that $\mathbf{v}_s$ is the velocity of the source as measured by the observer. We also write $\gamma_s = (1 - v_s^2)^{-1/2}$.) If the frequency of the starlight is ω_s in the star frame, then the frequency ω_o measured in K_o, will be different. Further, if the star appears at an orientation $\hat{\mathbf{n}}_s$ to an observer comoving with the star, then to an earthbound observer, the star appears in the sky at a different orientation $\hat{\mathbf{n}}_o$. To find ω_o and $\hat{\mathbf{n}}_o$, we consider a photon emitted by the star and absorbed at the earth. The four-momentum of the photon in the star and the earth frames are given by

$$k_s^\alpha = \omega_s(1, -\hat{\mathbf{n}}_s), \quad k_o^\alpha = \omega_o(1, -\hat{\mathbf{n}}_o). \tag{166.1}$$

We have used the facts that for light, $\omega = |\mathbf{k}|c$ in any frame and that the space part of the wave vector, $\mathbf{k}$, is directed opposite to the direction of observation, $\hat{\mathbf{n}}$. We now relate k_o^α to k_s^α using the basic LT. It turns out to be convenient to take $\mathbf{v}_s$ to be along the $-\hat{\mathbf{x}}$ direction. Writing $\mathbf{v}_s = -v_s\hat{\mathbf{x}}$, we get $t_o = \gamma_s(t_s - v_s x_s)$, $x_o = \gamma_s(x_s - v_s t_s)$, $y_o = y_s$. Since the four-momentum transforms similarly, we have

$$\omega_o = \gamma_s(\omega_s + \mathbf{v}_s \cdot \mathbf{k}_s) = \gamma_s\omega_s(1 - \mathbf{v}_s \cdot \hat{\mathbf{n}}_s),$$

$$k_{o,\parallel} = \gamma_s(k_{s,\parallel} - v_s\omega_s), \tag{166.2}$$

$$\mathbf{k}_{o,\perp} = \mathbf{k}_{s,\perp}.$$

Let us consider the frequency first. The formula above entails $\hat{\mathbf{n}}_s$, which is not the measured direction on earth. This problem is overcome by using the inverse LT, which gives

$$\omega_s = \gamma_s\omega_o(1 + \mathbf{v}_s \cdot \hat{\mathbf{n}}_o), \tag{166.3}$$

or

$$\omega_o = \frac{\omega_s}{\gamma_s(1 + \mathbf{v}_s \cdot \hat{\mathbf{n}}_o)}. \tag{166.4}$$

This is the formula for the Doppler shift of light. If the source is moving closer to the observer ($\mathbf{v}_s \cdot \hat{\mathbf{n}}_o < 0$), the observed frequency is higher than emitted, while if it is moving away ($\mathbf{v}_s \cdot \hat{\mathbf{n}}_o > 0$), the observed frequency is lower than emitted. We say that the light is "blueshifted" or "redshifted" in the two cases, respectively. When expanded in powers of v_s/c, the terms in the shift of order $1/c$ and $1/c^2$ are called the first-order and second-order Doppler effects. Note that even for a source moving at right angles to the direction of observation, there is a second-order shift in the frequency, toward the red.[2]

We can now solve for the apparent direction $\hat{\mathbf{n}}_o$. From eq. (166.2), we have

$$\mathbf{k}_o = \gamma_s\left(\frac{(\mathbf{k}_s \cdot \mathbf{v}_s)\mathbf{v}_s}{v_s^2} + \mathbf{v}_s\omega_s\right) + \frac{\mathbf{v}_s \times (\mathbf{k}_s \times \mathbf{v}_s)}{v_s^2}, \tag{166.5}$$

where we have combined the components parallel and perpendicular to $\mathbf{v}_s$ to give the full vector. Writing $\mathbf{k}_s = -\omega_s\hat{\mathbf{n}}_s$, $\mathbf{k}_o = -\omega_o\hat{\mathbf{n}}_o$, and substituting for ω_s from eq. (166.2), we get

$$\hat{\mathbf{n}}_o = \frac{\gamma_s\left((\hat{\mathbf{n}}_s \cdot \mathbf{v}_s)\mathbf{v}_s - v_s^2\mathbf{v}_s\right) + \mathbf{v}_s \times (\hat{\mathbf{n}}_s \times \mathbf{v}_s)}{\gamma_s v_s^2(1 - \mathbf{v}_s \cdot \hat{\mathbf{n}}_s)}. \tag{166.6}$$

It also pays to write this result in component form. Taking $\hat{\mathbf{n}}_o$ and $\hat{\mathbf{n}}_s$ to make an angle θ_o and θ_s with $\hat{\mathbf{x}}$, and recalling that $\mathbf{v}_s = -v_s\hat{\mathbf{x}}$, we get

$$\cos\theta_o = \frac{\cos\theta_s + v_s}{1 + v_s\cos\theta_s}, \quad \sin\theta_o = \frac{\sin\theta_s}{\gamma_s(1 + v_s\cos\theta_s)}, \quad \tan\theta_o = \frac{\tan\theta_s}{\gamma_s(1 + v_s\sec\theta_s)}. \tag{166.7}$$

This is identical to eq. (152.28) with $V' = 1$. We note once again the headlight effect; since $\theta_o < \theta_s$, the observer sees the light as emitted in a direction closer to that in which the source is moving.

---

[2] The factor $\gamma_s(1 + \mathbf{v}_s \cdot \hat{\mathbf{n}}_o)$ is sometimes written as $1 + z$, where z is known as the redshift. We shall not use this notation.

Let us apply our discussion to the phenomenon of aberration. The result is now more usefully written in terms of $\mathbf{v}_e = -\mathbf{v}_s$, the velocity of the earth relative to the stars. Expanding in powers of $\mathbf{v}_e/c$, we get

$$\hat{\mathbf{n}}_o = \hat{\mathbf{n}}_s + \frac{1}{c}\hat{\mathbf{n}}_s \times (\mathbf{v}_e \times \hat{\mathbf{n}}_s) + \frac{1}{2c^2}\left(2(\mathbf{v}_e \cdot \hat{\mathbf{n}}_s)^2\hat{\mathbf{n}}_s - (\mathbf{v}_e \cdot \hat{\mathbf{n}}_s)\mathbf{v}_e - v_e^2\hat{\mathbf{n}}_s\right) + \cdots . \tag{166.8}$$

Neglecting the terms of $O(c^{-2})$, we find that as the earth orbits the sun and $\mathbf{v}_e$ changes, the apparent orientation of the star changes according to

$$\delta\hat{\mathbf{n}}_o = \frac{1}{c}\hat{\mathbf{n}}_o \times (\delta\mathbf{v}_e \times \hat{\mathbf{n}}_o). \tag{166.9}$$

The maximum angular deviation is $\delta v_e/c \simeq 10^{-4}$, or 20.6″, taking $\delta v_e \simeq 3 \times 10^6$ cm/s, the speed of the earth around the sun. This is just what is seen. Over one year, all the stars describe ellipses with major axes of 41″ of arc; those near the celestial poles move in near circles, while those in the ecliptic move along near-straight lines.

Exercise 166.1 Derive the formulas for aberration using the result for addition of velocities (exercise 152.7).

It is instructive to derive the formula for Doppler shift in a second way that sidesteps the use of explicit LTs. We first derive a Lorentz-invariant expression for the energy of any moving particle as measured by an observer. Let us write the four-momentum of the particle as $p = (\mathcal{E}_{obs}, \mathbf{p}_{obs})$ in the observer's frame. Since his four-velocity is given in his own frame by $u_o = (1, \mathbf{0})$, it follows that (abbreviating "obs" further to just "o")

$$\mathcal{E}_o = u_o^\alpha p_\alpha. \tag{166.10}$$

For the right-hand side is a dot product and hence a Lorentz invariant, and it has the correct value in the observer's frame.

We now return to the Doppler shift. Let the four-velocities of the source and observer be written as u_s and u_o, and let the four-momentum of a photon emitted by the source be k. Since the frequency of light is the photon energy divided by $\hbar$, we have

$$\frac{\omega_o}{\omega_s} = \frac{u_o^\alpha k_\alpha}{u_s^\alpha k_\alpha}. \tag{166.11}$$

It remains to evaluate the dot products in some convenient frame. We use that of the observer. In this frame, $u_s = (\gamma_s, \gamma_s\mathbf{v}_s)$, $u_o = (1, \mathbf{0})$, and $k = \hbar\omega_o(1, -\hat{\mathbf{n}}_o)$. (Recall that $\hat{\mathbf{n}}_o$ is the direction in which the observer sees the source.) Then,

$$u_o^\alpha k_\alpha = \hbar\omega_o, \quad u_s^\alpha k_\alpha = \hbar\omega_o\gamma_s(1 + \mathbf{v}_s \cdot \hat{\mathbf{n}}_o). \tag{166.12}$$

Taking the ratio, we recover eq. (166.4)

The invariant expression (166.10) permits easy analysis of a high-precision test of special relativity using the Mossbauer effect.[3] An ^{57}Fe source and a similar absorber are

---

[3] D. C. Champeney, G. R. Isaak, and A. M. Khan, *Phys. Lett.* **7**, 241 (1963).

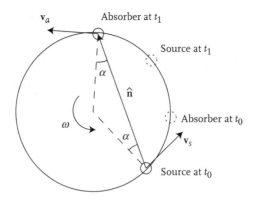

FIGURE 24.3. Test of the relativistic Doppler effect.

mounted on the edge of a disk rotating at a high angular velocity ω. A γ-ray is emitted by the source at some time t_0 in the lab frame and absorbed at a later time t_1 (see fig. 24.3). Because of the extreme narrowness of the Mossbauer line, the absorption would not occur at all if the Doppler shifts did not cancel exactly. We will show that this is indeed so, i.e., the energies of the γ-ray in the respective rest frames of both the absorber and emitter are the same. Let us denote the four-velocity of an inertial observer who is momentarily comoving with the source at t_0 by u_s, and of an observer comoving with the absorber at t_1 by u_a. In the lab frame, we have

$$u_s = \gamma(1, \mathbf{v}_s), \quad u_a = \gamma(1, \mathbf{v}_a). \tag{166.13}$$

Here $\gamma = (1 - \omega^2 r^2)^{-1/2}$ is the same for both source and absorber, since they are traveling at the same speed. Let the photon four-momentum (also in the lab frame) be written as $p_{\mathrm{ph}} = \mathcal{E}_{\mathrm{ph}}(1, \hat{\mathbf{n}})$. Then,

$$\frac{E_a}{E_s} = \frac{u_a^\mu p_{\mathrm{ph},\mu}}{u_s^\mu p_{\mathrm{ph},\mu}} = \frac{\gamma \mathcal{E}_{\mathrm{ph}}(1 - \hat{\mathbf{n}} \cdot \mathbf{v}_a)}{\gamma \mathcal{E}_{\mathrm{ph}}(1 - \hat{\mathbf{n}} \cdot \mathbf{v}_s)} = \frac{1 - \sin \alpha}{1 - \sin \alpha} = 1, \tag{166.14}$$

where the penultimate equality follows by simple geometry.

As the third property, let us consider how the energy density of the light transforms. It is more convenient to formulate the problem in terms of $I(\omega, \hat{\mathbf{n}})$, the energy flux of radiation traveling along the direction $\hat{\mathbf{n}}$ per unit frequency per unit solid angle. We can relate this to $N(\mathbf{r}, \mathbf{p})$, the phase space density of photons. Writing $\mathcal{E}(\mathbf{p})$ for the energy of a photon with momentum $\mathbf{p}$, the energy in a spatial range $d^3 x$ and momentum range $d^3 p$ is given by

$$d\mathcal{E} = \mathcal{E}(\mathbf{p}) N(\mathbf{r}, \mathbf{p}) d^3 x d^3 p. \tag{166.15}$$

Since light travels at the speed c, and $\mathcal{E}(\mathbf{p}) = pc = \hbar \omega$, we can write $d^3 x = c dt \, d(\text{Area})$, and since $d^3 p = p^2 dp d\Omega = (\hbar/c)^3 \omega^2 d\omega d\Omega$, we can write

$$d\mathcal{E} = \frac{\hbar^4}{c^2} \omega^3 N dt \, d(\text{Area}) \, d\omega \, d\Omega. \tag{166.16}$$

Dividing by $dt\,d(\text{Area})$ gives the flux, so it follows that

$$I(\omega, \hat{\mathbf{n}}) = \frac{\hbar^4}{c^2}\omega^3 N(\mathbf{r}, -(\hbar\omega/c)\hat{\mathbf{n}}). \tag{166.17}$$

But since the phase space density is a Lorentz invariant, we find

$$\frac{1}{\omega^3}I(\omega, \hat{\mathbf{n}}) = \text{invariant.} \tag{166.18}$$

This is the desired answer.

It is interesting to apply the above result to the 2.7 K cosmic microwave background. This radiation has a blackbody spectrum, for which

$$I(\omega) = \text{const}\frac{\omega^3}{e^{\hbar\omega/k_B T} - 1}, \tag{166.19}$$

with $T = 2.7$ K. This radiation is isotropic in its own frame, but not in the earth frame, which is moving with respect to it with some velocity $\mathbf{v}$. Denoting the radiation frame and the earth frame quantities with suffixes s and o as before, we have $I(\omega_o, \hat{\mathbf{n}}_o)/\omega_o^3 = I(\omega_s, \hat{\mathbf{n}}_s)/\omega_s^3$, where $\hat{\mathbf{n}}_o$ and $\hat{\mathbf{n}}_s$ are related by the formula for aberration, and ω_o and ω_s are related by the formula for the Doppler shift. As a result, in the earth's frame, the spectrum still has a blackbody form,

$$I(\omega_o, \hat{\mathbf{n}}_o) = \text{const}\frac{\omega_o^3}{e^{\hbar\omega_o/k_B T_o} - 1}, \tag{166.20}$$

but with $T_o = T_s(\omega_o/\omega_s)$. However, since the Doppler shift depends on the angle between the direction of observation and the earth's velocity, the observed temperature distribution will be anisotropic, with

$$T_o(\hat{\mathbf{n}}_o) = T_s\frac{(1 - v^2)^{1/2}}{1 - \mathbf{v}\cdot\hat{\mathbf{n}}_o}. \tag{166.21}$$

The measured anisotropy amounts to a temperature variation of ± 3.3 mK. This indicates that our solar system is moving at 370 km/sec relative to the cosmic rest frame.

Finally, let us examine the polarization of the light. Consider a plane wave in the star frame with four-potential

$$A'^\mu = a'^\mu e^{-ik'^\alpha x'_\alpha}. \tag{166.22}$$

Let us work in the Lorenz gauge, wherein $\partial'_\mu A'^\mu = 0$, and further impose the transversality condition $\nabla'\cdot\mathbf{A}' = 0$. Finally, let us choose directions of propagation exactly as in the discussion of aberration and Doppler shift. The wave vector and four-potential amplitude may then be written as

$$k'^\mu = \omega'(1, -\cos\theta', -\sin\theta', 0), \quad a'^\mu = (0, \zeta\sin\theta', -\zeta\cos\theta', \eta), \tag{166.23}$$

where ζ and η are arbitrary. One may now obtain the electric field three-vector:

$$\mathbf{E}' = i(k'^0\mathbf{a}' - \mathbf{k}'a'^0) = i\omega'(\zeta\sin\theta', -\zeta\cos\theta', \eta). \tag{166.24}$$

Note that we may write

$$\mathbf{E}' = i\omega'(\eta\hat{\mathbf{z}} + \zeta\hat{\mathbf{z}} \times \hat{\mathbf{k}}'), \tag{166.25}$$

From this, we see that ζ and η have the meaning of the complex amplitudes of the electric field along two orthogonal polarization directions.

We now find the electric field, $\mathbf{E} = i(\omega\mathbf{a} - \mathbf{k}a^0)$, in the earth frame. The goal will be to represent it in a form similar to eq. (166.25). The four-potential in the earth frame is given by an LT as

$$a^{\mu} = (-\gamma v\zeta\sin\theta', \gamma\zeta\sin\theta', -\zeta\cos\theta', \eta). \tag{166.26}$$

The earth-frame expression for the wave vector is already known. We now calculate $\mathbf{E}$ component by component and express E_x and E_y in terms of star-frame variables:

$$\begin{aligned} E_x &= i\omega\gamma\zeta\sin\theta' + i\gamma v\zeta\sin\theta'(-\gamma\omega')(\cos\theta' + v) \\ &= i\gamma^2\omega'\zeta\sin\theta'(1 + v\cos\theta') - i\gamma^2\omega'v\zeta\sin\theta'(\cos\theta' + v) \\ &= i\omega'\zeta\sin\theta'. \end{aligned} \tag{166.27}$$

$$\begin{aligned} E_y &= -i\omega\zeta\cos\theta' + i\gamma v\zeta\sin\theta'(-\omega'\sin\theta') \\ &= -i\gamma\omega'\zeta\cos\theta'(1 + v\cos\theta') - i\gamma\omega'\zeta v\sin^2\theta' \\ &= -i\gamma\omega'\zeta(\cos\theta' + v). \end{aligned} \tag{166.28}$$

$$E_z = i\omega\eta. \tag{166.29}$$

If we now examine the expressions for the earth-frame wave vector, we see that we may write

$$\mathbf{E} = i(-k^y\zeta, k^x\zeta, \omega\eta) = i\omega(\eta\hat{\mathbf{z}} + \zeta\hat{\mathbf{z}} \times \hat{\mathbf{k}}). \tag{166.30}$$

In other words, the relative amplitudes and phases of the two polarization components are given by the same two complex numbers, ζ and η, as in the star frame. Hence, all polarization properties are unchanged. In particular, the degree of polarization of light is Lorentz invariant, as are the degrees of linear and circular polarization, separately (see section 45).[4]

167 Motion of charge in an electromagnetic plane wave*

We now study how a charge moves in an external electromagnetic plane wave, ignoring the radiation produced as a result of the resulting accelerated motion. The problem is interesting in itself, and it is also relevant to the study of undulators and wigglers, which we undertake in section 175.

---

[4] We may mention that in quantum field theory, the polarization of a particle is often expressed in terms of the *helicity*, the projection of its spin onto the direction of the three-momentum. The helicity of a photon, indeed of any massless particle, is a Lorentz invariant.

We take the wave to be traveling along the z axis and write the fields in the wave as

$$\mathbf{E}(z, t) = (E_x(\xi),\, E_y(\xi),\, 0), \quad \mathbf{B}(z, t) = (-E_y(\xi),\, E_x(\xi),\, 0), \tag{167.1}$$

where

$$\xi = t - z. \tag{167.2}$$

The equation of motion of the charge is

$$\frac{dp^\mu}{d\tau} = m\frac{d^2 x^\mu}{d\tau^2} = q\, F^{\mu\nu}\frac{dx_\nu}{d\tau}, \tag{167.3}$$

where, using the properties of the wave fields, we get

$$F^{\mu\nu} = \begin{pmatrix} 0 & -E_x & -E_y & 0 \\ E_x & 0 & 0 & E_x \\ E_y & 0 & 0 & E_y \\ 0 & -E_x & -E_y & 0 \end{pmatrix}. \tag{167.4}$$

It is now a matter of solving the various components of the Lorentz force equation in the right order. Using dots to denote derivatives with respect to the proper time τ, we first note that

$$\dot{p}^0 = \dot{p}^z = q(E_x\dot{x} + E_y\dot{y}). \tag{167.5}$$

Hence,

$$p^0 - p^z = \text{const} \equiv \epsilon. \tag{167.6}$$

Integrating this equation once and choosing the zero of τ appropriately, we get

$$t - z = \alpha\tau, \quad (\alpha = \epsilon/m). \tag{167.7}$$

This is an important result, for it relates the independent variable ξ on which the fields depend to the proper time and also allows the equations for p^x and p^y to be integrated. We have

$$\dot{p}^{x,y} = q\, E_{x,y}(\dot{t} - \dot{z}) = q\alpha\, E_{x,y}, \tag{167.8}$$

so

$$p^{x,y}(\tau) = p^{x,y}(0) + q\alpha \int_0^\tau E_{x,y}(\xi)d\tau. \tag{167.9}$$

This expression can be written more revealingly in terms of the vector potential:

$$p^{x,y}(\tau) = p^{x,y}(0) - q\, A_{x,y}(\xi). \tag{167.10}$$

We have taken $\mathbf{A}(0) = 0$, which is always possible.

With the solutions for p^x and p^y in hand, the equation for p^z can also be integrated. Let us denote

$$\mathbf{p}_\perp = p^x \hat{\mathbf{x}} + p^y \hat{\mathbf{y}}. \tag{167.11}$$

Then, using eq. (167.8), we get

$$\dot{p}^z = q(E_x \dot{x} + E_y \dot{y}) = \frac{q}{m}(E_x p^x + E_y p^y)$$

$$= \frac{1}{m\alpha}(\dot{p}^x p^x + \dot{p}^y p^y)$$

$$= \frac{1}{2m\alpha} \frac{d}{d\tau} p_\perp^2. \tag{167.12}$$

Hence,

$$p^z(\tau) = p^z(0) + \frac{1}{2m\alpha}\left(p_\perp^2(\tau) - p_\perp^2(0)\right). \tag{167.13}$$

Substituting eq. (167.10), this becomes

$$p^z(\tau) = p^z(0) + \frac{1}{2m\alpha}\left(q^2 A^2(\xi) - 2q\mathbf{p}_\perp(0) \cdot \mathbf{A}(\xi)\right). \tag{167.14}$$

Equations (167.10) and (167.14) constitute an explicit solution for the momentum for any given initial conditions and any given waveform. They can be integrated once more to obtain the spatial coordinates of the particle, x, y, and z. Equations (167.6) and (167.7) give the energy and the time coordinate. All these variables are given parametrically in terms of the proper time. The constant ϵ can be related to the initial momentum as follows. We have

$$m^2 = (p^0)^2 - (p^z)^2 - p_\perp^2. \tag{167.15}$$

Using eqs. (167.6) and (167.13) and $\epsilon = m\alpha$, this becomes

$$m^2 = \epsilon^2 + 2\epsilon p^z(0) - p_\perp^2(0), \tag{167.16}$$

a quadratic that can be solved for ϵ.

Let us now specialize to the case of a monochromatic plane wave field. We take the vector potential as

$$A_{x,y}(\xi) = -\frac{E_{x0,y0}}{\omega}[\sin(\omega\xi + \phi_{x,y}) - \sin(\phi_{x,y})], \tag{167.17}$$

so that

$$E_{x,y}(\xi) = E_{x0,y0}\cos(\omega\xi + \phi_{x,y}). \tag{167.18}$$

We see from the general solution that the momentum varies periodically about an average value. It is most convenient to take the lab frame to be that in which this average value is zero, so that the particle is at rest on the average. The averaging procedure can be deduced by integrating the equation of motion $m(dz/d\tau) = p^z$ over one complete period. Since the particle must return to the same position in the average rest frame, we conclude that

$$\int_0^{\tau_0} p^z(\tau)\, d\tau = 0, \tag{167.19}$$

where τ_0 is the proper period. The important point that has emerged is that the integration variable in the averaging is not the lab time, but the proper time. It then follows from eqs. (167.17) and (167.7) that $\tau_0 = 2\pi/\alpha\omega$ and that

$$\overline{A}_{x,y} = \frac{E_{x0,y0}}{\omega} \sin\phi_{x,y}, \tag{167.20}$$

$$\overline{A^2}_{x,y} = \frac{E^2_{x0,y0}}{\omega^2} \left(\tfrac{1}{2} + \sin^2\phi_{x,y} \right), \tag{167.21}$$

where the overbars indicate averages over one period. If we now average eqs. (167.10) and (167.14), and use the conditions $\overline{p^{x,y}} = \overline{p^z} = 0$, we obtain the initial momenta:

$$p^{x,y}(0) = \frac{q\,E_{x0,y0}}{\omega} \sin\phi_{x,y}, \tag{167.22}$$

$$p^z(0) = \frac{q^2}{2m\omega^2\alpha} \left[(E^2_{x0} \sin^2\phi_x + x \leftrightarrow y) - \overline{E^2} \right], \tag{167.23}$$

with

$$\overline{E^2} = \frac{1}{2}(E^2_{x0} + E^2_{y0}). \tag{167.24}$$

Finally,

$$\epsilon^2 = m^2 + \frac{q^2}{\omega^2} \overline{E^2}. \tag{167.25}$$

The physical meaning of this quantity can be seen from eq. (167.6): it is the average energy of the particle in the average rest frame.

For completeness, we also give the explicit solutions for the momenta:

$$p^{x,y}(\tau) = \frac{q\,E_{x0,y0}}{\omega} \sin(\omega\xi + \phi_{x,y}), \tag{167.26}$$

$$p^z(\tau) = \frac{q^2}{2m\omega^2\alpha} \left[(E^2_{x0} \sin^2(\omega\xi + \phi_x) + x \leftrightarrow y) - \overline{E^2} \right]. \tag{167.27}$$

Let us further specialize to the case of linear polarization. We take

$$E_x = E_0 \cos\omega\xi, \quad E_y = 0. \tag{167.28}$$

With

$$\epsilon^2 = m^2 + \frac{q^2 E_0^2}{2\omega^2}, \tag{167.29}$$

and $\alpha = \epsilon/m$ as before, the momentum components are found to be

$$p^x = \frac{q\,E_0}{\omega} \sin(\alpha\omega\tau), \quad p^y = 0, \quad p^z = -\frac{q^2 E_0^2}{4\epsilon\omega^2} \cos(2\alpha\omega\tau), \tag{167.30}$$

while the spatial coordinates are

$$x = -\frac{q\,E_0}{\epsilon\omega^2} \cos(\alpha\omega\tau), \quad y = 0, \quad z = -\frac{q^2 E_0^2}{8\epsilon^2\omega^2} \sin(2\alpha\omega\tau). \tag{167.31}$$

These solutions are given parametrically in terms of the proper time τ, which is related to t by the implicit equation

$$t = \alpha\tau - \frac{q^2 E_0^2}{8\epsilon^2\omega^2}\sin(2\alpha\omega\tau). \qquad (167.32)$$

When t varies from 0 to $2\pi/\omega$, τ varies from 0 to $2\pi/\omega\alpha$. The particle moves in a figure-eight shape in the plane perpendicular to the magnetic field, similar to fig. 25.5b on page 598.

Exercise 167.1 Solve for the motion of a charge in a circularly polarized monochromatic plane wave.

168 Thomas precession*

In chapter 11 we saw how the spin of a particle located at a fixed point in space precesses in a magnetic field. If the particle is undergoing accelerated motion, however, the spin undergoes an additional precession, discovered by Thomas in 1927. This is a purely kinematic effect that is present even where there are no dynamical torques on the spin. It originates in the fact that the least noninertial coordinate system used by an accelerated observer must rotate in accordance with the law (167.24) for Fermi-Walker transport. Another way of viewing the effect is to consider a particle in a circular orbit. The velocity of the particle also executes a closed circuit in velocity space. Since relativistic velocity space is curved (see eq. (154.20)) and surrounding discussion), parallel transport of the spin vector around this circuit automatically entails a rotation. Indeed, from this point of view, the effect is much like the following illustration.[5]

A Roman legionary sets out from the North Pole, with orders to run down the prime meridian to the equator, turn left, run along the equator to the 90°E meridian, turn left again, and run up the meridian back to the pole. He holds a lance, which at the start is pointing straight ahead and parallel to the ground. The lance is not to be allowed to turn under penalty of death. That is, when the legionary turns his body, the lance must continue to point the same way as before. Therefore, on the leg along the equator, he must hold the lance pointing to his right, and along the last leg, he must hold it pointing straight back. When he returns to the pole, the hapless man is fed to the lions because the lance is pointing 90° to the left of its original direction!

The particle's spin is much like the legionary's lance, and after a complete orbit, it fails to return to its initial direction (that is, it precesses) for much the same reason.

To derive the precession quantitatively, let us consider an accelerating particle without any external torques on its spin. In doing this, it will pay to carefully distinguish the components of the spin four-vector, $\mathcal{S}^\alpha$, in several frames:

$$\mathcal{S}^\alpha = \begin{cases} (S^0, \mathbf{S}), & \text{(lab frame)}, \\ (0, \mathbf{s}_{\text{co}}), & \text{(comoving frame)}, \\ (0, \mathbf{s}), & \text{(boosted rest frame)}. \end{cases} \qquad (168.1)$$

---

[5] This marvelous example is due to Professor John Baez. See his website www.math.ucr.edu/home/baez/gr/parallel.transport.html.

By the comoving frame, we mean a local inertial frame instantaneously comoving with the particle, while by the boosted rest frame we mean a frame obtained by boosting from the lab frame by the lab velocity of the particle at any given instant. The distinction between these frames is key to the phenomenon. Let us consider the particle at proper time $\tau = 0$, and $\tau = \Delta\tau$, infinitesimally later. The comoving frame at $\Delta\tau$ is obtained from that at $\tau = 0$ by boosting by the velocity $\mathbf{a}_{co}\Delta\tau$, where $\mathbf{a}_{co}$ is the acceleration of the particle in the comoving frame at $\tau = 0$. On the other hand, the boosted rest frame at $\Delta\tau$ is obtained from the lab frame by a boost at velocity $\mathbf{v}_{lab}(\Delta\tau)$. The point is that even though the particle is at rest in both the comoving and boosted rest frames at $\Delta\tau$, these frames need not be identical, since the condition of being at rest does not fix the orientation of the spatial axes. In particular, even if the frames happen to be identical at $\tau = 0$, they may (and do) differ by a rotation at $\Delta\tau$.[6]

We caution the reader to pay close attention to the typography, since the differences between the various frames are denoted purely by the font and the use of upper- versus lowercase letters.

Let us now consider how the comoving observer sees the world. She sees no precession, since there are no torques. To her, therefore, $d\mathbf{s}_{co}/d\tau = 0$. She also sees, however, that $d\mathcal{S}^0/d\tau \neq 0$, and since $u^\alpha = (1, \mathbf{0})$ in her frame, she writes

$$\frac{d\mathcal{S}^\alpha}{d\tau} = ku^\alpha, \tag{168.2}$$

where k is a constant. This equation is in covariant form, so it must hold in the lab frame too. To find k, we contract both sides with u_α and use the results $u_\alpha u^\alpha = 1$ and $u_\alpha \mathcal{S}^\alpha = 0$. This yields

$$k = u_\alpha \frac{d\mathcal{S}^\alpha}{d\tau} = -\frac{du_\alpha}{d\tau}\mathcal{S}^\alpha = -a_\alpha \mathcal{S}^\alpha, \tag{168.3}$$

where a^α is the acceleration. Hence,

$$\frac{d\mathcal{S}^\alpha}{d\tau} = -(u^\alpha a^\beta)\mathcal{S}_\beta. \tag{168.4}$$

Invoking $u^\beta \mathcal{S}_\beta = 0$ once again, we can also write this as

$$\frac{d\mathcal{S}^\alpha}{d\tau} = -(u^\alpha a^\beta - u^\beta a^\alpha)\mathcal{S}_\beta, \tag{168.5}$$

which is the equation for Fermi-Walker transport.

We now wish to write eq. (168.4) in the lab frame. Using eq. (154.12) for the lab acceleration,

$$a^\beta = \gamma^2(0, \dot{\mathbf{v}}) + \dot{\gamma}u^\beta, \tag{168.6}$$

we get

$$a^\beta \mathcal{S}_\beta = -\gamma^2 \dot{\mathbf{v}} \cdot \mathbf{S}. \tag{168.7}$$

---

[6] This rotation is known as a Wigner or Thomas-Wigner rotation, and from a purely mathematical point of view, it originates in the noncommutativity of LTs. Finding it is equivalent to finding the Thomas precession.

With this equation and $d/d\tau = \gamma d/dt$, eq. (168.4) reduces to

$$\frac{dS^0}{dt} = \gamma^2(\dot{\mathbf{v}} \cdot \mathbf{S}), \tag{168.8}$$

$$\frac{d\mathbf{S}}{dt} = \gamma^2\mathbf{v}(\dot{\mathbf{v}} \cdot \mathbf{S}). \tag{168.9}$$

Because $S^\alpha S_\alpha = -\mathbf{s} \cdot \mathbf{s}$ is a constant, these two equations are not independent and it is more revealing to combine them into a single equation for $d\mathbf{s}/dt$. The most direct if somewhat laborious way to do this is to employ the Lorentz transformation directly. The transformation simplifies somewhat due to the fact $S^\alpha u_\alpha = 0$, but the algebra is still lengthy. We tabulate the resulting formulas for quick reference:

$$S^0 = \mathbf{v} \cdot \mathbf{S}, \tag{168.10}$$

$$S^0 = \gamma\mathbf{v} \cdot \mathbf{s}, \tag{168.11}$$

$$\mathbf{S} = \mathbf{s} + \frac{\gamma^2}{\gamma+1}\mathbf{v}(\mathbf{v} \cdot \mathbf{s}), \tag{168.12}$$

$$\mathbf{s} = \mathbf{S} - \frac{\gamma}{\gamma+1}\mathbf{v}(\mathbf{v} \cdot \mathbf{S}). \tag{168.13}$$

The proof is given below.

The procedure now is as follows. We find $d\mathbf{s}/dt$ using eq. (168.13). On the right-hand side we get $d\mathbf{S}/dt$ (for which we use eq. (168.9)), and $\mathbf{v} \cdot \mathbf{S}$ and $\dot{\mathbf{v}} \cdot \mathbf{S}$ (for which we use eq. (168.12)). In this way, everything is expressed in terms of $\mathbf{v} \cdot \mathbf{s}$ and $\dot{\mathbf{v}} \cdot \mathbf{s}$ times the vectors $\mathbf{v}$ and $\dot{\mathbf{v}}$. We also get the quantity $\dot{\gamma}$, for which the relation

$$\dot{\gamma} = \gamma^3\mathbf{v} \cdot \dot{\mathbf{v}} \tag{168.14}$$

proves useful. Various combinations of γ and v^2 can be simplified using eq. (152.7), and after two pages of algebra, one gets

$$\frac{d\mathbf{s}}{dt} = \frac{\gamma^2}{\gamma+1}\mathbf{v}(\dot{\mathbf{v}} \cdot \mathbf{s}) - \frac{\gamma^2}{\gamma+1}\dot{\mathbf{v}}(\mathbf{v} \cdot \mathbf{s}). \tag{168.15}$$

The right-hand side is a triple vector product, so the final result can be written as (restoring c)

$$\frac{d\mathbf{s}}{dt} = \boldsymbol{\omega}_T \times \mathbf{s}, \tag{168.16}$$

$$\boldsymbol{\omega}_T = \frac{\gamma^2}{\gamma+1}\frac{1}{c^2}(\dot{\mathbf{v}} \times \mathbf{v}). \tag{168.17}$$

The motion is clearly a precession (length of $\mathbf{s}$ stays fixed), and $\boldsymbol{\omega}_T$ is the Thomas precession frequency.

Proof of eqs. (168.10)–(168.13): Equation (168.10) follows directly from $u^\alpha S_\alpha = 0$. Equation (168.11) comes from the LT from the comoving to the lab frame if we remember that in the former $S^\alpha = (0, \mathbf{s})$:

$$S^0 = \gamma(0 + \mathbf{v} \cdot \mathbf{s}) = \gamma\mathbf{v} \cdot \mathbf{S}. \tag{168.18}$$

Similarly, equation (168.12) follows from the LT for the space part,

$$\mathbf{S} = \mathbf{s} + \frac{\gamma - 1}{v^2}\mathbf{v}(\mathbf{v}\cdot\mathbf{s}) + \gamma\mathbf{v}\times 0, \tag{168.19}$$

and the formula $\gamma - 1 = \gamma^2 v^2/(\gamma + 1)$. Lastly, eq. (168.13) follows from the inverse LT:

$$\mathbf{s} = \mathbf{S} + \frac{\gamma - 1}{v^2}\mathbf{v}(\mathbf{v}\cdot\mathbf{S}) - \gamma\mathbf{v}S^0. \tag{168.20}$$

But $S^0 = \mathbf{v}\cdot\mathbf{S}$, so the last two terms combine with a net coefficient of $\mathbf{v}(\mathbf{v}\cdot\mathbf{S})$ given by

$$\frac{\gamma - 1}{v^2} - \gamma = \frac{\gamma^2}{\gamma + 1} - \gamma = -\frac{\gamma}{\gamma + 1}. \tag{168.21}$$

This leads to eq. (168.13).

Exercise 168.1 A good way to understand Thomas precession is to consider a particle in circular motion in the xy plane,

$$x = r\cos\omega t, \quad y = r\sin\omega t, \tag{168.22}$$

and to solve eq. (168.4) for S^x and S^y in the lab frame directly. Although this can be done exactly for any ω, it is better to work in the limit $v = \omega r \ll 1$. Then we may set γ to 1 and neglect the difference between $\mathbf{s}$ and $\mathbf{S}$, as it is of order v^2 (see eq. (168.13)). Show that with these approximations,

$$\frac{ds^x}{dt} = \omega^3 r^2 \sin\omega t(\cos\omega t\, s^x + \sin\omega t\, s^y), \tag{168.23}$$

$$\frac{ds^y}{dt} = -\omega^3 r^2 \cos\omega t(\cos\omega t\, s^x + \sin\omega t\, s^y). \tag{168.24}$$

The rate of change of $\mathbf{s}$ is of order $\omega^3 r^2 = v^2\omega \ll \omega$, so $\mathbf{s}$ changes little over one orbital period $2\pi/\omega$. It is therefore valid to average the equations of motion over several such periods. Show that when this is done, you get the same result as eq. (168.16). Convince yourself that the precession is retrograde, i.e., opposite the orbital motion, and that $\omega_T = v^2\omega/2$.

We conclude this section with a discussion of spin–orbit coupling in atoms. This topic is covered in all intermediate and advanced texts on quantum mechanics,[7] all of which note the "mysterious Thomas half," which reduces the coupling by a factor of two. We show how this correction arises. Since electron speeds v in atoms are generally small, our treatment will be nonrelativistic, i.e., it will include only the lowest-order terms in v/c needed to see the effect.

Let us consider one of the electrons in the atom. Its spin feels a dynamic torque[8] (see section 75)

$$\frac{ge}{2mc}\mathbf{B}'\times\mathbf{s}. \tag{168.25}$$

We have written the magnetic field as $\mathbf{B}'$, the field in the comoving frame, since the expression $\mathcal{M}\times\mathbf{B}$ for the torque is valid only when the spin is at a fixed point in space.

---

[7] See, e.g., Baym (1969). This reference also contains an excellent discussion of the self-consistent field.
[8] In this discussion, we display all factors of c and $\hbar$ explicitly. Thus, $\mathbf{s}$ and $\mathbf{L}$ are dimensionful angular momenta.

For low speeds, we have from eq. (38.9)

$$\mathbf{B}' = \mathbf{B} - \frac{\mathbf{v} \times \mathbf{E}}{c}. \tag{168.26}$$

In addition, there is also a kinematic torque $\boldsymbol{\omega}_T \times \mathbf{s}$ from the Thomas precession. The complete equation of motion for the spin is, therefore,

$$\frac{d\mathbf{s}}{dt} = \frac{ge}{2mc}\mathbf{B} \times \mathbf{s} + \left(\frac{ge}{2mc^2}\mathbf{E} \times \mathbf{v} + \boldsymbol{\omega}_T \right) \times \mathbf{s}. \tag{168.27}$$

The last two terms constitute the total spin–orbit torque. We can view them as arising from a term in the Hamiltonian

$$\mathcal{H}_{so} = \left(\frac{ge}{2mc^2}\mathbf{E} \times \mathbf{v} + \boldsymbol{\omega}_T \right) \cdot \mathbf{s}. \tag{168.28}$$

We proceed to simplify $\mathcal{H}_{so}$. In the low speed limit, $\boldsymbol{\omega}_T = (\dot{\mathbf{v}} \times \mathbf{v})/2c^2$ and $\dot{\mathbf{v}} = -e\mathbf{E}/m$, so

$$\mathcal{H}_{so} = \frac{(g-1)e}{2mc^2}(\mathbf{E} \times \mathbf{v}) \cdot \mathbf{s}. \tag{168.29}$$

To write the electric field on the electron, which consists of the field from the nucleus plus the field from the other electrons, we employ the self-consistent field approximation,

$$-e\mathbf{E} = -\nabla V(r) = -\frac{\mathbf{r}}{r}\frac{dV(r)}{dr}, \tag{168.30}$$

in which $V(r)$ is the (spherically symmetric) self-consistent potential. Then, writing $\mathbf{v} = \mathbf{p}/m$, where $\mathbf{p}$ is the electron momentum, we get

$$e\mathbf{E} \times \mathbf{v} = \frac{1}{mr}\frac{dV}{dr}(\mathbf{r} \times \mathbf{p}). \tag{168.31}$$

But $\mathbf{r} \times \mathbf{p} = \mathbf{L}$, the orbital angular momentum of the electron, so we can write

$$\mathcal{H}_{so} = \frac{(g-1)}{2m^2c^2}\left(\frac{1}{r}\frac{dV}{dr} \right)\mathbf{L} \cdot \mathbf{s}. \tag{168.32}$$

This is the standard form of the spin–orbit coupling, the name for which is now clear.

The reader will undoubtedly have noted that the Thomas correction has the effect of changing the g factor to $(g-1)$ in eq. (168.32). Since $g = 2$ to high accuracy for the electron, $g - 1 = 1$, and so this correction is often viewed as halving the dynamic contribution to the energy. It is conceptually better, however, to think of the correction as subtractive rather than as multiplicative.

Exercise 168.2 Show that the covariant form of the equation of motion eq. (168.27) is

$$\frac{dS^\alpha}{d\tau} = -\frac{ge}{2mc}\left(F^{\alpha\beta}S_\beta - F^{\gamma\delta}u_\gamma S_\delta u^\alpha \right) - u^\alpha\frac{du^\beta}{d\tau}S_\beta. \tag{168.33}$$

Show further that if the magnetic gradient force $\nabla(\mathcal{M} \cdot \mathbf{B})$ can be neglected, this simplifies to

$$\frac{dS^\alpha}{d\tau} = -\frac{ge}{2mc}F^{\alpha\beta}S_\beta + \frac{(g-2)e}{2mc}F^{\gamma\delta}u_\gamma S_\delta u^\alpha. \tag{168.34}$$

This is known as the *Bargmann-Michel-Telegdi* or *BMT equation* (1959), although Thomas found essentially the same equation in 1927.

Radiation from relativistic sources

In this chapter we consider the radiation from charges moving at speeds close to that of light. The multipole expansion employed in chapter 10 is no longer valid, and the effects of retardation are much more pronounced. They lead to a number of new features, such as a dramatic increase in the power radiated, a collimation of the radiation, and a stretching of the power spectrum to very high frequencies.

169 Total power radiated

The fields produced by a charge carrying out an arbitrary motion were found in section 57. We reproduce the radiative parts of these fields for ready reference:

$$\mathbf{E} = \frac{q}{c^2(R_a - \boldsymbol{\beta}_r \cdot \mathbf{R}_a)^3} \left(\dot{\mathbf{v}}_r \times (\mathbf{R}_a - \boldsymbol{\beta}_r R_a) \right) \times \mathbf{R}_a, \tag{169.1}$$

$$\mathbf{B} = \frac{1}{R_a} \mathbf{R}_a \times \mathbf{E}. \tag{169.2}$$

Here $\mathbf{R}_a$ and $\mathbf{v}_r$ $(= c\boldsymbol{\beta}_r)$ denote the retarded position and velocity of the particle.[1]

When we studied radiation from slow charges, we were able to ignore the variation in the time delay over the spatial region of the radiator. Thus, there was no great need to differentiate between the times of emission and observation of radiation, since they were related by a single delay. This is not so when we consider ultrarelativistic particles. To formulate carefully the question of how much power is radiated, let us consider a charge moving along some trajectory, such that it is at position $\mathbf{r}_0$ at time t_0, and let us focus on the radiation emitted at this time, by which we mean the radiative part of the electromagnetic fields that can be regarded as having arisen at the time t_0. Imagine that we have a 4π solid-angle detector at a large distance R from $\mathbf{r}_0$. How much total power

---

[1] A word about notation. Recall from section 56 that the suffix "a" in $\mathbf{R}_a$ denotes the apparent position. While the apparent velocity $\mathbf{v}_a = d\mathbf{R}_a/dt$ is not the same as the retarded velocity $\mathbf{v}_r$, $\mathbf{R}_a = \mathbf{R}_r$.

falls on this detector at a time $t = t_0 + R/c$? When the charge is moving slowly, this power is given by the Larmor formula (58.12). For fast charges, this formula is no longer valid. A direct attack on this problem starting from the electric and magnetic fields of an arbitrarily moving charge, eqs. (57.19) and (57.14), is very cumbersome. It is simpler to consider the radiation in the frame that is momentarily comoving with the particle, and then Lorentz transform to the lab frame.

In the comoving frame, since the particle is momentarily still, the Larmor formula *is* valid. Denoting quantities in this frame by primes, the energy emitted by the particle in a time interval $\Delta t'$ is given by

$$\Delta \mathcal{E}' = \frac{2q^2}{3c^3}(\dot{\mathbf{v}}' \cdot \dot{\mathbf{v}}')\Delta t'. \tag{169.3}$$

To find the energy in the lab frame, we also need $\Delta \mathbf{G}'$, the momentum radiated in the same interval in the comoving frame. However, because of the symmetry of dipole radiation, $\Delta \mathbf{G}' = 0$ and

$$\Delta \mathcal{E} = \gamma(\Delta \mathcal{E}' + \mathbf{v} \cdot \Delta \mathbf{G}') = \gamma \Delta \mathcal{E}'. \tag{169.4}$$

To find the power radiated, P, we divide by Δt, and since $\Delta t = \gamma \Delta t'$, we get

$$P = P' = \frac{2q^2}{3c^3}a_{\text{co}}^2. \tag{169.5}$$

We have written the last result in terms of $\mathbf{a}_{\text{co}}$ ($\equiv \dot{\mathbf{v}}'$), the acceleration in the comoving frame. Using eq. (154.19) for a_{co}^2, we get

$$P = \frac{2q^2}{3c^3}\gamma^6\big(\dot{\mathbf{v}}^2 - (\dot{\mathbf{v}} \times \mathbf{v}/c)^2\big). \tag{169.6}$$

This is the relativistic generalization of the Larmor formula, found by Lienard in 1898.

The reader will note that, as a by-product, we have shown that the total power radiated is a Lorentz invariant. We can make this explicit by writing

$$P = -\frac{2q^2}{3c^3}a^\mu a_\mu. \tag{169.7}$$

If the charge is accelerated by passing through an EM field, the power can be written directly in terms of the fields. Since $\mathbf{a}_{\text{co}} = q\mathbf{E}_{\text{co}}/m$, we need E_{co}^2. This is obtained using the Lorentz transformation laws for the electric field:

$$
\begin{aligned}
E_{\text{co}}^2 &= \frac{1}{v^2}(\mathbf{E} \cdot \mathbf{v})^2 + \gamma^2\Big(\mathbf{E}_\perp + \frac{\mathbf{v}}{c} \times \mathbf{B}\Big)^2, \\
&= \gamma^2\Big[\Big(\mathbf{E} + \frac{\mathbf{v}}{c} \times \mathbf{B}\Big)^2 - \Big(\frac{\mathbf{v}}{c} \cdot \mathbf{E}\Big)^2\Big].
\end{aligned} \tag{169.8}
$$

Hence,

$$P = \frac{2}{3}\frac{q^4}{m^2c^3}\gamma^2\Big[\Big(\mathbf{E} + \frac{\mathbf{v}}{c} \times \mathbf{B}\Big)^2 - \Big(\frac{\mathbf{v}}{c} \cdot \mathbf{E}\Big)^2\Big]. \tag{169.9}$$

The two limiting cases of acceleration parallel and perpendicular to the velocity are of special interest. In the first case, which occurs in linear accelerators (linacs), the $\dot{\mathbf{v}} \times \mathbf{v}$ term drops out and

$$P = \frac{2q^2}{3c^3}\gamma^6 \dot{v}^2 = \frac{2q^2}{3c^3}\left(\frac{\dot{\mathcal{E}}}{mv}\right)^2, \tag{169.10}$$

where $\dot{\mathcal{E}} = d\mathcal{E}/dt$ is the rate at which the energy of the particle increases because of the acceleration. The dimensionless figure of merit is clearly the ratio $P/\dot{\mathcal{E}}$. It turns out to be best to express this in terms of $d\mathcal{E}/dx$, the gain in particle energy per unit distance along its trajectory. Since $d\mathcal{E}/dx = \dot{\mathcal{E}}/v$, we get

$$\frac{P}{\dot{\mathcal{E}}} = \frac{2q^2}{3m^2c^3v}\frac{d\mathcal{E}}{dx} = \frac{2}{3}\frac{c}{v}\frac{d\tilde{\mathcal{E}}}{d\tilde{x}}. \tag{169.11}$$

In the last form, $\tilde{\mathcal{E}} = \mathcal{E}/mc^2$, i.e., the energy in units of the rest energy, and $\tilde{x} = x/(q^2/mc^2)$, i.e., the position in units of the classical charge radius. If $v/c = O(1)$, this result shows that radiative losses will be insignificant unless the particle gains as much as its rest energy in a distance equal to one classical radius. For electrons this would mean gaining 0.5 MeV in $\sim 3 \times 10^{-13}$ cm. Needless to say, such accelerations are preposterous, and for more typical gains such as 0.5 MeV in 1 cm, radiative losses in linacs are negligible.

Matters are rather different when the acceleration is perpendicular to the velocity, as for example, in a circular synchrotron orbit. Now, $\dot{\mathbf{v}} \perp \mathbf{v}$, so $(\dot{\mathbf{v}} \times \mathbf{v})^2 = v^2\dot{v}^2$, and $|\dot{\mathbf{v}}| = v^2/r_0$, where r_0 is the radius of the particle orbit. Hence,

$$P = \frac{2q^2}{3c^3}\gamma^4 \dot{v}^2 = \frac{2q^2}{3c^3}\frac{\gamma^4 v^4}{r_0^2}. \tag{169.12}$$

The figure of merit now is the ratio of the energy loss per orbit, $\delta\mathcal{E} = (2\pi r_0/v)P$, to the energy of the particle, $\mathcal{E} = \gamma mc^2$. We get

$$\frac{\delta\mathcal{E}}{\mathcal{E}} = \frac{4\pi}{3}\frac{q^2}{mc^5}\frac{\gamma^3 v^3}{r_0} = \frac{4\pi}{3}\gamma^3\left(\frac{v}{c}\right)^3\frac{1}{\tilde{r}_0}. \tag{169.13}$$

Once again, $\tilde{r}_0$ is the orbit radius in units of the classical charge radius. The factor $1/\tilde{r}_0$ is very small, but it is counterweighed by the γ^3 factor. For the Cornell electron synchrotron, e.g., $\mathcal{E} \sim 10$ GeV, $r_0 = 100$ m, so $\gamma \sim 2 \times 10^4$, $1/\tilde{r}_0 \sim 3 \times 10^{-17}$, and $\delta\mathcal{E}/\mathcal{E} \sim 10^{-3}$. This is small compared to unity, but still significant. One loses the beam in a mere 1000 turns unless power is fed into it to maintain the energy. But one person's loss is another's gain. As we shall see in subsequent sections, synchrotron radiation is highly collimated and extends to frequencies vastly beyond the orbit frequency, making synchrotrons the X-ray source of choice for many investigations.

Exercise 169.1 Write the four-momentum radiated by a charge moving in an arbitrary trajectory in covariant four-dimensional form.

Answer:

$$\Delta P^\mu = -\frac{2q^2}{3c^3} a^\nu a_\nu u^\mu \Delta\tau. \tag{169.14}$$

Exercise 169.2 Derive eqs. (169.10) and (169.11) for linear acceleration.

Exercise 169.3 Derive eqs. (169.12) and (169.13) for circular acceleration.

Exercise 169.4 Show that the radiative energy lost in a betatron is given by

$$\Delta\mathcal{E} = \frac{\pi}{8}\frac{\omega}{\omega_0}\left(\frac{\mathcal{E}}{mc^2}\right)^4 \frac{e^2}{R}, \tag{169.15}$$

where ω and ω_0 are the frequencies of orbital motion and the driving magnetic field, $\mathcal{E}$ is the nominal energy the electrons would have in the absence of losses, and R is the radius of the electron orbit in the betatron.

Solution: Let $B(R, t) = B_0 \sin\omega_0 t$. Then $E(R, t) = -(\omega_0 R/c) B_0 \cos\omega_0 t$. Since $\mathbf{E} \parallel \mathbf{v} \perp \mathbf{B}$,

$$E_{\text{co}}^2 = E^2(R, t) + \frac{\gamma^2 v^2}{c^2} B^2(R, t) = \frac{\gamma^2 v^2}{c^2} B^2(R, t)\left(1 + \frac{1}{\gamma^2}\frac{\omega_0^2}{\omega^2}\right), \tag{169.16}$$

where we have used the result $v = \omega R$. Since $\gamma \gg 1$ and $\omega_0 \ll \omega$, the electric field contribution to E_{co}^2 is negligible. Writing $\gamma v = p/m$, where p is the electron momentum, and further using the result $p = (e R/c) B(R, t)$, the instantaneous power lost is found as

$$P = \frac{2}{3}\frac{e^6}{m^4 c^7} R^2 B^4(R, t). \tag{169.17}$$

To find the total energy lost, we integrate over time from 0 to $\pi/2\omega_0$ (a quarter-cycle), trade in B_0 for $\mathcal{E}$ using $\mathcal{E} = e R B_0$, and again use $\omega R = v \approx c$. This leads to eq. (169.15).[2]

170 Angular distribution of power

Let us now consider the angular distribution of the power. The energy flux is $(c/4\pi)|\mathbf{E}|^2$, which is directly obtained from the expression for the electric field. Hence, the energy radiated in a small time interval Δt and a small solid angle $\Delta\Omega$ is given by[3]

$$\Delta\mathcal{E} = \frac{q^2}{4\pi c^3} \frac{[\hat{\mathbf{n}} \times ((\hat{\mathbf{n}} - \mathbf{v}) \times \dot{\mathbf{v}})]^2}{(1 - \hat{\mathbf{n}} \cdot \mathbf{v})^6}\bigg|_{\text{ret}} \Delta\Omega\Delta t_{\text{obs}}. \tag{170.1}$$

The suffix in Δt_{obs} indicates that this is the time elapsed at the fixed location of the observer. Because of this, it would be a mistake to regard the coefficient of $\Delta\Omega\Delta t_{\text{obs}}$ as

---

[2] If this formula is applied to the Urbana-Champaign betatron described in section 72, the loss is obtained as 17% (57 MeV), which is about 2.5 times too big. It can be reduced to 43 MeV by extracting the beam slightly before a quarter-cycle, but that is still double the actual loss. The reason for this discrepancy is not clear to the author.

[3] We again measure velocities in units of c.

the angular power distribution. Because the source is moving, the photons collected at the detector are emitted at different locations and over an interval that is not the same as Δt_{obs}. The rate at which energy arrives in different directions is therefore different, and if we integrated the naive distribution (170.1) over all angles, we would not get the total power as found in the previous section. To get the correct result, we must refer the power to the retarded time or the time of emission, t_r, not the time of observation. In other words,

$$\frac{dP}{d\Omega} = \frac{\Delta\mathcal{E}}{\Delta\Omega\Delta t_r}. \tag{170.2}$$

Fortunately, the relation between Δt_{obs} and Δt_r is simple. We showed way back in eq. (56.14) that

$$\Delta t_{\text{obs}} = (1 - \mathbf{v} \cdot \hat{\mathbf{n}})_{\text{ret}} \Delta t_r. \tag{170.3}$$

Hence,

$$\frac{dP}{d\Omega} = \frac{q^2}{4\pi c^3} \left. \frac{\left[\hat{\mathbf{n}} \times \left((\hat{\mathbf{n}} - \mathbf{v}) \times \dot{\mathbf{v}}\right)\right]^2}{(1 - \hat{\mathbf{n}} \cdot \mathbf{v})^5} \right|_{\text{ret}}. \tag{170.4}$$

We leave it as an exercise for the reader to expand out the triple cross product, and square, and thus show that

$$\frac{dP}{d\Omega} = \frac{q^2}{4\pi c^3} \left[\frac{\dot{\mathbf{v}}^2}{(1 - \mathbf{v} \cdot \hat{\mathbf{n}})^3} + 2\frac{(\dot{\mathbf{v}} \cdot \hat{\mathbf{n}})(\dot{\mathbf{v}} \cdot \mathbf{v})}{(1 - \mathbf{v} \cdot \hat{\mathbf{n}})^4} - \gamma^{-2} \frac{(\dot{\mathbf{v}} \cdot \hat{\mathbf{n}})^2}{(1 - \mathbf{v} \cdot \hat{\mathbf{n}})^5} \right]. \tag{170.5}$$

To avoid misunderstanding, we stress that the difference between Δt_{obs} and Δt_r is not a time dilation effect. It arises solely because the interval Δt_r comprises events at different spatial locations. The point can be seen very clearly by considering a fast-moving charge that emits a burst of radiation for only 1 ns, say, as measured in the lab frame. Distant observers (also in the lab frame) located along different directions will see different burst durations. Consider an observer toward whom the charge is moving. The radiation emitted toward the end of the burst has a shorter distance to travel and reaches her after a shorter delay than the radiation emitted at the start of the burst. The net result is that the burst will be compressed in time for this observer, who might measure the burst duration as 0.9 ns, while one from whom the charge is receding might measure it as 1.1 ns. If all the observers recorded the energy they received for the same 1-ns interval of lab time, they would not account for all the energy emitted. To do this correctly, each observer must adjust her observation interval so that it corresponds to the same interval of emission time. This is what eq. (170.3) achieves.[4]

It is useful to derive the above result in another way. We again consider the four-momentum radiated in the comoving frame, this time, into a small solid angle around the direction $\hat{\mathbf{n}}'$. We have

$$\Delta\mathcal{E}' = \frac{q^2}{4\pi c^3} (\dot{\mathbf{v}}' \times \hat{\mathbf{n}}')^2 \Delta\Omega' \Delta t'. \tag{170.6}$$

---

[4] It should be apparent that the interval Δt of the previous section is Δt_r. Also, it is but a short step to extend this argument to obtain the Doppler shift formula.

We now note that

$$\Delta\mathcal{E}' = \gamma(\Delta\mathcal{E} - \mathbf{v}\cdot\Delta\mathbf{G}) = \gamma(1 - \mathbf{v}\cdot\hat{\mathbf{n}})\Delta\mathcal{E}, \tag{170.7}$$

since $\Delta\mathbf{G} = \Delta\mathcal{E}\hat{\mathbf{n}}$. Hence,

$$\Delta\mathcal{E} = \frac{q^2}{4\pi c^3}\frac{1}{\gamma(1 - \mathbf{v}\cdot\hat{\mathbf{n}})}(\dot{\mathbf{v}}'\times\hat{\mathbf{n}}')^2\Delta\Omega'\Delta t'. \tag{170.8}$$

Next, let us evaluate the squared cross product. We write

$$(\dot{\mathbf{v}}'\times\hat{\mathbf{n}}')^2 = (\dot{\mathbf{v}}')^2 - (\dot{\mathbf{v}}'\cdot\hat{\mathbf{n}}')^2. \tag{170.9}$$

We already know that

$$(\dot{\mathbf{v}}')^2 = \gamma^6\big(\dot{\mathbf{v}}^2 - (\mathbf{v}\times\dot{\mathbf{v}})^2\big) = \gamma^4\dot{\mathbf{v}}^2 + \gamma^6(\dot{\mathbf{v}}\cdot\mathbf{v})^2. \tag{170.10}$$

To evaluate $(\dot{\mathbf{v}}'\cdot\hat{\mathbf{n}}')$, we note that if k^μ is the four-wave vector of the light,

$$\dot{\mathbf{v}}'\cdot\hat{\mathbf{n}}' = -\frac{1}{\omega'}a_\mu k^\mu, \tag{170.11}$$

as may be verified by evaluating the dot product in the comoving frame. We now use the Doppler shift formula and evaluate $a_\mu k^\mu$ in the lab frame, using $k^\mu = \omega(1, \hat{\mathbf{n}})$, and eq. (154.12) for a^μ. This yields

$$\dot{\mathbf{v}}'\cdot\hat{\mathbf{n}}' = -\frac{1}{\gamma(1 - \mathbf{v}\cdot\hat{\mathbf{n}})}\left[\gamma^2(-\dot{\mathbf{v}}\cdot\hat{\mathbf{n}}) + \gamma\dot{\gamma}(1 - \mathbf{v}\cdot\hat{\mathbf{n}})\right]$$

$$= \gamma\frac{\dot{\mathbf{v}}\cdot\hat{\mathbf{n}}}{(1 - \mathbf{v}\cdot\hat{\mathbf{n}})} - \gamma^3(\dot{\mathbf{v}}\cdot\mathbf{v}), \tag{170.12}$$

where in the last term we have written $\dot{\gamma} = \gamma^3(\dot{\mathbf{v}}\cdot\mathbf{v})$. Collecting the last few results together, we get

$$(\dot{\mathbf{v}}'\times\hat{\mathbf{n}}')^2 = \gamma^4\dot{\mathbf{v}}^2 + 2\gamma^4\frac{(\dot{\mathbf{v}}\cdot\hat{\mathbf{n}})(\dot{\mathbf{v}}\cdot\mathbf{v})}{(1 - \mathbf{v}\cdot\hat{\mathbf{n}})} - \gamma^2\frac{(\dot{\mathbf{v}}\cdot\hat{\mathbf{n}})^2}{(1 - \mathbf{v}\cdot\hat{\mathbf{n}})^2}. \tag{170.13}$$

It remains to rewrite $\Delta\Omega'$. Using spherical polar angles with $\mathbf{v}$ along the pole, we have

$$\cos\theta' = \frac{\cos\theta - v}{1 - v\cos\theta}, \quad \varphi' = \varphi. \tag{170.14}$$

Hence,

$$\frac{d\Omega'}{d\Omega} = \frac{d(\cos\theta')}{d(\cos\theta)} = \frac{1 - v^2}{(1 - v\cos\theta)^2} = \frac{1}{\gamma^2(1 - \mathbf{v}\cdot\hat{\mathbf{n}})^2}. \tag{170.15}$$

Feeding eqs. (170.13) and (170.15) into eq. (170.8), we obtain

$$\Delta\mathcal{E} = \frac{q^2}{4\pi c^3}\gamma\left[\frac{\dot{\mathbf{v}}^2}{(1 - \mathbf{v}\cdot\hat{\mathbf{n}})^3} + 2\frac{(\dot{\mathbf{v}}\cdot\hat{\mathbf{n}})(\dot{\mathbf{v}}\cdot\mathbf{v})}{(1 - \mathbf{v}\cdot\hat{\mathbf{n}})^4} - \gamma^{-2}\frac{(\dot{\mathbf{v}}\cdot\hat{\mathbf{n}})^2}{(1 - \mathbf{v}\cdot\hat{\mathbf{n}})^5}\right]\Delta\Omega\Delta t'. \tag{170.16}$$

If we now note that $\gamma\Delta t' = \Delta t_r$, the emission time interval, and divide by $\Delta\Omega\Delta t_r$, we recover eq. (170.5).

The various powers of $(1 - \mathbf{v}\cdot\hat{\mathbf{n}})$ in the denominators in eq. (170.5) mean that the power is greatly enhanced along the forward direction of motion. Once again, we investigate

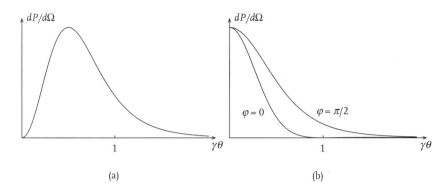

FIGURE 25.1. Angular power distribution for a relativistic particle for (a) $\dot{\mathbf{v}}\|\mathbf{v}$, (b) $\dot{\mathbf{v}} \perp \mathbf{v}$. The vertical scale is not the same for the two cases.

this for the cases of linear and circular motion separately. For linear motion, eq. (170.5) reduces to

$$\frac{dP}{d\Omega} = \frac{q^2\dot{v}^2}{4\pi c^3} \frac{\sin^2\theta}{(1 - v\cos\theta)^5},\tag{170.17}$$

where θ is the angle between $\mathbf{v}$ and $\hat{\mathbf{n}}$. For $v \simeq c$ and small angles,

$$1 - v\cos\theta \approx (1 - v) + \tfrac{1}{2}v\theta^2 \approx \tfrac{1}{2}(\gamma^{-2} + \theta^2).\tag{170.18}$$

Hence,

$$\frac{dP}{d\Omega} \approx \frac{8q^2\dot{v}^2}{\pi c^3} \frac{\theta^2}{(\theta^2 + \gamma^{-2})^5}.\tag{170.19}$$

We plot this in fig. 25.1a. The maximum power is attained for

$$\theta_{\max} = 1/2\gamma,\tag{170.20}$$

and the angular width of the peak is of the same order. Note, though, that there is a hole in the beam, for the power vanishes precisely along the direction of motion.

To analyze the case of circular motion, $\dot{\mathbf{v}} \perp \mathbf{v}$, let us choose spherical polar coordinates with $\mathbf{v}\|\hat{\mathbf{z}}$, $\dot{\mathbf{v}}\|\hat{\mathbf{x}}$. Then

$$\frac{dP}{d\Omega} = \frac{q^2\dot{v}^2}{4\pi c^3} \frac{1}{(1 - v\cos\theta)^3} \left[1 - \frac{\sin^2\theta\cos^2\varphi}{\gamma^2(1 - v\cos\theta)^2}\right],\tag{170.21}$$

$$\approx \frac{2q^2\dot{v}^2}{\pi c^3} \frac{1}{(\theta^2 + \gamma^{-2})^3} \left[1 - \frac{4\gamma^{-2}\theta^2\cos^2\varphi}{(\theta^2 + \gamma^{-2})^2}\right], \quad (\theta \ll 1).\tag{170.22}$$

This time the maximum power is attained at $\theta = 0$, i.e., in the forward direction, and the angular width of the peak is of order γ^{-1} for all φ (see fig. 25.1b). Once again, there are two holes in the power distribution at $\theta = \gamma^{-1}$, for $\varphi = 0$ and $\varphi = \pi$.

The analysis for other relative orientations of $\mathbf{v}$ and $\dot{\mathbf{v}}$ is not particularly illuminating, but it is clear that the radiation will always be concentrated in a cone of angular

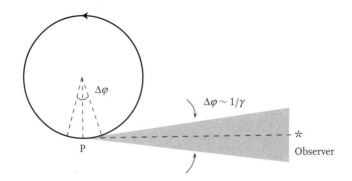

FIGURE 25.2. Diagram for finding the spectral distribution of synchrotron radiation.

width $\sim\gamma^{-1}$ around the forward direction and that there will always be two directions in which the power vanishes.

Exercise 170.1 Verify that integrating eqs. (170.19) and (170.22) over all angles leads to the correct expressions for the total power emission.

Exercise 170.2 For a charge moving in a circle, find the fraction of the power emitted in a cone of half-angle γ^{-1} around the forward direction.

171 Synchrotron radiation—qualitative discussion

Let us now consider a charge moving in a circular orbit, under the influence of a magnetic field, for example. When the charge is ultrarelativistic, the resulting radiation is called *synchrotron radiation*. In addition to man-made synchrotrons there are several astronomical sources of such radiation. The best known, perhaps, is in the Crab nebula, in which a central neutron star produces a strong magnetic field, in which electrons move, emitting radiation up to $\sim25\,\mathrm{GeV}$ in energy. In man-made sources, instead of just one electron, one typically has many bunches simultaneously circulating in a storage ring.

The total power and its angular distribution have been found in the previous two sections. It remains to find its spectral distribution. We shall investigate this in detail in the following sections, but its essential aspects can be found by the following argument. Knowing the answer in advance will make the quantitative analysis easier.

Since the power emitted by the charge is concentrated along the direction of motion, we obtain a narrow beam of angular width $\sim\gamma^{-1}$ that sweeps around the full 360° in the orbital plane as the charge moves in its orbit. An observer located in the plane of the orbit, as shown in fig. 25.2, will thus see short pulses of radiation separated by a time $2\pi/\omega_0$, where ω_0 is the orbital frequency. The radiation in one pulse is emitted from those points on the orbit that are within an angular width $\Delta\varphi \sim \gamma^{-1}$ of the point of tangency, P. Since the charge has angular speed ω_0, the emission occurs over a time interval $\Delta t_e \sim \Delta\varphi/\omega_0$.

This is received by the observer in an interval $\Delta t_{\text{obs}} = (1 - \mathbf{v} \cdot \hat{\mathbf{n}})\Delta t_e$. Since $\mathbf{v}$ is essentially parallel to the direction of emission for the entire duration of the pulse, we can write $(1 - \mathbf{v} \cdot \hat{\mathbf{n}}) \simeq (1 - v) \simeq \frac{1}{2}(1 - v^2) = 1/2\gamma^2$. Thus, $\Delta t_{\text{obs}} \sim 1/\gamma^3 \omega_0$. The spread in frequencies $\Delta \omega$ is given by the inverse of this pulse width, since $\Delta \omega \Delta t_{\text{obs}} \sim 1$. We thus discover that the spectrum extends up to a characteristic frequency

$$\omega_c \sim \gamma^3 \omega_0, \tag{171.1}$$

which is scaled up from ω_0 by the remarkable factor of γ^3.

172 Full spectral, angular, and polarization distribution of synchrotron radiation*

To analyze the spectrum quantitatively, it is convenient to begin with eq. (60.10) for the electric field in the frequency domain:

$$\mathbf{E}_\omega(\mathbf{R}) = \frac{ike^{ikR}}{cR} \mathbf{j}_{\mathbf{k}\omega}^\perp. \tag{172.1}$$

The quantity $\mathbf{j}_{\mathbf{k}\omega}^\perp$ is the projection of $\mathbf{j}_{\mathbf{k}\omega}$ onto the plane perpendicular to the line of sight from the observer to the apparent position of the source, with $\mathbf{j}_{\mathbf{k}\omega}$ being the spatial and temporal Fourier transform of the current distribution at the special wave vector $\mathbf{k} = (\omega/c)\hat{\mathbf{n}}$, with $\hat{\mathbf{n}} = \mathbf{R}/R$. That is, $\mathbf{j}_{\mathbf{k}\omega}^\perp = -\hat{\mathbf{n}} \times (\hat{\mathbf{n}} \times \mathbf{j}_{\mathbf{k}\omega})$, where

$$\mathbf{j}_{\mathbf{k}\omega} = \int\int d^3x \, dt \, \mathbf{j}(\mathbf{r}, t) e^{-i(\mathbf{k}\cdot\mathbf{r} - \omega t)}. \tag{172.2}$$

If the trajectory of the particle is denoted by $\mathbf{r}_0(t)$,

$$\mathbf{j}(\mathbf{r}, t) = q\dot{\mathbf{r}}_0(t)\delta(\mathbf{r} - \mathbf{r}_0(t)). \tag{172.3}$$

Hence,

$$\mathbf{j}_{\mathbf{k}\omega} = q \int_{-\infty}^{\infty} \dot{\mathbf{r}}_0(t) e^{i\Phi(t)} dt, \tag{172.4}$$

where

$$\Phi(t) = \omega\left[t - \frac{1}{c}\hat{\mathbf{n}} \cdot \mathbf{r}_0(t)\right]. \tag{172.5}$$

To evaluate $\mathbf{j}_{\mathbf{k}\omega}$, let us take the orbit to be of radius r_0, and let it lie in the xy plane, with the center at the origin, as shown in fig. 25.3. It suffices to take the observer in the xz plane at a distance $R \gg r_0$. Then, choosing the particle to be at $-r_0\hat{\mathbf{y}}$ at $t = 0$, we have

$$\mathbf{r}_0(t) = (r_0 \sin \omega_0 t, -r_0 \cos \omega_0 t, 0),$$

$$\dot{\mathbf{r}}_0(t) = v(\cos \omega_0 t, \sin \omega_0 t, 0),$$

$$\hat{\mathbf{n}} = (\cos \theta, 0, \sin \theta). \tag{172.6}$$

The angle θ is the same as that in section 170, and $v = \omega_0 r_0$ is the speed of the charge.

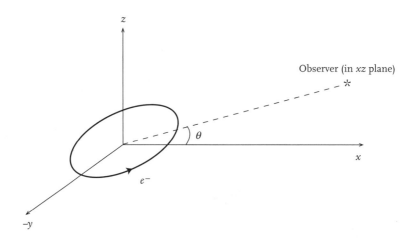

FIGURE 25.3. Perspective drawing to show synchrotron orbit and observer in the xz plane.

We now note that $\mathbf{j}_{\mathbf{k}\omega}$ is the Fourier transform of a periodic function with period $2\pi/\omega_0$, so it will be nonzero only at harmonics of ω_0, as one should expect for a periodic source. It is possible to evaluate these harmonics in closed form in terms of the Bessel function J_1. We have seen, however, that when $\gamma \gg 1$, the spectrum extends to harmonics of order γ^3. It is thus more useful to regard the spectrum as quasicontinuous and find its envelope. This is tantamount to considering only one pulse of radiation seen by the observer. We therefore consider only one passage of the particle through the point $-r_0\hat{\mathbf{y}}$, and cut the time integrals off at $\pm T$, where T is an arbitrary time much shorter than $2\pi/\omega_0$ but much greater than $1/\gamma\omega_0$.[5] In the interval $(-T, T)$, we can approximate

$$\dot{\mathbf{r}}_0(t) \approx v(\hat{\mathbf{x}} + \omega_0 t\hat{\mathbf{y}}). \tag{172.7}$$

The components of $\mathbf{j}_{\mathbf{k}\omega}$ are then given by

$$j_{\mathbf{k}\omega x} = qv \int_{-T}^{T} e^{i\Phi(t)} dt, \tag{172.8}$$

$$j_{\mathbf{k}\omega y} = qv\omega_0 \int_{-T}^{T} t e^{i\Phi(t)} dt. \tag{172.9}$$

In the same way, we can also expand $\Phi(t)$ in powers of t. In doing this, we limit ourselves to very small values of θ, since the power is insignificant otherwise, and replace v by c wherever possible. Then, with some algebra that is now familiar, we obtain

$$\Phi(t) \approx (1 - \frac{v}{c}\cos\theta)\omega t + \frac{v}{6c}\cos\theta\,\omega_0^2\omega t^3$$

$$\approx \frac{1}{2}(\theta^2 + \gamma^{-2})\omega t + \frac{1}{6}\omega_0^2\omega t^3. \tag{172.10}$$

---

[5] Our single-pulse analysis may in fact be more relevant to real synchrotrons. A strictly periodic trajectory would require the phase of the orbiting charge to have no drifts or fluctuations, which is difficult to achieve in practice.

Let us now analyze the integrals (172.8) and (172.9). For $\theta \sim \gamma^{-1}$, the two terms in the phase $\Phi(t)$ are of equal magnitude when $t \sim 1/\gamma\omega_0$, which we recognize as the duration over which one radiation pulse is emitted.[6] For times much larger than this, the integrand oscillates very rapidly, since

$$\frac{d\Phi}{dt} = \frac{1}{2}(\theta^2 + \gamma^{-2})\omega + \frac{1}{2}\omega_0^2\omega t^2, \tag{172.11}$$

and the second term is much greater than the first if $\omega_0 t \gg 1/\gamma$. The resulting positive and negative contributions to the integrals essentially cancel each other. Hence, almost the entire contribution to the integrals comes from times of order $1/\omega_0\gamma$, and the exact value of the cutoff T is irrelevant as long as $\gamma^{-1} \ll \omega_0 T \ll 1$.

To proceed further, let us abbreviate

$$a = \frac{1}{2}\omega_0^2\omega, \quad b = \frac{1}{2}\omega(\theta^2 + \gamma^{-2}), \tag{172.12}$$

so that

$$\Phi(t) = bt + \frac{1}{3}at^3. \tag{172.13}$$

Then

$$j_{\mathbf{k}\omega y} = -i\omega_0 \frac{\partial}{\partial b} j_{\mathbf{k}\omega x}, \tag{172.14}$$

and it suffices to focus on $j_{\mathbf{k}\omega x}$. To evaluate this, it is permissible to extend the integration limits to $\pm\infty$, as the added contribution from large t is exponentially small. The resulting integral is expressible in terms of the standard *Airy function*, defined by[7]

$$\mathrm{Ai}(z) = \frac{1}{\pi} \int_0^\infty \cos\left(\frac{t^3}{3} + zt\right) dt. \tag{172.15}$$

From this, we obtain the useful formula

$$\int_{-\infty}^\infty \exp\left[i\left(\frac{1}{3}at^3 + bt\right)\right] dt = \frac{2\pi}{a^{1/3}}\mathrm{Ai}\left(\frac{b}{a^{1/3}}\right). \tag{172.16}$$

It follows that

$$j_{\mathbf{k}\omega x} = 2\pi q v \left(\frac{2}{\omega\omega_0^2}\right)^{2/3} \mathrm{Ai}(\chi), \tag{172.17}$$

where

$$\chi = (\theta^2 + \gamma^{-2})\left(\frac{\omega}{2\omega_0}\right)^{2/3}. \tag{172.18}$$

The y component of $\mathbf{j}_{\mathbf{k}\omega}$ can now be expressed in terms of the derivative of the Airy function using eq. (172.14). Putting the two pieces together, we get

$$\mathbf{j}_{\mathbf{k}\omega} = \frac{2\pi q v}{\omega_0}\left[\left(\frac{2\omega_0}{\omega}\right)^{1/3}\mathrm{Ai}(\chi)\hat{\mathbf{x}} - i\left(\frac{2\omega_0}{\omega}\right)^{2/3}\mathrm{Ai}'(\chi)\hat{\mathbf{y}}\right]. \tag{172.19}$$

---

[6] The higher-order terms in $\Phi(t)$ are easily seen to be smaller by a factor of γ^2.
[7] Airy functions are discussed in more detail in appendix E.

Let us now find the electric field. For this, we need to project $\mathbf{j}_{k\omega}$ onto the plane perpendicular to $\hat{\mathbf{n}}$. Let $\hat{\mathbf{e}}_\parallel$ be the linear polarization in the plane of the orbit, and $\hat{\mathbf{e}}_\perp$ be orthogonal to it. For small θ, $\hat{\mathbf{e}}_\perp$ is essentially along $\hat{\mathbf{z}}$, perpendicular to the orbital plane. We have

$$-\hat{\mathbf{n}} \times (\hat{\mathbf{n}} \times \hat{\mathbf{y}}) = \hat{\mathbf{e}}_\parallel, \tag{172.20}$$

$$-\hat{\mathbf{n}} \times (\hat{\mathbf{n}} \times \hat{\mathbf{x}}) = -\sin\theta\,\hat{\mathbf{e}}_\perp. \tag{172.21}$$

Hence, specializing to small θ once more, we have

$$\mathbf{E}_\omega = -\frac{2\pi i k q v}{\omega_0 c R}\left[\left(\frac{2\omega_0}{\omega}\right)^{1/3}\theta\,\mathrm{Ai}(\chi)\hat{\mathbf{e}}_\perp + i\left(\frac{2\omega_0}{\omega}\right)^{2/3}\mathrm{Ai}'(\chi)\hat{\mathbf{e}}_\parallel\right]. \tag{172.22}$$

If we take the absolute square of this quantity and multiply by $c R^2/4\pi^2$, we will obtain the angular and spectral distribution of the energy radiated in one pulse. The power distribution is obtained by multiplying by a further factor of $\omega_0/2\pi$, the rate at which pulses are received. Hence,

$$\frac{d^2 P}{d\Omega d\omega} = \frac{q^2\omega^2}{2\pi\omega_0 c}\left[\left(\frac{2\omega_0}{\omega}\right)^{2/3}\theta^2[\mathrm{Ai}(\chi)]^2 + \left(\frac{2\omega_0}{\omega}\right)^{4/3}[\mathrm{Ai}'(\chi)]^2\right], \quad (\parallel, \perp). \tag{172.23}$$

The two terms in the square brackets give the power in the perpendicular and in-plane polarizations. The final parentheses show the order in which the two contributions are written.

The large body of knowledge about the Airy functions makes the results (172.22) and (172.23) quite practical.

Exercise 172.1 From the asymptotic behavior of the Airy functions, estimate the low- and high-frequency distribution of the integrated power in a cone of angle γ^{-1} around the forward direction of the beam. By low frequency we mean $\gamma^3\omega_0 \gg \omega \gg \omega_0$, and by high frequency we mean $\omega \sim \gamma^3\omega_0$.

173 Spectral distribution of synchrotron radiation*

To find the spectral distribution without regard to the emission direction, we must integrate eq. (172.23) over all orientations. This turns out to be involved. It is better to return to the formula (61.17) for the power in terms of the current–current and charge–charge autocorrelation functions,

$$\frac{d^2 P}{d\omega d\Omega} = \frac{\omega^2}{4\pi^2 c^3}\int_{-\infty}^{\infty} d\tau\left[\langle\mathbf{j}_k(t+\tfrac{1}{2}\tau)\cdot\mathbf{j}_k^*(t-\tfrac{1}{2}\tau)\rangle - c^2\langle\rho_k(t+\tfrac{1}{2}\tau)\rho_k^*(t-\tfrac{1}{2}\tau)\rangle\right]e^{i\omega\tau}. \tag{173.1}$$

The angular brackets now denote an average over one time period, $2\pi/\omega_0$. Further,

$$\mathbf{j}_k(t) = q\mathbf{v}(t)e^{-i\mathbf{k}\cdot\mathbf{r}_0(t)}, \quad \rho_k(t) = q e^{-i\mathbf{k}\cdot\mathbf{r}_0(t)}. \tag{173.2}$$

To save writing, we denote

$$t \pm \tfrac{1}{2}\tau = t_{1,2}, \quad \mathbf{r}_0(t \pm \tfrac{1}{2}\tau) = \mathbf{r}_{1,2}, \quad \mathbf{r}_{12} = \mathbf{r}_1 - \mathbf{r}_2. \tag{173.3}$$

Then,

$$\frac{d^2 P}{d\omega d\Omega} = \frac{\omega^2 q^2}{4\pi^2 c^3} \int_{-\infty}^{\infty} d\tau \left\langle \left(\mathbf{v}(t_1) \cdot \mathbf{v}(t_2) - c^2 \right) e^{-i(\omega/c)\hat{\mathbf{n}} \cdot \mathbf{r}_{12}} \right\rangle e^{i\omega\tau}. \tag{173.4}$$

The advantage of using the difference of the current–current and charge–charge correlation functions over the autocorrelation of the transverse part of the current is now apparent. We can perform the integration over angles, although it pays to not do so completely just yet. We denote the azimuthal angle about $\mathbf{r}_{12}$ by φ, and

$$\mu = \hat{\mathbf{n}} \cdot \hat{\mathbf{r}}_{12}. \tag{173.5}$$

Then $d\Omega = d\varphi\, d\mu$, the integration over φ is immediate, and we obtain

$$\int d\Omega\, e^{-i(\omega/c)\hat{\mathbf{n}} \cdot \mathbf{r}_{12}} = 2\pi \int_{-1}^{1} d\mu\, e^{-i\mu\omega r_{12}/c}. \tag{173.6}$$

The next step is to average over one period. We observe (by simple geometry) that[8]

$$\mathbf{v}(t_1) \cdot \mathbf{v}(t_2) = v^2 \cos(\omega_0\tau), \tag{173.7}$$

$$r_{12} = |\mathbf{r}(t_1) - \mathbf{r}(t_2)| = 2r_0 \sin(\omega_0\tau/2). \tag{173.8}$$

These expressions are both independent of t, so the averaging is simple, and we get

$$\frac{dP}{d\omega} = \frac{\omega^2 q^2}{2\pi c} \int_{-\infty}^{\infty} d\tau \int_{-1}^{1} d\mu \left(\frac{v^2}{c^2} \cos\omega_0\tau - 1 \right) e^{i\omega[\tau - \mu r_{12}(\tau)/c]}. \tag{173.9}$$

Since r_{12} and $\cos\omega_0\tau$ are periodic in τ with a frequency ω_0, the integral over τ is nonzero only for integer ω/ω_0. As before, however, we are interested in the case $v \simeq c$, when the spectrum extends to very high harmonics, and we may seek it in quasicontinuous form. Exactly as in the evaluation of the electric field, for $\omega \gg \omega_0$, only the very small values of τ contribute to the integral. We can thus expand the $\cos\omega_0\tau$ factor and the phase factor in the exponent in powers of τ and retain only the first two terms. We define

$$\Phi(\tau) \equiv \omega\left(\tau - \mu \frac{r_{12}(\tau)}{c} \right) \approx b\tau + \frac{1}{3}a\tau^3, \tag{173.10}$$

where

$$b = \left(1 - \mu \frac{v}{c} \right)\omega, \quad a = \frac{1}{8c}\mu r_0 \omega_0^3 \omega. \tag{173.11}$$

(Note that Φ and the a and b parameters are not the same as in the previous section.) Similarly,

$$\left(\frac{v^2}{c^2} \cos\omega_0\tau - 1 \right) \approx -\left(\gamma^{-2} + \frac{1}{2}\omega_0^2\tau^2 \right), \tag{173.12}$$

---

[8] Strictly speaking, we should put absolute value signs on the right-hand side in the formula for r_{12}, but this is unnecessary, as the integral over μ is an even function of r_{12}.

where we have put $v = c$ in the coefficient of the τ^2 term. In this way, we obtain

$$
\frac{dP}{d\omega} = \frac{\omega^2 q^2}{2\pi c} \int_{-1}^{1} d\mu \int_{-\infty}^{\infty} d\tau \left(\gamma^{-2} + \frac{1}{2} \omega_0^2 \tau^2 \right) \exp \left[i \left(\frac{1}{3} a \tau^3 + b\tau \right) \right].
\tag{173.13}
$$

The integral over τ can again be expressed in terms of Airy functions. Before we do so, however, we further simplify it by noting that the phase will continue to oscillate rapidly unless μ is very close to unity. We therefore write

$$
\zeta = 1 - \mu
\tag{173.14}
$$

and make a second expansion in ζ, keeping only the leading term. At the same time, we put $v = c$ wherever possible. With these simplifications, we have

$$
a \approx \frac{1}{8} \omega_0^2 \omega, \quad b \approx \frac{1}{2} (2\zeta + \gamma^{-2}) \omega,
\tag{173.15}
$$

and the argument of the Airy function is given by

$$
s \equiv \frac{b}{a^{1/3}} = (2\zeta + \gamma^{-2}) \left(\frac{\omega}{\omega_0} \right)^{2/3}.
\tag{173.16}
$$

To evaluate the τ integral in eq. (173.13), we need eq. (172.16) and, in addition, the integral of $\tau^2 e^{i\Phi}$. The latter is found by noting that[9]

$$
\int_{-\infty}^{\infty} (a\tau^2 + b) \exp \left[i \left(\frac{1}{3} a\tau^3 + b\tau \right) \right] d\tau = -i \int_{-\infty}^{\infty} \frac{d}{d\tau} \exp \left[i \left(\frac{1}{3} a\tau^3 + b\tau \right) \right] d\tau = 0.
\tag{173.17}
$$

Using this result and the basic formula (172.16) for the Airy function in eq. (173.13), we obtain

$$
\frac{dP}{d\omega} = \frac{2\omega q^2}{c} \left(\frac{\omega}{\omega_0} \right)^{2/3} \int_0^{\infty} [2(2\zeta + \gamma^{-2}) - \gamma^{-2}] \, \text{Ai}(s) \, d\zeta.
\tag{173.18}
$$

In this equation, s is given by eq. (173.16), and we have extended the upper limit of the ζ integral to infinity, as it is exponentially convergent at that limit. We now change the variable of integration to s, and define

$$
u = s(\zeta = 0) = \left(\frac{\omega}{\gamma^3 \omega_0} \right)^{2/3}.
\tag{173.19}
$$

Then, making use of the result (which follows directly from the differential equation for $\text{Ai}(s)$)

$$
\int_u^{\infty} s \, \text{Ai}(s) \, ds = \int_u^{\infty} \text{Ai}''(s) \, ds = -\text{Ai}'(u),
\tag{173.20}
$$

---

[9] We are being cavalier with the convergence of our integrals in fine physics tradition. The procedures can be justified by first deforming the contour of the τ integral so that it approaches infinity along the directions at $\pi/6$ and $5\pi/6$ to the real axis and then performing the small τ expansion of the factor multiplying $e^{i\Phi}$.

we obtain

$$\frac{dP}{d\omega} = \frac{\omega_0 q^2}{c} \left(\frac{\omega}{\omega_0}\right)^{1/3} \int_u^\infty (2s - u) \text{Ai}(s) ds$$

$$= -\frac{2\omega_0 q^2}{c} \left(\frac{\omega}{\omega_0}\right)^{1/3} \left[\text{Ai}'(u) + \frac{1}{2} u \int_u^\infty \text{Ai}(s) ds\right]. \tag{173.21}$$

This is the final form of the answer.

From eq. (173.21), we can obtain limiting forms at low and high frequencies. For $\gamma^3 \omega_0 \gg \omega \gg \omega_0$, $u \ll 1$, and the quantity in square brackets reduces to $\text{Ai}'(0) \simeq -0.259$. Hence,

$$\frac{dP}{d\omega} \approx 0.52 \frac{\omega_0 q^2}{c} \left(\frac{\omega}{\omega_0}\right)^{1/3}, \quad 1 \ll \omega/\omega_0 \ll \gamma^3. \tag{173.22}$$

In fact, numerics show that this expression has an error of only 15% even for $\omega = \omega_0$. At high frequencies, $\omega \sim \gamma^3 \omega_0$. Using the asymptotic forms for Ai and Ai', we get

$$\frac{dP}{d\omega} \approx \frac{\omega_0 q^2}{2c} \left(\frac{\omega}{\pi \gamma \omega_0}\right)^{1/2} \exp\left[-\frac{2}{3} \frac{\omega}{\gamma^3 \omega_0}\right]. \tag{173.23}$$

We thus see that the characteristic scale of the spectrum is $\gamma^3 \omega_0$, as deduced by the earlier qualitative argument.

Exercise 173.1 Using Eqs. (E.16), (E.19), and (E.18), show that we can also write

$$\frac{dP}{d\omega} = \frac{\sqrt{3} \gamma \omega_0 q^2}{2\pi c} F(\omega/\omega_c); \quad F(\xi) = \xi \int_\xi^\infty K_{5/3}(z) dz, \tag{173.24}$$

with

$$\omega_c = \tfrac{3}{2} \gamma^3 \omega_0. \tag{173.25}$$

This form of the answer is also commonly found and is better for numerical purposes. Figure 25.4 shows a graph of $F(\xi)$.

174 Angular distribution and polarization of synchrotron radiation*

Let us now integrate the full power distribution, eq. (172.23), with respect to the frequency, to obtain its angular distribution. To do this, we write $\text{Ai}(\chi)$ and $\text{Ai}'(\chi)$ in terms of time integrals such as eq. (172.16) and further define quantities $\tilde{\Phi}$, $\tilde{a}$, and $\tilde{b}$ that are obtained from the corresponding quantities without tildes (eqs. (172.10) and (172.12)) by dividing by ω. That is,

$$\tilde{\Phi} = \frac{1}{3} \tilde{a} t^3 + \tilde{b} t, \quad \tilde{a} = \frac{1}{2} \omega_0^2, \quad \tilde{b} = \frac{1}{2}(\theta^2 + \gamma^{-2}). \tag{174.1}$$

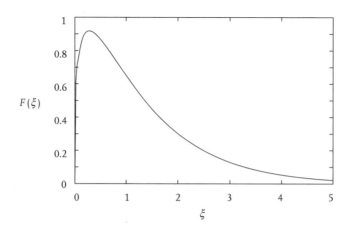

FIGURE 25.4. The function $F(\xi)$, which gives the scaled synchrotron frequency spectrum.

We further denote $\tilde{\Phi}(t_1)$ by $\tilde{\Phi}_1$, etc. Then, it is straightforward to show that

$$\frac{d\,P_\parallel}{d\Omega} = \alpha_\parallel \int_{-\infty}^{\infty}\int_{-\infty}^{\infty} dt_1\,dt_2\,t_1 t_2 \int_0^{\infty} \frac{d\omega}{\pi}\omega^2 e^{i(\tilde{\Phi}_1 - \tilde{\Phi}_2)\omega}, \tag{174.2}$$

$$\frac{d\,P_\perp}{d\Omega} = \alpha_\perp \theta^2 \int_{-\infty}^{\infty}\int_{-\infty}^{\infty} dt_1\,dt_2 \int_0^{\infty} \frac{d\omega}{\pi}\omega^2 e^{i(\tilde{\Phi}_1 - \tilde{\Phi}_2)\omega}, \tag{174.3}$$

where

$$\alpha_\parallel = \frac{q^2\omega_0^3 v^2}{8\pi^2 c^3}, \qquad \alpha_\perp = \frac{q^2\omega_0 v^2}{8\pi^2 c^3}. \tag{174.4}$$

The frequency integrals can now be evaluated:

$$\int_0^{\infty} \frac{d\omega}{\pi}\omega^2 e^{i(\tilde{\Phi}_1 - \tilde{\Phi}_2)\omega} = -\delta''(\tilde{\Phi}_1 - \tilde{\Phi}_2). \tag{174.5}$$

This function is nonvanishing only when $t_1 = t_2$, since $\tilde{\Phi}(t)$ is a monotonic function of t. We can use it to evaluate the t_1 integral by employing the identity (9.44). We have

$$\tilde{\Phi}' = \tilde{a}t^2 + \tilde{b}, \quad \tilde{\Phi}'' = 2\tilde{a}t, \quad \tilde{\Phi}''' = 2\tilde{a}. \tag{174.6}$$

Hence,

$$\delta''(\tilde{\Phi}_1 - \tilde{\Phi}_2) = \frac{1}{(\tilde{a}t_2^2 + \tilde{b})^3}\delta''(t_1 - t_2) + \frac{6\tilde{a}t_2}{(\tilde{a}t_2^2 + \tilde{b})^4}\delta'(t_1 - t_2) + \left[\frac{12\tilde{a}^2 t_2^2}{(\tilde{a}t_2^2 + \tilde{b})^5} - \frac{2\tilde{a}}{(\tilde{a}t_2^2 + \tilde{b})^4}\right]\delta(t_1 - t_2). \tag{174.7}$$

It follows that

$$\int_{-\infty}^{\infty} dt_2 \int_{-\infty}^{\infty} dt_1 \delta''(\tilde{\Phi}_1 - \tilde{\Phi}_2) = \int_{-\infty}^{\infty} dt_2 \left[\frac{12\tilde{a}^2 t_2^2}{(\tilde{a}t_2^2 + \tilde{b})^5} - \frac{2\tilde{a}}{(\tilde{a}t_2^2 + \tilde{b})^4}\right]$$

$$= -\frac{\tilde{a}}{\tilde{b}^4}\left(\frac{\tilde{b}}{\tilde{a}}\right)^{1/2}\frac{5\pi}{32}. \tag{174.8}$$

Likewise,

$$
\int_{-\infty}^{\infty} dt_2 t_2 \int_{-\infty}^{\infty} dt_1 t_1 \delta''(\tilde{\Phi}_1 - \tilde{\Phi}_2) = \int_{-\infty}^{\infty} dt_2\, t_2 \left(\frac{-6\tilde{a}t_2}{(\tilde{a}t_2^2 + \tilde{b})^4} + \left[\frac{12\tilde{a}^2 t_2^2}{(\tilde{a}t_2^2 + \tilde{b})^5} - \frac{2\tilde{a}}{(\tilde{a}t_2^2 + \tilde{b})^4} \right] t_2 \right)
$$

$$
= \int_{-\infty}^{\infty} dt_2 \left[\frac{12\tilde{a}^2 t_2^4}{(\tilde{a}t_2^2 + \tilde{b})^5} - \frac{8\tilde{a}t_2^2}{(\tilde{a}t_2^2 + \tilde{b})^4} \right]
$$

$$
= -\frac{1}{\tilde{b}^3} \left(\frac{\tilde{b}}{\tilde{a}} \right)^{1/2} \frac{7\pi}{32}. \tag{174.9}
$$

Collecting together all these results, we finally obtain

$$
\frac{dP}{d\Omega} = \frac{q^2 \omega_0^2 v^2}{32\pi c^3} \left[\frac{7}{(\theta^2 + \gamma^{-2})^{5/2}} + \frac{5\theta^2}{(\theta^2 + \gamma^{-2})^{7/2}} \right], \quad (\|, \perp). \tag{174.10}
$$

Again, the two terms in the square brackets correspond to $\|$ and $\perp$ polarizations, respectively.

From the last result, we can obtain the total power in the two polarizations separately by integrating over angles. In doing this, we must recall that θ is the latitude, not the polar angle, so in the small θ limit, the measure is just $d\theta$, and we can make the replacement

$$
\int d\Omega = \int_{-\infty}^{\infty} d\theta \int_0^{2\pi} d\varphi. \tag{174.11}
$$

Performing the integrations, we obtain

$$
P = 2\pi \frac{q^2 \omega_0^2 v^2}{32\pi c^3} \left[\frac{28}{3} \gamma^4 + \frac{4}{3} \gamma^4 \right], \quad (\|, \perp). \tag{174.12}
$$

Recalling that $\omega_0 v = \dot{v}$, we obtain

$$
P = \frac{2q^2}{3c^3} \gamma^4 \dot{v}^2, \tag{174.13}
$$

exactly as found before. Further, we see that the light is heavily polarized in the plane of the orbit, as

$$
\frac{P_\|}{P_\perp} = 7. \tag{174.14}
$$

175 Undulators and wigglers*

If a relativistic charge is placed in a magnetic field transverse to its velocity, and the magnetic field alternates in sign along the direction of the particle's velocity, the charge will undergo a sinuous motion in which it snakes to the left and right as it moves forward (fig. 25.5). The resulting acceleration will be essentially transverse to this forward direction, and the particle will emit radiation that is highly collimated along this same direction and strongly polarized in the plane of motion. Devices that produce such

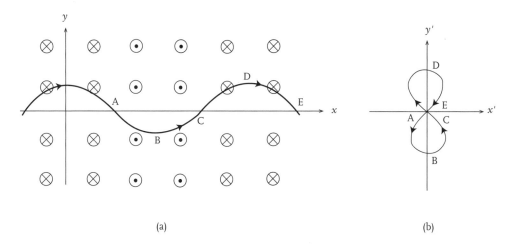

FIGURE 25.5. Particle trajectory in a periodic undulator or wiggler magnetic field, as seen in (a) the lab frame; (b) the average rest frame. Successive points along the trajectory are marked A through E. The excursions from the forward direction of motion are greatly exaggerated.

radiation are made by inserting a periodic array of alternating magnets in a synchrotron, and are known as undulators and wigglers. They can increase the brilliance of the X-rays over the bare synchrotron radiation significantly. The distinction between these two devices lies in the so-called K parameter. If the maximum angular deviation of the particle velocity from its forward direction is φ_0, then

$$K \equiv \gamma\varphi_0. \tag{175.1}$$

Since the radiation is concentrated in a cone of angle $\sim\gamma^{-1}$, this quantity is a scaled angular deviation. In a wiggler, $K \gg 1$, so an observer along the forward direction will see the beam of radiation rapidly flick to the left and right with a period determined by the spatial period or wavelength of the magnet structure. In an undulator, by contrast, $K \lesssim 1$, so the beam direction will be essentially constant, and the observer will see a coherent superposition of the radiation from different spatial locations along the particle beam.

The first step in analyzing the radiation is to understand the motion of the particle. We take the forward direction to be $\hat{\mathbf{x}}$, and approximate the magnetic field by

$$\mathbf{B} = B_u \cos k_u x\hat{\mathbf{z}}, \tag{175.2}$$

with $k_u = 2\pi/\lambda_u$, where λ_u is the spatial period. (The $\nabla \cdot \mathbf{B} = 0$ condition then requires $\mathbf{B}$ to have a z-dependent x component, but this is assumed to vanish in the xy plane, to which the motion is confined. We also neglect the higher spatial harmonics of the field.) Let the average velocity of the particle be $\bar{v}\hat{\mathbf{x}}$, and let $\bar{\gamma} = (1 - \bar{v}^2)^{-1/2}$. The motion is most simply analyzed in a frame traveling with this velocity, in which the particle is at rest on the average. Denoting quantities in this frame by primes, the electromagnetic fields are easily found to be

$$\mathbf{E}' \approx -\bar{\gamma}\bar{v}\,B_u \cos k_u x\hat{\mathbf{y}}, \quad \mathbf{B}' = \bar{\gamma}B_u \cos k_u x\hat{\mathbf{z}}. \tag{175.3}$$

Since, $\bar{v} \simeq 1$, $|\mathbf{E}'| \simeq |\mathbf{B}'|$ to very good approximation. Further, since

$$k_u x \approx k_u \bar{\gamma}(x' + t') \tag{175.4}$$

by the same token, and $\mathbf{E}' \perp \mathbf{B}'$, the fields are essentially those of an electromagnetic plane wave of wave number and frequency

$$\bar{k} = \bar{\omega} = \bar{\gamma} k_u, \tag{175.5}$$

traveling along the $-x$ direction. We can therefore use the results of section 167 by relabeling the directions thus: $\hat{\mathbf{z}} \to -\hat{\mathbf{x}}$, $\hat{\mathbf{x}} \to -\hat{\mathbf{y}}$, and $\hat{\mathbf{y}} \to \hat{\mathbf{z}}$. With ϵ as the constant of motion defined by

$$\epsilon = p'^0 + p'^x, \tag{175.6}$$

we have

$$x' = \frac{q^2 B_u^2}{8\epsilon^2 \bar{\gamma} k_u^3} \sin(2\epsilon\bar{\omega}\tau/m), \tag{175.7}$$

$$y' = \frac{q B_u}{\epsilon \bar{\gamma} k_u^2} \cos(\epsilon\bar{\omega}\tau/m), \tag{175.8}$$

$$t' = \frac{\epsilon}{m}\tau - \frac{q^2 B_u^2}{8\epsilon^2 \bar{\gamma} k_u^3} \sin(2\epsilon\bar{\omega}\tau/m). \tag{175.9}$$

The motion is confined to the xy plane: $z' = 0$ at all times. In the average rest frame, the particle executes a figure-eight curve, with the lobes of the 8 transverse to the main direction of motion (see fig. 25.5).

In the lab frame, the motion is now found to be (setting $\bar{v} = 1$ wherever possible)

$$x = t \approx \frac{\epsilon}{m}\bar{\gamma}\tau, \tag{175.10}$$

$$y = \frac{q B_u}{\epsilon \bar{\gamma} k_u^2} \cos(k_u t), \tag{175.11}$$

where we used the first equation to transform the argument of the cosine in the second. The trajectory is a sinusoid to first approximation, and, as expected, $y(t)$ has the same period as the magnetic field.

Let us now express the energy-like quantities ϵ and $\bar{\gamma}$, and the K parameter in terms of the parameters of the problem. Using eq. (167.29), we get

$$\epsilon^2 = m^2 + \frac{q^2}{2\bar{\omega}^2}(\bar{\gamma} B_u)^2 = m^2 + \frac{q^2 B_u^2}{2k_u^2}. \tag{175.12}$$

Next, we note that since $t = \gamma\tau$, eq. (175.10) for t implies

$$\bar{\gamma} = \frac{m\gamma}{\epsilon}. \tag{175.13}$$

Finally, from eqs. (175.10) and (175.11), we get

$$K = \gamma \left(\frac{dy}{dx}\right)_{y=0} = \frac{\gamma q B_u}{\epsilon \bar{\gamma} k_u} = \frac{q B_u}{mc^2 k_u}, \tag{175.14}$$

where we have restored c in the last form. Thus, K is proportional to both the magnetic field strength B_u and its spatial period $\lambda_u = 2\pi/k_u$. Typical values for wigglers are $B_u = 1\,\mathrm{T}$, $\lambda_u = 20\,\mathrm{cm}$, leading to $K \sim 10$. Undulators have smaller periods. Taking $B_u = 1\,\mathrm{T}$, $\lambda_u = 3\,\mathrm{cm}$ as typical, we get $K \sim 2$.

We can express ϵ and $\bar{\gamma}$ even more simply in terms of K:

$$\epsilon = m \left(1 + \tfrac{1}{2} K^2\right)^{1/2}, \tag{175.15}$$

$$\bar{\gamma} = \gamma \left(1 + \tfrac{1}{2} K^2\right)^{-1/2}. \tag{175.16}$$

If we recall that ϵ is the average energy of the particle in the average rest frame, it follows from eq. (175.15) that the motion in this frame will be nonrelativistic or only weakly relativistic if $K = O(1)$ (undulators), while it will be highly relativistic if $K \gg 1$ (wigglers).

We now discuss the spectrum and angular distribution of the radiation. We consider wigglers first. In this case, as already noted, an observer will see an X-ray beam that flicks back and forth. If there are N pairs of oppositely oriented magnets in the wiggler, the observer will see $2N$ pulses of radiation, each of which will have the essentially the same power spectrum as a synchrotron, and the effect of the magnets will be to increase the intensity by a factor $\sim 2N$. To find the frequency scale more quantitatively, let us write the particle trajectory as

$$y(t) = a \cos(k_u x), \tag{175.17}$$

where

$$a = \frac{q\,B_u}{\epsilon \bar{\gamma} k_u^2} = \frac{K}{\gamma}\frac{\lambda_u}{2\pi} \tag{175.18}$$

is the amplitude of the transverse motion. Denoting the minimum radius of curvature of trajectory by R, we have

$$\frac{1}{R} = k_u^2 a = \frac{q\,B_u}{m\gamma}. \tag{175.19}$$

The characteristic frequency scale of the radiation is therefore

$$\omega_c \simeq \gamma^3 \frac{1}{R} = \gamma^2 K(k_u c) = \gamma^2 \frac{q\,B_u}{mc}. \tag{175.20}$$

Since the curvature is determined entirely by the strength of the magnetic field and the particle's γ factor, it is quite comparable to the value for the bare synchrotron and so is the frequency scale.

For an undulator, the above argument is invalid, as the radiation beam direction is essentially unchanging, and one must coherently superpose the radiation fields from different points along the particle path. This superposition can shift the spectral weight around, and, a priori, the resulting power spectrum could be rather different from that of the synchrotron.

To understand the spectrum further, let us transform to the average rest frame of the particle. In this frame, the amplitude of the longitudinal motion relative to the transverse motion is given by

$$
\frac{x'_{max}}{a} = \frac{q \, B_u}{8\epsilon k_u} = \frac{1}{8} K \left(1 + \frac{1}{2} K^2\right)^{-1/2}.
\tag{175.21}
$$

This ratio is very small for $K \ll 1$ (it equals ~ 0.1 for $K = 1$). Thus, for an undulator, the longitudinal motion can be neglected to first approximation. Since $t' \approx \epsilon\tau/m$ in the same approximation, the remaining transverse motion is simple harmonic and generates pure dipole radiation at a frequency $\overline{\omega}$. In the next order, the longitudinal motion gives a small quadrupole contribution at $2\overline{\omega}$. For a wiggler, however, since $K \gg 1$, all multipoles are important, and the problem must be analyzed as that of a periodically moving charge. The radiation in the average rest frame will consist of harmonics of $\overline{\omega}$, extending up to $\sim K^3 \overline{\omega}$.[10] However, because the motion is relativistic, the long-wavelength multipole expansion is inapplicable, and one must use the exact Fourier decomposition of the current to find the power in a given harmonic. The analysis involved is unilluminating, so we omit it and make do with a qualitative understanding.

We can be more quantitative for undulators, as the dipole approximation is often quite adequate. The radiation from a linearly oscillating charge was found in section 62. Transcribing the results to the present problem, we find that in the average rest frame the angular power distribution is given by

$$
\frac{d P'}{d\Omega'} = \frac{q^2 a^2 \overline{\omega}^4}{8\pi c^3} (\hat{\mathbf{n}}' \times \hat{\mathbf{y}})^2.
\tag{175.22}
$$

In addition, the radiation is monochromatic, at frequency $\overline{\omega}$.

The angular distribution in the lab frame can now be found by applying a boost. From eq. (175.22), we see that the number of photons emitted in a time interval $\Delta t'$ and solid angle $\Delta\Omega'$ around the direction $\hat{\mathbf{n}}'$ is given by

$$
\Delta N_{ph} = \frac{1}{\hbar\overline{\omega}} \frac{q^2 a^2 \overline{\omega}^4}{8\pi c^3} (\hat{\mathbf{n}}' \times \hat{\mathbf{y}})^2 (\Delta t')(\Delta\Omega').
\tag{175.23}
$$

We now use the results that $\Delta t' = \Delta t/\overline{\gamma}$, and eq. (170.15) for $\Delta\Omega'/\Delta\Omega$, to write

$$
\Delta N_{ph} = \frac{q^2 a^2 \overline{\omega}^3}{8\pi\hbar c^3} (\hat{\mathbf{n}}' \times \hat{\mathbf{y}})^2 \frac{1}{\overline{\gamma}^3 (1 - \overline{\mathbf{v}} \cdot \hat{\mathbf{n}})^2} (\Delta t)(\Delta\Omega).
\tag{175.24}
$$

To think of these photons entirely in the lab frame, two points must be kept in mind. First, at an observation direction $\hat{\mathbf{n}}$, their frequency is Doppler shifted to

$$
\omega_0(\hat{\mathbf{n}}) = \frac{\overline{\omega}}{\overline{\gamma}(1 - \overline{\mathbf{v}} \cdot \hat{\mathbf{n}})}.
\tag{175.25}
$$

Second, the emission directions in the two frames, $\hat{\mathbf{n}}'$ and $\hat{\mathbf{n}}$, are related by the formula for aberration, which is more easily written in terms of spherical polar coordinates. If we take the polar direction as $\hat{\mathbf{x}}$ and measure the azimuth from $\hat{\mathbf{y}}$ (in the horizontal plane),

---

[10] Note that precisely on axis ($\theta = 0$), only odd harmonics are present.

we have

$$1 - \overline{\mathbf{v}} \cdot \hat{\mathbf{n}} = 1 - \overline{v} \cos \theta, \tag{175.26}$$

and

$$\cos \theta' = \frac{\cos \theta - \overline{v}}{1 - \overline{v} \cos \theta}, \qquad \sin \theta' = \frac{\sin \theta}{\overline{\gamma}(1 - \overline{v} \cos \theta)}, \qquad \varphi' = \varphi. \tag{175.27}$$

With these points in mind, the angular power distribution in the lab frame is given by

$$\frac{dP}{d\Omega} = \frac{\hbar \omega_0(\hat{\mathbf{n}}) \Delta N_{\mathrm{ph}}}{\Delta t \, \Delta \Omega}. \tag{175.28}$$

That is,

$$\frac{dP}{d\Omega} = \frac{q^2 a^2 \overline{\omega}^4}{8 \pi c^3} \frac{(\hat{\mathbf{n}}' \times \hat{\mathbf{y}})^2}{\overline{\gamma}^4 (1 - \overline{\mathbf{v}} \cdot \hat{\mathbf{n}})^3}. \tag{175.29}$$

Since the radiation is highly concentrated along the $\hat{\mathbf{x}}$ direction, we find it only for small θ. Writing $\cos \theta = 1 - \theta^2/2$, $\overline{v} = 1 - \overline{\gamma}^{-2}/2$, and keeping only lowest orders in θ and $\overline{\gamma}^{-1}$, we have

$$1 - \overline{v} \cos \theta = \tfrac{1}{2}(\overline{\gamma}^{-2} + \theta^2). \tag{175.30}$$

Similarly,

$$\begin{aligned} (\hat{\mathbf{n}}' \times \hat{\mathbf{y}})^2 &= \cos^2 \theta' + \sin^2 \theta' \sin^2 \varphi' \\ &= \frac{(1 - \overline{\gamma}^2 \theta^2)^2 + 4\overline{\gamma}^2 \theta^2 \sin^2 \varphi}{(1 + \overline{\gamma}^2 \theta^2)^2}, \end{aligned} \tag{175.31}$$

and the final result for the lab frame angular distribution is

$$\frac{dP}{d\Omega} = \frac{q^2 a^2 \overline{\gamma}^2 \overline{\omega}^4}{\pi c^3} \left[\frac{(1 - \overline{\gamma}^2 \theta^2)^2 + 4\overline{\gamma}^2 \theta^2 \sin^2 \varphi}{(1 + \overline{\gamma}^2 \theta^2)^2} \right]. \tag{175.32}$$

This small-angle form correctly shows that there are two holes in the radiation pattern: at $\theta = \overline{\gamma}^{-1}$, $\varphi = 0, \pi$.

The total power is now found by integrating eq. (175.32). We write $d\Omega = \theta \, d\theta \, d\varphi$, and extend the upper limit on θ to infinity. Then,

$$P_{\mathrm{tot}} = \frac{1}{3} \frac{q^2 a^2 \overline{\omega}^4}{c^3} = \frac{1}{3} c q^2 k_u^2 \gamma^2 K^2 \left(1 + \frac{1}{2} K^2 \right)^{-2}. \tag{175.33}$$

The same result is found by integrating the angular distribution in the average rest frame.

Finally, let us find the frequency spectrum in the lab frame. To do this, we first write the full frequency and angular distribution of the power. Since the radiation in the direction $\hat{\mathbf{n}}$ takes place at the single frequency $\omega_0(\hat{\mathbf{n}})$, eq. (175.32) implies that

$$\frac{d^2 P}{d\Omega d\omega} = \frac{q^2 a^2 \overline{\gamma}^2 \overline{\omega}^4}{\pi c^3} \left[\frac{(1 - \overline{\gamma}^2 \theta^2)^2 + 4\overline{\gamma}^2 \theta^2 \sin^2 \varphi}{(1 + \overline{\gamma}^2 \theta^2)^2} \right] \delta \left(\omega - \omega_0(\hat{\mathbf{n}}) \right). \tag{175.34}$$

For small θ,

$$\omega_0(\hat{\mathbf{n}}) = 2k_u c \frac{\bar{\gamma}^2}{1 + \bar{\gamma}^2 \theta^2}. \tag{175.35}$$

Thus, the spectrum extends from zero frequency up to a maximum

$$\omega_{\text{max}} = 2(k_u c)\bar{\gamma}^2, \tag{175.36}$$

which is somewhat different from the expression for the synchrotron frequency scale.

If we feed eq. (175.35) into eq. (175.34), the integration over angles is easy to perform. We leave it as an exercise for the reader to do this and show that

$$\frac{dP}{d\omega} = \frac{3 P_{\text{tot}}}{\omega_{\text{max}}} \xi(1 - 2\xi + 2\xi^2) \, \Theta(1 - \xi), \tag{175.37}$$

where Θ is the step function, and

$$\xi = \frac{\omega}{\omega_{\text{max}}}. \tag{175.38}$$

In contrast to synchrotron radiation, the undulator power spectrum is sharply cut off at ω_{max}. (The higher harmonics must be remembered, however.) Half the power is emitted from $\sim 0.77 \omega_{\text{max}}$ to ω_{max}.

To avoid misunderstanding, we must add a small caveat to the above discussion. The radiation is not purely monochromatic even at a given observation direction. In the average rest frame, the charge does not oscillate forever. Rather, if there are $2N$ magnets, so that one has N periods of oscillation, the total duration of the oscillation is $\sim 2\pi N/\bar{\omega}$. Thus, the radiation is spread out into a narrow band of relative width $1/N$. The same relative width is therefore seen in the lab frame. Since $N \sim 100$ in a typical undulator, this broadening is much smaller than that due to the θ-dependent Doppler shift and may be neglected. In fact, in any experiment, the radiation must be collected over a nonzero solid angle, and the resulting broadening is much greater than the $1/N$ broadening even in experiments with the smallest angular acceptances.

The same argument shows that at a fixed observation angle, and especially for $\theta = 0$, the intensity (not the power) of the radiation scales as N^2, provided one can maintain coherence over the entire extent of the magnet array.

A second corrective point is that the radiation from a wiggler is not just a more intense version of synchrotron radiation. For a definite direction of observation $\hat{\mathbf{n}}$, one sees a series of sharply defined harmonics of $\omega_0(\hat{\mathbf{n}})$, as opposed to the broad synchrotron spectrum. Only the angle-integrated spectra are similar.

We conclude by mentioning that users of X-ray light sources generally use two other figures of merit instead of $d^2 P/d\Omega d\omega$. The first, known as the *brightness*, is defined as the number of photons emitted per unit time per unit solid angle per unit relative frequency range. That is,

$$\text{Brightness} = \frac{N_{\text{ph}}}{\Delta t \, \Delta \Omega \, \Delta \omega/\omega}. \tag{175.39}$$

The second figure of merit is the *brilliance*, defined as

$$\text{Brilliance} = \text{Brightness}/\Sigma_x \Sigma_y, \tag{175.40}$$

where Σ_x and Σ_y are the transverse dimensions of the source. The unit conventions for the various intervals are seconds for Δt, $(\text{mrad})^2$ for $\Delta \Omega$, 10^{-3} for $\Delta \omega / \omega$, and mm^2 for the source area. The brightness is basically the same as $d^2 P / d\Omega d\omega$ times $n_b/\hbar$, where n_b is the number of electrons in one bunch. For the 7-GeV Advanced Photon Source at Argonne, the peak brightness is about 10^4 times greater than would be obtained without an undulator, while the corresponding X-ray energy is about 10 times less, 3 keV versus ~30 keV.

Brilliance is relevant when focusing devices are used to narrow the X-ray beam to a spot, as it is invariant if the same ratio is evaluated in the image plane. A 10^3–10^4 enhancement over bare synchrotron sources is commonly obtained. Unfortunately, it is beyond our scope to discuss this point further, as the source dimensions are determined by many engineering considerations in addition to purely physical ones.

Appendix A | Spherical harmonics

In this appendix we define and derive the principal properties of the spherical harmonics. Most of these results may be found in any text on mathematical physics, with one difference. The spherical harmonics are usually presented as functions of θ and φ, the spherical polar angular coordinates. We shall find it preferable to regard them as functions of Cartesian coordinates x, y, and z.[1] We shall label the succeeding arguments with subheadings in order to indicate when important parts are complete.

Solutions of Laplace's equation: Let us rewrite eq. (19.17) as

$$\frac{1}{|\mathbf{r} - \mathbf{r}'|} = \sum_{\ell=0}^{\infty} \frac{r_<^\ell}{r_>^{\ell+1}} P_\ell(\hat{\mathbf{r}} \cdot \hat{\mathbf{r}}'). \tag{A.1}$$

We now use the fact that $\nabla^2 |\mathbf{r} - \mathbf{r}'|^{-1} = 0$, and regard $\mathbf{r}'$ as a parameter. Since the result must hold for all $\mathbf{r}'$, the Laplacian of the right-hand side must vanish term by term for each power of r'. From the case $r' < r$, we conclude that

$$\nabla^2 \frac{1}{r^{\ell+1}} P_\ell(\hat{\mathbf{r}} \cdot \hat{\mathbf{r}}') = 0, \tag{A.2}$$

while from the case $r' > r$, we conclude that

$$\nabla^2 r^\ell P_\ell(\hat{\mathbf{r}} \cdot \hat{\mathbf{r}}') = 0. \tag{A.3}$$

---

[1] The author learned this point of view, and much of the subsequent analysis, in lectures by N. D. Mermin at Cornell University. A similar approach is taken in Schwinger et al. (1998). The centerpiece of the analysis is the marvelous generating function (A.51). The provenance of this formula is uncertain. It appears in Courant and Hilbert (1953), and in Erdeleyi et al. (1953), (chap. XI), both of whom cite unpublished notes by Herglotz as its source, and this author has not been able to find a reference to a publication by Herglotz. Further, both Courant and Hilbert and Erdeleyi et al. give the formula only in terms of spherical polar coordinates. The much more revealing vectorial form appears to be due to Schwinger. Mermin, however, does not recall seeing it in Schwinger's lectures at Harvard University, and credits lectures by J. H. Van Vleck as a more likely source.

Next, let us define

$$F_\ell(x, y, z; x', y', z') = r^\ell r'^\ell P_\ell(\hat{\mathbf{r}} \cdot \hat{\mathbf{r}}'). \tag{A.4}$$

Regarding $\mathbf{r}'$ as a constant, F_ℓ is a polynomial in x, y, and z, of degree ℓ. Further, it is a *homogeneous* polynomial, i.e., each term in it is of the form $x^a y^b z^c$, where a, b, and c are nonnegative integers with $a + b + c = \ell$. These assertions are easily verified by going back to the definition of P_ℓ. It follows from eq. (A.3) that

$$\nabla^2 F_\ell = 0. \tag{A.5}$$

In this way, we are led to consider solutions to Laplace's equation that are homogeneous polynomials of x, y, and z, of a given degree, ℓ. It is in this form that we shall determine the spherical harmonics. (The functions F_ℓ are sometimes called *solid harmonics*.)

The eigenvalue problem for $\mathbf{L}_{\mathrm{op}}^2$: As in eq. (6.25), we can write the Laplacian as the sum of a radial and an angular operator:

$$\nabla^2 = \frac{1}{r^2} \frac{\partial}{\partial r}\left(r^2 \frac{\partial}{\partial r}\right) - \frac{1}{r^2} \mathbf{L}_{\mathrm{op}}^2. \tag{A.6}$$

We now use this form in eq. (A.3). Since

$$\frac{1}{r^2} \frac{\partial}{\partial r}\left(r^2 \frac{\partial r^\ell}{\partial r}\right) = \frac{\ell(\ell+1)}{r^2} r^\ell, \tag{A.7}$$

we obtain

$$\mathbf{L}_{\mathrm{op}}^2 P_\ell(\hat{\mathbf{r}} \cdot \hat{\mathbf{r}}') = \ell(\ell+1) P_\ell(\hat{\mathbf{r}} \cdot \hat{\mathbf{r}}'). \tag{A.8}$$

This is an eigenvalue equation. It states that $P_\ell(\hat{\mathbf{r}} \cdot \hat{\mathbf{r}}')$ is an eigenfunction of $\mathbf{L}_{\mathrm{op}}^2$ with the eigenvalue $\ell(\ell+1)$. By itself, however, it does not define the eigenfunction completely. In other words, the solution to the eigenvalue problem

$$\mathbf{L}_{\mathrm{op}}^2 X_\ell = \ell(\ell+1) X_\ell \tag{A.9}$$

is not uniquely given by P_ℓ. The nonuniqueness is not just the trivial one of multiplication by a constant. To see this, let us consider the first few polynomials. We have, for $\ell = 1$,

$$rr' P_1(\hat{\mathbf{r}} \cdot \hat{\mathbf{r}}') = xx' + yy' + zz'. \tag{A.10}$$

Since $\nabla^2 x = \nabla^2 y = \nabla^2 z = 0$, it follows that

$$\mathbf{L}_{\mathrm{op}}^2 \frac{x}{r} = 2\frac{x}{r}, \quad \mathbf{L}_{\mathrm{op}}^2 \frac{y}{r} = 2\frac{y}{r}, \quad \mathbf{L}_{\mathrm{op}}^2 \frac{z}{r} = 2\frac{z}{r}. \tag{A.11}$$

Thus, for $\ell = 1$, there are at least three independent eigenfunctions. Likewise, for $\ell = 2$, we have

$$2r^2 r'^2 P_2(\hat{\mathbf{r}} \cdot \hat{\mathbf{r}}') = 3(xx' + yy' + zz')^2 - r^2 r'^2. \tag{A.12}$$

Consider the coefficient of z'^2. This is $3z^2 - r^2 = 2z^2 - (x^2 + y^2)$. Since

$$\nabla^2\left(2z^2 - x^2 - y^2\right) = 0 \tag{A.13}$$

(directly verified using Cartesian coordinates), it follows that the function $3(z/r)^2 - 1$ is an eigenfunction of $\mathbf{L}_{op}^2$ with eigenvalue $2(2+1)$. Other eigenfunctions with the same eigenvalue are obtained by replacing z with x and y, and yet more eigenfunctions are obtained by looking at coefficients of $x'z'$, $y'z'$, etc.; these are xz, yz, etc.

The second eigenoperator, L_z: Clearly, what is needed is a way to organize and separate out independent functions from the polynomial F_ℓ. We can do this by finding functions that are simultaneously eigenfunctions of $\mathbf{L}_{op}^2$ and one other operator that commutes with it. The requirement of commutativity is essential. Suppose that f is an eigenfunction of two operators A_{op} and B_{op} with eigenvalues a and b, respectively. Then, $A_{op}f = af$, $B_{op}f = bf$, and $A_{op}B_{op}f = A_{op}bf = bA_{op}f = baf$. Likewise, $B_{op}A_{op}f = abf$, implying that $A_{op}B_{op}f = B_{op}A_{op}f$. But this cannot be true in general unless A_{op} and B_{op} commute.

The second operator is commonly taken to be L_{op}^z. From eq. (6.24), we see that this does commute with $\mathbf{L}_{op}^2$ and, further, that no other operator commutes with both L_{op}^z and $\mathbf{L}_{op}^2$. The joint eigenfunction of these two operators is denoted $Y_{\ell m}(\hat{\mathbf{r}})$:

$$\mathbf{L}_{op}^2 Y_{\ell m}(\hat{\mathbf{r}}) = \ell(\ell+1)Y_{\ell m}(\hat{\mathbf{r}}), \tag{A.14}$$

$$L_{op}^z Y_{\ell m}(\hat{\mathbf{r}}) = mY_{\ell m}(\hat{\mathbf{r}}). \tag{A.15}$$

We already know that the eigenvalues of $\mathbf{L}_{op}^2$ are $\ell(\ell+1)$. We shall find the eigenvalues of L_{op}^z, which we have denoted by m, shortly. These two equations determine $Y_{\ell m}$ completely up to a constant factor.

It is more convenient at this point to use the linear combinations

$$\xi = x + iy, \quad \eta = x - iy, \tag{A.16}$$

instead of x and y. Clearly, if $r^\ell Y_{\ell m}$ is a homogeneous polynomial of degree ℓ in x, y, and z, it is a similar polynomial in ξ, η, and z. We now observe that in terms of η and ξ,

$$L_{op}^z = \xi\partial_\xi - \eta\partial_\eta. \tag{A.17}$$

Acting with this form of the operator on a general term in the polynomial $r^\ell Y_{\ell m}$, $\xi^a\eta^b z^c$, we obtain

$$L_{op}^z \xi^a\eta^b z^c = (a-b)\xi^a\eta^b z^c. \tag{A.18}$$

But this is a solution to the eigenvalue problem for L_{op}^z! For given ℓ, the largest value of $a - b$ is attained when $a = \ell$, $b = 0$, and the least when $a = 0$, $b = \ell$. Thus, the eigenvalues of L_{op}^z are integers, given by

$$m = -\ell, -\ell+1, \ldots, \ell. \tag{A.19}$$

Further, $r^\ell Y_{\ell m}$ is a sum of terms of the form $\xi^a\eta^b z^c$ in which the power of ξ exceeds that of η by m. Accordingly, we may write

$$r^\ell Y_{\ell m} = \sum_{\substack{a+b+c=\ell, \\ a-b=m}} C_{abc}\xi^a\eta^b z^c, \tag{A.20}$$

and the problem remains to find the coefficients C_{abc}. This is done in the next several steps.

Orthonormality: First, let us fix the overall scale of $Y_{\ell m}$. We do this via the orthonormality requirement

$$\langle Y_{\ell' m'}, Y_{\ell m}\rangle \equiv \int Y_{\ell' m'}^*(\hat{\mathbf{r}}) Y_{\ell m}(\hat{\mathbf{r}}) d\Omega = \delta_{\ell\ell'}\delta_{mm'}. \tag{A.21}$$

When $\ell = \ell'$ and $m = m'$, this equation fixes $Y_{\ell m}$ up to an overall phase factor. When $\ell \neq \ell'$, or $m \neq m'$, however, it asserts much more. It states that two different $Y_{\ell m}$'s are orthogonal, i.e., their inner product vanishes. The proof for this is classic. By definition,

$$L_{\text{op}}^z Y_{\ell m} = m Y_{\ell m},$$

$$L_{\text{op}}^z Y_{\ell' m'} = m' Y_{\ell' m'}.$$

Taking the inner product of the first equation with $Y_{\ell' m'}$ from the left, of the second with $Y_{\ell m}$ from the right, and subtracting, we get (making use of the reality of m')

$$\langle Y_{\ell' m'}, L_{\text{op}}^z Y_{\ell m}\rangle - \langle L_{\text{op}}^z Y_{\ell' m'}, Y_{\ell m}\rangle = (m - m')\langle Y_{\ell' m'}, Y_{\ell m}\rangle. \tag{A.22}$$

But, by hermiticity, eq. (8.26), the left-hand side vanishes. It follows that if $m \neq m'$, $Y_{\ell m}$ and $Y_{\ell' m'}$ are orthogonal.

By replacing L_{op}^z with $\mathbf{L}_{\text{op}}^2$ in the above argument, we deduce orthogonality for $\ell \neq \ell'$.

Raising and lowering operators: We now find a recursion relation that connects $Y_{\ell m}$ and $Y_{\ell, m-1}$. For this, we first write the operators $L_{\text{op}}^\pm$, defined in eq. (6.23), in the η-ξ variables:

$$L_{\text{op}}^+ = -\xi\partial_z + 2z\partial_\eta, \quad L_{\text{op}}^- = \eta\partial_z - 2z\partial_\xi. \tag{A.23}$$

We next note that

$$L_{\text{op}}^- \xi^a \eta^b z^c = c\, \xi^a \eta^{b+1} z^{c-1} - 2a\xi^{a-1}\eta^b z^{c+1}. \tag{A.24}$$

Suppose $a - b = m$. The monomials on the right-hand side are both of total degree ℓ, and in both, the power of ξ exceeds the power of η by $m - 1$. Accordingly, the right-hand side is an eigenfunction of L_{op}^z with eigenvalue $m - 1$. Since the left-hand side has eigenvalue m, L_{op}^- is called a *lowering operator* for L_z or m. We have shown that

$$L_{\text{op}}^- Y_{\ell m} \propto Y_{\ell, m-1}. \tag{A.25}$$

To find the constant of proportionality, we note that

$$\langle L_{\text{op}}^- Y_{\ell m}, L_{\text{op}}^- Y_{\ell m}\rangle = \langle Y_{\ell m}, L_{\text{op}}^+ L_{\text{op}}^- Y_{\ell m}\rangle = \langle Y_{\ell m}, \left(\mathbf{L}_{\text{op}}^2 - (L_{\text{op}}^z)^2 + L_{\text{op}}^z\right) Y_{\ell m}\rangle, \tag{A.26}$$

where in the first equality we used eq. (8.26c), and in the second the result (6.24f). But $Y_{\ell m}$ is an eigenfunction of both L_{op}^z and $\mathbf{L}_{\text{op}}^2$, so

$$\langle L_{\text{op}}^- Y_{\ell m}, L_{\text{op}}^- Y_{\ell m}\rangle = \left(\ell(\ell+1) - m(m-1)\right)\langle Y_{\ell m}, Y_{\ell m}\rangle = \left(\ell(\ell+1) - m(m-1)\right). \tag{A.27}$$

We choose the relative phases of the $Y_{\ell m}$'s so that

$$L_{op}^- Y_{\ell m} = \sqrt{\left(\ell(\ell+1) - m(m-1)\right)} Y_{\ell, m-1}. \tag{A.28}$$

Exercise A.1 Show that with this phase choice,

$$L_{op}^+ Y_{\ell m} = \sqrt{\left(\ell(\ell+1) - m(m+1)\right)} Y_{\ell, m+1}. \tag{A.29}$$

Explicit form for $Y_{\ell\ell}$: The next step is to determine $Y_{\ell\ell}$ explicitly. From eq. (A. 20), it follows that $Y_{\ell\ell} \propto \xi^\ell$. The normalization integral is

$$K_\ell \equiv \langle \xi^\ell, \xi^\ell \rangle = \int (\xi^* \xi)^\ell d\Omega = \int (1 - z^2)^\ell d\Omega, \tag{A.30}$$

since $\xi^* \xi = x^2 + y^2 = (1 - z^2)$ on the unit sphere. To evaluate the integral, we use spherical polar coordinates. In these, $z = \cos\theta$, $\xi = \sin\theta e^{i\varphi}$, $\eta = \sin\theta e^{-i\varphi}$, and the area element $d\Omega = \sin\theta \, d\theta \, d\varphi$, with $0 \le \theta \le \pi$, $0 \le \varphi \le 2\pi$. The integration over φ is trivial, and so

$$K_\ell = 2\pi \int_0^\pi \sin^{2\ell+1}\theta \, d\theta. \tag{A.31}$$

Integrating by parts, we get

$$K_\ell = 2\pi \sin^{2l}\theta(-\cos\theta)\Big|_0^\pi + 2\pi \int_0^\pi (\cos\theta).2\ell \sin^{2\ell-1}\theta \cos\theta \, d\theta$$

$$= 4\pi\ell \int_0^\pi \sin^{2\ell-1}(1 - \sin^2\theta) \, d\theta$$

$$= 2\ell(K_{\ell-1} - K_\ell). \tag{A.32}$$

Since $K_0 = 4\pi$, we find

$$K_\ell = \frac{2\ell}{2\ell+1} K_{\ell-1} = \frac{2\ell}{2\ell+1}\frac{2\ell-2}{2\ell-3} K_{\ell-2} = \cdots = \frac{(2\ell)!!}{(2\ell+1)!!}4\pi, \tag{A.33}$$

where $n!! = n(n-2)(n-4)\cdots$, with the last factor being 2 if n is even and 1 if n is odd. Also, for consistency, we define $0!! = 1!! = 1$. A simple rearrangement gives the alternative formula

$$K_\ell = \frac{2^{2\ell}(\ell!)^2}{(2\ell+1)!}4\pi. \tag{A.34}$$

Finally, we choose the phase so that

$$Y_{\ell\ell}(\hat{\mathbf{r}}) = \frac{(-1)^\ell}{2^\ell \ell!}\sqrt{\frac{(2\ell+1)!}{4\pi}}\left(\frac{\xi}{r}\right)^\ell. \tag{A.35}$$

The generating function: At this point, the problem is, in principle, solved, since we could find all the $Y_{\ell m}$'s by repeated application of L_{op}^- on $Y_{\ell\ell}$. A more compact answer is obtained by an indirect approach. We consider the polynomial $(\mathbf{a}\cdot\mathbf{r})^\ell$, where $\mathbf{a}$ is a constant vector. This is clearly homogeneous in x, y, and z and will be a sum of spherical harmonics of order ℓ if we can make it obey Laplace's equation. Now,

$$\nabla^2(\mathbf{a}\cdot\mathbf{r})^\ell = \nabla\cdot\left[\ell(\mathbf{a}\cdot\mathbf{r})^{\ell-1}\mathbf{a}\right] = \ell(\ell-1)(\mathbf{a}\cdot\mathbf{r})^{\ell-2}(\mathbf{a}\cdot\mathbf{a}), \tag{A.36}$$

which will vanish (for $\ell > 1$) only if $\mathbf{a}\cdot\mathbf{a} = 0$. This condition cannot be met by a nonzero real vector $\mathbf{a}$, but it can if we allow $\mathbf{a}$ to be complex, i.e., let its components be complex numbers. Up to an overall scale factor and a rotation, the most general form for a complex $\mathbf{a}$ is

$$\mathbf{a} = \hat{\mathbf{z}} - \frac{1}{2}\lambda(\hat{\mathbf{x}} + i\hat{\mathbf{y}}) + \frac{1}{2}\mu(\hat{\mathbf{x}} - i\hat{\mathbf{y}}). \tag{A.37}$$

Then

$$\mathbf{a}\cdot\mathbf{a} = 1 + \frac{1}{4}(\mu - \lambda)^2 - \frac{1}{4}(\mu + \lambda)^2 = 1 - \mu\lambda, \tag{A.38}$$

which vanishes if $\mu = 1/\lambda$. Accordingly, we take

$$\mathbf{a} = \hat{\mathbf{z}} - \frac{\lambda}{2}(\hat{\mathbf{x}} + i\hat{\mathbf{y}}) + \frac{1}{2\lambda}(\hat{\mathbf{x}} - i\hat{\mathbf{y}}). \tag{A.39}$$

Then

$$\mathbf{a}\cdot\mathbf{r} = z - \frac{\lambda}{2}\xi + \frac{1}{2\lambda}\eta, \tag{A.40}$$

By construction, $\nabla^2(\mathbf{a}\cdot\mathbf{r})^\ell = 0$ for any λ.

We now note that if we expand $(\mathbf{a}\cdot\mathbf{r})^\ell$ completely, the monomial $\xi^a\eta^b z^c$ will be multiplied by λ^{a-b}. Thus, the coefficient of λ^m in this expansion meets all the requirements for $Y_{\ell m}$, and we can write

$$(\mathbf{a}\cdot\mathbf{r})^\ell = \sum_{m=-\ell}^{\ell} c_m \lambda^m r^\ell Y_{\ell m}(\hat{\mathbf{r}}). \tag{A.41}$$

The only problem remaining is to find the constants c_m. This we do by using the formulas (A.23) to find the action of L_{op}^- and L_{op}^z on $(\mathbf{a}\cdot\mathbf{r})^\ell$:

$$L_{\text{op}}^-(\mathbf{a}\cdot\mathbf{r}) = [\eta + \lambda z], \tag{A.42}$$

$$L_{\text{op}}^z(\mathbf{a}\cdot\mathbf{r}) = \left[-\frac{\lambda}{2}\xi - \frac{1}{2\lambda}\eta\right]. \tag{A.43}$$

Hence,

$$\begin{aligned}
\left(\frac{1}{\lambda}L_{\text{op}}^- + L_{\text{op}}^z\right)(\mathbf{a}\cdot\mathbf{r})^\ell &= \ell(\mathbf{a}\cdot\mathbf{r})^{\ell-1}\left[\frac{1}{\lambda}\eta + z - \frac{\lambda}{2}\xi - \frac{1}{2\lambda}\eta\right] \\
&= \ell(\mathbf{a}\cdot\mathbf{r})^{\ell-1}\left[z - \frac{\lambda}{2}\xi + \frac{1}{2\lambda}\eta\right] \\
&= \ell(\mathbf{a}\cdot\mathbf{r})^\ell, \tag{A.44}
\end{aligned}$$

or

$$\frac{1}{\lambda}L_{\text{op}}^-(\mathbf{a}\cdot\mathbf{r})^\ell = \left(\ell - L_{\text{op}}^z\right)(\mathbf{a}\cdot\mathbf{r})^\ell. \tag{A.45}$$

We now feed the expansion (A.41) into both sides of this equation, which yields

$$\sum_{m=-\ell}^{\ell} c_m \lambda^{m-1} r^\ell \left(L_{op}^- Y_{\ell m} \right) = \sum_{m=-\ell}^{\ell} (\ell - m) c_m \lambda^m r^\ell Y_{\ell m}. \tag{A.46}$$

Equating equal powers of λ on both sides and using eq. (A.28), we obtain (for $m \neq \ell$)

$$c_m = \frac{1}{\ell - m} \sqrt{\ell(\ell + 1) - m(m + 1)} c_{m+1} = \left(\frac{\ell + m + 1}{\ell - m} \right)^{1/2} c_{m+1}. \tag{A.47}$$

Iterating this formula, we get

$$c_m = \sqrt{\frac{(\ell + m + 1)(\ell + m + 2) \cdots (2\ell)}{(\ell - m)(\ell - m - 1) \cdots 1}} c_\ell$$

$$= \sqrt{\frac{(2\ell)!}{(\ell + m)! \, (\ell - m)!}} c_\ell \tag{A.48}$$

By equating the coefficients of λ^ℓ on the two sides of eq. (A.41) and using eq. (A.35) for $Y_{\ell\ell}$, we obtain

$$c_\ell = \sqrt{\frac{4\pi}{(2\ell + 1)!}} \ell!, \tag{A.49}$$

so that

$$c_m = \sqrt{\frac{4\pi}{2\ell + 1}} \cdot \frac{\ell!}{\sqrt{(\ell + m)! \, (\ell - m)!}}. \tag{A.50}$$

Substituting this expression in eq. (A.41) and writing it out in full, we obtain, finally, our generating function for spherical harmonics,

$$\frac{1}{\ell!} \left(z - \frac{\lambda}{2} \xi + \frac{1}{2\lambda} \eta \right)^\ell = r^\ell \sqrt{\frac{4\pi}{2\ell + 1}} \sum_{m=-\ell}^{\ell} \frac{\lambda^m Y_{\ell m}(\hat{\mathbf{r}})}{\sqrt{(\ell + m)! \, (\ell - m)!}}. \tag{A.51}$$

Second solution of Laplace's equation: Using eqs. (A.6) and (A.14) it is easy to show that

$$\nabla^2 \frac{1}{r^{\ell+1}} Y_{\ell m}(\hat{\mathbf{r}}) = 0. \tag{A.52}$$

These are the solutions used to represent the far field of an electrostatic multipole.

Special values and transformations: Let us pause to derive some useful special properties of the spherical harmonics. First, consider the inversion or parity transformation $\mathbf{r} \to -\mathbf{r}$. Since $r^\ell Y_{\ell m}(\hat{\mathbf{r}})$ is a polynomial in x, y, and z of degree ℓ,

$$Y_{\ell m}(-\hat{\mathbf{r}}) = (-1)^\ell Y_{\ell m}(\hat{\mathbf{r}}). \tag{A.53}$$

Second, let us show the result (19.23) for $Y_{\ell m}^*(\hat{\mathbf{r}})$. In the generating function (A.51), let us replace λ by $-1/\lambda^*$ and take the complex conjugate. Since, $\xi^* = \eta$ and $\eta^* = \xi$, the left-hand side is reproduced as before. The right-hand side is changed to

$$r^\ell \sqrt{\frac{4\pi}{2\ell+1}} \sum_{m=-\ell}^{\ell} \frac{(-1)^m \lambda^{-m} Y_{\ell m}^*(\hat{\mathbf{r}})}{\sqrt{(\ell+m)!\,(\ell-m)!}}. \tag{A.54}$$

But this expression must equal the right-hand side of eq. (A.51). Changing the summation index in that sum to $-m$ and equating equal powers of λ, we obtain

$$Y_{\ell m}^*(\hat{\mathbf{r}}) = (-1)^m Y_{\ell,-m}(\hat{\mathbf{r}}). \tag{A.55}$$

This result shows that for a given ℓ, the set of $Y_{\ell m}$ consists of $2\ell+1$ real linearly independent functions.

Third, let us obtain $Y_{\ell m}(\hat{\mathbf{r}})$ for some special orientations. For $\hat{\mathbf{r}} = \hat{\mathbf{z}}$, we put $\xi = \eta = 0$ in eq. (A.51). The left-hand side is independent of λ. Thus, only the $m = 0$ term can survive in the sum on the right, and we conclude that

$$Y_{\ell m}(\hat{\mathbf{z}}) = \sqrt{\frac{2\ell+1}{4\pi}} \delta_{m0}. \tag{A.56}$$

Likewise, to obtain $Y_{\ell m}(\hat{\mathbf{x}})$, we put $z=0$, and $\xi = \eta = x$ in the generating function. The left-hand side becomes

$$\frac{1}{\ell!} \frac{1}{2^\ell \lambda^\ell} \left(1-\lambda^2\right)^\ell. \tag{A.57}$$

Expanding this by the binomial theorem and reading off the coefficient of λ^m, we find

$$Y_{\ell m}(\hat{\mathbf{x}}) = \begin{cases} \dfrac{(-1)^{(\ell+m)/2}}{2^\ell} \sqrt{\dfrac{2\ell+1}{4\pi}} \dfrac{\sqrt{(\ell+m)!\,(\ell-m)!}}{\left[\frac{1}{2}(\ell+m)\right]!\,\left[\frac{1}{2}(\ell-m)\right]!}; & \text{if } (\ell-m) \text{ is even,} \\[1em] 0 & \text{otherwise.} \end{cases} \tag{A.58}$$

Exercise A.2 Let $\mathbf{r}'$ be the vector obtained by rotating the vector $\mathbf{r}$ by an angle α about $\hat{\mathbf{z}}$. Show that

$$Y_{\ell m}(\hat{\mathbf{r}}') = e^{im\alpha} Y_{\ell m}(\hat{\mathbf{r}}). \tag{A.59}$$

Expansion of the inverse separation: We now prove the important result (19.16) for the expansion of the inverse separation. Let us imagine expanding $1/|\mathbf{r} - \mathbf{r}'|$ as in eq. (A.1), but think of the terms in the expansion as functions of $\hat{\mathbf{r}}$ and $\hat{\mathbf{r}}'$ separately. Since $\nabla^2 |\mathbf{r} - \mathbf{r}'|^{-1}$ vanishes, the terms involving r^ℓ must be of the type $r^\ell Y_{\ell m}(\hat{\mathbf{r}})$. Similarly, since $\nabla'^2 |\mathbf{r} - \mathbf{r}'|^{-1}$ also vanishes, the terms involving $1/r'^{(\ell+1)}$ must in fact be of the type $Y_{\ell m}(\hat{\mathbf{r}}')/r'^{\ell+1}$. These requirements can be met only by an expansion of the type

$$\frac{1}{|\mathbf{r}-\mathbf{r}'|} = \sum_{\ell=0}^{\infty} \frac{r^\ell}{r'^{\ell+1}} \sum_{m,m'} B_{\ell m m'} Y_{\ell m}(\hat{\mathbf{r}}) Y_{\ell m'}^*(\hat{\mathbf{r}}'). \tag{A.60}$$

The use of $Y_{\ell m}^*(\hat{\mathbf{r}}')$ instead of $Y_{\ell m}(\hat{\mathbf{r}}')$ is inconsequential and is done for later convenience. The task at hand is to find the coefficients $B_{\ell m m'}$.

To this end, let us note that

$$(\mathbf{L}_{op} + \mathbf{L}'_{op}) \frac{1}{|\mathbf{r} - \mathbf{r}'|} = -i(\mathbf{r} \times \nabla + \mathbf{r}' \times \nabla') \frac{1}{|\mathbf{r} - \mathbf{r}'|}. \tag{A.61}$$

On the right, we can replace ∇' by $-\nabla$, which yields

$$(\mathbf{L}_{op} + \mathbf{L}'_{op}) \frac{1}{|\mathbf{r} - \mathbf{r}'|} = -i(\mathbf{r} - \mathbf{r}') \times \nabla \frac{1}{|\mathbf{r} - \mathbf{r}'|}. \tag{A.62}$$

But the final gradient is parallel to $(\mathbf{r} - \mathbf{r}')$. Hence, the cross product vanishes, i.e.,

$$(\mathbf{L}_{op} + \mathbf{L}'_{op}) \frac{1}{|\mathbf{r} - \mathbf{r}'|} = 0. \tag{A.63}$$

Let us now consider the z component of eq. (A.63) and feed in eq. (A.60). Since $Y_{\ell m}$ is an eigenfunction of L^z_{op}, we get

$$0 = \sum_{\ell=0}^{\infty} \frac{r^\ell}{r'^{\ell+1}} \sum_{m,m'} B_{\ell mm'}(m - m') Y_{\ell m}(\hat{\mathbf{r}}) Y^*_{\ell m'}(\hat{\mathbf{r}}'). \tag{A.64}$$

The double sum over m and m' must therefore vanish, which implies that $B_{\ell mm'}$ vanishes if $m \neq m'$. We can therefore simplify eq. (A.60) to

$$\frac{1}{|\mathbf{r} - \mathbf{r}'|} = \sum_{\ell=0}^{\infty} \frac{r^\ell}{r'^{\ell+1}} \sum_{m} b_{\ell m} Y_{\ell m}(\hat{\mathbf{r}}) Y^*_{\ell m}(\hat{\mathbf{r}}'). \tag{A.65}$$

Next, let us feed eq. (A.65) into the $-$ component of eq. (A.63). The left-hand side vanishes. The right side will vanish only if the operator $L^-_{op} + L^{-\,\prime}_{op}$ annihilates the sum over m. Writing $Y^*_{\ell m}$ in terms of $Y_{\ell,-m}$, this condition reads

$$0 = \sum_{m=-\ell+1}^{\ell} (-1)^m b_{\ell,m} \sqrt{\ell(\ell+1) - m(m-1)}\, Y_{\ell,m-1}(\hat{\mathbf{r}}) Y_{\ell,-m}(\hat{\mathbf{r}}')$$

$$+ \sum_{m=-\ell}^{\ell-1} (-1)^m b_{\ell,m} \sqrt{\ell(\ell+1) - (-m)(-m-1)}\, Y_{\ell,m}(\hat{\mathbf{r}}) Y_{\ell,-m-1}(\hat{\mathbf{r}}'). \tag{A.66}$$

Shifting the summation index in the second sum from m to $m+1$, we obtain

$$0 = \sum_{m=-\ell+1}^{\ell} (-1)^m (b_{\ell,m} - b_{\ell,m-1}) \sqrt{\ell(\ell+1) - m(m-1)}\, Y_{\ell,m-1}(\hat{\mathbf{r}}) Y_{\ell,-m}(\hat{\mathbf{r}}'). \tag{A.67}$$

This equation can be satisfied for general $\hat{\mathbf{r}}$ and $\hat{\mathbf{r}}'$ only if $b_{\ell,m} = b_{\ell,m-1}$ for all applicable values of m, i.e., if

$$b_{\ell m} = b_{\ell 0}, \quad -\ell \leq m \leq \ell. \tag{A.68}$$

This simplifies eq. (A.65) even further.

Only one number, $b_{\ell 0}$, now remains to be found. To find this, let us take $\mathbf{r}$ and $\mathbf{r}'$ both along $\hat{\mathbf{z}}$. The left-hand side of eq. (A.65) is then a simple geometric series, and using eq. (A.56), we obtain

$$\sum_{\ell=0}^{\infty} \frac{r^\ell}{r'^{\ell+1}} = \sum_{\ell=0}^{\infty} \frac{r^\ell}{r'^{\ell+1}} b_{\ell 0} \frac{2\ell+1}{4\pi}. \tag{A.69}$$

This yields $b_{\ell 0} = 4\pi/(2\ell+1)$. Substituting in eq. (A.65) along with eq. (A.68), we obtain the result quoted in chapter 3:

$$\frac{1}{|\mathbf{r}-\mathbf{r}'|} = \sum_{\ell=0}^{\infty} \sum_{m=-\ell}^{\ell} \frac{4\pi}{2\ell+1} \frac{r_<^\ell}{r_>^{\ell+1}} Y_{\ell m}(\hat{\mathbf{r}}) Y_{\ell m}^*(\hat{\mathbf{r}}'). \tag{A.70}$$

We have used the symmetry of the left-hand side in $\mathbf{r}$ and $\mathbf{r}'$ to write the right-hand side more generally, indicating that the expansion may be in powers of either ratio, r/r' or r'/r, whichever is smaller. As in chapter 3, $r_<$ stands for the lesser of the distances r and r', and $r_>$ for the greater.

Connection with Legendre polynomials: The polynomials $P_\ell(\hat{\mathbf{r}} \cdot \hat{\mathbf{r}}')$ were first introduced by Legendre via eq. (A.1). However, they are also special cases of the spherical harmonics. To see this, let us take $\mathbf{r}' \parallel \hat{\mathbf{z}}$ in eq. (A.70) and use eq. (A.56) for $Y_{\ell m}(\hat{\mathbf{z}})$. Comparing with eq. (A.1), we get

$$P_\ell(\hat{\mathbf{r}} \cdot \hat{\mathbf{z}}) = \sqrt{\frac{4\pi}{2\ell+1}} Y_{\ell 0}(\hat{\mathbf{r}}). \tag{A.71}$$

It is also useful to rewrite eq. (A.1) as follows. Let us denote $u = \hat{\mathbf{r}} \cdot \hat{\mathbf{z}}$, and $t = r/r'$. Then, we get

$$(1+t^2-2tu)^{-1/2} = \sum_{\ell=0}^{\infty} t^\ell P_\ell(u). \tag{A.72}$$

This formula may be regarded as another generating function. Using it, the first few polynomials may be found to be

$$P_0(u) = 1,$$
$$P_1(u) = u,$$
$$P_2(u) = \tfrac{1}{2}(3u^2 - 1), \tag{A.73}$$
$$P_3(u) = \tfrac{1}{2}(5u^3 - 3u),$$
$$P_4(u) = \tfrac{1}{8}(35u^4 - 30u^2 + 3).$$

Among the properties of these polynomials, we mention the standardization condition

$$P_\ell(1) = 1, \tag{A.74}$$

which follows by putting $u = 1$ in eq. (A.72), and the orthonormality relation

$$\int_{-1}^{1} P_\ell(u) P_{\ell'}(u) \, du = \frac{2}{2\ell+1} \delta_{\ell \ell'}. \tag{A.75}$$

The addition theorem: If we compare the right-hand sides of eqs. (A.1) and (A.70), we obtain the addition theorem for spherical harmonics:

$$P_\ell(\hat{\mathbf{r}}_1 \cdot \hat{\mathbf{r}}_2) = \frac{4\pi}{2\ell+1} \sum_{m=-\ell}^{\ell} Y_{\ell m}^*(\hat{\mathbf{r}}_2) Y_{\ell m}(\hat{\mathbf{r}}_1). \tag{A.76}$$

This powerful result is of wide utility in many areas of physics. Its proof in classical texts on analysis entails some fairly intricate integrations.[2] The deepest insight into it comes from group theoretic methods, but to discuss that here would be too digressive.

The addition theorem immediately allows a number of interesting deductions, among which are the integral

$$\int P_\ell(\hat{\mathbf{r}}_1 \cdot \hat{\mathbf{r}}_2) \, P_{\ell'}(\hat{\mathbf{r}}_1 \cdot \hat{\mathbf{r}}_3) \, d\Omega_1 = \delta_{\ell\ell'} \frac{4\pi}{2\ell+1} \, P_\ell(\hat{\mathbf{r}}_2 \cdot \hat{\mathbf{r}}_3) \tag{A.77}$$

and the sum rule

$$\sum_{m=-\ell}^{\ell} |Y_{\ell m}(\hat{\mathbf{r}})|^2 = \frac{2\ell+1}{4\pi}. \tag{A.78}$$

The proofs are left to the reader. The orthonormality relation (A.75) can be seen as a special case of eq. (A.77).

We conclude this appendix with two exercises on recursion relations for the Legendre polynomials and spherical harmonics.

Exercise A.3 Show that

$$(2\ell+1)u \, P_\ell(u) = (\ell+1) \, P_{\ell+1}(u) + \ell \, P_{\ell-1}(u). \tag{A.79}$$

Solution: Apply to both sides of the generating formula (A.72), the operator

$$(1+t^2-2tu)\frac{\partial}{\partial t}, \tag{A.80}$$

use the generating function to express $(1+t^2-2tu)^{-1/2}$ in terms of the P_ℓ's once more, and equate equal powers of t on both sides. The result follows.

Exercise A.4 Derive the recursion relations

$$(2\ell+1)\frac{z}{r}\tilde{Y}_{\ell m} = \sqrt{(\ell+m)(\ell-m)}\,\tilde{Y}_{\ell-1,m} + \sqrt{(\ell+1+m)(\ell+1-m)}\,\tilde{Y}_{\ell+1,m},$$

$$(2\ell+1)\frac{\xi}{r}\tilde{Y}_{\ell m} = \sqrt{(\ell-1-m)(\ell-m)}\,\tilde{Y}_{\ell-1,m+1} - \sqrt{(\ell+1+m)(\ell+2+m)}\,\tilde{Y}_{\ell+1,m+1},$$

$$(2\ell+1)\frac{\eta}{r}\tilde{Y}_{\ell m} = -\sqrt{(\ell+m)(\ell-1+m)}\,\tilde{Y}_{\ell-1,m-1} + \sqrt{(\ell+1-m)(\ell+2-m)}\,\tilde{Y}_{\ell+1,m-1},$$

$$\tag{A.81}$$

---

[2] See, e.g., Copson (1970), sec. 11.7.

where

$$\tilde{Y}_{\ell m}(\hat{\mathbf{n}}) = \sqrt{\frac{4\pi}{2\ell+1}}\, Y_{\ell m}(\hat{\mathbf{n}}). \tag{A.82}$$

Solution: By using the analogs of eqs. (A.42) and (A.45) for $L_{\rm op}^+$, we obtain

$$\ell(z - \lambda\xi)(\mathbf{a}\cdot\mathbf{r})^{\ell-1} = (\ell + L_{\rm op}^z)(\mathbf{a}\cdot\mathbf{r})^{\ell}. \tag{A.83}$$

If we use the generating function for the $Y_{\ell m}$'s on both sides for $(\mathbf{a}\cdot\mathbf{r})^{\ell}$, shift from ℓ to $\ell+1$, and equate equal powers of λ on both sides, we obtain

$$\frac{z}{r}\tilde{Y}_{\ell m} - \left[\frac{\ell+m}{\ell+1-m}\right]^{1/2}\frac{\xi}{r}\tilde{Y}_{\ell,m-1} = \left[\frac{\ell+1+m}{\ell+1-m}\right]^{1/2}\tilde{Y}_{\ell+1,m}. \tag{A.84}$$

This gives us one linear combination of $z\tilde{Y}_{\ell m}$ and $\xi\tilde{Y}_{\ell,m-1}$. A second one can be obtained by starting from eq. (A.79). Using eq. (A.71), we can write this as

$$(2\ell+1)z\tilde{Y}_{\ell 0} = (\ell+1)\tilde{Y}_{\ell+1,0} + \ell\,\tilde{Y}_{\ell-1,0}. \tag{A.85}$$

We now operate on it with $L_{\rm op}^+$ m times, using the relations

$$(L_{\rm op}^+)^m\tilde{Y}_{\ell 0} = \left[\frac{(\ell+m)!}{(\ell-m)!}\right]^{1/2}\tilde{Y}_{\ell m}, \quad L_{\rm op}^+ z = -\xi, \quad \text{and} \quad L_{\rm op}^+ \xi = 0. \tag{A.86}$$

After some simplification, we obtain

$$\frac{z}{r}\tilde{Y}_{\ell m} - m\left[\frac{1}{(\ell+m)(\ell+1-m)}\right]^{1/2}\frac{\xi}{r}\tilde{Y}_{\ell,m-1}$$

$$= \frac{\ell+1}{2\ell+1}\left[\frac{\ell+1+m}{\ell+1-m}\right]^{1/2}\tilde{Y}_{\ell+1,m} + \frac{\ell}{2\ell+1}\left[\frac{\ell-m}{\ell+m}\right]^{1/2}\tilde{Y}_{\ell-1,m}. \tag{A.87}$$

Solving eqs. (A.84) and (A.87) for $z\tilde{Y}_{\ell m}$ and $\xi\tilde{Y}_{\ell,m-1}$, we obtain the first two of the recursion relations (A.81). The last relation follows by complex conjugation of the second with $m \to -m$.

Our approach to this problem has been purely analytic. In fact, much more powerful methods from group theory exist, which yield the product $Y_{\ell m}(\hat{\mathbf{n}})Y_{\ell'm'}(\hat{\mathbf{n}})$ of two spherical harmonics as a sum over other $Y_{\ell m}$'s.[3] Equation (A.81) is a special case of this result.

---

[3] See, e.g., Gottfried (1966), sec. 34.

Appendix B | Bessel functions

Bessel functions arise naturally in situations with cylindrical symmetry in the solution to many types of problems, Laplace and wave equations, for example. A knowledge of their principal properties from this viewpoint is therefore very useful, and the development herein is designed to obtain these as briefly as possible. More detailed expositions can be found in all texts on mathematical physics.[1]

Generating function: The function $\exp[u(t - t^{-1})/2]$ is analytic in u for all finite u, and for all $0 < |t| < \infty$. It can therefore be expanded as a Laurent series in t:

$$e^{\frac{1}{2}u(t-1/t)} = \sum_{m=-\infty}^{\infty} t^m J_m(u).$$ (B.1)

The coefficient $J_m(u)$ in this expansion is a function of u, known as the *Bessel function of the first kind of order m*. It is clearly nonsingular at all u, i.e., it is an entire function.

The left-hand side of eq. (B.1) is therefore a generating function for the Bessel functions, known as *Schlomlich's generating function*. We can cast it in another useful form by writing $t = ie^{i\varphi}$. This yields

$$e^{iu\cos\varphi} = \sum_{m=-\infty}^{\infty} e^{im\varphi} i^m J_m(u).$$ (B.2)

Parity: All the properties of the Bessel functions can be systematically derived from the generating function. The simplest is the behavior under reflection. By successively

---

[1] For most physicists, "all there is to be knowed" about Bessel functions may be found in Watson (1944). See especially chaps. 2, 3, 6, and 7. We shall be interested largely in the case of integer order and real argument, for which our pedestrian analysis suffices. A very useful compilation of properties is in Abramowitz and Stegun (1972), chap. 9.

writing $-\varphi$ and $\varphi + \pi$ for φ in eq. (B.2), we see that

$$J_{-m}(u) = (-1)^m J_m(u), \quad J_m(-u) = (-1)^m J_m(u). \tag{B.3}$$

Thus, J_{-m} is not independent of J_m, and it is even or odd in u accordingly as the order m is even or odd.

Integral representations: Since eq. (B.1) is a Laurent series, Cauchy's theorem yields

$$J_m(u) = \frac{1}{2\pi i} \int_C e^{\frac{1}{2}u(t-1/t)} \frac{dt}{t^{m+1}}, \tag{B.4}$$

where C is any contour that winds once around the origin in the counterclockwise direction. This result is even simpler from the viewpoint of eq. (B.2). Now, $J_m(u)$ is simply a Fourier coefficient,

$$J_m(u) = \frac{(-i)^m}{2\pi} \int_0^{2\pi} e^{i(u\cos\varphi - m\varphi)} d\varphi. \tag{B.5}$$

Behavior at zero and infinity: We can obtain a power series in u for $J_m(u)$ by expanding the integrand in eq. (B.5) in powers of u. If we consider, say, $m = 3$, the terms in this expansion of order u^0, u^1, and u^2 will be multiplied with 1, $\cos\varphi$, and $\cos^2\varphi$, which will vanish when multiplied with $e^{-3i\varphi}$ and integrated. The first nonzero term will be of order u^3. Extending this idea to general m, we obtain

$$J_m(u) \approx \frac{1}{2^m m!} u^m + \cdots, \quad u \approx 0. \tag{B.6}$$

With a little more effort, the full series can be obtained. Note in particular that

$$J_0(0) = 1, \quad J_m(0) = 0 \ (m \neq 0). \tag{B.7}$$

Next, we can obtain the leading asymptotic behavior at infinity by using the method of stationary phase in eq. (B.5). There are two points of stationary phase, one near $\varphi = 0$ and the other near $\varphi = \pi$, whose contributions are complex conjugates of each other. Hence,

$$J_m(u) \approx \frac{1}{2\pi} 2 \operatorname{Re} e^{-im\pi/2} \int_{-\infty}^{\infty} e^{iu(1-\frac{1}{2}\varphi^2)-im\varphi} d\varphi$$

$$\approx \frac{1}{2\pi} 2 \operatorname{Re} e^{i(u-m\pi/2)} \left(\frac{2\pi}{iu}\right)^{1/2} e^{im^2/2u}. \tag{B.8}$$

We have not kept track of all terms of order $1/u$, and so keeping only the leading term, we get

$$J_m(u) \approx \sqrt{\frac{2}{\pi u}} \cos\left(u - \frac{m\pi}{2} - \frac{\pi}{4}\right), \quad u \to +\infty. \tag{B.9}$$

The full asymptotic series can be obtained by systematic application of the steepest-descent method.

Together, eqs. (B.6) and (B.9) show that $J_m(u)$ is finite at $u = 0$ and that it oscillates with u, with an amplitude that decreases as u increases. A graph of the first few $J_m(u)$ is shown in fig. B.1.

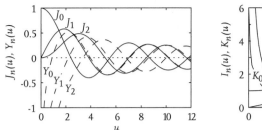 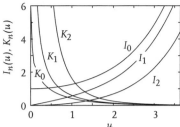

FIGURE B.1. The first few Bessel functions and modified Bessel functions of the first and second kind.

Differential equation: From the generating function (B.1), we can see that the Bessel functions obey a differential equation of the second order. Let us denote

$$F(u, t) = \tfrac{1}{2}u(t - t^{-1}).$$

(B.10)

Differentiating with respect to t and u, we obtain

$$\frac{\partial e^F}{\partial u} = \frac{1}{2}\left(t - \frac{1}{t}\right)e^F,$$

(B.11)

$$t\frac{\partial e^F}{\partial t} = \frac{1}{2}u\left(t + \frac{1}{t}\right)e^F,$$

(B.12)

$$\frac{\partial^2 e^F}{\partial u^2} = \frac{1}{4}\left(t^2 - 2 + \frac{1}{t^2}\right)e^F,$$

(B.13)

$$t\frac{\partial}{\partial t}t\frac{\partial}{\partial t}e^F = \left[\frac{1}{2}u\left(t - \frac{1}{t}\right) + \frac{1}{4}u^2\left(t^2 + 2 + \frac{1}{t^2}\right)\right]e^F.$$

(B.14)

Therefore,

$$\left(\frac{\partial^2}{\partial u^2} + \frac{1}{u}\frac{\partial}{\partial u} + 1 - \frac{1}{u^2}t\frac{\partial}{\partial t}t\frac{\partial}{\partial t}\right)e^F = 0.$$

(B.15)

Feeding in the expansion (B.1) for e^F, we obtain

$$\sum_m t^m\left(\frac{d^2}{du^2} + \frac{1}{u}\frac{d}{du} + 1 - \frac{m^2}{u^2}\right)J_m(u) = 0.$$

(B.16)

The sum must vanish term by term in t. Hence,

$$\left(\frac{d^2}{du^2} + \frac{1}{u}\frac{d}{du} + 1 - \frac{m^2}{u^2}\right)J_m(u) = 0.$$

(B.17)

This is Bessel's differential equation. Readers will recognize that this is the form that arises when we separate Laplace's equation in cylindrical coordinates, as, for example, in section 90.

Exercise B.1 Show that

$$J_{m+1}(u) = -u^m\frac{d}{du}\left(u^{-m}J_m(u)\right), \quad J_m(u) = u^{-(m+1)}\frac{d}{du}\left(u^{m+1}J_{m+1}(u)\right).$$

(B.18)

An important special case of these results, worth remembering separately, is

$$J_0'(u) = -J_1(u). \tag{B.19}$$

Bessel functions of the second kind: As a differential equation of second order, eq. (B.17) has two independent solutions. A common choice for the second solution is the so-called Bessel function of the second kind, also called Weber's function or Neumann's function, $Y_m(u)$. We do not bother to define this completely and give only limiting behaviors. As $u \to 0$,

$$Y_m(u) \approx \begin{cases} -\dfrac{(m-1)!}{\pi} \left(\dfrac{1}{2}u\right)^{-m}, & m \neq 0, \\ \dfrac{2}{\pi} \ln\left(\dfrac{1}{2}ue^C\right), & m = 0, \end{cases} \quad u \to 0, \tag{B.20}$$

where C is Euler's constant. And, as $u \to \infty$,

$$Y_m(u) \approx \sqrt{\frac{2}{\pi u}} \sin\left(u - \frac{m\pi}{2} - \frac{\pi}{4}\right), \quad u \to +\infty. \tag{B.21}$$

The first few $Y_m(u)$ are shown in fig. B.1.

Exercise B.2 The Wronskian of two functions $f(u)$ and $g(u)$ is defined as

$$W(f, g) = fg' - f'g. \tag{B.22}$$

Let f and g be two linearly independent solutions to Bessel's equation. Obtain a first-order differential equation for $W(f, g)$ and thus show that $W \propto 1/u$. Next, by examining $W(J_m, Y_m)$, show that Y_m must be singular as $u \to 0$ in accordance with eq. (B.20),[2] and verify the behavior as $u \to \infty$. Of course, the multiplicative constants cannot be found in this way. One way to fix them is by the convention

$$W(J_m, Y_m) = 2/\pi u. \tag{B.23}$$

Expansion of the inverse separation: Let us now apply the above results to the physical problem of expanding the inverse separation $|\mathbf{r} - \mathbf{r}'|^{-1}$ (which is the Green function of Laplace's equation) using Bessel functions. We will denote a three-dimensional vector by $\mathbf{R}$, and its projection on the xy plane by $\mathbf{r}$. Likewise, we will denote the corresponding reciprocal space quantities by $\mathbf{Q}$ and $\mathbf{q}$.

Recalling that the FT of $1/R$ is $4\pi/Q^2$, we have

$$\frac{1}{|\mathbf{R} - \mathbf{R}'|} = \int \frac{d^3Q}{(2\pi)^3} \frac{4\pi}{Q^2} e^{i\mathbf{Q}\cdot(\mathbf{R}-\mathbf{R}')}. \tag{B.24}$$

We write the right-hand side as

$$\int \frac{d^2q}{(2\pi)^2} e^{i\mathbf{q}\cdot(\mathbf{r}-\mathbf{r}')} \int \frac{dQ_z}{2\pi} \frac{4\pi}{Q_z^2 + q^2} e^{iQ_z(z-z')}. \tag{B.25}$$

---

[2] This result can also be found by seeking a power series solution.

The integral over Q_z can now be done easily, yielding

$$\frac{1}{|\mathbf{R}-\mathbf{R}'|} = \frac{1}{\left[|\mathbf{r}-\mathbf{r}'|^2 + (z-z')^2\right]^{1/2}} = \int \frac{d^2q}{(2\pi)^2} \frac{2\pi}{q} e^{i\mathbf{q}\cdot(\mathbf{r}-\mathbf{r}')} e^{-q|z-z'|}. \tag{B.26}$$

We now evaluate the $\mathbf{q}$ integral in polar coordinates. Denoting the angles that $\mathbf{r}$, $\mathbf{r}'$, and $\mathbf{q}$ make with the x axis by φ, φ', and α, we have

$$e^{i\mathbf{q}\cdot\mathbf{r}} = e^{iqr\cos(\alpha-\varphi)} = \sum_m i^m J_m(qr) e^{im(\alpha-\varphi)}, \tag{B.27}$$

$$e^{-i\mathbf{q}\cdot\mathbf{r}'} = e^{-iqr\cos(\alpha-\varphi')} = \sum_{m'} (-i)^{m'} J_{m'}(qr') e^{-im'(\alpha-\varphi')}, \tag{B.28}$$

using eq. (B.2) and its complex conjugate. Feeding these expansions into eq. (B.26) and performing the integration over α, we see that terms with $m \neq m'$ vanish, leaving us with

$$\frac{1}{|\mathbf{R}-\mathbf{R}'|} = \frac{1}{\left[|\mathbf{r}-\mathbf{r}'|^2 + (z-z')^2\right]^{1/2}} = \sum_{m=-\infty}^{\infty} \int_0^\infty dq\, J_m(qr) J_m(qr') e^{im(\varphi-\varphi')} e^{-q|z-z'|}. \tag{B.29}$$

This is the desired expansion.

Exercise B.3 Show that $\nabla^2 J_m(qr) e^{im\varphi} e^{\pm qz} = 0$.

Completeness and orthonormality: Since Bessel's equation is of the Sturm-Liouville type, Bessel functions can be used as a basis in which to expand an arbitrary function of r. To see how this works, let us first give a sloppy completeness proof, i.e., use them to represent a delta function. Let us operate on both sides of eq. (B.29) with the Laplacian. On the left, we get $\delta(\mathbf{R}-\mathbf{R}')$. On the right, making use of Bessel's equation, we see that all terms vanish, except that arising from the discontinuity in the slope of $|z-z'|$. Writing the left-hand side in terms of the one-dimensional delta functions of the separate coordinates, we get

$$-4\pi\frac{1}{r}\delta(r-r')\delta(\varphi-\varphi')\delta(z-z') = \sum_{m=-\infty}^{\infty} \int_0^\infty dq\, J_m(qr) J_m(qr') e^{im(\varphi-\varphi')} (-2q)\delta(z-z'). \tag{B.30}$$

Multiplying by $e^{im'\varphi}$ and integrating over φ leaves only one term in the sum on the right. Renaming m' as m, and canceling off $\delta(z-z')$, we obtain

$$\int_0^\infty q\, J_m(qr) J_m(qr')\, dq = \frac{1}{r}\delta(r-r'). \tag{B.31}$$

By interchanging the letters q and r in this equation, we obtain the statement of orthonormality. Note that the functions for any m are separately complete.

Two-dimensional Fourier transforms: Given a function $f(\mathbf{r})$, its Fourier transform in two dimensions is given by

$$f_\mathbf{q} = \int f(\mathbf{r}) e^{-i\mathbf{q}\cdot\mathbf{r}} d^2r. \tag{B.32}$$

Denoting the azimuthal angles in **r** and **q** space by φ_r and φ_q, and using the generating function in the form (B.2), we can write this transform as

$$f_{\mathbf{q}} = \sum_m (-i)^m \int_0^{2\pi} \int_0^\infty f(\mathbf{r}) J_m(qr) e^{im(\varphi_q - \varphi_r)} r \, dr \, d\varphi_r. \tag{B.33}$$

The inverse transform is written down in the same way and is seen to be correct on account of the completeness/orthonormality formula (B.31). Equation (B.33) is sometimes known as the *Fourier-Bessel expansion*. In particular, for a function that depends only on r, the radial distance, we get

$$f_{\mathbf{q}} = 2\pi \int_0^\infty f(r) J_0(qr) r \, dr. \tag{B.34}$$

Exercise B.4 Show that

$$\frac{1}{r} = \int_0^\infty J_0(qr) dq, \qquad \frac{1}{\sqrt{r^2 + z^2}} = \int_0^\infty J_0(qr) e^{-q|z|} dq. \tag{B.35}$$

Modified Bessel functions of the first kind: Let us return to Laplace's equation and solve it by separation of variables in cylindrical coordinates. If we choose the function of z as $e^{\pm ikz}$, and of φ as $e^{\pm im\varphi}$, the radial function will obey the differential equation (with $u = kr$)

$$\left(\frac{d^2}{du^2} + \frac{1}{u} \frac{d}{du} - 1 - \frac{m^2}{u^2} \right) F(u) = 0. \tag{B.36}$$

Solutions of this equation are known as *modified Bessel functions* or *Bessel functions of imaginary argument*, since this equation is the same as eq. (B.17) with u replaced by iu. However, the functions $J_m(iu)$ and $Y_m(iu)$ are impractical to use, so new linear combinations are employed.

The modified Bessel function of the first kind, denoted $I_m(u)$, is defined as

$$I_m(u) = e^{-im\pi/2} J_m(iu). \tag{B.37}$$

Writing $-iu$ for u in the integral (B.5) and recalling that $J_m(-u) = (-1)^m J_m(u)$, we obtain

$$I_m(u) = \frac{1}{2\pi} \int_0^{2\pi} e^{(u \cos \varphi - im\varphi)} d\varphi. \tag{B.38}$$

It follows that $I_m(u)$ is real for real u, that $I_{-m}(u) = I_m(u)$, and that $I_m(-u) = (-1)^m I_m(u)$.

Modified Bessel functions of the second kind: Let us now seek a second linearly independent solution of the differential equation (B.36). To do this, we first ask how I_m behaves as $u \to \infty$. Expanding the integrand in eq. (B.38) in powers of u and integrating, we get

$$I_m(u) = \sum_{n=0}^\infty \frac{\left(\frac{1}{2}u\right)^{m+2n}}{n!(n+m)!} \qquad (m \geq 0). \tag{B.39}$$

This is a series with all positive coefficients, so $I_m(u) \to \infty$ as $u \to +\infty$. There must therefore exist a second solution which vanishes as $u \to \infty$.[3]

Next, let us show that I_m obeys the differential equation directly from the integral form (B.38). We have

$$m^2 I_m(u) = -\frac{1}{2\pi} \int_0^{2\pi} e^{u \cos \varphi} \frac{d^2}{d\varphi^2} e^{-im\varphi} d\varphi. \tag{B.40}$$

Integrating twice by parts, we get

$$m^2 I_m(u) = \frac{1}{2\pi} \int_0^{2\pi} \left[-\frac{d^2}{d\varphi^2} e^{u \cos \varphi} \right] e^{-im\varphi} d\varphi$$

$$= \frac{1}{2\pi} \int_0^{2\pi} (-u^2 \sin^2 \varphi + u \cos \varphi) e^{u \cos \varphi} e^{-im\varphi} d\varphi. \tag{B.41}$$

On the other hand, differentiating directly under the integral sign, we get

$$\left(u^2 \frac{d^2}{du^2} + u \frac{d}{du} - u^2 \right) I_m(u) = \frac{1}{2\pi} \int_0^{2\pi} (u^2 \cos^2 \varphi + u \cos \varphi - u^2) e^{u \cos \varphi} e^{-im\varphi} d\varphi. \tag{B.42}$$

The differential equation is therefore satisfied.

For the second solution to eq. (B.36), let us, therefore, look for an integral representation that will vanish as $u \to \infty$ and that will obey the differential equation. A little bit of trial and error shows that (for $u > 0$)

$$K_m(u) \equiv \frac{1}{2} \int_{-\infty}^{\infty} e^{-u \cosh t - mt} dt = \int_0^{\infty} e^{-u \cosh t} \cosh(mt) dt, \tag{B.43}$$

meets both requirements. The behavior at $u \to \infty$ is immediate, and the differential equation can be seen to be obeyed in exactly the same way as for I_m. K_m is known as the *modified Bessel function of the second kind*. In fact, eq. (B.43) extends to noninteger values of m. We define $K_m(u) = K_m(u)$, $K_m(-u) = (-1)^m K_m(u)$.

The first few $I_m(u)$ and $K_m(u)$ are shown in fig. B.1.

Exercise B.5 Show the recursion relations (which also hold for noninteger m)

$$K_{m+1}(x) - K_{m-1}(x) = \frac{2m}{x} K_m(x), \qquad K_{m+1}(x) + K_{m-1}(x) = -2 K'_m(x). \tag{B.44}$$

Exercise B.6 Use the method of steepest descent to find the leading asymptotic behavior of $K_m(x)$ as $x \to \infty$ for fixed m.

Two-dimensional Green function revisited: There is a beautiful connection between K_0 and the Green function for Laplace's equation, which can be seen from eq. (B.43). The

---

[3] The definition (B.37) and the asymptotic form (B.9) imply that $I_m(u) \sim u^{-1/2} e^u$. The solution that vanishes as $u \to \infty$ must behave as $u^{-1/2} e^{-u}$. This can be seen either by means of the Wronskian or by looking at the differential equation at large u directly.

substitution $w = u \sinh t$ yields

$$K_0(u) = \frac{1}{2} \int_{-\infty}^{\infty} e^{-u \cosh t} dt = \frac{1}{2} \int_{-\infty}^{\infty} \frac{e^{-\sqrt{u^2 + w^2}}}{\sqrt{u^2 + w^2}} dw. \tag{B.45}$$

We now "undo" the square roots by recalling a basic integral identity:

$$K_0(u) = \int_{-\infty}^{\infty} dw \int_{-\infty}^{\infty} \frac{dv}{2\pi} \frac{e^{iv}}{v^2 + (u^2 + w^2)}. \tag{B.46}$$

This is the formula we seek. We can give it a physical look and feel by writing $u = \kappa r$, $v = q_x r$, and $w = q_y r$, and thinking of q_x and q_y as the components of a two-dimensional momentum space vector $\mathbf{q}$. Then,

$$K_0(\kappa r) = 2\pi \int \frac{d^2 q}{(2\pi)^2} \frac{e^{iq_x r}}{q^2 + \kappa^2}. \tag{B.47}$$

If we think of $\mathbf{r}$ as a two-dimensional vector, then by rotational invariance, we can write

$$K_0(\kappa r) = 2\pi \int \frac{d^2 q}{(2\pi)^2} \frac{e^{i\mathbf{q} \cdot \mathbf{r}}}{q^2 + \kappa^2}. \tag{B.48}$$

Equation (B.48) is an important two-dimensional Fourier transform identity. If we act on both sides by $(-\nabla^2 + \kappa^2)$, we get the $\mathbf{q}$-space integral of just $e^{i\mathbf{q} \cdot \mathbf{r}}$, and so

$$(-\nabla^2 + \kappa^2) K_0(\kappa r) = 2\pi \delta^{(2)}(\mathbf{r}). \tag{B.49}$$

This shows that K_0 is a two-dimensional Green function.

Exercise B.7 Show that

$$K_0(\kappa r) = \int_0^{\infty} \frac{J_0(qr)}{q^2 + \kappa^2} q \, dq = \int_0^{\infty} \frac{\cos qr}{\sqrt{q^2 + \kappa^2}} dq. \tag{B.50}$$

Exercise B.8 (addition theorems) Show that

$$J_0(k|\mathbf{r} - \mathbf{r}'|) = \sum_{m=-\infty}^{\infty} J_m(kr) J_m(kr') e^{im\varphi}, \tag{B.51}$$

$$K_0(k|\mathbf{r} - \mathbf{r}'|) = \sum_{m=-\infty}^{\infty} K_m(kr) I_m(kr') e^{im\varphi}, \tag{B.52}$$

where φ is the angle between $\mathbf{r}$ and $\mathbf{r}'$.

The time dependence of fields in electromagnetism is much more conveniently expressed or Fourier analyzed using complex exponentials such as $e^{-i\omega t}$ rather than real trigonometric expressions such as $\sin\omega t$ and $\cos\omega t$, since linear operations of calculus and algebra are then simpler. However, physical quantities such as the energy density and the Poynting vector are bilinear functions of $\mathbf{E}$, $\mathbf{B}$, $\mathbf{D}$, etc., and it then becomes important to be able to find these quantities efficiently from complex expressions. This is especially so when we want their time averages over durations that are long compared to the characteristic time periods. We discuss how to do this in this appendix.

Suppose $\mathbf{u}(t)$ and $\mathbf{v}(t)$ are two quantities (the fields at a given space point, e.g.) varying at frequency ω. We write

$$\mathbf{u}(t) = \mathbf{u}_0 e^{-i\omega t}, \quad \mathbf{v}(t) = \mathbf{v}_0 e^{-i\omega t}, \tag{C.1}$$

where $\mathbf{u}_0$ and $\mathbf{v}_0$ are complex. It is implicitly understood that we must take the real part of these expressions, and the whole point of this technique is not to burden ourselves with notations to this effect.

Suppose next that we need to consider a product like $\mathbf{u}\cdot\mathbf{v}$ and that we only want its average over many cycles of oscillation. Writing out the bilinear product in full, we have

$$\mathbf{u}(t)\cdot\mathbf{v}(t) = \frac{1}{4}(\mathbf{u}_0 e^{-i\omega t} + \mathbf{u}_0^* e^{i\omega t})\cdot(\mathbf{v}_0 e^{-i\omega t} + \mathbf{v}_0^* e^{i\omega t}) \tag{C.2}$$

$$= \frac{1}{4}(\mathbf{u}_0\cdot\mathbf{v}_0^* + \mathbf{u}_0^*\cdot\mathbf{v}_0) + \text{terms varying as } e^{\pm 2i\omega t}. \tag{C.3}$$

When we average over many cycles, the terms varying at frequency 2ω will drop out. Denoting the average by an overbar, we get

$$\overline{\mathbf{u}(t)\cdot\mathbf{v}(t)} = \frac{1}{2}\text{Re}\,(\mathbf{u}_0\cdot\mathbf{v}_0^*). \tag{C.4}$$

This is the basic formula.

We can also state the answer for the cross product:

$$\overline{\mathbf{u}(t) \times \mathbf{v}(t)} = \frac{1}{2} \operatorname{Re} (\mathbf{u}_0 \times \mathbf{v}_0^*). \tag{C.5}$$

As examples, for the time-averaged energy density and the Poynting vector of the monochromatic plane wave (42.1)–(42.3), we have

$$\bar{u} = \frac{1}{8\pi} \boldsymbol{\mathcal{E}}_0 \cdot \boldsymbol{\mathcal{E}}_0^*, \tag{C.6}$$

$$\overline{\mathbf{S}} = \frac{c}{8\pi} \operatorname{Re} \boldsymbol{\mathcal{E}}_0 \times \boldsymbol{\mathcal{B}}_0^*. \tag{C.7}$$

Note that there is no need for a notation to take the real part of the first expression.

Appendix D | Caustics

We discuss in this appendix two common caustics, the rainbow, and the number 3–shaped curve seen in a teacup when light from a compact source falls on it. We showed in section 43 that a caustic is the locus of the centers of curvature of the wave front. An equivalent characterization is that the caustic is the envelope of all the rays originating on that wave front, the so-called evolute of the wave front. It is this second characterization that we shall use to actually find the caustic in both our examples.

We also present a basic discussion of the intensity profile across a caustic, at the level of the Huygens principle and scalar diffraction theory.

The rainbow: A geometrical optics theory of the rainbow requires only a knowledge of the law of refraction. We discuss this law in section 136, which readers already familiar with the law may skip. Wave optics is needed for even the simplest understanding of the intensity and the polarization of the light, but if one only wants to see why one gets a rainbow in the first place and where it is located in the sky, the geometrical theory is enough.

Figure D.1 shows how a spherical water droplet scatters sunlight. The primary rainbow is formed by rays of class 3, the second bow by rays of class 4, and so on. Let us now find the path of these rays. If the initial angle of incidence is α, then at the first refraction, the refracted angle is β, where, with n being the refractive index of the water,

$$\sin \beta = \frac{1}{n} \sin \alpha. \tag{D.1}$$

Further, we see that the ray is turned through an angle $(\alpha - \beta)$ at each refraction and $(\pi - 2\beta)$ at each internal reflection. For a ray of the mth bow, which suffers two refractions and m internal reflections, the scattering angle is, therefore,

$$\chi_m = 2(\alpha - \beta) + m(\pi - 2\beta) = m\pi + 2\alpha - 2(m+1)\beta. \tag{D.2}$$

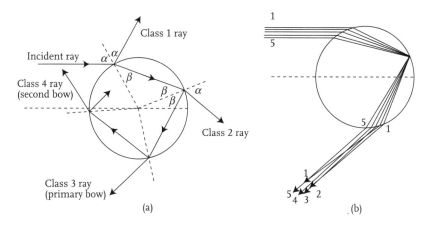

FIGURE D.1. Formation of the rainbow. Part (a) shows the labeling of the rays. In (b) we draw a bundle of rays belonging to the primary bow, showing that the scattering angle is nonmonotonically dependent on the impact parameter. In the diagram, the minimum scattering angle is achieved close to ray 3. One can also see a caustic surface *inside* the droplet, just before the first internal reflection.

This will have an extremum as a function of α where $d\chi_m/d\alpha = 0$. By making use of eq. (D.1), this condition can be solved to give

$$\sin\alpha_m = \left[\frac{(m+1)^2 - n^2}{(m+1)^2 - 1}\right]^{1/2}, \quad \sin\beta_m = \frac{1}{n}\sin\alpha_m. \tag{D.3}$$

At this extremum, the scattering cross section diverges, which is what leads to a caustic. More precisely, since the divergence happens at a certain angle, the caustic in this case is said to be *directional*.

Equation (D.3) has no solution for $m = 0$—the class 2 rays. These rays do not form a bow. Taking $n = 1.33$ for visible light in water, for the first three bows, we get

m	α_m	β_m	χ_m
1	$59.6°$	$40.4°$	$137.5°$
2	$71.9°$	$45.6°$	$129.9°$
3	$76.9°$	$47.1°$	$42.8°$

(D.4)

In making this table we have made use of the facts that χ_m only makes sense modulo 2π. Secondly, if χ_m exceeds π, we can equally well regard the scattering angle as $2\pi - \chi_m$. Neither of these two redefinitions affects the derivation of the extremum.

The first two bows form in a direction almost opposite to the sun and are most commonly seen when the sun is relatively low in the sky. They take the form of circular arcs of angles $\sim 42°$ $(180 - 137.5)$ and $\sim 50°$ $(180 - 129.9)$ about the antisolar point in the sky. Thus, if the sun is at $20°$ to the horizon in the west, the primary and secondary bows will have their maximum heights at about $23°$ and $30°$ to the horizon in the east. (It should also be kept in mind that the angular diameter of the sun is $0.5°$.) The geometry of observation is shown in fig. D.2.

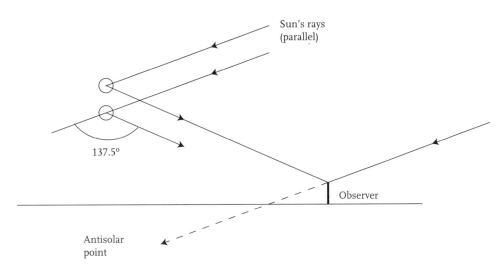

FIGURE D.2. Figure showing where in the sky an observer sees the rainbow. Only the primary bow is illustrated. Note that for a particular observer, the condition for scattering through the rainbow angle is satisfied by only one set of raindrops, and she will see no bow unless raindrops are present at the correct angle.

Second, let us consider the sequence of colors in each bow. Since violet light has a higher refractive index than red, it (violet) is refracted the most. In other words, the scattering angle for the caustic in violet light is larger, and the condition for this light to reach an observer will be satisfied by raindrops that are somewhat lower in the sky. Hence, the primary bow is violet on the inside of the arch and red on the outside. The sequence is reversed in the secondary bow. Under good conditions, one can discern a few pink and green *supernumerary bows* inside the primary bow under the violet. The space between the first two bows (which would be in the geometrical shadow were it not for other scattering) is noticeably darker than the surrounding sky and is known as *Alexander's dark space*. These two bows are fairly common. There are no reliable reports of observations of the third bow, which forms close to the direction of the sun.

Caustic in a teacup: Next, we will find the equation of the caustic in a teacup in the shape of a right circular cylinder, regarding the walls of the cup as a specularly reflecting surface and supposing the light to be incident normal to the axis of the cup. Since the light is incident normal to the axis of the cup, the problem is two dimensional. Let us choose coordinates with origin at the center of the cup, and the x axis along the direction of incidence. Let the cup have a radius R, and consider a ray that strikes it at $x = R\cos\theta$, $y = R\sin\theta$. The angle of incidence is then θ (see fig. D.3), and the reflected ray makes an angle 2θ with the x axis. Hence, it has the equation

$$(y - R\sin\theta) = \tan 2\theta(x - R\cos\theta) \tag{D.5}$$

or

$$\sin 2\theta\, x - \cos 2\theta\, y = R\sin\theta. \tag{D.6}$$

Similarly, the equation of the reflected ray with incidence angle θ' is

$$\sin 2\theta'\, x - \cos 2\theta'\, y = R\sin\theta'. \tag{D.7}$$

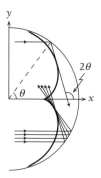

FIGURE D.3. The caustic in a teacup (heavy line). Also depicted is a bundle of rays showing how the caustic is formed.

If we solve eqs. (D.6) and (D.7), we obtain the point of intersection of the two reflected rays:

$$\begin{pmatrix} x \\ y \end{pmatrix} = -\frac{R}{\sin 2(\Delta\theta)} \begin{pmatrix} -\cos 2\theta & \cos 2\theta' \\ -\sin 2\theta & \sin 2\theta' \end{pmatrix} \begin{pmatrix} \sin \theta' \\ \sin \theta \end{pmatrix}, \tag{D.8}$$

where $\Delta\theta = \theta' - \theta$. If we now let $\Delta\theta \to 0$, this point of intersection clearly moves closer and closer to the envelope of the reflected rays, i.e., the caustic, until eventually, it lies on the caustic. In this way we obtain the coordinates of this point:

$$\begin{pmatrix} x \\ y \end{pmatrix} = \frac{R}{2} \begin{pmatrix} \cos\theta\cos 2\theta + 2\sin\theta\sin 2\theta \\ \sin 2\theta\cos\theta - 2\cos 2\theta\sin\theta \end{pmatrix}. \tag{D.9}$$

After some simplification, we obtain

$$x = \frac{R}{2}(3 - 2\cos^2\theta)\cos\theta, \quad y = R\sin^3\theta. \tag{D.10}$$

This is a parametric equation for the caustic. The curve, plotted in fig. D.3, is a classic plane curve known as a *nephroid*.

Intensity profile across a caustic: Let us now see how the wave nature of light softens the intensity singularity at a directional caustic. To keep the discussion simple, we use scalar diffraction theory and limit it to a two-dimensional treatment.

Let us consider an almost plane wave front, propagating along the $+x$ direction, with an equation $y(x)$, as shown in fig. D.4. In the geometrical theory, rays are normal to the wave front. The ray from a point (x, y) on the front reaches a distant observer at an angle $\theta_g = -\tan^{-1}(dx/dy)$, where θ_g is measured from the x axis, as shown in the figure. To form a caustic, the angle θ_g must attain a minimum (or maximum) as we move across the front. If we take $\theta_g = -\alpha y^2$, then since $\theta_g \approx -dx/dy$, the equation for the front is

$$x = \frac{1}{3}\alpha y^3, \tag{D.11}$$

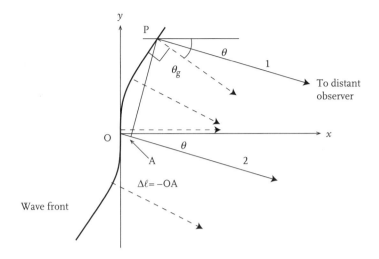

FIGURE D.4. A wave front in the shape of a cubic parabola that gives rise to a directional caustic at the observation angle $\theta_g = 0$. The dashed lines show geometrical optical rays emanating from different points on the wave front, while the lines marked 1 and 2 show the path lengths for secondary wavelets to a distant observer at a general observation angle θ.

where α is a positive constant with dimensions of (length)$^{-2}$. For rainbows formed by water droplets of radius a, we expect $\alpha \sim a^{-2}$. To obtain the EM wave amplitude at the distant observer, we apply the Huygens principle and add the amplitudes from all the wavelets originating at different points on the front. The distance from different parts of the front to the observer may be regarded as a constant as far as the attenuation is concerned, but its variation is essential to finding the phase of each wavelet. It follows from simple geometry that the path length difference for the rays 1 and 2 shown in the figure is

$$\Delta \ell = y \sin \theta - x \cos \theta, \tag{D.12}$$

where θ is now the angle from the observer to a general point on the front and not just the angle of the geometrical ray. Feeding in the form of the front and noting that $\theta \ll 1$, we get

$$\Delta \ell \approx y\theta - \frac{\alpha}{3} y^3. \tag{D.13}$$

If the wave vector is k, then the intensity at an angle θ is given by

$$I(\theta) \sim \left| \int_{-\infty}^{\infty} e^{-ik\Delta\ell(y)} dy \right|^2. \tag{D.14}$$

The integral can be written in terms of an *Airy function*, which is defined as

$$\mathrm{Ai}(x) = \frac{1}{2\pi} \int_{-\infty}^{\infty} \exp\left[-i\left(\frac{1}{3}t^3 + xt \right) \right] dt. \tag{D.15}$$

We obtain

$$I(\theta) = \mathrm{Ai}^2 \left[-\left(\frac{k^2}{\alpha}\right)^{1/3} \theta \right].$$
(D.16)

We discuss the Airy function in more detail in appendix E. $\mathrm{Ai}(x)$ is exponentially decaying for $x > 0$, and an oscillatory function for $x < 0$. The first three maxima of $\mathrm{Ai}^2(x)$ are located at $x = -1.02$, -3.25, and -4.83. Thus, $I(\theta)$ drops rapidly but continuously to zero for $\theta < 0$, replacing the singularity and discontinuous jump in the ray theory. The successive maxima for $\theta > 0$ correspond to the supernumerary bows inside the primary bow.

Appendix E | Airy functions

The Airy function and its natural associates are another set of standard special functions that arise frequently in physics. They can be related to modified Bessel functions of the second kind of order 1/3, 2/3, etc., and though these relationships provide the most convenient route to numerical computation, they also obscure many of the analytic properties of Airy functions. It is therefore advantageous to study the Airy functions in their own right, and they possess a host of beautiful functional properties.

Defining integral: The Airy function is defined for real x via the integral

$$\text{Ai}(x) = \frac{1}{\pi} \int_0^\infty \cos\left(\frac{1}{3}t^3 + xt\right) dt. \tag{E.1}$$

Equivalently,

$$\text{Ai}(x) = \frac{1}{2\pi} \int_{-\infty}^\infty \exp\left[i\left(\frac{1}{3}t^3 + xt\right)\right] dt. \tag{E.2}$$

Differential equation: If we think of x as the frequency ω, we can regard eq. (E.2) as giving us the Fourier transform of $\text{Ai}(\omega)$, which we denote by $\tilde{\text{Ai}}(t)$. We have

$$\tilde{\text{Ai}}(t) = \frac{1}{2\pi} e^{it^3/3}. \tag{E.3}$$

Hence,

$$i\frac{d}{dt}\tilde{\text{Ai}}(t) = -\frac{t^2}{2\pi} e^{it^3/3} = -t^2 \tilde{\text{Ai}}(t). \tag{E.4}$$

The right-hand side is the FT of $\text{Ai}''(\omega)$, while the left-hand side is the FT of $\omega\text{Ai}(\omega)$. It follows that $\text{Ai}(x)$ obeys the differential equation

$$y'' = xy, \tag{E.5}$$

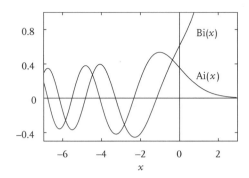

FIGURE E.1. The Airy functions Ai(x) and Bi(x).

where we have written x for ω once again. Ai(x) is that solution which is regular at $x = 0$ and vanishes as $x \to \infty$. A second solution of this equation, which diverges as $x \to \infty$, is denoted Bi(x). Graphs of Ai(x) and Bi(x) are shown in fig. E.1.

Equivalence with Bessel functions: In eq. (E.1), put

$$t = 2\sqrt{x}\sinh\phi. \tag{E.6}$$

Then,

$$\frac{1}{3}t^3 + xt = \frac{8}{3}x^{3/2}\sinh^3\phi + 2x^{3/2}\sinh\phi = \frac{2}{3}x^{3/2}\sinh 3\phi, \tag{E.7}$$

where we have used an identity for $\sinh 3\phi$ in the last step. We now define

$$\zeta(x) = \frac{2}{3}x^{3/2}, \quad u = 3\phi. \tag{E.8}$$

Changing the integration variable to ϕ and then to u, we obtain

$$\text{Ai}(x) = \frac{2}{3\pi}\sqrt{x}\int_0^\infty \cos[\zeta(x)\sinh u]\cosh\left(\frac{1}{3}u\right)du. \tag{E.9}$$

We can now connect eq. (E.9) to $K_{1/3}(\zeta)$. We begin with the defining integral representation (B.43) for $K_\nu(z)$:

$$K_\nu(z) = K_{-\nu}(z) = \frac{1}{2}\int_{-\infty}^\infty e^{-z\cosh t - \nu t}dt. \tag{E.10}$$

The substitution $e^t = s$ gives

$$K_\nu(z) = \frac{1}{2}\int_0^\infty \exp\left[-\frac{1}{2}z(s + s^{-1})\right]s^{-\nu-1}ds. \tag{E.11}$$

Next, we rotate the contour of integration in the s plane onto the positive imaginary axis. The contribution from the large quarter-circle vanishes for appropriately chosen ν. We then make the second substitution

$$s = ie^u. \tag{E.12}$$

Then the range of u is $(-\infty, \infty)$, and

$$\frac{1}{2}(s + s^{-1}) = i \sinh u, \quad s^{-\nu-1} ds = e^{-i\nu\pi/2} e^{-\nu u} du, \tag{E.13}$$

yielding

$$K_\nu(z) = \frac{1}{2} e^{-i\nu\pi/2} \int_{-\infty}^{\infty} e^{-iz\sinh u} e^{-\nu u} du. \tag{E.14}$$

Multiplying this equation by $e^{i\nu\pi/2}$, adding the corresponding equation with $-\nu$ instead of ν, using the fact $K_{-\nu}(z) = K_\nu(z)$, and dividing by $\cos(\nu\pi/2)$, we obtain

$$K_\nu(z) = \frac{1}{\cos(\nu\pi/2)} \int_0^{\infty} \cos(z \sinh u) \cosh(\nu u) du. \tag{E.15}$$

Comparing this result with eq. (E.9), we see that for $x > 0$,

$$\mathrm{Ai}(x) = \frac{1}{\pi} \left(\frac{x}{3}\right)^{1/2} K_{1/3}\left(\frac{2}{3} x^{3/2}\right). \tag{E.16}$$

This is the desired relationship. Without making explicit reference to $K_{1/3}$, we also have the representation

$$\mathrm{Ai}(x) = \frac{1}{\pi} \left(\frac{x}{3}\right)^{1/2} \int_0^{\infty} \exp\left[-\frac{2}{3} x^{3/2} \cosh t\right] \cosh\left(\frac{1}{3} t\right) dt. \tag{E.17}$$

The derivative $\mathrm{Ai}'(x)$: Using the recursion relations (B.44) for K_ν,

$$K_{\nu+1}(x) - K_{\nu-1}(x) = \frac{2\nu}{x} K_\nu(x), \quad K_{\nu+1}(x) + K_{\nu-1}(x) = -2 K_\nu'(x), \tag{E.18}$$

it is not difficult to show that

$$\mathrm{Ai}'(x) = -\frac{1}{\sqrt{3}\pi} x K_{2/3}\left(\frac{2}{3} x^{3/2}\right). \tag{E.19}$$

Formally, we can also differentiate under the integral sign in eq. (E.1) to obtain

$$\mathrm{Ai}'(x) = -\frac{1}{\pi} \int_0^{\infty} t \sin\left(\frac{1}{3} t^3 + xt\right) dt, \tag{E.20}$$

although one should, strictly speaking, first deform the integration contour to ensure convergence.

Behavior at zero and infinity: Both $\mathrm{Ai}(x)$ and $\mathrm{Ai}'(x)$ are regular at $x = 0$. Putting $x = 0$ in eqs. (E.1) and (E.20), and rotating the integration contour by $30°$, we obtain expressions in terms of $\Gamma(1/3)$ and $\Gamma(2/3)$. We content ourselves here by giving the numerical values:

$$\mathrm{Ai}(0) = 0.355028, \quad \mathrm{Ai}'(0) = -0.258819. \tag{E.21}$$

To find the asymptotic behavior of $\mathrm{Ai}(x)$ as $x \to +\infty$, we evaluate the integral (E.2) using the steepest-descent method. We define

$$f(t) = i\left(\frac{t^3}{3} + tx\right). \tag{E.22}$$

From $f'(t) = 0$, the saddle point is seen to be at $t = i\sqrt{x}$ (the other solution leads to a steepest-ascent contour). Since $f(ix^{1/2}) = -(2/3)x^{3/2}$, $f''(ix^{1/2}) = -2x^{1/2}$, and the steepest-descent contour through the saddle point is locally parallel to the real axis, the leading asymptotic behavior is

$$\mathrm{Ai}(x) \sim \frac{1}{2\pi} e^{-2x^{3/2}/3} \int_{-\infty}^{\infty} e^{-\sqrt{x}u^2}\, du$$

$$\sim \frac{1}{2\sqrt{\pi}x^{1/4}} e^{-2x^{3/2}/3} \quad (x \to +\infty). \tag{E.23}$$

The leading asymptotic behavior of $\mathrm{Ai}'(x)$ can be found either from one of the integral representations or by differentiating eq. (E.23):

$$\mathrm{Ai}'(x) \sim -\frac{x^{1/4}}{2\sqrt{\pi}} e^{-2x^{3/2}/3} \quad (x \to +\infty). \tag{E.24}$$

The behavior as $x \to -\infty$ can be given as follows[1]:

$$\mathrm{Ai}(x) \sim \frac{1}{\sqrt{\pi}(-x)^{1/4}} \sin\left(\frac{2}{3}(-x)^{3/2} + \frac{\pi}{4}\right) \quad (x \to -\infty). \tag{E.25}$$

The function Bi(x): For completeness, we discuss the function $\mathrm{Bi}(x)$ briefly. A second solution to Airy's differential equation must necessarily diverge as $x \to +\infty$. $\mathrm{Bi}(x)$ is chosen so as to be 90° out of phase with $\mathrm{Ai}(x)$ as $x \to -\infty$. (More precisely, $\mathrm{Bi}(z)$ is defined for complex argument z as a linear combination of $\mathrm{Ai}(ze^{\pm 2\pi i/3})$ with coefficients $e^{\pm \pi i/6}$.) Its asymptotic behaviors are

$$\mathrm{Bi}(x) \sim \begin{cases} \dfrac{1}{\sqrt{\pi}x^{1/4}} e^{2x^{3/2}/3} & (x \to +\infty), \\[2em] \dfrac{1}{\sqrt{\pi}(-x)^{1/4}} \cos\left(\dfrac{2}{3}(-x)^{3/2} + \dfrac{\pi}{4}\right) & (x \to -\infty). \end{cases} \tag{E.26}$$

The Wronskian with $\mathrm{Ai}(x)$ is

$$W(\mathrm{Ai}, \mathrm{Bi}) = \pi^{-1}. \tag{E.27}$$

The special values at zero are

$$\mathrm{Bi}(0) = 0.614927, \quad \mathrm{Bi}'(0) = 0.448288. \tag{E.28}$$

---

[1] This is not a true asymptotic relation in that it fails near the zeros. For a full discussion of this point, see, e.g., Bender and Orszag (1978), sec. 3.7.

Appendix F | Power spectrum of a random function

Suppose $f(t)$ is a function that extends from $t = -\infty$ to $t = +\infty$, fluctuating in such a way that its mean is zero. The classic example is the noise current in a resistor. The square $f^2(t)$ can then be interpreted as the instantaneous power dissipated, if the units are suitably chosen. The function $f(t)$ can be regarded as a superposition of many Fourier components, and it is of interest to ask how much of the power can be regarded as being dissipated in a given range of frequencies. The analysis that makes this question meaningful, and provides an answer to it, is standard and goes under the name the Wiener-Khinchin[1] theorem.

We shall say that $f(t)$ is *a stationary process* if the integrals

$$\langle f(t) \rangle \equiv \lim_{T \to \infty} \frac{1}{T} \int_{t_0}^{t_0+T} f(t)dt, \tag{F.1}$$

$$\langle f(t)f(t+\tau) \rangle \equiv \lim_{T \to \infty} \frac{1}{T} \int_{t_0}^{t_0+T} f(t)f(t+\tau)dt, \tag{F.2}$$

and similarly defined integrals of products of more than two f's are independent of t_0. In what follows, we will only consider processes in which $\langle f(t) \rangle = 0$. This is not a material restriction, for were it not true, we would simply take the function $f(t) - \langle f(t) \rangle$.

Equation (F.2) defines the *autocorrelation function*. By stationarity, it depends only on τ, so we write

$$G(\tau) = \langle f(t)f(t+\tau) \rangle. \tag{F.3}$$

By writing $t - \tau$ for t, we see that

$$G(\tau) = G(-\tau). \tag{F.4}$$

---

[1] Khinchin is also commonly transliterated as Khintchine and Khintchin.

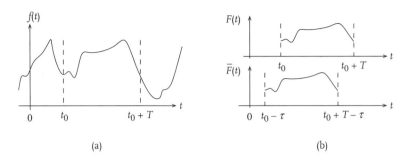

FIGURE F.1. Illustration of a stochastic process showing (a) one realization of $f(t)$ and (b) its unshifted and shifted truncations $F(t)$ and $\bar{F}(t)$.

Further, $G(\tau) \to 0$ as $\tau \to \pm\infty$ (recall that $\langle f(t) \rangle = 0$). In most physical processes, the decay is fairly rapid and is often exponential. We shall also require the integral $\int_{-\infty}^{\infty} |G(\tau)| \, d\tau$ to be finite.

The Wiener-Khinchin theorem states that the average power can be written as

$$P = \int_0^{\infty} d\omega \, W(\omega), \tag{F.5}$$

where W, the *power spectrum* of f, is given by

$$W(\omega) = \frac{1}{\pi} \int_{-\infty}^{\infty} G(\tau) e^{i\omega\tau} d\tau \quad (\omega > 0). \tag{F.6}$$

Some workers use the property that $G(\tau)$ is real and even to define a two-sided power spectrum that extends to negative frequencies also: $W(-\omega) = W^*(\omega) = W(\omega)$.

To show this result, we define (see fig. F.1)

$$F(t) = \begin{cases} f(t), & t_0 \le t \le t_0 + T, \\ 0, & \text{otherwise.} \end{cases}$$

$$\bar{F}(t) = F(t + \tau). \tag{F.7}$$

Note that $\bar{F}(t)$ vanishes outside the interval $t_0 - \tau \le t \le t_0 + T - \tau$. The FTs of $F(t)$ and $\bar{F}(t)$ are given by

$$F_\omega = \int_{t_0}^{t_0+T} f(t) e^{i\omega t} dt, \tag{F.8}$$

$$\bar{F}_\omega = \int_{t_0-\tau}^{t_0+T-\tau} f(t+\tau) e^{i\omega t} dt = F_\omega e^{-i\omega\tau}. \tag{F.9}$$

Secondly, since $F(t)$ and $\bar{F}(t)$ are both real,

$$F_{-\omega} = (F_\omega)^*, \qquad \bar{F}_{-\omega} = (\bar{F}_\omega)^*. \tag{F.10}$$

Let us now suppose that $\tau > 0$ and observe that we may clip an interval of width τ from the integral in eq. (F.3) without making any significant error:

$$G(\tau) = \lim_{T \to \infty} \left[\frac{1}{T} \int_{t_0}^{t_0+T-\tau} f(t)f(t+\tau)\,d\tau + O\left(\frac{f^2\tau}{T}\right) \right]. \tag{F.11}$$

The second term vanishes as $T \to \infty$.[2] We may now write

$$G(\tau) = \lim_{T \to \infty} \frac{1}{T} \int_{-\infty}^{\infty} F(t)\bar{F}(t)\,dt, \tag{F.12}$$

since the integrand vanishes outside the limits t_0 and $(t_0 + T - \tau)$, and equals $f(t)f(t+\tau)$ inside them. We now use Parseval's theorem and obtain

$$G(\tau) = \lim_{T \to \infty} \frac{1}{T} \int_{-\infty}^{\infty} \frac{d\omega}{2\pi} F_{-\omega} \bar{F}_\omega = \lim_{T \to \infty} \frac{1}{T} \int_{-\infty}^{\infty} \frac{d\omega}{2\pi} |F_\omega|^2 e^{-i\omega\tau}. \tag{F.13}$$

The last result follows from eq. (F.9). It allows us to identify the Fourier transform of $G(\tau)$ and thus write

$$W(\omega) = \lim_{T \to \infty} \frac{1}{T} \frac{|F_\omega|^2}{\pi}. \tag{F.14}$$

The theorem itself now follows very simply. The average power is nothing but

$$P = \lim_{T \to \infty} \frac{1}{T} \int_{t_0}^{t_0+T} f^2(t)\,dt = G(0). \tag{F.15}$$

Putting $\tau = 0$ in eq. (F.13) and using eq. (F.14), we obtain

$$P = \int_0^{\infty} d\omega\, W(\omega), \tag{F.16}$$

as desired.

Exercise F.1 Show that $|G(\tau)| \le G(0)$.

As an example of a stationary process, let us consider the so-called telegraph noise. The signal $f(t)$ takes on values $+a$ and $-a$, and switches between them at a mean rate γ. By this we mean that if we take a short interval $\Delta t \ll 1/\gamma$, the probability that the sign of f will change in this interval is $\gamma(\Delta t)$.

---

[2] This argument can be sharpened as follows. Call the part of the integral that has been cut out ΔG:

$$\Delta G = \frac{1}{T} \int_{t_0+T-\tau}^{t_0+T} f(t)f(t+\tau)\,dt.$$

By the Schwartz inequality,

$$(\Delta G)^2 \le \frac{1}{T^2} \int_{t_0+T-\tau}^{t_0+T} f^2(t)\,dt \int_{t_0+T}^{t_0+T+\tau} f^2(t)\,dt.$$

Each one of these integrals is proportional to τ for large τ by the assumptions under which G exists, so $\Delta G \sim \tau/T$.

Let us now find the probability $p(j; \tau)$ that we have j sign changes in an interval τ that may be large compared to $1/\gamma$. We divide the interval into N microscopic slices ϵ such that

$$\tau = N\epsilon, \tag{F.17}$$

and assume that ϵ is so small that the odds of two sign changes in one slice are negligible. The probability of a sign change in one slice is $\gamma\epsilon$; the probability of j sign changes in N slices is

$$p(j; \tau) = \binom{N}{j} (1 - \gamma\epsilon)^{N-j} (\gamma\epsilon)^j. \tag{F.18}$$

Letting $N \to \infty$ and $\epsilon \to 0$ such that the product remains fixed at τ, we get

$$p(j; \tau) = \frac{1}{j!} (\gamma\tau)^j e^{-\gamma\tau}. \tag{F.19}$$

Next, let us calculate the autocorrelation function for such noise. Consider the average

$$\frac{1}{T} \int_{t_0}^{t_0+T} f(t) f(t + \tau) dt \tag{F.20}$$

for $\tau > 0$ and very large T. The integrand will be a^2 for those values of t such that an even number of sign changes take place in the interval $(t, t + \tau)$, and it will be $-a^2$ for those values in which an odd number of sign changes occur. In other words, the average is

$$a^2[p(0; \tau) + p(2; \tau) + \cdots] + (-a^2)[p(1; \tau) + p(3; \tau) + \cdots]. \tag{F.21}$$

The summation is easy, and we get

$$G(\tau) = a^2 e^{-2\gamma|\tau|}, \tag{F.22}$$

where we have allowed for both signs of τ. Note that G is continuous and smooth (except at $\tau = 0$) even though the signal may have discontinuities.

The power spectrum of telegraph noise is therefore given by

$$W(\omega) = \frac{a^2}{\pi} \int_{-\infty}^{\infty} e^{-2\gamma|\tau|} e^{i\omega\tau} d\tau = \frac{a^2}{\pi} \frac{4\gamma}{\omega^2 + 4\gamma^2}. \tag{F.23}$$

The total power is

$$P = \int_0^{\infty} W(\omega) d\omega = a^2, \tag{F.24}$$

exactly as expected.

Although we began by requiring $f(t)$ to be a random process and to have zero mean, the Wiener-Khinchin theorem also applies to some signals for which these conditions are not met, as long as $G(\tau)$ exists. Let us suppose, e.g., that we have a signal that is perfectly harmonic at a single frequency ω_0:

$$f(t) = A \cos \omega_0 t. \tag{F.25}$$

Then its correlation function is easily shown to be

$$G(\tau) = \tfrac{1}{2}A^2 \cos(\omega_0 \tau), \tag{F.26}$$

and the power spectrum is therefore

$$W(\omega) = \frac{A^2}{2\pi} \int_{-\infty}^{\infty} \left[\frac{e^{-i\omega_0\tau} + e^{i\omega_0\tau}}{2} \right] e^{i\omega\tau} d\tau$$

$$= \frac{1}{2}A^2 \delta(\omega - \omega_0). \tag{F.27}$$

This is, of course, exactly what we want: the power spectrum of a purely monochromatic signal should be concentrated entirely at a single frequency. Moreover, the coefficient of the delta function, $\tfrac{1}{2}A^2$, is exactly the time average of $f^2(t)$.

A closely related example is that of a periodic signal. Let the period be T_0, so that

$$f(t) = f(t + T_0). \tag{F.28}$$

We may then expand f in a Fourier series,

$$f(t) = \sum_{n=-\infty}^{\infty} f_n e^{-i\omega_n t}, \tag{F.29}$$

where

$$\omega_n = \frac{2\pi}{T_0}n. \tag{F.30}$$

The Fourier coefficients are given by

$$f_n = \frac{1}{T_0} \int_0^{T_0} f(t) e^{i\omega_n t} dt. \tag{F.31}$$

It is not difficult to show that

$$G(\tau) = f_0^2 + 2\sum_{n\geq 1} |f_n|^2 \cos \omega_n \tau. \tag{F.32}$$

The Fourier transform gives the power spectrum as[3]

$$P(\omega) = f_0^2 \delta_{\text{full}}(\omega) + 2\sum_{n\geq 1} |f_n|^2 \delta(\omega - \omega_n). \tag{F.33}$$

The factor of 2 in the coefficient $2|f_n|^2$ may be unsettling to the reader, but if we write $f(t)$ in real form as

$$f(t) = f_0 + \sum_{n\geq 1} A_n \cos(\omega_n t + \delta_n), \tag{F.34}$$

---

[3] We attach a suffix "full" to $\delta(\omega)$ to indicate that it is to be integrated over a range $\omega = [-\epsilon, \epsilon]$, so that when this integration is done, the dc power, i.e., the component at zero frequency will be f_0^2. The literal application of eq. (F.6) would yield a dc part $2f_0^2\delta(\omega)$, but closer examination would reveal that the $\delta(\omega)$ would then be a "one-sided delta function," which is to be integrated only over the frequency range $[0, \epsilon]$. Said integration would yield an answer of $1/2$, not 1. We choose this nonstandard way of writing the result since the usual tendency is to immediately think of the coefficient multiplying a delta function as the weight at that frequency.

we discover that

$$A_n = 2|f_n|, \tag{F.35}$$

so that the power at frequency ω_n is $A_n^2/2$, as one would immediately conclude from the argument that the time average of $\cos^2(\omega_n t + \delta_n)$ is $1/2$.

In fact, further reflection shows that taking the Fourier transform of the correlation function is the most physical way of discovering any periodicities in a signal. It is therefore the most physical and operationally well-defined procedure for defining the power spectrum. After all, what does one mean by saying that a signal is periodic, and how does one make this meaning operational? One waits to see if the signal repeats, and one waits for longer times the more precisely one wants to find the frequency. The correlation function $G(\tau)$ is the perfect vehicle for extracting the repeating part of a signal. If $f(t)$ consists of a small-amplitude periodic component—the signal—superimposed on a large and random component—the noise, the $G(\tau)$ function will show a sinusoidal behavior at large τ, after any accidental correlations due to the noise have died away. The practical issues of how to calculate power spectral densities of a signal or data stream and how to remove noise from this signal are beyond our scope. A good introductory discussion is given in *Numerical Recipes* by Press et al. (1986), chapter 12.

The earth's magnetosphere provides a wonderful theater for the study of charged particle motion in an inhomogeneous magnetic field. A model that captures many basic features consists of treating the earth's field as a pure dipole field, omitting higher-order multipoles, the influence of the sun's magnetic field, the solar wind, etc. This problem is sometimes named after Carl Stormer, who studied it intensively around 1910.[1]

Let us take the dipole moment to be of magnitude $\mathcal{M}_0$ and pointing to the south, i.e., $-\hat{\mathbf{z}}$. ($\mathcal{M}_0 = 8.1 \times 10^{25}$ G cm^3 for the earth—see exercise 73.1.) For a charge q, which we will regard as positive, the relativistically correct equation of motion is

$$\dot{\mathbf{p}} = -\frac{q\mathcal{M}_0}{cr^5}\dot{\mathbf{r}} \times (3z\mathbf{r} - r^2\hat{\mathbf{z}}). \tag{G.1}$$

It should be stated at the outset that this equation is nonintegrable and is believed to show chaos. The reason is that we know only two constants of motion, which is not enough. One constant is the energy, or v, or p, equivalently. The second is a variable like the z component of the angular momentum, as we shall see below. To find individual trajectories, therefore, one must integrate the equations numerically. We shall focus instead on trying to understand the problem qualitatively, in particular, the conditions under which one can get trapped orbits, i.e., orbits where the particle does not escape to infinity. We will then see how we can understand these same features using the guiding center approximation.

Our first step is to cast eq. (G.1) into dimensionless form. It is easy to check that $q\mathcal{M}_0/c$ has dimensions of momentum times the square of length. Let the energy, momentum,

---

[1] Stormer, *The Polar Aurora* (1955). In light of the title of Stormer's book, it should be noted that a full understanding of the aurora entails several other phenomena, in particular, the solar wind.

and speed of the particle be denoted $\mathcal{E}_0$, p_0, and v_0, respectively. The quantities

$$\ell_{St} = \left(\frac{q\mathcal{M}_0}{p_0 c}\right)^{1/2}, \quad t_{St} = \frac{\ell_{St}}{v_0}, \tag{G.2}$$

will have dimensions of length and time, respectively. If we express all lengths in units of ℓ_{St} and times in units of t_{St}, and remember that $\mathbf{p} = \mathcal{E}_0 \mathbf{v}/c^2$ and that $\mathcal{E}_0$ is a constant, eq. (G.1) becomes

$$\ddot{\mathbf{r}} = -\frac{\dot{\mathbf{r}} \times (3z\mathbf{r} - r^2\hat{\mathbf{z}})}{r^5}. \tag{G.3}$$

Next, let us resolve eq. (G.3) into components in cylindrical coordinates (ξ, φ, z), where ξ is the distance from the axis of symmetry. We obtain

$$\ddot{z} = \frac{z\xi^2}{(\xi^2 + z^2)^{5/2}}\dot{\varphi}, \tag{G.4}$$

$$(\ddot{\xi} - \xi\dot{\varphi}^2) = -\frac{(2z^2 - \xi^2)}{(\xi^2 + z^2)^{5/2}}\xi\dot{\varphi}, \tag{G.5}$$

$$(\xi\ddot{\varphi} + 2\dot{\xi}\dot{\varphi}) = -\frac{3\xi z\dot{z} + (\xi^2 - 2z^2)\dot{\xi}}{(\xi^2 + z^2)^{5/2}}. \tag{G.6}$$

The last equation can be integrated once. If we multiply across by ξ, the left-hand side is $d(\xi^2\dot{\varphi})/dt$, while the right-hand side can be seen to be $(d/dt)(\xi^2/r^3)$. Thus,

$$\xi^2\dot{\varphi} - \frac{\xi^2}{r^3} = 2\gamma, \tag{G.7}$$

where γ is a constant (Stormer's original notation). This is the angular-momentum-like constant of motion mentioned above. The other constant of motion can be expressed in terms of the square of the velocity. In terms of our dimensionless variables, this condition is

$$\dot{\xi}^2 + \xi^2\dot{\varphi}^2 + \dot{z}^2 = 1. \tag{G.8}$$

If we solve eq. (G.7) for $\dot{\varphi}$ and substitute the answer in eqs. (G.4) and (G.5), we will get equations of motion for the particle in the ξ-z, or the meridianal, plane. It is simpler, however, to proceed as follows. Using eq. (G.7) to eliminate $\dot{\varphi}$ from eq. (G.8), we get

$$\dot{\xi}^2 + \dot{z}^2 + V(\xi, z) = 1, \tag{G.9}$$

where

$$V(\xi, z) = \left(\frac{2\gamma}{\xi} + \frac{\xi}{r^3}\right)^2. \tag{G.10}$$

(Recall that $r^2 = \xi^2 + z^2$.) Equation (G.9) is exactly the equation of energy conservation for a particle moving in two dimensions, ξ and z, in a potential energy $\frac{1}{2}V(\xi, z)$. Hence, it must be the case that

$$2\ddot{\xi} = -\frac{\partial V}{\partial \xi}, \quad 2\ddot{z} = -\frac{\partial V}{\partial z}. \tag{G.11}$$

We leave it to the reader to check this.

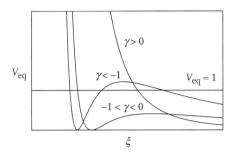

FIGURE G.1. Effective potential for motion in the equatorial plane.

The meridianal plane problem cannot be integrated any further. To progress further, it is useful to look at the special case of motion lying entirely in the equator, i.e., at trajectories with $z = \dot{z} = 0$. Then, $r = \xi$, and eq. (G.9) reduces to

$$\dot{\xi}^2 + V_{eq}(\xi) = 1, \tag{G.12}$$

with

$$V_{eq}(\xi) = V(\xi, 0) = \left(\frac{2\gamma}{\xi} + \frac{1}{\xi^2}\right)^2. \tag{G.13}$$

The problem is now one dimensional. We plot the potential V_{eq} in fig. G.1. It is obvious that the motion must be restricted to values of ξ such that $V_{eq} \leq 1$. If $\gamma > 0$, ξ is always greater than some value ξ_0. A particle starting at very large ξ with $\dot{\xi} < 0$ will approach the earth until it reaches $\xi = \xi_0$, where it will turn around and escape to infinity. Thus, we cannot have trapping for $\gamma > 0$. If $\gamma < 0$, the same behavior will arise if the maximum in V_{eq}, V_0, is less than 1. Trapping is possible only if $V_0 > 1$, since the particle can then bounce back and forth between ξ_1 and ξ_2. Since the maximum occurs at $\xi = -1/\gamma$, and $V_0 = \gamma^4$, we must have $\gamma \leq -1$ for equatorial plane trapping.

The full orbit can be found by first integrating eq. (G.12) to get $\xi(t)$ and then integrating eq. (G.7) with this result to get $\varphi(t)$. For trapped orbits, $\xi_1 < -1/2\gamma < \xi_2 < -1/\gamma$. Thus, at ξ_2, $\dot{\varphi} < 0$, and at ξ_1, $\dot{\varphi} > 0$. This is exactly what we expect from the combination of cyclotron and guiding center motion. The net drift is westward, since $q > 0$. For negative charges, all orbits can be obtained by reflecting in a meridianal plane.

Exercise G.1 Analyze the motion in the equatorial plane when γ is large and negative.

Solution: The motion in ξ is restricted to a small range near the bottom of the potential $V_{eq}(\xi)$, which may therefore be approximated as a parabola. Writing $\xi = \xi_0 + \eta$, where $\xi_0 = -1/2\gamma$, we get $V_{eq}(\xi) = 64\gamma^6\eta^2$. Thus, $\eta(t) = \eta_1 \cos \omega_\xi t$, where $\omega_\xi = 8|\gamma|^3$, and $64\gamma^6\eta_1^2 = 1$. From eq. (G.7), we get

$$\dot{\varphi} = \frac{2\gamma}{\xi^2} + \frac{1}{\xi^3} \approx -16\gamma^4\eta - 96\gamma^5\eta^2 + \cdots. \tag{G.14}$$

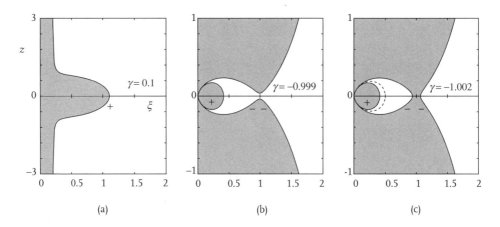

FIGURE G.2. The three characteristic patterns of allowed and forbidden regions in the Stormer problem. The forbidden regions are shaded. The full three-dimensional picture may be visualized by rotating the figure about the z axis. The $+$ and $-$ signs indicate the sign of v_φ on the boundary of the allowed region. In (c), the dashed line is the locus where v_φ vanishes.

To first order in η_1, $\varphi = \varphi_0 + 2\gamma\eta_1 \sin \omega_\xi t$—along with the solution for η, this is the cyclotron motion. Keeping the second-order term, we obtain an average azimuthal angular velocity $\langle \dot{\varphi} \rangle = -48\gamma^5\eta_1^2$. Hence, $\langle \dot{\varphi} \rangle / \omega_\xi = 6\gamma^2\eta_1^2 = \frac{3}{2}(\eta_1/\xi_0)^2$. Since η_1 and ξ_0 are, respectively, the dimensionless cyclotron radius and distance of the particle from the center of the earth, this is the same result found in exercise 73.1. We leave it to the reader to restore dimensional units and show that ω_ξ and η_1 are the cyclotron frequency and radius in the B field at a distance ξ_0 for a particle with momentum p_0.

Let us now consider the three-dimensional case. It follows from eq. (G.9) that the region of the ξ-z plane where $V(\xi, z) > 1$ is energetically forbidden. We can map this region by solving the equation $V = 1$ and using the results from the equatorial case. This equation has two branches,

$$\left(\frac{2\gamma}{\xi} + \frac{\xi}{r^3} \right) = \pm 1, \tag{G.15}$$

which may be rewritten as

$$z^2 = \left(\frac{\xi^2}{\pm\xi - 2\gamma} \right)^{2/3} - \xi^2. \tag{G.16}$$

As might be expected, the nature of the answer depends on γ. If $\gamma > 0$, only the branch with the $+$ sign matters. The forbidden region is shown in fig. G.2(a). As found earlier, when $z = 0$, only values of ξ greater than some ξ_0 are allowed. For such γ, the orbit can never approach the dipole, and the particle is never trapped. Nevertheless, such orbits are relevant to cosmic rays.

If $-1 < \gamma \leq 0$, the allowed region of the ξ axis is similar to the previous case, but now the $-$ sign branch in eq. (G.15) also enters (see fig. G.2(b)). The low and high ξ regions are connected by a narrow neck near $\xi = -1$ in the interesting case of γ close to -1. Particles can escape to infinity through this neck. Of course, the possibility of escape does not mean that escape is certain, and indeed there are infinitely many isolated periodic orbits that are trapped. However, work by Stormer and others shows that these are unstable, i.e., they become untrapped if perturbed even slightly. Conversely, particles can approach the dipole from infinity. These orbits are important for the auroral particles.

Stable trapped orbits appear when $\gamma \leq -1$. This is evident from fig. G.2(c), since the allowed region has two disconnected parts, an inner one and an outer one. Particles in the inner region can never reach the outer one. The full three-dimensional region can be visualized by rotating the figure about the z axis. We can understand the trajectories qualitatively as follows. First, we see from eq. (G.7) that

$$v_\varphi = \xi\dot\varphi = \frac{2\gamma}{\xi} + \frac{\xi}{r^3}. \tag{G.17}$$

Thus, $v_\varphi = \pm 1$ on the line (or surface in three dimensions) $V(\xi, z) = 1$. In fact, we can see that v_φ equals -1 on the outer line and $+1$ on the inner line. Further, on these lines $\dot\xi^2 + \dot z^2$ vanishes, i.e., $\dot\xi$ and $\dot z$ both vanish, so these lines are turning points of the meridianal motion, and on them the particle moves purely in the azimuthal direction. Next let us consider the line $V(\xi, z) = 0$. This is the locus on which $v_\varphi = 0$, so the particle's velocity lies entirely in the meridianal plane. The ξ and z coordinates of the particle oscillate back and forth across this line. All the while, the meridianal plane containing the particle spins about the z axis. However, every time the line $V = 0$ is crossed, we have a turning point of the azimuthal motion, and the direction of spin of the meridianal plane reverses. In spherical polar coordinates with θ as the polar angle (colatitude), $\xi = r \sin\theta$, and the condition $V = 0$ is

$$r = \frac{-1}{2\gamma} \sin^2\theta. \tag{G.18}$$

But this is precisely the equation of a line of force for a magnetic dipole (see exercise 21.2)! And so we arrive at exactly the picture formed on the basis of the guiding center approximation. The particle spirals up and down a line of force while executing cyclotron motion around it. At the same time, the particle drifts from one line of force to another.

It is difficult, however, to analyze the motion in the direction of the field lines quantitatively. Let us try and proceed as in exercise G.1, taking $\gamma \ll -1$ and assuming that the motion is limited to small values of z. Defining $\xi = -(2\gamma)^{-1} + \eta$ and expanding in powers of η and z, we get

$$V(\xi, z) \simeq 64|\gamma|^6(\eta + 3|\gamma|z^2)^2 + \cdots. \tag{G.19}$$

The equation of motion for z is now $\ddot z \simeq -384|\gamma|^7\eta z$, keeping only lowest-order terms. Even if we assume that z and $\dot z$ are small initially, the structure of this equation does not permit us to conclude that z will remain small forever. If we simply compare the two terms in V, we see that $\eta \sim |\gamma|^{-3}$, while $z \sim |\gamma|^{-2}$. This suggests that z will even become larger than η.

It is possible to analyze the trajectories much more completely in the guiding center approximation. This approximation holds provided $r_c \nabla B \ll B$. For a particle at a distance

R from the earth's center, $\nabla B \simeq B/R$, and $r_c \simeq cp_0/qB \simeq (cp_0/q\mathcal{M}_0)R^3 = R^3/\ell_{St}^2$, so the condition can be expressed as $r_c \ll R$, or $R^2 \ll \ell_{St}^2$, or

$$p_0 c \ll \frac{q\mathcal{M}_0}{R^2}. \tag{G.20}$$

The combination $e\mathcal{M}_0/R_e^2$ equals 60 GeV. (R_e is the radius of the earth.) There are two radiation belts surrounding the earth. The outer belt contains electrons with energies up to 1 MeV and extends up to 6–8 R_e. For all these particles, the condition (G.20) is comfortably satisfied. The inner belt contains protons with energies of 10–100 MeV and extends up to $\sim 2R_e$. Again, the condition is well satisfied for all the particles in this belt.

We have already analyzed the drift across lines of longitude in exercise G.1, at least for particles confined to the equatorial plane. When the latter restriction is lifted, the qualitative aspects of this drift will not change. The more interesting behavior is the bounce motion along lines of B, which coincide with lines of longitude. To study this, it is better to use spherical polar coordinates, but instead of the polar angle, θ, to use the latitude $\lambda = \pi/2 - \theta$. The equation of a line of B is then

$$r = r_{eq}\cos^2\lambda, \tag{G.21}$$

where r_{eq} is the distance to the line in the equatorial plane. The field at this distance is $B_{eq} = \mathcal{M}_0/r_{eq}^3$. Either one of the quantities B_{eq} or r_{eq} specifies a line of B. Along the line (see eq. (21.23)),

$$B(\lambda) = B_{eq}\frac{(1+3\sin^2\lambda)^{1/2}}{\cos^6\lambda}. \tag{G.22}$$

It is also convenient to define the arc length s along the line, measuring it from the equatorial point. Since

$$(ds)^2 = (dr)^2 + r^2(d\lambda)^2, \tag{G.23}$$

we obtain, upon using eq. (G.21),

$$\frac{ds}{d\lambda} = r_{eq}\cos\lambda(1+3\sin^2\lambda)^{1/2}. \tag{G.24}$$

Let us suppose that the velocity of a particle at the equator makes an angle α_{eq} with the line of force. Then, eq. (74.19) gives

$$v_\parallel = \frac{ds}{dt} = v_0\left[1 - \sin^2\alpha_{eq}\frac{B(s)}{B_{eq}}\right]^{1/2}. \tag{G.25}$$

This velocity will vanish, and the particle will mirror at a latitude λ_m, such that $B(\lambda_m) = B_{eq}/\sin^2\alpha_{eq}$. Hence,

$$\frac{\cos^6\lambda_m}{(1+3\sin^2\lambda_m)^{1/2}} = \sin^2\alpha_{eq}. \tag{G.26}$$

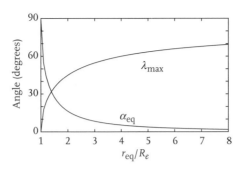

FIGURE G.3. Maximum latitude (λ_{max}) and equatorial loss cone angle (α_{eq}) for particles in the earth's magnetic field as a function of equatorial distance.

The particle will bounce back and forth between latitudes $\pm\lambda_m$. The time period of this bouncing is

$$T_B = \int_0^{s(\lambda_m)} \frac{4}{v_\parallel(s)} ds. \tag{G.27}$$

Changing the variable of integration from s to λ and using eqs. (G.22), (G.24), and (G.25), we get

$$T_B = \frac{4r_{eq}}{v_0} \int_0^{\lambda_m} \left[1 - \sin^2 \alpha_{eq} \frac{(1 + 3\sin^2 \lambda)^{1/2}}{\cos^6 \lambda}\right]^{-1/2} (1 + 3\sin^2 \lambda)^{1/2} \cos \lambda \, d\lambda. \tag{G.28}$$

The integral is a number of order 1, depending only on α_{eq}. For a 10-keV electron at $4 R_e$, $4r_{eq}/v_0$ is 1.7 sec, and the cyclotron orbital time is 7.3×10^{-5} sec. This confirms the overall picture of the motion—a very rapid cyclotron orbit, a much slower bounce motion between lines of latitude, and an even slower longitudinal drift around the earth.

The mirroring latitude λ_m is a decreasing function of α_{eq} [eq. (G.26)] and is independent of r_{eq}. However, a line with a given r_{eq} hits the surface of the earth at a latitude given by $\cos^2 \lambda_{max} = (R_e/r_{eq})$; λ_{max} is an increasing function of r_{eq}. Obviously, particles with orbits with λ_m greater than $\lambda_{max}(r_{eq})$ are not trapped; they hit the earth before they can mirror, and are lost. The corresponding loss cone in the angle α_{eq} is greater for particles that cross the equator nearer to the earth. The dependence of λ_{max} and α_{eq} on r_{eq} is shown in fig. G.3.

The bounce motion clearly becomes simple harmonic if λ_m is small. In this case, all quantities can be expanded in powers of λ. We obtain

$$B(\lambda) = B_{eq}(1 + \tfrac{9}{2}\lambda^2), \tag{G.29}$$

$$\alpha_{eq} = \tfrac{\pi}{2} - 3\lambda_m, \tag{G.30}$$

$$v_\parallel(\lambda) = \tfrac{9}{2}v_0(\lambda_m^2 - \lambda^2), \tag{G.31}$$

$$(ds/d\lambda) = r_{eq}(1 + \lambda^2). \tag{G.32}$$

Hence,

$$\frac{d\lambda}{dt} = \frac{3}{\sqrt{2}} \frac{v_0}{r_{eq}} (\lambda_m^2 - \lambda^2)^{1/2}. \tag{G.33}$$

The solution *is* simple harmonic. The bounce time period in this limit is given by

$$T_B = \frac{2\sqrt{2}\pi}{3} \frac{r_{eq}}{v_0}. \tag{G.34}$$

Since the particle velocity lies almost entirely in the equatorial plane, the result (73.12) for the time to make one drift orbit around the earth is applicable. The ratio of the drift time to the bounce time is easily seen to be

$$\frac{T_D}{T_B} = \sqrt{2}\frac{r_{eq}}{r_c} = \sqrt{2}\left(\frac{\ell_{St}}{r_{eq}}\right)^2. \tag{G.35}$$

It is also apparent that we can integrate the equation for the longitudinal drift of the guiding center in the general case. For this, and for a comparison between the actual particle trajectories and the guiding center motion, we refer the reader to the text by Alfven and Falthammar (1963).

In section 130, we discussed the Maxwell receding image construction for the eddy currents induced in a thin metal sheet by a monopole that materializes abruptly above it at some instant. In this appendix, we give an alternative, constructive proof of this solution using Fourier transform methods. The notation follows section 130. We use $\mathbf{R}$ and $\mathbf{r}$ to denote points in the full three-dimensional space and on the metal sheet, respectively, and $\mathbf{Q}$ and $\mathbf{q}$ for the corresponding reciprocal space quantities.

Our starting point is the result

$$\frac{1}{|\mathbf{R}|} = \int \frac{d^3Q}{(2\pi)^3} \frac{4\pi}{Q^2} e^{i\mathbf{Q}\cdot\mathbf{R}}. \tag{H.1}$$

We write the right-hand side as

$$\int_{\mathbf{q}} e^{i\mathbf{q}\cdot\mathbf{r}} \int \frac{dq_z}{2\pi} \frac{4\pi}{q_z^2 + q^2} e^{iq_z z}. \tag{H.2}$$

(We have resorted to the shorthand $\int_{\mathbf{q}}$ for $\int (d^2q/4\pi^2)$, and we will use a similar shorthand $\int_{\mathbf{r}}$ for $\int d^2r$.) The integral on q_z can now be evaluated easily, yielding

$$\frac{1}{|\mathbf{r} + z\hat{\mathbf{z}}|} = \int_{\mathbf{q}} e^{i\mathbf{q}\cdot\mathbf{r}} \left(\frac{2\pi}{q} e^{-q|z|} \right). \tag{H.3}$$

This is a very useful result. We may think of it as a two-dimensional Fourier transform of the function $|\mathbf{r} + z\hat{\mathbf{z}}|^{-1}$ with respect to $\mathbf{r}$, with z as a parameter. By differentiating with respect to z, another useful identity is obtained:

$$\frac{z}{(r^2 + z^2)^{3/2}} = 2\pi \int_{\mathbf{q}} (\text{sgn }z) e^{i\mathbf{q}\cdot\mathbf{r}} e^{-q|z|}. \tag{H.4}$$

The basic strategy is to use the boundary conditions (130.2) and (130.7) to find the magnetic scalar potential. Instead of the current $\mathbf{K}$, however, it is easier to use the equivalence of a current loop and a dipole sheet as established in section 22, and find the scalar potential produced by this dipole sheet. Consider the annulus between radii r and $r + dr$ on the metal sheet. The current flowing through this annular loop is $K_\varphi dr$. Hence, the magnetic moment density $\mathbf{m}(\mathbf{r}, t) = m(\mathbf{r}, t)\hat{\mathbf{z}}$ on the equivalent dipole sheet is

$$m(\mathbf{r}, t) = \frac{1}{c} \int_r^\infty K_\varphi(\mathbf{r}', t) dr'. \tag{H.5}$$

The scalar potential at a point $\mathbf{R} = \mathbf{r} + z\hat{\mathbf{z}}$ is then given by

$$\phi^{\text{ind}}(\mathbf{R}, t) = \int_{\mathbf{r}'} \frac{\mathbf{m}(\mathbf{r}', t) \cdot (\mathbf{R} - \mathbf{r}')}{|\mathbf{R} - \mathbf{r}'|^3} = \int_{\mathbf{r}'} m(\mathbf{r}', t) \frac{z}{[(\mathbf{r} - \mathbf{r}')^2 + z^2]^{3/2}}. \tag{H.6}$$

The right-hand side is in the form of a convolution integral over $\mathbf{r}'$. It is therefore easy to Fourier transform it with respect to $\mathbf{r}$. Making use of the identity (H.4), we obtain

$$\phi^{\text{ind}}(\mathbf{q}, z, t) = 2\pi(\text{sgn } z)m(\mathbf{q}, t)e^{-q|z|}. \tag{H.7}$$

Since $\mathbf{B}^{\text{ind}}(\mathbf{R}, t) = -\nabla\phi^{\text{ind}}(\mathbf{R}, t)$, we get

$$\mathbf{B}^{\text{ind}}(\mathbf{q}, z, t) = -(i\mathbf{q} + \hat{\mathbf{z}}\partial_z)\phi^{\text{ind}}(\mathbf{q}, z, t)$$
$$= -[i\mathbf{q} - q(\text{sgn } z)\hat{\mathbf{z}}]2\pi(\text{sgn } z)m(\mathbf{q}, t)e^{-q|z|}. \tag{H.8}$$

In particular, the z component of the magnetic field has the FT

$$B_z^{\text{ind}}(\mathbf{q}, z, t) = 2\pi q e^{-q|z|} m(\mathbf{q}, t). \tag{H.9}$$

We now enforce the boundary condition (130.7), which we repeat for convenience:

$$\left(\frac{\partial}{\partial t} \mp v_0 \frac{\partial}{\partial z} \right) B_z^{\text{ind}} = -\frac{\partial}{\partial t} B_z^{\text{dir}} \quad (z = 0\pm). \tag{H.10}$$

Let us take the two-dimensional Fourier transform of both sides. Using eq. (H.9) on the left-hand side, we get

$$2\pi q \left(\frac{\partial}{\partial t} + v_0 q \right) m(\mathbf{q}, t). \tag{H.11}$$

For the right-hand side, we need the field due to the source monopole. We have

$$B_z^{\text{dir}}(\mathbf{r}, t) = -q_m \frac{h}{(r^2 + h^2)^{3/2}} \theta(t). \tag{H.12}$$

We now make use of the identity (H.4) with h for z to Fourier transform the spatial part. Further, since $d\theta(t)/dt = \delta(t)$, the right-hand side of the boundary condition is

$$-\frac{\partial}{\partial t}\text{F.T.}\left[B^{\text{dir}}(\mathbf{r}, t) \right] = 2\pi q_m e^{-qh}\delta(t), \tag{H.13}$$

and the condition itself becomes

$$\left(\frac{\partial}{\partial t} + v_0 q \right) m(\mathbf{q}, t) = q_m \frac{e^{-qh}}{q}\delta(t). \tag{H.14}$$

This is an easy-to-solve differential equation for $m(\mathbf{q}, t)$. We rewrite it as

$$\frac{\partial}{\partial t}\left[e^{qv_0 t}m(\mathbf{q}, t)\right] = q_m \frac{e^{-qh}}{q} e^{qv_0 t}\delta(t),$$ (H.15)

and integrate both sides from $-\infty$ to t. Thus,

$$m(\mathbf{q}, t) = q_m \frac{e^{-q(h+v_0 t)}}{q}\theta(t).$$ (H.16)

Inverting the Fourier transform with the aid of eq. (H.3), and omitting the $\theta(t)$ factor with the understanding that all quantities are given for $t \geq 0$ only, we have

$$m(\mathbf{r}, t) = \frac{q_m}{2\pi}\frac{1}{\sqrt{r^2 + h^2(t)}},$$ (H.17)

where we have abbreviated $h(t) = h + v_0 t$.

The scalar potential is now easily found. Equation (H.7) gives

$$\phi^{\text{ind}}(\mathbf{q}, z, t) = 2\pi q_m(\text{sgn } z)\frac{e^{-q\left(|z|+h(t)\right)}}{q}.$$ (H.18)

Using eq. (H.3) one last time, we obtain, finally,

$$\phi^{\text{ind}}(\mathbf{R}, t) = \pm\frac{q_m}{|\mathbf{R} \pm h(t)\hat{z}|} \quad (z \gtrless 0).$$ (H.19)

This is the potential due to Maxwell's image monopoles as found in section 130. The current distribution is given by eq. (H.5) as

$$K_\varphi(\mathbf{r}, t) = -c\frac{\partial}{\partial r}m(\mathbf{r}, t) = \frac{q_m c}{2\pi}\frac{r}{\left(r^2 + (h + v_0 t)^2\right)^{3/2}},$$ (H.20)

also as found before.

Bibliography

This bibliography lists books that are cited in the text, with the section numbers where they are cited in square brackets. References to journal articles appear in the text, in footnotes.

Abramowitz, M., and Stegun, I. A., *Handbook of Mathematical Functions, with Formulas, Graphs, and Mathematical Tables*, 10th printing, National Bureau of Standards, U. S. Govt. Printing Office, Washington, DC (1972) [secs. 49, 90, appendix B].

Aharoni, A., *Theory of Ferromagnetism*, Cambridge University Press, Cambridge, UK (2000) [sec. 106].

Alfven, H., and Falthammar, C. G., *Cosmical Electrodynamics*, 2nd edition, Oxford University Press, Oxford, UK (1963) [sec. 73, appendix G].

Arnold, V. I., *Mathematical Methods of Classical Mechanics*, Springer-Verlag, New York (1978) [sec. 74].

Ashcroft, N. A., and Mermin, N. D., *Solid State Physics*, Thomson Learning, New York (1976) [secs. 94, 123].

Balian, R., *From Microphysics to Macrophysics*, Springer-Verlag, Berlin (1991) [sec. 97].

Baym, G., *Lectures on Quantum Mechanics*, Benjamin/Cummings, Menlo Park, CA (1969) [secs. 25, 26, 47, 147, 168].

Bender, C. M., and Orszag, S. A., *Advanced Mathematical Methods for Scientists and Engineers*, McGraw-Hill, New York (1978) [chap. 2, appendix E].

Berestetskii, V. B., Lifshitz, E. M., and Pitaevskii, L. P., *Relativistic Quantum Theory* (Vol. 4 of the *Course of Theoretical Physics* by L. D. Landau and E. M. Lifshitz), Pergamon Press, Oxford, UK (1964) [sec. 47].

Boas, M. L., *Mathematical Methods in the Physical Sciences*, 3rd edition, Wiley, New York (2005) [chap. 2].

Born, M., and Wolf, E., *Principles of Optics*, Pergamon Press, Oxford, UK (1975) [sec. 43].

Brillouin, L., *Wave Propagation and Group Velocity*, Academic Press, New York (1960) [sec. 134].

Brown, W. F., *Magnetoelastic Interactions*, Springer, Berlin (1966) [sec. 4].

Copson, E. T., *Introduction to the Theory of Functions of a Complex Variables*, Oxford University Press, Oxford, UK (1970) [chap. 2, secs. 90, 112, appendix A].

Courant and Hilbert, *Methods of Mathematical Physics*, Vols. I & II, 1st English edition, Interscience, New York (1953) [chap. 2, appendix A].

Darrigol, O., *Electrodynamics from Einstein to Ampere*, Oxford University Press, Oxford, UK (2000) [sec. 2].

Dell, R. M., and Rand, D. A. J., *Understanding Batteries*, Royal Society of Chemistry, Cambridge, UK (2001) [sec. 115].

Erdelyi, A., Magnus, W., Oberhettinger, F., and Tricomi, F. G., *Higher Transcendental Functions*, Vols. I & II, based, in part, on notes left by Harry Bateman and compiled by the Staff of the Bateman Manuscript Project, McGraw-Hill, New York (1953) [appendix A].

Eyges, L., *The Classical Electromagnetic Field*, Addison-Wesley, Reading, MA (1972) [sec. 29].

Feynman, R. P., *Statistical Mechanics*, Benjamin/Cummings, Reading, MA (1972) [sec. 139].

Feynman, R. P., Leighton, R. B., and Sands, M., *The Feynman Lectures in Physics*, Vols. I, II, and III, Addison-Wesley, Reading, MA (1964) [secs. 6, 8, 14, 25, 29, 116, 151].

Fitzpatrick, R., *Plasma Physics: A Graduate Level Course*, self-published by the author (2006) [sec. 132].

Flanders, H., *Differential Forms with Applications to the Physical Sciences*, Dover, New York (1989) [sec. 8].

Goldstein, H., Poole, C., and Safko, J., *Classical Mechanics*, Addison-Wesley, New York (1992) [sec. 34].

Gottfried, K., *Quantum Mechanics*, Benjamin, NY (1966) [sec. 47, appendix A].

Hardy, G. H., *A Course of Pure Mathematics*, 10th edition, Cambridge University Press, Cambridge, UK (1952) [chap. 2].

Heitler, W., *The Quantum Theory of Radiation*, Dover, New York (1984) [sec. 79].

Jackson, J. D., *Classical Electrodynamics*, 3rd edition, Wiley, New York (1999) [secs. 2, 4, 36, 62, 64, 87, 92, 97, 152].

Jeans, J., *The Mathematical Theory of Electricity and Magnetism*, 5th edition, Cambridge University Press, Cambridge, UK (1966) [preface, chap. 14, sec. 130].

Jona, F., and Shirane, G., *Ferroelectric Crystals*, Dover, New York (1993) [sec. 99].

Jose, J. V., and Saletan, E. J., *Classical Dynamics: A Contemporary Approach*, Cambridge University Press, Cambridge, UK (1998) [sec. 2, chap. 12].

Kittel, C., *Introduction to Solid State Physics*, 2nd edition, Wiley, New York (1975) [secs. 14, 94].

Kittel, C., *Quantum Theory of Solids*, 2nd revised printing, Wiley, New York (1987) [sec. 94].

Kraus, J. D. *Antennas*, 2nd edition, McGraw-Hill, New York (1988) [sec. 64].

Kreyszig, E., *Advanced Engineering Mathematics*, 9th edition, Wiley, New York (2005) [chap. 2].

Landau, L. D., and Lifshitz, E. M., *Course of Theoretical Physics*, Vol. 1, *Mechanics*, 3rd edition, Pergamon Press, Oxford, UK (1976) [secs. 36, 74].

Landau, L. D., and Lifshitz, E. M., *Course of Theoretical Physics*, Vol. 2, *Classical Theory of Fields*, 4th revised English edition, Pergamon Press, Oxford, UK (1975) [sec. 71].

Landau, L. D., and Lifshitz, E. M., *Statistical Physics, Course of Theoretical Physics* Vol. 5, 2nd English edition, Pergamon Press, Oxford, UK (1969) [sec. 121].

Landau, L. D., and Lifshitz, E. M., *Course of Theoretical Physics* Vol. 8, *Electrodynamics of Continuous Media*, 1st English edition, Pergamon Press, Oxford, UK (1960) [secs. 53, 90, 96, 97, 127].

Lighthill, M. S., *Introduction to Fourier Analysis and Generalized Functions*, Cambridge University Press, Cambridge, UK (1958) [sec. 9].

Mahan, B., *University Chemistry*, Addison-Wesley, Palo Alto, CA (1965) [sec. 115].

Mandel, L., and Wolf, E., *Optical Coherence and Quantum Optics*, Cambridge University Press, New York (1995) [sec. 51].

Maxwell, J. C., *A Treatise on Electricity and Magnetism*, 3rd edition, reprinted by Dover, New York (1954) [sec. 21].

Morse, P. M., and Feshbach, H., *Methods of Theoretical Physics*, Vols. 1 and 2, McGraw-Hill, New York (1953) [sec. 11].

Northrop, T. G., *The Adiabatic Motion of Charged Particles*, Interscience, New York (1963) [sec. 73].

Panofsky, W. K. H., and Phillips, M., *Classical Electricity and Magnetism*, Addison-Wesley, Reading, MA (1955) [sec. 97].

Peskin, M. E., and Schroeder, D. V., *An Introduction to Quantum Field Theory*, Addison-Wesley, Reading, MA (1995) [sec. 26].

Press, W. H., Flannery, B. P., Teukolsky, S. A., and Vetterling, W. T., *Numerical Recipes*, Cambridge University Press, Cambridge, UK (1986) [sec. 92, appendix F].

Pugh, E. M., and Pugh, E. W., *Principles of Electricity and Magnetism*, Addison-Wesley, Reading, MA (1970) [sec. 4].

Purcell, E. M., *Electricity and Magnetism*, McGraw-Hill, New York (1985) [preface, chap. 9, secs. 57, 92].

Roederer, J. G., *Dynamics of Geomagnetically Trapped Radiation*, Springer-Verlag, Berlin (1970) [sec. 73].

Sakurai, J. J., *Advanced Quantum Mechanics*, Benjamin/Cummings, Menlo Park, CA (1967) [sec. 26].

Schwinger, J., DeRaad, L. L., Jr., Milton, K. A., and Tsai, W.-Y., *Classical Electrodynamics*, Perseus Books, Reading, MA (1998) [sec. 48, appendix A].

Shankar, R., *Principles of Quantum Mechanics*, Plenum, New York (1980) [secs. 79, 89].

Smythe, W. R., *Static and Dynamic Electricity*, 3rd revised edition, McGraw-Hill, New York (1968) [chap. 14].

Stoer, J., and Bulirsch, R., *Introduction to Numerical Analysis*, Springer-Verlag, New York (1980) [sec. 92].

Stormer, C., *The Polar Aurora*, Oxford University Press, London (1955) [appendix G].

Terletskii, Y. P., *Paradoxes in the Theory of Relativity*, Plenum, New York (1968) [sec. 1].

Thomson, J. J., *Electricity and Matter*, Scribner, New York (1904) [chap. 9].

Vincent, C. A., Bonino, F., Lazzari, M., and Scrosati, B., *Modern Batteries*, Edward Arnold, London (1984) [sec. 115].

Watson, G. N., *Theory of Bessel Functions*, 2nd edition, Cambridge University Press, Cambridge, UK (1944) [sec. 90, appendix B].

White, R. M., *Quantum Theory of Magnetism*, 2nd corrected and updated edition, Springer-Verlag, Berlin (1983) [sec. 124].

Whittaker, E. T., *A History of the Theories of Aether and Electricity*, revised and enlarged edition, Nelson, London (1951) [sec. 2].

Whittaker, E. T., and Watson, G. N., *A Course of Modern Analysis*, 4th edition, Cambridge University Press, Cambridge, UK (1927) [sec. 90].

Williams, W. S. C., *Nuclear and Particle Physics*, Clarendon Press, Oxford, UK (1991) [sec. 14].

Index